T0329309

CHEETAHS: BIOLOGY AND CONSERVATION

This is a Volume in the

Biodiversity of the World:
Conservation from Genes to Landscapes
Edited By Philip J. Nyhus

CHEETAHS: BIOLOGY AND CONSERVATION

BIODIVERSITY OF THE WORLD:
CONSERVATION FROM GENES
TO LANDSCAPES

Series editor

PHILIP J. NYHUS

Environmental Studies Program, Colby College, Waterville, ME, United States

Edited by

LAURIE MARKER

Cheetah Conservation Fund, Otjiwarongo, Namibia

LORRAINE K. BOAST

Cheetah Conservation Botswana, Gaborone, Botswana

ANNE SCHMIDT-KÜNTZEL

Cheetah Conservation Fund, Otjiwarongo, Namibia

ELSEVIER

ACADEMIC PRESS

An imprint of Elsevier

Academic Press is an imprint of Elsevier
125 London Wall, London EC2Y 5AS, United Kingdom
525 B Street, Suite 1800, San Diego, CA 92101-4495, United States
50 Hampshire Street, 5th Floor, Cambridge, MA 02139, United States
The Boulevard, Langford Lane, Kidlington, Oxford OX5 1GB, United Kingdom

Notices
Knowledge and best practice in this field are constantly changing. As new research and experience broaden
our understanding, changes in research methods, professional practices, or medical treatment may become
necessary.

Practitioners and researchers must always rely on their own experience and knowledge in evaluating and
using any information, methods, compounds, or experiments described herein. In using such information or
methods they should be mindful of their own safety and the safety of others, including parties for whom they
have a professional responsibility.

To the fullest extent of the law, neither the Publisher nor the authors, contributors, or editors, assume any
liability for any injury and/or damage to persons or property as a matter of products liability, negligence
or otherwise, or from any use or operation of any methods, products, instructions, or ideas contained in the
material herein.

Library of Congress Cataloging-in-Publication Data
A catalog record for this book is available from the Library of Congress

British Library Cataloguing-in-Publication Data
A catalogue record for this book is available from the British Library

ISBN: 978-0-12-804088-1

For information on all Academic Press publications visit our
website at https://www.elsevier.com/books-and-journals

  Working together
to grow libraries in
developing countries

www.elsevier.com • www.bookaid.org

Publisher: John Fedor
Acquisition Editor: Anna Valutkevich
Editorial Project Manager: Pat Gonzalez
Production Project Manager: Mohana Natarajan
Designer: Matthew Limbert

Typeset by Thomson Digital

To Caitlin and Chloe, as representatives
of the future generations, and to all those helping
the cheetah win its race for survival.

Photo Credits
Cover photograph by Angela Scott

Section break photographs by the following artists:
1. The Cheetah – Peter Scheufler
2. Threats – Nejat Said/Cheetah Conservation Fund
3. Solutions – Elena Chelysheva
4. Captivity – Peter Scheufler
5. Techniques – Remote camera trap/Cheetah Conservation Botswana
6. The Future – Angela Scott

Companion Site
Protocols and forms (which were provided by authors) can be found at the
Cheetahs: Conservation and Biology companion site:
https://www.elsevier.com/books-and-journals/book-companion/9780128040881

Contents

1
THE CHEETAH

1. A Brief History of Cheetah Conservation

LAURIE MARKER, JACK GRISHAM, BRUCE BREWER

2. History of the Cheetah–Human Relationship

BENISON PANG, BLAIRE VAN VALKENBURGH,
KENNETH F. KITCHELL, JR., AMY DICKMAN,
LAURIE MARKER

3. The Cheetah: Evolutionary History and Paleoecology

BLAIRE VAN VALKENBURGH, BENISON PANG,
MARCO CHERIN, LORENZO ROOK

4. Cheetah Rangewide Status and Distribution

LAURIE MARKER, BOGDAN CRISTESCU,
TESS MORRISON, MICHAEL V. FLYMAN,
JANE HORGAN, ETOTÉPÉ A. SOGBOHOSSOU,
CHARLENE BISSETT, VINCENT VAN DER MERWE,
IRACELMA B. DE MATOS MACHADO,
EZEQUIEL FABIANO, ESTHER VAN DER MEER,
ORTWIN ASCHENBORN, JOERG MELZHEIMER,
KIM YOUNG-OVERTON, MOHAMMAD S. FARHADINIA,
MARY WYKSTRA, MONICA CHEGE,
ABDOULKARIM SAMNA, OSMAN G. AMIR,
AHMED SH MOHANUN, OSMAN D. PAULOS,
ABEL R. NHABANGA, JASSIEL L.J. M'SOKA,
FARID BELBACHIR, ZELEALEM T. ASHENAFI,
MATTI T. NGHIKEMBUA

2

CONSERVATION THREATS

11. The Status of Key Prey Species and the Consequences of Prey Loss for Cheetah Conservation in North and West Africa

LAURIE MARKER, THOMAS RABEIL, PIERRE COMIZZOLI,
HAYLEY CLEMENTS, MATTI T. NGHIKEMBUA,
MATT W. HAYWARD, CRAIG J. TAMBLING

12. The Impact of Climate Change on the Conservation and Survival of the Cheetah

MATTI T. NGHIKEMBUA, FLAVIO LEHNER, WILBUR
OTTICHILO, LAURIE MARKER, STEVEN C. AMSTRUP

13. The Costs and Causes of Human-Cheetah Conflict on Livestock and Game Farms

AMY DICKMAN, NIKI A. RUST, LORRAINE K. BOAST,
MARY WYKSTRA, LOUISA RICHMOND-COGGAN,
REBECCA KLEIN, MOSES SELEBATSO,
MAURUS MSUHA, LAURIE MARKER

14. Pets and Pelts: Understanding and Combating Poaching and Trafficking in Cheetahs

PATRICIA TRICORACHE, KRISTIN NOWELL,
GÜNTHER WIRTH, NICHOLAS MITCHELL,
LORRAINE K. BOAST, LAURIE MARKER

3

CONSERVATION SOLUTIONS

15. Use of Livestock Guarding Dogs to Reduce Human-Cheetah Conflict

AMY DICKMAN, GAIL POTGIETER, JANE HORGAN,
KELLY STONER, REBECCA KLEIN, JEANNINE MCMANUS,
LAURIE MARKER

16. Improved and Alternative Livelihoods:
Links Between Poverty Alleviation,
Biodiversity, and Cheetah Conservation

MARY WYKSTRA, GUY COMBES, NICK OGUGE,
REBECCA KLEIN, LORRAINE K. BOAST,
ALFONS W. MOSIMANE, LAURIE MARKER

17. Coordination of Large Landscapes
for Cheetah Conservation

LARKIN A. POWELL, REINOLD KHARUXAB, LAURIE MARKER,
MATTI T. NGHIKEMBUA, SARAH OMUSULA, ROBIN S. REID,
ANDREI SNYMAN, CHRIS WEAVER, MARY WYKSTRA

18. Cheetah Conservation
and Educational Programs

COURTNEY HUGHES, JANE HORGAN, REBECCA KLEIN,
LAURIE MARKER

19. Protected Areas for Cheetah Conservation

BOGDAN CRISTESCU, PETER LINDSEY, OLIVIA MAES,
CHARLENE BISSETT, GUS MILLS, LAURIE MARKER

20. Cheetah Translocation
and Reintroduction Programs:
Past, Present, and Future

LORRAINE K. BOAST, ELENA V. CHELYSHEVA,
VINCENT VAN DER MERWE, ANNE SCHMIDT-KÜNTZEL,
ELI H. WALKER, DEON CILLIERS, MARKUS GUSSET,
LAURIE MARKER

21. Global Cheetah Conservation Policy:
A Review of International Law
and Enforcement

KRISTIN NOWELL, TATJANA ROSEN

4

CAPTIVE CHEETAHS

22. History of Cheetahs in Zoos and Demographic Trends Through Managed Captive Breeding Programs

LAURIE MARKER, KATE VANNELLI, MARKUS GUSSET,
LARS VERSTEEGE, KAREN ZIEGLER MEEKS,
NADJA WIELEBNOWSKI, JAN LOUWMAN,
HANNEKE LOUWMAN, LAURIE BINGAMAN LACKEY

23. The Role of Zoos in Cheetah Conservation: Integrating *Ex Situ* and *In Situ* Conservation Action

KARIN R. SCHWARTZ, MARKUS GUSSET,
ADRIENNE E. CROSIER, LARS VERSTEEGE,
SIMON EYRE, AMANDA TIFFIN,
ANTOINETTE KOTZÉ

24. Clinical Management of Captive Cheetahs

ANA MARGARITA WOC COLBURN,
CARLOS R. SANCHEZ, SCOTT CITINO,
ADRIENNE E. CROSIER, SUZANNE MURRAY,
JACQUES KAANDORP, CHRISTINE KAANDORP,
LAURIE MARKER

25. Diseases Impacting Captive and Free-Ranging Cheetahs

KAREN A. TERIO, EMILY MITCHELL, CHRIS WALZER,
ANNE SCHMIDT-KÜNTZEL, LAURIE MARKER,
SCOTT CITINO

26. Nutritional Considerations for Captive Cheetahs

KATHERINE WHITEHOUSE-TEDD, ELLEN S. DIERENFELD,
ANNE A.M.J. BECKER, GEERT HUYS, SARAH DEPAUW,
KATHERINE R. KERR, J. JASON WILLIAMS,
GEERT P.J. JANSSENS

6
THE FUTURE

List of Contributors

Osman G. Amir   Somalia Wildlife and Natural History Society, Mogadishu, Somalia

Steven C. Amstrup   Polar Bears International, Bozeman, MT; University of Wyoming, Laramie, WY, United States

Leah Andresen   Nelson Mandela University, Port Elizabeth, South Africa

Ortwin Aschenborn   University of Namibia, Windhoek, Namibia

Zelealem T. Ashenafi   Born Free Foundation Ethiopia, Addis Ababa, Ethiopia

Jonathan D. Ballou   Smithsonian Conservation Biology Institute, Washington, DC, United States

Anne A.M.J. Becker   Ghent University, Ghent, Belgium

Annie Beckhelling   Cheetah Outreach, Cape Town, South Africa

Farid Belbachir   University of Béjaïa, Béjaïa, Algeria

Laurie Bingaman Lackey   World Association of Zoos and Aquariums, Gland, Switzerland

Charlene Bissett   South African National Parks, Kimberley, South Africa

Lorraine K. Boast   Cheetah Conservation Botswana, Gaborone, Botswana

Birgit Braun   Action Campaign for Endangered Species, Korntal-Münchingen, Germany

Christine Breitenmoser   IUCN/SSC Cat Specialist Group; KORA, Bern, Switzerland

Bruce Brewer   Cheetah Conservation Fund, Otjiwarongo, Namibia

Femke Broekhuis   Mara Cheetah Project, Kenya Wildlife Trust, Nairobi, Kenya; University of Oxford, Tubney, Abingdon, United Kingdom

Rox Brummer   Green Dogs Conservation, Alldays, South Africa

Tim Caro   University of California Davis, Davis, CA, United States

Linda Castaneda   Cincinnati Zoo & Botanical Garden, Cincinnati, OH, United States

Pauline Charruau   University of Veterinary Medicine Vienna, Austria

Monica Chege   Kenya Wildlife Service, Nairobi, Kenya

Elena V. Chelysheva   Mara-Meru Cheetah Project, Kenya Wildlife Services, Nairobi, Kenya

Marco Cherin   University of Perugia, Perugia, Italy

Deon Cilliers   Cheetah Outreach Trust, Cape Town, South Africa

Scott Citino   White Oak Conservation Center, Yulee, FL, United States

Hayley Clements   Nelson Mandela University, Port Elizabeth, South Africa; Monash University, Melbourne, VIC, Australia

Guy Combes   Guy Combes Studio, Antioch, CA, United States

Pierre Comizzoli   Smithsonian Conservation Biology Institute, Washington, DC, United States

Bogdan Cristescu   Cheetah Conservation Fund, Otjiwarongo, Namibia

Adrienne E. Crosier   Smithsonian Conservation Biology Institute, Front Royal, VA, United States

Desiré L. Dalton   National Zoological Gardens of South Africa, Pretoria; University of Venda, Thohoyandou, South Africa

Harriet T. Davies-Mostert   Endangered Wildlife Trust, Johannesburg; University of Pretoria, Pretoria, South Africa

Jacqueline T. Davis   University of Cambridge, Cambridge, United Kingdom

Iracelma B. de Matos Machado   Ministry of Agriculture, Luanda, Angola

Sarah Depauw   Odisee College University, Sint Niklaas, Belgium

Amy Dickman   University of Oxford, Tubney, Abingdon, United Kingdom

Ellen S. Dierenfeld   Ellen S. Dierenfeld, LLC, St. Louis, MO, United States

Sarah M. Durant   Zoological Society of London, London, United Kingdom; Wildlife Conservation Society, New York, NY, United States

Susie Ekard   San Diego Zoo Safari Park, San Diego, CA, United States

Simon Eyre   Wellington Zoo Trust, Wellington, New Zealand

Ezequiel Fabiano   University of Namibia, Katima Mulilo, Namibia

Mohammad S. Farhadinia   Future4Leopards Foundation, Tehran, Iran; University of Oxford, Tubney, Abingdon, United Kingdom

Michael V. Flyman   Department of Wildlife and National Parks, Gaborone, Botswana

Katherine Forsythe   Anchor Environmental Consultants, Tokai, South Africa

Angela K. Fuller   Cornell University, Ithaca, NY, United States

Kyle Good   Cheetah Conservation Botswana, Bulawayo, Zimbabwe

Jack Grisham   Saint Louis Zoo, St. Louis, MO, United States

Rosemary Groom   Zoological Society of London, London, United Kingdom; Wildlife Conservation Society, New York, NY, United States

Markus Gusset   World Association of Zoos and Aquariums, Gland, Switzerland

Holly Haefele   Fossil Rim Wildlife Center, Glen Rose, TX, United States

Axel Hartmann   Bicornis Veterinary Consulting, Bicornis Conservation Trust, Otjiwarongo, Namibia

Matt W. Hayward   Nelson Mandela University, Port Elizabeth, South Africa; Bangor University, Bangor, United Kingdom; University of Newcastle, Callaghan, NSW, Australia

Cathryn Hilker   Cincinnati Zoo & Botanical Garden, Cincinnati, OH, United States

Jane Horgan   Cheetah Conservation Botswana, Maun, Botswana

Courtney Hughes   University of Alberta, Edmonton; Alberta Environment and Parks, Edmonton, AB, Canada

Luke T.B. Hunter   Panthera, New York, NY, United States

Geert Huys   Ghent University, Ghent, Belgium

Audrey Ipavec   Zoological Society of London, London, United Kingdom; Wildlife Conservation Society, New York, NY, United States

Geert P.J. Janssens   Ghent University, Ghent, Belgium

Richard M. Jeo   The Nature Conservancy in Montana, Helena, MT, United States

Douglas W. Johnson   University of Queensland, Brisbane, QLD, Australia

Sandra Johnson   Queensland University of Technology, Brisbane, QLD, Australia

Warren E. Johnson   Smithsonian Conservation Biology Institute, Front Royal, VA, United States

Houman Jowkar   Persian Wildlife Heritage Foundation; Conservation of the Asiatic Cheetah Project, Tehran, Iran

Christine Kaandorp   GaiaZOO, Kerkrade, The Netherlands

Jacques Kaandorp   Safaripark Beekse Bergen, Hilvarenbeek, The Netherlands

Katherine R. Kerr   San Diego Zoo Global, San Diego, CA, United States

Reinold Kharuxab   Namibia University of Science and Technology, Windhoek, Namibia

Kenneth F. Kitchell, Jr.   University of Massachusetts Amherst, Amherst, MA; Louisiana State University, Baton Rouge, LA, United States

Rebecca Klein   Cheetah Conservation Botswana, Gaborone, Botswana

Diana C. Koester   Cleveland Metroparks Zoo, Cleveland, OH, United States

Antoinette Kotzé   National Zoological Gardens of South Africa, Pretoria; University of Free State South Africa, Bloemfontein, South Africa

Imke Lüders   GEOlife's Animal Fertility and Reproductive Research, Hamburg, Germany

Flavio Lehner National Center for Atmospheric Research, Boulder, CO, United States

Kristin Leus IUCN SSC Conservation Planning Specialist Group—Europe, Copenhagen Zoo, Merksem, Belgium

Peter Lindsey Panthera, New York, NY; University of Pretoria, Pretoria, South Africa; Wildlife Network, Palo Alto, CA, United States

Michelle Lloyd Monarto Zoo, Monarto, SA, Australia

Hanneke Louwman Wassenaar Wildlife Breeding Centre, Wassenaar, The Netherlands

Jan Louwman Wassenaar Wildlife Breeding Centre, Wassenaar, The Netherlands

Jassiel L.J. M'soka Ministry of Tourism and Arts, Chilango, Lusaka, Zambia

David W. Macdonald University of Oxford, Tubney, Abingdon, United Kingdom

Olivia Maes Cheetah Conservation Fund, Toddington, United Kingdom

Laurie Marker Cheetah Conservation Fund, Otjiwarongo, Namibia

Nikki Marks Queen's University Belfast, Belfast, United Kingdom

Aaron Maule Queen's University Belfast, Belfast, United Kingdom

Natasha McGowan Queen's University Belfast, Belfast, United Kingdom

Jeannine McManus Landmark Foundation Leopard Project, Riversdale, South Africa

Julie Meachen Des Moines University, Des Moines, IA, United States

Karen Ziegler Meeks White Oak Conservation, Yulee, FL, United States

Joerg Melzheimer Leibniz Institute for Zoo and Wildlife Research, Berlin, Germany

Kerrie Mengersen Queensland University of Technology, Brisbane, QLD, Australia

Marilyn Menotti-Raymond Laboratory of Genomic Diversity, Frederick, MD, United States

Gus Mills University of Oxford, Tubney, Abingdon, United Kingdom; The Lewis Foundation, Johannesburg, South Africa

Emily Mitchell National Zoological Gardens of South Africa, Pretoria, South Africa

Nicholas Mitchell Zoological Society of London, London, United Kingdom; Wildlife Conservation Society, New York, NY, United States

Ahmed Sh Mohanun Somalia Wildlife and Natural History Society; Somalia Ministry of Wildlife and Rural Development, Mogadishu, Somalia

Tess Morrison Cheetah Conservation Fund, Otjiwarongo, Namibia

Alfons W. Mosimane University of Namibia, Neudamm, Namibia

Maurus Msuha Tanzania Wildlife Research Institute, Arusha, Tanzania

Suzanne Murray Smithsonian Biology Institute, National Zoological Park, Washington, DC, United States

Matti T. Nghikembua Cheetah Conservation Fund, Otjiwarongo, Namibia

Abel R. Nhabanga Banhine National Park, Gaza, Mozambique

Kristin Nowell Cat Action Treasury and World Conservation Union (IUCN) Red List Programme, Cape Neddick, ME, United States

Stephen J. O'Brien St. Petersburg State University, St. Petersburg, Russia; Nova Southeastern University, Fort Lauderdale, FL, United States

Nick Oguge University of Nairobi, Nairobi, Kenya

Sarah Omusula Action for Cheetahs in Kenya Project, Nairobi, Kenya

Stephane Ostrowski Wildlife Conservation Society, New York, NY, United States

Wilbur Ottichilo Republic of Kenya, Parliament; Parliament Conservation Caucus; Environment Committee, Nairobi, Kenya

Benison Pang University of California at Los Angeles, Los Angeles, CA, United States

Osman D. Paulos Ethiopian Wildlife Conservation Authority, Addis Ababa, Ethiopia

Ruben Portas Leibniz Institute for Zoo and Wildlife Research, Berlin, Germany

Gail Potgieter Tau Consultants, Maun, Botswana

Larkin A. Powell University of Nebraska-Lincoln, Lincoln, NE, United States

Thomas Rabeil Sahara Conservation Fund, Bussy St. Georges, France

Marcela Randau University College London, London, United Kingdom

Suzi Rapp Columbus Zoo and Aquarium, Powell, OH, United States

Robin S. Reid Colorado State University, Fort Collins, CO, United States

Louisa Richmond-Coggan Cheetah Conservation Fund, Otjiwarongo, Namibia

James M. Robinson F.R.C.S Royal College of Surgeons, London, United Kingdom

Lorenzo Rook University of Florence, Florence, Italy

Janet Rose-Hinostroza San Diego Zoo Safari Park, San Diego, CA, United States

Tatjana Rosen Panthera and IISD Earth Negotiations Bulletin, Khorog, GBAO, Tajikistan

Niki A. Rust University of Kent, Canterbury; WWF-UK, Woking, United Kingdom

Abdoulkarim Samna Ministry of Environment and Sustainable Development, Niamey, Niger

Alicia Sampson Cincinnati Zoo & Botanical Garden, Cincinnati, OH, United States

Carlos R. Sanchez Fort Worth Zoo, Fort Worth, TX, United States

M. Sanjayan Conservation International, Arlington, VA, United States

David M. Scantlebury Queen's University Belfast, Belfast, United Kingdom

George B. Schaller Panthera; Wildlife Conservation Society, New York, NY, United States

Anne Schmidt-Küntzel Cheetah Conservation Fund, Otjiwarongo, Namibia

Martin Schulman University of Pretoria, Pretoria, South Africa

Karin R. Schwartz Cheetah Conservation Fund, Otjiwarongo, Namibia

Moses Selebatso Kalahari Research and Conservation, Gaborone, Botswana

Andrei Snyman University of Nebraska-Lincoln, Lincoln, NE, United States; Northern Tuli Predator Project and Research Mashatu, Mashatu Game Reserve, Botswana

Etotépé A. Sogbohossou University of Abomey-Calavi Cotonou, Cotonou, Benin

Simone Sommer University of Ulm, Ulm, Germany

Linda Stanek Columbus Zoo and Aquarium, Powell, OH, United States

Gerhard Steenkamp University of Pretoria, Onderstepoort, South Africa

Kelly Stoner Ruaha Carnivore Project, Iringa, Tanzania

Chris Sutherland University of Massachusetts-Amherst, Amherst, MA, United States

Craig J. Tambling Nelson Mandela University, Port Elizabeth; University of Fort Hare, Alice, South Africa

Karen A. Terio University of Illinois, Brookfield, IL, United States

Amanda Tiffin Wellington Zoo Trust, Wellington, New Zealand

Carl Traeholt Copenhagen Zoo, Frederiksberg, Denmark

Kathy Traylor-Holzer IUCN SSC Conservation Planning Specialist Group, Apple Valley, MN, United States

Patricia Tricorache Cheetah Conservation Fund, Islamorada, FL, United States

Linda van Bommel University of Tasmania, Hobart, TAS; Australian National University, Canberra, ACT, Australia

Esther van der Meer Cheetah Conservation Project Zimbabwe, Victoria Falls; National University of Science and Technology, Bulawayo, Zimbabwe

Vincent van der Merwe Endangered Wildlife Trust, Modderfontein, Gauteng, South Africa

Leanne Van der Weyde Cheetah Conservation Botswana, Gaborone, Botswana

Kate Vannelli Cheetah Conservation Fund, Otjiwarongo, Namibia

Blaire Van Valkenburgh University of California at Los Angeles, Los Angeles, CA, United States

Lars Versteege Safaripark Beekse Bergen, Hilvarenbeek, The Netherlands

Bettina Wachter Leibniz Institute for Zoo and Wildlife Research, Berlin, Germany

Eli H. Walker Cheetah Conservation Fund, Otjiwarongo, Namibia

Chris Walzer University of Veterinary Medicine, Vienna, Austria

Chris Weaver World Wildlife Fund, Windhoek, Namibia

Katherine Whitehouse-Tedd Nottingham Trent University, Southwell, United Kingdom

Nadja Wielebnowski Oregon Zoo, Portland, OR, United States

J. Jason Williams Indianapolis Zoological Society, Indianapolis, IN, United States

Günther Wirth Independent Researcher, Hargeisa, Somaliland

Ana Margarita Woc Colburn Nashville Zoo at Grassmere, Nashville, TN, United States

Rosie Woodroffe Zoological Society of London, London, United Kingdom

Claudia Wultsch American Museum of Natural History, New York, NY, United States

Mary Wykstra Action for Cheetahs in Kenya Project, Nairobi, Kenya

Kim Young-Overton Zambian Carnivore Programme, Panthera, New York, NY, United States

Foreword

I shall never tire of cheetahs—watching them, studying their grace and agility, researching their natural history, and uncovering anecdotes of human fascination with them. Cheetahs are so special among cats, mammals, living evolutionary creations, and threatened wildlife. Their aerodynamic body and breath-taking speed is heralded as a genetic adaptation driven by an endless pursuit of faster and faster-prey species. The ancestors of the cheetahs were sculpted by evolutionary forces at work on the plains where the earliest cheetah-relatives first appeared. Cheetahs hunted ungulates and grew to become the fastest running species on the planet. The second fastest land animal today is the American pronghorn, a distant giraffe relative resembling African antelopes that likely served as prime cut for the cheetah's early relatives for millennia. By what is termed "coevolution," cheetahs and pronghorns or the like, likely stimulated each other to develop agility for high-speed chases across vast prairies. Then around 12,000 years ago, as the last ice age receded during the late Pleistocene epoch, several species of large vertebrates (including American lions, saber-tooth tigers, mastodons, giant sloths, and dire-wolves) disappeared rather abruptly from the Americas. All were victims of a global cataclysm that nearly extirpated the cheetah. The majority of living cheetahs today are found in Africa and are derived from an ancestral migration of cheetahs from Asia.

In more recent times, the cheetah population grew to hundreds of thousands of individuals across Africa and southern Asia until an inevitable human development reduced their habitat and future considerably. Cheetahs today, remain highly vulnerable and are protected by international agreements, to preserve these treasures and to reverse the deliberate march to extinction the species has endured. Conservation programs for cheetahs have appeared and, as is seen in this new work, scientific research has blossomed around this highly specialized animal. As a consequence, the cheetah is probably more thoroughly researched in terms of scientific rigor than nearly any other threatened species. This monograph is a testament to the depth of science enquiry pursued to assist conservation management of the African cheetah. The subjects of the 40 chapters are myriad and all around cheetahs—ecology, genetics, reproduction, evolution, paleontology, behavior, morphology, nutrition, infectious disease, habitat assessment, poaching, and translocation. Each chapter offers an in-depth glimpse of the discoveries made in studying cheetahs, not only by initial patient behavioral observation, but marshaling state-of-the-art scientific technologies. The composition of the whole is unprecedented in cheetah lore and perhaps in breadth of any threatened species. Conservation action must be science-driven and this collection offers a healthy and rigorous dose of research, results, and interpretation.

These cheetah studies provide a definitive workbook for science advances, analyses, and conservation applications. Early on cheetahs became the poster-model for the medical and reproductive perils of historic purges of a species' genetic variability. Crippled with ~90%–99% depletion of endemic variation, cheetahs still became a success story as diverse science approaches were applied. By consensus of zoos and breeders the captive population was designated as a "research population" in 1987, to assure that the field research would inform the captive managers with insight into the behavioral ecology to provide the best management practices available. The cheetahs' isolationist behavior requires enormous territories, rendering the species hypersensitive to habitat loss. However, behavioral isolation in nature likely protected them, as a genetically depauperate species, from numerous deadly infectious-disease outbreaks, which would spread far more rapidly in a crowded captive setting.

In 1990, a fledgling new democracy in Namibia would pledge in their constitution, a national respect and continued protection for native wildlife including the dwindling cheetah population, numbering less than 15,000 across Africa. That same year the Cheetah Conservation Fund (CCF) was established by committed citizenry led by Laurie Marker in Namibia, a country with the highest density of cheetahs in Africa. Over the last decades, CCF and other conservation organizations have massaged attitudes about cheetahs in Namibia and across Africa and Iran from a "problem animal" to a cherished symbol of these countries' natural resources. These conservation organizations encouraged and facilitated governments, to develop natural plans to stop the reduction and find ways to restore cheetah habitats where they presently and formerly existed. Cheetahs rescued from traps on livestock farms were offered for reintroduction in Namibia and other African countries; their modest goal—to eventually restore cheetah populations across their former and recently truncated natural range.

Cheetahs have dwindled worldwide to numbers fewer than 8000 individuals in natural habitats today. Successful conservation programs invariably heed the balance of people and wildlife in developing communities and cultures in some cases more accustomed to destruction than construction. They do this with calm and deliberate science, education outreach, and international sponsorship. In an era where science in so many areas is often mistrusted and denied, how refreshing to see the fruits of science-based action described by these authors in the important realm of cheetah and wildlife conservation.

I personally believe that the cheetah can be saved and conserved even in the face of its skirmishes with extinction in its past. However, science is the new currency of effective conservation. The bright glow of hard research data in any setting will outshine the most raucous of unfounded opinions on the way forward for wildlife management. In my own pursuits across conservation communities, I have met countless dedicated, concerned, and determined conservationists. Their commitment and experience often makes them convinced of their own perceptions and pet ideas. Their resolve centers upon a universal agreement that saving the species is really our only goal. Reliable validated data go a long way to winning the day in such conflicts. This rich conservation science compendium should be absorbed and relished by saviors for cheetahs, as well as by the many dedicated champions who have chosen to facilitate effective conservation programs across the planet.

I shall never tire of cheetahs.

Stephen J. O'Brien
Chairman Emeritus,
Cheetah Conservation Fund

Acknowledgments

The authors and editors wish to thank the large number of organizations and individuals who provided data, information, editorial assistance, and advice during the production of this book. In addition, special thanks goes to the Cheetah Conservation Fund for its commitment to and financial contribution to the success of this publication.

We acknowledge the international team of experts who contributed to this publication by peer reviewing chapters within this book: Laurie Bingaman Lackey (World Association of Zoos and Aquarium, European Association of Zoos and Aquaria, and Asociación Colombiana de Parques Zoológicos y Acuarios); Christopher Bonar (Dallas Zoo); Urs Breitenmoser (IUCN/SSC Cat Specialist Group); Abigail Breuer (Wildlife Friendly Enterprise Network); Tim Caro (University of California Davis); Scott Citino (White Oak Conservation Center); Raymond Coppinger (Hampshire College); Adrienne E. Crosier (Smithsonian Conservation Biology Institute); Melanie Culver (US Geological Survey and University of Arizona); Amy Dickman (University of Oxford); Cees van Duijn (INTERPOL); Mary Duncan (Saint Louis Zoo); Richard Edwards (University of Nebraska); Brian Gerber (Colorado State University); Ashwell Glasson (Southern African Wildlife College); Markus Gusset (WAZA); Holly Haefele (Fossil Rim Wildlife Center); Adam Hartstone-Rose (University of South Carolina); Matt W. Hayward (Bangor University); Oliver Höner (Leibniz Institute for Zoo and Wildlife Research); Luke T.B. Hunter (Panthera); Chloe Inskip (Chester Zoo); Rodney Jackson (Snow Leopard Conservancy); Julie S. Johnson-Pynn (Berry College); Derek Keeping (University of Alberta); Andrew C. Kitchener (National Museums Scotland); Rebecca Klein (Cheetah Conservation Botswana); Klaus-Peter Koepfli (Smithsonian Conservation Biology Institute); Miha Krofel (University of Ljubljana); Robert Lacy (Chicago Zoological Society); Debbie Luke (Society for Conservation Biology); David Mallon (Manchester Metropolitan University); Michael Manfredo (Colorado State University); Michael Maslanka (Smithsonian Institution's National Zoological Park and Conservation Biology Institute); Roland Melisch (TRAFFIC); Gus Mills (The Lewis Foundation and University of Oxford); Robin Naidoo (WWF-US); James Nichols (US Geological Survey); Kristin Nowell (Cat Action Treasury and IUCN Red List Program); Stephen O'Brien (St. Petersburg State University and Nova Southeastern University); Cynthia Olson (Cheetah Conservation Fund Scientific Advisory Board); Karen Povey (Point Defiance Zoo and Aquarium); Larkin Powell (University of Nebraska-Lincoln); David Raubenheimer (University of Sydney); Paul Schuette (University of Alaska and Zambian Carnivore Programme); Karin Schwartz (Cheetah Conservation Fund); Philip Seddon (University of Otago); Julie Stein (Wildlife Friendly Enterprise Network); Todd Steury (Auburn University); Bill Swanson (Cincinnati Zoo & Botanical Garden); Karen Terio (University of Illinois Zoological Pathology Program); Julie Thomson (TRAFFIC); Linda Van Bommel (Australian National University); Lars Werdelin (Swedish Museum of Natural History); Florian Weise (CLAWS Conservancy); David Wildt (Smithsonian Institution's National Zoological Park and Conservation Biology Institute);

Christiaan Winterbach (Tau Consultants); Susan Yannetti (Cheetah Conservation Fund); and a number of reviewers who wish to remain anonymous.

In addition, the authors of the relevant chapters wish to acknowledge: the many Iranian experts who shared their knowledge and records since mid-1990s, Iranian Department of Environment, UNDP Iran, Conservation of Asiatic Cheetah Project, Iranian Cheetah Society (ICS), Persian Wildlife Heritage Foundation (PWHF), and Panthera and Wildlife Conservation Society (WCS) for establishing and supporting conservation efforts to save the Asiatic cheetahs across the country (Chapter 5); contributions by Peter Barber, Elena Chelysheva, and Darcy Ogada (Chapter 16); Dawn Glover (Cheetah Outreach), Esther van der Meer (Cheetah Conservation Project Zimbabwe), Mary Wykstra (Action for Cheetahs in Kenya), and Mohammad Farhadinia (Iranian Cheetah Society) for providing information about their education programs (Chapter 18); Earthwatch volunteers, Working Abroad volunteers, and all other citizen scientists who participate in cheetah conservation programs worldwide, Leah Andresen (Mozambique Carnivore Project), Charlene Bissett (Rhodes University), Femke Broekhuis (Mara Cheetah Project), Tim Caro (University of California), Elena Chelysheva (Mara Meru Cheetah Project), Amy Dickman (Ruaha Carnivore Project), Ezequiel Fabiano (Angola Carnivore Project), Mohammad Farhadani (Iranian Cheetah Society), Dada Gottelli (Zoological Society London), Donna Hanssen (Africat), Jane Horgan and Rebecca Klein (Cheetah Conservation Botswana), Laurie Marker (Cheetah Conservation Fund), Esther van der Meer (Cheetah Conservation Project Zimbabwe), Joerg Melzheimer and Bettina Wachter (IZW Cheetah Research Project), Gus Mills (Kgalagadi Cheetah Project), Nick Mitchell (Range Wide Conservation Program for Cheetah and African Wild Dogs), Niki Rust (University of Kent), Paul Schuette (Zambian Carnivore Programme), Craig Tambling (Nelson Mandela Metropolitan University), Atie tak Tehrani (Iranian Cheetah Society), Mary Wykstra (Action for Cheetahs in Kenya), and Kim Young-Overton (Kafue Carnivore Coalition) for providing an insight into the use of citizen science in cheetah research (Chapter 34); the Howard G. Buffett Foundation for supporting the work relative to the chapter, AZA SAFE for their support of the southern African review workshop in 2015, all participants of the strategic planning workshops and national action planning workshops for providing information on cheetah distribution, status, and conservation, and Karen Minkowski and Lisanne Petracca for providing invaluable assistance with the distributional mapping (Chapter 39).

SECTION 1

THE CHEETAH

1

A Brief History of Cheetah Conservation

Laurie Marker*, Jack Grisham**, Bruce Brewer*

*Cheetah Conservation Fund, Otjiwarongo, Namibia
**Saint Louis Zoo, St. Louis, MO, United States

INTRODUCTION

The cheetah (*Acinonyx jubatus*), the most unique of the 41 species of felids, (Kitchener et al., 2016) is at the crossroads of its survival. With an estimated population of 7100 adult and adolescent cheetahs in their natural habitat (Durant et al., 2017), long-term conservation research programs are collectively working on strategies to ensure their survival. Over the past 4 decades, a small but prolific group of international researchers and conservation biologists has emerged, all dedicated to solving the problems that threaten cheetah survival. Their collective research is presented in the chapters of this book and brings together what we currently know about the cheetah, the challenges it is facing, and the solutions that have been developed. Many of the cheetah conservation strategies that currently are being undertaken have a unique and interesting history of how they began. Here we endeavor to offer a historical overview and timeline that ties together the information presented in the chapters of this book.

HISTORICAL CONTEXT

Ancient History—Challenges for the Cheetah Populations

The cheetah is a survivor; its challenging evolutionary history has shaped a unique physiology, optimized for speed (Chapter 7). The first fossil records of cheetah (*Acinonyx*) date from approximately 4 million years ago and evidence of related species was retrieved in America, Europe, Asia, and Africa (Chapter 3). Following a founder effect approximately 100,000 years ago, the cheetah escaped extinction in the Pleistocene, which left the species with both reduced numbers and diminished genetic diversity (Chapter 6).

Modern History—Human Pressure on the Cheetah Populations

For the past 5000 years, humans throughout Asia, Europe, and Africa have revered the cheetah (Chapter 2); however, humans' fascination with the cheetah has manifested in ways that have contributed to the species' near demise (Wrogemann, 1975; Chapters 2, 14, and 22).

The first recognized use of the word cheetah was in 1610, but it wasn't until 1775 that a German naturalist, Johann Christian Daniel von Schreber, published the first description of the species, which at that time was commonly found throughout Asia and Africa. Cheetahs were appreciated as hunting companions in India, and in Hindu, cheetahs were known as "Chita" or the "Spotted One," and were often referred to as "hunting leopard." The pressure on the wild cheetah populations due to the harvest of cheetahs for Maharajahs' hunting parties, and later for safari hunting, contributed to the decline and eventual extinction of cheetahs from the wild in India (Chapters 4, 5, and 22).

In Africa, the first nationally protected area, Kruger National Park, was established in the 1890s. Measures to protect wildlife were in part motivated by plummeting wildlife populations, as over exploitation, habitat loss, and human persecution (Schreber, 1775) arose as a result of a shift from traditional lifestyle to agriculture under colonial influences. However, large carnivores were viewed as vermin, and until the 1970s, cheetahs and other predators were killed in many African national parks to protect game (in addition to their persecution on farmland) (Linnell et al., 2001; Woodroffe and Ginsberg, 1997).

In addition, during the 1960s, cheetahs, along with other exotic wildlife species, were in high demand to stock the world's zoos. Due to poor captive breeding success, most zoo-bound cheetahs were captured from wild populations, putting big pressure on populations in East Africa and Namibia (Chapter 22). As a consequence, game dealers showed farmers how to catch cheetahs with cage traps. From the late 1960s through the 1970s several thousand cheetahs were caught for the world's zoos (Marker-Kraus et al., 1996). Cage traps typically were set at cheetah marking trees, visited primarily

by males; but zoos preferred females. For every female cheetah caught, up to 20 males were captured, many of which were killed by the farmers (Marker-Kraus et al., 1996).

As a result of human development, cheetah numbers are estimated to have dropped from 100,000 in 1900 to 7,100 in 2016 (Durant et al., 2017; Myers, 1975; Chapters 4 and 29).

1960s—THE BEGINNING OF KNOWLEDGE

In the 1960s, the first reports of concern over declining populations of cheetahs emerged when the East African Wildlife Society began an investigation into the species' status (Graham, 1966). A couple of years later the results of the first study of cheetahs in the wild was published, where George Schaller shared his findings from Tanzania's Serengeti National Park (Schaller, 1968). His work described the unique hunting style of the cheetah, illuminating traits and strategies (Chapter 9). In 1969, Joy Adamson wrote about raising an orphaned cheetah, which she reintroduced into the wild in Kenya's Meru National Park (Adamson, 1969). Adamson's reports included the first close observations of birthing, cub behavior, and development (Adamson, 1972).

1970s—THE NEED FOR CONSERVATION ACTION IS RECOGNIZED

Learning About the Cheetah—Early Research Studies in Africa

In the 1970s, Randal Eaton shared additional insight into cheetah behavior, describing breeding and hunting behaviors, social structure, and prey preferences he observed in Kenya (Eaton, 1974; Chapters 8 and 9). In the Serengeti, George and Lory Frame studied behavioral ecology to evaluate the status and survival of cheetahs. They recorded hunting methods, social interactions, maternal behavior, family structure, mother–cub interactions, coexistence with other predators, and developed a methodology for identifying cheetahs by spot patterns on the face (Chapter 32). Their method of weighing carcasses after meals and noting the portion consumed became the norm for many cheetah feeding ecology studies (Frame and Frame, 1981).

In 1975, Norman Myers was the first to publish about the range wide decline in cheetah due to habitat loss and human-wildlife conflict (Myers, 1975). While there were perhaps 40,000 wild cheetahs in 1960, there were reportedly fewer than 20,000 by 1975, and of those, fewer than 3,000 in Africa's protected areas (Myers, 1975; Chapter 19). The reduced numbers were attributed to conflict with larger predators inside protected areas and conflict with the growing human populations outside protected areas (Myers, 1975). It was recognized that the mere existence of protected areas was insufficient to guarantee the long-term survival of this wide-ranging carnivore. Myers reported that cheetahs occurred at low densities with a limited distribution in only a small portion of sub-Saharan Africa and was the first to call for action "in the near future to reverse this decline."

Myers voiced his concerns over a growing African human population and the human disturbances impacting Africa's wildlife. In particular, carnivores like the cheetah were under pressure due to the potential threat they posed to livestock. In 1975, Africa's population was 450 million people and growing by 3.5%–4.5% per year, exerting unsustainable pressures on wild lands and wildlife. He reported that livestock farmers in Kenya, Namibia, and Zimbabwe were motivated to engage in cheetah persecution as they were receiving "compensation revenue" through the sales of skins (Myers, 1975). The countries Myers considered having the greatest potential for cheetah conservation initiatives were Botswana, Kenya, Namibia, Tanzania, and Zimbabwe, and sustainable land management was encouraged to balance the needs of wildlife, people, livestock, and the land (Myers, 1975).

Conflict with African Farmers Acknowledged

When CITES (1975) put an end to the export of wild cheetahs, Namibian farmers no longer had a market for trapped cheetahs (Marker-Kraus et al., 1996). As a result, trapped cheetahs generally were killed, but a few were translocated to national parks and game reserves (Marker-Kraus et al., 1996; Chapter 20). To provide care for cheetahs captured in conflict with farmers, the Pretoria Zoo, in partnership with Anne Van Dyk, developed the De Wildt Cheetah and Wildlife Centre in South Africa in 1971 (renamed Anne Van Dyk Cheetah Centre in 2014). In the years to follow, they also became the most successful breeding center in the world, providing captive bred cheetahs to the world's zoos (Marker-Kraus, 1990; Chapter 22).

Conflict between the farmers and cheetahs continued (Chapter 13). During the 1980s, Namibian game and livestock farmers reported killing over 800 cheetahs per year (CITES, 1992). The problems facing wild cheetahs were brought forward to the US conservation community in 1977, when Marker conducted research in Namibia on cheetah rehabilitation, and learned firsthand about the cheetah and farmer conflict (Marker-Kraus and Kraus, 1990). Cheetahs in South Africa and Zimbabwe were confronted with similar issues (Marker-Kraus and Kraus, 1990), while those in Kenya and Tanzania faced habitat loss, illegal snaring, and poaching (Myers, 1975).

Cooperative Captive Cheetah Programs Begin

In response to the world's declining biodiversity, the United States (US) Endangered Species Act (ESA, 1973) and the World Conservation

Union's (IUCN) Convention on International Trade in Endangered Species of Wild Fauna and Flora (CITES, 1975) were passed (Chapter 21). As a result, sourcing wild cheetahs for zoo exhibits stopped, and zoos began to collaborate through managed breeding programs to maximize genetic diversity, promote conservation, and educate the public (Chapters 22 and 23).

By the early 1970s, after returning from Kenya, Eaton began developing safari parks in the United States. His recommendation was for large habitats to foster better environments for breeding cheetahs. With this plan, he helped develop several safari parks in the United States, including the Wildlife Safari in Oregon, where Laurie Marker began working with cheetahs in 1974. Wildlife Safari became one of the first successful breeding facilities in the United States. During the same period, the San Diego Wild Animal Park in California and Whipsnade Safari Park in the United Kingdom also were developed. These facilities were some of the few facilities in the world to begin breeding cheetahs successfully.

1980s—GENETIC RESEARCH AND CAPTIVE MANAGEMENT

Reproductive Difficulties and Genetic Discovery

In the 1980s, zoos were struggling to breed cheetahs (Chapter 22). In 1983, groundbreaking research identifying the remarkable genetic uniformity of cheetahs was published by Drs. Stephen O'Brien and David Wildt from the US National Institute of Health, and Dr. Mitch Bush from the Smithsonian Institution's National Zoo. The cheetah's loss of genetic diversity was attributed to a historic bottleneck that was postulated to have occurred over 10,000 years earlier (O'Brien et al., 1983; Chapter 6). During this initial study, sperm samples that were collected showed 70% abnormalities and were considered an effect from the limited genetic diversity

(Wildt et al., 1983; Chapter 27). Since then, multiple studies, ranging from skin graft acceptance to whole genome analysis, have supported the findings of low genetic diversity (Chapter 6). And extensive research on reproductive physiology has since tackled major breeding-related questions (Chapter 27).

In 1983, a male cheetah traded from another zoo introduced a feline corona virus causing feline infectious peritonitis (FIP) to the population of the Wildlife Safari in Oregon (Chapters 6 and 25). The resulting high morbidity and mortality was believed to be facilitated by the genetic vulnerability (O'Brien et al., 1985). This alerted the captive managers to additional risks for the species and a need for increased clinical management (Chapter 24).

Captive Cheetah Management and the Cheetah Species Survival Plan® (SSP)

In 1982, Wildlife Safari hosted the first US national cheetah meeting, bringing together zoos willing to cooperate on breeding. The foundation for managing demographic structure in captive cheetahs began with the first regional cheetah studbook for North America (Marker, 1983), followed in 1988 with the first International Cheetah Studbook (Marker-Kraus, 1990). The studbook is a registry that lists all known animals belonging to zoos and private facilities, thus creating the preconditions for selecting breeding animals. Marker has maintained the International Cheetah Studbook since its inception (Chapter 22).

In 1984, the North American Cheetah Species Survival Plan® (SSP) of the American Association of Zoos and Aquariums (AZA) was developed (Chapter 23). The SSP brought collaborative research and management to the forefront and helped develop comprehensive plans to conserve captive and wild cheetahs.

In 1987, under the leadership of the Cheetah SSP Species Coordinator, Jack Grisham, and Dr. Ulysses Seal, founding chair of IUCN's

FIGURE 1.1 (A) Dr. Stephen O'Brien, geneticist, National Cancer Institute; Dr. Linda Munson, veterinarian pathologist, University of California at Davis; Jack Grisham, AZA SSP Cheetah Coordinator, St. Louis Zoological Gardens; Dr. Laurie Marker, International Cheetah Studbook Keeper and Director, Cheetah Conservation Fund, Namibia; Dr. David Wildt, reproductive physiologist, Smithsonian Institution's National Zoological Park worked closely in collaborative cheetah research. (B) Participants at the first Global Cheetah Meeting in South Africa. (C) Cheetah census workshop meeting participants in Tanzania. (D) Participants at the Southern Africa Rangewide Cheetah and Wild Dog Regional Meeting, Botswana, 2007.

Conservation Breeding Specialist Group (CBSG), a national meeting on cheetahs convened. Here, the Cheetah SSP designated the US captive cheetah population as a research population (Grisham and Lindburg, 1989), and the first systematic research plan was designed and implemented. The initial 3-year, multidisciplinary research project provided a basic understanding of cheetah biology and was a critical step in forming conservation strategies for *ex situ* populations. Its results appeared in a special edition of *Zoo Biology* (Wildt and Grisham, 1993). Fig. 1.1A shows some of the research collaborators involved in designing and implementing the research plan.

Cheetah Studies in Africa

In 1980, photographer and tour guide David Drummond reintroduced 3 orphaned cheetah cubs in the Maasai Mara National Reserve (Drummond, 2005). The female cub started a long lineage of cheetahs that were resident at the Governor's Camp area of the Maasai Mara. Drummond was one of the first to present the issues of poaching, wildlife interactions, and challenges of pastoral communities in terms of predators.

Tanzania—The Serengeti Research Project

In 1980, Dr. Tim Caro and his students continued the longitudinal study of cheetah behavior initiated by the Frames in the Serengeti National Park. Today, the research program is overseen by the Zoological Society of London (ZSL) through Dr. Sarah Durant, making it the longest-running cheetah research project. Caro's team also was interested in understanding wild versus captive behavior and used the Serengeti population to provide baseline data on breeding behavior and mother cub interactions (Chapter 9). *Cheetahs of the Serengeti Plains: Group Living in an Asocial Species* (Caro, 1994) has been the primary reference book on cheetah behavioral ecology since its publication.

During the same time frame, Hamilton studied the ecology of cheetahs in sub-Saharan Africa. His findings showed that, albeit in low densities, cheetahs were persisting even in areas where they were predicted to be extinct due to rising human populations. He reported cheetahs to be remarkably successful predators, well adapted to coexistence with nomadic pastoralists in arid and semiarid lands over large areas (Hamilton, 1986).

1990s—POPULATION RESEARCH STUDIES, AFRICAN CONSERVATION PROGRAMS, HEALTH ANALYSES OF WILD CHEETAHS, AND POPULATION VIABILITY ANALYSES

Range-Wide Population Studies

At the end of the 1980s there was little understanding of the population distribution or the ecology and biology of healthy, free-ranging cheetahs outside of protected areas. In the 1990s, Paula Gros undertook cheetah population surveys in East Africa (Gros, 1996, 1998, 2002; Gros and Rejmanek, 1999) and provided a rich insight into population distribution. During the same period, the cheetah's global status was summarized using country specialist information gathered in the 1980s from all range countries (Marker, 1998).

The Beginning of African Conservation Programs

Namibia—The Cheetah Conservation Fund (CCF)

In response to accumulating evidence about the cheetah's habitat and population decline and to gain knowledge on cheetahs in nonprotected areas, Laurie Marker founded the Cheetah Conservation Fund (CCF) in 1990 and set up an international field research and education center along with a model farm in Namibia. Marker began working with Dieter Morsbach, research scientist from the Namibian Ministry of the Environment (MET), and the livestock farming communities who were trapping and killing high numbers of cheetahs each year (Chapter 13). In 1991 an extensive survey of Namibian rural farming communities provided a better understanding of the threats to the cheetah and the techniques employed to prevent livestock depredation by cheetahs (Marker-Kraus et al., 1996; Marker et al., 2005). This community survey—along with health, disease, reproduction, genetic, and ecological surveys on Namibian cheetahs—provided the groundwork for many cheetah-range country programs (Marker et al. 2010).

CCF also launched national and international outreach and education programs to raise awareness of the cheetah's vulnerable status (Marker and Boast, 2015; Chapter 18). In 1994, CCF launched the first of many African livestock guarding dog programs in cooperation with Dr. Ray Coppinger and the Livestock Guarding Dog Association of America (Chapter 15).

South Africa—Cheetah Outreach

In 1997, Cheetah Outreach (CO), near Cape Town, was founded by Annie Beckheling and Mandy Schumann. Launched as an educational cheetah encounter facility, its focus is conserving South Africa's wild cheetahs. Adapting CCF's educational outreach programs, Cheetah Outreach began to educate school children and the public and developed a cheetah ambassador program (Chapter 28) working closely to develop nutritional guidelines for captive cheetahs (Chapter 26). In addition, they developed a livestock guarding-dog program and work closely with livestock farmers to help reduce conflict.

Systematic Biobanking and Reproductive Research

In 1994, an international research collaboration between the Namibian MET, a SSP research team, and CCF initiated the investigation of links between reproductive traits, nutrition, and diet. CCF's Genome Resource Bank, which included the first "field banked" sperm samples, was initiated (Crosier et al., 2007; Chapter 27). At the same time, reproductive research on cheetah males was being conducted at a few zoos in the United States (Wildt and Grisham, 1993) and in South Africa at the De Wildt Cheetah and Wildlife Center (Bertschinger et al., 2008; Chapter 27). Data on the basic biology of the female cheetah took longer to complete, with the first results in 2011 bringing insight into ovarian development and reproductive cycling (Crosier et al., 2011; Wachter et al., 2011; Chapter 27).

In 1996, Dr. Linda Munson, cheetah SSP veterinarian pathologist, trained Namibian veterinarians and field biologists in systematic sample collection and assisted in developing sample collection research protocols that led to long term, collaborative global disease studies (Chapter 25). Further training brought together pathologists from the United States, Europe, and South Africa.

Population, Habitat, and Viability Analysis (PHVA)

In 1996, the cheetah SSP sponsored a Population, Habitat and Viability Assessment (PHVA; Chapter 38) for the Namibian Cheetah and Lion. It was hosted by CCF and gathered the IUCN CBSG, the IUCN Cat Specialist Group, the MET, and members of the Cheetah SSP. The workshop provided a platform for Namibian farmers, wildlife officials, international scientists, and other stakeholders and set forth a strategy for managing cheetahs in Namibia while addressing issues affecting neighboring cheetah-range countries (Berry et al., 1997). The first comprehensive conservation plan for managing the wild Namibian cheetah (Berry et al., 1997), and the first National Cheetah Plan followed (Nowell, 1996).

2000s—CHEETAH CONSERVATION PROGRAMS, RANGE-WIDE WORKSHOP, AND PROGRAMS, OFF THE BEATEN TRACK

Expansion of Cheetah-Specific Conservation Programs Across the Cheetah's Range

The new millennium saw the development and expansion of cheetah conservation programs and new areas of research.

Kenya—CCF, Kenya/Action for Cheetah, Kenya (ACK)

In 2001, the Kenyan government and its citizens voiced concern about the decline of their cheetah population. The decline was attributed to a reduction in wild prey caused by poaching and the transition from large, collective ranches to smaller farms (Chapter 11), and also to habitat fragmentation (Chapter 10). Although cheetahs had been photographed extensively in the Maasai Mara (Ammann and Ammann, 1984; Scott and Scott, 1998), there was no cheetah conservation work being undertaken in the country. To address the void, Mary Wykstra and Laurie Marker developed CCF Kenya in 2001, which later became Action for Cheetahs in Kenya (ACK).

Iran—Iranian Cheetah Society (ICS)

In 2001, Mohammad Farhadinia, Kaveh Hobeali and Morteza Eslami founded the Iranian Cheetah Society (ICS), an NGO at the forefront of cheetah conservation in Iran (Chapter 5).

Zimbabwe—Cheetah Conservation Program and Cheetah Conservation Project

In 2002, following long-term cheetah studies from Vivian Wilson, in Zimbabwe, a cheetah conservation program was developed under the leadership of Netty Purchase, the carnivore project coordinator from the Marwell Zimbabwe Trust, and Verity Bowman from the Dambari Trust. In 2012, Dr. Esther van der Meer launched the Cheetah Conservation Project Zimbabwe.

Namibia—Cheetah Research Project

In 2002, the Leibniz Institute of Zoo and Wildlife Research (IZW) from Germany established the Cheetah Research Project in the southeast of Namibia under the direction of Dr. Bettina Wachter.

Botswana—Cheetah Conservation Botswana (CCB)

In 2004, Rebecca Klein, Anne-Marie Houser, and Dr. Kyle Good established Cheetah Conservation Botswana (CCB) with the assistance of the Mokolodi Wildlife Foundation and CCF. Their first research camp, at Jwaneng Diamond Mine's game reserve, monitored the reserve's cheetah population, and conducted outreach, education, and research on nearby farmland. CCB developed an administration and education base at Mokolodi Nature Reserve near Gaborone, followed by a field station, model farm, and education center on farmland near Ghanzi.

Carnivore Projects

In addition to the afore-mentioned cheetah projects, several carnivore projects have put a strong emphasis on cheetah conservation. These programs include: Endangered Wildlife Trust (EWT, South Africa), The Africat Foundation and N/a'an ku sê (Namibia), Botswana Predator Conservation Trust, Tanzanian Wildlife Research Institute, and Ruaha carnivore project (Tanzania).

Range-Wide Workshops and Collaboration

In 2001, the Cheetah SSP joined with *in situ* cheetah conservation efforts and sponsored the first Global Cheetah Action Plan Workshop in cooperation with the CBSG, CBSG South Africa, and CCF. Over 50 people from 11 countries attended the workshop in South Africa presenting on *in situ* and *ex situ* cheetah research (Fig. 1.1B). Working groups convened to discuss census research, protection of cheetahs outside protected areas, education and communication, and health and viability of the *ex situ* population. The findings provided the basis for the first Global Cheetah Action Plan that helped link research initiatives and enhance collaborations (Bartels et al., 2002a). This group called itself the "Global Cheetah Forum" (GCF).

Keeping the momentum, a second GCF took place in 2002. Participants included collaborators working on the Asiatic cheetah and members of the IUCN Cat Specialist Group. The highest priority was completing a census of free-ranging cheetahs to determine how and where range-wide conservation efforts could be implemented (Bartels et al., 2002b). In addition, the Forum members determined that conservation education and training programs should continue to be a top priority in range countries (Chapter 18).

Following this meeting in 2002, cheetah conservation organizations and representatives of the South African farming community met formally for the first time and developed the National Cheetah Conservation Forum (NCCF). Led by a team from the De Wildt Cheetah and Wildlife Centre, EWT, several universities, the National Research Foundation, the Agricultural Research Council and other governmental institutions, new research and conservation initiatives were launched, including a census of the South African cheetah populations.

Part of South Africa's approach to reducing conflict with livestock and game farmers was to begin a large-scale reintroduction program into private game reserves (Chapter 20). The cheetah populations on these game reserves needed to be artificially connected through animal movement. To achieve this, the cheetah metapopulation strategy was launched in 2009, and is managed by the EWT. In addition, these private reserves allowed for additional ecological studies on cheetah in these protected areas (Chapter 8).

The next GCF meeting took place in 2004, in the Serengeti in Tanzania with over 30 delegates from 7 countries (Fig. 1.1C). The SSP-sponsored "Cheetah Census Technique Development Workshop" was hosted by Dr. Sarah Durant from the ZSL and Wildlife Conservation Society (WCS). Census techniques for acquiring reliable, quantitative information on cheetah distribution and their numbers across Africa were discussed, and a cheetah census technique manual was developed to standardize "best practice" guidelines (Bashir et al., 2004).

Building on this collaboration, in 2005 a southern African Cheetah Regional Workshop brought over 30 people from 6 countries to CCF in Namibia (Dickman et al., 2006). This workshop was facilitated by IUCN CBSG southern Africa and moderated by cochair of IUCN's Cat Specialist Group, Dr. Christine Breitenmoser. The aims were to assess and evaluate accomplishments in the southern African region and to set new objectives. Key determinations appeared in a special issue of Cat News (Breitenmoser and Breitenmoser, 2007).

Following the 2005 workshop, the Cat Specialist Group in Switzerland undertook the Cheetah Compendium (http://www.catsg.org/cheetah/20_cc-compendium/index.htm). It is a web-based communication tool that houses a library of information, data, documents, maps, and other material relevant to the conservation of the cheetah.

Range Wide Cheetah Program

The Range Wide Conservation Program for Cheetah and African Wild Dogs (RWCP) was established in 2007 through a collaboration of Canid and Cat Specialist Groups of IUCN, and led by Drs. Sarah Durant and Rosie Woodruff. Under this program, regional cheetah workshops for southern and eastern Africa took place in 2007 (IUCN/SSC, 2007a,b; Figs. 1.1D and 1.2A), with initial meetings for central, northern, and western Africa in 2012 (IUCN/SSC, 2012; Fig. 1.2B) and a follow up meeting for southern Africa in 2015 (RWCP and IUCN/SSC, 2015; Fig. 1.2C). Range-wide priority conservation plans were developed with government officials for both the cheetah and African wild dog, drawing on the similarities of these species' conservation requirements (Chapter 39).

Drawing on regional plans, national workshops developed country-specific plans in many cheetah-range states (Chapter 39). These workshops led to increased government awareness and support throughout the cheetah's range—as well as ongoing census research allowing for mapping of cheetah populations by national and regional experts—and an understanding of the threats the species was facing and will likely face in the future (Durant et al., 2017).

The lack of local capacity was a key finding of regional plans. To address this challenge, CCF, in cooperation with the Howard G. Buffett Foundation and the RWCP, trained more than 300 government wildlife officials, university professors, scientists, conservation managers, conservation NGO officers, and community extension officers from 15 cheetah-range countries between 2008 and 2011. The aim of the courses was to teach research techniques (Chapters 29–38) and to promote a unified and systematic approach to cheetah research and conservation (Marker and Boast, 2015).

FIGURE 1.2 (A) Participants at the East African Rangewide Cheetah and Wild Dog Regional meeting, Kenya, 2007. (B) North, West, and Central Rangewide Cheetah and Wild Dog Regional meeting participant in Niger, 2012. (C) Southern Africa Rangewide Regional Cheetah and Wild Dog meeting participants in South Africa, 2015. (D) International researchers and government officials at the International Workshop on the Asiatic Cheetah in Iran.

Off the Beaten Track—Other Cheetah Populations

Iran

Cheetahs were known to still be present in Iran after the Iranian Revolution ended in 1979 (Joslin, 1984), although little information was available on their population size or distribution before 2000. The current population is estimated at less than 50 adult and adolescent individuals (Durant et al., 2017; Chapter 5).

In 2001 the Iranian government arranged two separate meeting one headed by Dr. George Schaller from the Wildlife Conservation Society (WCS) and the other headed by Dr. Laurie Marker from the CCF to discuss options to save Iran's small population of Asiatic cheetah (*Acinonyx jubatus venaticus*). Attendees included government officials from the Department of Environment (DOE), the United Nations Development Program (UNDP) and various Iranian researchers involved with the Iranian cheetah. The outcome of the meetings provided support for Iran to begin working under a multiyear UNDP grant aimed at saving the critically endangered Asiatic cheetah population.

Two important meetings followed. In 2004, the Iranian Centre for Sustainable Development (CENESTA) hosted an International Workshop on the Conservation of Asiatic Cheetah, with participation by Asiatic cheetah conservation partners and local communities to discuss conservation strategies with stakeholders throughout the cheetah's Iranian range (Fig. 1.2D); and in 2010, an Iranian Cheetah Strategic Planning meeting reviewed the previous decade of work and planned Iranian's cheetah survival strategies for the following 5 years (Breitenmoser et al., 2010). An overview of the Iranian cheetah situation is found in Chapter 5.

NW Africa—Algeria

The first survey of cheetahs in Algeria was undertaken in 2005 in the Ahaggar National Park, Central Sahara (Busby et al. 2009; Wacher et al., 2005). Interviews with nomadic herders helped assess the nature of interactions between nomads, cheetahs, and other wildlife. A year later, the North African Region Cheetah Organization (Observatoire du Guépard en Régions d'Afrique du Nord (OGRAN) met in Tamanrasset, Algeria, for a 3-day conference to discuss conservation strategies in Algeria highlighting data collection, census techniques, training and education needed to conserve this critically endangered cheetah population. As a result, PhD fieldwork began in the Ahaggar National Park, collecting the region's first camera trap evidence of cheetahs (Belbachir et al., 2015).

India

The cheetah disappeared from India in 1956 (Divyabhanusinh, 1999). For some time, there have been discussions on reintroducing cheetahs to India (Ranjitsinh and Jhala, 2010). In 2009, the Wildlife Trust of India (WTI), headed by Chairman Dr. M.K. Ranjitsinh, hosted a team of experts including representatives from the IUCN (including its Cat Specialist Group, Reintroduction Specialist Group, and Veterinary Specialist Group), Oxford University's WildCRU, Cheetah Outreach, and CCF, along with Indian authorities and forestry directors from various regions.

They concluded the following: The original cause of the extinction of the cheetah in India had been adequately addressed; a network of protected areas had been established; and effective wildlife legislation, change in the conservation ethos, and nationwide awareness could lead to a successful cheetah reintroduction (Ranjitsinh and Jhala, 2010). However, the reintroduction project has been stalled indefinitely due to political issues concerning what subspecies could or should be used (Laing and Nelson, 2012; O'Brien, 2013).

International Attention to Illegal Wildlife Trafficking of Cheetahs—UAE/North Africa

By 2006, it was evident that there was need for a long-term plan for combatting illegal wildlife trafficking and for cheetah conservation awareness in Ethiopia. The Ethiopian Wildlife Conservation Authority assigned a task force to develop guidelines and recommendations for a sanctuary for wild orphan animals, and in 2010, the Born Free Foundation in Ethiopia built a sanctuary to hold cheetahs, lions, and other confiscated animals.

The first solicited report on illegal trade was compiled by Nowell to CITES in 2015 (Chapter 14). In 2016, at the CITES Convention of the Parties (CoP17), several resolutions were accepted to work toward the reduction of supply and demand for illegally trafficked cheetah cubs (CITES, 2016; Chapter 14). Work continues between the Horn of Africa and the Gulf States.

2015 ONWARDS—THE RECENT YEARS

In 2015, the AZA launched their Saving Animals From Extinction (SAFE) program to focus on conservation of the cheetah along with nine other species. The goal of SAFE is to restore healthy populations in the wild by connecting scientists with stakeholders and to identify threats, launch action plans, find new resources and engage the public (Chapter 23).

In mid 2015, the RWCP met again to update the southern African regional cheetah plan and population maps (RWCP and IUCN/SSC, 2015). This time the meeting was supported by SAFE with representatives from the AZA community, IUCN SSC Cat Specialist Group, seven national governments, and a multitude of cheetah research and conservation NGOs.

In December 2016, a 54 coauthored paper was published presenting the current estimate on cheetah numbers and distribution (Durant et al., 2017; Chapter 39). It highlighted that the cheetah populations continue to decline in range and number in addition to the need for further investment into their conservation. With 77% of the remaining 7100 adult and adolescent free-ranging population found outside protected areas, it is being recommended the species be uplisted to endangered status on the IUCN red list (Durant et al., 2017; Chapter 39).

CONCLUSIONS

The survival of the cheetah needs to be the responsibility of everyone, not just governments and conservationists. In light of largely human-caused global changes in the environment (Chapter 12), people have the responsibility to help ensure the availability of wide-ranging spaces for cheetah conservation (Chapter 17). This is only possible if the livelihoods (Chapter 16) of local populations living in the same habitat as the cheetah are also taken into consideration in a holistic conservation approach. We hope that the passion and cooperative efforts of active cheetah conservationists will be augmented by the world at large—everyday people who care about wildlife and cheetahs—to help the cheetah reverse indefinitely its fragile march toward extinction. The last chapters of this book Chapters, 39 and 40) will help define the way forward to help secure a future for the cheetah.

References

Adamson, 1969. The Spotted Sphinx. Collins; Harvill P, New York.

Adamson, 1972. Pippa's Challenge. Collins; Harvill P, New York.

Ammann, K., Ammann, K., 1984. Cheetah. Camerapix Publishing International, Nairobi.

Bartels, P., Berry, H.H., Cilliers, D., Dickman, A., Durant, S.M., Grisham, J., Marker, L., Munson, L., Mulama, M., Schoeman, B., Tubbesing, U., Venter, L., Wildt, D.E., Ellis, S., Freidmann, Y. (Eds.), 2002a. Global Cheetah Conservation Action Plan—final report from the workshop. Global Cheetah Conservation Action Plan—Workshop held at Shumba Valley Lodge in South Africa from the 27th to the 30th of August 2001.

Bartels, P., Bouwer, V., Crosier, A., Cilliers, D., Durant, S.M., Grisham, J., Marker, L., Wildt, D.E., Friedmann, Y. (Eds.), 2002. Global Cheetah Action Plan Review Final Workshop Report. (SSC/IUCN) Conservation Breeding Specialist Group, Apple Valley, MN.

Bashir, S., Daly, B., Durant, S.M., Förster, H., Grisham, J., Marker, L., Wilson, K., Friedmann, Y. (Eds.), 2004. Global Cheetah (Acinonyx jubatus) Monitoring Workshop. Final Workshop Report. (SSC/IUCN) Conservation Breeding Specialist Group. Endangered Wildlife Trust, Apple Valley, MN.

Belbachir, F., Pettorelli, N., Wacher, T., Belbachir-Bazi, A., Durant, S.M., 2015. Monitoring rarity: the critically endangered Saharan cheetah as a flagship species for a threatened ecosystem. PLoS ONE 10 (1), e0115136.

Berry, H., Bush, M., Davidson, B., Forge, O., Fox, B., Grisham, J., Howe, M., Hurlbut, S., Marker-Kraus, L., Martenson, J., Munson, L., Nowell, K., Schumann, M., Shille, T., Stander, F., Venzke, K., Wagener, T., Wildt, D., Ellis, S., Seal, U. (Eds.), 1997. 1996 Population Habitat Viability Assessment for the Namibian Cheetah and Lion. IUCN/SSC Conservation Breeding Specialist Group, Apple Valley, MN.

Bertschinger, H.J., Meltzer, D.G.A., van Dyk, A., 2008. Captive breeding of cheetahs in South Africa—30 years of data from the de Wildt Cheetah and Wildlife Centre. Reprod. Domest. Anim. 43 (Suppl. 2), S66–S73.

Breitenmoser, C., Breitenmoser, U. (Ed), 2007. The Status and Conservation of the Cheetah in Southern Africa. CAT News Special Issue NO 3. KORA, Switzerland.

Breitenmoser, U., Breitenmoser-Würsten, Ch., von Arx, M., 2010. Workshop on the Conservation of the Asiatic Cheetah. Cat News (No. 52).

Busby, G.B.J., Gottelli, D., Wacher, T., Marker, L., Belbachir, F., De Smet, K., Belbachir-Bazi, A., Fellous, A., Belghoul, M., Durant, S.M., 2009. Leopards and cheetahs in southern Algeria. Oryx 43 (3), 412–415.

Caro, T.M., 1994. Cheetahs of the Serengeti Plains: Group Living in an Asocial Species. University of Chicago Press, Chicago.

CITES, 1975. Convention on International Trade in Endangered Species of Wild Fauna and Flora Signed at Washington, DC (on 3 March 1973).

CITES, 1992. Quotas for trade in specimens of cheetah. Eighth Meeting of the Convention of International Trade in Endangered Species of Wild Fauna and Flora, pp. 1–5.

CITES, 2016, Species-specific matters: illegal trade in Cheetahs Acinonyx jubatus. COP17 Doc. 49, Seventeenth Meeting of the Conference of the Parties Johannesburg (South Africa), September 24–October 5, 2016.

Crosier, A.E., Comizzoli, P., Baker, T., Davidson, A., Munson, L., Howard, J., Marker, L.L., Wildt, D.E., 2011. Increasing age influences uterine integrity, but not ovarian function or oocyte quality in the cheetah (Acinonyx jubatus). Biol. Reprod. 85, 243–253.

Crosier, A.E., Marker, L., Howard, J., Pukazhenthi, B.S., Henghali, J.N., Wildt, D.E., 2007. Ejaculate traits in the Namibia cheetah (Acinonyx jubatus): influence of age, season and captivity. Reprod. Fertil. Dev. 19, 370–382.

Dickman, A., Marnewick, K., Daly, B., Good, K., Marker, L., Schumann, B., Ezequiel, F., de Jonge, M., Stein, A., Hengali, J., Eddins, S., Cilliers, D., Selebatso, M., Klein, R., Melzheimer, J., Beckhelling, A., Schulze, S., Carlisle, G. and Friedmann, Y. (Eds.), 2006. Southern African Cheetah (Acinonyx jubatus) Conservation Planning Workshop. Final Report. (SSC/IUCN) Conservation Breeding Specialist Group. Endangered Wildlife Trust, Johannesburg.

Divyabhanusinh, 1999. The End of a Trail: The Cheetah in India, Banyan Books, New Delhi.

Drummond, D., 2005. Queen of the Mara. Troubador Publishing Ltd, Leicester UK.

Durant, S.M., Mitchell, N., Groom, R., Pettorelli, N., Ipavec, A., Jacobson, A., Woodroffe, R., Bohm, M., Hunter, L., Becker, M., Broekuis, F., Bashir, S., Andresen, L., Aschenborn, O., Beddiaf, M., Belbachir, F., Belbachir-Bazi, A., Berbash, A., Brandao de Matos Machado, I., Breitenmoser, C., Chege, M., Cilliers, D., Davies-Mostert, H., Dickman, A., Fabiano, E., Farhadinia, M., Funston, P., Henschel, P., Horgan, J., de Iongh, H., Jowkar, H., Klein, R., Lindsey, P., Marker, L., Marnewick, K., Melzheimer, J., Merkle, J., Msoka, J., Msuha, M., O'Neill, H., Parker, M., Purchase, G., Saidu, Y., Samaila, S., Samna, A., Schmidt-Küntzel, A., Selebatso, E., Sogbohossou, E., Soultan, A., Stone, E., van der Meer, E., van Vuuren, R., Wykstra, M., Young-Overton, K., 2017. The global decline of cheetah and what it means for conservation. Proc. Natl. Acad. Sci. USA 114, 528–533.

Eaton, R.L., 1974. The Cheetah: Biology, Ecology and Behavior of an Endangered Species. Van Nostrand Reinholt, New York.

ESA, 1973. Endangered species act of 1973. As amended through the 108th Congress, Department of the Interior US Fish and Wildlife Service Washington, DC.

Frame, G., Frame, L., 1981. Swift and Enduring Cheetahs and Wild Dogs of the Serengeti. E.P. Dutton, New York.

Graham, A.D., 1966. East African Wildlife Services Survey—Extracts from the Report by Wildlife Services, Nairobi, Kenya.

Grisham, J., Lindburg, D. (Eds.), 1989. Cheetah Species Survival Plan Research Master Plan. American Zoo and Aquarium Association, Apple Valley, MN.

Gros, P.M., 1996. Status of the cheetah in Malawi. Nyala 19, 33–36.

Gros, P.M., 1998. Status of the cheetah Acinonyx jubatus in Kenya: a field-interview assessment. Biol. Conserv. 85, 137–149.

Gros, P.M., 2002. The status and conservation of the cheetah Acinonyx jubatus in Tanzania. Biol. Conserv. 106, 177–185.

Gros, P.M., Rejmanek, M., 1999. Status and habitat preferences of Uganda cheetahs: an attempt to predict carnivore occurrence based on vegetation structure. Biodivers. Conserv. 8, 1561–1583.

Hamilton, P.H., 1986. Status of the cheetah in Kenya, with reference to sub-Saharan Africa. In: Miller, S.D., Everett, D.D. (Eds.), Cats of the World: Biology, Conservation and Management. National Wildlife Federation, Washington DC.

IUCN/SSC, 2007a. Regional Conservation Strategy for the Cheetah and African Wild Dog in Eastern Africa. IUCN/SSC, Gland, Switzerland.

IUCN/SSC, 2007b. Regional Conservation Strategy for the Cheetah and African Wild Dog in Southern Africa. IUCN/SSC, Gland, Switzerland.

IUCN/SSC, 2012. Regional Conservation Strategy for the Cheetah and African Wild Dog in Western, Central, and Northern Africa. IUCN/SSC, Gland, Switzerland.

Joslin, P., 1984. Cited in Divyabhanusinh, The origin, range and status of the Asiatic (or Indian) cheetah or hunting leopard (Acinonyx jubatus venaticus). In: Proceedings of Cat Specialist Group Meeting, pp. 183–185 (Unpublished report).

Kitchener, A.C., Breitemoser-Würsten, Ch., Eizirik, E., Gentry, A., Werdelin, L., Wilting, A., Yamaguchi, N., Abramov, A., Christiansen, P., Driscoll, C., Duckworth, W., Johnson, W., Luo, S.-J., Meijaard, E., O'Donoghue, P., Sanderson, J., Seymour, K., Bruford, M., Groves, C., Hoffmann, M., Nowell, K., Timmons, Z., Tobe, S., 2016. A revised taxonomy of the Felidae. The Final Report of the Cat Classification Task Force of the IUCN Cat Specialist Group. Cat News Special Issue 11, p. 80.

Laing, A., Nelson, D., 2012. Project to ship cheetahs from Africa to India 'Totally misconceived'. The Telegraph, 9 May, 2011.

Linnell, J.D.C., Swenson, J.E., Anderson, R., 2001. Predators and people: conservation of large carnivores is possible at high human densities if management policy is favourable. Anim. Conserv. 4 (4), 345–349.

Marker, L.L., 1983. 1982 North American Regional Cheetah Studbook. Wildlife Safari, Winston.

Marker, L., 1998. Current status of the cheetah (Acinonyx jubatus). In: Penzhorn, B.L. (Ed.), A Symposium on Cheetahs as Game Ranch Animals. Onderstepoort, South Africa, pp. 1–17.

Marker, L., Boast, L., 2015. Human wildlife conflict 10 years later—lessons learnt and their application to cheetah conservation. Hum. Dimens. Wildl. 20 (4), 1–8.

Marker, L., Dickman, A., Macdonald, D., 2005. Perceived effectiveness of livestock guarding dogs placed on Namibian farms. J. Rangeland Manage. 58 (4), 329–336.

Marker, L., Dickman, A.J., Mills, M.G.L., Macdonald, D.W., 2010. Cheetahs and ranches in Namibia: a case study.

In: Macdonald, D.W., Loveridge, J. (Eds.), Biology andConservation of Wild Felids. Oxford University Press, Oxford, (Chapter 15).

Marker-Kraus, L., 1990. 1988 International Cheetah Studbooks. Smithsonian Press, Washington, DC.

Marker-Kraus, L., Kraus, D., 1990. Investigative trip to Zimbabwe and Namibia. Cat News 12, 16–17.

Marker-Kraus, L., Kraus, D., Barnett, D., Hurlbut, S., 1996. Cheetah Survival on Namibian Farmlands. Cheetah Conservation Publication, Windhoek.

Myers, N., 1975. The Cheetah, *Acinonyx jubatus*, in Africa. IUCN Monograph, Morges, Switzerland.

Nowell, K., 1996. Draft Namibian Cheetah Conservation Strategy. IUCN/SSC Cat Specialist Group. Ministry of Environment and Tourism, Namibia.

O'Brien, S.J., 2013. The "Exotic Aliens" controversy: a view from afar. J. Bombay Nat. Hist. Soc. 110 (2), 108–113.

O'Brien, S.J., Roelke, M.E., Marker, L., Newman, A., Winkler, C.A., Meltzer, D., Colly, L., Evermann, J.F., Bush, M., Wildt, D.E., 1985. Genetic basis for species vulnerability in the cheetah. Science 227, 1428–1434.

O'Brien, S.J., Wildt, D.E., Goldman, D., Merril, C.R., Bush, M., 1983. The cheetah is depauperate in genetic variation. Science 221, 459–462.

Ranjitsinh, M.K., Jhala, Y.V., 2010. Assessing the Potential for Reintroducing the Cheetah in India. Wildlife Trust of India; the Wildlife Institute of India, Noida; Dehradun (India), TR2010/001.

RWCP & IUCN/SSC, 2015, Regional Conservation Strategy for the Cheetah and African Wild Dog in Southern Africa (Revised and Updated, August 2015).

Schaller, G.B., 1968. Hunting behaviour of the cheetah in the Serengeti National Park, Tanzania. Afr. J. Ecol. 6 (1), 95–100.

Schreber, J.C.D., 1775. Die Säugethiere in Abbildungen Nach der Natur mit Beschreibungen. Wolfgang Walther, Erlangen, vol. 2(14).

Scott, J., Scott, A., 1998. Jonathan Scott's Safari Guide to East African Animals. Newpro UK Ltd, Farringdon.

Wacher, T., De Smet, K., Belbachir, F., Belbachir-Bazi, A., Fellous, A., Belghoul, M. & Marker, L. 2005. Sahelo–Saharan Interest Group Wildlife Surveys. Central Ahaggar Mountains (March 2005) (iv + p. 34).

Wachter, B., Thalwitzer, S., Hofer, H., Lonzer, J., Hildebrandt, T.B., Hermes, R., 2011. Reproductive history and absence of predators are important determinants of reproductive fitness: the cheetah controversy revisited. Conserv. Lett. 4, 47–54.

Wildt, D.E., Bush, M., Howard, J.G., O'Brien, S.J., Meltzer, D.J., 1983. Unique seminal quality in the south African cheetahs and a comparative evaluation in the domestic cat. Biol. Reprod. 29, 1019–1025.

Wildt, D.E., Grisham, J., 1993. Basic research and the cheetah SSP® program. Zoo Biol. 12, 3–4.

Woodroffe, R., Ginsberg, J.R., 1997. Past and future causes of wild dogs' population decline. In: Woodroffe, R., Ginsberg, J.R., Macdonald, D.W. (Eds.), The African Wild Dog Status Survey and Conservation Action Plan. IUCN, Gland.

Wrogemann, N., 1975. Cheetah Under the Sun. McGraw-Hill Book Company, Johannesberg.

2

History of the Cheetah–Human Relationship

Benison Pang*, Blaire Van Valkenburgh*,
Kenneth F. Kitchell, Jr.**,†, Amy Dickman‡,
Laurie Marker§

*University of California at Los Angeles, Los Angeles, CA, United States
**University of Massachusetts Amherst, Amherst, MA, United States
†Louisiana State University, Baton Rouge, LA, United States
‡University of Oxford, Tubney, Abingdon, United Kingdom
§Cheetah Conservation Fund, Otjiwarongo, Namibia

The relationship between cheetahs (*Acinonyx jubatus*) and humans has persisted over millennia, across the entire world. In Europe and even more so in Asia and Africa, cheetahs have been significant components of national culture.

CHEETAHS IN PREHISTORY

It has been suggested that *Acinonyx pardinensis*, the extinct "giant cheetah," of which the modern-day cheetah is the closest living relative, may have been an important carcass provider to Early Pleistocene humans (2.6–0.8 million years ago), both because modern cheetahs are easy to steal from and do not fully consume their kills (Hemmer et al., 2011). However, the anatomy, and consequently the predatory habits of *A. pardinensis*, were most likely different from those of the modern cheetah, and there are no data on how *A. pardinensis* consumed carcasses (Chapter 3). Additionally, the complexity of the carnivore guild in the Early Pleistocene means that using the behavior of modern *A. jubatus* to inform our view of *A. pardinensis* may be somewhat premature (Cherin et al., 2014). Although Europe contains the earliest-known representations of cheetahs in the Chauvet caves of France (32000–26000 BCE; Bradshaw Foundation, 1999), outlines of late Neolithic (5300–4500 BCE) felines can be found on the desert floor at 'Awja 1 near the Jordan/Saudi border (Fujii, 2014). Fujii makes a good case for these outlines being those of spotted felines, despite the relative crudeness of the outlines.

CHEETAHS IN AFRICA

Much like other felids, the cheetah was of religious significance to ancient Egyptian mythology. As early as the 1st dynasty (around 3rd millennium BCE) the ancient Egyptian pantheon included the first feline goddess Mafdet, who was portrayed at times as a cheetah. Mafdet's varied roles included being a representation of justice and execution, a protector against snakes, as well as a guardian of the pharaoh's chambers. Even though Mafdet was eventually supplanted by the better-known cat goddess Bast (also translated into Bastet), her image as a protector remained. Ancient Egyptians also believed that cheetahs were the animals that would carry away the spirits of the pharaohs when they passed away.

Egyptian tombs commonly displayed illustrations of cheetahs from the 3rd dynasty (mid-3rd millennium BCE) onwards, and they seem to have been extremely popular during the 18th and 19th Dynasties (early to mid-2nd millennium BCE), based on the sheer prevalence of artistic depictions during that time (Guggisberg, 1975). For example, King Tutankhamen's tomb included a statue of the king on a cheetah, among other cheetah-related artifacts (Wrogemann, 1975). A funerary couch in the antechamber of King Tutankhamun's tomb was initially identified as a cheetah (Wrogemann, 1975). However, even though the felid appears to have tear marks, the overall appearance (head shape, vestigial mane, and absence of spots) has a much higher resemblance with a lion (*Panthera leo*). Misidentification of big cats in art has been a long-standing problem (Castel, 2002), in particular in cultures which do not have specific names to distinguish the species referred to; so it is not surprising to find that the cheetah is also vulnerable to this issue.

Another early representation of a relationship between cheetahs and humans can be found in the reliefs from the Deir el-Bahari temple in northern Africa (c.1500 BCE) (Allsen, 2006). The imagery is that of an expedition in which cheetahs are illustrated with collars and leashes, suggesting taming of some sort (Fig. 2.1).

WAR TROPHIES
DEIR EL BAHARI.

FIGURE 2.1 **Cheetahs in Egyptian art.** A depiction of cheetahs from the Deir el-Bahari temple in Egypt. *Source: Wellcome Library, London under a CC-BY 4.0 license.*

Cheetahs displaced hunting dogs as the hunting animal of choice in ancient Egypt (Bodenheimer, 1960), and were also used for hunting in Libya during the time of the pharaohs (Harper, 1945). Cheetahs were not necessarily used for hunting to obtain food, but for the sporting challenge, an activity known as coursing (Guggisberg, 1975; Kingdon 1977). This prioritization of sport over food acquisition is similar to falconry, in which the thrill is in the hunt rather than the carcass, and was likely limited to royalty.

Relative to northern Africa, the relationship between cheetahs and humans in southern Africa does not seem to have carried as much social significance. Bushmen, left behind comparatively little artwork of cheetahs or other felids relative to other animals, although rock etchings of cheetahs that are 2000–6000-years old have been found at Twyfelfontein in Namibia (Marker, personal observation). Other uses of the cheetah in southern Africa include divination, the ritual practice of gaining insight into the unknown. In divination, different animals typically represent different traits. With the cheetah's reputation for speed, it is unsurprising that the foot bones of a cheetah represent "fleet-footedness" (Wrogemann, 1975). It is interesting to note that modern San hunters of southern Africa have been depicted as hunting alongside cheetahs; however, San hunters are not known to use cheetahs as hunting companions, but rather follow behind cheetahs and take their kills away.

CHEETAHS IN ASIA

Cheetahs were widely popular in Asia, particularly as hunting animals. During the 16th century, Akbar the Great of the Mughal Empire (covering the Indian subcontinent) used cheetahs to hunt antelope (Fig. 2.2) and owned as many as 1000 at a time and over 9000 in total, during his 49-year reign (Allsen, 2006). They were referred to as Khasa, or Imperial cheetahs (Allsen, 2006). While Akbar the great's captive cheetah numbers are the most striking example,

the Mughal Empire was not alone in owning cheetahs. Allsen (2006) notes that the Persian epic "Shanameh," written around the first millennium CE, refers to cheetahs in a hunting context, suggesting their use in Persia in pre-Islamic times, centuries prior to their use in the Mughal Empire (Umayyad Caliphate:7th century AD). During the Crusades (11–13th century), tamed wild cheetahs, trained to hunt, were also seen pursuing gazelles in Syria and Palestine (Guggisberg, 1975). Azerbaijan khans, as well as Armenian and Kartlian royalty owned hunting cheetahs up to the 14th century (Nowell and Jackson, 1996). The Ali Qapu palace in Isfahan, Iran also contains a depiction of cheetahs among its wall paintings, dating from around the 16th century (Fig. 2.3A). To this day, Iran is the only remaining haven for the critically endangered Asiatic cheetah, whose numbers have plummeted in recent years (Farhadinia et al., 2012; Chapter 5).

As described in Allsen (2006), the training for hunting cheetahs during the Mughal empire (16–19th century) was complicated, taking up to a year. In the Arabic tradition, hunting with cheetahs required training that was broken down into several stages. The first stage of training involved using means, such as food and sleep deprivation to create submission in the cheetah. The submissive state of the cheetah was required to acclimatize the potential hunting animal to human presence; this stage involved frequent human contact, whether by walking it around on a leash or simply by having constant human interaction. Acclimatization to humans was followed by horseback riding, where the cheetah was trained to leap onto a pillion (a cushioned seat attached to the back of a saddle). Around this time, the cheetah was then slowly allowed to indulge in its natural instincts, by being given the opportunity to lap up the blood of a newly slaughtered animal for instance. In the final stages of the training, the cheetah was released upon a prey animal that had previously been driven to exhaustion. Both Akbar's cheetahs and cheetahs of the Arabic caliphs received

FIGURE 2.2   **Emperor Akbar hunting with cheetahs (1602).** *Source: Scanned by AshLin, 2005. In: Kothari, A.S., Chhapgar, B.F. (Eds.), Treasures of Natural History. Bombay Natural History Society and Oxford University Press, Mumbai, p. 16.*

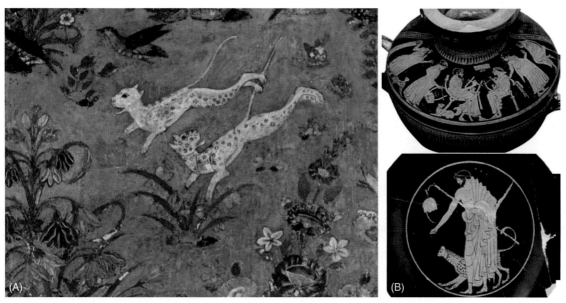

FIGURE 2.3 **Art depicting cheetah in early cultures.** (A) A wall painting of cheetahs at the Ali Qapu Place in Isfahan, Iran. (B) Early representations of cheetah with humans on pottery. Top: c.490 BC, made in Attica (historical region, part of modern-day Greece). Note the presence of a monkey to the right, probably another exotic pet. Bottom: c.480 BC, made in Attica. *Source: Part A, Adina L. Savin; part B (top) attributed to the Apollodoros Painter. Information taken from the British Museum, catalog number 1836,0224.230, part B (bottom) attributed to The Pig Painter. Information taken from the British Museum, catalog number 1864,1007.85. Trustees of the British Museum.*

substantial food rations. Akbar's cheetahs were apparently ranked in a system and food was rationed accordingly, with higher-ranked cheetahs being given larger portions. All hunting cheetahs were drawn from wild populations, leading to a sharp decline in the number of Asiatic cheetahs (Chapters 5 and 22). By 1927, the cheetah was said to be rare in India, such that most cheetahs had to be imported from Africa (Finn, 1929).

In China, there appears to have been two peaks of popularity in owning cheetahs for human use, with the first being as early as the 7th century. Allsen (2006) describes mural paintings in tombs of the Tang Dynasty (618–907 CE) that depict hunting cheetahs, and Turkestan emissaries were reported to have sent gifts of cheetahs to China as part of an appeal for military support. The second surge of popularity occurred around the 13th century CE and apparently persisted through to the Ming Dynasty (1368–1644 CE).

During this time, many live cheetahs were gifted or used as barter beyond their natural range to both China and western Europe (Allsen, 2006). Sources for these gifted cheetahs into China included Persia (Allsen, 2006) and Sultan Abu Said Mirza of the Timurid Dynasty.

Outside China and subsequent to the Tang Dynasty, the Liao Dynasty (907–1125 CE), an empire in eastern Asia including parts of present-day China, Korea, Mongolia, and Russia, was also said to have received gifts of cheetahs. The Mongol Empire under Kublai Khan (1215–1294 CE) used cheetahs to hunt antelope and fallow deer (Wrogemann, 1975), as evidenced by the tomb figurines portraying both cheetahs and caracals sitting on horseback, as was commonplace for hunting cats. It has also been suggested that cheetahs were used by ancient Mesopotamians for hunting (Bodenheimer, 1960). Cited as evidence is a

Mesopotamian seal from 3000 BCE that shows a hunting party with an animal, thought to be a cheetah (Guggisberg 1975; Van Buren 1939). However, as was the case with the funerary couch of King Tutankhamun, the morphology of the animal in question is notably different from that of a cheetah long muzzle and pointed, erect ears) and is more likely that of a hunting dog (Van Buren, 1936).

CHEETAHS IN EUROPE

European use of cheetahs was popular over a short period of time and appears to have been limited to just a few civilizations, most notably the Greeks, who were apparently the earliest adopters of cheetahs in European history. During the golden age of Athens, around the 5th century BCE, a wealthy noble class arose. The young, leisured males of this class were frequently depicted on vases, often showing off their pets, which were status symbols and often cost a great deal of money. A handful of vases depicted these youths with spotted cats, sometimes leading them on leashes, holding them in their laps, bringing them to music lessons (Fig. 2.3B, top), and introducing them to pet dogs (Ashmead, 1978). Despite the ancients' lack of clarity in differentiating spotted cats (Kitchell, 2014, pp. 28, 29, 107), these cats are unlikely to be leopards (*Panthera pardus*), as leopards on Greek vase paintings were usually depicted as decorative or heraldic animals as opposed to pets. Also, some well-rendered drawings show the characteristic head shape, rounded ears, and long legs of a cheetah (Fig. 2.3B, bottom). The cats were probably not used for hunting, since there is no such mention in ancient Greek literature. It is also worth noting that most of the animals depicted on the vases are rather small. It may be that they were acquired as kittens to be hand raised and tamed; the lifespan of these wild animals in a city like Athens may have been short.

The Greek artwork provides hints of the role cheetahs played in society. Notable examples include two vases (Ashmead, 1978, ill. 8 and 9), which seem to indicate bids being offered for the cheetah on display, and a pelike by the Tyszkiewicz Painter, in the Musée Boulogne-sur-mer (n. 134, Ashmead, 1978, ill. 12), which depicts the cheetah's role as an exotic gift between male lovers.

In 1231, Frederick II, a Holy Roman Emperor, was also observed to have owned cheetahs as part of his court menagerie (Sunquist and Sunquist, 2002). The sport of hunting with cheetahs then spread to Italy, becoming a status symbol among the elite (Harper, 1945). In 1479, the Italian Duke of Ferrara gifted cheetahs to King Louis XI of France, and by the end of the Renaissance, hunting cheetahs were fashionable in both countries. Further east, tame cheetahs were recorded in Prague in the 16th century, and Emperor Leopold of Austria attempted the use of hunting cheetahs around 1700, although without much success (Guggisberg, 1975).

MODERN DAY

Since the early 1900s the use of cheetahs for hunting declined, save for sporadic instances, with social media outlets, such as Facebook and YouTube, providing evidence of cheetahs being used to hunt deer on rare occasions. In some regions cheetahs are still kept as exotic pets, especially in the Arab states of the Persian Gulf (CITES, 2014) and are often wild-born animals, illegally sourced from the wild (Chapter 14).

In the modern day, there are also some myths about cheetahs that are associated with local cultures. For example, in Tanzania, some villagers mentioned that cheetahs "eat rats and drink milk" (Dickman, 2009), and in Namibia, some farmers reported that cheetahs were vampires and sucked the prey's blood, as the point of their relatively short canine teeth only left small marks in the neck during strangulation (Marker-Kraus et al., 1996). However, there are far fewer stories and myths associated with cheetahs than with some of the other conflict-causing carnivores, such as lions and spotted hyenas (*Crocuta crocuta*), possibly because the myths are most frequently associated with fear linked to attacks on humans (Knight, 2000). Among those larger carnivores, myths about witchcraft can have very damaging impacts in terms of local perceptions and persecutions (Dickman, 2010).

Cheetahs fall victim to some tribes for use in ceremonies—such as the Sonjo, Maasai, and Sukuma tribes in Tanzania (TAWIRI, 2007). Similarly, Turkana respondents in Kenya reported that they sold cheetah skins to tribes in South Sudan because fathers needed to wear spotted skins during ceremonies to celebrate their daughters' reaching maturity (Wykstra, unpublished data). In addition, cheetah skins are used to make traditional man-slippers and are sought after by rich Sudanese (Marker, 2000). It is also reported that the cheetah has been a prized target animal during shooting practice in Turkana, while Mara respondents said that cheetahs were often used to prepare for hunting lions, as cheetahs could be easily tracked until they tired, and were much less likely to attack the hunters (Wykstra, unpublished data).

CONCLUSIONS

Today, the cheetah can still be seen in traditional batiks, carvings, beadwork, and embroidery across its range, indicating its continued cultural significance. Cheetahs have had a long history with people, but competition over land and resources poses one of the greatest threats to cheetahs today, including real and perceived conflict over livestock depredation (Chapters 10, 11, and 13). Within Africa, the problem is well-known; cheetah depredation has been recorded throughout its remaining range with varying

degrees of severity (Chapter 13). Northeastern Iran is a curious example where cheetahs prey on livestock, but are not significantly persecuted by herders; it has been suggested that this is partly due to herders misattributing the loss of their livestock to wolves rather than cheetahs (Farhadinia et al., 2012; Chapter 5). With over 1700 individuals in captivity (Marker, 2016), compared with around 7100 adult and adolescent individuals in the wild (Durant et al., 2017), combined with the cryptic behavior characteristic of the cheetah in the wild, it is likely that the majority of human–cheetah interactions today come in the form of zoo encounters (Chapters 23 and 28).

References

Allsen, T.T., 2006. The Royal Hunt in Eurasian History. University of Pennsylvania Press, Philadelphia.

Ashmead, A., 1978. Greek cats. Expedition 20, 38–48.

Bodenheimer, F.S., 1960. Animal and Man in Bible Lands. E.J. Brill, Leiden.

Bradshaw Foundation, 1999. Visiting the Chauvet Cave. Available from: http://www.bradshawfoundation.com/chauvet/page1.php.

Castel, E., 2002. Panthers, leopards and cheetahs. Notes on identification. Trabajos de Egiptología 1, 17–28.

Cherin, M., Iurino, D.A., Sardella, R., Rook, L., 2014. *Acinonyx pardinensis* (Carnivora, Felidae) from the early Pleistocene of Pantalla (Italy): predatory behavior and ecological role of the giant Plio-Pleistocene cheetah. Quat. Sci. Rev. 87, 82–97.

CITES, 2014. Convention on International Trade in Endangered Species of Wild Fauna and Flora. Sixty-fifth meeting of the Standing Committee Geneva, Switzerland, Doc 39, Rev 2.

Dickman, A.J., 2009. Key determinants of conflict between people and wildlife, particularly large carnivores, around Ruaha National Park, Tanzania. PhD thesis, University College London, United Kingdom.

Dickman, A.J., 2010. Complexities of conflict: the importance of considering social factors for effectively resolving human-wildlife conflict. Animal Conserv. 13, 458–466.

Durant, S.M., Mitchell, N., Groom, R., Pettorelli, N., Ipavec, A., Jacobson, A., Woodroffe, R., Bohm, M., Hunter, L., Becker, M., Broekuis, F., Bashir, S., Andresen, L., Aschenborn, O., Beddiaf, M., Belbachir, F., Belbachir-Bazi, A., Berbash, A., Brandao de Matos Machado, I., Breitenmoser, C., Chege, M., Cilliers, D., Davies-Mostert, H., Dickman, A., Fabiano,

E., Farhadinia, M., Funston, P., Henschel, P., Horgan, J., de Iongh, H., Jowkar, H., Klein, R., Lindsey, P., Marker, L., Marnewick, K., Melzheimer, J., Merkle, J., Msoka, J., Msuha, M., O'Neill, H., Parker, M., Purchase, G., Saidu, Y., Samaila, S., Samna, A., Schmidt-Küntzel, A., Selebatso, E., Sogbohossou, E., Soultan, A., Stone, E., van der Meer, E., van Vuuren, R., Wykstra, M., Young-Overton, K., 2017. The global decline of cheetah and what it means for conservation. Proc. Natl. Acad. Sci. USA 114, 528–533.

Farhadinia, M.S., Hosseini-Zavarei, F., Nezami, B., Harati, H., Absalan, H., Fabiano, E., Marker, L., 2012. Feeding ecology of the Asiatic cheetah *Acinonyx jubatus venaticus* in low prey habitats in northeastern Iran: implications for effective conservation. J. Arid Environ. 87, 206–211.

Finn, F., 1929. Sterndale's Mammalia of India. Thacker & Co. Limited, Bombay.

Fujii, S., 2014. Slab-lined feline representations: new finding at a late neolithic open-air sanctuary in southernmost Jordan. In: Stucky, R.A., Kaelin, O., Mathys, H.P. (Eds.), Proceedings, 9th ICAANE, Basel, 2014. Harrassowitz Verlag, Wiesbaden.

Guggisberg, C.A.W., 1975. Wild Cats of the World. Taplinger, New York.

Harper, F., 1945. Extinct and Vanishing Mammals of the Old World. Special Publication no. 12. Committee for International Wildlife Protection, New York.

Hemmer, H., Kahlke, R.D., Vekua, A.K., 2011. The cheetah *Acinonyx pardinensis* (Croizet et Joubert, 1828) s. l. at the hominin site of Dmanisi (Georgia)—a potential prime meat supplier in early Pleistocene ecosystems. Quat Sci. Rev. 30, 2703–2714.

Kingdon, J., 1977. Cheetah (*Acinonyx jubatus*). East African Mammals, Vol 3A. Academic Press, New York.

Kitchell, K., 2014. Animals in the Ancient World from A to Z. Routledge, New York.

Knight, J., 2000. Natural Enemies: People-Wildlife Conflicts in Anthropological Perspective. Routledge, London.

Marker, L., 2000. International Cheetah Studbook, 1999. Cheetah Conservation Fund, Namibia.

Marker, L., 2016. International Cheetah Studbook, 2015. Cheetah Conservation Fund, Namibia.

Marker-Kraus, L., Kraus, D., Barnett, D., Hurlbut, S., 1996. Cheetah Survival on Namibian Farmlands. Cheetah Conservation Fund, Windhoek.

Nowell, K., Jackson, P., 1996. Wild Cats. Burlington Press, Cambridge.

Sunquist, M., Sunquist, F., 2002. Wild Cats of the World. University of Chicago Press, Chicago and London.

TAWIRI, 2007. Proceedings of the First Tanzania Cheetah Conservation Action Plan Workshop, TAWIRI, Arusha.

Van Buren, E.D., 1936. Mesopotamian fauna in the light of the monuments. Archiv für Orientforschung 11, 1–37.

Van Buren, E.D., 1939. The fauna of ancient Mesopotamia as represented in art. Pontificum Institutum Biblicum, Rome.

Wrogemann, N., 1975. Cheetah Under the Sun. McGraw Hill, Johannesburg.

3

The Cheetah: Evolutionary History and Paleoecology

Blaire Van Valkenburgh*, Benison Pang*,
Marco Cherin**, Lorenzo Rook†

*University of California at Los Angeles, Los Angeles, CA, United States
**University of Perugia, Perugia, Italy
†University of Florence, Florence, Italy

INTRODUCTION

Cheetahs, *Acinonyx jubatus*, are remarkable cats and among the most unusual of felids, living or extinct. Their elongate limbs, slender body, and small head distinguish them easily from other similar sized felids, such as the leopard or jaguar. These features and others, such as not having fully retractile claws, are related to their ability to run at speeds that exceed 100 km/h (Chapter 7). Their swiftness enhances their hunting success, but comes at a price. Their slender frame makes them an easy target for other, more robust carnivores, and consequently they often lose their kills to lions, spotted hyenas, or packs of wild dogs.

Given the many distinguishing features of cheetahs, it might be expected that their evolutionary history and relationships would be easy to unravel. However, this has not been the case. Based on analyses of morphological characters, cheetahs were often placed in a group with the pantherines (e.g., leopard, lion; Salles, 1992), but occasionally aligned with the North American mountain lion (*Puma concolor*) and jaguarundi (*Herpailurus yagouaroundi*) instead (Adams, 1979; Herrington, 1986; Van Valkenburgh et al., 1990). In addition, the when and where of their origins have been debated, with some arguing for a New World (the Americas) origin with subsequent migration to the Old World (Eurasia and Africa) (Adams, 1979; Dobrynin et al., 2015; Johnson et al., 2006), and others arguing for the opposite scenario (Barnett et al., 2005; Li et al., 2016; Werdelin et al., 2010). The disagreement stems in large part from the discovery of cheetah-like felids in North America approximately 2.5 million years ago (ma), at a time that was nearly coincident with that of the then oldest known representatives of *Acinonyx* in the Old World.

Much of this controversy has been resolved over the past 2 decades due to new fossil discoveries, the development of a robust felid phylogeny based on nuclear and mitochondrial DNA data, and advances in ancient DNA (aDNA) technology. As a result, it is now clear that (1) *Acinonyx* is part of a clade that includes the mountain lion, jaguarundi (Chapter 6), and extinct North American cheetah-like cats in the genus *Miracinonyx*, (2) *Acinonyx* originated in Africa circa 4 ma and was not derived from any New World felid, and (3) the evolution of highly cursorial, cheetah-like cats in the New World is an example of convergent or parallel evolution.

In this chapter, we review current knowledge concerning the phylogenetic position of *Acinonyx*, its fossil record and origins, as well as its relationship to the New World cheetah-like cats.

EVOLUTIONARY HISTORY

The cheetah is a member of the family Felidae within the order Carnivora. Felids first appear in the fossil record approximately 35–30 ma in Europe (Werdelin et al., 2010). The oldest known species, *Proailurus lemanensis*, was similar in size to a bobcat (*Lynx rufus*) and is easily recognizable as a felid. Like living felids, it has a skull that is more rounded than elongate in profile and short jaws that contain a reduced number of teeth that are all specialized for carnivory. *Proailurus* is followed by an evolutionary radiation of felids that eventually sorts itself into two recognizable lineages, the subfamilies Machairodontinae (sabertooth cats) and Felinae (conical-toothed cats), approximately 13–10 ma (Werdelin et al., 2010). The machairodonts are characterized by elongate, knife-like upper canine teeth, and dominate the felid fossil record in terms of body mass and diversity until the Pleistocene. The earliest members of the Felinae appear about 9 ma in western Eurasia, and are similar in size to wildcats (*Felis lybica*), and hence much smaller than the contemporaneous machairodonts (Werdelin et al., 2010). All of the

living felids, including the cheetah, descend from these first felines.

Due to a general similarity of form in felines and their poor fossil record, it has been difficult to tease apart evolutionary relationships within the subfamily based on skeletal and dental features. However, molecular data have been much more successful, and divide the living species into eight distinct lineages or clades (Chapter 6). As noted earlier, *Acinonyx* belongs to the Puma lineage, which diverged from the stem felid lineage approximately 8–6 ma, followed by a split between New World and Old World branches within the clade about 5 ma (Johnson et al., 2006; Li et al., 2016). Thus, the molecular data would predict the first appearance of *Acinonyx* to be about 5 ma, which is close to the date of 4 ma estimated for the earliest *Acinonyx* specimens in the fossil record.

The genus *Acinonyx* first appears in the fossil record in both eastern and southern Africa circa 4 ma. The fossils are fragmentary and cannot be definitively assigned to *A. jubatus*, but appear to be larger than the extant species (Werdelin and Lewis, 2005). This early cheetah migrated out of Africa, and possibly evolved into *Acinonyx pardinensis*, the giant cheetah of the Old World around 3.5 ma (Werdelin and Peigné, 2010). Fossils of *A. pardinensis* are widespread between 2.5 and 0.5 ma, ranging from China to India, North Africa and Europe (Fig. 3.1) (Cherin et al., 2014). Given the large geographic range of the giant cheetah, it is not surprising to find morphological differences between fossil specimens separated by large distances, and this has led to the naming of additional species and/or subspecies. However, the current consensus (Cherin et al., 2014; Geraads, 1997, 2014; Hemmer et al., 2008; Werdelin et al., 2010) appears to be that they all belong to *A. pardinensis*, including the early Pleistocene *Sivapanthera linxianensis* from China (Qiu et al., 2004). An initial report on a skull found in China claimed that it was a primitive species of *Acinonyx* but this was retracted later because the analysis was based on a composite specimen (Deng, 2011; Mazak, 2012).

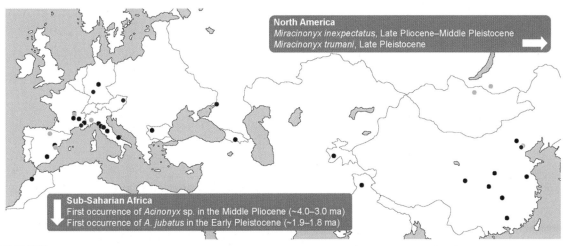

North America
Miracinonyx inexpectatus, Late Pliocene–Middle Pleistocene
Miracinonyx trumani, Late Pleistocene

Sub-Saharian Africa
First occurrence of *Acinonyx* sp. in the Middle Pliocene (~4.0–3.0 ma)
First occurrence of *A. jubatus* in the Early Pleistocene (~1.9–1.8 ma)

FIGURE 3.1   **Map of the Plio-Pleistocene records of *Acinonyx pardinensis* in the Old World.** The approximate ages are indicated by color as follows: *green*, 3.5–2 ma; *red*, 2–0.8 ma; *blue*, 0.8–0.5 ma. See Cherin et al. (2014) for details.

Fossils of the living species, *A. jubatus*, appeared first in southern Africa at least 1.8 ma, followed by slightly younger fossils from eastern Africa (Werdelin and Peigné, 2010). Although it is not possible to be certain, it seems unlikely that *A. pardinensis* is ancestral to *A. jubatus*, instead they represent separate descendants from a common African ancestor (see further). Interestingly, fossils of *A. jubatus* are rare and perhaps unknown outside Africa, despite historical records that indicate the presence of the cheetah in arid and semi-arid regions of the Middle East, southwestern Asia, and India (Krausman and Morales, 2005). Consequently, it is not possible to determine more precisely when *A. jubatus* left Africa.

Although the history of the New World cheetah-like cats in the genus *Miracinonyx* is now known to be largely separate from that of the Old World cheetahs, it is of interest because they share common ancestry based on both genetic and morphological data. Ancient DNA sequences from specimens of the youngest of the New World cheetah-like cats, *Miracinonyx trumani*, confirm that *Miracinonyx* and *Acinonyx* are part of the Puma lineage and also suggest that *Miracinonyx* is more closely related to the New World

Puma (mountain lion) and *Herpailurus* (jaguarundi), than to *Acinonyx* (Barnett et al., 2005). However, the *Miracinonyx–Puma* + *Herpailurus* clade is based on a single mitochondrial marker and needs to be confirmed with more detailed aDNA analyses (O'Brien et al., 2016). Nevertheless, the grouping of all New World members of the Puma lineage was also supported by cladistic analyses of postcranial and craniodental morphology (Herrington, 1986; Van Valkenburgh et al., 1990). These more extensive studies contradicted a previous study by Adams (1979) who had argued for *Miracinonyx* and *Acinonyx* as sister taxa exclusive of *Puma* and *Herpailurus* that descended from an as yet unknown common New World ancestor.

The confusion concerning origins is understandable given the remarkable convergence in overall skeletal and cranial morphology between *M. trumani* and *A. jubatus* (Adams, 1979; Martin et al., 1977). Both evolved adaptations for extreme speed, including relatively long, slender limbs, small heads, and an elongated lumbar region of their spines. However, a comparison of their entire skeletons (Fig. 3.2), rather than isolated elements, reveals that the North American cat was not so extreme in its morphology as

1. THE CHEETAH

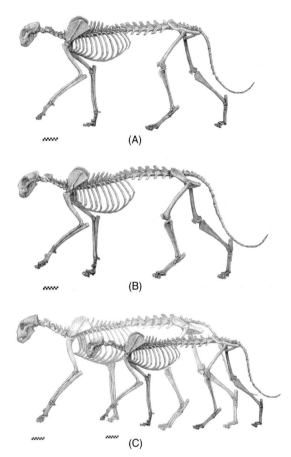

FIGURE 3.2 Skeletal reconstructions of (A) *Acinonyx jubatus*, (B) *Miracinonyx trumani*, and (C) *A. jubatus* in front of *A. pardinensis* to show the much larger size of the Pleistocene cheetah. Scale bars = 10 cm. Although the convergence between *A. jubatus* and *M. trumani* is remarkable, note the relatively larger head and more robust limb bones of *M. trumani*. *Source: Illustration by Mauricio Anton.*

A. jubatus. Despite being similar in overall body size, *M. trumani* had more robust limb bones, a relatively larger skull, and stronger neck (larger cervical vertebrae) than *A. jubatus*. Interestingly, the extant cheetah has a relatively bigger scapula, which may be largely an adaptation to increase stride length and hence speed. This and its overall more gracile build suggests that the extant species could have outrun its North American vicar were they to have met.

The fact that *Acinonyx* is present in Africa at 4 ma, at least 1 million years prior to the appearance of *Miracinonyx* in the New World, strongly suggests that the origin of the Puma lineage lies in the Old World. A possible scenario would begin 7–6 ma in the Old World with the appearance of a medium sized feline that had a generalized mountain lion-like morphology (e.g., *Puma pardoides*; Cherin et al., 2013; Hemmer et al., 2004; Madurell-Malapeira et al., 2010). This ancestral cat could have then produced two descendant lineages, one that evolved into *Acinonyx* in Africa, and another that emigrated to the New World at least once about 3 ma and produced *Miracinonyx, Puma, and Herpailurus*. However, we have yet to discover a fossil felid of the appropriate age (7–6 ma) and morphology that can serve as the probable ancestor in the Old World. Based on its morphology, *P. pardoides* could be ancestral but it does not appear in the record until about 3 ma, 500,000 years or more after the first fossils of *Acinonyx* in Africa. This could change with additional discoveries.

An alternative scenario posits a New World origin of the Puma lineage with subsequent migration of cheetah ancestors into the Old World. This is based on historical reconstructions of biogeography generated from genetic data (Dobrynin et al., 2015; Johnson et al., 2006; O'Brien et al., 2016). These reconstructions have been challenged recently on two fronts. First, the fossil and morphological evidence summarized earlier strongly support an Old World origin (Faubry et al., 2016). Second, a new, potentially more robust reconstruction of historical biogeography that relies on Bayesian analysis of genetic data supports an Old World rather than New World origin of the Puma lineage (Li et al., 2016). The debate over the geographic location of the origin of the Puma lineage is likely to continue until more fossils are found and/or a more definitive aDNA analysis is completed on fossil members of the lineage.

MORPHOLOGICAL EVOLUTION AND PALEOECOLOGY OF *ACINONYX*

There are a number of distinguishing features of extant cheetahs that make it possible to recognize their ancestors in the fossil record. Many of these features are craniodental and skeletal specializations for cursoriality (Chapter 7). For example, cheetahs have elongate limbs in which the radius exceeds the length of the humerus and the tibia is near equal in length to the femur. Their skulls are lighter and smaller relative to their body size than in other big cats and the teeth are specialized for slicing skin and meat.

They display multiple piercing cusps on their premolars (Hartstone-Rose, 2011) and a nearly completely blade-like upper carnassial that lacks the blunt, crushing protocone of other felids. The skull of cheetahs has an unusual domed appearance in lateral view (Fig. 3.3) caused by expansion of the frontal bone and downward flexion of the face on the braincase (Cherin et al., 2014; Geraads, 2014). It is not clear why the frontal bone and sinus are expanded in cheetahs. It has been suggested to relate to respiratory demands and possible brain-cooling during sprinting (Chapter 7), but this has yet to be supported with physiological data on respiratory function in felids.

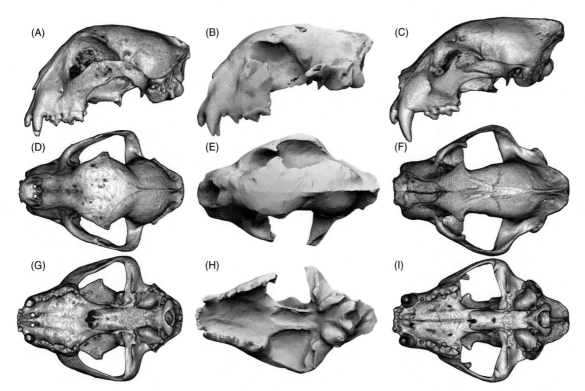

FIGURE 3.3 Comparative cranial morphology of *A. jubatus* (A, D, G), *A. pardinensis* from Pantalla, Italy (B, E, H), and the extant jaguar, *Panthera onca* (C, F, I). The 3D models are shown in left lateral (A, B, C), dorsal (D, E, F), and ventral (G, H, I) views. All crania are shown at the same size to facilitate shape comparisons. Cranial length is 177 mm in *A. jubatus*, 190 mm in *A. pardinesis*, and 224 mm in *P. onca*. *Source: Artwork by David A. Iurino.*

How many of these features were present in the earliest known representatives of *Acinonyx*? Unfortunately the oldest specimens from Africa are too fragmentary to provide much information, but seem to suggest that they were not as specialized for cursoriality as the living cheetah, especially in their forelimb skeleton, which is intermediate in form between that of extant leopards and cheetahs (Werdelin and Lewis, 2013). There are many more fossils of *A. pardinensis*, including both craniodental material and limb bones, and it is clear that it was much larger than *A. jubatus* and not quite so specialized in its skull and teeth (Figs. 3.2 and 3.3). The skull is not as domed and the downward flexion of the face on the braincase is not as extreme (Fig. 3.3) (Cherin et al., 2014; Geraads, 2014). Moreover, the teeth are not quite so slender and specialized for pure meat eating (Cherin et al., 2014; Van Valkenburgh et al., 1990). In terms of limb proportions, *A. pardinensis* was similar to *A. jubatus* in having slightly longer radii and tibiae than humeri and femora, respectively (Van Valkenburgh et al., 1990), although this is based on at most two specimens, if the species assignments for the fossils are correct. The giant cheetah's limb bones also appear to have been relatively thicker than those of *A. jubatus* as expected given the much larger body size (Fig. 3.2). Although there was a recent thorough review of the craniodental morphology of *A. pardinensis* (Cherin et al., 2014), a similar review of all known postcranial material of the giant cheetah is sorely needed and would help clarify its adaptations.

The large size of the giant cheetah is intriguing and has implications for both its paleoecology and predatory behavior. Based on linear regressions of tooth dimensions versus body mass in living felids, *A. pardinensis* had a body mass between 60 and 121 kg, which is about double that of the living species (35–40 kg) (Cherin et al., 2014; Spassov, 2011). This near doubling is similar to that observed for Pleistocene spotted hyenas and lions relative to their modern counterparts (Van Valkenburgh et al., 2016). The giant cheetah was a member of a very diverse guild of formidable predators that included sabertooth cats, as well as large hyenas and lions. Undoubtedly, its large size was key to its ability to coexist with these species and defend itself against kleptoparasitism. Prey size tends to increase with predator size (Sinclair et al., 2003), and consequently, it seems likely that the giant cheetah take larger prey on an average than *A. jubatus*, perhaps in the 50–100 kg range as opposed to the 23–56 kg range that is typical for the extant cheetah (Hayward et al., 2006; Turner and Anton, 1997).

In view of the elongate limbs of the giant cheetah, it likely hunted in a similar style to the living species, using extreme speed to overtake prey and knock them to the ground, before killing with a suffocating bite to the throat (Chapter 7). However, without more study of the postcranial material, it is difficult to determine how closely the behavior of the giant cheetah mirrored that of *A. jubatus*, and Cherin et al. (2014) suggest caution in interpretations of its behavior. In any case, it is likely that *A. pardinensis* frequented both savannah and thicket habitats as cheetahs do today. Although cheetahs are most often associated with grasslands, they are known to forage in woodland or thickets to avoid losing their kills to larger predators (e.g., Bissett and Barnard, 2007), and it is likely that the giant cheetahs would have behaved similarly, given the presence of even larger hyenas, lions, and sabertooth cats.

The giant cheetah has its last occurrence in the fossil record in the middle Pleistocene, approximately 500 thousand years ago. The cause of its extinction is unclear and not coincident with the disappearance of other large carnivores, but the record is far from complete. Although Hemmer et al. (2008) suggested that *A. pardinensis* evolved into *A. jubatus*, this seems unlikely. *A. jubatus* appears in Africa approximately 1.9 ma, well before the extinction of the giant cheetah, and has no fossil record outside of Africa, suggesting that its immediate ancestry lies fully within that continent.

CONCLUSIONS

Although the cheetah appears to be such a highly specialized carnivore that it might not easily survive major environmental perturbations and biotic change, the fossil record indicates otherwise. The genus *Acinonyx* has persisted for more than 5 million years in the Old World, and its sister taxon, *Miracinonyx*, persisted for nearly as long in North America. The end members of both these lineages, *A. jubatus* and *M. trumani*, appeared about 2 million years ago with similarly extreme cursorial morphology. Subsequently, they both lived through multiple climatic fluctuations during the Pleistocene, although ultimately, *M. trumani* went extinct approximately 18,000 years ago. Their persistence through these oscillations suggests that, despite being relatively specialized, cheetah-like cats are capable of adapting to significant environmental and biotic change, including shifts in prey availability and the intensity of competition with co-existing carnivores. This evolutionary tenacity can give us hope for the continued survival of cheetahs, but without serious conservation efforts, the present-day threats of climate change (Chapter 12), human persecution (Chapter 13), and small population size (Chapter 10) are likely to overwhelm this exceptional felid.

References

Adams, D.B., 1979. The cheetah: native American. Science 205, 1155–1158.

Barnett, R., Barnes, I., Phillips, M.J., Martin, L.D., Harington, C.R., Leonard, J.A., Cooper, A., 2005. Evolution of extinct sabre-tooths and the American cheetah-like cat. Curr. Biol. 15, R589–R590.

Bissett, C., Barnard, R.T.F., 2007. Habitat selection and feeding ecology of the cheetah (*Acinonyx jubatus*) in thicket vegetation: is the cheetah a savanna specialist? J. Zool. 271, 310–317.

Cherin, M., Iurino, D.A., Sardella, R., 2013. Earliest occurrence of *Puma pardoides* (Owen, 1946) (Carnivora, Felidae) at the Plio-Pleistocene transition in western Europe: new evidence from the Middle Villafranchian assemblage of Montopoli, Italy. C. R. Palevol 12, 165–171.

Cherin, M., Iurino, D.A., Sardella, R., Rook, L., 2014. *Acinonyx pardinensis* (Carnivora, Felidae) from the early Pleistocene of Pantalla (Italy): predatory behavior and ecological role of the giant Plio-Pleistocene cheetah. Quat. Sci. Rev. 87, 82–97.

Deng, T., 2011. *Acinonyx kurteni* based on a fossil composite. Vertebrat. Palasiatic. 49, 362–364.

Dobrynin, P., Liu, S., Tamazian, G., Xiong, Z., Yurchenko, A.A., Krasheninnikova, K., Kliver, S., Schmidt-Küntzel, A., Koepfli, K.P., Johnson, W., Kuderna, L.F., García-Pérez, R., de Manuel, M., Godinez, R., Komissarov, A., Makunin, A., Brukhin, V., Qiu, W., Zhou, L., Li, F., Yi, J., Driscoll, C., Antunes, A., Oleksyk, T.K., Eizirik, E., Perelman, P., Roelke, M., Wildt, D., Diekhans, M., Marques-Bonet, T., Marker, L., Bhak, J., Wang, J., Zhang, G., O'Brien, S.J., 2015. Genomic legacy of the African cheetah *Acinonyx jubatus*. Genome Biol. 16, 277.

Faubry, S., Werdelin, L., Svenning, J.C., 2016. The difference between trivial and scientific names: there were never any true cheetahs in North America. Genome Biol. 17, 89.

Geraads, D., 1997. Carnivores du Pliocene terminal de Ahl al Oughlam (Casablanca, Moroc). Geobios 30, 127–164.

Geraads, D., 2014. How old is the cheetah skull shape? The case of *Acinonyx pardinensis* (Mammalia, Felidae). Geobios 47, 39–44.

Hartstone-Rose, A., 2011. Reconstructing the diets of extinct South African carnivorans from premolar 'intercuspid notch' morphology. J. Zool. 285, 119–127.

Hayward, M.W., Hofmeyr, M., O'Brien, J., Kerley, G.I.H., 2006. Prey preferences of the cheetah *Acinonyx jubatus*: morphological limitations or the need to capture rapidly consumable prey before kleptoparasites arrive? J. Zool. 270, 615–627.

Hemmer, H., Kahlke, R.D., Keller, T., 2008. Cheetahs in the middle Pleistocene of Europe: *Acinonyx pardinensis* (sensu lato) *intermedius* (Thenius, 1954) from the Mosbach Sands (Wiesbaden, Hesse, Germany). Neues Jahrbuch Geol. Palaeontol. Abhandlungen 249, 349–356.

Hemmer, H., Kahlke, R.D., Vekua, A.K., 2004. The old world puma—*Puma pardoides* (Owen, 1846) (Carnivora: Felidae)—in the Lower Villafranchian (Upper Pliocene) of Kvabebi (East Georgia, Transcaucasia) and its evolutionary and biogeographical significance. Neues Jahrbuch Geol. Palaeontol. Abhandlungen 233, 197–231.

Herrington, S.J., 1986. Phylogenetic Relationships of the Cats of the World. PhD thesis, University of Kansas, US.

Johnson, W.E., Ezirik, E., Pecon-Slattery, J., Murphy, W.J., Antunes, A., Teeling, E., O'Brien, S.J., 2006. The late Miocene radiation of modern Felidae: a genetic assessment. Science 311, 73–77.

Krausman, P.R., Morales, S.M., 2005. *Acinonyx jubatus*. Mamm. Sp. Acc. 771, 1–6.

Li, G., Davis, B.W., Eizirik, E., Murphy, W.J., 2016. Phylogenomic evidence for ancient hybridization in the genomes of living cats (Felidae). Genome Res. 26, 1–11.

Madurell-Malapeira, J., Alba, D.M., Moya-Sola, S., Aurell-Garrido, J., 2010. The Iberian record of the puma-like cat *Puma pardoides* (Owen, 1946) (Mammalia, Felidae). C. R. Palevol 9, 55–62.

Martin, L.D., Gilbert, B.M., Adams, D.B., 1977. A cheetah-like cat in the North American Pleistocene. Science 195, 981–982.

Mazak, J., 2012. Retraction. Proc. Natl. Acad. Sci. 109, 15072.

O'Brien, S.J., Koepfli, K.P., Eizirik, E., Johnson, W., Driscoll, C., Antunes, A., Schmidt-Küntzel, A., Marker, L., Dobrynin, P., 2016. Response to comment by Faurby, Werdelin, and Svenning. Genome Biol. 17, 90.

Qiu, Z., Deng, T., Wang, B., 2004. Early Pleistocene mammalian fauna from Longdan, Dongxiang, Gansu, China. Palaeontol. Sin. 191, 1–198.

Salles, L.O., 1992. Felid phylogenetics: extant taxa and skull morphology (Felidae, Aeleuroidea). Am. Mus. Novit. 3047, 1–67.

Sinclair, A.R.E., Mduma, S., Brashares, J.S., 2003. Patterns of predation in a diverse predator-prey system. Nature 425, 288–290.

Spassov, N., 2011. *Acinonyx pardinensis* (Croizet et Joubert) remains from the middle Villafranchian locality of Varshets (Bulgaria) and the Plio-Pleistocene history of cheetahs in Eurasia. Estud. Geol. 67, 245–253.

Turner, A., Anton, M., 1997. The Big Cats and Their Fossil Relatives. Columbia University Press, New York.

Van Valkenburgh, B., Grady, F., Kurten, B., 1990. The Plio-Pleistocene cheetah-like cat *Miracinonyx inexpectatus* of North America. J. Vert. Paleontol. 10, 434–454.

Van Valkenburgh, B., Hayward, M.W., Ripple, W.J., Meloro, C., Roth, V.L., 2016. The impact of large terrestrial carnivores on Pleistocene ecosystems. Proc. Natl. Acad. Sci. 113, 862–867.

Werdelin, L., Lewis, M., 2005. Plio-Pleistocene Carnivora of eastern Africa: species richness and turnover patterns. Zool. J. Linn. Soc. 144, 121–144.

Werdelin, L., Lewis, M., 2013. Koobi Fora Research Project. The Carnivora. Califfornia Academy of Sciences, San Francisco.

Werdelin, L., Peigné, S., 2010. Carnivora. In: Werdelin, L., Sanders, W.J. (Eds.), Fossil Mammals of Africa. University of California Press, Berkeley, California, pp. 603–658.

Werdelin, L., Yamaguchi, N., Johnson, W.J., O'Brien, S.J., 2010. Phylogeny and evolution of cats (Felidae). In: Macdonald, D.W., Loveridge, A. (Eds.), The Biology and Conservation of Wild Felids. Oxford University Press, Oxford, pp. 59–82.

Cheetah Rangewide Status and Distribution

Laurie Marker*, Bogdan Cristescu*,
Tess Morrison*, Michael V. Flyman**,
Jane Horgan[†], Etotépé A. Sogbohossou[‡],
Charlene Bissett[§], Vincent van der Merwe[¶],
Iracelma B. de Matos Machado[††], Ezequiel Fabiano[‡‡],
Esther van der Meer[¶¶], Ortwin Aschenborn[§§],
Joerg Melzheimer***, Kim Young-Overton[†††],
Mohammad S. Farhadinia[‡‡‡,¶¶¶], Mary Wykstra[§§§],
Monica Chege****, Abdoulkarim Samna[††††],
Osman G. Amir[‡‡‡‡], Ahmed Sh Mohanun[‡‡‡‡,¶¶¶¶¶],
Osman D. Paulos[¶¶¶¶], Abel R. Nhabanga[§§§§],
Jassiel L.J. M'soka*****, Farid Belbachir[†††††],
Zelealem T. Ashenafi[‡‡‡‡‡], Matti T. Nghikembua*

*Cheetah Conservation Fund, Otjiwarongo, Namibia
**Department of Wildlife and National Parks, Gaborone, Botswana
[†]Cheetah Conservation Botswana, Maun, Botswana
[‡]University of Abomey-Calavi Cotonou, Cotonou, Benin
[§]South African National Parks, Kimberley, South Africa
[¶]Endangered Wildlife Trust, Modderfontein, Gauteng, South Africa
[††]Ministry of Agriculture, Luanda, Angola
[‡‡]University of Namibia, Katima Mulilo, Namibia
[¶¶]Cheetah Conservation Project Zimbabwe, Victoria Falls, Zimbabwe
[§§]University of Namibia, Windhoek, Namibia

Cheetahs: Biology and Conservation
http://dx.doi.org/10.1016/B978-0-12-804088-1.00004-6

***Leibniz Institute for Zoo and Wildlife Research, Berlin, Germany
†††Zambian Carnivore Programme, Panthera, New York, NY, United States
‡‡‡Future4Leopards Foundation, Tehran, Iran
¶¶¶University of Oxford, Tubney, Abingdon, United Kingdom
§§§Action for Cheetahs in Kenya Project, Nairobi, Kenya
****Kenya Wildlife Service, Nairobi, Kenya
††††Ministry of Environment and Sustainable Development, Niamey, Niger
‡‡‡‡Somalia Wildlife and Natural History Society, Mogadishu, Somalia
¶¶¶¶Ethiopian Wildlife Conservation Authority, Addis Ababa, Ethiopia
§§§§Banhine National Park, Gaza, Mozambique
*****Ministry of Tourism and Arts, Chilango, Lusaka, Zambia
†††††University of Béjaïa, Béjaïa, Algeria
‡‡‡‡‡Born Free Foundation Ethiopia, Addis Ababa, Ethiopia
¶¶¶¶¶Somalia Ministry of Wildlife and Rural Development, Mogadishu, Somalia

INTRODUCTION

The cheetah (*Acinonyx jubatus*) is listed as vulnerable by the World Conservation Union (IUCN) Red List of Threatened Species with two subspecies, the Asiatic cheetah (*Acinonyx jubatus venaticus*) (Griffith, 1821) and the Northwest African cheetah (*Acinonyx jubatus hecki*) Hilzheimer, 1913, listed as critically endangered (Durant et al., 2015). The cheetah's historical distribution in Africa covered a substantial proportion of the continent, but as a result of range contraction in the last century the cheetah is now found in only 9% of its historical range, of which 77% is outside protected areas (Durant et al., 2017; Ray et al., 2005). The species is extinct in nearly its entire Asian range, with the exception of a remnant population in Iran. Population declines are largely attributed to habitat loss, land fragmentation, and conflict with humans (Marker et al., 2003; Myers, 1975). In addition, the decline in Asia was heightened owing to the Asian aristocracy capturing cheetahs to use them for hunting (Divyabhanusinh, 1995; Chapter 2). Reduced numbers and habitat fragmentation threaten to further reduce the cheetah's already

low genetic diversity (Chapters 6 and 10). Rangewide cheetahs have been confirmed in 33 populations in 20 countries with a population estimate of 7100 adult and adolescent cheetahs (Durant et al., 2017).

Obtaining reliable population estimates for cheetahs is difficult due to the species' wide ranging movements, low population densities (Chapter 8), and probable behavioral avoidance of people, especially outside protected areas, where cheetahs are often persecuted (Durant et al., 2007; RWCP and IUCN/SSC, 2015). The first status survey for cheetahs was conducted in the early 1970s (Myers, 1975); the population in sub-Saharan Africa was estimated at 14,000. Later surveys of selected countries were conducted in the 1980s (Gros, 1996, 1998, 2002; Gros and Rejmanek, 1999) and a summary of knowledge of global status was collated in 1998 estimating 15,000 adult individuals (Marker, 1998).

In 2007 the Rangewide Conservation Program for Cheetah and African Wild Dogs, spearheaded by Dr. Sarah Durant and Dr. Rosie Woodroffe, was initiated to enable landscape-level planning for cheetah and wild dog conservation across

TABLE 4.1   Percentage of Historical Cheetah Range Falling Into Each Range Category

	Resident (%)	Possible (%)	Connecting (%)	Unknown (%)	Recoverable (%)	Unrecoverable (%)
Southern Africa	22.8[a]	6.6	1.0	26.1	2.7	40.7
Eastern Africa	6.2	17.7	0.2	62.8	0.0	13.2
Western, Central, and Northern Africa	8.8	7.7	0.5	24.5	1.6	56.9

Range categories: resident, land where cheetahs are known to be still resident; *possible*, land where cheetahs may still be resident, but where residency has not been confirmed in the last 10 years; *connecting*, land where cheetahs may not be resident, but which dispersing animals may use to move between occupied areas or to recolonize extirpated range; *unknown*, land where the cheetahs' status is currently unknown and cannot be inferred using knowledge of the local status of habitat and prey; *recoverable*, land where habitat and prey remain over sufficiently large areas that either natural or assisted recovery of cheetahs might be possible within the next 10 years, if reasonable conservation action were to be taken; *unrecoverable*, land where habitat has been so heavily modified or fragmented as to be uninhabitable by cheetahs for the foreseeable future.

[a] *Includes 1.2% of historical range considered transient. Transient is defined as habitat used intermittently by cheetahs, but where cheetahs are known not to be resident and which does not connect to other resident range. Does not include historical range where cheetahs are now in fenced reserves as a metapopulation (equivalent to 0.1% of historical range).*

Source: RWCP and IUCN/SSC, 2015; IUCN/SSC, 2007a; IUCN/SSC, 2012.

their ranges in Africa. The regional planning framework involved the IUCN/SSC Cat and Canid Specialist Groups, Wildlife Conservation Society (WCS), and Zoological Society of London (ZSL). As part of this initiative three regional strategies were devised to compartmentalize the species' distribution into (1) southern Africa (IUCN/SSC, 2007b; RWCP and IUCN/SSC, 2015), (2) eastern Africa (IUCN/SSC, 2007a), and (3) western, central, and northern Africa (IUCN/SSC, 2012; Table 4.1).

The regional strategies, including a cheetah distribution map for each region, were drafted at planning workshops held in Botswana 2007; updated at a subsequent meeting in 2015 in South Africa, Kenya (2007), and Niger (2012). Since these regional range-wide strategies were developed, to date, 17 cheetah range countries have developed national strategies (Chapter 39).

A range wide distribution map based on best available expert knowledge collated by Durant et al. (2017) is presented in Fig. 4.1, including map units to cross reference populations. Corresponding population estimates are shown in Table 4.2. The status and distribution of cheetahs

in each of the 19 African countries, where they are resident, will be briefly discussed. The last remaining population of Asiatic cheetah, in Iran, is briefly mentioned, but will be discussed in detail in Chapter 5.

SOUTHERN AFRICA

Southern Africa is the cheetah's regional stronghold on the continent, holding over 4000 adults and adolescents, the majority of which are in a uniquely large contiguous transfrontier landscape, covering southern Angola, Botswana, Namibia, and northern South Africa, (Durant et al., 2017; Fig. 4.1, Table 4.2; S1, S13, S14, S15, S17, S18, S19). Cheetahs in southwestern Mozambique and the adjoining population in Kruger National Park in South Africa are considered part of this landscape (Durant et al., 2017) but it has not been confirmed if the populations are connected (Endangered Wildlife Trust [EWT] unpublished data).

Historically, the southern African cheetah subspecies (*A. j. jubatus*) (Schreber, 1775)

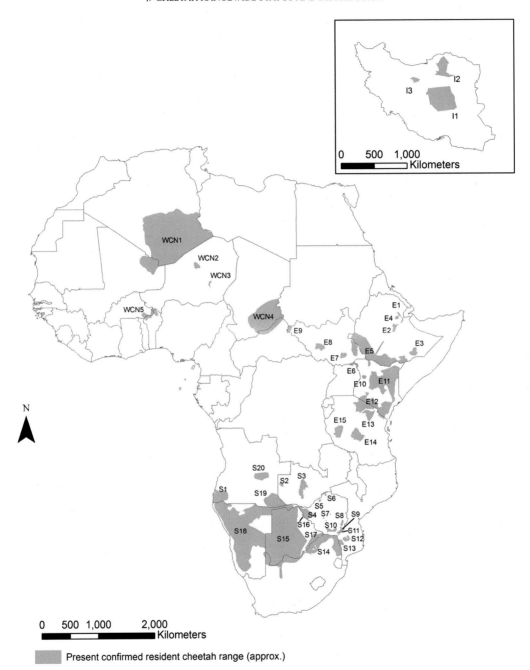

FIGURE 4.1 **Global distributions of resident free-ranging cheetahs.** Main map shows distribution in Africa, whereas inset map shows distribution in Iran. Cheetah population polygons are depicted in brown. Details on populations by map unit (black font) are provided in Table 4.2. *Source: Modified from Durant et al., 2017; with updated data from Farhadinia et al., 2016; and Van der Meer, 2016.*

TABLE 4.2   Global Distribution of Free-Ranging Cheetah Populations in 2016

Region	Area name	Countries	Resident range (km²)	Population size	Map unit
Southern Africa	Southern Africa five-country polygon	Angola/Botswana/ Mozambique/ Namibia/South Africa[a]	1,212,179	4,021[b]	S1, S13, S14, S15, S17, S18, S19
	Kafue	Zambia	26,222	65	S3
	Pandamatenga/Hwange/ Victoria Falls	Botswana/Zimbabwe	24,796	46[c]	S4, S16
	Moxico	Angola	25,717	26	S20
	Malilangwe/Save/ Gonarezhou/Maunge	Mozambique/ Zimbabwe	9,241	44[d]	S8, S9, S11
	Bubiana/Nuanetsi/Bubye Valley Conservancy and neighboring farms	Zimbabwe	9,143	47	S10
	Banhine	Mozambique	7,266	10	S12
	Zambezi Valley Complex	Zimbabwe	5,034	15[e]	S5, S6
	Liuwa	Zambia	3,170	20	S2
	Midlands Rhino Conservancy/Debshan and neighboring farms	Zimbabwe	1,685	8	S7
	Region-wide		*1,324,453*	*4,302*	
Eastern Africa	Serengeti/Mara/Tsavo/ Laikipia/Samburu	Kenya/Tanzania	280,114	1,362	E11, E12
	Ethiopia/Kenya/South Sudan	Ethiopia/Kenya/South Sudan	191,180	191	E5
	Ruaha ecosystem	Tanzania	30,820	200	E14
	Katavi-Ugalla	Tanzania	23,955	60	E15
	Maasai Steppe	Tanzania	20,409	51	E13
	Southern national park (NP)	South Sudan	14,680	147	E8
	Ogaden	Ethiopia	12,605	32	E3
	Badingilo NP	South Sudan	8,517	85	E7
	Blen-Afar	Ethiopia	8,170	20	E2
	Radom NP	South Sudan	6,821	68	E9
	Kidepo/southern South Sudan/northwest Kenya	Kenya/South Sudan/ Uganda	6,694	19	E6
	Afar	Ethiopia	4,480	11	E1
	South Turkana	Kenya	3,580	36	E10
	Yangudi Rassa	Ethiopia	3,046	8	E4
	Region-wide		*615,071*	*2,290*	

(*Continued*)

1. THE CHEETAH

TABLE 4.2 Global Distribution of Free-Ranging Cheetah Populations in 2016 (*cont.*)

Region	Area name	Countries	Resident range (km²)	Population size	Map unit
Western, Central and Northern Africa	Adrar des Ifoghas/Ahaggar/Ajjer and Mali	Algeria/Mali	762,871	191	WCN1
	Central African Republic (CAR)/Chad	CAR/Chad	238,234	238	WCN4
	W-Arly and Pendjari protected area complex	Benin/Burkina Faso/Niger	25,345	25	WCN5
	Air and Air and Ténéré	Niger	8,052	2	WCN2
	Termit Massif	Niger	2,820	1	WCN3
	Region-wide		1,037,322	457	
Asia	Southern landscape	Iran	107,566	NA	I1
	Northern landscape	Iran	33,445	NA	I2
	Kavir	Iran	5,856	NA	I3
	Region-wide		146,867	<40[f]	
Global Total			3,123,713,	7,089	

Population estimates were derived using various methodologies, including primarily expert-based opinion, and thereby likely having wide margins of errors. Estimates refer to adult and adolescent animals only. For a spatial layout of populations refer to Fig. 4.1.

[a] *Cheetahs managed as part of the South African metapopulation program are not included in this table.*

[b] Population breakdown, *Namibian resident range (S18), 1,498 (506,980 km²); Botswana resident range (S15), 1,547 (429,622 km²); Northern South Africa and Kruger NP (S14), 696 (142,303 km²); Iona NP and surrounds (S1); 39 (44,966 km²); Tuli Block (S17), 142 (23,204 km²); Limpopo NP, Lebombo and Sabie (S13), 41 (6,823 km²) data from RWCP and IUCN/SSC, 2015; Luengue-Luiana NP and Mavinga NP and surrounds (S19); 58 (58,281 km²) data from PANTHERA, unpublished data and Fabiano, unpublished data.*

[c] Population breakdown, *Greater Hwange to Victoria Falls (S4), 41 (23,340 km²) data from Van der Meer, 2016; Pandamatenga (S16), 5 (1,456 km²) data from RWCP and IUCN/SSC, 2015.*

[d] Population breakdown: *Gonarezhou NP and Malilangwe (S9), 28 (5,733 km²); Save Valley Conservancy (S8), 10 (2,664 km²) data from Van der Meer, 2016; Maunge (S11), 6 (844 km²) data from RWCP and IUCN/SSC, 2015.*

[e] Population breakdown: *Mana Pools National Park, part of Sapi and Chewore (S6), 12 (3612 km²), Matusadona National Park (S5), 3 (1,422 km²) data from Van der Meer, 2016.*

[f] *Population estimates by Durant et al., 2017 were Southern landscape (I1), 20; Northern landscape (I2), 22; Kavir (I3), 1. However, recent data analyses estimates the overall population to be less than 40 adults (Farhadinia et al., 2016).*

Modified from Modified from Durant et al., 2017; with updated data from Farhadinia et al., 2016; and Van der Meer, 2016

ranged over most Savanna ecosystems in the region (Shortridge, 1934). Resident or transient populations are now known to persist in only 22.9% of this historical range (RWCP and IUCN/SSC, 2015; Table 4.1), of this resident range 75% is located outside of protected areas (Durant et al., 2017). Most of this loss has taken place in the heavily utilized and populated areas of South Africa, Malawi (believed extinct), and Zimbabwe (RWCP and IUCN/SSC, 2015). The decline from 6260 adults and adolescents in 2007

(IUCN/SSC, 2007a) to todays' numbers may be partially attributed to an actual loss in numbers, as well as to the availability of greater and more detailed data (RWCP and IUCN/SSC, 2015). Data on cheetah distribution and conservation status are lacking in 26.1% of the cheetahs historic range in the southern region (RWCP and IUCN/SSC, 2015; Table 4.1). An additional 6.6% is considered possible range, where the cheetah may still be resident, but has not been sighted in the last 10 years (Table 4.1).

Major threats to cheetah persistence in the region are generally shared throughout sub-Saharan Africa (Table 4.3). However, conflict with game farmers over the loss of valuable game animals to cheetah predation is a distinct and important challenge to cheetah conservation in southern Africa and is on the rise (RWCP and IUCN/SSC, 2015).

Also unique to the southern Africa region (specifically Namibia and Zimbabwe) is a cheetah trophy hunting program aimed at providing a financial incentive for landowners to tolerate cheetah presence. Export of these cheetah trophies has been regulated under quota since 1992 by the Convention on International Trade in Endangered Species of Wild Fauna and Flora (CITES, 1992; Chapter 21).

Angola

Map unit: S1, S19, S20 (Table 4.2; Fig. 4.1).

The cheetah in Angola is classified as endangered by the Angolan Wildlife Authority (van Vuuren et al., 2010). Angola has different

TABLE 4.3 Major Threats to Cheetah Populations

Region	Country	Threats	Source
Southern Africa	Region-wide	Habitat loss and fragmentation, conflict with livestock and game farmers, prey loss, accidental snaring, road accidents, small population size, infectious disease[a], illegal hunting, illegal trade, irresponsible tourism[a], increased use of poison, poor coexistence with communities, detrimental land use policies, insufficient political commitment, military land mines, corruption in law enforcement agencies, loss of resilience in cheetah populations due to climate change	RWCP and IUCN/SSC (2015)
	Angola	Habitat loss, illegal killing of cheetahs, illegal killing of prey	This chapter
	Botswana	Habitat loss and fragmentation, conflict with livestock and game farmers, prey loss, accidental snaring, road kills, infectious disease[b], illegal hunting of cheetahs, illegal trade	DWNP (2009)
	Mozambique	Habitat loss and fragmentation, illegal killing of prey, illegal killing of cheetahs for traditional medicine uses and lack of effective implementation of strategy and management plans focusing on cheetahs	Fusari et al. (2010); this chapter
	Namibia	Habitat loss, conflict with livestock and game farmers, indiscriminate killing of cheetahs, land use changes, legal hunting[b]	Ministry of Environment and Tourism (2013); this chapter
	South Africa	Habitat loss, conflict with livestock and game farmers, game farm fencing, illegal trade	Lindsey et al. (2009)
	Zambia	Habitat loss, illegal killing of prey, illegal killing of cheetahs	This chapter
	Zimbabwe	Habitat loss and fragmentation, poverty, resource constraints, poor governance, prey loss, snaring, dominant competitors, and to a lesser extent conflict with livestock and game farmers and unsustainable offtake	Van der Meer (2016)

(Continued)

TABLE 4.3   Major Threats to Cheetah Populations (*cont.*)

Region	Country	Threats	Source
Eastern Africa	*Region-wide*	Habitat loss and fragmentation, conflict with livestock farmers, prey loss, accidental snaring, road accidents, illegal hunting, infectious disease[b], illegal trade, poorly managed tourism	IUCN/SSC (2007b)
	Ethiopia	Habitat loss and fragmentation, conflict with livestock farmers, prey loss, poorly managed tourism[a], infectious disease[b], illegal trade	EWCA (2012)
	Kenya	Habitat loss and fragmentation, conflict with livestock farmers, prey loss, incidental snaring, road accidents, poorly managed tourism[a], infectious disease[b], illegal hunting of cheetahs, illegal trade[b]	KWS (2010)
	South Sudan	Illegal trade, lack of management plan, competition with livestock for space, competition between prey and livestock for grazing, habitat loss, absence of policy, poor infrastructure, lack of strict laws for carnivore protection, legislation not clear about utilization, insufficient community involvement, underfunded ministry, inadequate budget distribution, weak law enforcement, illegal killing, snaring, conflict with livestock farmers, poor capacity, prey depletion, dominant competitors	South Sudan Wildlife Service (2010)
	Tanzania	Habitat loss and fragmentation, conflict with livestock farmers, prey loss, infectious disease[b], accidental snaring, road accidents, poorly managed tourism, illegal trade, dominant competitors	Ministry of Natural Resources and Tourism (2014)
	Uganda	Habitat loss, conflict with livestock farmers, depleted wildlife prey, poaching	Uganda Wildlife Authority (2010)
Western, Central, and Northern Africa	*Region wide*	Habitat loss and fragmentation, conflict with livestock farmers, prey loss, accidental snaring, road accidents, illegal hunting, infectious disease[a], small population size	IUCN/SSC (2012)
Asia	Iran	Habitat loss, road accident, conflict with livestock farmers, illegal killing of cheetah and its prey	Chapter 5

[a] *Low threat.*
[b] *Potential threat of unconfirmed significance.*

biomes, with a vast biodiversity from north to south, and has the potential to be a stronghold for cheetahs. However, a 27-year armed conflict (1975–2002) had negative impacts on Angolan wildlife, including the cheetah population. After 14 years of peace, researchers are now conducting field projects with the aim of updating data on cheetah distribution in Angola.

Historically, cheetahs occurred across 11 provinces (Bengo, Bié, Cunene, Huambo, Huíla, Cuando Cubango, Luanda, Lunda Sul, Malanje, Moxico, and Namibe), but today they are only confirmed at 3 of these locations: Namibe (mainly in the region of the National Park of Iona; S1), Cuando Cubango (particularly in Mavinga and Luengue-Luiana National Parks; S19), and Moxico (outside of protected areas; S20) (Durant et al., 2017; Marker et al., 2010). With regards to other provinces the status is unknown, with additional research needed. There is possibly free movement of animals between these localities and cheetah range in Namibia. While the total size of the cheetah population in Angola is unknown, the number of animals in the 3 confirmed locations is estimated at approximately 123 adult and adolescent cheetahs (RWCP and IUCN/SSC, 2015).

Despite the armed conflict ceasing in 2002, the cheetah population in Angola is thought to be declining due to anthropogenic factors. Poaching of prey species continues in parks and reserves for internal and external commercialization (Bersacola et al., 2014). Cheetah meat is consumed as a source of protein and the species' teeth, skin, bones, and paws can be used for crafts, religious props, and superstitious purposes (Fabiano, unpublished data). Additionally, cheetah habitat is threatened by expanding urbanization and agriculture, and indiscriminate landscape burning (Table 4.3).

In order to reverse the current probable declining status of the cheetah in Angola, the Angolan Ministry of Environment (National Institute of Biodiversity and Conservation Areas) and the Ministry of Agriculture (Institute of Veterinary Services and Forestry Development Institute) are pursuing a number of actions including mitigating human-wildlife conflict through environmental education at all levels, and drafting and updating wildlife legislation. Antipoaching efforts are being augmented by increasing the number of rangers and investing in crossborder security to reduce illegal trafficking of wildlife.

Botswana

Map unit: S15, S16, S17 (Table 4.2; Fig. 4.1).

With an estimated 1694 cheetahs mainly in one large contiguous population, Botswana is one of the most important range states (RWCP and IUCN/SSC, 2015). The cheetah has historically been highly regarded in Botswana. In 1967 the cheetah's status was changed from "Royal Game" to "Conserved Animal" by the colonial forces, and

"tribally conserved animal" by Bangwato and Batawana tribes (Spinage, 1991). Currently, the cheetah is considered a "Protected Game Animal" in the Wildlife Conservation and National Parks Act of 1992, the highest level of species protection and as such cannot be legally hunted or captured except in accordance with a permit issued by the Director of Wildlife and National Parks (Botswana Government, 1992). Despite its high level of protection under the law, Botswana's cheetahs are faced with a number of threats, primarily, human-wildlife conflict (Table 4.3).

Cheetahs are still widely distributed in Botswana, with most of the country consisting of resident range. The exceptions are the north–east district, and parts of Chobe and Central Districts (RWCP and IUCN/SSC, 2015). Cheetahs are concentrated in the south-western parts of the country, where population densities of other large carnivores and humans are relatively low. A large part of the cheetah's distribution is outside protected areas with the highest population densities recorded in agricultural zones (Klein, 2007). Despite significant threats to cheetahs in the country, Botswana's population is one of the few in the region to have remained stable, with little change since the estimated 1768 cheetahs in 2007 (Klein, 2007) and the 1000–2000 stated by Myers in 1975. Botswana is a priority area in southern Africa, as it links cheetah populations with all of its neighboring countries. Transboundary conservation is therefore of vital importance. Transboundary protected areas like the Kgalagadi Transfrontier Park are important areas linking South Africa with Botswana. Similarly, northern areas of Botswana are now incorporated in the Kavango-Zambezi Transfrontier Conservation Area (KAZA-TFCA), which focuses on building corridor areas for wildlife populations in southern Zambia and Angola, with those in northern Namibia, Zimbabwe, and Botswana. However, with key populations of cheetahs primarily in western Botswana, similar conservation areas linking with eastern Namibia are necessary.

Mozambique

Map unit: S11, S12, S13 (Table 4.2; Fig. 4.1).

Cheetahs were once present throughout the west, north, and south of Mozambique (Purchase, 2007a). Habitat loss, decreasing numbers of prey and persecution hunting for skins and traditional medicine resulted in a substantial population decline by the mid-1900s (Purchase, 2007a). This was further exacerbated by the lack of wildlife law enforcement during the 16-year civil war (1977–92), which allowed the illegal hunting and exportation of cheetahs to continue. The wildlife sector now has a high-level of political support and cheetahs are a protected species. A 1975 report estimated the population size to be 200 individuals, with cheetahs located in three regions: South of the Zambezi River, between the Gorongosa National Park and the Pungue River, and in and around the Limpopo Valley National Park (Myers, 1975).

The population has since declined and is currently estimated at 57 individuals (RWCP and IUCN/SSC, 2015). Of the areas identified by Myers (1975) as cheetah strongholds, the species is now considered locally extinct in Gorongosa National Park, as well as Niassa Game Reserve, Gile Game Reserve, Maputo Game Reserve (Purchase et al., 2007), and Zinave National Park (Nhabanga et al., 2011). Cheetahs are considered transient in the Limpopo Valley National Park (S13; Andresen et al., 2012), possibly as a result of the connection with Kruger National Park (South Africa) as part of the Greater Limpopo Transfrontier Park. Survey and camera-trap records also revealed that cheetahs are present at Sabie Game Reserve (S13; Nhabanga et al., 2011). Banhine National Park (S12) has a small, confirmed population of cheetahs, which naturally recolonized from South Africa following an increase in conservation efforts in the area (Andresen et al., 2015). A small, transboundary population of cheetahs is present in the Maunge complex (S11) and neighboring national park and conservancy land in Zimbabwe (S8, S9), this threatened population needs special attention.

Cheetahs persist outside protected areas in Mozambique, albeit in very low numbers. They have been observed in Chicualacuala area. Additionally, two confirmed cheetah sightings suggest a small population of cheetahs in the Tete Province. Cheetah data for Mozambique are scarce and mostly outdated as large areas of Mozambique have not been surveyed (Purchase, 2007a).

Namibia

Map unit: S18 (Table 4.2; Fig. 4.1).

Cheetahs are classified as protected game in Namibia under the Nature Conservation Ordinance of 1975 (SWA, 1975), but may be killed or captured without a permit in defense of human life and livestock when "the life of such livestock is actually being threatened." Historically, cheetahs occupied all of Namibia, with population densities varying primarily based on rainfall and ecosystem productivity. Besides a reduced number in the southern part of Namibia, the overall national distribution of cheetahs appears to have changed little since the earliest accounts (Joubert and Mostert, 1975; Marker-Kraus et al. 1996; Shortridge, 1934); cheetahs are still widely distributed especially in the central and northwestern parts of the country.

The majority (over 80%) of cheetahs live outside protected reserves on commercial livestock farms, game ranches, and communal conservancies (Marker et al. 2010; Morsbach, 1987). As a result, the persistence of Namibia's cheetah population depends on the strategic management of these areas with best-practice livestock protection mechanisms in place. The number of game fenced farms is increasing in Namibia, resulting in greater conflict with game farmers (Marker et al., 2010; Chapter 13). Currently the Namibian population is estimated at 1498 adults and adolescents (RWCP and IUCN/SSC, 2015). Long-term abundance and distribution studies in north central Namibia reported cheetah densities of 0.6/100 km² (Fabiano, 2013). However, further research is needed on the population density of cheetahs on farmlands, as well as in the vast communal conservancies of northwest Namibia.

Comparison of estimated cheetah resident range in 2007 (Purchase et al., 2007) to 2015 (Durant et al., 2017) suggests the cheetah's range has expanded to the south and to the west of the country. This range expansion is encouraging; however, cheetah densities in this arid landscape are low (0.2 individuals/100 km²; RWCP and IUCN/SSC, 2015). This can be attributed to the reintroduction of cheetahs to this region and to changing land uses from smallstock farming to more wildlife-based land use.

South Africa

Map unit: S14 (Table 4.2; Fig. 4.1).

The cheetah in South Africa is listed as vulnerable under the country's National Red List (Van der Merwe et al., 2016). Legislation regarding the protection of cheetahs varies by province, but in all provinces the cheetah receives some degree of protection and a permit is required for its removal (Purchase et al., 2007). Historically, cheetahs were widely distributed throughout South Africa with the exception of forests (Marker, 1998). Owing to threats, including direct human persecution, habitat loss and illegal trade, the distribution of cheetahs in South Africa has shrunk dramatically. Cheetahs are considered resident in 12% of their historical range equivalent to 142,303 km² (RWCP and IUCN/SSC, 2015), of which 20% of the resident range occurs inside protected areas (Marnewick et al., 2007; RWCP and IUCN/SSC, 2015).

The South African cheetah population can be divided into; free-roaming/naturally occurring cheetahs (both, inside and outside large protected areas) and cheetahs found in small to medium-sized reserves (managed as a metapopulation; Marnewick et al., 2009). Protected areas are conservatively estimated to support 412 cheetahs (RWCP and IUCN/SSC, 2015), this number is likely to be an underestimate as the Kruger National Park alone has been reported to hold 412 cheetahs (Marnewick et al., 2014) and the South African portion of the Kgalagadi Transfrontier Park has an estimated population of 80 cheetahs (Mills et al., 2017). The only free-roaming cheetahs outside of protected areas, persist on the border with Botswana and Zimbabwe (IUCN/SSC, 2007a; Marnewick et al., 2007). In the 1990s the number of cheetahs on non-protected land increased as many cattle farms in the area switched to game ranching, thereby increasing the available prey for cheetahs. Conservatively, a population of 284 individuals is estimated to occur outside protected areas, with the total free-ranging population of cheetahs in South Africa estimated to be 696 adults and adolescents (RWCP and IUCN/SSC, 2015).

This estimate excludes the cheetah held in the metapopulation. The metapopulation management program was established in 2009 following an increase in the number of small game reserves holding cheetahs (Castley et al., 2001; Smith and Wilson, 2002; Wells, 1996), in part due to a compensation-translocation program (Chapter 20). In August 2017, the managed metapopulation comprised of 325 cheetahs, held in 54 fenced reserves (16 state-owned and 38 private game reserves; Endangered Wildlife Trust, unpublished data, 2017).

Despite human persecution, the free-ranging cheetah population in South Africa continues to persist, but as the number of game ranches with electrified "predator-proof" fencing increases, blocking predator access to this habitat, it is feared the population will decline.

Zambia

Map unit: S2, S3 (Table 4.2, Fig. 4.1).

In Zambia, cheetahs are specially protected, but can be killed with special permission from the government wildlife authority to control predation on livestock (Purchase, 2007b). Like elsewhere in Africa, cheetahs in Zambia have suffered a major reduction in range over the last half-century. Cheetahs have disappeared not just across unprotected lands, but also in eight of the national parks. The species is now mostly confined to a handful of protected areas in the west of the country. The only relatively large cheetah population occurs in the Greater Kafue Ecosystem, a 66,000 km² protected-area complex with Kafue National Park at its center (S3). In 2015, through a citizen science program (Chapter 34), cheetah sightings were collated, mostly from inside the park boundary and in the adjacent Namwala, Mumbwa West, and Shishifulo game management areas. The population size in the Kafue Ecosystem is not known with certainty, but it is currently estimated at 65 adults and adolescents (Durant et al., 2017); a research program to estimate population size using mark-recapture and photographic sighting is underway. Owing to human occupancy in the area, cheetahs are considered locally extinct outside most of the Greater Kafue Ecosystem, including Lochinovar and Blue Lagoon National Parks (Myers, 1975).

A small population of cheetahs also remains in northwest Zambia in Liuwa National Park (S2). The park includes expansive cheetah habitat, comprising open grassland plains with relatively abundant prey. However, cheetah population densities appear low for this park, while spotted hyena (Crocuta crocuta) population densities are unusually high (Zambian Carnivore Program, unpublished data).

Historically, cheetahs were also present across the Bangweulu Basin in northeast Zambia (Grimsdell and Bell, 1975). Since the 1980s, cheetahs were considered, no longer present, although a cheetah sighting in this region was recorded in Kasanka National Park, in 2000 (Purchase, 2007b). In southeast Zambia

cheetahs were once present in low abundance in the Lower Zambezi complex, including Lower Zambezi National Park. In 1994 an attempt was made to reintroduce cheetahs to this area, but two of the three released cheetahs were caught in snares shortly after release (Phiri, 1996). Until 2000 cheetahs were also recorded sporadically in the Luangwa Valley in Zambia's eastern province and the Munyamadzi Corridor between North and South Luangwa Parks (Purchase, 2007b).

The only additional area where cheetahs have been confirmed in the last decade is Zambia's southwestern corner in Sioma Ngwezi National Park, where they are potentially connected to populations in Namibia and southeast Angola. During a large carnivore survey in 2015, cheetahs were only detected once in this area, suggesting a small remaining population (Zambian Carnivore Program, unpublished data).

Overall, a minimum population of 85 cheetahs was reported across Zambia (RWCP and IUCN/SSC, 2015); however, 60% of Zambia is classified as unknown status and requiring survey (RWCP and IUCN/SSC, 2015). While there is a need to survey some areas where cheetahs may still remain, for example, West Lunga National Park in the very northern portion of Zambia, an apparent absence of cheetah-livestock conflict is consistent with cheetah being largely absent outside protected areas.

Cheetah range reduction seems to have coincided with increased human population and agricultural development competing with cheetah for habitat, together with a substantial decline in prey species (Myers, 1975; Purchase, 2007b). Today the main threats to cheetahs in Zambia are continued habitat reduction and bush-meat poaching, which substantially depletes prey and causes cheetah mortality in snares (Table 4.3). Many of Zambia's national parks are classified as recoverable habitat and there are plans to reintroduce cheetahs to some of these sites. However, for the foreseeable future securing the remaining populations in the

Greater Kafue Ecosystem and the Liuwa National Park is crucially important for ensuring the persistence of this species in Zambia.

Zimbabwe

Map unit: S4, S5, S6, S7, S8, S9, S10 (Table 4.2; Fig. 4.1).

In Zimbabwe the cheetah is a specially protected species under the 1975 Parks and Wildlife Act (Heath, 1997). No person is allowed to hunt, possess, or sell live cheetahs or parts of cheetahs, unless a permit is issued by the Zimbabwe Parks and Wildlife Management Authority (Parliament of Zimbabwe, 2001).

Cheetahs occurred historically, throughout Zimbabwe (ZPWMA, 2009), with population estimates ranging from 400 in 1975 (Myers, 1975) to more than 1500 cheetahs in 1999 (Davison, 1999a). The apparent increase in cheetah numbers was thought to have occurred on commercial farmlands partly in response to many Zimbabwean farmers stocking game in the 1980s (Davison, 1999b; Wilson, 1988). In 1992, 80% of the cheetah population was found on commercial farmlands (CITES, 1992; Davison, 1999a) and in an attempt to reduce human-cheetah conflict CITES granted Zimbabwe an annual export quota of 50 animals, which helped to establish a trophy hunting program and allow landowners to derive income from cheetahs (Chapter 21). In 2000 with introduction of Phase II of the government's land reform program, vast areas of large-scale commercial farming were converted into indigenous small-scale commercial and subsistence farming (du Toit, 2004), and the game ranching industry in Zimbabwe virtually collapsed (Lindsey et al., 2009). The changes in land use caused by the land reforms, the economic depression that followed and an increase of the human population resulted in overexploitation of natural resources, and degradation of wildlife populations and habitat, including that of cheetahs (du Toit, 2004; Lindsey et al., 2011).

A nationwide cheetah assessment showed that these developments had a severe impact on Zimbabwe's cheetah population. Resident cheetah range is thought to have declined by 61% and the population has reduced to ca. 150–170 adult and adolescent cheetahs (Van der Meer, 2016). The majority of this remaining population (80%) is found in protected areas (Van der Meer, 2016). As a result human-cheetah conflict is minimal and no longer considered the main conservation threat (Van der Meer, 2016). The largest, free-roaming, cheetah populations are found in Greater Hwange (S4), and Gonarezhou National Park and neighboring concessions (S9 and S8; Table 4.2). These populations are part of transfrontier populations with Botswana and Mozambique, respectively (Durant et al., 2017). In addition, there are a few relatively isolated cheetah populations in fenced conservancies (S7, S10). With the shift from historically free-ranging populations outside of protected areas to populations limited to protected areas, it is necessary to revise cheetah management in Zimbabwe. Conservation efforts should focus on securing the main free-ranging populations and facilitate their connectivity with the larger transfrontier populations.

EASTERN AFRICA

The range of the 3 cheetah subspecies in eastern Africa (Smithers, 1975) has contracted severely since the 1970s (IUCN/SSC, 2007a). Resident populations of cheetahs now occur in Ethiopia, Kenya, South Sudan, Tanzania, and Uganda with a population estimate of 2290 individuals (Table 4.2).

A. j. raineyi Heller, 1913 (for which the name *ngorongorensis* Hilzheimer, 1913 has priority) is still resident in Kenya, northern Tanzania, and Uganda; this subspecies was merged with *A. j. jubatus* (Kitchener et al., 2017) due to genetic similarity (Charruau et al., 2011; see Chapter 6 for more detail). *A. j. soemmeringii* (Fitzinger, 1855) is found in South Sudan and Ethiopia. Cheetah distribution is unknown in some parts of the region, particularly in Somalia, where human safety concerns in the context of decades of political instability and border disputes make field surveys unfeasible. While it is unconfirmed whether cheetahs are still present in Somalia, they might be persisting at the southern end of the country in Lag Badana National Park (Amir et al., 2016). In the north, cheetah presence is possible in the Nugal Valley area (Amir unpublished data), which spans across the Puntland State of Somalia and eastern Somaliland, a self-declared autonomous region. Cheetahs might also be present along most of the Ethio-Somali border of western Somaliland, either as resident or transboundary populations, based on illegal cheetah trade confiscations and personal interviews (CCF unpublished data). Somalia (and Somailand) have a high incidence of illegal wildlife trafficking of cheetahs, but trafficked cheetahs are thought to primarily originate from the neighboring countries of Ethiopia and Kenya (Chapter 14).

Resident cheetah populations in eastern Africa are known to persist in just 6% of the species regional-historical range (pre 1900; IUCN/SSC, 2007a). Close to 75% of this resident range is located outside of protected areas (Durant et al., 2017). Cheetahs are protected under national law in all the countries with extant populations in the region. In addition to being subjected to similar threats, as some of the other cheetah populations in sub-Saharan Africa, poorly managed tourism, is a unique threat to cheetahs in eastern Africa (Table 4.3; KWS, 2010; Chapter 39).

Ethiopia

Map unit: E1, E2, E3, E4, E5 (Table 4.2; Fig. 4.1).

Once widespread throughout Ethiopia, excluding the country's high mountain areas (IUCN/SSC, 2007a), the cheetah population has since experienced a substantial decline in its geographical range. Five subpopulations are still found in the country (Durant et al., 2017; Table 4.2), with cheetahs mainly persisting at low densities in the southern plains; namely neighboring the southern and south–western borders to Kenya and South Sudan (E5). Cheetahs are also found in three populations in central northeastern Ethiopia (E1, E2, E4) and one population close to the border with Somalia (E3) (Ethiopian Wildlife Conservation Authority; EWCA, 2012). The population size in this Ogaden region (E3) is uncertain. The region is ecologically distinct (being more arid) from other areas where the countries resident cheetah populations are known to persist (EWCA, 2012). A high number of cheetah cubs have been confiscated from illegal trade in the area in recent years (Chapter 14), which probably reflects that there are a large number of cheetahs there. It is important that surveys are conducted in this area. However, there is increased political and military strife in the region, which puts increased pressure on wildlife and the environment, and continues to prevent conservation work.

Cheetahs have been recorded in 10 of Ethiopia's protected areas, namely the Yabello Wildlife Sanctuary, Garhaile National Park, Chebera Churchura National Park, Mago National Park, Mazie National Park, Nechisar National Park, Omo National Park, Senkele Wildlife Sanctuary, Awash National Park, and Babille Elephant Sanctuary (EWCA, 2012). The Dawa River valley on the Ethiopia–Kenya border, and the Borena region north of Moyale contain important cheetah habitats outside protected areas.

Half of the cheetahs' range in Ethiopia is outside of parks and reserves, and cheetah population densities within protected areas is low (EWCA, 2012). Transfrontier cheetah conservation management, both inside, as well as outside protected areas, is therefore important for cheetah survival, as well as for the region's economy through tourism (EWCA, 2012). Areas that have extirpated cheetah populations, but are suitable for recovery, have also been identified (EWCA, 2012).

Kenya

Map unit: E5, E6, E10, E11, E12 (Table 4.2; Fig. 4.1).

Cheetahs occurred across most of Kenya before human activity modified substantial proportions of the natural habitat (Myers, 1975). Cheetahs are confirmed resident in about 23% of their historical range; another 56% (most of northern and eastern Kenya) of their former range may support nonconfirmed populations (IUCN/SSC, 2007a). Recent surveys have shown cheetah occurrence in some of the "unknown" and "possible" range (Wykstra, Action for Cheetahs, unpublished data), but it is uncertain to what extent individuals in these areas are resident versus transitory. Over 80% of resident range and over 95% of possible range fall outside protected areas (IUCN/SSC, 2007a).

Two large and globally important resident cheetah populations are found in Kenya and Tanzania. These populations are comprised of Laikipia/Samburu (E11), which falls entirely in Kenya, and Serengeti, Maasai Mara, Maasailand, and Tsavo NP (E12), which spans the Kenya–Tanzania border (Table 4.2). The Laikipia/Samburu population may also be connected to the small South Turkana cheetah population (Wykstra, unpublished data; Table 4.2; E10). Additional small populations might possibly persist in Meru and Kora National Parks, and Boni National Reserve (eastern Kenya), the latter potentially connected to Somalia. A small population might still be residing in north-western Kenya bordering South Sudan and Uganda (E6), and another small population in Northern Kenya is thought to link to populations in Ethiopia and South Sudan (E5).

Conservation activities outside protected areas are crucial to safeguard Kenya's cheetah population, both to maintain the range and allow for connectivity. Measures, such as the proposed designation of "carnivore conservation zones" on private and community lands (IUCN/SSC, 2007a) will be vital for the long-term survival of this species in Kenya.

South Sudan

Map unit: E5, E6, E7, E8, E9 (Table 4.2; Fig. 4.1).

Cheetahs were historically widespread, in low population densities, across South Sudan. Today, most of South Sudan's known cheetah ranges are close to or on its international boundaries (South Sudan Wildlife Service, 2010). One transfrontier population is found in Boma National Park on the eastern border with Ethiopia (E5). Populations bordering other countries include those in Radom National Park on the western border with Central African Republic (E9), as well as in the Elemi region, which borders Omo National Park, Ethiopia and Turkana plains, Kenya (E5). Smaller populations occur in Kidepo Game Reserve, which is contiguous with Uganda's Kidepo Valley National Park (E6), as well as possibly in Lantoto National Park, which is contiguous with Congo's Garamba National Park. Other important cheetah populations are located in Southern National Park (E8) and Badingilo National Park (Table 4.2; E7). Because much of the cheetah range in South Sudan overlaps with neighboring countries, transfrontier cooperation on cheetah monitoring and management is critical.

Most of the cheetah survey work in South Sudan has focused on protected areas. However, like elsewhere in Africa, a large proportion of South Sudan's cheetahs may be located outside protected areas, with reported sightings in Pibor, Kapoeta, and Elemi (South Sudan Wildlife Service, 2010). Identifying unprotected cheetah habitat and providing connectivity between cheetah populations that might inhabit areas outside parks and reserves will be important for ensuring the long-term viability of cheetah in South Sudan (South Sudan Wildlife Service, 2010). Such considerations will need to be taken into account as South Sudan rebuilds its infrastructure in the wake of the past civil war (1983–2005).

Tanzania

Map unit: E12, E13, E14, E15 (Table 4.2; Fig. 4.1).

Historically cheetahs were distributed across most of the country, with the exception of a coastal strip of up to 160 km wide from the north–east of the country and reaching southwards to 40 km north of Lindi, were cheetah residence is unconfirmed (Ministry of Natural Resources and Tourism, 2014). As in other parts of Africa, cheetahs in Tanzania have experienced a significant decline in range. Cheetahs are resident across much of the northern part of the country along large sections of the border with Kenya, and in the Maasai Steppe (Table 4.2; E13). They are also resident within the Ruaha-Rungwa (E14) and Katavi-Ugalla (E15) areas (Ministry of Natural Resources and Tourism, 2014).

More than half (56%) of the cheetah's resident range in Tanzania, and 37% of the possible range fall within protected areas (Ministry of Natural Resources and Tourism, 2014). The area between Ruaha-Rungwa and Katavi-Ugalla represents a significant expanse of mostly unprotected possible cheetah range and one of strategic importance, considering its potential to connect the two resident populations. Cheetahs were present in the Selous Game Reserve, but the last confirmed sighting there was in the late 1990s (Siege, personal communication).

Given the cheetah's low-population density, the populations inside protected areas are almost certainly dependent on adjoining unprotected lands for their long-term viability (IUCN/SSC, 2007a). Hence, conservation activities both inside and outside of reserves are important, as well as transfrontier cooperation.

Uganda

Map unit: E6 (Table 4.2; Fig. 4.1).

Cheetahs have historically inhabited the north–eastern semiarid part of Uganda. In the 1970s cheetahs were reported to occur in several regions in this area (Gros and Rejmanek, 1999), with a population of possibly 100–250 individuals (Myers, 1975). More recently the species is thought to only be found in Kidepo Valley National Park (Uganda Wildlife Authority, 2010; E6). It is possible that some cheetahs exist further south in the Karenga Community Wildlife Area or elsewhere in Karamoja, but there have been no recent sightings to confirm residence (Uganda Wildlife Authority, 2010). The current cheetah range in Uganda is contiguous with small sections of south Sudan and north–west Kenya (E6). This transfrontier population is thought to contain 19 adult and adolescent cheetahs (Durant et al., 2017), with less than 10 individuals resident in Uganda (Uganda Wildlife Authority, 2010).

The decline in cheetah distribution was associated with several human-mediated factors, and habitat loss might be a lesser concern to cheetah conservation in Uganda than direct persecution of cheetahs and their prey by humans (Gros and Rejmanek, 1999). Cheetahs have been illegally hunted for their skins, shot as a preemptive means to limit livestock predation, or because of confusion with more dangerous animals (e.g., leopards Panthera pardus) (Gros and Rejmanek, 1999; Table 4.3). Cheetah prey populations have also been depleted through poaching for meat and trophies (Edroma, 1984), potentially curtailing seasonal migrations of herbivores (Gros and Rejmanek, 1999).

Owing to the transboundary distribution, crossborder cooperation in addition to mitigating human threats is necessary to safeguard this last-known resident cheetah population in Uganda.

WESTERN, CENTRAL, AND NORTHERN AFRICA

The cheetah was historically widespread in western, central, and northern Africa, with the exception of the coastal area and the rainforest belt. The species range has experienced significant decline, with resident populations known to persist in only 9% of historical range and considered irretrievably lost from 57% of the region (IUCN/SSC, 2012; Table 4.1). Today the cheetah population is highly fragmented and the species persists in only five populations in seven countries in the region (IUCN/SSC, 2012). Eighty-three percent of the resident range is situated in unprotected lands (Durant et al., 2017). Cheetahs in the region are confronted with many of the same major threats as those in other areas of Africa (Table 4.3), but prey loss (Chapter 11) and small population sizes (Chapter 10) are probably more critical limiting factors here than for southern and eastern African populations.

Although there is no full agreement on the number of cheetah subspecies in western, central, and northern Africa, two subspecies are generally thought to remain in the region (Smithers, 1975). In northwest Africa the critically endangered subspecies, *A. j. hecki*, is still found in one contiguous transfrontier population that includes northern Benin, southern Burkina Faso, and southern Niger (Belbachir 2008; Chapter 11; Table 4.2; WCN5). In central Africa *A. j. soemmeringii* is found in a contiguous population in Chad and northern Central African Republic (CAR; WCN4).

Algeria, Benin, Burkina Faso, Central African Republic (CAR), Chad, Mali, and Niger

Map unit: WCN1, WCN2, WCN3, WCN4, WCN5 (Table 4.2; Fig. 4.1).

Cheetahs are protected under national law in all countries in the region, but in practice this protection has not prevented cheetah populations from declining. Today the cheetah population in the region is small (457; Table 4.2) and highly fragmented (Durant et al., 2017; IUCN/ SSC, 2012). Overall, the conservation status of the cheetah is poorly documented (Caro, 2013; Claro et al., 2006; Di Silvestre, 2003; Newby and Grettenberger, 1986; Nowell and Jackson, 1996; Wacher et al., 2005; Chapter 11) and research into cheetah range dynamics and population trends in the region is urgently needed.

Central Africa only harbors 1 resident cheetah population. This transfrontier population (WCN4) in south-eastern Chad and northern CAR (IUCN/SSC, 2012; Table 4.2) has a population estimate of 238 individuals and occurs predominantly (81.4% of cheetahs) on unprotected land (Durant et al., 2017). In contrast, almost the entire cheetah population in western Africa is found in protected areas (IUCN/SSC, 2012).

Two additional transfrontier populations exist in the region. Northern Africa's transfrontier population is located in south-central Algeria and north-eastern Mali (estimated at 191 cheetahs: Table 4.2; WCN1), and possibly extends into western Libya (Belbachir et al., 2015). Western Africa's transfrontier population is located in the W-A-P ecosystem (W transfrontier Biosphere Reserve in Benin, Burkina Faso and Niger, Arly National Park in Burkina Faso, and Pendjari Biosphere Reserve in Benin; estimated at 25 individuals: Table 4.2; WCN5). The region also hosts two subpopulations in the Termit and Tin-Toumma (WCN2) and the Aïr Tenere (WCN3) protected areas in Niger, estimated to contain 3 individuals collectively (Durant et al., 2017, Table 4.2).

While population estimates in the region are speculative, they are based on recent observations. Cheetahs have been sighted recently in all areas except in Mali, where the last sighting occurred pre-2000; on the other hand, sightings in Pendjari Biosphere Reserve (Benin) have increased since 2010 (Berzins et al., 2007; Claro et al., 2006). Camera-trap photos of cheetahs have also been reported in the south-central Saharan Ahaggar in Algeria (Belbachir, 2015), the Termit massif in Niger (Claro and Sissler, 2003; Claro et al., 2006; Rabeil et al., 2008; Sillero-Zubiri et al. 2015), and the Aïr mountains (Dragesco-Joffć, 1993; Newby, 1992).

Considering that protected areas in northern, western, and central Africa do not guarantee the safeguarding of wildlife and other biodiversity, because of illegal uses and anthropogenic pressure, special attention should be paid to areas surrounding protected areas and the education of local communities. Moreover, the conservation of cheetahs in these areas requires integrated transfrontier management of protected areas, because more than 95% of the population resides in transfrontier ecosystems (IUCN/SSC, 2012).

ASIA

Historically, the Asiatic cheetah (*A. j. venaticus*) ranged across 16 countries in southwest and central Asia, from the Arabian Peninsula, through Iran and Afghanistan, to the Indian subcontinent almost to the border with Bangladesh (Jackson, 1998; Nowell and Jackson, 1996). By the late 1800s, much earlier than the other cheetah subspecies, the Asiatic cheetah was starting to vanish from most of it's range due to habitat loss, land fragmentation, conflict with humans and its capture and use in the sport of "coursing" (Chapter 2). Between the 1940s and 1980s the cheetah became extinct across most of it's Asian range, with the exception of Iran (Divyabhanusinh, 1984, 1995; Nowell and Jackson, 1996). The species persisted sporadically until 1980 in the Arabian Peninsula (Harrison and Bates, 1991) and disappeared from central Asia by the mid-1980s, though a few individuals may have persisted longer (Bannikov, 1984; Kaczensky and Linnell, 2015; Lukarevskii, 2001; Mallon, 2007).

Although considered extinct in the countries bordering Iran (e.g., Afghanistan; Habibi, 2003, Pakistan, and Iraq; Hatt, 1959), there have been occasional reports of cheetahs in these areas in the past. A cheetah skin was reported as originating in the province of Samangan in Afghanistan (Manati and Nogge, 2008), but is more likely to have been imported, and there has been no recent evidence of cheetahs near the Iran border. A recent survey in Pakistan also produced no evidence of cheetahs in the Iran–Pakistan borderlands (Husain, 2001). In Iraq cheetahs used to be seen in Basra, close to the Iranian territory (Corkill, 1929), but are no longer thought to occur there.

Southern Turkmenistan and areas in Uzbekistan retain suitable habitat and prey to host a small cheetah population, if recolonization or reintroduction were possible (Kaczensky and Linnell, 2015; Marker, 2012). This region also offers a corridor to the central Asian steppes in the vicinity of Lake Aral, where the species once resided (Breitenmoser, 2002).

Iran

Map unit: I1, I2, I3 (Table 4.2; Fig. 4.1).

Iran is the only country in the region where cheetahs still remain, although in small populations (Farhadinia et al., 2017), with the total number estimated to be less than 40 adults (Farhadinia et al., 2016; Table 4.2). Chapter 5 focuses on this unique remnant population.

CONCLUSIONS

Cheetah distribution has decreased dramatically over the past century and will likely continue to do so as growing human populations and their livestock and farmed game continue to encroach on cheetah habitat. Small fragmented populations are most vulnerable to this growing pressure as they have reduced ability to adapt to environmental and ecological changes, leaving them at a high risk of extinction (Chapters 6 and 10).

Throughout the species range there is a greater need to work closely with governments and communities to enable the remaining cheetah populations to persist, and if possible thrive. While protected areas remain important for cheetah survival, they are not large enough on their own to maintain viable populations (Chapter 19). With 67% of the world's remaining cheetah populations living outside protected areas, community projects (Chapter 16) and the reduction of human-cheetah conflict are key throughout most of the species' range (Durant et al., 2017). Developing effective policies and management tools to mitigate conflict with livestock farmers are necessary to promote coexistence (Chapters 13 and 15).

With the development of range-wide and national planning meetings for cheetahs, beginning in 2007, the implementation of strategic plans needs support, including working with communities and government officials at regional and national levels. These plans need to be integrated into developmental goals for each of the cheetah range countries to recognize the contribution of this species to terrestrial biodiversity. This would assist range-wide conservation planning and also raise awareness of the importance of improving cheetah and predator management strategies.

With these increasing pressures, viable cheetah populations can only be secured if the remaining, relatively large, free-ranging cheetah populations (and their habitats) can be linked in transfrontier populations. Thus, the importance of approaching cheetah conservation at an international landscape level is critical.

References

Amir, O.G., de Leeuw, J., Koech, G., 2016. Assessment of the biodiversity in terrestrial and marine landscapes of the proposed Lag Badana National Park and surrounding areas, Jubaland, Somalia. A report prepared for the IGAD Biodiversity Management Program. World Agroforestry Centre (ICRAF) Nairobi and Somali Wildlife and Natural History Society (SWNHS), Mogadishu, Somalia, pp. 75.

Andresen, L, Everatt, K, Kerley, G., 2015. Preliminary report on the apex predators of Banhine National Park and the potential Limpopo-Banhine corridor. Centre for African Conservation Ecology, Nelson Mandela Metropolitan University, Report No. 61. Available from: http://ace.nmmu.ac.za/ace/media/Store/documents/Technical%20reports/ACE-Report-61-Banhine-Corridor.pdf.

Andresen, L., Everatt, K.T., Somers, M.J., Purchase, G.K., 2012. Evidence for a resident population of cheetah in the Parque Nacional do Limpopo, Mozambique. S. Afr. J. Wildl. Res. 42, 144–146.

Bannikov, A.G., 1984. Cheetah. In: Borodin, A.M. (Ed.), Red Book of the USSR. Vol. 1. Animals, Lesnaya Promyshlennost', Moscow, pp. 48–49 (in Russian).

Belbachir, F. 2008. *Acinonyx jubatus ssp. hecki*. The IUCN Red List of Threatened Species 2008: T221A13035738. http://dx.doi.org/10.2305/IUCN.UK.2008.RLTS. T221A13035738.en.

Belbachir, F., Pettorelli, N., Wacher, T., Belbachir-Bazi, A., Durant, S.M., 2015. Monitoring rarity: the critically endangered Saharan cheetah as a flagship species for a threatened ecosystem. PLoS One 10, e0115136.

Bersacola, E., Svensson, M., Bearder, S., Mills, M., Nijman, V., 2014. Hunted in Angola-surveying the bushmeat trade. SWARA 38, 31–36.

Berzins, R., Claro, F., Akpona, A.H., Alfa Gambari Imorou, S., 2007. Conservation du guépard et développement durable dans les aires protégées du nord Bénin-Mission d'enquête auprès des villageois et des agents d'aires protégées (16/12/2005–26/02/2006). Société Zoologique de Paris, Paris, France.

Botswana Government, 1992. Wildlife Conservation and National Parks Act, Chapter 38:01. Gaborone, Botswana.

Breitenmoser, U. 2002. Feasibility study on cheetah reintroduction in Turkmenistan. Cat News 36, 13–15.

Caro, T., 2013. Big 5 and Conservation. Anim. Conserv. 16 (3), 261–262.

Castley, J.G., Boshoff, A.F., Kerley, G.I.H., 2001. Compromising South Africa's natural biodiversity-inappropriate herbivore introduction. S. Afr. J. Sci. 97, 344–348.

Charruau, P., Fernandes, C., Orozco-TerWengel, P., et al., 2011. Phylogeography, genetic structure and population divergence time of cheetahs in Africa and Asia: evidence for long-term isolates. Mol. Ecol. 4, 706–724.

CITES, 1992. Quotas for trade in specimens of cheetah. Eighth meeting of the Convention of International Trade in Endangered Species of Wild Fauna and Flora, pp. 1–5.

Claro, F., Leriche, H., van Syckle, S.J., Rabeil, T., Hergueta, S., Fournier, A., Alou, M., 2006. Survey of the cheetah in W National Park and Tamou Fauna Reserve, Niger. Cat News 45, 4–7.

Claro, F., Sissler, C., 2003. Saharan cheetahs in the Termit region of Niger. Cat News 38, 23–24.

Corkill, N.L., 1929. On the occurrence of the cheetah (*Acinonyx jubatus*) in Iraq. J. Bom. Nat. Hist. Sot. 33, 760–762.

Davison, B., 1999a. An Estimate of the Status of the Cheetah in Zimbabwe. Zimbabwe Department of National Parks and Wildlife Management, Harare.

Davison, B., 1999b. Zimbabwe's Cheetah Policy and Management Plan. Zimbabwe Department of National Parks and Wildlife Management, Harare.

Di Silvestre, I., 2003. Distribution et abundance des grands carnivores dans le Parc Régional WMoussa A: Dénombrement des Guépards dans le Parc W Niger.

Divyabhanusinh, 1984. The origin, range and status of the Asiatic (or Indian) cheetah or hunting leopard (*Acinonyx jubatus venaticus*). In: Proceeding of Cat Specialist Group Meetingpp. 182–195.

Divyabhanusinh, 1995. The end of a Trail: The Cheetah in India. Banyan Books, pp. 248.

Dragesco-Joffé), A., 1993. La vie sauvage au Sahara. In: Gillet Hubert. Journal d'agriculture traditionnelle et de botanique appliquée, 36e année, bulletin no. 2,1994. Phytogéographie tropicale: réalités et perspectives. Propos d'ethnobiologie, pp. 298–300. Available from: http://www.persee.fr/doc/jatba_0183-5173_1994_num_36_2_3556_t1_0298_0000_2.

du Toit, R., 2004. Review of Wildlife Issues Associated With the Land Reform Programme in Zimbabwe. WWF-SARPO Occasional Paper No. 10. Harare, Zimbabwe.

Durant, S.M., Bashir, S., Maddox, T., Laurenson, M.K., 2007. Relating long-term studies to conservation practice: the case of the Serengeti cheetah project. Conserv. Biol. 21, 602–611.

Durant, S.M., Mitchell, N., Groom, R., Pettorelli, N., Ipavec, A., Jacobson, A., Woodroffe, R., Bohm, M., Hunter, L.T.B., Becker, M.S., Broekuis, F., Bashir, S., Andresen, L., Aschenborn, O., Beddiaf, M., Belbachir, F., Belbachir-Bazi, A., Berbash, A., Brandao de Matos Machado, I., Breitenmoser, C., Chege, M., Cilliers, D., Davies-Mostert, H., Dickman, A.J., Ezekiel, F., Farhadinia, M.S., Funston, P., Henschel, P., Horgan, J., de Iongh, H.H., Jowkar, H., Klein, R., Lindsey, P.A., Marker, L., Marnewick, K., Melzheimer, J., Merkle, J., M'soka, J., Msuha, M., O'Neill, H., Parker, M., Purchase, G., Sahailou, S., Saidu, Y., Samna, A., Schmidt-Küntzel, A., Selebatso, E., Sogbohossou, E.A., Soultan, A., Stone, E., Van der Meer, E., van Vuuren, R., Wykstra, M., Young-Overton, K., 2017. The global decline of cheetah *Acinonyx jubatus* and what it means for conservation. Proc. Natl. Acad. Sci. USA 114, 528–533.

Durant, S., Mitchell, N., Ipavec, A., Groom, R., 2015. *Acinonyx jubatus*. The IUCN Red List of Threatened Species 2015: e.T219A50649567. Available from: http://dx.doi.org/10.2305/IUCN.UK.2015-4.RLTS.T219A50649567.en.

DWNP, 2009. National Conservation Action Plan for cheetahs and wild dog in Botswana. Department of Wildlife and National Parks, Gaborone.

Edroma, E.L., 1984. Drastic decline in numbers of animals in Uganda National Parks. In: Joss, P.J., Lynch, P.W., Williams, O.B. (Eds.), Rangelands. A Resource under Siege. Cambridge University Press, New York.

EWCA, 2012. National Action Plan for the Conservation of Cheetahs and African Wild Dogs in Ethiopia. Ethiopian Wildlife Conservation Authority, Addis Ababa, Ethiopia.

Fabiano, E., 2013. Demografia histórica e contemporânea de guepardos (*Acinonyx jubatus*) na Namíbia, África Austral. Pontifícia Universidade Católica do Rio Grande do Sul, Rio Grande do Sul, Brazil, PhD thesis.

Farhadinia, M.S., Akbari, H., Eslami, M., Adibi MA, 2016. A review of ecology and conservation status of Asiatic cheetah in Iran. Cat News Special Issue 10, 18–26.

Farhadinia, M.S., Hunter, L.T.B., Jourabchian, A.R., Hosseini-Zavarei, F., Akbari, H., Ziaie, H., Schaller, G.B., Jowkar, H., 2017. The critically endangered Asiatic cheetah A*cinonyx jubatus venaticus* in Iran: a review of recent distribution, and conservation status. Biodivers. Conserv. 26, 1027, doi:10.1007/s10531-017-1298-8.

Fitzinger, H., 1855. Bericht an die kaiserl. Academie der Wissenschaften uber die von dem Herrn Consulatsverweser Dr. Theodor v. Heuglin fur die kaiserliche Menagerie zu Schonbrunn mitgebrachten lebenden Tiere. Mathematische-Naturwisenschaftliche Classe 17, 242–253.

Fusari, A., Mahumane, M.C., Cuambe, E.O., Cumbi, R., Barros, P., 2010. Plano de Acção Nacional para a Conservação da Chita (*Acinonyx jubatus*) e Mabeco (*Lycaon pictus*) em Moçambique. Ministério do Turismo e Ministério da Agricultura, Maputo, Moçambique.

Griffith, E., 1821. General and particulated descriptions of the vertebrated animals, arranged conformably to the modern discoveries and improvements in zoology. Baldwin, Cradock and Joy, London, UK. Grimsdell.

Grimsdell, J.J.R., Bell, R.H.V., 1975. Black Lechwe Research Project, Final Report: Ecology of the Black Lechwe in the Bangweulu Basin of Zambia. NCSR, Falcon Press, Ndola, Zambia.

Gros, P.M., 1996. Status of the cheetah in Malawi. Nyala 19, 33–36.

Gros, P.M., 1998. Status of the cheetah *Acinonyx jubatus* in Kenya: a field-interview assessment. Biol. Conserv. 85, 137–149.

Gros, P.M., 2002. The status and conservation of the cheetah *Acinonyx jubatus* in Tanzania. Biol. Conserv. 106, 177–185.

Gros, P.M., Rejmanek, M., 1999. Status and habitat preferences of Uganda cheetahs: an attempt to predict carnivore occurrence based on vegetation structure. Biodivers. Conserv. 8, 1561–1583.

Habibi, K., 2003. Mammals of Afghanistan. Zoo Outreach Organization, Coimbatore, India.

Harrison, D.L., Bates, P.J., 1991. Mammals of Arabia, 2nd ed. Harrison Zoological Museum, Sevenoaks, UK.

Hatt, R.T., 1959. The Mammals of Iraq. Miscellaneous Publications, Museum of Zoology, University of Michigan, USA, pp. 113.

Heath, D., 1997. Leopard and Cheetah Management in Zimbabwe. Zimbabwe Department of National Parks and Wildlife Management, Harare.

HHELLER, E. 1913. *Acinonyx jubatus raineyi* In: New races of carnivores and baboons from equatorial Africa and Abyssinia. Smithsonian Miscellaneous Collections 61 (19), 1–12.

Hilzheimer, M., 1913. Über neue Gepparden nebst Bemerkungen uber die Nomenklatur dieser Tiere. Sitzungsber. Ges. Naturf. Freunde Berlin 5, 283–292.

Husain, T., 2001. Survey for the Asiatic cheetah, *Acinonyx jubatus*, in Balochistan province, Pakistan Barbara Delano Foundation, Balochistan, pp. 39.

IUCN/SSC, 2007a. Regional Conservation Strategy for the Cheetah and African Wild Dog in Eastern Africa. IUCN/SSC, Gland, Switzerland.

IUCN/SSC, 2007b. Regional Conservation Strategy for the Cheetah and African Wild Dog in Southern Africa. IUCN/SSC, Gland, Switzerland.

IUCN/SSC, 2012. Regional Conservation Strategy for the Cheetah and African Wild Dog in Western, Central and Northern Africa. IUCN/SSC, Gland, Switzerland.

Jackson, P., 1998. Asiatic Cheetah in Iran. Cat News 28, 2–3.

Joubert, E., Mostert, P.K.N., 1975. Distribution patterns and status of some mammals in South West Africa. Madoqua 9.

Kaczensky, P., Linnell, J.D., 2015. Rapid assessment of the mammalian community in the Badhyz Ecosystem, Turkmenistan, NINA Report 1148, pp. 38.

Kitchener, A.C., Breitemoser-Wq/4rsten, C., Eizirik, E., Gentry, A., Werdelin, L., Wilting, A., Yamaguchi, N., Abramov, A., Christiansen, P., Driscoll, C., Duckworth, W., Johnson, W., Luo, S.-J., Meijaard, E., O'Donoghue, P., Sanderson, J., Seymour, K., Bruford, M., Groves, C., Hoffmann, M., Nowell, K., Timmons, Z., Tobe, S., 2017. A revised taxonomy of the Felidae. The final report of the Cat Classification Task Force of the IUCN Cat Specialist Group. Cat News Special Issue 11, 80.

Klein, R., 2007. Status report for the cheetah in Botswana. Cat News Special Issue 3, 14–21.

KWS, 2010. Conservation and Management Strategy for Cheetah and Wild Dogs in Kenya. Kenya Wildlife Service, Kenya.

Lindsey, P., Romañach, S., Davies-Mostert, H., 2009. The importance of conservancies for enhancing the conservation value of game ranch land in southern Africa. J. Zool. 277, 99–105.

Lindsey, P.A., Romanãch, S.S., Matema, S., Matema, C., Mupamhadzi, I., Muvengwi, J., 2011. Dynamics and underlying causes of illegal bushmeat trade in Zimbabwe. Oryx 45, 84–95.

Lukarevskii, V.S., 2001. Leopard, Striped Hyena and Wolf in Turkmenistan. Signar, Moscow (in Russian).

Mallon, D.P., 2007. Cheetahs in central Asia: a historical summary. Cat News 46, 4–7.

Manati, A.R., Nogge, G., 2008. Cheetahs in Afghanistan. Cat News 49, 17–18.

Marker-Kraus, L., Kraus, D., Barnett, D., Hurlbut, S., 1996. Cheetah Survival on Namibian Farmlands. Cheetah Conservation Fund, Windhoek.

Marker, L., 1998. Current status of the cheetah (*Acinonyx jubatus*). In: Penzhorn, B.L. (Ed.), A symposium on cheetahs as game ranch animals. South African Veterinary Association, Onderstepoort, pp. 1–17.

Marker, L., 2012. Reintroduction of Cheetah to Uzbekistan. Feasibility study. Field Trip Report 10 August 2012. Cheetah Conservation Fund. Otjiwarongo, Namibia.

Marker, L.L., Fabiano, E., Nghikembua, M., 2010. A rapid ecological survey in the Iona National Park, March 2010, Namibe, Angola, Cheetah Conservation Fund, Otjiwarongo, Namibia.

Marker, L., Mills, M.G.L., Macdonald, D.W., 2003. Factors influencing perceptions and tolerance towards cheetahs on Namibian farmlands. Conserv. Biol. 17 (5), 1–9.

Marnewick, K., Bekhelling, A., Cilliers, D., Lane, E., Mills, G., Herring, K., Caldwell, P., Hall, R., Meintjes, S., 2007. The status of cheetah in South Africa. Cat News Special Issue 3, 22–31.

Marnewick, K., Ferreira, S.M., Grange, S., Watermeyer, J., Maputla, N., et al., 2014. Evaluating the status of and African wild dogs *Lycaon pictus* and cheetahs *Acinonyx jubatus* through tourist-based photographic surveys inthe Kruger National Park. PLoS One 9 (1), e86265.

Marnewick, K., Hayward, M.W., Cilliers, D., Somers, M.J., 2009. Survival of cheetahs relocated from ranchland to fenced protected areas in South Africa. In: Hayward, M.W., Somers, M.J. (Eds.), Reintroduction of Top-Order Predators. 1st ed. Wiley-Blackwell, Oxford, UK, pp. 282–306.

Mills, M.G.L., Mills, M.E.J., Edwards, C.T., Gottelli, D., Scantelbury, D.M., 2017. Kalahari cheetahs: adaptations to an arid region. Oxford University Press, Oxford.

Ministry of Environment and Tourism, Government of Namibia, 2013. National Conservation Action Plan for Cheetahs in Namibia. Ministry of Environment and Tourism, Government of Namibia, Namibia.

Ministry of Natural Resources and Tourism, 2014. National Action Plan for the Conservation of Cheetahs & African Wild Dogs in Tanzania. Goverment of Tanzania. Dar es Salaam, Tanzania.

Morsbach, D., 1987. Cheetah in Namibia. Cat News 6, 25–26.

Myers, N., 1975. The Cheetah *Acinonyx jubatus* in Africa. IUCN Monograph No. 4 IUCN, Morges, Switzerland.

Newby, J.E., 1992. Parks for people-a case study from the Aïr Mountains of Niger. Oryx 26 (1), 19–28.

Newby, J.E., Grettenberger, J.F., 1986. The human dimension in natural resource conservation: a Sahelian example from Niger. Environ. Conserv. 13 (3), 249–256.

Nhabanga, A., Purchase, G., Groom, R., Minkowski, K., 2011. Regional Conservation Planning for Cheetah and African Wild Dog in Southern Africa: Mozambique as a Success Story. Presentation at AHEAD-GLTFCA working group - 11th Meeting, March 2011, South Africa.

Nowell, K., Jackson, P., 1996. Wild Cats: Status Survey and Conservation Action Plan. IUCN/SSC Cat Specialist Group, Gland and Cambridge.

Parliament of Zimbabwe, 2001. Parks and Wild Life Act (Chapter 20:14) Act 22 of 2001. Zimbabwe Parks and Wildlife Management Authority, Harare, Zimbabwe.

Phiri, C.M., 1996. Cheetah translocation project in Lower Zambezi National Park, Zambia. Report. National parks and wildlife service, Zambia.

Purchase, N., 2007a. Mozambique: preliminary assessment of the status and distribution of cheetah. Cat News, Special Issue 3, 3–39.

Purchase, N., 2007b. Status and distribution of cheetah in Zambia: a preliminary. assessment. Cat News, Special Issue 3, 40–42.

Purchase, N., Marker, L., Marnewick, K., Klein, R., Williams, S., 2007. Regional assessment of the status, distribution and conservation needs of cheetahs in southern Africa. Cat News, Special Issue 3, 44–46.

Rabeil, T, Harouna, A, Newby, J., 2008. Avant projet de classement d'une aire protégée dans le Termit-Tin Toumma, Niger, Zinder, Niger: Antilopes Sahélo-Sahariennes.

Ray, J.C., Hunter, L., Zigouris, J., 2005. Setting Conservation and Research Priorities for Larger African Carnivores. Wildlife Conservation Society, New York, p. 203.

RWCP and IUCN/SSC, 2015. Regional Conservation Strategy for the Cheetah and African Wild Dog in Southern Africa; Revised and Updated, August 2015, London, United Kindgom.

Schreber J.C.D., 1775. Die Säugethiere in Abbildungen Nach der Natur mit Beschreibungen, vol. 2(14). Wolfgang Walther, Erlangen.

Shortridge, G.C., 1934. The mammals of South West Africa. Heinemann, London.

Sillero-Zubiri, C., Rostro-García, S., Burruss, D., Matchano, A., Harouna, A., Rabeil, T., 2015. Saharan cheetah *Acinonyx jubatus hecki*, a ghostly dweller on Niger's Termit massif. Oryx 49, 591–594.

Smith, N., Wilson, S.L., 2002. Changing land use trend in the thicket biome: pastoralism to game farming. Unpublished Report No. 38. Centre for African Conservation Ecology, Nelson Mandela Metropolitan University, Port Elizabeth, South Africa, pp. 22.

Smithers, R.H.N., 1975. Family Felidae, part 8.1. In: Meester, J., Setzer, H.W. (Eds.), The Mammals of Africa: An Identification Manual. Smithsonian Institution Press, Washington, DC, pp. 1–10.

South Sudan Wildlife Service, 2010. National Action Plan for the Conservation of Cheetahs and African Wild Dogs in South Sudan. South Sudan Wildlife Service, South Sudan.

Spinage, C., 1991. History and Evolution of the Fauna Conservation Laws of Botswana, occasional paper No. 3, Botswana Society, Gaborone.

SWA, 1975. Nature Conservation Ordinance. Official Gazette Extraordinary, No. 4, pp. 75.

Uganda Wildlife Authority, 2010. Strategic Action Plan for Large Carnivore Conservation in Uganda, 2010–2020. Uganda Wildlife Authority, Kampala, Uganda.

Van der Meer, E., 2016. The cheetahs of Zimbabwe. Distribution and population status 2015. Cheetah Conservation Project Zimbabwe, Victoria Falls, Zimbabwe. Available from: cheetahzimbabwe.org.

Van der Merwe, V., Marnewick, K., Bissett, C., Groom, R., Mills, M.G.L., Durant, S.M., 2016. A conservation assessment of *Acinonyx jubatus*. In: Child, M.F., Roxburgh, L., Do Linh San, E., Raimondo, D., Davies-Mostert, H.T. (Eds.), The Red List of Mammals of South Africa, Swaziland and Lesotho. South African National Biodiversity Institute and Endangered Wildlife Trust, South Africa.

van Vuuren, J.B., Robinson, T.J., VazPinto, P., Estes, R., Matthee, C.A., 2010. Western Zambian sable: are they a geographic extension of the giant sable antelope? Southern Africa Journal of Wildlife Research 40, 35–42, doi: 10.3957/056.040.0114.

Wacher, T., de Smet, K., Belbachir, F., Belbachir-Bazi, A., Fellous, A., Belghoul, M., Marker, L., 2005. Sahelo-Saharan Interest Group Wildlife Surveys, Central Ahaggar Mountains. Sahelo Saharan Interest Group, Zoological Society of London, London.

Wells, M.P., 1996. The social role of protected areas in the new South Africa. Environ. Conserv. 23, 322–331.

Wilson, V.J., 1988. Distribution and status of cheetah in Zimbabwe. Chipangali Wildlife Trust, Bulawayo, Zimbabwe.

ZPWMA, 2009. National Conservation Action Plan for cheetahs and wild dogs in Zimbabwe. Zimbabwe Parks and Wildlife Management Authority, Harare, Zimbabwe.

5

Asiatic Cheetahs in Iran: Decline, Current Status and Threats

Mohammad S. Farhadinia*,¶, Luke T.B. Hunter**,
Houman Jowkar†,‡, George B. Schaller**,§,
Stephane Ostrowski§

*University of Oxford, Tubney, Abingdon, United Kingdom
**Panthera, New York, NY, United States
†Persian Wildlife Heritage Foundation, Tehran, Iran
‡Conservation of the Asiatic Cheetah Project, Tehran, Iran
§Wildlife Conservation Society, New York, NY, United States
¶Future4Leopards Foundation, Tehran, Iran

INTRODUCTION

Despite being popularly considered an African felid, the cheetah (*Acinonyx jubatus*) once occurred across large extents of dry Asia (Nowell and Jackson, 1996). In the last 100 years, the Asiatic cheetah *A. j. venaticus* has experienced dramatic reductions in numbers and geographic range (Divyabhanusinh, 2007; Jackson, 1998) and is now confined to a small population in the central plateau of Iran (Farhadinia, 2004).

Asiatic cheetahs began to disappear from most of their historic range much earlier than other cheetah subspecies (Nowell and Jackson, 1996). Historically, they ranged across southwest and central Asia to India (Durant et al., 2015), but they became extinct from the majority of their Asian range between the 1940s and 1980s (Nowell and Jackson, 1996). Currently, the last remaining population of Asiatic cheetah is categorized as Critically Endangered on the IUCN Red List (Durant et al., 2015) and is included in CITES Appendix I (Nowell, 2014). The Asiatic cheetah has been considered a symbol for wildlife conservation in Iran since the 1950s, after being officially declared a protected species by Iranian law.

In this chapter, we present conservation information for the Asiatic cheetah and outline the cheetah's current range in Iran. Additionally, we review the main threats to the subspecies and describe the actions required to ensure the survival of the last population of cheetahs in Asia.

LAST STRONGHOLD FOR ASIATIC CHEETAHS: PAST AND PRESENT RANGE IN IRAN

Before World War II, the cheetah population in Iran was believed to be around 400 individuals, living across the steppes and desert areas of the eastern half of the country and in western terrains near the Iraqi border (Harrington, 1971). Since the late 1950s, the newly created Department of Environment established protection measures for cheetahs, their prey, and habitat to halt poaching (Firouz, 1974). Subsequently, cheetah sightings increased in some regions, suggesting that conservation measures were effective at saving the cheetah from an irreversible decline (Firouz, 1974; Mowlavi, 1985). In the 1970s, the cheetah's range was thought to still include most of the arid lands of the eastern half of Iran and some borderlands with Iraq (Firouz, 1974). The population was estimated to be 200–300 individuals (Firouz cited in Goodwin and Holloway, 1974). However, Joslin (1984) considered this estimate to be too high and proposed 100 cheetahs as a more realistic number. None of these past estimates were based on comprehensive surveys or robust counting methods.

The Iranian civil revolution in 1979 followed by the war with Iraq between 1980 and 1988 effectively ceased organized wildlife conservation activities in Iran. Livestock occupied many protected areas and cheetahs and their prey were heavily poached. During this period, the cheetah is believed to have disappeared from a number of historical strongholds and became confined to the remotest areas in the eastern half of the country, which still offered prey populations and relative safety (Asadi, 1997; Farhadinia, 2004).

Between 1980 and 2001, the cheetah was confirmed from 11 sites, Ariz, Bafq, Dareh Anjir, Dorouneh, Kamki Bahabad, Kavir National Park (NP), Naybandan, and Touran Biosphere Reserve (BR) (Jourabchian and Farhadinia, 2008), as well as Khabr NP, Shahdad and Bahram-e Gour Protected Area (PA) (Jowkar, 1999). Between 2001 and 2016, the cheetah was confirmed unequivocally from an additional seven sites, Abbasabad PA, Boshrouyeh, Kalmand PA, Khosh Yeilagh Wildlife Refuge (WR), Miandasht WR, Darband PA, and Siahkouh NP and PA (Farhadinia et al., 2016b). Based on photographs of footprints, the cheetah was additionally considered to occur in Rafsanjan County, Kerman Province, Takhti Iran No Hunting Area (NHA), North Khorasan Province, and Chah Shirin NHA, Semnan Province during 2000s, that is, a total number of 21 sites confirmed with cheetah presence since 1980.

More recent surveys in Khabr NP, Shahdad, and Bahram-e Gour PA (Ghoddousi et al., 2007; Jourabchian and Farhadinia, 2008) where cheetahs formerly occurred have failed to find recent evidence of the species. Accordingly, since 2001 cheetah presence is unequivocally known in Iran in only 18 areas (Fig. 5.1). All current sites apart from Boshrouyeh and Rafsanjan Counties have received some level of official protection from the Department of Environment (Farhadinia et al., 2016b). The apparent expansion of the known range of the Asiatic cheetah over the 2000s compared to the period 1980–2001 is likely due to greater survey effort and the increased use of camera traps rather than actual range recovery or expansion.

The present range of the cheetah in Iran is difficult to quantify accurately as protected areas with cheetah records have been only partially surveyed and few surveys have been carried out outside protected areas. Using occurrence data, the cheetahs' "Extent of Occurrence (EOO)" was calculated to be approximately 240,000 km^2 (Farhadinia et al., 2017). Furthermore, using presence data, an ensemble model of habitat

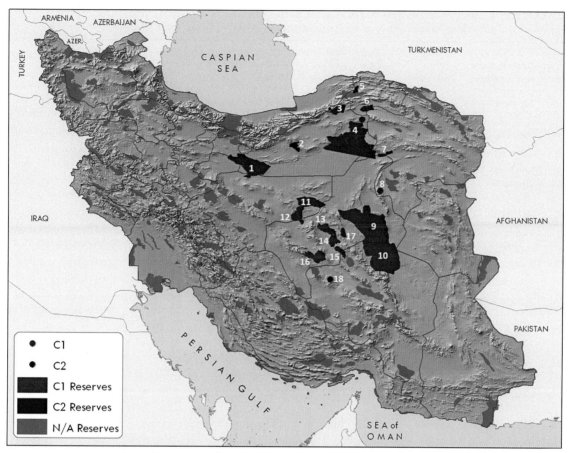

FIGURE 5.1 **A map of Iran showing protected areas** *(patches)* **and nonprotected localities** *(plain circle)* **that reported at least one confirmed Asiatic cheetah record since 2001.** *Red patches* (C1) are areas with hard evidences of cheetah presence (reliable field reports with photographs, videos, or dead specimens). *Blue patches* (C2) show areas with soft evidences of cheetah presence (reliable field observations but without the aforementioned material). *Dark gray* areas are protected areas without cheetah record since 2001: (1) Kavir, (2) Chah Shirin, (3) Khosh Yeilaq, (4) Touran, (5) Takhti Iran, (6) Miandasht, (7) Dorouneh, (8) Boshrouyeh, (9) Naybandan, (10) Darband, (11) Abbas Abad, (12) Siahkouh, (13) Dareh Anjir, (14) Ariz, (15) Bafq, (16) Kalmand, (17) Kamki Bahabad, and (18) Rafsanjan. *Source: Map obtained from Farhadinia et al., 2016b.*

suitability based on seven species distribution models suggested that ca. 320,000 km^2 is still suitable habitat to host the cheetahs in Iran (Ahmadi et al., 2017). The spatial configuration of the cheetah records since 2001 suggests that there are three main landscapes for cheetahs in Iran, with possible movement of individuals between them (Farhadinia et al., 2016b; Moqanaki and Cushman, 2016).

The Northern Landscape

The northern Landscape includes six protected areas where the cheetah has been reported: Touran BR and five smaller areas, Chah Shirin NHA, Dorouneh PA, Khosh Yeilaq WR, Miandasht WR, and possibly Takhti Iran NHA (Table 5.1). Touran BR is one of the largest reserves in the country and has been known for

TABLE 5.1   Characteristics of Areas in Iran Where Cheetah Occurrence has Been Confirmed Since 2001 (Based on Hard Evidence or Footprint Identification Approved by Experts)

No.	Reserve name and category	Province	Area (km²)	Year officially protected	Wild ungulate prey species[a]	Sympatric large carnivores[a]
1	Naybandan WR	S. Khorasan	15169	2001	Wild goat, wild sheep, chinkara	Striped hyena, gray wolf[e], Persian leopard[e]
2	Bafq PA	Yazd	885	1996	Wild goat, wild sheep, chinkara	Persian leopard, gray wolf[e]
3	Kavir NP and PA	Semnan	6911	1967	Wild goat, wild sheep, chinkara	Striped hyena, gray wolf, Persian leopard
4	Touran BR	Semnan	14414	1972	Wild goat, wild sheep, goitered gazelle, chinkara	Striped hyena, gray wolf, Persian leopard
5	Ariz NHA	Yazd	1313	1999	Wild goat, wild sheep chinkara	Persian leopard[e]
6	Kamki Bahabad NHA	Yazd	650	2011	Wild goat, wild sheep, chinkara	Striped hyena, gray wolf Persian leopard
7	Dareh Anjir WR	Yazd	1753	2002	Wild goat, wild sheep, chinkara	Striped hyena[e]
8	Miandasht WR	N. Khorasan	844	1975	Goitered gazelle wild sheep[e]	Striped hyena, gray wolf
9	Kalmand PA	Yazd	2291	1990	Wild goat, wild sheep, goitered gazelle	Striped hyena, gray wolf, Persian leopard
10	Takhti Iran NHA[b]	N. Khorasan	230	2006	Wild goat, wild sheep, goitered gazelle	Gray wolf, Persian leopard, striped hyena[e]
11	Ranfsanjan County[b]	Kerman	NA	NA	Wild goat, wild sheep	Striped hyena, gray wolf
12	Chah Shirin NHA[b]	Semnan	NA	NA	Wild goat, wild sheep, chinkara	Striped hyena, gray wolf
13	Darband WR	Kerman	13640	2010	Wild goat, wild sheep, chinkara	Striped hyena, gray wolf, Persian leopard
14	Siahkouh NP and PA	Yazd	2057	2001	Wild goat, wild sheep, chinkara	Gray wolf
15	Abbas Abad WR	Esfahan	3050	2009	Wild goat, wild sheep, chinkara	Striped hyena, gray wolf, Persian leopard[g]
16	Khosh Yeilagh WR	Semnan	1500	1965	Wild goat, wild sheep, goitered gazelle	Striped hyena, gray wolf, Persian leopard
17	Boshrouyeh[c]	S. Khorasan	NA	NA	NA	Striped hyena, gray wolf
18	Dorouneh PA[d]	Razavi Khorasan	750	2007	Wild goat, wild sheep, chinkara[f]	Striped hyena, gray wolf, Persian leopard
	Total		65370			

BR, Biosphere reserve; N, north NHA, no hunting area; NP, national park; PA, protected area; S, south WR, wildlife refuge.

[a]*Species scientific names: chinkara (Gazella bennettii), goitered gazelle (Gazella subgutturosa), gray wolf (Canis lupus), Persian leopard (Panthera pardus saxicolor), striped hyena (Hyaena hyaena), wild sheep (Ovis orientalis), and wild goat (Capra aegagrus).*

[b]*The cheetah's existence was confirmed based on fresh footprints. There is a reserve nearby, but cheetah sign was detected outside the reserve.*

[c]*A cheetah was confirmed to be poached; there is no protected area.*

[d]*It is connected to Dasht-e-Laghari WR (established in 2012 with an area of 240 km² where cheetahs are occasionally reported).*

[e]*rarely.*

[f]*all in extremely low density.*

[g]*to be confirmed.*

decades as a cheetah stronghold (Asadi, 1997; Etemad, 1985; Farhadinia, 2004; Hajji, 1986; Ziaie, 2008). A recent camera-trap survey in 2012 detected 5 adult cheetahs in the area (Ashayeri et al., 2013). In addition to Touran BR, Miandasht WR is also known as a breeding area for cheetahs, with at least six litters recorded between 2002 and 2016. In February 2016, an adult male cheetah first recorded as a cub in Miandasht WR was found dead North of Touran BR. Khosh Yeilaq WR was once a stronghold for the species, with a population of over 30 adult and adolescent cheetahs (Joslin, 1984). Since the early 1980s, cheetahs were assumed to have disappeared from the area with no records until 2011, when two individuals were filmed by game guards. In 2013, an adult female cheetah was photographed by game guards in Dorouneh PA, an area adjacent to Touran BR, feeding on a domestic goat. Finally, Takhti Iran NHA, located around 100 km from the international border with Turkmenistan, might have hosted transient cheetahs, based on sporadic records during the 2000s (Farhadinia et al., 2008).

The Southern Landscape

The southern Landscape includes 11 areas that have recorded cheetah occurrence since 2001, including Abbas Abad WR, Ariz NHA, Bafq PA, Boshruyeh County, Darband WR, Dareh Anjir WR, Kalmand PA, Kamki Bahabad NHA, Naybandan WR, Rafsanjan County, and Siahkouh NP and PA (Table 5.1). The first camera trap photograph of a cheetah in Iran was taken in Naybandan WR in October 2001 (Jourabchian and Farhadinia, 2008), and this male was found dead by game guards in January 2011, at an estimated age of at least 13 years. Cheetahs have also been reported from Darband WR based on mortality records of three males in 2009, including two adult males that died from poisoning (Saeid,

personal communication). Dareh Anjir WR is thought to play an important role for cheetahs in the southern Landscape as the majority of cheetah individuals detected in the South have used the area. Finally in 2011, an adult cheetah was killed by local herders near Boshrouyeh, around 90 km North of Naybandan WR, between the northern and southern Landscapes.

There is no evidence of reproduction from the southern Landscape since 2011. Game guards have reported cheetah cubs in Naybandan WR in 2014 and 2015, respectively, but without confirmatory photographs.

Kavir National Park

There have been sporadic cheetah sightings outside of the aforementioned landscapes, predominantly of single, adult individuals. Cheetahs have been sighted irregularly in Kavir NP, located approximately 50 km southeast of Tehran (Asadi, 1997; Bayat, 1984; Hajji, 1986; Jowkar, 1999; Mowlavi, 1985). Three camera trapping surveys in 2003, 2005, and 2009–10, resulting in a cumulated effort of ca. 5300 trap nights, captured only two different individuals, one adult of unknown gender in 2003 and one adult male in 2009–10 (Ghadirian et al., 2010). There are no recent records of reproduction from Kavir and no cheetah has been reported from Kavir NP since December 2015.

Other Areas

Recent field investigations yielded no evidence of cheetahs in numerous areas within the historical range of the cheetahs in Iran, including Bidouyeh PA (Allahgholi et al., 2007), Bahram-e-Gour PA (Ghoddousi et al., 2007), and Bajestan (Cheraghi et al., 2007). In western Iran, the cheetah was historically present in both the eastern and western Zagros

Mountain (Ziaie, 2008) but there is no evidence of its occurrence in these regions for more than 40 years. Additional surveys are still needed to confirm whether cheetahs might still occur in Hormozgan and Sistan-va-Baluchestan Provinces.

ASIATIC CHEETAH'S SMALL POPULATION IN IRAN

While a robust and precise estimation of the population size of cheetahs in Iran is still lacking, several crude population estimates have been proposed, all agreeing to fewer than 100 adult cheetahs for the entire country (<60: Schaller and O'Brien, 2001; Farhadinia, 2004; <40: Jourabchian, 1999; 50–100: Asadi, 1997; 70–100: Ziaie, 2008; 60–100: Jowkar et al., 2008). In combination with the large EOO (ca. 240,000 km^2), such low numbers would translate into extremely low densities (<1/2, 500 km^2).

As a commonly used tool for monitoring elusive species, camera trapping has been utilized to determine the abundance and density of cheetah populations (Chapter 29). In areas with high cheetah detectability photographic surveys have also been used to monitor cheetah population trends (Marnewick et al., 2014). However, such surveys commonly detect only a proportion of the cheetah individuals in an area (Belbachir et al., 2015), which may not represent the true demographic composition of the population. Even within intensively studied areas, abundance estimates derived through camera trapping are strongly male-biased, especially when cameras are placed at scent marking posts (Marker et al., 2008; Marnewick et al., 2008) which are visited more frequently by males (Chapter 9). Therefore, these estimates are unlikely to reflect true parameters to be used for long-term population monitoring.

It is even more difficult to obtain accurate population estimates in Iran, where there is a very low rate of female detection and where cheetahs are wide-ranging (Farhadinia et al., 2013, 2016a). Additionally, obtaining a proper understanding about population size is more challenging at reserve level, due to high interreserve mobility of the cheetahs (Fig. 5.2; Farhadinia et al., 2016a), which may lead to underestimation of animal numbers. Indeed, unless surveys are conducted simultaneously in neighboring reserves, camera trapping of nearby reserves might yield no picture of cheetahs, simply because a given individual could be in adjacent reserves at the time of the camera trapping survey. Thus, rigorous monitoring of such a rare species as the Asiatic cheetah over an area the size of Italy remains a major challenge for its conservation.

Nevertheless, as a result of the first country-scale assessment based on an intensive camera trapping survey across more than half of the known cheetah reserves between 2010 and 2013, a minimum of 20 adult cheetah individuals were identified (Farhadinia et al., 2014). Therefore, it was concluded that Iran likely hosts a smaller population than previously believed, estimated to comprise fewer than 40 adult cheetahs (Farhadinia et al., 2016b).

Between 2001 and 2011, photographs of female cheetahs with cub(s) were obtained from 10 cheetah reserves whereas between 2011 and 2016 confirmed records of cheetah breeding was obtained only from Miandasht WR and Touran BR, both in the northern part of the current range (Fig. 5.3; Farhadinia et al., 2016b). To the best of our knowledge, apart from an ambiguous record of cheetah breeding in Naybandan in 2015, no adult female accompanied with cub(s) has been detected in other cheetah protected areas since 2011, suggesting that the country's population has been experiencing a decline in productivity. If true, this is a major concern for the population's long-term persistence.

FIGURE 5.2 **A coalition of two male cheetahs which have been detected in five different reserves between 2009 and 2017 in central Iran.** *Source: ICS/DoE/CACP/UNDP/Panthera.*

CURRENT THREATS AND CHALLENGES FOR CHEETAH SURVIVAL IN IRAN

Direct Threats

We compiled 50 records of cheetah mortality between 2001 and 2016, of which 34 (68.0%) were confirmed by photographs or collected carcasses; 16 (32.0%) were reported by reliable sources but without collected evidence, and were thus not included in the analyses. Out of the 34 confirmed cases 7 (20.1%) cheetahs died of unknown causes, in contrast to the majority of casualties ($n = 27$), which were claimed to be human-mediated.

Between 2001 and 2016, car collisions were responsible for 15 cheetah casualties, including 8 (6 males; 2 females) in Yazd Province, 1 male in Darband WR and 6 (1 male; 5 females) on one primary road bordering to the North Touran BR. The growing network of roads is a major emerging problem for cheetahs in Iran

8/18/2012   8:39 PM

FIURE 5.3   **A family of Asiatic cheetah first photographed in August 2012 in Miandasht WR.** At least two individuals of this family died due to human mediated factors by 2017. *Source: ICS/DoE/CACP/UNDP/Panthera.*

(Mohammadi and Kaboli, 2016; Moqanaki and Cushman, 2016).

Most of the cheetah's range does not host high densities of livestock (Jourabchian and Farhadinia, 2008), with the exception of Touran BR and Miandasht WR, which supported ca. 42,000 and 6,000 sheep and goats, respectively, in authorized grazing areas during winter 2016–17. Generally, herders have a positive attitude toward cheetahs compared with other large predators (Hamidi and Nezami, 2009) possibly due to a low encounter rate, and few conflicts.

However, between 2001 and 2016, herders and their guard dogs have been responsible for seven confirmed mortality cases. In addition, 13 mortality cases that could have resulted from a conflict with herders and their dogs were reported by trained experts and rangers, but with no collected evidence.

Intentional shooting or poisoning of cheetahs is rarely documented in Iran, possibly because encountering cheetahs in the wild is a rare, accidental event. However, we believe that the five confirmed cases of poaching and poisoning detected between 2001 and 2016, underestimate the actual level of this threat. There have been occasional rumors of cheetahs being shot in remote areas, which have never been confirmed, possibly as cases are not reported to avoid incurring the very high penalty (currently 1 billion IRR equal to US $ 28,570). Nevertheless, in the context of a very small and declining population size, any poaching is extremely concerning.

As mortality in wildlife is notoriously difficult to estimate, and cheetahs are killed illicitly and/or at remote sites our figures underestimate cheetah mortality in Iran. In contrast, there is no evidence of trade in captive cheetahs from Iran to the Arab states of the Persian Gulf, unlike in north east and east Africa (Nowell, 2014; Chapter 14).

Indirect Threats

As seen in cheetahs in North and West Africa (Chapter 11) and for many large carnivores, the decrease of main prey is possibly the most important threat to cheetah survival in Iran (Farhadinia, 2004; Hunter et al., 2007; Ziaie, 2008). The Asiatic cheetah lives in arid environments supporting a low density of medium-sized wild ungulates, such as chinkara (*Gazella bennettii*), goitered gazelle (*Gazella subgutturosa*), wild sheep (*Ovis orientalis*), and wild goat (*Capra aegagrus*), which in general suffer chronic poaching (Farhadinia and Hemami, 2010; Ziaie, 2008). Illegal hunting of prey in and around protected areas is widespread and continues to be a main limiting factor for the cheetahs (Breitenmoser et al., 2009; Hunter et al., 2007; Schaller and O'Brien, 2001). It has been proposed that the cheetahs can survive by preying on small sized mammals, particularly hares (Karami, 1992; Ziaie, 2008), but hares may be too small to sustain cheetahs (especially females with cubs; Hunter et al., 2007; Chapters 8

and 11) and recent fecal analyses (based only on phenotypic identification of scats) suggest that small mammals constitute a minority of cheetah diet composition in the surveyed areas (Farhadinia et al., 2012; Rezaie, 2014; Zamani, 2010). Depletion of medium-sized wild ungulates as a main prey base can increase cheetah predation on livestock and the risk of retaliatory destruction of cheetahs (Farhadinia et al., 2012).

Equally important, habitat loss is a major threat to cheetah and prey survival (Chapter 10). In Iran, habitat loss is mostly due to overgrazing, development (e.g., road construction and mining), and drought (Asadi, 1997; Farhadinia, 2004; Hunter et al., 2007; Karami, 1992; Moqanaki and Cushman, 2016; Marashi et al., 2017).

Most of the cheetah's range in Iran overlaps with mineral deposits, which results in pressure to allow exploitation, particularly in Abbas Abad WR, Naybandan WR, Darband WR, and in most reserves in Yazd Province. Fortunately, a significant proportion of the cheetah's range is protected by the Department of Environment which has successfully blocked many requests for new mining or expansion of existing exploitations. Between 2012 and 2016, three such requests have been rejected, in Dareh Anjir WR, Touran BR, and Naybandan WR (Conservation of the Asiatic Cheetah Project/DoE, unpublished).

MANAGEMENT IMPLICATIONS

The cheetah in Iran is presently restricted to arid regions where it is reliant on nomadic and/or highly dispersed ungulate prey. It still occurs in 18 areas, which are estimated to cover less than 20% of the species' suitable habitat (Ahmadi et al., 2017). Therefore, protecting a higher proportion of cheetah habitat is urgently needed. In addition, direct observations (Farhadinia et al., 2016a,b) and connectivity models (Moqanaki and Cushman, 2016) support that cheetahs also use large tracks of land between and beyond protected areas. Recently models of linkage areas that could serve as corridors have been developed to help optimize conservation actions outside protected areas and inform future targeted land planning (Ahmadi et al., 2017).

In view of the very small population size of cheetahs in Iran, any mortality matters, and it is urgent that the main sources of human-caused mortality are better controlled. Herding practices and herder behavior within cheetah range must be addressed, ideally through participatory and sustainable processes. Guard dogs accompanying authorized livestock herds into protected areas during grazing seasons should be reduced in numbers. Cheetahs continue to be killed by herders opportunistically and out of ignorance of the cheetah's protected status, including in protected areas. Requiring a mandatory attendance to education workshops could be linked to obtaining permission for grazing in protected areas.

Increasing the tolerance toward predator presence should be another priority (Chapter 13). An insurance program for compensating livestock predation by cheetahs has been developed and facilitated by the Conservation of the Asiatic Cheetah Project. Although the insurance-based compensation program is useful to prevent the retaliatory killing of cheetahs, it may lack sustainability beyond the period of agreement with the donor (i.e., 4 years for the current agreement). As a supplementary solution, it would also be interesting to assess the viability of rewarding herders for cheetah sightings and information. Performance-payment type programs, such as those run for wolverines (*Gulo gulo*) in Sweden (Persson et al., 2015) may improve coexistence

between cheetahs and communities. An effective vaccination program of livestock could also help gain herders' engagement in cheetah conservation and at the same time address the risk of spillover of infectious agents between infected livestock and wild ungulates, such as in the case of Peste des Petits Ruminants outbreak which has previously resulted in mortalities of cheetah prey in the Kavir NP (Schaller and O'Brien, 2001).

Considering the high mobility of cheetahs between reserves (Farhadinia et al., 2013, 2016a) and the growing network of roads, particular focus must be placed on providing safe road crossings between reserves to diminish road collisions (Ahmadi et al., 2017; Moqanaki and Cushman, 2016; Mohammadi and Kaboli, 2016). Efforts to reduce road collisions involving cheetahs have focused on road signage. Unfortunately because it was installed far from the road or they were poorly visible at night, its effectiveness at reducing speed and increasing drivers' vigilance was limited (Mohammadi and Kaboli, 2016). In some selected areas, wildlife crossings, overpasses, and underpasses may be feasible, such features have reduced mortality and increased connectivity for large carnivores elsewhere (see Grilo et al., 2015 for a review). Most documented cheetah road kills have occurred on a small number of highway stretches passing through hilly terrain (Mohammadi and Kaboli, 2016).

Controlling anthropogenic threats to wild ungulates, particularly poaching, is also critical (Breitenmoser et al., 2009). Overgrazing by livestock, in particular within protected areas, must be addressed by introducing sustainable practices of rangeland management, combined with revising and controlling existing grazing permissions. These permissions are based on historical right of access rather than on periodic assessments of the carrying capacity, and as a result livestock numbers far exceed sustainable levels in many areas (Jowkar et al., 2016). In addition it has been proposed that the regime of artificial water supplementation in protected areas often benefits livestock more than wild ungulates, and has an unknown impact on the rangeland used by cheetah prey, a hypothesis that requires further examination.

The rapid pace of development in Iran exacerbates pressures on the natural habitat of cheetahs and their prey (Jowkar et al., 2016). Within cheetah range, the most rapid and extensive development is driven by the mining sector's exploration and exploitation activities. For instance, large-scale extraction and ongoing exploration efforts particularly for iron ore occur in the vicinity of Dareh Anjir WR, Kalmand PA, and Bafq PA in Yazd Province (Fig. 5.1) and more are planned. Compounding the immediate loss of habitat, mining requires the construction of new roads and railways, increasing the vulnerability of cheetahs to accidents and enabling easier access to remote areas by those poaching wild ungulates (Chapter 11).

Sound monitoring of the trend of the Asiatic cheetah population is central to its management and conservation. Unfortunately the development of a robust monitoring system that would combine and synergize available methodologies [e.g., camera trapping (Chapter 29), molecular tools (Chapters 6 and 31), opportunistic photos, and citizen science (Chapter 33)] has so far failed to emerge, owing in part to the scale of the task. Additional challenges are presented by the lack of human resources and sometimes capacity, financial constraints, and limitations to transfer funds, assets, and samples internationally. Past camera trap studies have been extremely useful at identifying important sites for cheetahs and have helped direct conservation efforts. But a robust, coordinated, long-term monitoring program for cheetahs in the main protected

areas is still needed. Such a program should establish long-term, fixed monitoring sites (i.e., camera stations) to be surveyed during the same survey window each year, thus providing a consistent, minimum annual count of cheetahs and hence information on population trends. To control for the low camera-capture probability for female cheetahs (Chapter 29); equipping scarce water resources with camera traps during warm months should be tested, as lactating females are likely to make an increased use of these sites because of their higher water requirements (Laurenson, 1995).

CONSERVATION PROGRAM TO SAFEGUARD THE ASIATIC CHEETAH

Since the 1950s the Iranian cheetah has been subjected to regular reassessments of its status, yet largely based on anecdotal information on population size (Dareshuri and Harrington, 1976; Firouz, 1974; Harrington, 1971; Joslin, 1984; Jowkar 1999; Karami, 1992; Khalili, 1984; Mowlavi, 1985). Efforts have been made to support cheetah populations and their main prey with the creation of PAs, such as Touran BR and Kavir NP during the 1960s and 1970s (Makhdoum, 2008) without which cheetahs would have possibly gone extinct.

Since the early 1990s, multiple attempts were made to establish a comprehensive baseline of protection measures for the Asiatic cheetah (e.g., Asadi, 1997; Dareshuri, 1997; Farhadinia, 2004; Jowkar, 1999; Jourabchian, 1999; Karami, 1992) but the most comprehensive effort was achieved with the creation in September 2001 of the Conservation of Asiatic Cheetah Project (CACP) of the DoE with the support of the Global Environment Facility and the United Nations Development Programme (UNDP) in Iran. The program also involved national and international non-governmental organizations, which worked at increasing social mobilization, developed measures to improve local incomes and community well-being, initiated co-management activities between communities and local DoE operations, and implemented science and monitoring activities (Breitenmoser et al., 2009). The goal of the project was to "secure the conservation of the Asiatic cheetah in the I.R. of Iran and the related complex of rare and endangered wild species, and their natural habitats with the support and collaboration of local communities" (Conservation of the Asiatic Cheetah Project, 2008).

In addition to extensive awareness raising activities that targeted international, national, and local stakeholders, the CACP implemented measures to elevate the protection level of confirmed and potential cheetah habitats and increase the number of game guards in protected areas used by cheetahs. Darband was elevated to a WR in 2011 and the core of Miandasht to a National Park in 2014. In 2017 there were 126 full-time employed DoE game guards in charge of protection and law-enforcement across the cheetah range. Strong deterrents have also been established for killing of cheetahs, including imprisonment and the highest fine for a wildlife species destruction in Iran. Efforts also included the collection and analysis of data, particularly sighting records by trained game guards.

To improve rangeland rehabilitation, CACP in collaboration with provincial staff of the DoE and non-governmental organizations have developed and tested new rangeland management practices aimed at improving rangeland conditions in protected areas where large numbers of livestock coexist seasonally with wild ungulates. As a likely outcome, the population of goitered gazelles, a key prey species for cheetahs, in Miandasht WR and NP experienced a threefold increase between 2005 and 2016.

Since 2012, CACP succeeded to raise significant funds from corporate sources for cheetah conservation. These funds have supported compensatory measures for livestock killed by cheetahs, as well as incentive measures to improve the engagement of game guards (e.g., free health insurance coverage of game guards and their families) in protected areas used by cheetahs. However, these corporate engagements are subjected to short-term agreements (e.g., 3–5 years) that require regular renegotiations.

The Department of Environment with CACP has also played a fundamental role as "watch dog" and has successfully limiting encroachment of the mining industries and linear infrastructures into cheetah protected areas, such as the planned construction of a road in the middle of Bafq PA or a project of large-scale mining development in Dareh Anjir WR. These achievements were the result of assiduous vigilance and strong political will from the central authority of the Department of Environment.

Public awareness campaigns were established in communities in and around cheetah habitats to increase people's knowledge about the cheetah and its ecosystem, and to dispel misconceptions and myths. Also, following a proposition of the Department of Environment and in agreement with the Federation of International Football Associations the Iranian football federation selected the Asiatic cheetah as the official emblem for the football kit for the national team competing in the World Cup 2014. This generated considerable attention over the conservation of the cheetah and supported with success national fund raising campaigns.

The Asiatic cheetah has provided Iran with the opportunity to engage in modern wildlife conservation. The first phase evaluation of CACP has stated that "the conservation of the Asiatic cheetah has definitely created more national and international awareness than any other wildlife conservation project in the region. In Iran, it has generated wide interest among young researchers on big cats, carnivores, wildlife conservation, and research in general, and it has the potential to help spread this interest across the national borders to the whole region" (Breitenmoser et al., 2009). A conservation strategy and an action plan were developed in 2010 and now require revision to outline a new strategic framework, and provide the necessary impetus to conservation actions that will enable the long-term survival of the Asiatic cheetah in Iran. Despite all conservation efforts the Asiatic cheetah remains one of the rarest felids in the world, and will be critically dependent on continuing and effective conservation support for the foreseeable future.

References

Ahmadi, M., Balouchi, B.N., Jowkar, H., Hemami, M.-R., Fadakar, D., Malakouti-Kah, S., Ostrowski, S., 2017. Combining landscape suitability and habitat connectivity to conserve the last surviving population of cheetah in Asia. Divers. Distrib. 23 (6), 592–603.

Allahgholi M.A., Yusefi H., Khalatbari L., Mobargha M., 2007. Status of Asiatic cheetah in Bidouyeh Protected Area, Conservation of Asiatic Cheetah, final report, Tehran, Iran (in Persian).

Asadi H., 1997. The environmental limitations and future of the Asiatic cheetah in Iran. Project Progress Report, IUCN/SSC Cat SG, Tehran.

Ashayeri, D., Hamidi, A.H.Kh., Ashayeri, S., Abolghasemi, H., Ghadirian, H., Ajami, A., 2013. Conservation of Asiatic cheetah and its sympatric species in Touran. Persian Wildlife Heritage Found. Newslett. 3 (7), 5–6.

Bayat, H.R., 1984. Gazelles of Iran (in Persian). Environ. Q. 12, 13–15.

Belbachir, F., Pettorelli, N., Wacher, T., Belbachir-Bazi, A., Durant, S.M., 2015. Monitoring rarity: the critically endangered Saharan cheetah as a flagship species for a threatened ecosystem. PloS One 10 (1), e0115136.

Breitenmoser U., Alizadeh A., Breitenmoser-Würsten Ch., 2009. Conservation of the Asiatic Cheetah, its Natural Habitat and Associated Biota in the I.R. of Iran. Unpublished typescript report, p. 74.

Cheraghi S., Almasi M., Satey N., 2007. Status of Asiatic cheetah in Bajestan, Conservation of Asiatic Cheetah, final report, Tehran, Iran (in Persian).

Conservation of the Asiatic Cheetah Project, 2008. Conservation of the Asiatic cheetah, its Natural Habitats and Associated Biota in the I.R. of Iran. Final Report Project Number IRA/00/G35 (GEF/UNDP/DoE).

Dareshuri B., 1997 cited in Rezaie F., 2002. Cheetah in Iran. Donyayeh Vahsh Q. 1(3), 14–18 (in Persian).

Dareshuri B.F., Harrington F.A., 1976. A guide to the mammals of Iran, Department of the Environment, Tehran, p. 92 (in Persian).

Divyabhanusinh C., 2007. The End of a Trail, third ed. Oxford University Press, Oxford, p. 307.

Durant S., Mitchell N., Ipavec A., Groom R., 2015. *Acinonyx jubatus*. The IUCN Red List of Threatened Species 2015: e.T219A50649567. Available from: http://dx.doi.org/10.2305/IUCN.UK.2015-4.RLTS.T219A50649567.en.

Etemad, E., 1985. Mammals of Iran. Iranian Department of the Environment, Iran, (in Persian).

Farhadinia, M.S., 2004. The last stronghold: Cheetah in Iran. Cat News 40, 11–14.

Farhadinia, M.S., Akbari, H., Eslami, M., Adibi, M.A., 2016b. A review of ecology and conservation status of Asiatic cheetah in Iran. Cat News Special Issue 10, 18–26.

Farhadinia, M.S., Akbari, H., Musavi, S.J., Eslami, M., Azizi, M., Shokouhi, J., Gholikhani, N., Hosseini-Zavarei, F., 2013. Movements of Asiatic cheetah *Acinonyx jubatus venaticus* across multiple arid reserves in central Iran. Oryx 47 (3), 427–430.

Farhadinia M.S., Eslami M., Hobeali K., Hosseini-Zavarei F., Gholikhani N., Taktehrani A., 2014. Status of Asiatic cheetah in Iran: a country-scale assessment. Project Final Report, Iranian Cheetah Society (ICS), Tehran, Iran. p. 26.

Farhadinia, M.S., Gholikhani, N., Behnoud, P., Hobeali, K., Taktehrani, A., Hosseini-Zavarei, F., Eslami, M., Hunter, L.T.B., 2016a. Wandering the barren deserts of Iran: illuminating high mobility of the Asiatic cheetah with sparse data. J. Arid Environ. 30 (134), 145–149.

Farhadinia, M.S., Hemami, M.R., 2010. Prey selection by the critically endangered Asiatic cheetah in central Iran. J. Nat. Hist. 44, 1239–1249.

Farhadinia, M.S., Hosseini, F., Nezami, B., Harati, H., Marker, L., Fabiano, F., 2012. Feeding ecology of the Asiatic cheetah *Acinonyx jubatus venaticus* in low prey habitats in northeastern Iran: implications for effective conservation. J. Arid Environ. 87, 1–6.

Farhadinia, M.S., Hunter, L.T.B., Jourabchian, A.R., Hosseini-Zavarei, F., Akbari, H., Ziaie, H., Schaller, G.B., Jowkar, H., 2017. The critically endangered Asiatic cheetah *Acinonyx jubatus venaticus* in Iran: a review of recent distribution, and conservation status. Biodivers. Conserv. 26, 1027–1046.

Farhadinia, M.S., Nezami, B., Hosseini-Zavarei, F., 2008. Status of the Asiatic cheetah in North Khorasan Province. Department of Environment (in Persian).

Firouz, E., 1974. Environment Iran. National Society of the Conservation of Natural Resources and Human Environment, Tehran.

Ghadirian T., Eslami M., Hamidi A. Kh., Moqanaki E. M., 2010. Minimum population assessment of the Asiatic cheetah *Acinonyx jubatus venaticus* using camera photo-trapping in Kavir National Park, Semnan, Iran. Final report to the Conservation of the Asiatic Cheetah Project (CACP), Tehran, p. 63 (in Persian).

Ghoddousi A., Habibi Moghaddam A., Ashayeri D., Fahimi H., 2007. Status of Asiatic cheetah in Bahram-e-Gour Protected Area, Conservation of Asiatic Cheetah, Final report, Tehran, Iran (in Persian).

Goodwin, H., Holloway, C., 1974. Red Data Book. IUCN, Switzerland.

Grilo, C., Smith, D.J., Klar, N., 2015. Carnivores: struggling for survival in roaded landscapes. In: van der Ree, R., Smith, D.J., Grilo, C. (Eds.), Handbook of Road Ecology. Wiley, Chichester, pp. 300–312.

Hajji A., 1986. An introduction to the cheetahs of Iran. BSc thesis, Shahid Beheshti University, Iran (in Persian).

Hamidi A. Kh., Nezami B., 2009. Community attitudes towards the Asiatic cheetah in Touran and Miandasht. Final report, Conservation of Asiatic Cheetah Project (CACP), Tehran (in Persian).

Harrington, F.A., 1971. Present status of the cheetah in Iran, Unpublished typescript report.

Hunter, L., Jowkar, H., Ziaie, H., Schaller, G., Balme, G., Walzer, C., Ostrowski, S., Zahler, P., Robert-Charrue, N., Kashiri, K., Christie, S., 2007. Conserving the Asiatic cheetah in Iran: launching the first radio-telemetry study. Cat News 46, 8–11.

Jackson, P., 1998. Asiatic cheetah in Iran. Cat News 28, 2–3.

Joslin, P., 1984 cited in Divyabhanusinh., 1984. The origin, range and status of the Asiatic (or Indian) cheetah or hunting leopard (*Acinonyx jubatus venaticus*)—A tentative position paper. In : Jackson, P. (Ed.). The Plight of the Cats. Proceedings of the Meeting and Workshop of the IUCN/SSC Cat Specialist Group at Kanha National Park, Madhya Pradest, India, 9–12 April, pp. 186–198.

Jourabchian A.R., 1999. Cheetah status in Khorasan Province, Khorasan Provincial Department of the Environment, Unpublished report. p. 34 (in Persian).

Jourabchian A.R., Farhadinia M.S., 2008. Final report on Conservation of the Asiatic cheetah, its Natural Habitats and Associated Biota in Iran. Project Number IRA/00/G35 (GEF/UNDP/DoE), Tehran, Iran (in Persian with English summary).

Jowkar H., 1999. The preliminary study on Asiatic cheetah and its status in Iran. BSc thesis, Islamic Azad University, Iran. p. 176 (in Persian).

Jowkar, H., Ostrowski, S., Hunter, L., 2008. Asiatic cheetah cub recovered from a poacher in Iran. Cat News 48, 13.

Jowkar, H., Ostrowski, S., Tahbaz, M., Zahler, P., 2016. The conservation of biodiversity in Iran: threats challenges and hopes. Iran. Stud. 49 (6), 1065–1077.

Karami, M., 1992. Cheetah distribution in Khorasan Province, Iran. Cat News 16, 4.

Khalili, S.M., 1984. State of the cheetah in Iran. Environ. Q. 28, 2–11, (in Persian).

Laurenson, M.K., 1995. Behavioral costs and constraints of lactation in free-living cheetahs. Anim. Behav. 50 (3), 815–826.

Makhdoum, M.F., 2008. Management of protected areas and conservation of biodiversity in Iran. Integr. J. Environ. Stud. 65, 563–585.

Marashi, M., Masoudi, S., Moghadam, M.K., Modirrousta, H., Marashi, M., Parvizifar, M., Dargi, M., Saljooghian, M., Homan, F., Hoffmann, B., Schulz, C., Starick, E., Beer, M., Fereidouni, S., 2017. Peste des Petits Ruminants virus in vulnerable wild small ruminants, Iran, 2014–2016. Emerg. Infect. Dis. 23, 704–706.

Marker, L., Fabiano, E., Nghikembua, M., 2008. The use of remote camera traps to estimate density of free-ranging Cheetahs in North-Central Namibia. Cat News 49, 22–24.

Marnewick, K., Ferreira, S.M., Grange, S., Watermeyer, J., Maputla, N., Davies-Mostert, H.T., 2014. Evaluating the status of African wild dogs *Lycaon pictus* and cheetahs *Acinonyx jubatus* through tourist-based photographic survey in the Kruger National Park. PloS One 9 (1), e86265.

Marnewick, K., Funston, P.J., Karanth, K.U., 2008. Evaluating camera trapping as a method for estimating cheetah abundance in ranching areas. South Afr. J. Wildl. Res. 38 (1), 59–65.

Mohammadi, A., Kaboli, M., 2016. Evaluating wildlife–vehicle collision hotspots using kernel-based estimation: a focus on the endangered Asiatic cheetah in central Iran. Hum. Wildl. Interact. 10 (1), 103–109.

Moqanaki, E.M., Cushman, S.A., 2016. All roads lead to Iran: predicting landscape connectivity of the last stronghold for the critically endangered Asiatic cheetah. Animal Conserv. 20 (1), 29–41.

Mowlavi, M., 1985. Cheetah in Iran. Cat News 2, 7.

Nowell, K., 2014. An assessment of the conservation impacts of legal and illegal trade in cheetahs *Acinonyx jubatus*. IUCN SSC Cat Specialist Group report prepared for the CITES Secretariat, 65th meeting of the CITES Standing Committee, Geneva, 7-11 July. CITES SC65 Doc. 39.

Nowell, K., Jackson, P., 1996. Wild Cats: Status Survey and Conservation Action Plan. IUCN, Gland.

Persson, J., Rauset, G.R., Chapron, G., 2015. Paying for an endangered predator leads to population recovery. Conserv. Lett. 8 (5), 345–350.

Rezaie A., 2014. Trophic niche partitioning between Asian Cheetah *(Acinonyx jubatus venaticus)* and Persian Leopard *(Panthera pardus saxicolor)* in the Bafq Protected Area. MSc thesis, University of Tehran, Iran (In Persian).

Schaller G.B., O'Brien T., 2001. A preliminary survey of the Asiatic cheetah and its prey in the I.R. of Iran. Report to WCS, Iran DoE and UNDP-GEF.

Zamani N., 2010. Food habits of Asiatic cheetah in Naybandan and Dareh Anjir Wildlife Refuges. MSc thesis, University of Tehran, Iran, (in Persian).

Ziaie H., 2008. A field guide to mammals of Iran, second ed. Tehran (Iran): Wildlife Center Publication. p. 432 (in Persian).

Conservation Genetics of the Cheetah: Genetic History and Implications for Conservation

Anne Schmidt-Küntzel, Desiré L. Dalton**,†,
Marilyn Menotti-Raymond‡, Ezequiel Fabiano§,
Pauline Charruau¶, Warren E. Johnson††, Simone
Sommer‡‡, Laurie Marker*, Antoinette Kotzé**,§§,
Stephen J. O'Brien***,†††*

*Cheetah Conservation Fund, Otjiwarongo, Namibia
**National Zoological Gardens of South Africa, Pretoria,
South Africa
†University of Venda, Thohoyandou, South Africa
‡Laboratory of Genomic Diversity, Frederick, MD, United States (Retired)
§University of Namibia, Katima Mulilo, Namibia
¶University of Veterinary Medicine Vienna, Austria
††Smithsonian Conservation Biology Institute, Front Royal,
VA, United States
‡‡University of Ulm, Ulm, Germany
§§University of Free State South Africa, Bloemfontein, South Africa
***St. Petersburg State University, St. Petersburg, Russia
†††Nova Southeastern University, Fort Lauderdale, FL,
United States

INTRODUCTION

The cheetah (*Acinonyx jubatus*) is one of the most recognized examples of the important links between evolutionary history, genetic variation, and conservation. Its value to the biodiversity of the world is not only warranted by its unique physical characteristics, such as being the fastest land mammal (Chapter 7), but also its unique evolutionary lineage as the only extant representative of its genus, *Acinonyx*. Concerns over levels of genetic variation among cheetahs were first raised as captive programs grappled with difficulties in breeding cheetahs (Chapter 27). These observations led to research investigating the biological basis of the low rates of captive breeding success (10%–15%) and the concurrent high rate of infant mortality (29%) (O'Brien et al., 1985). This research led to the discovery of low levels of genetic diversity in the cheetah, which were attributed to one or several severe population bottlenecks. As a consequence, debates arose regarding the impact of low genetic diversity on the survival of the species, and the cheetah has been featured in genetic textbooks since the 1980s. Early research on cheetah represented one of the first studies in the new field of conservation genetics.

In the last century, cheetah numbers have declined drastically due to loss of habitat and prey, persecution due to real or perceived livestock depredation, and removal from the wild to supply captive facilities and private individuals (Chapters 10, 11, 13, and 14). The reduction in numbers and fragmented distribution add to the urgency to preserve the genetic diversity left today.

In this chapter, we review the current status of cheetah genetics and its impact on the species' conservation. While publications on cheetah genetics may appear contradictory at times, the fundamental conclusions have been consistent for over 30 years, with various measures confirming low genetic diversity (section "Genetic Diversity"), which was shown to have originated thousands of years ago (section "Historic Demography"), and with a relatively recent divergence

of extant published subspecies (section "Subspecies Definition and Divergence"). The differences debated among geneticists only affect the interpretation of genetic results regarding precise timing of events and extent of reduced genetic diversity. This chapter also covers the cheetah's phylogenetic (evolutionary relation based on genetic data) position among other felids (section "Species-Level Taxonomy"), the genetic structure of the subspecies and within geographical regions (section "Phylogeography"), an overview of additional genetic studies including kinship (section "Additional Insights Into Cheetah Genetics"), and implications of genetic findings for cheetah conservation (section "Discussion").

SPECIES-LEVEL TAXONOMY

The cheetah is a member of the family Felidae (Fig. 6.1), which comprises 41 living species that are distributed throughout the world, with the exception of Australasia and the polar regions (Kitchener et al., 2017). One of the most striking aspects of molecular genetic studies in the Felidae was how rapidly felids evolved into eight different lineages (over a 6-million-year period), each with unique biogeographical histories. Earlier groupings of felid lineages were largely based on morphological features and life-history patterns. The cheetah was generally considered to be an early divergence from the felid radiation due to some of its unique adaptations, including its incompletely retractile claws (Chapter 7). However, the advent of genetic approaches has provided clarity to more confidently reconstruct felid evolutionary history, and today the cheetah is included in the Puma lineage, which was the sixth of eight lineages to branch off during felid evolution [7 million years ago (MYA); Johnson et al., 2006; Li et al., 2016; Werdelin et al., 2010; Fig. 6.1]. The cheetah's closest living relatives are known to be the puma (*Puma concolor*) and the jaguarundi (*Herpailurus yagouaroundi*)

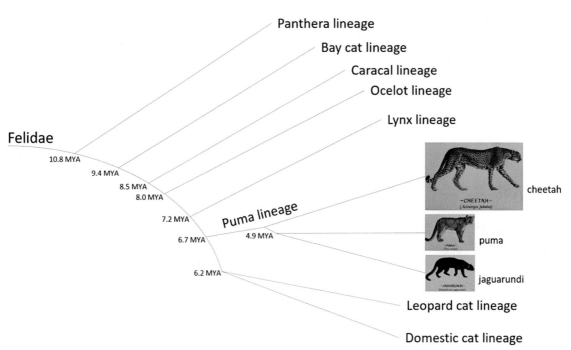

FIGURE 6.1 **Phylogenetic tree depicting the cheetah (*Acinonyx jubatus*) within the *Puma* lineage, relative to the other lineages of extant felid species.** Time of divergence for each lineage is indicated at the base of the branch in million years ago (MYA). The figure is based on the molecular data presented in Johnson et al. (2006).

(Johnson et al., 2006; Li et al., 2016; Werdelin et al., 2010; Fig. 6.1), with whom the cheetah likely shared a common ancestor and evolutionary history prior to their divergence. Today the cheetah is found in the Old World, and the puma and jaguarundi in the New World (Chapter 3). The cheetah has 19 chromosomal pairs (i.e., 38 chromosomes) like most felid species (O'Brien et al., 2006). The cheetah is the only extant representative of its genus.

GENETIC DIVERSITY

Genetic variation (polymorphism) forms the raw material of evolution. Novel variation rises from DNA mutations. DNA mutation rates that alter the amino acid sequence and may be detectable in allozyme migration rates (section

"Allozymes") are several orders of magnitude slower than those observed in the mitochondrial DNA (section "Mitochondrial DNA") or in repetitive elements in nuclear DNA, such as microsatellites (section "Microsatellites"). While significant gain of genetic variation is limited by mutation rates and takes numerous generations, variation can be lost within a single generation if only a subset of the population reproduces due to either high death rates or a high number of nonreproducing individuals.

Allozymes

The first study to indicate reduced genetic diversity of the cheetah documented low variation at protein-based markers compared to other felids and mammals (O'Brien et al., 1983; O'Brien et al., 1985; Table 6.1). Allozymes are

TABLE 6.1   Overview of Studies on Genetic Diversity in the Cheetah

Marker detail[a]	N (region)[b]	Polymorphism[c]	Diversity[d]	Study	Comparison
ALLOZYMES/SOLUBLE PROTEINS					
47 allozyme loci	55 (S)	0 (0%) pol loci		O'Brien et al. (1983)	8%–12%[e]
155 sol. prot. loci	55 (S)		0.013 Ho	O'Brien et al. (1983)	0.03–0.07 Ho[e]
52 allozyme loci	55 (S)	0 (0%) pol loci		O'Brien et al. (1985)	
49 allozyme loci	30 (E)	2 (4%) pol loci	0.014 Ho	O'Brien et al. (1987)	
	43 (S)	1 (2%) pol locus	0.0004 Ho		
MITOCHONDRIAL MARKERS (mtDNA)					
505 nt (28 RFLP)	39 (E), 35 (S)	6 variants, 7 hapl	0.182% div	Menotti-Raymond and O'Brien (1993)	
525 nt (all CR)	1 (E), 17 (S), 2 (NE)	15 variants	1.31% div	Freeman et al. (2001)	4.16–7.45%[f]
915 nt (221 CR)	29 (S), 26 (NE), 11 (E), 1 (N), 11 (Asia)	29 pol sites, 18 hapl	0.66% div	Charruau et al. (2011)	N/A[g]
Whole genome	4 (S)		0.071% div	Dobrynin et al. (2015)	
	3 (E)		0.008%div		
MICROSATELLITE MARKERS					
Whole genome (2 probes; RFLP 3 ez)	15–17 (E and S)	167 pol fragments	0.435 Ho	Menotti-Raymond and O'Brien (1993)	
10 loci (random)	5 (S), 5 (E)		0.39 Ho	Menotti-Raymond and O'Brien (1995)	0.61–0.77 Ho[h]
82 loci (random)	10 (NE)		0.44 Ho	Driscoll et al. (2002)	0.08–0.63 Ho[i]
	20 (S)		0.44–0.46 Ho		
38 loci	98 unrel (S)		0.64–0.70 He	Marker et al. (2008)	
13 loci	147 (E)		0.65 He	Gottelli et al. (2007)	
18 loci	27 (S)		0.70 He	Charruau et al. (2011)	
	25 (NE)		0.67 He		
14 loci	32 (S)		0.62 He	Dalton et al. (2013)	
MAJOR HISTOCOMPATIBILITY COMPLEX (MHC)					
MHC I (RFLP 4 ez)	9 (S)		0.05 Ho	Yuhki and O'Brien (1990)	0–0.51 Ho[j]
	13 (E)		0.07 Ho		
MHC I (~1100 nt seq)	2 (S?)	2 alleles		Yuhki and O'Brien (1994)	
MHC I (SSCP)	108 (S)	10 alleles		Castro-Prieto et al. (2011)	
MHC I (seq)	4 (S), 3 (E)	11 alleles		Dobrynin et al. (2015)	136 alleles[f]
MHC II-DRB (RSCA)	25	5 alleles		Drake et al. (2004)	

TABLE 6.1 Overview of Studies on Genetic Diversity in the Cheetah (*cont.*)

Marker detail[a]	N (region)[b]	Polymorphism[c]	Diversity[d]	Study	Comparison
MHC II-DRB (SSCP)	139 (S)	4 alleles		Castro-Prieto et al. (2011)	14/52 alleles[k]
MHC II-DRB (seq)	4 (S), 3 (E)	7 alleles		Dobrynin et al. (2015)	54 alleles[f]
WHOLE GENOME SEQUENCING					
Whole genome	4 (S), 3 (E)		0.02% div	Dobrynin et al. (2015)	

Studies are organized by marker type. Details about the marker and the number of animals tested from each region (Columns 1, 2) as well as the measured outcome (Columns 3, 4) are shown. Values from other species are indicated for comparison (Column 6).

[a] *CR, control region; ez, restriction enzyme; nt, nucleotide; RFLP, restriction fragment length polymorphism; RSCA, reference strand mediated conformational analysis; seq, sequencing; sol. prot., soluble proteins; SSCP, single strand conformation polymorphism.*

[b] *E, eastern Africa; NE, northeastern Africa; S, southern Africa; unrel, unrelated.*

[c] *hapl, Haplotypes; pol, polymorphic.*

[d] *He, expected heterozygosity; Ho, observed heterozygosity; % div, % nucleotide diversity.*

[e] *25 loci in caracal, Caracal caracal; leopard, Panthera pardus; lion, Panthera leo; ocelot, Leopardus pardalus; tiger, Panthera tigris; Newman et al. (1985).*

[f] *In domestic cat, Felis catus; same study.*

[g] *Not comparable to other studies due to other studies covering different mitochondrial regions.*

[h] *In domestic cat, lion, puma, Puma concolor; same study.*

[i] *0.08–0.16 Ho in lions from the Gir forest, pumas from Florida; 0.33–0.47 in lions from Serengeti and Ngorongoro, pumas from Idaho; 0.63 in domestic cat; same study.*

[j] *0–0.08 Ho in lions from the Gir forest and Ngorongoro; 0.17–0.51 in domestic cat, lions from the Serengeti, mole rat, human; same study*

[k] *14 alleles in 14 Bengal tigers; Pokorny et al. (2010); 52 in 25 Gir; Sachdev et al. (2005).*

variant forms of enzymes and can be detected with electrophoretic analyses. The observed differences of protein migration reflect differences at the amino acid level, which correspond to alleles (alternative genetic variants) of the corresponding gene(s) (locus/loci) at the DNA level. The percentage of polymorphic allozyme loci within several mammal species was estimated to range from 15% to 60% (O'Brien et al., 1987). Thus, it was surprising that analyses of 55 southern African cheetahs failed to identify genetic polymorphisms across a total of 52 allozyme loci (O'Brien et al., 1983; O'Brien et al., 1985; Table 6.1), and that additional analyses only identified 3 polymorphic loci in 73 eastern and southern African cheetahs [observed heterozygosity (differing alleles at a given locus/set of loci; measure of diversity) = 0.014 and 0.0004, respectively; O'Brien et al., 1987; Table 6.1]. Presence of multiple protein forms at 155 abundant soluble protein loci within individual cheetahs was also low (observed heterozygosity = 0.013) compared to 7 other felid species (O'Brien et al., 1983; Table 6.1).

However, insights from these studies were limited, because protein-based markers only assess amino acid changes that alter the electrophoretic mobility of the protein. DNA changes, such as silent/synonymous substitutions are masked in proteins, resulting in an underestimation of the amount of genetic variation in all species. In addition, protein-based markers are more likely to be linked to functional differences and as such allele frequency differences may be susceptible to selective pressure (environmental conditions affecting survival of organisms with a particular characteristic). The subsequent development of DNA-based markers provided more detailed information about the degree of genetic diversity in cheetahs.

Skin-Graft Acceptance

Concurrently with allozyme studies, functional studies demonstrated that reciprocal skin allografts between 12 unrelated cheetahs and 2 siblings showed no signs of acute graft rejection, whereas xenografts (from domestic cat, *Felis catus*) were rapidly rejected (O'Brien et al., 1985). These results were attributed to reduced *functional* allelic variation at the cheetah's major histocompatibility complex (MHC), an important immune gene family, which encodes cell surface proteins responsible for distinguishing foreign from self molecules. The inferred reduced functional variation was ultimately supported by molecular studies (section "Major Histocompatibility Complex").

Mitochondrial DNA

As sequencing techniques became available, low levels of genetic variation were also observed in mitochondrial DNA (mtDNA). mtDNA is independent from the nuclear genome, and represents the maternal demographic history. The mitochondrial genome evolves faster than the nuclear coding genome, with the control region (CR) being the most rapidly evolving region of the mitochondrial genome. mtDNA, in particular the CR, has been informative for investigations of diversity patterns, population structure, and phylogeography (phylogenetic structure in relation to location). The complete cheetah mtDNA genome has 17,047 bp (Burger et al., 2004) and has 91% similarity with the mtDNA genome of the domestic cat (Lopez et al., 1996).

In the 1990s, a study based on restriction fragment length polymorphism (RFLP) inferred low levels of nucleotide variation (0.18% diversity) in cheetah mtDNA relative to comparable studies in other species (Menotti-Raymond and O'Brien, 1993; Table 6.1). In 2001, nucleotide variation in the mtDNA-CR in 20 cheetahs was observed to be relatively low (1.31%; Freeman et al., 2001; Table 6.1). Nucleotide variation was even lower when a short sequence of the mtDNA coding region was included with the mtDNA-CR (Charruau et al., 2011), or when the entire mtDNA genome was evaluated (Dobrynin et al., 2015; Table 6.1). Dobrynin et al. (2015) found a 90% reduction in nucleotide variation across 7 cheetahs (4 Namibian, 3 Tanzanian)

relative to other mammals (Dobrynin et al., 2015; Table 6.1).

Microsatellites

Microsatellites (nuclear DNA markers consisting of variable numbers of tandem repetitions of 2–6 nucleotides; also called STR or short tandem repeats) accumulate new variation quickly as they have high mutation rates (several orders of magnitude higher than DNA coding for proteins), and are usually not associated with any function (i.e., they do not code for a protein) and thus are not subjected to selection pressure. These markers have been used widely in conservation genetics, population genetics, and wildlife forensics. Most non-domestic felid studies have selected a subset from 583 polymorphic microsatellite loci, which were mapped and characterized in the domestic cat (Menotti-Raymond et al., 2003).

Initial studies, with 10 microsatellite markers in 10 cheetahs, demonstrated low observed heterozygosity (0.39; Menotti-Raymond and O'Brien, 1995; Table 6.1). Expected heterozygosity was slightly higher (0.46–0.48) when measured in 82 randomly selected microsatellite markers in 30 cheetahs (Driscoll et al., 2002; Table 6.1). These estimates were comparable to the heterozygosity levels of other felids in small isolated populations (e.g., pumas from Idaho, lions from Serengeti and Ngorongoro crater), but lower than in domestic cats (Driscoll et al., 2002; Table 6.1). Subsequent studies of cheetahs only included a subset of these microsatellite markers, which were not selected randomly, but instead for their known relatively high level of polymorphism, leading to higher heterozygosity estimates. Hence, these increased heterozygosity estimates do not reflect the genetic diversity of the cheetah species *per se*, but are summarized in Table 6.1 for informational purposes.

Terrell et al. (2016) suggested a reduction in genetic diversity in the wild population from a dataset spanning 30 years. The study was based on 46 individuals born between 1976 and 2007 from South Africa and Namibia, which were genotyped (characterized on a genetic level) with 12 microsatellite markers.

While microsatellite markers demonstrated levels of heterozygosity in the cheetah that were not always significantly lower than for other species (Table 6.1), this does not contradict findings of low genetic diversity in cheetahs at other, more slowly evolving markers. It merely indicates that the variation at microsatellite loci is more recently evolved in origin. The length of time needed to accumulate this new microsatellite variation can inform estimates of the timing of events that led to the loss of genetic variation (section "Historic Demography").

Major Histocompatibility Complex

The MHC is one of the most polymorphic loci known in vertebrates and has important immune functions. An increasing number of studies have documented an association between the diversity of MHC genotypes or individual alleles with disease susceptibility in wildlife, thus confirming the importance of the selection pressure from pathogens on the MHC (reviewed in Sommer, 2005).

Most comprehensive studies of the MHC in felids (also known as the feline leucocyte antigen), have been conducted in the domestic cat (Winkler et al., 1989). The cat is a good model as the general architecture of the MHC appears relatively conserved within each class of vertebrates; the number of MHC class I or II genes, however, can vary substantially among species. The domestic cat MHC region is located on chromosome B2 and includes 19 MHC I and 8 MHC II genes (Yuhki et al., 2008). The cheetah MHC sequence resolved 278 genes with complete homology to the domestic cat for all MHC II and most MHC I genes. Its structural organization was also found to be highly similar to that of the domestic cat (Dobrynin et al., 2015).

An early study of cheetah MHC I based on RFLP markers showed reduced genetic diversity in cheetah (observed heterozygosity = 0.05–0.07) compared to other species, which was only comparable to that of lions from isolated populations (Gir Forest and Ngorongoro Crater; Yuhki and O'Brien, 1990; Table 6.1). Only 2 MHC I alleles were identified in 2 individual cheetahs through sequencing (Yuhki and O'Brien, 1994; Table 6.1) and 5 MHC II-DRB alleles were identified in 25 individuals through Reference Strand-Mediated Conformational Analysis (Drake et al., 2004; Table 6.1). Castro-Prieto et al. (2011) identified 10 unique MHC I and 4 MHC II-DRB alleles in 108 and 139 Namibian cheetahs, respectively (Table 6.1). While a 5-fold increase in the number of MHC I alleles identified in 2011 may appear like an increase in genetic diversity, the identification of only 8 additional alleles despite a 54-fold increase in the number of study animals is in fact further confirmation of low levels of allelic diversity in the cheetah. The low level of allelic diversity was further confirmed by comparison to other species, which harbor more alleles in fewer individuals (Table 6.1). More recently, a 95%–98% reduction in single nucleotide variants (SNVs) was observed when the complete MHC sequence of 7 cheetahs (4 Namibian, 3 Tanzanian) was compared to that of human (*Homo sapiens*), dog (*Canis familiaris*), and an outbred domestic cat; only 11 variants affecting the amino acid sequence were detected in the MHC I coding region of cheetahs (Dobrynin et al., 2015; Table 6.1).

Whole Genome Sequence Variants

Dobrynin et al. (2015) described patterns of diversity across the entire genome of the cheetah. Five commonly employed metrics confirmed the genic and genomic lack of diversity of the species: SNV incidence was 90% less than that observed in a feral domestic cat; SNV density was 8–15× less than in the domestic cat,

European wildcat (*Felis silvestris silvestris*), or human (however it was higher than in lions from the Gir Forest); regions of continuous homozygosity (identical alleles at a given locus/set of loci) were 10–15× longer than in the domestic cat; heterozygosity levels were 15%–61% of the levels observed in the domestic cat, tiger (*Panthera tigris*), and human; SNVs in coding genes were 98% reduced compared to the domestic cat or European wildcat (Dobrynin et al., 2015; Table 6.1; Fig. 6.2).

Since the initial discovery of reduced genetic diversity in the cheetah in the 1980s, conservation, and scientific interest have turned toward identifying its cause (section "Historic Demography") and assessing the impact of low genetic diversity on the cheetah's chances of long-term survival (section "Importance of Low Genetic Diversity on Cheetah Survival").

HISTORIC DEMOGRAPHY

The cumulative results, indicative of reduced genetic diversity in the cheetah, were consistent with a genetic bottleneck or a series of demographic reductions over time and space. Menotti-Raymond and O'Brien (1993) proposed a scenario of distant past, rather than recent, reduction in the global population size. The demographic event causing this drastic loss of diversity was estimated to have occurred during the end of the Pleistocene (10,000–12,000 years ago; Table 6.2). This proposal was based on the time estimated for the near-reconstitution of genetic variation at rapidly evolving minisatellite (nuclear DNA marker consisting of variable numbers of tandem repetitions of 5–50 nucleotides) loci. The authors obtained similar estimates with mtDNA RFLP data calibrated on estimates of divergence of the species from the *Panthera* genus. This estimated timeframe was later corroborated with 82 microsatellite markers (Driscoll et al., 2002), with the time needed for nuclear

alleles to reach fixation in the MHC (Castro-Prieto et al., 2011), and with whole genome data (Dobrynin et al., 2015) (Table 6.2). In the Asiatic cheetah population a more recent, independent, bottleneck was inferred based on a significant heterozygosity excess observed in the 18 microsatellite loci tested (Charruau et al., 2011).

Several alternative hypotheses to the single bottleneck scenario have been proposed to explain the severe loss of genetic variation. First, that the uniformity resulted from a persistent low effective population size (*Ne*; theoretical number which roughly reflects the number of animals genetically contributing to the population), possibly resulting from the high reproductive variance linked with the cheetah's presumed polygynous mating system (Pimm et al., 1989). Second, that low effective population sizes were maintained by a continuous cycle of extinction of subpopulations followed by recolonization, that is, metapopulation dynamics (Gilpin, 1991; Hedrick, 1996; Pimm et al., 1989).

The availability of SNVs derived from whole genome data of individuals from both eastern and southern Africa permitted more robust analyses of historical demographic patterns (Dobrynin et al., 2015). These analyses support the premise that cheetah populations expanded uniformly following a founder event 100,000 years ago. This was followed by a more recent split into an eastern and southern population, which were subjected to a bottleneck around 10,000–12,000 years ago. An alternative scenario of a gradual decline in the effective population size was supported by analyses of diploid whole genome sequence data to estimate past population sizes (Table 6.2).

Fabiano et al. (in preparation) suggested a gradual decline in population numbers, commencing at least 20,000 years ago, based on different coalescent-based approaches applied to published microsatellite profiles in the Namibian cheetah. While there was evidence

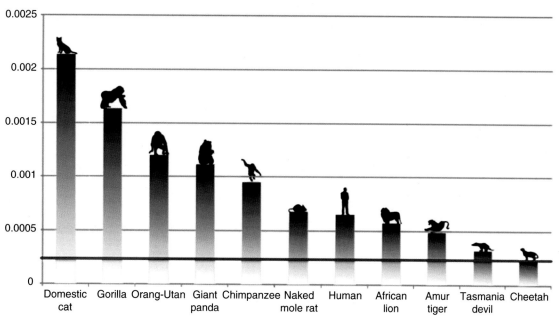

FIGURE 6.2 **Rates of single nucleotide variants (SNV) representing the diversity of the cheetah (*Acinonyx jubatus*) genome relative to other mammal genomes.** SNV rates were estimated for one individual per represented species using all variant positions, without filtering for repetitive regions. *Source: Reprinted from Dobrynin et al., 2015.*

of a continuous decline during this time period, some methods suggest an accelerated decline around 10,000 and 13,000 years ago (Table 6.2).

SUBSPECIES DEFINITION AND DIVERGENCE

Prior to the availability of genetic data, the species' taxonomy was based on morphological and geographical information. According to these taxonomic criteria, the extant cheetah populations were classified into four African and one Asiatic subspecies (Smithers, 1975), namely: *A. jubatus hecki* Hilzheimer, 1913 in northwest Africa; *A. j. raineyi* Heller, 1913 (for which the name *ngorongorensis* Hilzheimer, 1913 has priority) in east Africa; *A. j. jubatus* Schreber, 1775 in southern Africa; *A. j. soemmeringii* in

northeast Africa; and *A. j. venaticus* Griffith, 1821 from north Africa to central India.

A. j. jubatus and *A. j. raineyi* were the first two subspecies to be assessed with molecular tools. Initial allozyme analyses in 1987 detected some minor differences (O'Brien et al., 1987; Table 6.3). The separation of the two sub-Saharan populations was further supported with mtDNA-CR (Freeman et al., 2001), microsatellites (Driscoll et al., 2002), and whole genome variation (Dobrynin et al., 2015; Table 6.3). The time of divergence was estimated to be a minimum of 4500 years ago by both Driscoll et al. (2002) (Table 6.3) and O'Brien et al. (2017). In 2017, based on their interpretation of the published evidence, the International Union for Conservation of Nature (IUCN) Cat Specialist Group's Cat Classification Task Force has suggested that *A. j. raineyi* and *A. j. jubatus* be synonymized into a

TABLE 6.2   Studies Investigating the Historic Demography of the Cheetah With Molecular Genetic Methods

Marker[a]	N	Generation time (years)[b]	Mutation rate/ calibration[c]	Model[d]	Time range (years ago)	Conclusion
Menotti-Raymond and O'Brien (1993)						
mtDNA RFLP (505 nt; 28 ez)	74	6	Cal: Panthera ancestor: 1.6–2.0 MYA	Molecular clock (1 initial hapl)	28,000–36,000	
Minisat fing prt (2 probes, 3 ez)	7–16	6	μ: 4.7×10^{-4}– 1.7×10^{-3}	$1/\mu \times G$ (1 initial allele)	3,529–12,766	10,000–12,000
Driscoll et al. (2002)						
Microsatellite (82 loci)	20	6	μ: 5.6×10^{-4}– 2.05×10^{-3}	$1/\mu \times G$	2,928–10,716	Min 12,000
				SMM	4,631–16,950	
Castro-Prieto et al. (2011)						
MHC I/II-DRB	108/139	2.4		Fixation $4N_e$	2,976–14,880	
Dobrynin et al. (2015)						
Whole-genome (1,8 M var. sites)	7	3	μ: 0.3×10^{-8}	DaDi	11,084–12,589	12,000
		3		PSMC	Since 100,000	Gradual decline
Fabiano et al. (in preparation)						
Microsatellite (31 loci)	89	2.4 and 6		Coal (DIYABC-FDA, MSVAR1.3, VarEff)	Since >20,000	Gradual decline[e]

Marker information, number of animals, main parameters (Columns 1–5), as well as outcomes (Columns 6 and 7) are shown for each study.

[a] *ez, Restriction enzyme; M var. sites, Million variable sites; Minisat fing prt, minisatellite fingerprint; mtDNA, mitochondrial DNA; nt, nucleotide; RFLP, restriction fragment length polymorphism.*

[b] *2.4 years (Kelly et al.,1998); 3 years (modified from Kelly et al., 1998); 6 years (Marker and O'Brien, 1989).*

[c] *cal: Reference time point used as calibration; μ: mutation rate given per generation; MYA, million years ago.*

[d] *$1/\mu \times G$ (Nei, 1987); coal, coalescent methods, MSVAR1.3 (Storz and Beaumont, 2002), VarEff (Nikolic and Chevalet, 2013), DIYABC-FDA (Cornuet et al., 2010); DaDi, Diffusion approximation to the allele frequency spectrum (Gutenkunst et al., 2009); fixation $4N_e$, fixation of a neutral nuclear marker is expected after $4N_e$ generations (Nichols, 2001); PSMC, Pairwise Sequential Markovian Coalescent (Li and Durbin, 2011); SMM, Stepwise Mutation Model (Valdes et al., 1993).*

[e] *An accelerated decline may be present 10,000 and 13,000 years ago.*

single subspecies (Kitchener et al., 2017). Additionally, Kitchener et al. (2017) put forward that as additional data become available the four subspecies that the IUCN Cat Specialist Group currently recognizes may be further merged in the future. However, this does not affect the amount of diversity found between the populations identified to date, which we will present here.

A. j. soemmeringii was first compared to *A. j. jubatus* in 2001 based on mtDNA (Freeman et al., 2001), with additional support provided in 2011 (Charruau et al., 2011). Time of divergence was estimated to be between 1,600 and 72,296 years ago by Charruau et al. (2011) (Table 6.3) and approximately 5,000 years by O'Brien et al. (2017).

A. j. venaticus was only genetically assessed in 2011, when Charruau et al. (2011) analyzed the first samples from the subspecies in a range-wide study. This finding refuted the possibility that the current Asiatic cheetah population may have originated from individuals imported from eastern Africa for hunting purposes in the past century. The time of divergence between *A. j. venaticus* and *A. j. jubatus* was estimated at 4,700–67,400 years ago. The extent of the separation of *A. j. venaticus* from the African subspecies was not clear-cut. mtDNA data placed the split between *A. j. jubatus* and *A. j. venaticus* slightly more recently than that of *A. j. jubatus* with *A. j. soemmeringii*, while microsatellite data suggested that the divergence with *A. j. soemmeringii* was the more recent event (Charruau et al. 2011; Table 6.3). The pairwise genetic distance (F_{ST}: measure of difference between populations) of *A. j. venaticus* to the African subspecies appeared to be possibly slightly larger than distances within African populations (mtDNA F_{ST} values: 0.818–0.958 compared to 0.724–0.930; Charruau et al., 2011). However, it is important to keep in mind that divergence values between *A. j. venaticus* and the other subspecies could have been stochastically increased due to a postulated recent bottleneck in *A. j. venaticus* (section "Historic Demography"). O'Brien et al. (2017) estimates a time of divergence between *A. j. venaticus* and *A. j. jubatus* of approximately 6500 years ago.

A. j. hecki was not specifically assessed, as none of the studies was able to include confirmed *A. j. hecki* samples from west Africa.

PHYLOGEOGRAPHY

Range Wide

Historically, the distributions of each subspecies were defined based on morphological differences and presumed connectivity between populations. Once the subspecies were confirmed genetically, the distribution of each subspecies and their genetic structure could be verified (Fig. 6.3A). Charruau et al. (2011) obtained a sample collection of 94 cheetahs from 18 countries from the extant and historical cheetah range. These samples were expected to represent four of the five cheetah subspecies recognized at that time.

A. j. jubatus was confined to individuals from southern African countries, which included Botswana, South Africa, and Namibia. These samples consistently clustered (grouped) together, with both nuclear (microsatellite) and mtDNA data. Depending on the type of analyses, a single cheetah sample from the Democratic Republic of Congo grouped with *A. j. jubatus*, or slightly outside. The mtDNA haplotype (linked group of variants that is inherited together) group of *A. j. jubatus* was the most diverse (8 haplotypes) of the investigated sample collection and was centrally positioned in the mtDNA haplotype networks, with the haplotypes of the other subspecies radiating from it.

Haplotypes assigned to *A. j. raineyi* were confined to east African countries, which included Kenya and Tanzania. However, the maternal lineages (mtDNA) fell into 2 separate haplotype groups, one of which clustered with *A. j. jubatus*, separately from other *A. j. raineyi* haplotypes. As a consequence, *A. j. raineyi* has been included in *A. j. jubatus* by Kitchener et al. (2017)

A. j. venaticus was confined to extant samples from Iran. Historical samples of Asiatic cheetahs (Oman, Jordan, India, Iraq, medieval Iran) clustered with the extant Iranian cheetah samples, as observed with both nuclear and mtDNA data. Additionally, one sample from the extinct population in northeastern Egypt also clustered with the Asiatic cheetah samples of *A. j. venaticus*.

Historical samples from Libya appeared distinct based on mtDNA data. With microsatellite data the Libyan sample clustered more closely to, but separately from, the Asiatic cheetah (bootstrap support of 72% for the divergence between the branch of the Asiatic samples and the Libyan sample). Cheetahs from southern Egypt, western Sahara, and Algeria shared the mtDNA haplotype with the Libyan samples, and were separate from the *A. j. venaticus* haplotypes. Thus the mtDNA data supported the Asiatic subspecies

TABLE 6.3   Divergence Between Putative Cheetah Subspecies Estimated by Molecular Genetic Methods

Marker[a]	N_1/N_2[b]	Parameter[c]	Model[d]	Outcome	Conclusion
A. j. jubatus/A. j. raineyi (classified as consubspecific in 2017)					
Allozymes (49 loci)	43/30		D	0.004 distance (8× less than humans)	
mtDNA (505 nt; 28 ez)	35/39			100% unique hapl.	
mtDNA (525 nt)	17/1		K2P	15 substitutions	
Minisat (2 probes; 3 ez)	15–17		APD	Mean APD: 48.2 29% unique frag.	
Microsat (82 loci)	10/10	μ: 5.6×10^{-4}–2.05×10^{-3}	$(\delta\mu)^2$	4,253 ya	≥4,500 ya
		Cal: bottleneck 12,000 ya	Prop unique	4,514 ya	
A. j. jubatus/A. j. soemmeringii					
mtDNA (525 nt)	17/2		K2P	9 substitutions	
mtDNA (915 nt)	29/26	Clo: Puma-cheetah 4.92 MYA	Coal	32,200–244,000 ya	66,500 ya
			D_A	26,660–202,100 ya	55,085 ya
			IMa	43,928–379,317 ya	72,296 ya
			0 Migr	24,067–117,615 ya	66,698 ya
Microsat (18 loci)	27/25	μ: 2.05×10^{-4}–2.05×10^{-3}	$(\delta\mu)^2$	3,200–32,400 ya	
			D_{SW}	1,600–15,600 ya	
A. j. jubatus/A. j. venaticus					
mtDNA (915 nt)	29/11	Clo: Puma-cheetah 4.92 MYA	D_A	20,300–153,800 ya	41,900 ya
			Coal	15,570–118,020 ya	32,170 ya
			IMa	27,420–379,222 ya	44,403 ya
			0 migr	16,295–83,677 ya	42,120 ya
Microsat (18 loci)	27/8	μ: 2.05×10^{-4}–2.05×10^{-3}	$(\delta\mu)^2$	6,700–67,400 ya	
			D_{SW}	4,700–47,200 ya	

Studies are organized by subspecies pair. Marker information, animal numbers per subspecies and parameters (Columns 1–4) as well as outcomes (Columns 5 and 6) are indicated for each study. ya, Years ago.

Studies: *A. j. jubatus/A. j. raineyi*: Allozymes, O'Brien et al., 1987; mtDNA (505 nt), Menotti-Raymond and O'Brien, 1993; mtDNA (525 nt), Freeman et al., 2001; Minisat, Menotti-Raymond and O'Brien, 1993; Microsat, Driscoll et al., 2002.

A. j. jubatus/A. j. soemmeringii: mtDNA (525 nt), Freeman et al., 2001; mtDNA (915 nt), Charruau et al., 2011; Microsat, Charruau et al., 2011.

A. j. jubatus/A. j. venaticus: mtDNA, Charruau et al., 2011; Microsat, Charruau et al., 2011.

[a] *ez, Restriction site enzyme; Microsat, microsatellite; Minisat, minisatellite fingerprint; mtDNA, mitochondrial DNA; nt, nucleotide.*

[b] N_1/N_2 *is the no. of individuals N_1 of subspecies 1 followed by the no. N_2 of subspecies 2; subspecies 1 and 2 correspond to the two subspecies indicated for each section.*

[c] *Generation time used for all studies is 6 years (Marker and O'Brien, 1989); cal, reference time point used as calibration; clo, reference time point used for molecular clock; μ, mutation rate given per generation; MYA, million years ago.*

[d] *0 migr, Isolation with migration rate = 0, conservation method (Wakeley and Hey, 1997); $(\delta\mu)^2$, microsatellite genetic distance (Goldstein and Pollock, 1997; Zhivotovsky and Feldman, 1995); APD, average percent difference; Coal, coalescent method (Gaggiotti and Excoffier, 2000); D, Nei's raw number of nucleotide differences between populations (Nei and Li, 1979); D_A, net number of nucleotide differences between populations (Nei and Li, 1979); D_{SW}, stepwise weighted genetic distance (Shriver et al., 1995); IMa, isolation with migration, demographic method (Hey and Nielsen, 2007); K2P, Kimura's 2-parameter model (Kimura, 1980); Prop unique, proportion of unique alleles.*

boundaries suggested by Nowell and Jackson (1986) for *A. j. venaticus*, with cheetahs from the northern Sahara being of a different subspecies (Belbachir, 2007; Krausman and Morales, 2005). The cheetahs from the northern Sahara may be of the same subspecies as the west African cheetah (*A. j. hecki*), but this could not be confirmed, as no west African samples were available.

A. j. soemmeringii was confined to northeast African countries, which included Sudan, Djibouti, Ethiopia, and Somalia. Although these samples clustered together, some substructuring was identified. In Freeman et al. (2001), 1 of the *A. j. soemmeringii* mtDNA haplotypes was more similar to the *A. j. raineyi* haplotype (1 substitution apart) than to the other *A. j. soemmeringii* mtDNA haplotypes (9 substitutions apart). This likely reflects a more complex migration/colonization history, imperfect lineage sorting, or weakly defined boundaries of subspecies. All the *A. j. soemmeringii* mtDNA haplotypes shared a 1-amino acid deletion in the mtDNA-ND5 protein, which, if confirmed, could be used as diagnostic site to trace illegally traded specimens.

Globally, there was no evidence of gene flow between the subspecies in recent generations, as no admixture was detected with Bayesian Analysis of Population Structure. Genetic differentiation was further supported by the Neighbor-Joining (NJ) phylogenetic tree based on 18 microsatellite markers.

Southern Africa

Patterns of genetic variation identified 3 broad regional populations in southern Africa: Namibia, the Kalahari, and South Africa, when analyzing 13 microsatellite loci in 51 cheetahs from Botswana, Namibia, and the north-western parts of South Africa (Fig. 6.3B; Kotze et al., 2008). F_{ST} values between Namibia and South Africa (0.115) and Namibia and the Kalahari (0.108) were higher than those between South Africa and the Kalahari (0.059). However, these values remain below the F_{ST} values recommended for high statistical certainty of population assignment (0.15–0.20; Manel et al., 2002). Also, while the significance of the differentiation between populations ($P < 0.001$) suggests that accurate population assignment may be possible, attempts to assign 6 cheetahs of unknown origin to a specific region (from a Bayesian exclusion test method and a frequency-based method) were inconclusive. The efficacy of assignment testing in cheetahs will most likely be improved through more comprehensive sampling (more subpopulations and more individuals per region; Manel et al., 2002). mtDNA data may also help to resolve questions of origin within southern Africa, given that 8 haplotypes were identified in a 915 bp sequence of mtDNA in Botswana, Namibia, and South Africa (Charruau et al., 2011).

Namibia

The Namibian cheetah population was the first national population to be assessed genetically. Marker et al. (2008) analyzed 35 microsatellite markers in 89 unrelated individuals whose distribution covered most (over 90%) of the cheetah's natural distribution in the country (Fig. 6.3B). No regional structure was detected when analyzing samples individually. When individuals were grouped by region, very modest support was given to a tentative grouping of the north-western regions (Outjo, Grootfontein; NJ tree, bootstrap support of 98%) and of the southern regions (Windhoek and Gobabis; bootstrap support of 79%). Multiple component analysis distributed the regions somewhat according to their geographical location, with the most northern region (Outjo) appearing to be most distinct, but F_{ST} values for Okahanja and Outjo ($F_{ST} = 0.086$) were below the recommended threshold for separate population assignment (0.15–0.20; Manel et al., 2002). Overall it was concluded that the Namibian cheetah population is panmictic (without barrier to breeding and without population structure), with gene

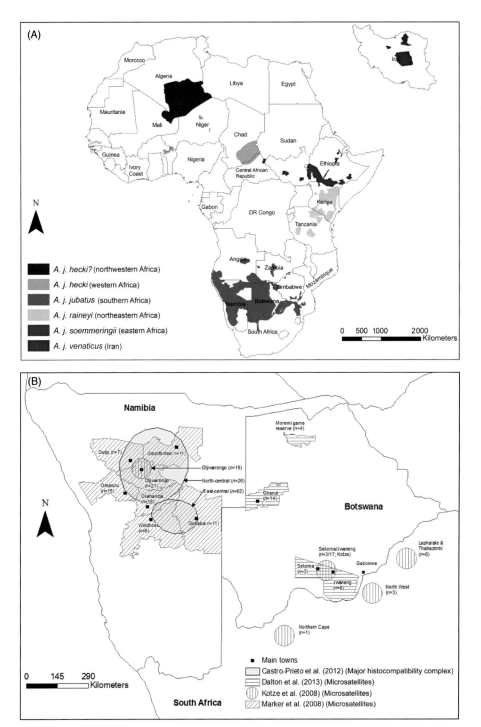

FIGURE 6.3 (A) Current confirmed resident range-wide distribution map (according to Durant et al., 2017), representing the putative subspecies according to genetic verification in Charruau et al. (2011). Subspecies genetically confirmed to date (*A. jubatus jubatus*; *A. j. soemmeringii*; *A. j. venaticus*) are represented with intense colors (*green, blue, red,* respectively); the northern African subspecies (which was confirmed genetically, and may be *A. j. hecki*) is in *brown,* and the expected *A. j. hecki* range (which has not yet been genetically tested) in *light brown.* *A. j. raineyi* was classified as consubspecific of *A. j. jubatus* in Kitchener et al. (2017) and is represented in lighter shade of *green.* (B) Geographical representation of the regions covered by the southern African phylogenetic studies.

flow maintained between the studied regions and precluding the emergence of any major structure. Connectivity was further supported by identification of two migrants to and from the Outjo region.

More recently, Castro-Prieto et al. (2012) reached similar conclusions through analyses of exon 2 of MHC class I, and class II-DRB genes that were genotyped with the single strand conformational polymorphism method in 26 individuals from the north-central region and 62 individuals from the east-central region (Fig. 6.3B). All identified alleles and resulting haplotypes were present in both regions. No regional differentiation was detected for MHC II-DRB, although haplotype frequency for MHC class I varied between north-central and east-central Namibia with a moderate F_{ST} support ($F_{ST} = 0.07$; $P < 0.01$). The difference in allele frequency at MHC class I (which targets intracellular pathogens) was attributed to different viral selective pressures in the two regions.

Botswana

Dalton et al. (2013) conducted an analysis of the Botswanan population (Fig. 6.3B) in 32 unrelated animals with 14 microsatellite loci. Although this study included a small sampling number, it still provided essential insights into the lack of population structure in Botswana. All animals were assigned to one unique group. Absence of substructure was further supported from the analysis of molecular variance (93% of variation shared among localities) and low genetic distance between the two largest sampled populations (Ghanzi vs. Jwaneng $F_{ST} = 0.035$; $P < 0.05$). Weak subdivision among the geographical populations suggests that gene flow occurs, which can be attributed to natural cheetah movements. The Moremi population appeared slightly "distinct" from the neighboring Ghanzi population ($F_{ST} = 0.079$), which was tentatively attributed to the presence of the Okavango delta as potential natural barrier between these two populations.

ADDITIONAL INSIGHTS INTO CHEETAH GENETICS

Kinship

Genetic analyses of fecal samples in the Serengeti identified multiple paternities for subsequent litters (8 females), as well as within litters (10/23 litters); 3 observed adoption events could also be confirmed genetically (Gottelli et al., 2007). In Namibia 21 of 23 females with cubs were confirmed as biological mothers, and 17 of 21 presumed sibling groups without a dam were confirmed to be related (Marker et al., 2008); in all exceptions (2 mothers and 4 sibling groups) individuals had been sampled from captive facilities with suspected human-induced animal grouping.

Coalition males appeared to be related in Namibia based on 23 of 26 male coalitions (Marker et al., 2008); while in Botswana, 3 of 4 wild-caught coalitions appeared to include at least one unrelated individual (Dalton et al., 2013). All three unrelated Namibian groups and one of the unrelated Botswanan groups dispersed at the time of release, suggesting that the grouping may have been an artifact of capture.

It was also shown that females with higher levels of genetic relatedness had greater home-range overlap (Marker et al., 2008), suggesting a matriarchal society.

Genetic Investigations of Infectious Diseases Affecting Cheetahs

The identification of the infectious agent responsible for one of the better documented viral outbreaks in the captive cheetah population (Chapter 25) was made possible due to samples properly stored since the outbreak in the early 1980s (Pearks Wilkerson et al., 2004). Using a phylogenetic approach, the virus was identified as a coronavirus, very similar to domestic cat coronavirus responsible for feline infectious peritonitis. However, in cheetahs, morbidity was 100% (symptoms included diarrhea, jaundice, and seizures) and overall mortality was 60% within 3 years (85% in cubs). This was significantly higher compared to the 5%–10% mortality in domestic cat or the corresponding coronavirus (SARS) outbreak in humans in 2002, making this the deadliest documented death toll of a coronavirus. The authors attributed the high death toll to the lack of genetic diversity and the naïveté of the cheetah population to this virus, which appeared to have jumped from the domestic cat into the cheetah species.

A new species of infectious blood-borne parasite, *Babesia lengau*, was characterized genetically in cheetah (Bosman et al., 2010). While, as opposed to its effect in domestic cats, *Babesia* is not currently known to cause any pathology in the cheetah, additional research is continuing to investigate the effect of *Babesia* on specific health parameters (Schmidt-Küntzel et al., in preparation).

Investigations of Potential Genetic Predisposition to Disease

A single nucleotide polymorphism (SNP), described in the *Serum Amyloid A* (*SAA*) gene, was genotyped in captive cheetahs to assess whether there was a correlation between amyloidosis disease status (Chapter 25) and the SNP genotype (Franklin et al., 2015). It was found that the SNP had a semidominant effect on the associated protein level within each study population (N = 58), but that the institution at which the animals were housed had an even larger effect. In addition, there was no significant association between genotype and disease status (N = 48). Thus, the genetic impact of *SAA* on amyloid levels in cheetahs is minimal and outcompeted by other factors.

No correlation could be detected between variants in the mtDNA genome and myelopathic pathology (Burger et al., 2004). And to date no correlation could be identified between oxalate nephrosis and the coding regions of published

candidate genes (Cheetah Conservation Fund and National Zoological Gardens of South Africa, unpublished data).

Investigations of the Molecular Basis for Heritable Traits

The cheetah has long been known for its poor sperm quality, with less than 20% viable sperm observed in reproductive studies (Chapter 27). In a whole genome study comparing the cheetah sequence to that of other species, several mutations affecting gene function were found in *A-kinase anchor protein 4* (*AKAP4*), a gene involved in spermatogenesis, and were shown to be likely fixed (only 1 allele present in the population) in the cheetah (Dobrynin et al., 2015). Those mutations may be in part responsible for the documented poor sperm quality.

Other phenotypes (heritable traits that can be seen/measured) of interest observed in the cheetah are kinked tails, crowded incisors, palatal depression, and coat variations (Chapter 7). While no molecular work has been performed on the morphological traits to date, insight was gained on several coat related phenotypes. Genes from the keratin-associated protein family were found to be expressed at higher levels in the yellow background of the cheetah fur (Kaelin et al., 2012), which is consistent with observations that fur of the black spots is softer relative to the coarser textured yellow background. Conversely, genes responsible for pigmentation were expressed at higher levels in the black spots relative to the less pigmented yellow background of cheetah fur (Hong et al., 2011). Another gene whose expression was increased in the black spots was a paracrine hormone, which was hypothesized to be involved in coordination of the spot pattern (Kaelin et al., 2012). The genetic basis for the king cheetah coat variant was determined to be a mutation in the *transmembrane aminopeptidase Q* gene (Kaelin et al., 2012). Rare cases of gross morphological deformities have occurred in the past, but no literature is available to substantiate whether they were based on low genetic diversity, inbreeding in a captive setting, or teratogenic influences.

Signatures of Selection, Copy Number Variation, and Changes in Gene Families

In a recent genome-wide analysis of the cheetah (Dobrynin et al., 2015) signatures of positive selection (by comparison with lion, tiger, cat, human, and mouse) were identified in close to a thousand genes. Ten of the genes were involved in muscle contraction (both cardiac and striated muscles), specifically the mitogen-activated protein kinase pathway (which is linked to stress), and in the regulation of catabolic processes, indicating that the cheetah underwent some degree of specialization in these pathways. In the same study over 10 million nucleotides of segmental duplications were identified, and affect genes that are believed to be involved in energy balance, nutrition, and sensory adaptation. In addition, gene expansion was observed in the MHC extended class I region, which includes vomeronasal receptors, as well as olfactory and G-coupled receptor genes; these gene expansions were tentatively linked to behavior (pheromones) and physiology (e.g., LDH-A and LDH-B are linked to a carnivorous diet). Both segmental duplications and gene expansions lead to temporary redundancy, allowing new gene functions to arise.

Evidence of historical positive selection on antigen binding sites that interact directly with pathogen-derived proteins was detected for both MHC classes, particularly MHC I (Castro-Prieto et al., 2011). Signatures of selection in the MHC were also identified in the whole genome study (Dobrynin et al., 2015). A study of the cytochrome P450 gene (CYP2D6), involved in drug metabolism, showed considerable genetic diversity and signs of relaxed selection pressure in felids, including cheetahs (Schenekar et al., 2011).

DISCUSSION

Importance of Low Genetic Diversity on Cheetah Survival

The most notable and still poorly understood feature of cheetah evolutionary history is how the cheetah has persisted in spite of remarkably low levels of genetic variation. Genomic variation is generally considered to be crucial for long-term survival of species as it provides potential for adaptive responses (natural selection of advantageous genetic variation) to environmental changes, such as climate change (Chapter 12), and adaptability of immunity to disease outbreaks (O'Brien and Evermann, 1988). Therefore, the initial discovery of genetic uniformity of the cheetah was quickly followed by concerns about the species' chances of long-term survival. However, the discovery that the event or events leading to the loss of diversity could be placed over 10,000 years ago and that cheetah numbers had recovered by the 19th century, indicate the cheetah's ability to survive and thrive, despite reduced levels of genetic diversity, over extended periods of time. However, this does not guarantee the cheetah's survival in the future, as lack of genetic diversity limits the ability to adapt and evolve, in particular in the light of major changes in environmental conditions or pathogenic pressure.

Reduced genetic variation, particularly at adaptively important MHC loci, has been associated with high susceptibility to infectious diseases in captive cheetahs (O'Brien et al., 1985; O'Brien et al., 1986; O'Brien and Evermann, 1988). A prime example is the high death toll caused by a coronavirus outbreak in a North American zoo (section "Genetic Investigations of Infectious Diseases Affecting Cheetahs," Chapter 25). Despite this, free-ranging cheetahs from eastern and southern Africa show robust health (Caro 1994; Munson et al., 2004; Munson et al., 2005; Thalwitzer et al., 2010) and do not seem to have compromised immunocompetence (Castro-Prieto et al., 2011). In addition, in the wild the cheetah's large home ranges (Chapter 8) reduce the risk of infectious disease transmission. However, this may not be sufficient to protect the species in the event of an emerging disease, especially given the low levels of MHC diversity.

At an individual level manifestation of deleterious traits caused by excessive levels of homozygosity (inbreeding depression) appear to be limited compared with the puma population in Florida (Florida panther), which suffered from atrial defects, poor sperm quality, cryptorchidism, and high disease load (Roelke et al., 1993). The cheetah is only known to suffer from poor sperm quality; none of the other traits observed in the cheetah (e.g., kinked tails, crowded incisors; section "Investigations of the Molecular Basis for Heritable Traits") are detrimental to individual health, or the capacity to survive and reproduce. This, and evidence of positive signatures of selection on genes involved in muscle contraction and stress metabolism (section "Signatures of Selection, Copy Number Variation, and Changes in Gene Families"), suggests that the low levels of genetic diversity were caused, or followed by, strong selective pressures, which perhaps purged deleterious alleles from the species (e.g., Hedrick and Garcia-Dorado, 2016).

Despite limited sperm quality, cheetah matings produce sufficient viable cubs (up to 6 cubs per litter every 2 years; Chapter 9) to maintain and even increase the population. However, further loss of genetic diversity could impair reproductive success, which is the ultimate requirement for species survival. Indeed, increased infant mortality was observed in captive inbred individuals (O'Brien et al., 1985).

Genetic Diversity and *In Situ* Cheetah Conservation

With the reduction of its natural range, cheetah numbers are declining (Durant et al., 2017; Chapter 4), and most cheetah populations today are fragmented with loss of connectivity between them (Chapter 10). Small populations,

such as the critically endangered Iranian cheetah (*A. j. venaticus*; Chapter 5), which has the lowest amount of genetic diversity of all the currently recognized cheetah subspecies (Charruau et al., 2011), are particularly at risk of losing further genetic diversity. Therefore, it is crucial to maintain or regain sufficiently large population sizes and connectivity, while preserving existing variation through viable long-term storage of sperm and oocytes (Chapter 27).

Whenever possible, animals should remain in, or if captured, be returned to the wild (Chapter 20). Captured animals that are not suitable for release back into the wild should be considered for breeding programs (section "Genetic Diversity and *Ex Situ* Cheetah Conservation"). Wild cheetah populations of the same subspecies and geographical region were generally panmictic; minor population structure was only observed in allele frequencies of the rapidly evolving immune response genes. As such, translocations of wild caught individuals performed as part of conservation actions within these populations, only mimic natural connectivity (which may be restricted by anthropogenic barriers to gene flow).

An additional level of complexity arises when populations are from different subspecies. While in principle exchange between populations of different subspecies should be avoided as they may be considered evolutionary significant units (Moritz, 1994), a compromise between preserving the existing structure and the urgency to rescue a small population at risk of disappearing, may have to be reached. Relatively recent times of subspecies divergence and the merging of two subspecies in 2017, are additional considerations during such decision making processes (section "Subspecies Definition and Divergence"). This dilemma may have to be addressed for the Iranian population at some point if the numbers remain below 100 individuals (Chapter 5). Additional information on the genetic health of the existing population is critically and urgently needed to determine if this population is likely to survive without management actions.

Genetic Diversity and *Ex Situ* Cheetah Conservation

Captive cheetah populations have been established in part to serve as a reservoir for the wild population (Chapter 23). Breeding decisions are guided by the genealogical data managed through the regional and international cheetah studbooks (Chapter 23). Reputable captive programs (Chapter 22) aim to retain 90% of genetic diversity over 100 years (Lacy, 2012). However, this may not be sufficient in the long term as the small number of founders only represents a subset of the genetic diversity found in the wild (founder effect) and the 10% genetic diversity lost is irreversible. This loss of diversity can only be compensated for by the recovery of lost breeding lines through "reinjection" of viably preserved reproductive material (e.g., gametes) of founders into the captive population, addition of new founder individuals (or gametes) obtained from the wild, or inclusion of unrelated captive individuals to the breeding pool. As with wild populations, interconnectivity of captive populations should be maintained as much as possible, through regional and interregional exchange of animals or reproductive material. As the cost of genetic research continues to decrease, it will provide the possibility to assess the genetic makeup of all captive cheetahs, and thus the integration of animals of unknown origin into the cheetah breeding pool, if genetic evaluation determines that they represent a new breeding lineage. Of note, inbred individuals (high homozygosity levels) can also be used for further breeding if they represent a unique breeding lineage, as homozygosity is not heritable.

CONCLUSIONS

The field of conservation genetics will see more advanced analyses emerging, including the evaluation of heritable traits, landscape genetics, and increased precision in assessing the extent and timing of the events that cause the loss of genetic diversity. Non invasive samples are increasingly employed to provide answers regarding populations that have not yet been intensely studied (Chapter 31). Additional research involving contemporary and museum samples is currently under way to fill existing knowledge gaps regarding the published subspecies (Léna Godsall Bottriell, personal communication). However, it is crucial to remember that data obtained from genetic studies published to date agree sufficiently in confirming the low genetic diversity of the species at nonrepetitive loci (section "Genetic Diversity"), dating the origin of the low diversity to more than 10,000 years ago (section "Historic Demography"), providing support for genetic differences, although short divergence times, between the populations corresponding to most published subspecies (section "Subspecies Definition and Divergence"), and showing only minimal population structure within geographical regions (section "Phylogeography"). This in turn enables a joint message in terms of recommendations for cheetah conservation, as well as *in situ* and *ex situ* management (sections "Genetic Diversity and *In Situ* Cheetah Conservation" and "Genetic Diversity and *Ex Situ* Cheetah Conservation"). We hope that by presenting all available data on cheetah genetics, this chapter provides clarity to the results and conclusions arising from the field of conservation genetics, and contribute to the global efforts for the cheetah's long-term survival.

References

Belbachir, F., 2007. Les grandes questions relatives à la conservation des grands félins d'Algérie: cas du guépard et du léopard. In: Compte-Rendu de la Deuxième Réunion de l'observatoire du Guépard en Régions d'Afrique du Nord—Tamanrasset (ed. OGRAN,). Société Zoologique de Paris, Paris, p. 42.

Bosman, A.M., Oosthuizen, M.C., Peirce, M.A., Venter, E.H., Penzhorn, B.L., 2010. *Babesia lengau* sp. nov., a novel Babesia species in cheetah (*Acinonyx jubatus*, Schreber, 1775) populations in South Africa. J. Clin. Microbiol. 48 (8), 2703–2708.

Burger, P.A., Steinborn, R., Walzer, C., Petit, T., Mueller, M., Schwarzenberger, F., 2004. Analysis of the mitochondrial genome of cheetahs (*Acinonyx jubatus*) with neurodegenerative disease. Gene 338 (1), 111–119.

Caro, T.M., 1994. Cheetahs of the Serengeti Plains: Group Living in an Asocial Species. The University of Chicago Press, Chicago.

Castro-Prieto, A., Wachter, B., Melzheimer, J., Thalwitzer, S., Hofer, H., Sommer, S., 2012. Immunogenetic variation and differential pathogen exposure in free-ranging cheetahs across Namibian farmlands. PLoS One 7 (11), e49129.

Castro-Prieto, A., Wachter, B., Sommer, S., 2011. Cheetah paradigm revisited: MHC diversity in the world's largest free-ranging population. Mol. Biol. Evol. 28 (4), 1455–1468.

Charruau, P., Fernandes, C., Orozco-TerWengel, P., Peters, J., Hunter, L., Ziaie, H., Jourabchian, A., Jowkar, H., Schaller, G., Ostrowski, S., Vercammen, P., Grange, T., Schlötterer, C., Kotze, A., Geigl, E.M., Walzer, C., Burger, P.A., 2011. Phylogeography, genetic structure and population divergence time of cheetahs in Africa and Asia: evidence for long-term geographic isolates. Mol. Ecol. 20 (4), 706–724.

Cornuet, J.M., Ravigné, V., Estoup, A., 2010. Inference on population history and model checking using DNA sequence and microsatellite data with the software DIYABC (v1. 0). BMC Bioinf. 11, 401.

Dalton, D.L., Charruau, P., Boast, L., Kotzé, A., 2013. Social and genetic population structure of free-ranging cheetah in Botswana: implications for conservation. Eur. J. Wildl. Res. 59 (2), 281–285.

Dobrynin, P., Liu, S., Tamazian, G., Xiong, Z., Yurchenko, A.A., Krasheninnikova, K., Kliver, S., Schmidt-Küntzel, A., Koepfli, K.P., Johnson, W., Kuderna, L.F., García-Pérez, R., de Manuel, M., Godinez, R., Komissarov, A., Makunin, A., Brukhin, V., Qiu, W., Zhou, L., Li, F., Yi, J., Driscoll, C., Antunes, A., Oleksyk, T.K., Eizirik, E., Perelman, E., Roelke, M., Wildt, D., Diekhans, M., Marques-Bonet, T., Marker, L., Bhak, J., Wang, J., Zhang, G., O'Brien, S.J., 2015. Genomic legacy of the African cheetah, *Acinonyx jubatus*. Genome Biol. 16, 277.

Drake, G.J.C., Kennedy, L.J., Auty, H.K., Ryvar, R., Ollier, W.E.R., Kitchener, A.C., Freeman, A.R., Radford, A.D., 2004. The use of reference strand-mediated conformational analysis for the study of cheetah (*Acinonyx jubatus*) feline leucocyte antigen class II DRB polymorphisms. Mol. Ecol. 13 (1), 221–229.

Driscoll, C.A., Menotti-Raymond, M., Nelson, G., Goldstein, D., O'Brien, S.J., 2002. Genomic microsatellites as evolutionary chronometers: a test in wild cats. Genome Res. 12 (3), 414–423.

Durant, S.M., Mitchell, N., Groom, R., Pettorelli, N., Ipavec, A., Jacobson, A., Woodroffe, R., Bohm, M., Hunter, L., Becker, M., Broekuis, F., Bashir, S., Andresen, L., Aschenborn, O., Beddiaf, M., Belbachir, F., Belbachir-Bazi, A.,

Berbash, A., Brandao de Matos Machado, I., Breitenmoser, C., Chege, M., Cilliers, D., Davies-Mostert, H., Dickman, A., Fabiano, E., Farhadinia, M., Funston, P., Henschel, P., Horgan, J., de Iongh, H., Jowkar, H., Klein, R., Lindsey, P., Marker, L., Marnewick, K., Melzheimer, J., Merkle, J., Msoka, J., Msuha, M., O'Neill, H., Parker, M., Purchase, G., Saidu, Y., Samaila, S., Samna, A., Schmidt-Küntzel, A., Selebatso, E., Sogbohossou, E., Soultan, A., Stone, E., van der Meer, E., van Vuuren, R., Wykstra, M., Young-Overton, K., 2017. The global decline of cheetah and what it means for conservation. Proc. Natl. Acad. Sci. USA 114, 528–533.

Franklin, A.D., Schmidt-Küntzel, A., Terio, K.A., Marker, L.L., Crosier, A.E., 2015. Serum amyloid A protein concentration in blood is influenced by genetic differences in the cheetah (*Acinonyx jubatus*). J. Hered. 107 (2), 115–121.

Freeman, A.R., Machugh, D.E., Mckeown, S., Walzer, C., Mcconnell, D.J., Bradley, D.G., 2001. Sequence variation in the mitochondrial DNA control region of wild African cheetahs (*Acinonyx jubatus*). Heredity 86 (3), 355–362.

Gaggiotti, O., Excoffier, L., 2000. A simple method of removing the effect of a bottleneck and unequal population sizes on pairwise genetic distances. Proc. R. Soc. B 267 (1438), 81–87.

Gilpin, M., 1991. The genetic effective size of a metapopulation. Biol. J. Linn. Soc. 42 (1–2), 165–175.

Goldstein, D.B., Pollock, D.D., 1997. Launching microsatellites: a review of mutation processes and methods of phylogenetic inference. J. Hered. 88 (5), 335–342.

Gottelli, D., Wang, J., Bashir, S., Durant, S.M., 2007. Genetic analysis reveals promiscuity among female cheetahs. Proc. R. Soc. B 274 (1621), 1993–2001.

Gutenkunst, R.N., Hernandez, R.D., Williamson, S.H., Bustamante, C.D., 2009. Inferring the joint demographic history of multiple populations from multidimensional SNP frequency data. PLoS Genet. 5 (10), e1000695.

Hedrick, P.W., 1996. Bottleneck(s) in cheetahs or metapopulation. Conserv. Biol. 10 (3), 897–899.

Hedrick, P.W., Garcia-Dorado, A., 2016. Understanding inbreeding depression, purging, and genetic rescue. Trends Ecol. Evol. 31 (12), 940–952.

Hey, J., Nielsen, R., 2007. Integration within the Felsenstein equation for improved Markov chain Monte Carlo methods in population genetics. PNAS 104 (8), 2785–2790.

Hong, L.Z., Li, J., Schmidt-Küntzel, A., Warren, W.C., Barsh, G.S., 2011. Digital gene expression for non-model organisms. Genome Res. 21 (11), 1905–1915.

Johnson, W.E., Eizirik, E., Pecon-Slattery, J., Murphy, W.J., Antunes, A., Teeling, E., O'Brien, S.J., 2006. The late Miocene radiation of modern Felidae: a genetic assessment. Science 311 (5757), 73–77.

Kaelin, C.B., Xu, X., Hong, L.Z., David, V.A., McGowan, K.A., Schmidt-Küntzel, A., Roelke, M.E., Pino, J., Pontius, J., Cooper, G.M., Manuel, H., 2012. Specifying and sustaining pigmentation patterns in domestic and wild cats. Science 337 (6101), 1536–1541.

Kimura, M., 1980. A simple method for estimating evolutionary rate of base substitutions through comparative studies of nucleotide sequences. J. Mol. Evol. 16 (2), 111–120.

Kitchener, A.C., Breitemoser-Würsten, Ch., Eizirik, E., Gentry, A., Werdelin, L., Wilting, A., Yamaguchi, N., Abramov, A., Christiansen, P., Driscoll, C., Duckworth, W., Johnson, W., Luo, S.-J., Meijaard, E., O'Donoghue, P., Sanderson, J., Seymour, K., Bruford, M., Groves, C., Hoffmann, M., Nowell, K., Timmons, Z., Tobe, S., 2017. A revised taxonomy of the Felidae. The final report of the Cat Classification Task Force of the IUCN Cat Specialist Group. Cat News Spec. Issue 11, 80.

Kotze, A., Ehlers, K., Cilliers, D., Grobler, J., 2008. The power of resolution of microsatellite markers and assignment tests to determine the geographic origin of cheetah (Acinonyx jubatus) in Southern Africa. Mamm. Biol. 73 (6), 457–462.

Krausman, P., Morales, S., 2005. Acinonyx jubatus. Mamm. Species 771, 1–6.

Lacy, R.C., 2012. Achieving true sustainability of zoo populations. Zoo Biol. 32 (1), 19–26.

Li, G., Davis, B.W., Eizirik, E., Murphy, W.J., 2016. Phylogenomic evidence for ancient hybridization in the genomes of living cats (Felidae). Genome Res. 26 (1), 1–11.

Li, H., Durbin, R., 2011. Inference of human population history from individual whole-genome sequences. Nature 475 (7357), 493–496.

Lopez, J.V., Cevario, S., O'Brien, S.J., 1996. Complete nucleotide sequences of the domestic cat (Felis catus) mitochondrial genome and a transposed mtDNA tandem repeat (Numt) in the nuclear genome. Genomics 33 (2), 229–246.

Kelly, M.J., Laurenson, M.K., FitzGibbon, C.D., Collins, D.A., Durant, S.M., Frame, G.W., Bertram, B.C.R., Caro, T.M., 1998. Demography of the Serengeti cheetah (Acinonyx jubatus) population: the first 25 years. J. Zool. 244 (4), 473–488.

Manel, S., Berthier, P., Lukart, G., 2002. Detecting wildlife poaching identifying the origin of individuals with bayesian assignment test and multilocus genotypes. Conserv. Biol. 16 (3), 650–659.

Marker, L., O'Brien, S.J., 1989. Captive breeding of the cheetah (Acinonyx jubatus) in North American zoos (1871-1986). Zoo Biol. 8 (1), 3–16.

Marker, L.L., Pearks Wilkerson, A.J., Sarno, R.J., Martenson, J., Breitenmoser-Würsten, C., O'Brien, S.J., Johnson, W.E., 2008. Molecular genetic insights on cheetah (Acinonyx jubatus) ecology and conservation in Namibia. J. Hered. 99 (1), 2–13.

Menotti-Raymond, M., David, V.A., Chen, Z.Q., Menotti, K.A., Sun, S., Schäffer, A.A., Agarwala, R., Tomlin, J.F., O'Brien, S.J., Murphy, W.J., 2003. Second-generation integrated genetic linkage/radiation hybrid maps of the domestic cat (Felis catus). J. Hered. 94 (1), 95–106.

Menotti-Raymond, M.A., O'Brien, S.J., 1993. Dating the genetic bottleneck of the African cheetah. Proc. Natl. Acad. Sci. USA 90 (8), 3172–3176.

Menotti-Raymond, M.A., O'Brien, S.J., 1995. Evolutionary conservation of ten microsatellite loci in four species of Felidae. J. Hered. 86 (4), 319–322.

Moritz, C., 1994. Defining 'evolutionarily significant units' for conservation. Trends Ecol. Evol. 9 (10), 373–375.

Munson, L., Marker, L., Dubovi, E., Spencer, J.A., Evermann, J.F., O'Brien, S.J., 2004. Serosurvey of viral infections in free-ranging Namibian cheetahs (Acinonyx jubatus). J. Wildl. Dis. 40 (1), 23–31.

Munson, L., Terio, K.A., Worley, M., Jago, M., Bagot-Smith, A., Marker, L., 2005. Extrinsic factors significantly affect patterns of disease in free-ranging and captive cheetah (Acinonyx jubatus) populations. J. Wildl. Dis. 41 (3), 542–548.

Nei, M., 1987. Molecular Evolutionary Genetics. Columbia University Press, New York, NY.

Nei, M., Li, W., 1979. Mathematical model for studying genetic variation in terms of restriction endonucleases. Proc. Natl. Acad. Sci. USA 76 (10), 5269–5273.

Newman, A., Bush, M., Wildt, D.E., Van Dam, D., Frankenhuis, M.T., Simmons, L., Phillips, L., O'Brien, S.J., 1985. Biochemical genetic variation in eight endangered or threatened felid species. J. Mamm. 66 (2), 256–267.

Nichols, R., 2001. Gene trees and species trees are not the same. Trends Ecol. Evol. 16 (7), 358–364.

Nikolic, N., Chevalet, C., 2013. Variation of effective population size. R package version (VarEff).

Nowell, K., Jackson, P., 1986. Wild cats. Status Survey and Conservation Action Plan. IUCN, Gland.

O'Brien, S.J., Evermann, J.F., 1988. Interactive influence of infectious disease and genetic diversity in natural populations. Trends Ecol. Evol. 3 (10), 254–259.

O'Brien, S.J., Johnson, W.E., Driscoll, C.A., Dobrynin, P., Marker, L., 2017. Conservation genetics of the cheetah-lessons learned and new opportunities. J. Hered. 108 (6), 671–677.

O'Brien, S.J., Menninger, J.C., Nash, W.G., 2006. An Atlas of Mammalian Chromosomes. John Wiley and Sons Publishers, New York, NY.

O'Brien, S.J., Roelke, M.E., Marker, L., Newman, A., Winkler, C.A., Meltzer, D., Colly, L., Evermann, J.F., Bush, M., Wildt, D.E., 1985. Genetic basis for species vulnerability in the cheetah. Science 227 (4693), 1428–1434.

O'Brien, S.J., Wildt, D.E., Bush, M., 1986. The cheetah in genetic peril. Sci. Am. 254 (5), 84–92.

O'Brien, S.J., Wildt, D.E., Bush, M., Caro, T.M., Fitzgibbon, C., Aggundey, I., Leakey, R.E., 1987. East-African cheetahs—evidence for two population bottlenecks. Proc. Natl. Acad. Sci. USA 84 (2), 508–511.

O'Brien, S.J., Wildt, D.E., Goldman, D., Merril, C.R., Bush, M., 1983. The cheetah is depauperate in genetic variation. Science 221 (4609), 459–462.

Pearks Wilkerson, A.J., Teeling, E.C., Troyer, J.L., Bar-Gal, G.K., Roelke, M., Marker, L., Pecon-Slattery, J., O'Brien, S.J., 2004. Coronavirus outbreak in cheetahs: lessons for SARS. Curr. Biol. 14 (6), R227–R228.

Pimm, S.L., Gittleman, J.L., McCracken, G.F., Gilpin, M., 1989. Plausible alternatives to bottlenecks to explain reduced genetic diversity. Trends Ecol. Evol. 4 (6), 176–178.

Pokorny, I., Sharma, R., Goyal, S.P., Mishra, S., Tiedemann, R., 2010. MHC class I and MHC class II DRB gene variability in wild and captive Bengal tigers (Panthera tigris tigris). Immunogenetics 62 (10), 667–679.

Roelke, M.E., Martenson, J.S., O'Brien, S.J., 1993. The consequences of demographic reduction and genetic depletion in the endangered Florida panther. Curr. Biology 3 (6), 340–350.

Sachdev, M., Sankaranarayanan, R., Reddanna, P., Thangaraj, K., Singh, L., 2005. Major histocompatibility complex class I polymorphism in Asiatic lions. Tissue Antigens 66 (1), 9–18.

Schenekar, T., Winkler, K.A., Troyer, J.L., Weiss, S., 2011. Isolation and characterization of the CYP2D6 gene in felidae with comparison to other mammals. J. Mol. Evol. 72 (2), 222–231.

Shriver, M.D., Jin, L., Boerwinkle, E., Deka, R., Chakraborty, R., 1995. A novel measure of genetic distance for highly polymorphic tandem repeat loci. Mol. Biol. Evol. 12 (5), 914–920.

Smithers, R.H.N., 1975. Family felidae. In: Meester, J., Sezter, H.W. (Eds.), The Mammals of Africa. An Identification Manual. Smithsonian Institution Press, Washington, DC.

Sommer, S., 2005. The importance of immune gene variability (MHC) in evolutionary ecology and conservation. Front. Zool. 2, 16.

Storz, J.F., Beaumont, M.A., 2002. Testing for genetic evidence of population expansion and contraction: an empirical analysis of microsatellite DNA variation using a hierarchical Bayesian model. Evolution 56 (1), 154–166.

Terrell, K.A., Crosier, A.E., Wildt, D.E., O'Brien, S.J., Anthony, N.M., Marker, L., Johnson, W.E., 2016. Continued decline in genetic diversity among wild cheetahs (Acinonyx jubatus) without further loss of semen quality. Biol. Conserv. 200, 192–199.

Thalwitzer, S., Wachter, B., Robert, N., Wibbelt, G., Müller, T., Lonzer, J., Meli, M.L., Bay, G., Hofer, H., Lutz, H., 2010. Seroprevalences to viral pathogens in free-ranging and captive cheetahs (Acinonyx jubatus) on Namibian farmland. Clin. Vacc. Immunol. 17 (2), 232–238.

Valdes, A.M., Slatkin, M., Freimer, N.B., 1993. Allele frequencies at microsatellite loci: the stepwise mutation model revisited. Genetics 133 (3), 736–749.

Wakeley, J., Hey, J., 1997. Estimating ancestral population parameters. Genetics 145 (3), 847–855.

Werdelin, L., O'Brien, S.J., Johnson, W.E., Yamaguchi, N., 2010. Felid phylogenetics and evolution. In: McDonald, D., Loverage, L. (Eds.), The Biology and Conservation of Wild Felids. Oxford University Press, United Kingdom, pp. 59–82.

Winkler, C., Schultz, A., Cevario, S., O'Brien, S.J., 1989. Genetic characterization of FLA, the cat major histocompatibility complex. Proc. Natl. Acad. Sci. USA 86 (3), 943–947.

Yuhki, N., Mullikin, J.C., Beck, T., Stephens, R., O'Brien, S.J., 2008. Sequences, annotation and single nucleotide polymorphism of the major histocompatibility complex in the domestic cat. PLoS One 3 (7), e2674.

Yuhki, N., O'Brien, S.J., 1990. DNA variation of the mammalian major histocompatibility complex reflects genomic diversity and population history. Proc. Natl. Acad. Sci. USA 87 (2), 836–840.

Yuhki, N., O'Brien, S.J., 1994. Exchanges of short polymorphic DNA segments predating speciation in feline major histocompatibility complex class I genes. J. Mol. Evol. 39 (1), 22–33.

Zhivotovsky, L.A., Feldman, M.W., 1995. Microsatellite variability and genetic distances. Proc. Natl. Acad. Sci. 92 (25), 11549–11552.

Cheetah Specialization: Physiology and Morphology

Julie Meachen, Anne Schmidt-Küntzel**,
Holly Haefele†, Gerhard Steenkamp§,
James M. Robinson‡, Marcela Randau¶,
Natasha McGowan††, David M. Scantlebury††,
Nikki Marks††, Aaron Maule††, Laurie Marker***

*Des Moines University, Des Moines, IA, United States
**Cheetah Conservation Fund, Otjiwarongo, Namibia
†Fossil Rim Wildlife Center, Glen Rose, TX, United States
‡F.R.C.S Royal College of Surgeons, London, United Kingdom
§University of Pretoria, Onderstepoort, South Africa
¶University College London, London, United Kingdom
††Queen's University Belfast, Belfast, United Kingdom

INTRODUCTION

Of the 41 extant species of cats, cheetahs (*Acinonyx jubatus*) are the most distinctive and unique hunters, chasing prey farther and faster than other cats (Ewer, 1973). First classified by early taxonomists as *Felis jubatus* (Schreber, 1775), the cheetah was soon differentiated from the genus *Felis* by placing it into the monospecific genus *Acinonyx* (Brookes, 1828), of which *A. jubatus* is the only living species. The scientific genus name *Acinonyx* is a reference to the species' semiretractile claws. In Greek, *a* means not, *kaina* means thorn, and *onus* means claw (Gotch, 1979). *Jubatus,* in Latin means maned, referring to the crest or mane on the shoulders and back of young cheetahs (Gotch, 1979). Other temporary genus names were *Cynailurus* and *Cynofelis* (Lesson, 1842; Wagler, 1830) referring to some of the canid-like features of the cheetah and *Guepardus* (Duvernoy, 1834).

Cheetahs have evolved for speed rather than power and aggression (flight vs. fight). The configuration of their muscles, slender body, and

FIGURE 7.1 **Cheetah model displaying skeletal proportions and coat pattern in a flexed and suspended running position.** *Source: Blue Rhino Studio and The Field Museum, Chicago.*

stride length of up to 6 m allow them to reach speeds of up to 113 km/h (Hildebrand, 1959; Hildebrand, 1961). When nearing full speed, they can complete 3.5 strides per second, including 2 airborne phases (Hildebrand, 1961). The cheetah's suite of morphological adaptations includes a small streamlined head; long, light limbs; powerful hind legs; flexible shoulders and spine; a long muscular tail; semiretractile claws; wide nostrils; and enlarged nasal passages (Fig. 7.1). All combine to make it the fastest land mammal over moderate distances (300–400 m; Gray, 1968).

The literature shows evidence that the cheetah body mass is related to prey size (Carbone et al., 1999; Caro, 1994; Sunquist and Sunquist, 2002). However, Wilson et al. (2015) found that prey maneuverability may be a more important factor than prey size for a cheetah. They found that different sized prey turn at different frequencies and that the cheetah's turning frequency correlates with optimal prey turning frequency. These prey selection constraints

appear to be more important in cheetahs than larger felid species like lions (*Panthera leo*), who only chase their prey for short bursts (approx. 22–100 m) and expend the most energy during a kill in the grappling/subduing phase, whereas cheetahs will chase prey to a distance up to 500 m (Hunter and Hamman, 2003; Schaller, 1972).

The cheetah also has a unique way to bring down its prey and the functional morphology of cheetahs has been primarily studied to better understand their hunting style, which sets them apart from other large cats (>25 kg), such as the pantherines (e.g., in the genus *Panthera*), the puma (*Puma concolor*), and sometimes the Eurasian lynx (*Lynx lynx*). The two key features of other large cats for prey-killing include strong teeth and jaws, as well as powerful forelimbs to tackle their prey. Instead, the cheetah uses a combination of its well-honed speed and its forepaws (Caro, 1994). They use the enlarged first digit, or dewclaw, like a sharp hook to trip the prey during a chase and throw it off balance (Londei, 2000). Once the prey has fallen to the ground the cheetah will bite the throat to suffocate the animal (Caro, 1994). Their unique hunting technique enables cheetahs to be efficient killers with shorter canines and a relatively light body mass and skull.

In this chapter, we briefly present the cheetah's overall appearance, and address its morphological specializations and some corresponding physiological adaptations. We also present data on the specializations found in the cheetah's organ systems.

BODY MASS

Wild-born cheetahs weigh on average 150–300 g at birth (Kingdon, 1977), while captive-born cheetah cubs are larger at around 500 g at birth (Wack et al., 1991). Body mass in adult cheetahs ranges widely from approximately 20 kg to greater than 60 kg, depending upon the animal's age, body condition, sex, subspecies, or location of origin (Marker and Dickman, 2003; Sunquist and Sunquist, 2002). Similar to other large carnivores, the cheetah's mass may continue to increase into adulthood, and stabilize between 49 and 96 months (Marker and Dickman, 2003).

Botswanan cheetahs were found to be heavier and longer (mass 42.2 ± 8.2 kg, total body length 204.1 ± 15.5 cm; Boast et al., 2013) than Namibian cheetahs (mass 40.4 ± 7.0 kg, total body length 197.6 ± 11.6 cm; Boast et al., 2013; Marker and Dickman, 2003). Kenyan cheetahs and a captive population in the UAE likely originating from northern Africa were found to be smaller than the southern African cheetahs; the cheetahs from the UAE also being the lightest group (Meachen et al., in preparation). Like other large felids cheetahs show marked sexual dimorphism in body size, with males being larger than females (Boast et al., 2013; Marker and Dickman, 2004; Meachen et al., in preparation).

OVERALL APPEARANCE, COAT, AND COLORATION

The cheetah's coat is tan, or buff colored, with solid black spots (Fig. 7.1), which are believed to provide camouflage (Marker, 2014). Each cheetah has a distinct spot pattern, which can be used for individual identification (Chapter 32). The tail can also be used for identification and usually has several black rings at the end; the tip can be black or white. Malar stripes (tear marks) on each side of the nose differentiate the cheetah from all the other cats and are hypothesized to reduce glare from the bright sun (Marker, 2014).

The cheetah's skin is pigmented under both tan and black hair (Marker, personal observation). The black hair is softer, which was associated with reduced expression of keratin producing genes (Kaelin et al., 2012). Black spots appear raised relative to the buff colored coat; this was attributed to an optical effect caused by a different structure rather than the hypothesized longer hair [Cheetah Conservation Fund (CCF), unpublished data]. A rare, naturally occurring coat color mutant, the king cheetah, has converging rather than individual spots (Kaelin et al., 2012). Cubs are born fully furred, with an indistinct spot pattern making them appear gray. Juveniles display a longer coat on the back of their neck and spine, called a mantel (Adamson, 1969).

CRANIAL MORPHOLOGY AND PHYSIOLOGY

Skull

Cheetahs have an aerodynamic build, with a small head and lightweight body (Fig. 7.1). Some important cranial modifications appear to make the skull lighter and more streamlined for a cursorial lifestyle.

A shorter muzzle with a large braincase and greater postorbital processes permits overall reduction in skull size without compromising

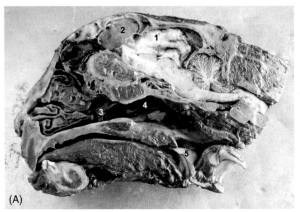
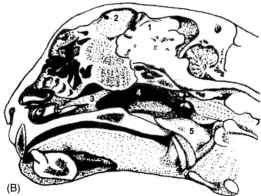

FIGURE 7.2 (A) Sagittal section of frozen cheetah head, and (B) corresponding drawing; the head sections show the cheetah's vaulted braincase (1), enlarged frontal sinuses (2), nasal passages (3), and internal nares (4); the epiglottis (5) is shown in the oral space (A) and in the nasal space (B).

brain volume (Emerson and Bramble, 1993; Meachen-Samuels and Van Valkenburgh, 2009a; O'Regan, 2002). This morphology is seen in the small cats (such as *Felis*) and cheetahs retain this shape, even with larger growth. Cheetahs have smaller canines, mediolaterally narrower zygomatic arches (cheek bones), and less robust skulls than the puma, a sister species (Meachen-Samuels and Van Valkenburgh, 2009a). The lighter skull may be linked to a slight difference in preferred prey size, as cheetahs tend to hunt slightly smaller prey than pumas (Caro, 1994; Meachen-Samuels and Van Valkenburgh, 2009a; Sunquist and Sunquist, 2002).

The braincase in the cheetah is vaulted to accommodate a larger volume of frontal sinuses (Fig. 7.2), which is unique to cheetahs, even when compared to the other members of its clade, the puma and the jaguarundi (*Herpailurus yagouaroundi*; Segura et al., 2013; Chapter 3). Interorbital breadth is larger than expected from skull size in cheetahs; this corresponds to wide nasal bones that facilitate rapid breathing while both hunting and strangling prey, which was hypothesized to allow the cheetah to cool down from the strenuous hunt (Kingdon, 1997; O'Regan, 2002; Taylor and Rowntree, 1973). Pocock (1916a) described a large opening of

the internal nares in 1 cheetah specimen, which would afford a larger area for nasal respiration and possible heat dissipation. This was confirmed when the height and width of the internal nares of the closely related puma ($N = 22$) and similarly sized leopards (*Panthera pardus*, $N = 24$) were compared to the cheetah ($N = 24$; Robinson, unpublished data). The corresponding enlarged postnasal space (Fig. 7.2) has been thought to crowd out the roots of the upper canines (Ewer, 1973), and cheetahs do indeed show a reduced canine and P^4 tooth size (O'Regan, 2002). While enlarged nasal passages and internal nares of cheetah have been confirmed, a difference in the size of the external nares of cheetah relative to leopard could not be supported when adjusting for the difference in body mass (Torregrosa et al., 2010).

Oral Cavity and Bite

The soft tissue morphology of the cheetah's mouth including the vomeronasal organ is comparable to the domestic cat. Steenkamp et al. (2017) showed that the cheetah has an average of 68% palatal pigmentation. Cheetahs have 7–9 rugae (folds in the palate, which may facilitate feeding and drinking), which is on average one more ruga than the domestic cat (Steenkamp et al., 2017).

Cheetahs have 30 permanent teeth when they reach adulthood and share the dental formula described for the Felidae: 3-1-3-1/3-1-2-1, upper/lower incisors-canines-premolars-molars (Krausman and Morales, 2005). A recent study compared the tooth indexes among cheetahs, lions, and leopards and found the relative height of the cheetah mandibular cheek teeth to be significantly higher-crowned than those of lions and leopards (Steenkamp et al., 2017). Steenkamp et al. (2017) hypothesized that these relatively larger teeth make cheetahs effective in consuming enough food in a shorter period [presumably to avoid kleptoparasitism (prey theft) by other predators; Chapter 9] despite their smaller size.

These relatively big teeth induce modifications, "palatal depressions," in the cheetah's palate. These palatal depressions are absent until the permanent mandibular cheek teeth have erupted and disappear if the mandibular cheek tooth corresponding to it is extracted (Steenkamp et al., 2017). The larger of the palatal depressions are situated next to the maxillary (upper) 4th premolar to accommodate the mandibular (lower) carnassial. A further two sets of depressions, when present, are situated palatal to the 2nd and 3rd premolars which would accommodate the mandibular 3rd and 4th premolars. Only the larger caudal depression has a bony depression accompanying it. This depression was previously inaccurately referred to as a mild (nonpathological) form of focal palatine erosion (Marker and Dickman, 2004). No erosion or inflammation is present in normal depressions (Steenkamp et al., 2017). Extreme depressions can become pathological (Chapters 24 and 25). Additional traits, such as crowded incisors and absence of the upper premolar were also reported in the captive and wild cheetah populations of Namibia (Marker and Dickman, 2004).

Cheetahs kill their prey through a bite to the neck which squeezes the trachea, causing the animal to suffocate (Chapter 9). This suffocation technique ideally requires the bite to crush the cartilage rings supporting the windpipe or trachea, and may cause death within 10 min, less if major blood vessels—carotid and jugular—are ruptured (Leyhausen, 1979). Using a dry skull method, a 2005 study found that the estimated bite force at the canines (CB_s) was lower in cheetahs ($CB_s = 472$) than that estimated in tigers (*Panthera tigris*) or jaguars (*Panthera onca*; $CB_s = 1525$ and 1014, respectively), and was comparable to the leopard ($CB_s = 467$). The cheetah's bite force quotient (BFQ = estimated absolute bite force relative to extrapolated mass of 30 kg) was comparable to lions (*P. leo* BFQ = 112, *A. jubatus* BFQ = 119; Wroe et al., 2005). These data contradict expectations based on predicted reduced mass of the masseter muscle given the relatively small sagittal crest and suggest that the adaptations for reduced weight in the cheetah skull may not compromise the cheetah's ability to deliver a powerful bite. Additionally, the foreshortened face of the cheetah may confer higher mechanical advantage to bites made with the canines during prey killing (Biknevicius and Van Valkenburgh, 1996).

Larynx and Vocalizations

Cheetahs, unlike the pantherines (e.g., lions, leopards), purr instead of roaring. In pantherines, the epihyoideum—one of the structures associated with the larynx and the hyoid—is an elastic ligament, whereas in all other cats, including the cheetah, it is ossified. An elastic epihyoideum allows for roaring, while an ossified epihyal allows other cats to purr (Pocock, 1916b; Weissengruber et al., 2002). In fact, much of the cheetah's laryngeal morphology is the same as that of a domestic cat. The act of purring is caused by the rapid twitching of the vocalis muscle, while long vocal folds and an elongated vocal tract in the pantherines make purring no longer possible (Weissengruber et al., 2002). Cheetahs are also able to emit a wide spectrum of other unique vocalizations (e.g., chirping, stutter call; Chapter 9) that are used for communication.

Eye Retina and Vision

Cheetah cubs first open their eyes around 2 weeks of age. Like many other cats, cheetahs have eyes with a round pupil, and simple dichromatic color vision (Jacobs, 1993), where rods outnumber cones 100:1 in the peripheral retina and 20:1 in the area centralis (Steinberg et al., 1973). However, unlike other large cats that live in mostly closed habitats and have a predominance of M (middle to long wavelength) cones, cheetahs have an equal number of M and S (short wavelength) cones, which have evolved for optimal visual acuity along the horizontal meridian during chases (Ahnelt et al., 2006).

POSTCRANIAL MORPHOLOGY

Limbs

Cheetah bodies are lightweight in comparison to the build of other cats and their only real defense is to run from danger and avoid confrontation. Speed is the driver of the cheetah's postcranial specialization. The cheetah is the fastest living land mammal (Sharp, 1997). The first record of a cheetah at full speed was measured to be 113 km/h (Hildebrand, 1959). Later measurements on 1 cheetah using three trials recorded an average speed of 105 km/h (Sharp, 1997). These speeds were measured using captive cheetahs that chased a lure along a straight track. Using GPS data from 367 hunts of five collared wild cheetahs, a top speed of 93 km/h was recorded with a mean high speed of 54 km/h (SD = 12.24). Top speeds were usually sustained for only 1–2 s (Wilson et al., 2013).

To achieve these high speeds the cheetah's limbs are specialized for running, at the expense of grappling (Fig. 7.1). The cheetah's long bones are dissimilar to those of other large cats.

Instead, their forelimbs are long, thin, and gracile, showing convergence with canids (Meachen-Samuels and Van Valkenburgh, 2009b). The reduced supinatory abilities of their humerus and radius bones also show convergence with canids (Andersson and Werdelin, 2003). Additionally, radiographs of cheetah humeri reveal that cheetahs have thinner, less robust cortical bone than other large cats, confirming that cheetah forelimbs are not subjected to the same twisting forces associated with grappling prey.

Cheetah forelimbs are modified to be distally elongated with a long radius and ulna, relative to humerus (Fig. 7.1). Like in other cats, the cheetah clavicle is vestigial, but unlike in other large cats, it has an "acromial hook" (i.e., lateral curve), which at this time is still unexplained but may be an adaptation for running (Hartstone-Rose et al., 2012). The exceptional mobility of their scapula allows cheetahs to extend their shoulder blade to further increase their stride length (Hildebrand, 1959). The cheetah's muscles are concentrated at the top of their limbs (proximally), and the muscle mass in their forelimbs is lower relative to hindlimbs (Hudson et al., 2011a). This muscle distribution is further indicative of a highly cursorial locomotor mode and is commonly found in canids, but rare in felids (Hudson et al., 2011a; Meachen-Samuels, 2010).

The major muscles of propulsion in the cheetah are located in the hindlimbs (i.e., hamstring and gastrocnemius; Wilson et al., 2013). Cheetahs also have an enlarged psoas muscle which stabilizes the hip during acceleration and long hindlimb bones which are adapted for resisting ground forces and accelerating (Hudson et al., 2011b). The cheetah's ischium is elongated, providing the extensor muscles with more leverage and propulsive power (Hudson et al., 2011b). In accordance with the relative muscle masses in the fore and hind limbs, the cheetah had less cortical bone (solid exterior part) relative to its length in the humerus, while the femur cortical bone was on par with other large cats (Meachen-Samuels and Van Valkenburgh, 2010).

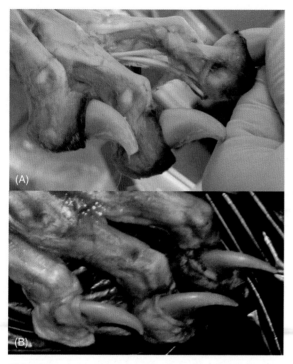

FIGURE 7.3 (A) Dissected cheetah paw with third digit pulled out to show partial retraction ability and blunt keratin sheaths, compared to (B) dissected puma paw showing claws fully retracted and sharp keratin sheaths.

The cheetah is the only cat with short, blunt claws, lacking skin sheaths, thus providing more traction when running (Ewer, 1973). Previous literature has stated that cheetahs have nonretractile claws (Hudson et al., 2011a; Wilson et al., 2013); however, more accurately, cheetah claws are semiretractile (Fig. 7.3). First, the claw core (bony portion of the claw) is longer in a cheetah than in other large cats, so even when retracted as far as possible, the claw still sticks out of the protective sheath. Second, the medial phalanx of cheetahs has a morphology halfway between a canid (mostly symmetrical) and other large felids (asymmetrical, with a grove to "store" the distal phalanx), which only allows partial retraction of the distal phalanx (Russell and Bryant, 2001). The foot has additional adaptations to speed, including less rounded paws

than the other large cats and hard digital pads that are pointed in the front. Their metacarpal/metatarsal pads are also very hard and have strong indentations in the back forming a characteristic angular "W" shape and may serve functionally, as antiskid devices, similar to tire treads (Marker, 2014).

Bony Spine and Tail

Differences in osteological measurements are also found in the vertebral column of cheetahs when compared to other cats (Randau et al., 2016a,b). The differences between cheetahs and other cats are concentrated in the posterior region of the vertebral column, from the 10th thoracic (i.e., the vertebra at the diaphragm) to the last lumbar vertebra, and are most conspicuously related to vertebral body length and height measurements (Randau et al., 2016a,b). Compared to the average morphology of lumbar vertebrae of other cats, the cheetah's lumbar vertebrae (L1–L7) have longer vertebral bodies and show less length variation (Randau et al., 2016a). Cheetah anterior lumbar vertebrae are also shorter in height than the average of other cats (Randau et al., 2016a). These differences in length and height are directly correlated to the degree of movement between two consecutive vertebra and overall passive lateral flexibility of the vertebral column (Koob and Long, 2000; Long et al., 1997; Pierce et al., 2011) and may facilitate the longer stride lengths observed in this species.

The cheetah's long, flat, muscular tail counteracts its body weight, and acts as a stabilizer for balance during quick turns in a high-speed chase (Marker, 2014).

RUNNING ENERGETICS

Muscles

Forty-five percent of a cheetah's body mass is concentrated in the locomotor muscles (Wilson et al., 2013). It has also been shown that cheetah muscles contain a high proportion of powerful, fast-twitch glycolytic fibers, compared to other large cats (Hyatt et al., 2010). This means that cheetah muscles are specially adapted for quick bursts of speed rather than sustained oxygen-consuming activity. Scantlebury et al. (2014) found that cheetahs were mobile for 2.86 h/day on average, which accounted for 42% of their daily energy budgets. However, because of the lower power costs of movement compared to cursorial predators, such as wild dogs (*Lycaon pictus*; Gorman et al., 1998), kleptoparasitism from other carnivores may not be as important an energetic cost to cheetahs as has been presumed in the past (Caro, 1994).

Thermoregulation

The average daily core body temperature (T_b) of cheetahs was measured at 38.3 ± 0.2°C (Hetem et al., 2013). During movement, respiration increases in muscles and heat is generated. When cheetahs were exercised on a treadmill (speed: 1–12 km/h; distance: 2 km), Taylor and Rowntree (1973) reported that T_b subsequently increased from 39°C to 40.5–41.0°C, and concluded that heat must be stored during exercise and subsequently dissipated. This result could not be replicated by Hetem et al. (2013) when measuring T_b during chases in an enclosure, leading them to conclude that sufficient heat dissipation must have had to occur during a hunt for T_b to remain relatively constant. Some preliminary evidence suggests that cheetahs may have the ability to adopt strategies of either evaporative water loss or increased T_b, depending on the environmental conditions (Queen's University Belfast and CCF, unpublished data).

Infrared thermography (QUB and CCF, unpublished data) indicates that the "normal" temperature of the nasal ridge was 30.7 ± 2.66°C (Fig. 7.4), but in 34% of facial thermal images the nasal ridge appeared significantly cooler (23.4 ± 2.75°C), regardless of exercise status. The "cool nose" may be indicative of selective brain cooling, a process which has been documented in other felids to prevent the brain from overheating (Sicuro and Oliveira, 2011; Taylor and Lyman, 1972). The lower temperature of the nasal ridge could signify transnasal cooling

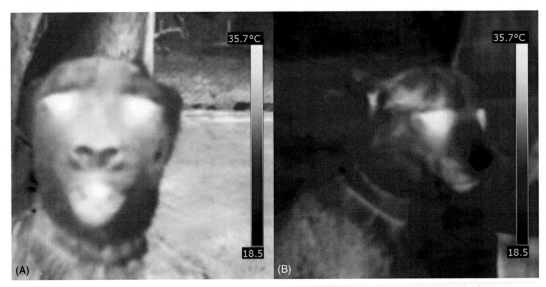

FIGURE 7.4   Thermal images of cheetah face showing (A) a warm and (B) a cool nose relative to the body temperature.

through evaporative water loss from the nasal passages; which is consistent with cheetahs' enlarged nasal passages and nasal bones relative to skull size. In addition, if cheetahs show the same trend as other short-snouted felids, a significant portion of their ethmoturbinals would be expected to be devoted to temperature regulation (Van Valkenburgh et al., 2014).

ORGAN SYSTEMS

Gastrointestinal: Stomach, Liver, and Gall Bladder

Like other felids, cheetahs are adapted to eating meat, which is reflected in a simple stomach, short intestines and a small, uncoiled cecum (Stevens and Hume, 1995). *Helicobacter* sp. bacteria is a common finding in the cheetah's stomach and has been associated with gastritis in captive cheetahs (Chapter 25).

Normal ultrasonographic liver and gall bladder anatomy is described in a study by Carstens et al. (2006). The relative echogenicity of the liver to the spleen and kidney cortex was similar to dogs and domestic cats. Gall bladders were single or bilobed. Age was found to significantly affect gall bladder length, with older cheetahs having larger gall bladders (Carstens et al., 2006). Cheetahs may have enlarged livers relative to other felids, but findings are not conclusive (Gray, 1968).

Immune System: Spleen and Blood

Fatty nodules in the spleen (and occasionally the liver), described as myelolipomatosis or splenic nodular lipomatosis, are a common finding in captive cheetahs on ultrasound and during necropsy, but they do not appear to have a negative impact on the health of the animal (Chapter 25).

Cheetahs have the same AB–erythrocyte antigens blood types as domestic cats, and the same

blood typing procedures can be applied (Griot-Wenk and Giger, 1999). Interestingly, despite the known decreased genetic diversity (Chapter 6), two different blood types were found in 8 cheetahs; 6 had blood type B, similar to other felids in the Puma group and African and Asian golden cats, while the other 2 cheetahs had blood type AB (Griot-Wenk and Giger, 1999).

Cardiopulmonary: Heart and Lungs

Cheetahs have been described as having relatively large hearts, bronchi, and lungs as an adaptation for high speed hunting (Eaton, 1974). Gray (1968) also described an enlarged heart and increased lung capacity. However, these findings are controversial. Spector (1956) did not detect increased heart and lung to body mass ratios; while the measurements were limited to 2 cheetahs, they were comparable to preliminary data obtained from 26 necropsies (lungs: 1.28 ± 0.55 g/100 g body mass; heart: 0.57 ± 0.18 g/100 g body mass; CCF, unpublished data). More recently, a radiographic study also found the relative size of cheetah hearts to be similar to domestic cats when the ratios of several measurements were compared (Schumacher et al., 2003).

Respiratory rates increase significantly after a chase or during high environmental temperatures, going from 9 breaths/min at rest to 100–206 breaths/min while panting (Frame and Frame, 1981; Taylor and Rowntree, 1973).

Endocrine: Adrenal Glands

Chronic stress, associated with some captive settings, has been hypothesized to contribute to cheetahs' susceptibility to multiple unusual diseases and poor reproductive success (Chapters 25 and 27), as such, an understanding of adrenal gland morphology is important. Corticomedullary ratios of captive (zoo) cheetahs were found to be greater than free living cheetahs, as measured at necropsy, with mean captive ratios of 2.07 and free-ranging 1.92 (Terio et al., 2004). When inferring relative adrenal size ultrasonographically, no difference could be detected between cheetahs living in captivity in Namibia and free-ranging cheetahs (Wachter et al., 2011). Adrenal weight, width, corticomedullary ratio, and corticomedullary hyperplasia correlated with age (Gillis-Germitsch et al., 2016; Kirberger and Tordiffe, 2016).

Genitourinary: Ovaries, Testes, and Kidneys

Historically, reproductive success in captivity has been poor (Chapter 27) and sustainability of the captive cheetah population continues to be challenged by a low reproduction rate (Chapter 22). Knowledge of reproductive anatomy, physiology, and morphology is therefore an important basis if increases in captive reproductive success are to take place. Ovarian, uterine, and testicular reference volumes are presented in Chapter 27. Ultrasonographic uterine metrics did not vary with age (Crosier et al., 2011; Wildt et al., 1993). Paraovarian and endometrial cysts are common findings in cheetah, but do not appear to negatively affect fertility (Chapter 27). Male cheetahs, like other felids, have spines on their penis that erect during copulation and are believed to induce ovulation (Bertschinger et al., 2006). Cryptorchidism is rare in cheetahs, but isolated cases have been documented in wild and captive Namibian (Crosier et al., 2007) and captive North American cheetahs (Haefele, unpublished data), while in the puma population found in Florida (Florida panther), another genetically compromised felid, cryptorchidism was found commonly (Roelke et al., 1993) prior to its genetic rescue.

Renal disease is a common finding in cheetahs (Chapter 25); it is therefore important to be able to differentiate a normal from an affected kidney. In a study of 15 captive cheetahs, the ratio of kidney length to the second lumbar vertebrae length was found to be 1.81 ± 0.14, a ratio smaller than

for domestic cat standards (Hackendahl and Citino, 2005). Average kidney length, width, and height at level of renal sinus, were 63.9 ± 5.7 mm, 42.1 ± 5 mm, and 38.1 ± 5.2 mm, respectively. Left and right kidneys were not significantly different in length (Carstens et al., 2006). While male kidneys were found to be 10.6% longer than females, the mass of animal appears to be a more important predictor of kidney length than sex or age (Carstens et al., 2006).

SUMMARY

With its characteristic tear marks, the cheetah is easily distinguished from other felids. Since cheetahs hunt mainly in the daytime, they also have adaptations for diurnal visual acuity, including more short wavelength cones in their retinas, a horizontal-oriented visual system. Cheetahs have a unique body shape among cats that has evolved for speed rather than power. As the fastest land mammal, the cheetah has many specializations that make it distinct among cats. Cheetah morphology includes adaptations in the skull, limbs, and spine. Cheetah physiology is also optimized for velocity, including adaptations in the muscle fibers, thermoregulatory processes, and many organs.

The adaptations that have made the cheetah so successful at hunting their prey have also made them more fragile in a changing ecosystem. All efforts should be taken to preserve these unique cats for centuries to come.

References

Adamson, J., 1969. The spotted sphinx. Collins, London.

Ahnelt, P.K., Schubert, C., Kübber-Heiss, A., Schiviz, A., Anger, E., 2006. Independent variation of retinal S and M cone photoreceptor topographies: a survey of four families of mammals. Vis. Neurosci. 23, 429–435.

Andersson, K., Werdelin, L., 2003. The evolution of cursorial carnivores in the tertiary: implications of elbow-joint morphology. Proc. R. Soc. Lond. 270, S163–S165.

Bertschinger, H.J., Jago, M., Nöthling, J., Human, A., 2006. Repeated use of the GnRH analogue deslorelin to down-regulate reproduction in male cheetahs (*Acinonyx jubatus*). Theriogenology 66, 1762–1767.

Biknevicius, A.R., Van Valkenburgh, B., 1996. Design for killing: craniodental adaptations of predators. Gittleman, J.L. (Ed.), Carnivore Behavior, Ecology, and Evolution, vol. 2, Cornell University Press, Ithaca, NY, pp. 393–428.

Boast, L.K., Houser, A.M., Good, K., Gusset, M., 2013. Regional variation in body size of the cheetah (*Acinonyx jubatus*). J. Mammal. 94, 1293–1297.

Brookes, J., 1828. A Catalogue of the Anatomical and Zoological Museum of Joshua Brookes, Esq., F.R.S. F.L.S. &c. Richard Taylor, London.

Carbone, C., Mace, G., Roberts, S., Macdonald, D.W., 1999. Energetic constraints on the diet of terrestrial carnivores. Nature 402, 286–288.

Caro, T., 1994. Cheetahs of the Serengeti Plains. University of Chicago Press, Chicago.

Carstens, A.N.N., Kirberger, R.M., Spotswood, T.I.M., Wagner, W.M., Grimbeek, R.J., 2006. Ultrasonography of the liver, spleen, and urinary tract of the cheetah (*Acinonyx jubatus*). Vet. Radiol. Ultrasound 47, 376–383.

Crosier, A.E., Comizzoli, P., Baker, T., Davidson, A., Munson, L., Howard, J., Marker, L.L., Wildt, D.E., 2011. Increasing age influences uterine integrity, but not ovarian function or oocyte quality, in the cheetah (*Acinonyx jubatus*). Biol. Reprod. 85, 243–253.

Crosier, A.E., Marker, L., Howard, J., Pukazhenthi, B.S., Henghali, J.N., Wildt, D.E., 2007. Ejaculate traits in the Namibian cheetah (*Acinonyx jubatus*): influence of age, season and captivity. Reprod. Fertil. Dev. 19, 370–382.

Duvernoy, G.L., 1834. Notice Critique sur les Espèces de Grands Chats Nommées, par Hermann: *Felis Chalybeate* et *Guttata*. Société d'Histoire Naturelle de Strasbourg, Strasbourg, France.

Eaton, R.L., 1974. The Cheetah: Biology, Ecology and Behavior of an Endangered Species. Van Nostrand Reinhold, New York.

Emerson, S., Bramble, D., 1993. Scaling, allometry and skull design. Hanken, J., Hall, B. (Eds.), The Skull: Functional and Evolutionary Mechanisms, vol. 3, University of Chicago Press, Chicago, pp. 384–421.

Ewer, R., 1973. The Carnivores. Cornell University Press, Ithaca, New York.

Frame, G., Frame, L., 1981. Swift & Enduring Cheetahs and Wild Dogs of the Serengeti. Elsevier-Dutton Publishing Co., New York, NY.

Gillis-Germitsch, N., Vybiral, P.R., Codron, D., Clauss, M., Kotze, A., Mitchell, E.P., 2016. Intrinsic factors, adrenal gland morphology, and disease burden in captive cheetahs (*Acinonyx jubatus*) in South Africa. Zoo Biol., 40–49.

Gorman, M.L., Mills, M.G., Raath, J.P., Speakman, J.R., 1998. High hunting costs make African wild dogs vulnerable to kleptoparasitism by hyaenas. Nature 391, 479–481.

Gotch, A.F., 1979. Mammals—their latin names explained. Blanford Press, Poole, Dorset.

Gray, J., 1968. Animal Locomotion. W.W. Norton, New York, NY.

Griot-Wenk, M., Giger, U., 1999. The AB blood group system in wild felids. Anim. Genet. 30, 144–147.

Hackendahl, N.C., Citino, S.B., 2005. Radiographic kidney measurements in captive cheetahs (*Acinonyx jubatus*). J. Zoo Wildl. Med. 36, 321–322.

Hartstone-Rose, A., Long, R.C., Farrell, A.B., Shaw, C.A., 2012. The clavicles of *Smilodon fatalis* and *Panthera atrox* (mammalia: Felidae) from Rancho La Brea, Los Angeles, California. J. Morphol. 273, 981–991.

Hetem, R.S., Mitchell, D., de Witt, B.A., Fick, L.G., Meyer, L.C.R., Maloney, S.K., Fuller, A., 2013. Cheetah do not abandon hunts because they overheat. Biol. Lett. 9, e0130472.

Hildebrand, M., 1959. Motions of the running cheetah and horse. J. Mammal. 40, 481–495.

Hildebrand, M., 1961. Further studies on locomotion of the cheetah. J. Mammal. 42, 84–91.

Hudson, P.E., Corr, S.A., Payne-Davis, R.C., Clancy, S.N., Lane, E., Wilson, A.M., 2011a. Functional anatomy of the cheetah (*Acinonyx jubatus*) forelimb. J. Anat. 218, 375–385.

Hudson, P.E., Corr, S.A., Payne-Davis, R.C., Clancy, S.N., Lane, E., Wilson, A.M., 2011b. Functional anatomy of the cheetah (*Acinonyx jubatus*) hindlimb. J. Anat. 218, 363–374.

Hunter, L., Hamman, D., 2003. Cheetah. Struik Publishers, Cape Town, SA.

Hyatt, J.-P.K., Roy, R.R., Rugg, S., Talmadge, R.J., 2010. Myosin heavy chain composition of tiger (*Panthera tigris*) and cheetah (*Acinonyx jubatus*) hindlimb muscles. J. Exp. Zool. 313A, 45–57.

Jacobs, G.H., 1993. The distribution and nature of colour vision among the mammals. Biol. Rev. Camb. Philos. Soc. 68, 413–471.

Kaelin, C.B., Xu, X., Hong, L.Z., David, V.A., McGowan, K.A., Schmidt-Küntzel, A., Roelke, M.E., Pino, J., Pontius, J., Cooper, G.M., Manuel, H., Swanson, W.F., Marker, L., Harper, C.K., van Dyk, A., Yue, B.S., Mullikin, J.C., Warren, W.C., Eizirik, E., Kos, L., O'Brien, S.J., Barsh, G.S., Menotti-Raymond, M., 2012. Specifying and sustaining pigmentation patterns in domestic and wild cats. Science 337, 1536–1541.

Kingdon, J., 1977. East African Mammals. University of Chicago Press, New York.

Kingdon, J., 1997. The Kingdon Field Guide to African Mammals. Academic Press, London, UK.

Kirberger, R.M., Tordiffe, A.S., 2016. Ultrasonographic adrenal gland findings in healthy semi-captive cheetahs (*Acinonyx jubatus*). Zoo Biol. 35, 260–268.

Koob, T.J., Long, J.H., 2000. The vertebrate body axis: evolution and mechanical function. Am. Zool. 40, 1–18.

Krausman, P.R., Morales, S.M., 2005. *Acinonyx jubatus*. Mamm. Species, 1–6.

Lesson, R.P., 1842. Nouveau Tableau du Règne Animal. Arthus Bertrand, Paris.

Leyhausen, P., 1979. Cat Behavior: The Predatory and Social Behavior of Domestic and Wild Cats. Garland STMP Press, New York.

Londei, T., 2000. The cheetah (*Acinonyx jubatus*) dewclaw: specialization overlooked. J. Zool 251, 535–537.

Long, Jr., J.H., Pabst, D.A., Shepherd, W.R., McLellan, W.A., 1997. Locomotor design of dolphin vertebral columns: bending mechanics and morphology of *Delphinus delphis*. J. Exp. Biol. 200, 65–81.

Marker, L., 2014. A Future For Cheetahs. Cheetah Conservation Fund, Virginia.

Marker, L.L., Dickman, A.J., 2003. Morphology, physical condition, and growth of the cheetah (*Acinonyx jubatus jubatus*). J. Mammal. 84, 840–850.

Marker, L.L., Dickman, A.J., 2004. Dental anomalies and incidence of palatal erosion in namibian cheetahs (*Acinonyx jubatus jubatus*). J. Mammal. 85, 19–24.

Meachen-Samuels, J., 2010. Comparative scaling of the forelimbs of felids and canids using radiographic images. J. Mamm. Evol. 17, 193–209.

Meachen-Samuels, J., Van Valkenburgh, B., 2009a. Craniodental indicators of prey-size preference in the Felidae. Biol. J. Linn. Soc. 96, 784–799.

Meachen-Samuels, J., Van Valkenburgh, B., 2009b. Forelimb indicators of prey-size preference in the Felidae. J. Morphology 270, 729–744.

Meachen-Samuels, J.A., Van Valkenburgh, B., 2010. Radiographs reveal exceptional forelimb strength in the Sabertooth cat, *Smilodon fatalis*. PLoS One 5, e11412.

O'Regan, H.J., 2002. Defining Cheetahs, a multivariate analysis of skull shape in big cats. Mamm. Rev. 32, 58–62.

Pierce, S.E., Clack, J.A., Hutchinson, J.R., 2011. Comparative axial morphology in pinnipeds and its correlation with aquatic locomotory behaviour. J. Anat. 219, 502–514.

Pocock, R.I., 1916a. On some of the cranial and external characters of the Hunting Leopard or Cheetah. F.R.S. Ann. Mag. Nat. Hist. 18, 419–429.

Pocock, R.I., 1916b. On the hyoidean apparatus of the lion (*F. leo*) and related species of Felidae. Annu. Mag. Nat. Hist. Zool. Bot. Geol. 8, 222–229.

Randau, M., Cuff, A.R., Hutchinson, J.R., Pierce, S.E., Goswami, A., 2016a. Regional differentiation of felid vertebral column evolution: a study of 3D shape trajectories. Org. Divers. Evol. 17, 305–319.

Randau, M., Goswami, A., Hutchinson, J.R., Cuff, A.R., Pierce, S.E., 2016b. Cryptic complexity in felid vertebral evolution: shape differentiation and allometry of the axial skeleton. Zool. J. Linn Soc. 178, 183–202.

Roelke, M.E., Martenson, J.S., O'Brien, S.J., 1993. The consequences of demographic reduction and genetic depletion in the endangered Florida panther. Curr. Biol. 3, 340–350.

Russell, A., Bryant, H., 2001. Claw retraction and protraction in the Carnivora: the cheetah (*Acinonyx jubatus*) as an atypical felid. J. Zool. 254, 67–76.

Scantlebury, D.M., Mills, M.G.L., Wilson, R.P., Wilson, J.W., Mills, M.E.J., Durant, S.M., Bennett, N.C., Bradford, P., Marks, N.J., Speakman, J.R., 2014. Flexible energetics of cheetah hunting strategies provide resistance against kleptoparasitism. Science 346, 79–81.

Schaller, G., 1972. The Serengeti Lion: A Study of Predator Prey Relationships. University of Chicago Press, Chicago.

Schreber, J.C.D., 1775. Die Säugthiere in Abbildungen nach der Natur mit Beschreibungen. Walther, Erlangen, Germany.

Schumacher, J., Snyder, P., Citino, S.B., Bennett, R.A., Dvorak, L.D., 2003. Radiographic and electrocardiographic evaluation of cardiac morphology and function in captive cheetahs (*Acinonyx jubatus*). J. Zoo Wildl. Med. 34, 357–363.

Segura, V., Prevosti, F., Cassini, G., 2013. Cranial ontogeny in the Puma lineage, *Puma concolor, Herpailurus yagouaroundi*, and *Acinonyx jubatus* (Carnivora: Felidae): a three-dimensional geometric morphometric approach. Zool. J. Linn. Soc. 169, 235–250.

Sharp, N., 1997. Timed running speed of a cheetah (*Acinonyx jubatus*). J. Zool. 241, 493–494.

Sicuro, F.L., Oliveira, L.F.B., 2011. Skull morphology and functionality of extant Felidae (Mammalia: Carnivora): a phylogenetic and evolutionary perspective. Zool. J. Linn. Soc. 161, 414–462.

Spector, W.S., 1956. Handbook of Biological Data. W.B. Saunders Company, Philadelphia and London.

Steenkamp, G., Boy, S.C., van Staden, P.J., Bester, M.N., 2017. How the cheetahs' specialized palate accommodates its abnormally large teeth. J. Zool 301, 290–300.

Steinberg, R.H., Reid, M., Lacy, P.L., 1973. The distribution of rods and cones in the retina of the cat (*Felis domesticus*). J. Comp. Neurol. 148, 229–248.

Stevens, C.E., Hume, I.D., 1995. Comparative Physiology of the Vertebrate Digestive System, second ed. Press Syndicate, New York.

Sunquist, M., Sunquist, F., 2002. Wildcats of the World. University of Chicago Press, Chicago.

Taylor, C.R., Lyman, C.P., 1972. Heat storage in running antelopes: independence of brain and body temperatures. Am. J. Physiol. 222, 114–117.

Taylor, C.R., Rowntree, V.J., 1973. Temperature regulation and heat balance in running cheetahs: a strategy for sprinters? Am. J. Physiol. 224, 848–851.

Terio, K.A., Marker, L., Munson, L., 2004. Evidence for chronic stress in captive but not free-ranging cheetahs (*Acinonyx jubatus*) based on adrenal morphology and function. J. Wildl. Dis. 40, 259–266.

Torregrosa, V., Petrucci, M., Pérez-Claros, J.A., Palmqvist, P., 2010. Nasal aperture area and body mass in felids: ecophysiological implications and paleobiological inferences. Geobios 43, 653–661.

Van Valkenburgh, B., Pang, B., Bird, D., Curtis, A., Yee, K., Wysocki, C., Craven, B.A., 2014. Respiratory and olfactory turbinals in feliform and caniform carnivorans: the influence of snout length. Anat. Rec. 297, 2065–2079.

Wachter, B., Thalwitzer, S., Hofer, H., Lonzer, J., Hildebrandt, T.B., Hermes, R., 2011. Reproductive history and absence of predators are important determinants of reproductive fitness: the cheetah controversy revisited. Conserv. Lett. 4, 47–54.

Wack, R.F., Kramer, L.W., Cupps, W., Currie, P., 1991. Growth rate of 21 captive-born, mother-raised cheetah cubs. Zoo Biol. 10, 273–276.

Wagler, J., 1830. Natürliches System der Amphibien, mit vorangehender Classification der Säugthiere und Vögel. J. G. Cotta'schen Buchhandlung, Munich, Germany.

Weissengruber, G.E., Forstenpointner, G., Peters, G., Kübber-Heiss, A., Fitch, W.T., 2002. Hyoid apparatus and pharynx in the lion (*Panthera leo*), jaguar (*Panthera onca*), tiger (*Panthera tigris*), cheetah (*Acinonyx jubatus*) and domestic cat (*Felis silvestris f. catus*). J. Anat. 201, 195–209.

Wildt, D., Brown, J., Bush, M., Barone, M., Cooper, K., Grisham, J., Howard, J., 1993. Reproductive status of cheetahs (*Acinonyx jubatus*) in North American zoos: the benefits of physiological surveys for strategic planning. Zoo Biol. 12, 45–80.

Wilson, R.P., Griffiths, I.W., Mills, M.G.L., Carbone, C., Wilson, J.W., Scantlebury, D.M., 2015. Mass enhances speed but diminishes turn capacity in terrestrial pursuit predators. eLife 4, e06487.

Wilson, A.M., Lowe, J.C., Roskilly, K., Hudson, P.E., Golabek, K.A., McNutt, J.W., 2013. Locomotion dynamics of hunting in wild cheetahs. Nature 498, 185–189.

Wroe, S., McHenry, C., Thomason, J., 2005. Bite club: comparative bite force in big biting mammals and the prediction of predatory behaviour in fossil taxa. Proc. R. Soc. Lond. B 272, 619–625.

Ecology of Free-Ranging Cheetahs

Laurie Marker, Bogdan Cristescu*, Amy Dickman**,
Matti T. Nghikembua*, Lorraine K. Boast†,
Tess Morrison*, Joerg Melzheimer‡, Ezequiel Fabiano§,
Gus Mills**,¶, Bettina Wachter‡, David W. Macdonald***

*Cheetah Conservation Fund, Otjiwarongo, Namibia
**University of Oxford, Tubney, Abingdon, United Kingdom
†Cheetah Conservation Botswana, Gaborone, Botswana
‡Leibniz Institute for Zoo and Wildlife Research, Berlin, Germany
§University of Namibia, Katima Mulilo, Namibia
¶The Lewis Foundation, Johannesburg, South Africa

INTRODUCTION

Cheetahs historically occurred throughout the Sahel and sub-Saharan Africa and south western Asia, but are now restricted to 10% of their historic range in Africa and the central deserts of Iran (Chapters 4 and 5). The ecology of the cheetah has been extensively studied in some areas, primarily in southern Africa and the Serengeti National Park (NP) in East Africa. This chapter provides a summary of the cheetah's ecology focusing on habitat, density, prey species consumed, intraguild interactions, home range sizes, movements, and activity patterns.

HABITAT TYPE

Cheetahs are adapted for high-speed chases (Chapter 7) and require good visibility to detect and pursue prey; as a result, grassland and open savannahs are a critical component of the cheetah's range (Caro, 1994). Cheetahs also benefit from high grass or bush areas that enable them to remain undetected while stalking prey for longer periods than in open areas, thereby reducing chase time and conserving energy (Mills et al., 2004; Rostro-García et al., 2015). Additionally, the denser cover associated with bushveld and woodland provides safety for young cubs from other predators and minimizes the risk of

Cheetahs: Biology and Conservation
http://dx.doi.org/10.1016/B978-0-12-804088-1.00008-3

kleptoparasitism, that is, the takeover of a kill by other large carnivore species (Broomhall, 2001; Durant, 1998; Mills et al., 2004; Purchase and du Toit, 2000; Rostro-García et al., 2015). Previously, the cheetah had been considered a species of open landscapes (Mills et al., 2004). This impression was probably due to the ease of sighting cheetahs in these open areas, and the long-term studies conducted on cheetahs in such habitats in East Africa (Caro, 1994; Eaton, 1974; Schaller, 1968). However, research has revealed that cheetahs use a wide variety of habitats from hyper- and semi-desert to savannah woodland, dry forest, and dense vegetation (Durant et al., 2015). Examples of the diversity of regions utilized by cheetahs include sections of the Sahara desert in Algeria, woodland and grassland in Kruger NP in South Africa, the Okavango Delta in Botswana, and thornbush savannah on Namibian farmlands (Belbachir et al., 2015; Broekhuis, 2012; Broomhall, 2001; Marker-Kraus et al., 1996). Cheetahs are absent from tropical and montane forests, but sightings have been reported at altitudes of 4000 m on Mt Kenya (Young and Evans, 1993). Cheetah habitat selection is based on a variety of factors, including visibility, prey density, and at a fine scale on the avoidance of larger predators (Hayward et al., 2006; Marker et al., 2008a,b; Muntifering et al., 2006; Rostro-García et al., 2015).

Cheetahs in eastern and southern Africa primarily occupy the Savannah Biome, characterized by a mix of grassland, bushveld, and woodland (Low and Rebelo, 1996). On Namibian farmland, females showed a preference for sparse bush, where ungulate biomass was higher (Marker et al., 2008a; Muntifering et al., 2006). In parts of southern and eastern Africa, landscapes with high visibility and grass cover are quickly disappearing due to the overgrazing by livestock, which has led to bush encroachment, an overgrowth of the native bush (Olaotswe et al., 2013). The disappearance of native grasses and the rapid spread of thorny thickets have reduced prey densities and visibility, and have led

to eye injuries in cheetahs that can inhibit hunting ability (Bauer, 1998; Muntifering et al., 2006).

In western, central, and northern Africa, much of the current cheetah habitat is characterized by wide valleys, high mountain ranges, and hyper- to subarid climate (Belbachir et al., 2015). Cheetahs are primarily located in the countries containing more productive habitats within the region, such as Benin, Burkina Faso, and southern Niger. In the central Sahara, which is classified as a hyperarid region, cheetahs occur in high mountain habitat that receives slightly higher rainfall than the surrounding desert. The higher precipitation results in richer vegetation and small permanent waterholes that support antelope populations (Nowell and Jackson, 1996).

In Iran, cheetahs typically live in hilly terrain, foothills, and rocky valleys within the desert ecosystem (Farhadinia et al., 2008; Hunter et al., 2007; Sarhangzadeh et al., 2015). They are also known to select mountainous habitats in gravelly desert ranges that rise to 3000 m (Sarhangzadeh et al., 2015). Rainfall in some of these arid regions is <100 mm per year and the sparse vegetation consists predominantly of shrubs <1 m in height (Durant et al., 2015).

While cheetahs use a wider range of habitats than previously thought (Bissett and Bernard, 2007; Mills et al., 2004; Rostro-García et al., 2015), their adaptability is unlikely to keep up with the rapid changes occurring across the cheetah's range today, which include road networks, habitat deterioration, fragmentation (Chapter 10), and bush encroachment.

CHEETAH DENSITIES

Wildlife management and conservation decisions often rely on information related to animal density. In comparison with other large African felids, cheetahs occur at relatively low densities, generally 10%–30% of the typical densities for lions (*Panthera leo*) or leopards (*Panthera pardus*) in prime habitat (Durant, 2007). Density estimates for cheetahs typically range between 0.3 and 3.0 adult cheetahs per 100 km^2 in free-ranging environments (e.g., Durant et al., 2011; Gros, 1998; Marker, 2002; Mills, 2015; Mills and Biggs, 1993). Generally, cheetahs are found at substantially lower densities in harsh hyperarid environments and densities as low as 0.021–0.055 per 100 km^2 have been recorded for the Saharan cheetahs in Algeria (Belbachir et al., 2015). In "predator-proof" fenced reserves that are intensively managed, high densities of cheetahs are possible when competing predators, such as lions and spotted hyenas (*Crocuta crocuta*) are absent and/or when the prey base is artificially supplemented. However, Swanson et al. (2014) suggests that mitigating mesopredator suppression may be possible through fine-scale spatial avoidance and that the impact on cheetahs by lions may be less severe than reported (Mills et al. 2017). Density data are not available longitudinally, with the exception of a few intensively studied populations, such as the Serengeti NP in Tanzania (Durant et al., 2011) and Namibian farmland (Marker, 2002; Marker et al., 2007; Fabiano, 2013). However, density and population estimates should be treated with caution due to the large ranging movements of cheetahs and their nonheterogeneous use of the landscape, which can result in "hotspots" of high cheetah density (Broekhuis and Gopalaswamy, 2016; Durant et al., 2007). Extrapolation of locally acquired data to other areas may thus not be appropriate.

Occupancy modeling or relative abundance indices represent attractive alternatives to density estimation (Chapters 30 and 37). Monitoring temporal trends in presence, or the variability in relative abundance, might be preferable to channeling limited resources into obtaining a snapshot of density. Monitoring cheetah occupancy across seasons is particularly important in areas where cheetahs shift locations to track migratory prey species (e.g., in the Serengeti NP; Durant et al., 1988).

PREY SPECIES CONSUMED

Cheetahs have been recorded to feed on 73 different prey species throughout their range (Table 8.1). The prey body size varies from rodents to adult ungulates. Based on observations of cheetah kills, the preferred prey species is typically a small to medium sized ungulate with a body mass between 14 and 56 kg, and most often, less than 40 kg (Clements et al., 2014; Hayward et al., 2006; Schaller, 1972). A large proportion of prey consumed consists of the juveniles of free-ranging ungulate species (Burney, 1980; McLaughlin, 1970; Mills et al., 2004). However, when medium to large prey is not available, cheetahs are able to subsist on smaller prey. Outside of birth peaks of ungulates, springhares (*Pedetidae capensis*) and hares (Leporidae) were the most frequently eaten species on Namibian farmlands as determined by scat analysis (Marker et al., 2003b; Wachter et al., 2006), and in Tanzania as determined by observations (Cooper et al., 2007). Similarly, using data from direct observations and tracking spoor, Mills (2015) found that in the southern Kalahari most cheetah kills (except for coalition males) fell outside the prey range, identified by Hayward et al. (2006), with many kills comprising smaller prey, such as steenbok (*Raphicerus campestris*), springbok (*Antidorcas marsupialis*) lambs, hares, and springhares. In the northern areas of Africa and some parts of Iran, hares are the most widely distributed and abundant wild prey species available, but while an adult cheetah can survive on such a diet, they are thought to be insufficient to feed family groups with growing litters of cubs (Farhadinia et al., 2012; Hunter et al., 2007). Of the 21 published studies reviewed by Hayward et al. (2006), only three studies (all in South Africa) considered hares as a potential prey item (Broomhall, 2001; Mills and Biggs, 1993; Pienaar, 1969). Whether the other studies failed to detect hares, as many of the studies were based on kill observation data that tend to be biased to large prey items, or a

TABLE 8.1   Prey Species Identified as Being Consumed by Cheetahs in Various Regions and Countries

Region	Country	Prey	Source, [method] (location, habitat)
Southern Africa	Botswana	Bird[6A], black-backed jackal[10A], blesbok[19A], blue wildebeest[16AB], bushbuck[72B], cattle[7A], eland[69AB], gemsbok[47A], goat[12A], grey duiker[67AB], impala[1AB], kudu[73AB], red hartebeest [2A], scrub hare[39A], sheep[51A], springbok[4A], springhare[54AB], steenbok[57AB], warthog[55B], waterbuck[34A], zebra[23AB]	A. Boast et al., 2016[a] (Fhs) B. Craig et al., 2017[ac] (NTGR)
	Namibia	Blesbok[19A], blue wildebeest[16B], bustards[48A], eland[69A], gemsbok[47AC], goat[12AC], grey climbing mouse[21C], gray duiker[67AC], ground squirrel[64C], guineafowl[45A], impala[1A], Kirk's dik-dik[43A], kudu[73AC], ostrich[65A], red hartebeest[2AC], scrub hare[39A], sheep[51A], springbok[4ABC], springhare[54A], steenbok[57A], warthog[55AC]	A. Marker-Kraus et al. (1996)[a] (Ft); Marker et al. (2003a,b)[a,c,b] (Ft); Morsbach (1987)[c] (Ft) B. Berry (1981)[e] (ENP) C. Wachter et al. (2006)[a] (Ft)
	South Africa	Aardvark[46A], bat-eared fox[49B], black-backed jackal[10AE], black wildebeest[15JK], blesbok[19AEIK], blue wildebeest[16ABCEFGI], bontebok[20J], buffalo[68A], bushbuck[72AFGHIK], bustards[48A], Cape porcupine[33A], cattle[7I], eland[73AEJK], gemsbok[47BDJK], giraffe[26AI], grey duiker[67AFGHIJK], guineafowl[45A], impala[1ACDEFGHIJK], kudu[73ACDEFGHIJK], Kirk's dik-dik[43A], mountain reedbuck[60DK], nyala[70AFGI], ostrich[65ALEFI], red duiker[13I], red hartebeest[2BDEK], roan antelope[31A], sable antelope[32A], scrub hare[39AFJK], Sharpe's grysbok[58A], southern reedbuck[59AFI], springbok[4BDEHJK], steenbok[57ABFGIJ], tsessebe[18A], warthog[55ADEFGIK], waterbuck[34ADFG], zebra[23AIK]	A. Broomhall (2001)[c,d] (KNP); Labuschagne (1979)[c] (KNP); Mills and Biggs (1993)[c], (KNP); Pienaar (1969)[f] (KNP); Mills et al. (2004)[c,d] (KNP); Owen-Smith and Mills (2008)[f] (KNP) B. Mills (1984)[f] (SK) C. Hirst (1969)[f,c] (TPNR) D. Hofmeyr and van Dykaver (1998)[c] (PNP) E. Hofmeyr and van Dykaver (1998)[c] (MGR) F. Radloff and Du Toit (2004)[c] (MMPGR) G. Tambling et al. (2014)[c] (KaPGR) H. Bissett and Bernard (2007)[c,d] (KwPGR) I. Hunter (1998)[c,d] (PRR) J. Vorster (2011)[a,c] (SWR) K. Cristescu (2006)[d] (SGR); O'Brien (2012)[c,d] (SGR)
	Zambia	Blue wildebeest[16], bushbuck[72], grey duiker[67], impala[1], kudu[73], Lichtenstein's hartebeest[2], oribi[50], puku[37], southern reedbuck[59], warthog[55], zebra[23]	Mitchell et al. (1965)[c,f] (KaNP)
	Zimbabwe	Bushbuck[72], grey duiker[67], guineafowl[45], impala[1], kudu[73], waterbuck[34]	Purchase and du Toit (2000)[a,c] (MNP)
Eastern Africa	Kenya	Blue wildebeest[16C], bushbuck[72C], cattle[7C], Cape hare[38C], Coke's hartebeest[2ABC], gerenuk[40B], giant rat[17C], giraffe[26C], goat[12C], Grant's gazelle[44ABC], grey duiker[67C], guineafowl[45B], Günther's dik-dik[42B], impala[1ABC], lesser kudu[71BC], rock hyrax[56C], sheep[51C], springhare[54C], steenbok[57C], Thomson's gazelle[25ABC], vervet monkey[14C], warthog[55C], waterbuck[34A], yellow baboon[53C], zebra[23C]	A. Burney (1980)[c] (NNP); Eaton (1974)[e] (NNP); Graham (1966)[b] (NNP); McLaughlin (1970)[e] (NNP) B. Hamilton (1986)[b] (NKKR) C. Mutoro (2015)[a] (FRSK)
	Tanzania	Blue wildebeest[16], bohor reedbuck[61] Coke's hartebeest[2A], Grant's gazelle[44], Kirk's dik-dik[43], scrub hare[39], Thomson's gazelle[25], topi[18], zebra[23]	Caro (1994)[c] (SNP); Fitzgibbon and Fanshawe (1989)[c] (SNP); Frame (1986)[c,d] (SNP); Kruuk and Turner (1967)[c] (SNP); Schaller (1968)[c] (SNP)

TABLE 8.1 Prey Species Identified as Being Consumed by Cheetahs in Various Regions and Countries (*cont.*)

Region	Country	Prey	Source, [method] (location, habitat)
Western, Central and Northern Africa	Algeria	Camel[8], Cape hare[38], donkey[22], Dorcas gazelle[28]	Wacher et al. (2005)[b,f] (ANP)
	Benin	Cape hare[38], grey duiker[67], guineafowl[45], oribi[42], red-fronted gazelle[24], warthog[55]	Rabeil and Comizzoli (personal communication)[b] (PBR)
	Central African Republic	Bubal hartebeest[2], Buffon's kob[36], oribi[50], warthog[55]	Ruggiero (1991)[c,f] (MGFNP)
	Egypt (extinct)	Camel[8], Cape hare[38], donkey[22], Dorcas gazelle[28], Egyptian golden jackal[9], goat[12], grey heron[5], rodents[63], sheep[51]	Saleh et al. (2001)[a,b,c] (EWD)
	Niger	Bustards[48A], camel[8B], Cape hare[38A], Dorcas gazelle[28B], gazelles[29A], grey duiker[67A], guineafowl[45A], livestock[41A], oribi[50A], red-fronted gazelle[24A], warthog[55A], western hartebeest[2A]	A. Poche (1973)[c] (PW); Claro et al. (2006)[b] (PW); Sogbohossou et al. (2011)[b] (PW); B. Claro and Sissler (2003)[f,b] (TR); Matchano (2011)[f,g] (TR)
Asia	Iran	Cape hare[38B], goat[12BC], Goitered gazelle[30BC], insect[35BC], Jebeer gazelle[27A], Persian ibex[11A], reptile[62B], rodents[63BC], scrub hare[39C], See-See partridge[3B], sheep[51BC], wild boar[66C], wild sheep[52A]	A. Farhadinia and Hemami (2010)[f] (DAWR); B. Farhadinia et al. (2012)[a,c] (MWR); C. Farhadinia et al. (2012)[a,c] (NHA)

Note: Instances of cannibalism are rare and were not listed.

Superscript uppercase letters ([A, B, C, etc.]) in column four refer to the data source; multiple sources documenting identical species are represented by one letter.

Superscript numbers ([1, 2, 3, etc.]) refer to the prey species scientific name:

[1]*Aepyceros melampus*, [2]*Alcelaphus buselaphus*, [3]*Ammoperdix griseogularis*, [4]*Antidorcas marsupialis*, [5]*Ardea cinerea*, [6]*Aves* spp., [7]*Bos taurus*, [8]*Camelidae* spp., [9]*Canis anthus*, [10]*Canis mesomelas*, [11]*Capra aegagrus*, [12]*Capra aegagrus hircus*, [13]*Cephalopus natalensis*, [14]*Chlorocebus pygerythrus*, [15]*Connochaetes gnou*, [16]*Connochaetes taurinus*, [17]*Crycetomis emini*, [18]*Damaliscus lunatus*, [19]*Damaliscus pygargus phillipsi*, [20]*Damaliscus pygargus pygargus*, [21]*Dendromus melanotis*, [22]*Equus africanus asinus*, [23]*Equus burchellii*, [24]*Eudorcas rufifrons*, [25]*Eudorcas thomsonii*, [26]*Giraffa camelopardalis*, [27]*Gazella bennetti*, [28]*Gazella dorcas*, [29]*Gazella* spp., [30]*Gazella subgutturosa*, [31]*Hippotragus equinus*, [32]*Hippotragus niger*, [33]*Hystrix africaeaustralis*, [34]*Kobus ellipsiprymnus*, [35]*Insecta* spp., [36]*Kobus kob*, [37]*Kobus vardonii*, [38]*Lepus capensis*, [39]*Lepus saxatilis*, [40]*Litocranius walleri*, [41] livestock, [42]*Madoqua guentheri*, [43]*Madoqua kirkii*, [44]*Nanger granti*, [45]*Numididae*, [46]*Orycteropus afer*, [47]*Oryx gazella*, [48]*Otididae*, [49]*Otocyon megalotis*, [50]*Ourebia ourebi*, [51]*Ovis aries*, [52]*Ovis orientalis*, [53]*Papio cynocephalus*, [54]*Pedetidae* spp., [55]*Phacochoerus africanus*, [56]*Procavia capensis*, [57]*Raphicerus campestris*, [58]*Raphicerus sharpei*, [59]*Redunca arundinum*, [60]*Redunca fulvorufula*, [61]*Redunca redunca*, [62]*reptilia* spp., [63]*rodentia* spp., [64]*Sciuridae* spp., [65]*Struthio camelus*, [66]*Sus scrofa*, [67]*Sylvicapra grimmia*, [68]*Syncerus caffer*, [69]*Taurotragus oryx*, [70]*Tragelaphus angasii*, [71]*Tragelaphus imberbis*, [72]*Tragelaphus scriptus*, [73]*Tragelaphus strepsiceros*.

Methods of data collection are coded by superscript lowercase letters ([a, b, c, etc.]):
[a]Scat analysis, [b]interviews, [c]direct observation from opportunistic monitoring, [d]direct observation from continuous monitoring, [e]direct observation (monitoring style is unknown), [f]carcass analysis, [g]livestock depredation.

List of abbreviations of study locations with associated habitat type:
ANP, Ahaggar National Park, desert; DAWR, Dar-Anjir Wildlife Refuge, desert; ENP, Etosha National Park, woodland savanna; EWD, Egyptian Western Desert, desert; Fhs, farmland cheetahs, hardveld, sandveld; Ft, farmland cheetahs, thornbush savanna; FRSK, free-ranging cheetahs in southern Kenya, open low shrubland, shrub savanna, crops, grassland; KaNP, Kafue National Park, woodland, scrub and grassland; KaPGR, Karongwe Private Game Reserve, savannah; KNP, Kruger National Park, woodland savannah, mopane veld, thornveld; KwPGR, Kwandwe Private Game Reserve, savanna, thicket; MGFNP, Manovo-Gounda St. Floris National Park, savanna, forest; MGR, Madikwe Game Reserve, bushveld, thornveld; MMPGR, Mala Mala Private Game Reserve, woodland savanna, bushveld, open grassland; MNP, Matusadona National Park, woodland, grassland; MWR, Miandasht Wildlife Refuge, desert; NHA, Behkadeh No Hunting Area, mountainous and plain Mediterranean climate; NKKR, Northern Kenya and Kora Reserve, bushland, woodland, shrubland, thicket; NNP, Nairobi National Park, grassland, dry forest; NTGR, Northern Tuli Game Reserve, mopane veld; PBR, Pendjari Biosphere Reserve, savanna grassland, savanna woodland, gallery forest; PNP, Pilansberg National Park, savanna, thicket; PRR, Phinda Resource Reserve, lowveld bushveld, coastal bushveld-grassland; PW, Park W, desert, woodland, shrubland savanna; SGR, Shamwari Game Reserve, thicket, grassland, disturbed land, savanna; SK, Southern Kalahari, open shrub semi desert, tree savannah semi desert; SNP, Serengeti National Park, open woodland, grassland; SWR, Sanbona Wildlife Reserve, Succulent Karoo, thicket, renosterveld; TPNR, Timbavati Private Nature Reserve, woodland savanna, lowveld grassland; TR, Termit Region, desert, wooded and grassy steppe.

true absence of hares in the diet of cheetahs is unclear.

In addition to wild prey, cheetahs have been recorded to prey on domestic animals, including smallstock (sheep and goats), calves (generally less than 6 months of age) (Marker-Kraus et al., 1996), and camels (Wacher et al., 2005). On Namibian farmlands, based on scat analyses, cheetahs primarily predated on wild prey and livestock was only found in 6.4% of scats, with 1.0 calf and 0.4 sheep killed per 100 km^2 by cheetahs per year (Marker et al., 2003b).

Kill rates (number of kills per unit time) are important for estimating effects of predation on prey populations, as well as determining prey base and energetic inputs required by cheetahs. The mean kill rate recorded in six studies was one kill per 3.8 days (SD = 2.4; Mills et al., 2004). Consumption rates have been estimated to be 0.1 kg eaten/kg cheetah body mass/day (Mills, 2015).

INTRAGUILD INTERACTIONS

The abundance of competitively superior carnivore species has a major impact on the cheetah's feeding ecology, behavior, reproductive success, and survival (see also Chapter 9). In turn, cheetahs have an impact on the ecology of mesopredators, influencing biodiversity in local carnivore communities (Terborgh et al., 1999).

Cheetahs may be affected by competition for food resources with larger predator species due to an overlap in prey use (Kruuk and Turner, 1967). They are subject to kleptoparasitism from species, such as lions, spotted hyenas, and leopards (Caro, 1994; Schaller, 1972; Stander, 1990). These competitors are thought to use descending vultures as a cue to locate kills (Kruuk and Turner, 1967). Both lions and spotted hyenas can displace groups of cheetahs, including female cheetahs with adolescent cubs and male coalitions from their kills (Caro, 1994). In areas with these large competitors, cheetahs have been estimated to lose between 2.2% and

11.8% of kills through kleptoparasitism (Bissett and Bernard, 2007; Hunter et al., 2007; Mills and Mills, 2014; Mills et al., 2017; Chapter 9). Kleptoparasitism may force cheetahs to repeat hunting, thus increasing daily energy expenditure (Scantlebury et al., 2014) and the probability for injuries sustained through hunting.

Cheetahs, particularly cubs, are also vulnerable to predation by other large carnivore species (Caro, 1994; Durant, 1998, 2000; Laurenson et al., 1995; Chapter 9). As a consequence, in areas with abundant lion and spotted hyena populations cheetahs have higher mortality rates (Laurenson, 1994; Marker et al., 2003a; Wachter et al., 2011) and regions outside of protected areas, where lions and spotted hyenas have largely been eradicated, often support higher densities of cheetahs than protected areas (Durant et al., 2015). In protected areas with lions, cheetahs are most likely only able to persist through fine-scale avoidance mechanisms (Cooper et al., 2007; Mills, 2015; Swanson et al., 2014) by altering habitat use and activity patterns to avoid larger carnivores. In these areas, cheetahs may select hunting areas with less prey but more protective cover from competitors (Bissett and Bernard, 2007; Broekhuis and Gopalaswamy, 2016; Durant, 2000; Rostro-García et al., 2015). Cheetahs use of space is suggested to be a hierarchical process, driven primarily by prey acquisition and further modulated by avoidance of other predators, that is, a reactive, rather than a predictive, response to risk (Broekhuis et al., 2013).

CHEETAH HOME RANGES

A home range is the area in which a species resides, reproduces, and interacts socially, and from which it derives its survival needs (Burt, 1943). Territorial male cheetahs defend their territory from potentially intruding conspecifics, whereas other males adopt a "floater" tactic characterized by large home ranges and lack of territorial defense (Caro, 1994). Body size, age, and group size likely interact in determining territorial versus "floater" tactics of male cheetahs (Caro and Collins, 1987; Caro et al., 1989; Chapter 9). Territorial males are typically in their prime age (48–96 months; Marker et al., 2008a).

Often, cheetahs will temporarily venture outside their normal home range due to natural or anthropogenic disturbances or for exploratory purposes (Laver, 2005; Marker, 2002). Dispersing individuals, females with cubs, or "floaters" engage in wide-ranging behavior which may cause the boundaries of their ranges to temporarily expand (Laver, 2005).

Cheetah home range sizes have been estimated in many geographical ranges and land uses, from national parks to mixed use livestock and wildlife farmlands, conservancies and "predator-proof" game fenced farms. Because "predator-proof" fencing limits cheetah movements by preventing cheetahs from leaving the reserve, only home range sizes of free-ranging cheetahs, that is, those outside "predator-proofed" reserves, are reported here (Table 8.2). Cheetahs can have very large ranges, for example, in Namibia, the mean home range for territorial males, floaters, and females was 1713 km^2 (SD ± 1154; n = 41) (calculated from Marker et al. (2008a); Table 8.2). Extensive ranges have also been reported in Algeria (Belbachir et al., 2015) and Iran (Farhadinia et al., 2016). In the short grass plains of the Serengeti NP, mean home ranges were 454 km^2 (SD ± 379; n = 50; calculated from Caro (1994); Table 8.2), less than a third of those in Namibia. Large home ranges might be a product of low carrying capacity (e.g., low prey availability; Chapter 11, human-wildlife conflict; Chapter 13, and habitat perturbation, such as bush encroachment) (Belbachir et al., 2015; Farhadinia et al., 2016; Marker et al., 2008a; Muntifering et al., 2006). Research methods might influence inferred home range size, with direct observations possibly yielding smaller areas than those derived from VHF and GPS collars.

The mean home range overlap between all cheetahs (territorial and floater males, and females) on north-central Namibian farmland is 15.8% ± 17% (Marker et al., 2008a). However, in systems with changing prey availability due to migration, the degree of home range overlap between sexes varies. In the Serengeti NP, some female cheetahs follow the prey herds, while territorial males remain stationary and defend small territories in areas that are attractive to females (Caro, 1994; Durant et al., 1988). Overlapping ranges facilitate social interactions, gene flow, and reproduction.

At independence, individuals seek to establish their own home ranges; young females remain in proximity to their maternal range, whereas young males have been known to disperse up to 200 km from their natal range (Marker et al., 2008a). Each home range has a core area, defined as an area of intensive use (Samuel et al., 1985) or most concentrated ranging (Seaman and Powell, 1990). The average size of the core area of territorial males, floaters, and females in north-central Namibia (determined using 50% adaptive kernel; n = 41) was 13.9% (SD ± 5.3) of their home range size (Marker et al., 2008a) as compared to 10.4% (SD ± 0.26, 50% fixed kernel; n = 8) in Botswana (Houser et al., 2009). Female home range size is possibly influenced by factors, including avoiding males when not in estrous, avoiding areas where other cheetah or predators hunt, and the need to familiarize cubs with the home range (Bekoff et al., 1984). Home ranges and daily movement of female cheetahs increase as dependent cubs become older (Marker, 2002).

CHEETAH MOVEMENT

Animal movement patterns are an output of behavioral choices based on environmental cues, experience, and memory (Nathan et al., 2008; Schick et al., 2008). The cheetah shows substantial capacity for long-range movements,

TABLE 8.2 Cheetah Home Range Sizes in Different Study Areas

Region	Country	Site	Field data	Home range type	Solitary[a] male, km² (n)	Coalition[a] male, km² (n)	Female, km² (n)	Mean[b] ± SE[c], km² (n)	Source
Southern Africa	Botswana	Farmland	GPS/GSM collar	95% Fixed kernel	579 (2)	849 (1)	274 (2)	567 ± 166 (5)	Houser et al. (2009)
				95% MCP	316 (2)	598 (1)	286 (2)	400 ± 99 (5)	
	Namibia	Farmland	VHF radiocollar	95% Adaptive kernel	1490 (15)	1344 (11)	2160 (15)	1651 ± 1594 (41)	Marker et al. (2008b)
				95% MCP	1697 (15)	1608 (11)	1836 (15)	1836 ± 2010 (41)	
	South Africa	Kruger NP	VHF radiocollar	95% MCP	—	332 (2)	160 (2)	246 ± 86 (4)	Mills (1998)
		Kruger NP	VHF radiocollar	95% Fixed kernel	250 (1)	188 (1)	212 (2)	217 ± 18 (4)	Broomhall (2001)
		Kruger NP	VHF radiocollar	95% MCP	195 (1)	126 (1)	161 (2)	161 ± 20 (4)	Broomhall et al. (2003)
		Southern Kalahari	Random observations	95% MCP	—	125 (3)	320 (4)	223 ± 98 (7)	Mills (1998)
		Farmland	GPS/GSM collar	95% Local convex hull	405 (4)	146 (2)	300 (3)	284 ± 75 (9)	Marnewick and Somers (2015)
				95% MCP	1597 (4)	515 (2)	698 (3)	937 ± 334 (9)	
Eastern Africa	Tanzania	Serengeti plains	VHF radiocollar, direct observation	95% MCP	407 (31)[d]	—	833 (19)	549 ± 166 (50)	Caro (1994)
Western, Central, and Northern Africa	Algeria	Ahaggar Cultural Park	Camera trapping[e]	Mean max. dist. moved (MMDM)	—	—	—	1583	Belbachir et al. (2015)
Asia	Iran	Protected area	Camera trapping[e]	100% MCP	2475 (11)[d]	—	1090 (6)	2105 ± 806 (17)	Farhadinia et al. (2016)
		Protected area	GPS/GSM collar	95% MCP	—	1700 (1)	—	1700 (1)	Hunter (2011)

Region	Country	Site	Field data	Home range type	Territorial males, km² (n)	Floater male, km² (n)	Mean[f] ± SE[c], km² (n)	Source
Southern Africa	Namibia	Farmland	VHF radiocollar	95% Adaptive kernel	531 (10)	2300 (10)	1663 ± 983 (20)	Marker (2002) (reanalyzed data)
	Namibia	Farmland	GPS/VHF collars	95% Adaptive kernel	368 (26)	1595 (28)	981 ± 867 (54)	Melzheimer et al. (in preparation)
	Botswana	Farmland	GPS collars	95% Fixed kernel	526 (3)	3248 (1)	1115 ± 1932 (4)	Van der Weyde et al. (2016)
				95% MCP	730 (3)	2270 (1)	2065 ± 2089 (4)	
Eastern Africa	Tanzania	Serengeti NP	VHF radiocollar, direct observation	95% MCP	37 (22)	777 (9)	407 ± 523 (31)	Caro (1994)

NP, National Park.

[a] Includes territorial and nonterritorial (floater) males. For a detailed accounting of territorial behavior refer to Chapter 9.
[b] Mean was calculated based on average home range sizes for solitary males, coalition males, and females.
[c] Standard error is based on the overall mean values per social group in this table.
[d] Solitary and coalition males pooled.
[e] Home ranges are likely underestimated because of the small number of locations derived from camera trapping data.
[f] Mean was calculated based on average home range sizes for territorial and floater males.

especially during dispersal, when following migratory prey or when seeking breeding opportunities (Durant et al., 1988). In the Kalahari, cheetahs predominantly travel at a speed of 2.5–3.8 km/h (Scantlebury et al., 2014), covering on average 11 km in a 24 h period (Mills, 2015). Daily movement distances recorded in Botswana are considerably larger for males (mean = 6.13 km/24 h, SD = 0.60) than females (mean = 2.16 km/24 h, SD = 0.16) (Houser et al., 2009). Generally, as prey availability decreases, cheetahs cover larger distances to obtain access to suitable prey (Caro, 1994). For example, in Iran, where prey availability is generally low, individual cheetahs were detected in multiple reserves that were spatially separated by up to 217 km (Farhadinia et al., 2016).

In some parts of the cheetah's range, movements of cheetahs and other wildlife are increasingly impeded by habitat fragmentation due to anthropogenic barriers, such as "predator-proof" game fence lines, roads, and expanding human settlements (Hayward and Kerley, 2009; Moqanaki and Cushman, 2016; Chapter 10). Long-term persistence of cheetah populations in these areas will be reliant on dispersal as a distinct type of long-range movement. We suggest that investigating cheetah movement ecology, such as through combining high resolution movement data and direct observations (Grünewälder et al., 2012), will become critical for understanding ecological requirements, individual, and population limiting factors, as well as for informing strategic decisions for cheetah conservation.

CHEETAH ACTIVITY PATTERNS

The activity patterns of cheetahs vary regionally as they follow a time plan tactic (Eaton, 1970) aimed at either minimizing direct contact with their main competitors (e.g., lions and spotted hyenas), which are predominantly active during the night (Durant, 1998;

Hayward and Slotow, 2009) or reducing interference competition, for example, interactions with other species (Durant, 2000), including humans.

Diurnal behavior is common in areas where cheetahs coexist with large competitors (Bissett et al., 2015). For example, in the Okavango Delta, Botswana, 73% of the feeding events of cheetahs occurred during the day, 14% at dawn and dusk and 12% took place at night (Broekhuis et al., 2014). In the absence of their main competitors, for example, on much of the farmland in Namibia and Botswana or in areas where cheetahs are the top predators, for example, Algeria, cheetahs are predominantly active at night and early morning/late evening (Belbachir et al., 2015; Fabiano, 2013; Houser et al., 2009; Marker, 2002; McVittie, 1979; Nghikembua et al., 2016).

Nocturnal activity patterns are also favored in order to avoid humans (e.g., livestock herders and farmers) and in hyperarid ecosystems (e.g., in the Sahara Mountains), where it most likely serves as a behavioral adaptation to minimize water loss and exposure to the extreme hot dry daytime conditions (Belbachir et al., 2015; Bothma et al., 1999).

Additionally, the cheetahs' activity pattern is influenced by the lunar cycle and possibly supports the "starvation hypothesis," whereby cheetahs increase their activity during moonlit nights to maximize hunting/feeding opportunities because of better visibility to hunt prey (Broekhuis et al., 2014; Cozzi et al., 2012). This is suggested despite cheetahs exposing themselves to a higher risk of detrimental interactions with lions and spotted hyenas.

CONCLUSIONS

Human populations are likely to continue to modify the landscapes of Africa and Iran. They are likely to change wildlife habitats, increase the number of livestock, and fence more land with "predator-proof" fences. The pace of these modifications might be higher than the speed with which cheetahs can adjust to cope with them. It is therefore crucial to understand how these changes affect the spatial movements of the cheetahs, their habitat preferences, activity pattern, and interactions with other large carnivore species, and most importantly their ability to persist in the presence of high human densities. The long-term persistence of cheetahs will depend on a good knowledge of these aspects of the cheetah's ecology and behavior, and mitigating these effects may become the most important factors in cheetah conservation ecology. This information can be used to plan science-based conservation solutions.

References

Bauer, G.A., 1998. Cheetah—running blind. In: Penzhorn, B.L., (Ed.) Proceedings of a Symposium on Cheetahs as Game Ranch Animals, South African Veterinary Association, Onderstepoort. pp. 106–108.

Bekoff, M., Daniels, T.J., Gittleman, J.L., 1984. Life history patterns and the comparative social ecology of carnivores. Ann. Rev. Ecol. Syst. 15, 191–232.

Belbachir, F., Pettorelli, N., Wacher, T., Belbachir-Bazi, A., Durant, S.M., 2015. Monitoring rarity: the critically endangered Saharan cheetah as a flagship species for a threatened ecosystem. PLoS One 10, e0115136.

Berry, H.H., Kerley, G.I.H., 1981. Abnormal levels of disease and predation as limiting factors for wildebeest in the Etosha National Park. Madoqua 12, 242–253.

Bissett, C., Bernard, R.T.F., 2007. Habitat selection and feeding ecology of the cheetah (Acinonyx jubatus) in thicket vegetation: is the cheetah a savanna specialist? J. Zool. 271, 310–317.

Bissett, C., Parker, D.M., Bernard, R.T.F., Perry, T.W., 2015. Management-induced niche shift? The activity of cheetahs in the presence of lions. Afr. J. Wildl. Res. 45, 197–203.

Boast, L., Houser, A.M., Horgan, J., Reeves, H., Phale, P., Klein, R., 2016. Prey preferences of free-ranging cheetahs on farmland: scat analysis versus farmers' perceptions. Afr. J. Ecol. 54 (4), 424–433.

Bothma, J., du, P., Walker, C., 1999. Larger Carnivores of the African Savannas. Springer, Berlin.

Broekhuis, F., 2012. Niche Segregation by Cheetah (Acinonyx jubatus) as a Mechanism for Co-Existence With Lion (Panthera leo) and Spotted Hyaena (Crocuta crocuta). DPhil thesis, University of Oxford, United Kingdom.

Broekhuis, F., Cozzi, G., Valeix, M., McNutt, J.W., Macdonald, D.W., 2013. Risk avoidance in sympatric large carnivores: reactive or predictive? J. Anim. Ecol. 82, 1098–1105.

Broekhuis, F., Gopalaswamy, A.M., 2016. Counting cats: spatially explicit population estimates of cheetah (Acinonyx jubatus) using unstructured sampling data. PLoS One 11, e0153875.

Broekhuis, F., Grünewälder, S., McNutt, J.W., Macdonald, D.W., 2014. Optimal hunting conditions drive circalunar behavior of a diurnal carnivore. Behav. Ecol. 25, 1268–1275.

Broomhall, L.S., 2001. Cheetah Acinonyx jubatus in the Kruger National Park: A Comparison with Other Studies Across the Grassland-Woodland Gradient in African Savannas. MSc thesis, University of Pretoria, South Africa.

Broomhall, L.S., Mills, M.G.L., du Toit, J.T., 2003. Home range and habitat use by cheetahs (Acinonyx jubatus) in the Kruger National Park. J. Zool. 261, 119–128.

Burney, D.A., 1980. The Effects of Human Activities on Cheetah (Acinonyx jubatus) in the Mara Region of Kenya. MSc thesis, University of Nairobi, Kenya.

Burt, H.B., 1943. Territoriality and home range concepts as applied to mammals. J. Mammal. 24, 346–352.

Caro, T.M., 1994. Cheetahs of the Serengeti Plains: Group Living in an Asocial Species. University of Chicago Press, Chicago and London.

Caro, T.M., Collins, D.A., 1987. Male cheetah social organisation and territoriality. Ethology 74, 52–64.

Caro, T.M., Fitzgibbon, C.D., Holt, M.E., 1989. Physiological costs of behavioural strategies for male cheetahs. Anim. Behav. 38, 309–317.

Claro, F., Leriche, C., van Syckle, S.J., Rabeil, T., Hergueta, S., Fournier, A., Alou, M., 2006. Survey of the Cheetah in W National Park and Tamou Fauna Reserve, Niger. Cat News 45, 4–7.

Claro, F., Sissler, C., 2003. Saharan cheetahs in the Termit region of Niger. Cat News 38, 23–24.

Clements, H.S., Tambling, C.J., Hayward, M.W., Kerley, G.I.H., 2014. An objective approach to determining the weight ranges of prey preferred by and accessible to the five large African carnivores. PLoS One 9, e101054.

Craig, C.A., Brassine, E.I., Parker, D.M., 2017. A record of cheetah (Acinonyx jubatus) diet in the Northern Tuli Game Reserve, Botswana. African Journal of Ecology. doi: 10.1111/aje.12374.

Cooper, A.B., Pettorelli, N., Durant, S.M., 2007. Large carnivore menus: factors affecting hunting decisions by cheetahs in the Serengeti. Anim. Behav. 73, 651–659.

Cozzi, G., Broekhuis, F., McNutt, J.W., Turnbull, L.A., Macdonald, D.W., Schmid, B., 2012. Fear of the dark or dinner by moonlight? Reduced temporal partitioning among Africa's large carnivores. Ecology 93, 2590–2599.

Cristescu, B., 2006. Space Use and Diet of Selected Large Carnivores in a South African Game Reserve. MSc thesis, University of Leeds, United Kingdom.

Durant, S., 1998. Competition refuges and coexistence: an example from Serengeti carnivores. J. Anim. Ecol. 67, 81–92.

Durant, S.M., 2000. Living with the enemy: avoidance of hyenas and lions by cheetahs in the Serengeti. Behav. Ecol. 11, 624–632.

Durant, S., 2007. Range-wide conservation planning for cheetah and wild dog. Cat News 46, 13.

Durant, S.M., Bashir, S., Maddox, T., Laurenson, M.K., 2007. Relating long-term studies to conservation practice: the case of the Serengeti Cheetah Project. Conserv. Biol. 21, 602–611.

Durant, S.M., Caro, T.M., Collins, D.A., Alawi, R.M., Fitzgibbon, C.D., 1988. Migration patterns of Thomson's gazelles and cheetahs on the Serengeti Plains. Afr. J. Ecol. 26, 257–268.

Durant, S.M., Craft, M.E., Hilborn, R., Bashir, S., Hando, J., Thomas, L., 2011. Long-term trends in carnivore abundance using distance sampling in Serengeti National Park, Tanzania. J. Appl. Ecol. 48, 1490–1500.

Durant, S., Mitchell, N., Ipavec, A., Groom, R., 2015. *Acinonyx jubatus*. The IUCN Red List of Threatened Species 2015: e.T219A50649567. Available from: http://dx.doi.org/10.2305/IUCN.UK.2015-4.RLTS.T219A50649567.en.

Eaton, R.L., 1970. Group interactions, spacing and territoriality in cheetahs. J. Tierpsychol. 27, 481–491.

Eaton, R.L., 1974. The Cheetah: Biology, Ecology and Behavior of an Endangered Species. Van Nostrand Reinholt, New York.

Fabiano, E.C., 2013. Demografia histórica e contemporânea de guepardos (*Acinonyx jubatus*) na Namíbia, África Austral. Tese de Doutorado, Pontifícia Universidade Católica do Rio Grande do Sul.

Farhadinia, M.S., Gholikhani, N., Behnoud, P., Hobeali, K., Taktehrani, A., Hosseini-Zavarei, F., Eslami, M., Hunter, L.T.B., 2016. Wandering the barren deserts of Iran: illuminating high mobility of the Asiatic cheetah with sparse data. J. Arid Environ. 134, 145–149.

Farhadinia, M.S., Hemami, M.-R., 2010. Prey selection by the critically endangered Asiatic cheetah in central Iran. J. Nat. Hist. 44, 1239–1249.

Farhadinia, M.S., Hosseini-Zavarei, F., Nezami, B., Harati, H., Absalan, H., Fabiano, E., Marker, L., 2012. Feeding ecology of the Asiatic cheetah *Acinonyx jubatus venaticus* in low prey habitats in northeastern Iran: implications for effective conservation. J. Arid Environ. 87, 1–6.

Farhadinia, M.S., Jourabchian, A., Eslami, M., Hosseini, F., Nezami, B., 2008. Is food availability a reliable indicator of cheetah presence in Iran? Cat News 49, 14–18.

Fitzgibbon, C.D., Fanshawe, J.H., 1989. The condition and age of Thomson Gazelles killed by Cheetahs and wild dogs. J. Zool. 218, 99–107.

Frame, G.W., 1986. Carnivore Competition and Range Use in the Serengeti Ecosystem in Tanzania. PhD thesis, Utah State University, United States.

Graham, A., 1966. East African wildlife society cheetah survey: extracts from the report by wildlife services. East Afr. Wildl. J. 4, 50–55.

Gros, P.M., 1998. Status of the cheetah *Acinonyx jubatus* in Kenya: a field-interview assessment. Biol. Conserv. 85, 137–149.

Grünewälder, S., Broekhuis, F., Macdonald, D.W., Wilson, A.M., McNutt, J.W., Shawe-Taylor, J., Hailes, S., 2012. Movement activity based classification of animal behaviour with an application to data from cheetah (*Acinonyx jubatus*). PLoS One 7, e49120.

Hamilton, P.H., 1986. Status of the cheetah in Kenya, with reference to sub-Saharan Africa. In: Miller, S.D., Everett, D.D. (Eds.), Cats of the World: Biology, Conservation and Management. National Wildlife Federation, Washington DC.

Hayward, M.W., Hofmeyr, M., O'Brien, J., Kerley, G.I.H., 2006. Prey preferences of the cheetah (*Acinonyx jubatus*) (Felidae: Carnivora): morphological limitations or the need to capture rapidly consumable prey before kleptoparasites arrive? J. Zool. 270, 615–627.

Hayward, M.W., Kerley, G.I.H., 2009. Fencing for conservation: restriction of evolutionary potential or a riposte to threatening processes? Biol. Conserv. 142, 1–13.

Hayward, M.W., Slotow, R., 2009. Temporal partitioning of activity in large African carnivores: tests of multiple hypotheses. S. Afr. J. Wildl. Res. 39, 109–125.

Hirst, S., 1969. Populations in a transvaal lowveld nature reserve. Zool. Afr. 4, 199–230.

Hofmeyr, M., van Dyk, G., 1998. Cheetah introductions to two north west parks: case studies from Pilanesberg National Park and Madikwe Game Reserve. In: Proceedings of a Symposium on Cheetahs as Game Ranch Animals, Onderstepoort, 23–24 October 1998, p. 71.

Houser, A.M., Somers, M.J., Boast, L.K., 2009. Home range use of free-ranging cheetah on farm and conservation land in Botswana. S. Afr. J. Wildl. Res. 39, 11–22.

Hunter, L.T.B., 1998. The Behavioural Ecology of Reintroduced Lions and Cheetahs in the Phinda Resource Reserve, Kwazulu-Natal, South Africa. PhD thesis, University of Pretoria, South Africa.

Hunter, L., 2011. Carnivores of the World. Princeton University Press, Princeton.

Hunter, J.S., Durant, S.M., Caro, T.M., 2007. To flee or not to flee: predator avoidance by cheetahs at kills. Behav. Ecol. Sociobiol. 61, 1033–1042.

Kruuk, H., Turner, M., 1967. Comparative notes on predation by lion, leopard, cheetah and wild dog in the Serengeti area, east Africa. Mammalia 31, 1–27.

Labuschagne, W., 1979. 'n Bio-Ekologiese en Gedragstudie Van de Jagluiperd *Acinonyx jubatus* jubatus (Schreber 1775). MSc thesis, University of Pretoria, South Africa.

Laurenson, M.K., 1994. High juvenile mortality in cheetahs (*Acinonyx jubatus*) and its consequences for maternal care. J. Zool. 234, 387–408.

Laurenson, M.K., Wielebnowski, N., Caro, T.M., 1995. Extrinsic factors and juvenile mortality in cheetahs. Conserv. Biol. 9, 1329–1331.

Laver, P.N., 2005. Cheetah of the Serengeti Plains: A Home Range Analysis. MSc thesis, Virginia Tech, United States.

Low, A.B., Rebelo, A.G. (Eds.), 1996. Vegetation of South Africa, Lesotho and Swaziland. Department of Environmental Affairs and Tourism, Pretoria.

Marnewick, K., Somers, M.J., 2015. Home ranges of cheetahs (*Acinonyx jubatus*) outside protected areas in South Africa. Afr. J. Wildl. Res. 45, 223–232.

Marker, L.L., 2002. Aspects of Cheetah (*Acinonyx jubatus*) Biology, Ecology and Conservation Strategies on Namibian Farmlands. PhD thesis, University of Oxford, United Kingdom.

Marker, L.L., Dickman, A.J., Mills, M.G.L., Jeo, R.M., Macdonald, D.W., 2008b. Spatial ecology of cheetahs (*Acinonyx jubatus*) on north-central Namibian farmlands. J. Zool. 274, 226–238.

Marker, L., Dickman, A., Wilkinson, C., Schumann, B., Fabiano, E., 2007. The Namibian cheetah: status report. Cat News Special Issue 3, pp. 4–13.

Marker, L., Fabiano, E., Nghikembua, M., 2008a. The use of remote camera traps to estimate density of free-ranging cheetahs in north-central Namibia. Cat News (49), 22–24.

Marker, L., Mills, M.G.L., Macdonald, D.W., 2003b. Factors influencing perceptions and tolerance towards cheetahs on Namibian farmlands. Conserv. Biol. 17 (5), 1–9.

Marker, L.L., Muntifering, J.R., Dickman, A.J., Mills, M.G.L., Macdonald, D.W., 2003a. Quantifying prey preferences of free-ranging Namibian cheetahs. S. Afr. J. Wildl. Res. 33, 43–53.

Marker-Kraus, L., Kraus, D., Barnett, D., Hurlbut, S., 1996. Cheetah Survival on Namibian Farmlands. Cheetah Conservation Fund, Windhoek.

Matchano, A., 2011. Mise en œuvre d'une stratégie dans le cadre du conflit homme/carnivore dans le massif de Termit et ses environs. Carnivore Project, Sahara Conservation Fund, Niameyp. 32, (in Rabeil & Comizzoli pers comm).

McLaughlin, R., 1970. Aspects of the Biology of the Cheetah (*Acinonyx jubatus*, Schreber) in Nairobi National Park. MSc thesis, University of Nairobi, Kenya.

McVittie, R., 1979. Changes in the social behaviour of South West African cheetah. Madoqua 11, 171–184.

Mills, M.G.L., 1984. Prey selection and feeding habits of large carnivores in the southern Kalahari. Koedoe Suppl. 27, 281–294.

Mills, M.G.L., 1986. Cheetah ecology and behaviour in East and South Africa. In: Penzhorn, B.L. (Ed.), Symposium on Cheetahs as Game Ranch, Animals. Onderstepoort, South Africa, pp. 18–22.

Mills, M.G.L., 2015. Living near the edge: a review of the ecological relationships between large carnivores in the arid Kalahari. Afr. J. Wildl. Res. 45, 127–137.

Mills, M.G.L., Biggs, H.C., 1993. Prey apportionment and related ecological relationships between large carnivores in Kruger National Park. Symp. Zool. Soc. Lond. 65, 253–268.

Mills, M.G.L., Broomhall, L.S., du Toit, J.T., 2004. Cheetah *Acinonyx jubatus* feeding ecology in the Kruger National Park and a comparison across African savanna habitats: is the cheetah only a successful hunter on open grassland plains? Wildl. Biol. 10, 177–186.

Mills, M.G.L., Mills, M.E.J., 2014. Cheetah cub survival revisited: a re-evaluation of the role of predation, especially by lions, and implications for conservation. J. Zool. 292, 136–141.

Mills, M.G.L., Mills, M.E.J., Edwards, C.T., Gottelli, D., Scantelbury, D.M., 2017. Kalahari cheetahs: adaptations to an arid region. Oxford University Press, Oxford.

Mitchell, B.L., Shenton, J.B., Uys, J.C.M., 1965. Predation on large mammals in the Kafue national park. Zambia. Zool. Afr. 1, 297–318.

Morsbach, D., 1987. Cheetah in Namibia. Cat News 6, 25–26.

Moqanaki, E.M., Cushman, S.A., 2016. All roads lead to Iran: predicting landscape connectivity of the last stronghold for the critically endangered Asiatic cheetah. Anim. Conserv. 20, 29–41.

Muntifering, J.R., Dickman, A.J., Perlow, L.M., Hruska, T., Ryan, P.G., Marker, L.L., Jeo, R.M., 2006. Managing the matrix for large carnivores: a novel approach and perspective from cheetah (*Acinonyx jubatus*) habitat suitability modelling. Anim. Conserv. 9, 103–112.

Mutoro, N.M., 2015. Assessment of Cheetah Prey Base Outside Protected Areas in Salama and Kapiti Plains of Southern Kenya. MSc thesis, University of Nairobi, Kenya.

Nathan, R., Getz, W.M., Revilla, E., Holyoak, M., Kadmon, R., Saltz, D., Smouse, P.E., 2008. A movement ecology paradigm for unifying organismal movement research. Proc. Natl. Acad. Sci. USA 105, 19052–19059.

Nghikembua, M., Harris, J., Tregenza, T., Marker, L., 2016. Spatial and temporal habitat use by GPS collared male cheetahs in modified bushland habitat. Open J. For. 6, 269–280.

Nowell, K., Jackson, P., 1996. Wild Cats. Status Survey and Conservation Action Plan. IUCN/SSC Cat Specialist Group, Gland and Cambridge.

O'Brien, J., 2012. The Ecology and Management of the Large Carnivore Guild on Shamwari Game Reserve, Eastern Cape. PhD thesis, Rhodes University, South Africa.

Olaotswe, E., Kgosikoma, I., Kabo Mpgots, I., 2013. Understanding the causes of bush encroachment in Africa: The key to effective management of savanna grasslands. Trop. Grasslands 1, 215–219.

Owen-Smith, N., Mills, M.G.L., 2008. Predator–prey size relationships in an African large-mammal food web. J. Anim. Ecol. 77, 173–183.

Pienaar, U.D.V., 1969. Predator-prey relationships amongst the larger mammals of the Kruger National Park. Koedoe 12, 109–176.

Poche, R., 1973. Niger's threatened Park W. Oryx 12, 216–222.

Purchase, G.K., du Toit, J.T., 2000. The use of space by cheetahs in Matusadona National Park, Zimbabwe. S. Afr. J. Wildl. Res. 30, 139–144.

Radloff, F.G.T., Du Toit, J.T., 2004. Large predators and their prey in a southern African savanna: a predator's size determines its prey size range. J. Anim. Ecol. 73, 410–423.

Rostro-García, S., Kamler, J.F., Hunter, L.T.B., 2015. To kill, stay or flee: the effects of lions and landscape factors on habitat and kill site selection of cheetahs in South Africa. PLoS One 10, e0117743.

Ruggiero, R.G., 1991. Prey selection of the lion (Panthera leo L.) in the Manovo-Gounda-St. Floris National Park, Central African Republic. Mammalia 55, 23–33.

Saleh, M.A., Helmy, I., Giegengack, R., 2001. The cheetah, *Acinonyx jubatus* (Schreber, 1776) in Egypt (Felidae, Acinonychinae). Mammalia 65, 177–194.

Samuel, M.D., Pierce, D.J., Garton, E.O., 1985. Identifying areas of concentrated use within the home range. J. Anim. Ecol. 54, 711–719.

Sarhangzadeh, J., Akbari, H., Esfandabad, B.S., 2015. Ecological niche of the Asiatic Cheetah (*Acinonyx jubatus venaticus*) in the arid environment of Iran (Mammalia: Felidae). Zool. Middle East 61, 109–117.

Scantlebury, D.M., Mills, M.G.L., Wilson, R.P., Wilson, J.W., Mills, M.E.J., Durant, S.M., Bennett, N.C., Bradford, P., Marks, N.J., Speakman, J.R., 2014. Flexible energetics of cheetah hunting strategies provide resistance against kleptoparasitism. Science 346, 79–81.

Schaller, G.B., 1968. Hunting behaviour of the Cheetah in the Serengeti National Park, Tanzania. East Afr. Wildl. J. 6, 95–100.

Schaller, G.B., 1972. The Serengeti Lion: A Study of Predator-Prey Relations. University of Chicago Press, Chicago.

Schick, R.S., Loarie, S.R., Colchero, F., Best, B.D., Boustany, A., Conde, D.A., Halpin, P.N., Joppa, L.N., McClellan, C.M., Clark, J.S., 2008. Understanding movement data and movement processes: current and emerging directions. Ecol. Lett. 11, 1338–1350.

Seaman, D.E., Powell, R.A., 1990. Identifying patterns and intensity of home range use. Inte. Conf. Bear Res. Manage. 8, 243–249.

Sogbohossou, E.A., de Iongh, H.H., Sinsin, B., de Snoo, G.R., Funston, P.J., 2011. Human–carnivore conflict around Pendjari Biosphere Reserve, northern Benin. Oryx 45, 569–578.

Stander, P.E., 1990. Notes on foraging habits of cheetah. S. Afr. J. Wildl. Res. 20, 130–132.

Swanson, A., Caro, T., Davies-Mostert, H., Mills, M.G., Macdonald, D.W., Borner, M., Masenga, E., Packer, C., 2014. Cheetahs and wild dogs show contrasting patterns of suppression by lions. J. Anim. Ecol. 83, 1418–1427.

Tambling, C.J., Wilson, J.W., Bradford, P., Scantlebury, M., 2014. Fine-scale differences in predicted and observed cheetah diet: does sexual dimorphism matter? S. Afr. J. Wildl. Res. 44, 90–94.

Terborgh, J., Estes, J.A., Paquet, P., Ralls, K., Boyd-Heger, D., Miller, B.J., Noss, R.F., 1999. The role of top carnivores in regulating terrestrial ecosystems. In: Soule, M.E., Terborgh, J. (Eds.), Continental Conservation: Scientific Foundations of Regional Reserve Networks. Island Press, Washington, pp. 39–64.

Van der Weyde, L.K., Hubel, T.Y., Horgan, J., Shotton, J., McKenna, R., Wilson, A.M., 2016. Movement patterns of cheetahs (*Acinonyx jubatus*) in farmlands in Botswana. Biol. Open 6, 118–124.

Vorster, P.H., 2011. The Feeding and Spatial Ecology of Cheetahs (*Acinonyx jubatus*) and Lions (*Panthera leo*) in the Little Karoo, South Africa. MSc thesis, Rhodes University, South Africa.

Wacher, T., De Smet, K., Belbachir, F., Belbachir-Bazi, A., Fellous, A., Belghoul, M., Marker, L., 2005. Sahelo-Saharan Interest Group Wildlife Surveys, Central Ahaggar Mountains (March 2005). Sahelo Saharan Interest Group.

Wachter, B., Jauernig, O., Breitenmoser, U., 2006. Determination of prey hair in faeces of free-ranging Namibian cheetahs with a simple method. Cat News 44, 8–9.

Wachter, B., Thalwitzer, S., Hofer, H., Lonzer, J., Hildebrandt, T.B., Hermes, R., 2011. Reproductive history and absence of predators are important determinants of reproductive fitness: the cheetah controversy revisited. Conserv. Lett. 4, 47–54.

Young, T.P., Evans, M.R., 1993. Alpine vertebrates of Mount Kenya, with particular notes on the rock hyrax. J. East Afr. Nat. Hist. Soc. Natl. Mus. 82, 55–79.

Behavior and Communication of Free-Ranging Cheetahs

Bettina Wachter*, Femke Broekhuis**,†,
Joerg Melzheimer*, Jane Horgan‡, Elena V. Chelysheva§,
Laurie Marker¶, Gus Mills†,††, Tim Caro‡‡

*Leibniz Institute for Zoo and Wildlife Research, Berlin, Germany
**Mara Cheetah Project, Kenya Wildlife Trust, Nairobi, Kenya
†University of Oxford, Tubney, Abingdon, United Kingdom
‡Cheetah Conservation Botswana, Maun, Botswana
§Mara-Meru Cheetah Project, Kenya Wildlife Services, Nairobi, Kenya
¶Cheetah Conservation Fund, Otjiwarongo, Namibia
††The Lewis Foundation, Johannesburg, South Africa
‡‡University of California Davis, Davis, CA, United States

INTRODUCTION

The study of the behavior of free-ranging cheetahs (*Acinonyx jubatus*) can provide valuable insights into the characteristics of life of this elusive species. Although the cheetah is a well-studied species, research is still discovering new information about their behavior. Data can be collected using direct observations in the field and/or indirect data collection with the help of technology, such as GPS collars (Chapter 32) or motion-triggered cameras (Chapter 29). Direct observation requires the habituation of the animals to the observer and therefore is predominantly limited to protected areas where cheetahs receive greater protection from human-mediated threats. This chapter presents the behavior of free-ranging cheetahs in the context of reproduction, social interactions, hunting, feeding, and communication.

BREEDING BEHAVIOR

In terrestrial mammal species, several mating systems have been described. Most species form monogamous pairs, harems, multimale–multifemale groups, males gathering in leks, or males roaming solitarily in search of solitary females (Clutton-Brock, 2016). In felids, males typically range over large areas that encompass several female ranges, allowing them access to these females when they are receptive (Sandell, 1989). Cheetahs do not fit in any of these mammalian mating systems, and one female has access to several males. Cub survival and reproductive success of cheetah females are highly variable.

Mating System and Behavior

Male cheetahs are either territory holders defending small territories or "floaters" roaming in large, overlapping, undefended home ranges. In the Serengeti National Park (NP) in Tanzania, the territories are on an average 48 km^2 (SD = 23, n = 9) and the home ranges of floaters are 777 km^2 (SD = 459, n = 9) in size (Caro, 1994), whereas in east-central Namibia, the territories are 379 km^2 (SD = 161, n = 28) and the home ranges of floaters are 1595 km^2 (SD = 1131, n = 28) in size (Melzheimer et al., in preparation). Both territory holders and floaters are either solitary or live in coalitions of two to three, or rarely four males (Caro, 1994; Frame, 1984). Females roam alone or with their dependent offspring in large overlapping home ranges (Serengeti NP: 833 km^2; SD = 370, n = 19; Caro, 1994, east-central Namibia: 857 km^2; SD = 760, n = 20; Melzheimer et al., in preparation, north-central Namibia: 1836 km^2; SD = 2010, n = 15; Marker et al., 2008), primarily overlapping with related individuals (Chapter 8). Female home ranges encompass several male territories and overlap with the home ranges of floaters, therefore allowing them access to several males (Caro, 1994).

When males encounter a female in their territory, they attempt to limit her movements (see later). A single male does not necessarily monopolize mating during estrus. In the Serengeti NP and the Kgalagadi Transfrontier Park (TP) in Botswana/South Africa, multiple paternity occurred in 43% (n = 23) and 29% (n = 24) of litters, respectively, and cubs did not necessarily originate from a single coalition of males, but could also result from males of another territory (Gottelli et al., 2007; Mills et al., 2017). Females may gain fitness benefits for their offspring by mating with multiple males.

Female adoption of abandoned cubs also occurs. Genetic analysis confirmed that cubs from three supposed adoptions were genetically unrelated to the adult female they were with (Gottelli et al., 2007). Adoptions might result in increased long-term survival probability for male offspring. For example, adolescent males have a higher chance of survival when they grow up with a sister because they rely strongly on their sisters to catch prey for the group (Caro, 1994; Durant et al., 2004). Additionally, adoptions of male cubs might increase the size of a future male coalition and thus the chances of taking over a territory (Caro, 1994).

Mating behavior of cheetahs in the wild is only poorly described. The breeding animals stay together for 2 or 3 days and mate several times with gaps of up to 8 h between mating events (Frame and Frame, 1981). The duration of a reproductive cycle of a captive female cheetah ranges between 5 and 30 days and once mating has been successful the gestation period lasts approximately 92 days (Chapter 27). Females in the Serengeti NP have their first litter when they are on average 2.4-years old (SD = 3.1 months, n = 22, Kelly et al., 1998) and in the Kgalagadi TP when they are on an average 3.1-years old (SD = 3.4 months, n = 5, Mills et al., 2017). Males in captivity are physiologically capable of breeding at less than 2 years (Crosier et al., 2007; Wildt et al., 1993), however, breeding in the wild is likely to be delayed until a male becomes territorial (Caro, 1994; Marker et al., 2003).

Breeding Months

Cheetahs do not have a particular breeding season but give birth throughout the year. However, there are some local differences. In the Serengeti NP, a similar number of litters are born in the wet and the dry season (Laurenson et al., 1992).

On farmland in north-central Namibia, female cheetahs have several birth peaks distributed throughout the year, that is, in the rainy season in February and March, in the cold-dry season in June and July, and in the hot-dry season in October and November (Marker et al., 2003; Marker-Kraus et al., 1996). In the Kgalagadi TP, few cubs are born from October to December, which is often the driest time of the year (Mills et al., 2017). The finding that cheetahs do not have a systematically defined breeding season is consistent with females quickly resuming cycling after the loss of a litter, which can happen at any time of the year. On average, females in the Serengeti NP conceive again 17.8 days (SD = 13.5, n = 9) after losing the previous litter (Laurenson et al., 1992). Interbirth intervals between successfully reared litters were on an average 20.1 months (SD = 3.0, n = 36) in the Serengeti NP (Kelly et al., 1998) and 24 months (SD = 3.0, n = 6) in north-central Namibia (Marker et al., 2003).

Reproductive Success of Females

Long-term studies in the Serengeti NP have identified several factors that affect yearly and lifetime female reproductive success in cheetahs. Females gave birth to litters with 2 to 6 cubs with a mean of 3.5 cubs (n = 25, Caro, 1994). This mean might be a slight underestimate because litters were not examined at birth but on average 15 days after the estimated day of birth (range 6–35 days) (Caro, 1994). On average, females in the Serengeti NP raised 1.7 cubs (SD = 2.4, n = 108 females) to independence in their lifetime, and mean annual reproductive success was 0.36 cubs (SD = 0.51, n = 103 females) reaching independence (Kelly et al., 1998). Annual reproductive success increased with the age of the mothers until they were 7-years old and then decreased again (Pettorelli and Durant, 2007). Thus, lifetime reproductive success increased with longevity in a nonlinear relationship, and leveled off at 11 years of age in the Serengeti NP (Pettorelli and Durant, 2007). The increased number of reproductive opportunities associated with a long life outcompeted the advantage of high annual reproductive success. Females with a high output early in life died younger, suggesting a trade-off between annual reproductive success and longevity (Pettorelli and Durant, 2007). Variance in reproductive success was considerable in the Serengeti NP; most females (61% of n = 108) raised 0 or 1 cub during their lifetime, whereas a few females (1.9%) raised 9 or 10 cubs during their lifetime (Kelly et al., 1998). In the Kgalagadi TP, the variance in female reproductive success was lower. The ratio of females that raised ≤5 cubs to those that raised ≥6 cubs was 3.0:1 and thus significantly lower than the ratio of 12.5:1 in the Serengeti NP (Mills et al., 2017).

Females that have higher reproductive success were found in areas with lower densities of lions (*Panthera leo*) and spotted hyenas (*Crocuta crocuta*), than less successful females (Durant, 2000). Cheetahs regularly avoid lions and spotted hyenas spatially and temporally (Durant, 1998). Annual and lifetime reproductive success of females in the Serengeti NP, as well as litter size at independence, was higher when lion abundance was low (from 1969 to 1979) compared to when lion population was high (from 1980 to 1994) (Kelly et al., 1998). However, on a population level, lion density does not inflict sufficient demographic consequences on cheetahs to have a negative effect on population size (Swanson et al., 2014).

Cub Survival

The main factor determining cub survival during the first 2 months of life, when cubs are hidden in the lair, is predation. Although mothers are very vigilant within 1 km of the lair (Caro, 1994) and adopt several predator avoidance behaviors at the lair, such as rarely sitting up while nursing the cubs, returning to the lair at night and being silent and cryptic, predators sometimes detect lairs (Caro, 1994; Laurenson, 1994). In case a predator detects the lair and the mother is present, the mother will try to drive it away or

TABLE 9.1   Number of Litters and Cubs Observed in the Lair, Upon Emergence From the Lair, at 4 Months of Age and at 14 Months of Age in Three Study Sites

	Serengeti NP (Laurenson, 1994)[a]				Kgalagadi TP (Mills et al., 2017)				Farmland in east-central Namibia (Wachter et al., 2011)			
	Litters		Cubs		Litters		Cubs		Litters		Cubs	
	#	%	#	%	#	%	#	%	#	%	#	%
In lair	36		125		25		83		n.a.		n.a.	
Survived to emergence	10	27.8	36	28.8	13	52.0	41	49.4	3	n.a.	14	n.a.
Survived to 4 months	5–6	15.3[b] 55.0[c]	10–12	8.8[b] 30.6[c]	11	44.0[b] 84.6[c]	29	34.9[b] 70.7[c]	3	n.a. 100.0[c]	11	n.a. 78.6[c]
Survived to 14 months	3–4	9.7[b] 35.0[c]	5–7	4.8[b] 16.7[c]	11	44.0[b] 84.4[c]	26	31.3[b] 63.4[c]	3	n.a. 100.0[c]	11	n.a. 78.6[c]

n.a., Not applicable; NP, National Park; TP, Transfrontier Park.

[a] Percentages of [b] and [c] are based on the mean of the minimum and maximum number of litters and cubs, respectively.

[b] Percentage compared with data in the lair.

[c] Percentage compared with data at emergence.

threaten it; however sometimes unsuccessfully (Laurenson, 1994). Of 12 litters in the Serengeti NP for which the cause of death was definitely or probably known, 7 entire litters (58.3%) were killed by predators, and an additional litter was thought to have died because the mother was killed by predators (Laurenson, 1994). Other causes of entire litter death were starvation following abandonment by the mother (16.7%), fire (8.3%), and pneumonia infection following exposure to bad weather (8.3%). Of those cubs emerging from the lair (n = 36), only 16.7% reached the age of 14-months old, that is, an age a few months before independence when offsprings are still with their mother and survival can be determined (Laurenson, 1994) (Table 9.1).

In the Kgalagadi TP, an ecosystem with only 26.3% of the lion density and 0.9% of the spotted hyena density of the Serengeti NP, survival of cubs observed in the lair to 14 months of age was 69.9%, 3.8 times higher than that recorded in the Serengeti NP (Table 9.1). Nevertheless, the primary cause of mortality in the lair was still predation, with 88.9% of cub deaths (n = 31) being attributed to predators. The predator species was not always identified, but leopards (*Panthera pardus*) and probably also smaller predator species, such as honey badgers (*Mellivora capensis*) and black-backed jackals (*Canis mesomelas*) contributed to cub deaths (Mills and Mills, 2013).

On farmland in east-central Namibia where there are no lions and spotted hyenas, cheetah cub survival after emergence from the lair was 4.7 times and 1.2 times higher than in the Serengeti NP and in the Kgalagadi TP, respectively (Wachter et al., 2011) (Table 9.1). Thus, the presence of large carnivore species affects cheetah cub survival.

PARENT-OFFSPRING BEHAVIOR

Evolutionary theory predicts that there will be a conflict between giving and receiving parental care and that the benefit will always be greater for offspring than for parents (Trivers, 1974). In cheetahs, males do not participate in raising offspring, thus parental care falls solely to females. The study of mother-offspring behavior can therefore reveal valuable insights into life history decisions of females, such as time allocated to weaning, vigilance, and predator defense, teaching hunting skills, and the timing of separation from the offspring.

FIGURE 9.1 **Cubs in the lair in two differing habitats.** (A) Cubs 1 week of age in the Kgalagadi Transfrontier Park, Southern Kalahari, (B) cubs 3 weeks of age in the Maasai Mara, Kenya. *Source: Part A, Gus Mills; part B, Elena Chelysheva.*

Weaning

Females in the Serengeti NP hide their cubs in lairs (Fig. 9.1) for approximately 2 months and move them on an average every 5.6 days to new lairs (Laurenson, 1993). During these first 2 months, mothers nurse the cubs mainly during the early morning hours between 0630 and 0930 (Caro, 1994). At 2 months old, the cubs emerge from the lair and are usually nursed between 0630 and 0800 and again an hour during midday. At that age, they are also introduced to solid food (Caro, 1994). First, the cubs approach and sniff the carcass but retreat with fear. After a couple of exposures, they lose their fear, become interested and start feeding. At the age of 3 months, the cubs spend more time during the day eating solid food than nursing. When cubs are 4-months old, the mother stops nursing them by covering its nipples with the hind-legs, rolling on its belly or sitting up (Caro, 1994; Mills et al., 2017). Once the cubs have left the lair, they follow the mother and feed and rest where she does.

Predator Defense

Mothers with young cubs that have left the lair increase their vigilance and notice predators at greater distances than mothers with cubs at the lair or with old cubs (8.5–14.0 months of age, Caro, 1994). Mothers are also more likely to behave aggressively toward predators and large herbivores when their cubs are young and vulnerable to predation. This behavior includes stalking, chasing, slapping, or attempting to bite the predator or large herbivore and decreases when cubs become older, probably because cubs become able to outrun predators (Laurenson, 1994). Once cubs are nearly fully grown, mothers can benefit from having large, almost independent cubs because they are, as a group, more vigilant and therefore less likely to be approached unnoticed by male cheetahs and other predator species (Caro, 1994).

Play

For carnivore species, it has been suggested that play behavior serves as a training for hunting (Caro, 1987, 1995). Four categories of play behavior have been identified: (1) locomotor play, which is mainly observed in very young cubs and might train their abilities to escape predation, (2) noncontact social play, which involves crouching and stalking family members, (3) contact social play, and (4) object play (Caro, 1995). The latter two categories (Fig. 9.2) are mostly seen in older cubs and include the mother capturing and releasing live prey for the cubs to train their hunting skills (Caro and

FIGURE 9.2　(A) Contact social play, (B) object play with a Thomson's gazelle fawn. *Source: Elena Chelysheva.*

Hauser, 1992). Cubs that often crouched and stalked family members also frequently did so toward prey animals; similarly, cubs that often showed contact and object play, frequently tapped and played with live prey provided by the mother (Caro, 1995).

Independence

In the Serengeti NP, cheetah females leave their cubs (i.e., cubs become independent) when the cubs are on average 17.1-months old (SD = 1.9, n = 70 litters, Kelly et al., 1998) and in the Kgalagadi TP, when they are on an average 18.9-months old (SD = 2.0, n = 8 litters, Mills et al., 2017). In the Serengeti NP and Kgalagadi TP, females were pregnant in 44% of 36 cases and 60% of 11 cases, respectively, when they left the cubs (Kelly et al., 1998; Mills et al., 2017).

After separation from their mother, sibling groups stay together for a few more months before females and males separate (Caro, 1994). Females might stay together for another few weeks before they split from each other and establish separate, but largely overlapping home ranges close to their natal range (Caro, 1994; Frame 1984; Marker et al., 2008). Males usually disperse and settle away from their natal area; in the Serengeti NP they disperse approximately 20 km from the natal range (Frame, 1984), while in Botswana and Namibia, males have been captured 200 km away from their natal range (Marker, 2002; Cheetah Conservation Botswana, unpublished data). Brothers stay together and form a coalition, whereas single males of a litter either remain alone or join other solitary or coalition males to become part of a coalition (Caro, 1994; Mills et al., 2017).

MALE BEHAVIOR

Cheetah males have a broad spectrum of behaviors that range from friendly to highly aggressive depending on the social context. Interactions within coalition males are predominantly friendly, whereas interactions with females and their cubs can be aggressive but are rarely harmful. In contrast, interactions with unfamiliar males are often aggressive and can lead to death.

Male Behavior Toward Males Within a Coalition

Males within coalitions are often related. However, because the size of a coalition increases the chance for the males to take over a territory, it seems beneficial for two singleton males to form a coalition or for two brothers to accept an additional singleton male as a coalition partner, and unrelated coalition members have been identified (Caro, 1994; Chapter 6).

Coalition partners are very tolerant to one another's proximity and mainly exchange friendly behavior (Caro, 1994). In the Serengeti NP, they rest in close proximity and groom each other regularly. Unrelated males joining two brothers suffered some aggression in the beginning, but once the coalitions were fully established, the unrelated male rested at the same distance to the brothers as the brothers did to each other. When one coalition partner loses contact with the other(s), they call continuously until they find each other again (Caro, 1994). In one incidence, one of three coalition males lost contact with the others and reunited with them after 31 days (Hubel et al., 2016). Males within a coalition of two or three males are egalitarian with respect to access to food or females. This is also true for coalitions that include unrelated partners (Caro, 1994). Thus, mating access might be relatively equal, but large-scale paternity analyses is needed to investigate this.

Male Behavior Toward Females and Cubs

When males meet a female, interactions are often aggressive and usually intimidating for the female. Male coalitions typically surround a female, and if the female tries to escape, the males will slap her down or bite her (Caro, 1994). Females might slap back but are generally not successful in breaking away from males. Encounters are terminated when males lose interest and move away. Males typically sniff the ground where the female has been sitting or lying, most likely to assess whether she is in estrus. Encounters between males and females in the Serengeti NP lasted between 3 min and 2 days, whereas in Phinda Resource Reserve, a 170 km^2 small fenced area in South Africa, they lasted between 21 min and 18 days (Caro, 1994; Hunter and Skinner, 2003).

In the Serengeti NP, males approached solitary females in 69% of 45 encounters, compared to 28% of 40 encounters when females were accompanied by cubs (Caro, 1994). This differs from Phinda Resource Reserve, where males approached females with cubs in all 19 observed encounters (Hunter and Skinner, 2003). Offspring usually stayed within 50 m of their mother during the encounter and were mostly ignored by the males (Caro, 1994; Hunter and Skinner, 2003). Sometimes the aggressive behavior of males was directed toward the cubs (observed in cubs aged 3–14-months old), but the cubs were never seriously harmed, and females never attempted to defend the cubs. In other felid species, such as lions, males unrelated to the cubs might attack and kill them, bringing the mother of the killed cubs quickly into estrus again (Pusey and Packer, 1987). In cheetahs, infanticide has not been documented so far. Reasons that cheetah males are unlikely to commit infanticide might include the difficulty for males to monopolize females in estrus, an important factor in species in which infanticide occurs (Lukas and Huchard, 2014). This lack of monopolization is based in the mating system of the cheetah that consists of females roaming in large areas and covering multiple male territories and home ranges (Wolff and Macdonald, 2004). Thus, paternity of cubs in cheetahs is uncertain and males would risk killing related offspring (Lukas and Huchard, 2014).

Male Behavior Toward Unfamiliar Males

Males fight heavily to take over a territory, indicating that territories contain valuable resources, such as access to receptive females (Caro, 1994; Caro and Collins, 1987). For floaters to successfully take over a territory, a group size larger than the one holding the territory is necessary (Caro, 1994; Melzheimer et al., in preparation). Such fights can be very violent and can result in death (Caro, 1994; Melzheimer et al., in preparation). In the Serengeti NP, territory holders were heavier (mean ± SD: 45.7 kg ± 3.0) and physiological measures showed that they had better body conditions than floaters (mean ± SD: 36.3 kg ± 5.0, Caro and Collins, 1987; Caro et al., 1989). Floaters that successfully take over a territory consequently improve their body condition (i.e., increase their body mass index; BMI = kg/m^2). Fully grown males that were captured and measured as floaters or as very early territorial males in east-central Namibia had a significantly lower BMI (mean ± SD: 27.3 ± 1.3) than when they were captured again as established territory holders (mean ± SD: 30.3 ± 2.6; Melzheimer et al., in preparation).

HUNTING AND FEEDING BEHAVIOR

Cheetahs hunt and kill a wide variety of small to medium sized prey animals, based primarily upon their availability (Chapter 8). Their hunting is characterized by their high speed (Chapter 7) and their success is influenced by intrinsic factors, such as the hunting technique and skills of the individual cheetah, as well as extrinsic factors, such as the prey species and the habitat.

Hunting Techniques

Cheetahs use speed to hunt and are the fastest land mammal on earth, reaching speeds of up to 93 km/h in the wild (Wilson et al., 2013a) (Chapter 7). Speed is only part of a successful hunt because the faster a cheetah runs the more difficult it is to turn. Thus, cheetahs have different hunting techniques that are dependent on prey type, the habitat, and the skills of the individual hunter (Hilborn et al., 2012; Wilson et al., 2013a,b). Cheetahs either hunt alone or in groups (Bailey et al., 2013; Stander, 1990). When hunting in groups, cheetahs either appear to coordinate their movements, possibly to intercept fleeing prey, or only one male is hunting while the other(s) observe or follow the hunter,

but does/do not actively participate in the hunt (Caro, 1994; Stander, 1990).

In the Serengeti NP, 80% of the 295 observed hunts involved a stalk that then turned into a chase, whereas in 17% of hunts cheetahs chased prey without a stalk and in the remaining 3% the behavior was not classifiable. A hunt without a stalk can occur when flushing hidden prey, such as hares or fawns or when prey comes into close proximity (Hilborn et al., 2012). The actual chase can be split into two components: the acceleration phase in which a cheetah catches up with its prey and a deceleration phase needed to take down the prey. High-speed chases are likely to be a good strategy for large prey unlikely to make quick turns. For small prey, such as steenbok (*Raphicerus campestris*) and gazelle fawns, that can turn quickly, cheetahs accelerate to catch up with the prey and then slow down 5–8 s before taking down the prey to accommodate their prey's quick turns (Fitzgibbon, 1990a; Wilson et al., 2013b). In less open habitats, such as the open woodland savannah and thick bush of the Kruger NP in South Africa, cheetahs use the vegetation as cover to approach the prey animals and shorten their hunting distance. They also hide at the edges of cleared areas and initiate a hunt from such edges into the open areas (Mills et al., 2004). This is consistent with the preferred habitat of cheetahs being the edges between open and thick areas in Namibia (Muntifering et al., 2006).

In contrast to other felids, cheetahs rely on their dewclaws (claws of the first digits of the forepaws) and a bite to the throat when taking down medium to large sized prey (Chapter 7). Smaller prey, such as hares are killed by a bite in the skull (Estes, 1991). If a coalition of males kills a prey animal larger than themselves, such as a calf of a gemsbok (*Oryx gazelle*), the cheetahs are not able to strangle the prey. In such cases, 1 cheetah holds the head of the prey down while the other(s) attack the body and inflict damage in the abdominal areas (Mills et al., 2017).

The main factors that influenced the cheetah's initiation of a hunt in the Serengeti NP were the abundance of their main prey species and the abundance of lions. The presence of Thomson's gazelles in medium-sized groups (approximately 135 animals) increased the likelihood that a cheetah initiated a hunt (compared to larger or smaller groups of gazelles), whereas the presence of lions decreased the likelihood of a hunt being initiated (Cooper et al., 2007). Further, when females had cubs, they were more likely to initiate a hunt compared to females without cubs, probably because the former have a higher demand on energy (Cooper et al., 2007).

A hunt can last from 1 s to 165 min (Caro, 1987). The stalk will determine the length of the hunt, as the actual chase tends to be short, lasting a mean of 37.9 s (SD = 11.6, n = 19, Scantlebury et al., 2014). When hunting, cheetahs will approach the prey to 20–70 m, before breaking into a sprint (Estes, 1991; Mills et al., 2017). They then run a mean distance of 173 m (SD = 116, n = 367 hunts of five cheetahs), but distances of over 559 m have been recorded (Wilson et al., 2013a).

Acquisition of Hunting Skills

The process of acquiring hunting skills is complex. It starts with play behavior of cubs, continues with observing the mother when hunting, improves with joint hunts with the mother or siblings, and is perfected with continuous attempts throughout life (Caro, 1994). Young cubs from the age of 2 months will follow their mothers when she is going hunting. In the beginning, cubs stay behind while the mother starts to stalk the prey from a crouched position. Sometimes they chase inappropriate prey species, such as birds or jackals (Caro, 1994). The majority of the learning takes place through prey captured by the mother, which she then lets go in the presence of the cubs. Between the age of 5.0 and 8.5 months, mothers release almost one-third of the prey that they catch (Caro, 1994). Cubs are left to play with the released prey, often chasing

it or tripping it and pinning it down. Particularly when cubs are young, the mothers will eventually come in and kill the prey (Caro and Hauser, 1992). From approximately 11 months of age, mothers decrease the number of prey that they release for the cubs and cubs start initiating hunts of appropriate prey species. At this stage, however, cubs will rarely capture or make a kill without the help from their mother. Only as offspring become older than 12 months will they start making kills themselves. Some hunting skills, such as deciding on the right distance at which to initiate a chase, can take up 3.5 years to master and even then, hunts are not always successful (Caro, 1994).

Hunting Success

Failed hunts were previously attributed to cheetahs overheating during hunts (Taylor and Rowntree, 1973), but this has recently been disproved (Hetem et al., 2013; Chapter 7). The reasons why a hunt is successful or not, is thought to be a complex interplay between the hunting skills of the individual cheetah, including its age and condition, its ability to approach close enough to the prey before chasing it, the prey species being hunted, and the habitat (Hilborn et al., 2012; Mills et al., 2017). Hunting success is not influenced by the hunger level of the cheetah, nor does hunger level influence the decision to initiate a hunt (Cooper et al., 2007; Hilborn et al., 2012). When hunting gazelle fawns, cheetahs are more successful when fawns run away as compared to when they drop down to hide (Fitzgibbon, 1990a). When hunting adult gazelles, cheetahs are more successful at catching male gazelles than females, because males are less vigilant than females and position themselves often on the periphery of a herd (Fitzgibbon, 1990b; Mills et al., 2004).

In the Serengeti NP, hunting success of cheetah mothers and single young females were 26.9% of 643 attempts and 26.7% of 101 attempts, respectively, whereas hunting success of single adult females was 43.3% of 67 attempts (Caro, 1994). Cubs caused at least 16.4% of hunting failures ($n = 478$) by cheetah mothers because the prey became aware of the cubs' presence (Caro, 1987). Hunting success of solitary males was 39.1% of 133 attempts, of coalitions with 2 males 30.4% of 115 attempts and of coalitions with three males 52.2% of 69 attempts (Caro, 1994). While cheetahs tend to be more successful in open habitat types (Mills et al., 2004), they can also hunt successfully in denser vegetation (Wilson et al., 2013a).

Feeding Behavior

Once a cheetah makes a successful kill, it often moves the kill to a protected location, perhaps to reduce the chance of the kill being discovered by other predators. In the Serengeti NP, approximately 65% of the 491 observed kills were moved away from the kill site (Hunter et al., 2007). One individual was observed to move its kill 712 m away, but on average cheetahs move their kill an estimated 64.5 m (SE = 5.8 m) before feeding (Hunter et al., 2007).

Cheetahs often open their kills on the abdominal side and first feed on muscle and viscera before chewing on skin and bones (Caro, 1994). The length of feeding events varies considerably across ecosystems and is likely related to prey size, hunger level, and the probability of kleptoparasitism by other large scavengers, such as spotted hyenas and lions (Broekhuis et al., 2014; Hunter et al., 2007; Mills et al., 2004). Cheetah males in coalitions and mothers with many cubs leave the leftovers of the kills earlier, after they have finished feeding, than solitary males and mothers with few cubs, respectively, probably to decrease the risk of being detected by large predators at their larger kills or as a larger foraging group (Hunter et al., 2007). Feeding time can also depend on weather conditions; in the Maasai Mara, Kenya cheetahs fed for up to 3 h from a carcass when it was not raining, while they stayed at the same spot feeding from

a carcass for up to 2 days when it was raining (Chelysheva, 2015). On average, cheetahs fed in the Maasai Mara for 116 min (SD = 74, $n = 136$; Chelysheva, unpublished data). In Moremi, Botswana, nocturnal feeding bouts ($n = 109$) lasted on average 62.3 min (SD = 3.5), which was significantly shorter than feeding bouts during the day ($n = 664$) of an average of 97.2 min (SD = 3.6). It is not known whether cheetahs select for smaller prey at night or reduce the time on a kill to minimize the chance of being detected by lions and spotted hyenas that are active at night (Broekhuis et al., 2014). Avoidance of large scavengers reduces the risk of kleptoparasitism and predation, particularly on cubs (Laurenson, 1994). In the Serengeti NP, kleptoparasitism occurred in 11.4% of the 605 observed cheetah kills (Hunter et al., 2007). Kleptoparasitism events were primarily caused by spotted hyenas (78%), followed by lions (15%), other cheetahs (3%), and disturbance by tourists and other wildlife species (4%). Similarly, in the Kruger NP, cheetahs lost 11.8% of their kills to other predators, whereas in the Kgalagadi TP, only 6.1% of 378 kills were lost, and in the Kwandwe Private Game Reserve, South Africa, only 2.2% of 224 kills were lost (Bisset and Bernard, 2007; Mills et al., 2004, 2017).

While it is rare, cheetahs occasionally scavenge (Caro, 1982). For example, in Etosha NP, Namibia, an adult male and two accompanying adolescent cubs were observed scavenging on the carcass of a giraffe (*Giraffa camelopardalis*) (Stander, 1990), and in the Maasai Mara, a cheetah mother and her three fully grown offspring were observed chasing an adult spotted hyena away from its freshly killed topi (*D. lunatus jimela*) (Broekhuis and Irungu, 2016).

COMMUNICATION

Acoustic and olfactory communication are important parts of social interactions of a species. They can be used when two or more individuals of a species meet but also when individuals receive acoustic information over a distance or olfactory information with a time delay. Communication is used in various contexts, such as territory defense, attraction of sexual partners, warning of conspecifics against predators, protection of offspring, social binding, and attraction of cubs by their mother to food sources (Bradbury and Vehrencamp, 1998).

Acoustic Communication

The vocal repertoire of cheetahs is rich and consists of several sounds used during social interactions. Different authors have categorized and named the sounds slightly differently. Here we present the ones that most studies reported (Table 9.2). Yelping is a staccato high-pitched growling sound used for long-distant calls between adults and sometimes by mothers calling their cubs. Mothers also emit a single soft chirrup or a series of chirrups when calling their cubs. A sharp chirrup sound elicits close-following when a mother is moving, and cubs produce the birdlike chirp when calling their mother and littermates (Estes, 1991). Males that

have lost their coalition partners yelp or chirp loudly and in long series, or yip with a short, high-pitched birdlike call. Males chirrup loudly when encountering females, particularly when the female is in estrus, while females yip fearfully in such situations. In aversive situations, cheetahs growl, bleat, hiss, cough/spit or yowl, which is a drawn-out moan of variable pitch (Caro, 1994; Estes, 1991). Adults and cubs often snort/snuffle when feeding on the same carcass in close proximity to each other (Chelysheva, unpublished data). Meow, a none-context-specific call, can best be used to identify individuals and sexes. Males meow with lower voices, that is, lower frequency, than females (Smirnova et al., 2016). Unlike other large cats, cheetahs are able to purr during friendly interactions, particularly during social grooming (Estes, 1991), and are unable to roar (Chapter 7).

Olfactory Communication

The cheetah is a solitary species and roams over large home ranges. Thus, cheetahs do not meet very often and use olfactory cues to communicate indirectly. Scent marking involves spraying urine and defecating on

TABLE 9.2 Vocal Repertoire of Cheetahs, Context When it is Used and Description of Calls

Sound	Context	Description
Yelp	Adult males calling coalition partners, mother calling cubs	Loud, long-distance call (< 2 km), staccato, high-pitched
Chirrup	Mother calling cubs, males encountering (estrus) females	Soft or loud call, single, or in series
Chirp	Adult males calling coalition partners, cubs calling mother or littermates	Loud call, in series
Yip	Adult males calling coalition partners, females encountering adult male(s)	Loud call, high-pitched, birdlike
Purr	Friendly interaction between cheetahs	Soft call, in series
Growl, bleat, hiss, cough, spit, yowl	Aversive interactions between cheetahs or cheetahs and other predator species	Short call, single, or in series
Snort, snuffle	Adults and cubs when feeding	Soft call, single, or in series
Meow	Not context specific	Soft call

FIGURE 9.3　**Territorial males scent marking on a prominent tree.** (A) In the Kgalagadi Transfrontier Park, Southern Kalahari, (B) on Namibian farmland. *Source: Part A, Gus Mills; part B, Joerg Melzheimer.*

prominent landmarks, such as trees (Fig. 9.3), rocks, termite mounds, bushes, or tall grass tuffs on the top of a dune (Caro, 1994; Marker-Kraus et al., 1996; Marnewick et al., 2006; Mills et al., 2017). Conspicuousness of the marking trees is important when cheetahs mark, thus tall trees with large canopy areas are preferred marking sites (Walker et al., 2016). Floaters do not defecate on prominent landmarks and urinate far less frequently than territorial males, both in their home ranges and in the territories of other males (Caro, 1994). Females communicate less by olfactory cues than males and use urine spraying mainly to advertise estrus (Caro, 1994; Mills et al., 2017; Wielebnowski and Brown, 1998).

In east-central Namibia, a camera trap study running for 812 days at 24 marking trees recorded 2820 visits by cheetahs, of which cheetahs marked in 1034 (36.7%) cases. Of these 1034 markings, 95.6% were made by territorial males, 3.1% by floaters, and 1.4% by females (Wachter & Melzheimer, unpublished data). The high proportion of marking by territorial males is likely to indicate their ongoing claim on the territory, whereas the marking by floaters might be a challenge to the territory owners.

CONCLUSIONS

Long-term studies of free-ranging cheetahs in southern and eastern Africa have answered many questions regarding cheetah behavior. Cheetahs have a unique social and mating system compared with other felids, with males having two spatial tactics, (i.e., defending small territories or roaming in large home ranges), and females having large home ranges encompassing several male territories. Being the fastest land mammal, they exhibit various hunting and feeding behaviors, while avoiding confrontation with larger carnivore species. In addition to long-term monitoring, technological advances, such as GPS collars that record aspects of the physiology and environment of the cheetahs, and motion-triggered cameras have advanced our understanding of cheetah behavior in previously less understood topics, such as spatial movements, hunting styles, and scent marking. As human populations further encroach on cheetah habitat, understanding the behavior of cheetahs in human-dominated landscapes may aid the development of new conservation solutions.

References

Bailey, I., Myatt, J., Wilson, A., 2013. Group hunting within the Carnivora: physiological, cognitive and environmental influences on strategy and cooperation. Behav. Ecol. Sociobiol. 67, 1–17.

Bisset, C., Bernard, R.T.F., 2007. Habitat selection and feeding ecology of the cheetah (*Acinonyx jubatus*) in thicket vegetation: is the cheetah a savanna specialist? J. Zool 271, 310–317.

Bradbury, J.W., Vehrencamp, S.L., 1998. Principles of Animal Communication. Sinauer Associates, Sunderland.

Broekhuis, F., Grünewälder, S., McNutt, J.W., Macdonald, D.W., 2014. Optimal hunting conditions drive circalunar behavior of a diurnal carnivore. Behav. Ecol. 25 (5), 1268–1275.

Broekhuis, F., Irungu, O., 2016. Role reversal: record of cheetahs (*Acinonyx jubatus*) kleptoparasitizing a kill from a spotted hyena (*Crocuta crocuta*). J. Afr. Ecol. 55 (1), 115–117.

Caro, T.M., 1982. A record of cheetah scavenging in the Serengeti. Afr. J. Ecol. 20, 213–214.

Caro, T.M., 1987. Indirect costs of play: cheetah cubs reduce maternal hunting success. Anim. Behav. 35, 295–297.

Caro, T.M., 1994. Cheetahs of the Serengeti: Group living in an asocial species. University of Chicago Press, Chicago.

Caro, T.M., 1995. Short-term costs and correlates of play in cheetahs. Anim. Behav. 49, 333–345.

Caro, T.M., Collins, D.A., 1987. Male cheetah social organisation and territoriality. Ethology 74, 52–64.

Caro, T.M., Fitzgibbon, C.D., Holt, M.E., 1989. Physiological costs of behavioural strategies for male cheetahs. Anim. Behav. 38, 309–317.

Caro, T.M., Hauser, M.D., 1992. Is there teaching in nonhuman animals? Q. Rev. Biol. 67, 151–174.

Chelysheva E.V., 2015. Losers or Survivors: The Mara's cheetah adapt to survive. Swara January–March, 2015, pp. 39–43.

Clutton-Brock, T.H., 2016. Mammal societies, first ed. Wiley-Blackwell, West- Sussex.

Cooper, A.B., Pettorelli, N., Durant, S.M., 2007. Large carnivore menus: factors affecting hunting decisions by cheetahs in the Serengeti. Anim. Behav. 73, 651–659.

Crosier, A.E., Marker, L., Howard, J., Pukazhenthi, B.S., Henghali, J.N., Wildt, D.E., 2007. Ejaculate traits in the Namibian cheetah (*Acinonyx jubatus*): influence of age, season and captivity. Reprod. Fertil. Dev. 19, 370–382.

Durant, S.M., 1998. Competition refuges and coexistence: an example from Serengeti carnivores. J. Anim. Ecol. 67, 370–386.

Durant, S.M., 2000. Predator avoidance, breeding experience and reproductive success in endangered cheetah, *Acinonyx jubatus*. Anim. Behav. 60, 121–130.

Durant, S.M., Kelly, M., Caro, T.M., 2004. Factors affecting life and death in Serengeti cheetahs: environment, age and sociality. Behav. Ecol. 15, 11–22.

Estes, R.D., 1991. The Behavior Guide of African mammals. University of California Press, Berkeley.

Fitzgibbon, C.D., 1990a. Anti-predator strategies of immature Thomson's gazelles: hiding and the prone response. Anim. Behav. 40, 846–855.

Fitzgibbon, C.D., 1990b. Why do hunting cheetahs prefer male gazelles? Anim. Behav. 40, 837–845.

Frame, G.W., 1984. Cheetah. Macdonald, D.W. (Ed.), The Encyclopedia of Mammals, vol. 1, Allen and Unwin, London, pp. 40–43.

Frame, G.W., Frame, L.H., 1981. Swift and enduring: cheetah and wild dogs in the Serengeti. E.P. Dutton, New York.

Gottelli, D., Wang, J., Bashir, S., Durant, S.M., 2007. Genetic analysis reveals promiscuity among female cheetahs. Proc. R. Soc. Lond. 274, 1993–2001.

Hetem, R.S., Mitchell, D., de Witt, B.A., Fick, L.G., Meyer, L.C.R., Maloney, S.K., Fuller, A., 2013. Cheetah do not abandon hunts because they overheat. Biol. Lett 9, 20130472.

Hilborn, A., Pettorelli, N., Orme, C.D.L., Durant, S.M., 2012. Stalk and chase: how hunt stages affect hunting success in Serengeti cheetah. Anim Behav. 84, 701–706.

Hubel, T.Y., Shotton, J., Wilshin, S., McKenna, R., Horgan, J.E., Wilson, A.M., 2016. Cheetah reunion—the challenge of finding your friends again. PLoS ONE 11 (12), e0166864.

Hunter, J.S., Durant, S.M., Caro, T.M., 2007. To flee or not to flee: predator avoidance by cheetahs at kills. Behav. Ecol. Sociobiol. 61, 1033–1042.

Hunter, L.T.B., Skinner, J.D., 2003. Do male cheetahs *Acinonyx jubatus* commit infanticide? Trans. R. Soc. S. Afr. 58, 79–82.

Kelly, M.J., Laurenson, M.K., Fitzgibbon, C.D., Collins, D.A., Durant, S.M., Frame, G.W., Bertram, B.C.R., Caro, T.M., 1998. Demography of the Serengeti cheetah (*Acinonyx jubatus*) population: the first 25 years. J. Zool. 244, 473–488.

Laurenson, M.K., 1993. Early maternal behavior of wild cheetahs: implications for captive husbandry. Zoo Biol. 12, 31–43.

Laurenson, M.K., 1994. High juvenile mortality in cheetahs (*Acinonyx jubatus*) and its consequences for maternal care. J. Zool. 234, 387–408.

Laurenson, M.K., Caro, T.M., Borner, M., 1992. Female cheetah reproduction. Natl. Geogr. Res. Explor. 8, 64–75.

Lukas, D., Huchard, E., 2014. The evolution of infanticide by males in mammalian societies. Science 346, 841–844.

Marker, L.L., 2002. Aspects of cheetah (*Acinonyx jubatus*) biology, ecology and conservation strategies on Namibian farmlands. PhD thesis, University of Oxfor, United Kingdom.

Marker, L.L., Dickman, A.J., Jeo, R.M., Mills, M.G.L., Macdonald, D.W., 2003. Demography of the Namibian cheetah, *Acinonyx jubatus jubatus*. Biol. Conser. 114, 413–425.

Marker, L.L., Dickman, A.J., Mills, M.G.L., Jeo, R.M., Macdonald, D.W., 2008. Spatial ecology of cheetahs (*Acinonyx jubatus*) on North-central Namibian farmlands. J. Zool. 274, 226–238.

Marker-Kraus, L., Kraus, D., Barnett, D., Hurlbut, S., 1996. Cheetah survival on Namibian farmlands. Cheetah Conservation Fund, Windhoek.

Marnewick, K.A., Bothma, J.d.P., Verdoorn, G.H., 2006. Using camera-trapping to investigate the use of a tree as a scent-marking post by cheetahs in the Thabazimbi district. S. Afr. J. Wildl. Res. 36, 139–145.

Mills, M.G.L., Broomhall, L.S., du Toit, J.T., 2004. Cheetah *Acinonyx jubatus* feeding ecology in the Kruger National Park and a comparison across African savanna habitats: Is the cheetah only a successful hunter on open grassland plains? Wildl. Biol. 10, 177–186.

Mills, M.G.L., Mills, M.E.J., 2013. Cheetah cub survival revisited: a re-evaluation of the role of predation, especially by lions, and implications for conservation. J. Zool. 292 (2), 136–141.

Mills, M.G.L., Mills, M.E.J., Edwards, C.T., Gottelli, D., Scantelbury, D.M., 2017. Kalahari cheetahs: Adaptations to an arid region. Oxford University Press, Oxford.

Muntifering, J.R., Dickman, A.J., Perlow, L.M., Hruska, T., Ryan, P.G., Marker, L.L., Jeo, R.M., 2006. Managing the matrix for large carnivores: a novel approach and perspective from cheetah (*Acinonyx jubatus*) habitat suitability modelling. Anim. Conser. 9, 103–112.

Pettorelli, N., Durant, S.M., 2007. Longevity in cheetahs: the key to success? Oikos 116, 1879–1886.

Pusey, A.E., Packer, C., 1987. The evolution of sex-based dispersal in lions. Behaviour 101, 275–310.

Sandell, M., 1989. The mating tactics and spacing patterns of solitary carnivores. In: Gittleman, J.L. (Ed.), Carnivore Behavior, Ecology and Evolution. Cornell University Press, New York, pp.164–182.

Scantlebury, D.M., Mills, M.G.L., Wilson, R.P., Wilson, J.W., Mills, M.E.J., Durant, S.M., Bennett, N.C., Bradford, P., Marks, N.J., Speakman, J.R., 2014. Flexible energetics of cheetah hunting strategies provide resistance against kleptoparasitism. Science 346, 79–81.

Smirnova, D.S., Volodin, I.A., Demina, T.S., Volodina, E.V., 2016. Acoustic structure and contextual use of calls by captive male and female cheetahs (*Acinonyx jubatus*). PLoS ONE 11 (6), e0158546.

Stander, P.E., 1990. Notes on foraging habits of cheetah. S. Afr. J. Wildl. Res. 20, 130–132.

Swanson, A., Caro, T., Davies-Mostert, H., Mills, M.G.L., Macdonald, D.W., Borner, M., Masenga, E., Packer, C., 2014. Cheetahs and wild dogs show contrasting patterns of suppression by lions. J. Anim. Ecol. 83, 1418–1427.

Taylor, C.R., Rowntree, V.J., 1973. Temperature regulation and heat balance in running cheetah: a strategy for sprinters? Am. J. Physiol. 224, 848–851.

Trivers, R.L., 1974. Parent-offspring conflict. Am. Zool. 14, 249–264.

Wachter, B., Thalwitzer, S., Hofer, H., Lonzer, J., Hildebrandt, T.B., Hermes, R., 2011. Reproductive history and absence of predators are important determinants of reproductive fitness: the cheetah controversy revisited. Conser. Lett. 4, 47–54.

Walker, E.H., Nghikembua, M., Bibles, B., Marker, L., 2016. Scent-post preference of free-ranging Namibian cheetahs. Glob. Ecol. Conser. 8, 55–57.

Wielebnowski, N., Brown, J.L., 1998. Behavioral correlates of physiological estrus in cheetahs. Zoo Biol. 17, 193–209.

Wildt, D.E., Brown, J.L., Bush, M., Barone, M.A., Cooper, K.A., Grisham, J., Howard, J.G., 1993. Reproductive status of cheetahs (*Acinonyx jubatus*) in North American zoos: the benefits of physiological surveys for strategic planning. Zoo Biol. 12, 45–80.

Wilson, A., Lowe, J., Roskilly, K., Hudson, P., Golabek, K., McNutt, J., 2013a. Locomotion dynamics of hunting in wild cheetahs. Nature 498, 185–189.

Wilson, J.W., Mills, M.G.L., Wilson, R.P., Peters, G., Mills, M.E.J., Speakman, J.R., Durant, S.M., Bennett, N.C., Marks, N.J., Scantlebury, M., 2013b. Cheetahs, *Acinonyx jubatus*, balance turn capacity with pace when chasing prey. Biol. Lett. 9, 20130620.

Wolff, J.O., Macdonald, D.W., 2004. Promiscuous females protect their offspring. Trends Ecol. Evol. 19, 127–134.

CONSERVATION THREATS

Drivers of Habitat Loss and Fragmentation: Implications for the Design of Landscape Linkages for Cheetahs

Richard M. Jeo*, Anne Schmidt-Küntzel**,
Jonathan D. Ballou[†], M. Sanjayan[‡]

*The Nature Conservancy in Montana, Helena, MT, United States
**Cheetah Conservation Fund, Otjiwarongo, Namibia
[†]Smithsonian Conservation Biology Institute, Washington, DC, United States
[‡]Conservation International, Arlington, VA, United States

INTRODUCTION

Understanding the patterns and drivers of habitat fragmentation and connectivity at multiple scales remains critical for the design of conservation strategies. The cheetah (*Acinonyx jubatus*) was historically widespread, ranging throughout nonforested areas of Africa, the Middle East, Central Asia, and the Indian subcontinent (Nowell and Jackson, 1996). Human-caused habitat loss and fragmentation are among the principal drivers of decline and remain major threats to long-term persistence (Crooks et al., 2011; Durant et al., 2017; Schipper et al., 2008). Large-bodied, wide-ranging

mammals are sensitive to habitat loss and fragmentation because they generally require large, connected tracts of habitat to support stable populations (Ripple et al., 2014). Exacerbating the problem, the majority of cheetahs live outside of protected areas in low densities (Durant et al., 2017). Human pressure in the cheetah's historical range has steadily increased over the last 25 years as the human population of sub-Saharan Africa doubled from 500 million people in 1990 to an estimated 1 billion in 2015 (United Nations, Department of Economic and Social Affairs, Population Division, 2015). With the human population expected to double again by 2050, additional habitat loss and fragmentation may further isolate and fragment existing cheetah core stronghold areas and increase the chance of species extinction. Populations can be connected via translocations (metapopulation management; Chapter 20) or natural movement (via landscape connectivity). Strategies that enhance or protect landscape connectivity can reduce the impact of habitat loss and fragmentation and increase population viability (Chetkiewicz et al., 2006; Crooks and Sanjayan, 2006) and are urgently needed particularly in the human-dominated matrix.

In this chapter, we explore the current status and impact of fragmentation on the viability of cheetah populations, what exactly defines cheetah habitat, and the primary factors that contribute to habitat fragmentation. We then discuss connectivity conservation strategies, including implications for the design of cheetah conservation linkage areas.

CHALLENGES OF A FRAGMENTED POPULATION: WHERE DOES THE CHEETAH CURRENTLY SURVIVE?

Assessing current cheetah status is difficult, as the species is cryptic, wide-ranging, and occurs at low densities throughout its range (Chapter 8). Although there remains considerable uncertainty, the consensus among experts is that the current cheetah distribution is at only 9% of its original range, with a total population estimate of approximately 7100 adult and adolescent individuals (Durant et al., 2017). The largest contiguous population resides in southern Africa (~4000 adult and adolescent animals, mostly composed of animals from Botswana and Namibia); and the second largest population resides in eastern Africa (~1400 animals) in the transfrontier areas of Kenya and Tanzania (Durant et al., 2017). Another five populations range between 150 and 250 adult and adolescent cheetahs (in descending order:

transfrontier populations of the Central African Republic and Chad; the Ruaha ecosystem in Tanzania; transfrontier population of Ethiopia, Kenya, and South Sudan; transfrontier population of Algeria and Mali; and the Southern National Park of South Sudan; Durant et al., 2017). All other populations, including the relict population of Asiatic cheetah in Iran (Chapter 5), are estimated to be less than 100 adult and adolescent animals, of which half number 20 or less (Durant et al., 2017).

All of these extant populations exist within a mosaic of wild lands, protected areas, rural and communal lands used for livestock production, game farms, agricultural croplands, and other human-altered landscapes. In addition to the factors that have been the direct causes of population declines (e.g., habitat loss, reduction in prey populations—Chapter 11, human-caused mortality—Chapter 13, and interspecific competition—Chapters 8 and 9), these small disjoined populations are particularly vulnerable and face increased risk from extinction due to stochastic events (Frankham et al., 2017). Cheetahs are challenged by their low genetic diversity (Chapter 6), and although additional inbreeding effects on already genetically similar populations are largely unknown, the potential consequences include reduced fitness and a reduced ability to adapt to future changes, such as the effects of climate change (Channell and Lomolino, 2000; Chapter 12).

Although the exact number of animals required for a small population to have the capacity to survive depends on many factors including genetic diversity, numbers of deleterious traits, and reproductive success, most cheetah populations are already smaller than the recommended minimum viable population size (Traill et al., 2010). A general estimate for an *effective population size* (i.e., N_e, a theoretical number which roughly reflects the number of animals genetically contributing to the population) is 50 to minimize inbreeding depression over 5 generations (Soule and

Wilcox, 1980). However, N_e tends to be 5–10 times smaller than the actual number of animals in the population (Frankham et al., 2014), thus an N_e of 50 would approximate a required minimum size of 250–500 adult individuals to ensure short term survival. While inbreeding depression may already occur in larger population sizes (as early as when N_e drops to 100), the increased efficiency of purging deleterious alleles from small populations hopefully compensates for inbreeding depression to some extent (García-Dorado, 2015). For long-term maintenance of genetic diversity, the recommended numbers are 10 times larger with N_e of 500 or 1,000 (corresponding to 2,500–5,000 and 5,000–10,000 adult individuals; Frankham et al., 2014; Traill et al., 2010). These numbers suggest that all of the small, isolated cheetah populations are already at an extreme risk of extinction and even the largest population in southern Africa remains tenuous.

However, rather than defining specific minimum targets for cheetah populations, perhaps the more important question is: what conservation strategies can best enhance the chances of the species' survival? If remaining populations can exchange migrants naturally or via translocation (Chapter 20), the risk of extinction due to stochastic events would be dramatically reduced. However, "one migrant per generation" (Mills and Allendorf, 1996) may not be sufficient as recent work has demonstrated that migrants must survive and breed to have an impact, therefore a more likely estimate of the number required is approximately 5–19 per generation (Frankham et al., 2017). In light of these numbers, conservation efforts that facilitate landscape connectivity are increasingly urgent. In the following sections, we explore how the interactions between habitat factors and cheetah biology can be used to identify linkage areas, and to guide conservation strategies that enhance animal movement and hopefully reduce the cheetah's extinction risk.

WHAT LIMITS AND FRAGMENTS CHEETAH HABITAT?

A number of studies have documented habitat features associated with cheetah presence (e.g., Caro, 1994; Marker et al., 2008; Nghikembua et al., 2016). Although these studies provide useful information, to address questions regarding habitat for the cheetah, the biotic factors that drive an animal's fitness must be explicitly considered (Mitchell and Hebblewhite, 2012). Following this guidance, we need to answer a number of critical questions regarding cheetah habitat—that is, what combinations of structural and functional characteristics of landscapes drive survival and reproduction? In the following sections, we briefly review factors that influence functional connectivity for the cheetah, in the context of their unique biology. We discuss evidence for fitness-driven variables that have been linked to cheetah habitat, including (1) prey availability, (2) interspecific impacts (i.e., competition, conflict, and displacement associated with dominant carnivores), (3) land-cover conversion, roads, and fences, and (4) human tolerance, conflicts, and persecution.

Prey Availability

Cheetahs are obligate carnivores as their relatively fragile teeth, skull, and jaw musculature largely restrict their diet to meat (Chapter 7). There is clear evidence that one of the primary factors related to density of carnivores is density of available prey (Carbone and Gittleman, 2002; Fuller and Sievert, 2001), and sufficient access to prey remains a fundamental energetic requirement for cheetahs and a key fitness variable.

Although medium-sized game (23–56 kg) is preferred (cheetahs have been documented to prey on a wide range of species, ranging in size from birds and rodents to large ungulates weighing up to 270 kg), small prey species (<23 kg), such as common duiker (*Sylvicapra grimmia*) or hares (*Lepus* spp.), contribute substantially to the cheetah's diet (Hayward et al., 2006; Chapters 8 and 11). Energetic models predict that cheetahs could theoretically subsist exclusively on small prey, but this strategy would make them energetically vulnerable and only low cheetah densities would be expected (Carbone et al., 2007). Nevertheless, this evidence suggests that marginal habitats could be valuable, even areas where only small prey species are available, and could be utilized for dispersal or travel between more productive habitats. In general, we conclude here that (1) the best habitat areas for cheetah are areas with abundance of medium-sized prey species, and (2) marginal, arid habitats with lower density of medium-sized prey may potentially serve as habitat for cheetah if sufficient smaller prey is available. In such areas, smaller prey species would have to make up a larger proportion of their diet than have been previously observed.

Interspecific Impacts: Competition, Conflict, and Displacement

The presence of dominant competitors, particularly lions (*Panthera leo*) and also leopards (*Panthera pardus*) and spotted hyaenas (*Crocuta crocuta*), restricts the distribution and density of subordinate species, such as cheetahs (Durant, 2000, 1998; Laurenson, 1994), by competing for resources and through interspecies conflict. High density of dominant competitors has been shown to increase local extinction risk of subordinate species (Hayward and Kerley, 2008) including the cheetah. Cheetahs avoid competitors by ranging widely and utilizing areas where few competitors occur (Creel, 2001; Durant, 2000; Mills and Gorman, 1997). Interspecific impacts thus represent key biotic factors that drive habitat suitability and influence fragmentation patterns for the cheetah. However, as lions and spotted hyenas are generally absent or present at low density outside of protected areas, interspecific effects perhaps represent a lesser concern for landscape scale connectivity (and is certainly a less important factor than

human conflict, prey density, and conversion to cropland). Nevertheless, increasing numbers of leopards in key places that are important for cheetah habitat connectivity may be a concern.

Land Conversion, Fences, and Roads

Widespread land-cover alterations over the past century, ranging from human development of urban and residential areas to the development of cultivated croplands and its associated road and fencing infrastructure, have drastically reduced land available to cheetahs and created barriers between and within populations (Crooks et al., 2011; Ray et al., 2005). The effects of wildlife fencing and land conversions have a direct impact on ungulate prey species which, in turn, impacts cheetahs and other carnivores that prey on the ungulates. For example, large-scale fences constructed to prevent the spread of disease from wildlife to livestock have prevented migratory species from reaching vital food and water resources (Williamson and Williamson, 1984), thus leading to their decline. Similarly, game-proof and predator-proof fences (used commonly around game farms in southern Africa) have been shown to both impede cheetah movements and disrupt movement of natural prey species (Marker et al., 2008). In general, it remains likely that the loss of productive habitat through crop conversion and its associated impacts will continue to play a major role in habitat fragmentation for the cheetah as human populations expand throughout the cheetah's range (Woodroffe, 2000).

Roads represent another primary cause of mortality for carnivores throughout the world. Vehicular collisions are the main documented human-caused mortalities of critically endangered Asiatic cheetahs in Iran (Farhadinia et al., 2016; Chapter 5). In Tanzania, roads have not only contributed to habitat fragmentation, but also caused sufficient levels of direct mortality of wildlife in Mikumi National Park to create population sinks for some species (Caro, 1994). This pattern also held in Zimbabwe's Hwange National Park, where half of all African wild dog deaths were attributed to roads (Woodroffe and Ginsberg, 1997). Roads, railroads, and navigable waterways also serve as starting points for land conversion and facilitate human encroachment on protected areas and lead indirectly to habitat loss and fragmentation (Watson et al., 2015). Large-scale infrastructure development and conversion of native vegetation into cultivated cropland remain important fitness-related variables for cheetahs, especially development impacts that fragment the few remaining population strongholds.

Human Conflict

The majority of the cheetah's current range (77%), along with the majority of animals (67%), are outside of protected areas (Durant et al., 2017). In these human-dominated landscapes, conflict with livestock farmers—both commercial and subsistence farmers—results in trapping and direct removal of cheetahs (Marker and Dickman, 2004; Chapter 13) and represents a clear biological driver of fitness. Conflict is driven by a variety of activities, including competition over a reduced prey base (see section, "Land Conversion, Fences, and Roads") and conflict with livestock production.

Actual and perceived livestock predation by large carnivores are the most widespread causes of conflict, and retaliatory killing by affected farmers is a major threat to all large carnivores including cheetahs (Marker et al., 2003a; Woodroffe et al., 2005; Chapter 13). There is a strong evidence that human tolerance levels and the levels of human persecution of carnivores are principal factors that drive habitat security and global carnivore population trends (Linnell et al., 2001; Marker et al., 2003b; Ripple et al., 2014; Woodroffe, 2000).

Finally, despite mounting evidence and a general recognition that human conflict is a principal factor in cheetah mortality, there is a paucity of data on the sociological drivers of conflict across the cheetah's range (Dickman, 2010) especially in remote areas. These areas may be important contributors to long-term habitat connectivity.

Understanding interactions between nomadic livestock producers, cheetah prey species, and cheetah mortality is needed to develop effective strategies to mitigate habitat fragmentation in such areas (Andresen et al., 2014; Homewood et al., 2001; Ogutu et al., 2005).

THE FOUR KINDS OF CHEETAH HABITAT: A CONCEPTUAL MODEL

If we consider cheetah habitat as being largely defined by human behavior—through impacts on prey species and direct impacts on the cheetah itself—a simple 2-axes conceptual model can be illustrative (Fig. 10.1). Suitable habitat (Fig. 10.1A) is found largely in existing cheetah stronghold population areas and has both sufficient levels of human tolerance and prey that allow cheetahs to persist.

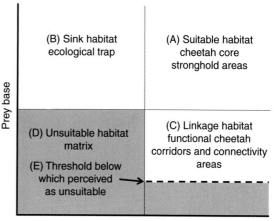

FIGURE 10.1 **Conceptual model illustrating four types of cheetah habitat, based on two major factors: human tolerance and prey availability.** (A) Suitable cheetah habitat based on acceptable levels of human tolerance and prey biomass that is sufficient to support resident cheetah populations. (B) Ecological trap or sink habitats, where high levels of prey biomass are combined with low tolerance resulting in human-caused cheetah mortality. (C) Linkage habitat or functional cheetah corridor areas where prey density is adequate for movement and human tolerance is high. (D) Unsuitable habitat. (E) Prey density threshold required for cheetah habitat selection.

Ecological traps (Fig. 10.1B) are areas where productive, prey-rich, or otherwise attractive areas are combined with low human tolerance, resulting in high levels of cheetah mortality (Battin, 2004; Hale et al., 2015). Private lands managed for trophy hunting, live game sales, or game meat production of wildlife species (a.k.a. "game farms") can be a particular problem for cheetahs and have been shown to be ecological traps on Namibian farmlands (Marker et al., 2003b). In southern Africa, highly productive areas with high levels of livestock biomass, coupled with low wild prey biomass, have been shown to be areas with high levels of conflict (Winterbach et al., 2014).

Linkage habitats (Fig. 10.1C) are areas that have sufficient habitat quality (i.e., appropriate land cover, prey density, and other factors) that allows cheetahs to select and utilize such areas for movement and dispersal combined with a high level of human tolerance (and thus high habitat security). We emphasize that such areas do not have to be linear or even continuous, but they do have to be spatially widespread enough across the landscape to allow animals to travel between suitable habitats. For linkage areas, habitat quality is important only insofar as

cheetahs are unlikely to utilize the corridor if the habitat is perceived to be unsuitable (represented conceptually by Fig. 10.1E). As such, linkage habitat can be marginally productive, requiring only enough prey and other factors to attract cheetah use. But how do we recognize and invest in these areas? This topic and implications for conservation strategies are discussed in the following sections.

FUNCTIONAL CHEETAH CORRIDOR DESIGN: STRATEGIES TO INCREASE LANDSCAPE PERMEABILITY

Corridors have been suggested as a mechanism to connect habitat, but both the conceptual and real world identification of corridors often neglect key ecological factors for the target organism (Chetkiewicz et al., 2006). The review of evidence presented in the aforementioned sections suggests that both human conflict and prey availability are principal drivers of fitness for the cheetah outside of protected areas, where interspecific competition is low (Table 10.1). Therefore, efforts to mitigate fragmentation

TABLE 10.1   Summary of Key Drivers of Habitat Loss and Fragmentation for Cheetahs

Key driver	Rate of impact	Magnitude	Notes
Human persecution/ conflict	Fast	High	Demonstrated immediate and direct impact on cheetah populations
Reduction in prey	Fast	Medium	Obligate carnivores but have considerable diet plasticity; historical large migrations already eliminated
Land conversion	Intermediate	High	Suitable land for cultivation already largely converted; large tracts of unoccupied habitat appear to exist; secondary impacts include roads, fences, domestic animals, and similar impact
Habitat degradation	Intermediate	Medium	Overgrazing in pastoral areas; desertification; direct impact on prey availability
Interspecific competition/ predation	Fast	High/low	Large competitors key factor in some protected areas, but largely absent outside
Climate change	Slow	Uncertain	Direct and indirect impacts include desertification, changes in prey distribution, and changes in land available for crop cultivation

patterns and delineate functional cheetah corridors should focus primarily on human tolerance and prey availability, consistent with recent studies and literature reviews (Winterbach et al., 2013, 2015).

Measuring and mapping connectivity and habitat fragmentation patterns have been greatly facilitated by the development of new analytical tools and easy access to high-resolution spatial data (Chapter 36). However, it is crucial not only to restrict analyses to physical characteristics of the landscape, but also to include the more difficult analyses of factors, such as human tolerance, to avoid misleading conclusions and, worse, misguided conservation investments.

Similarly, the concept of linear corridors as tools for conservation has been broadened to encompass landscape permeability or "linkages," where key ecological processes are supported without necessarily being linear, continuous, or structurally distinct across the landscape (Bennett, 1999; Chetkiewicz et al., 2006). As such, we suggest that functional cheetah corridors are areas that have acceptable levels of human tolerance and sufficient access to prey necessary to support movement. Functional corridors would provide population connectivity within and between cheetah core populations and allow natural movement and genetic exchange. We continue to explore the concept of functional cheetah corridors and implications for development of linkage areas in the subsequent sections.

CHEETAH BIOLOGY AND RESILIENCE FACTORS: IDENTIFYING POTENTIAL LINKAGE AREAS

Unlike larger-bodied predators, cheetahs have more flexibility in their energetics and arguably more plasticity in the behavior that underlies their ability to access food. Moreover, conservation efforts have provided successful models in the reduction of human conflict (Marker

and Dickman, 2004), and cheetahs are generally viewed as more benign than other large predators. We suggest that these factors contribute to the cheetah's resilience, and can provide insight into the design of functional cheetah corridors.

Cheetah Resilience Factors

Large carnivore species and populations have widely varying levels of ecological resilience to human-caused habitat fragmentation (Crooks, 2002; Gittleman, 2001). The cheetah's resilience is challenged by their large home range size, relatively large body size, and primarily diurnal hunting behavior largely because these traits tend to increase human conflict and persecution. Additional factors that reduce cheetah resilience include their sensitivity to interspecific competition and low population density. These combined factors have driven range-wide declines and cheetahs remain among the most vulnerable large carnivores worldwide (Ray et al., 2005).

On the positive side, the cheetah possesses a number of traits that bestow considerable ecological resilience, including their high mobility, secretive behavior, considerable habitat flexibility (Bissett and Bernard, 2006), diverse and flexible prey base (Hayward et al., 2006), and large litter size (Kelly et al., 1998). Cheetahs are habitat generalists, excluded only from forest and extremely rugged areas. They are able to exist in a wide range of land-cover types, ranging from thick bush to arid and hyperarid areas and once ranged widely throughout Africa and from the Arabian Peninsula, through Central Asia and India.

These traits enable cheetahs to utilize marginal habitats outside of protected areas, where there is little interspecific competition from dominant predators and relatively low risk of cropland conversion. Identification of marginal, human-unoccupied habitats used by cheetahs could serve to increase connectivity within and between existing core populations and reduce habitat fragmentation, by serving as functional cheetah corridors or linkage areas.

Identifying Potential Linkage Areas

Are there unoccupied areas, at low risk for cropland conversion, that could serve as functional linkages for cheetahs? One of the results of the last few centuries of human alteration has been an irreversible baseline shift that has eliminated once great migrations and left vast areas devoid of herds of ungulate prey and their predators including cheetah. Although most of the productive habitat areas have already been converted or otherwise lost, there remains potential in vast areas of marginal, arid habitat. In Fig. 10.2, we plot two human land-uses (cropland and nomadic livestock production) against rainfall to conceptually identify these areas of marginal habitat that may be candidates for cheetah habitat linkages. Rainfall is one of the principal requirements for cropland. Most of the areas that have been converted to crops have annual rainfall between approximately 300 and 1200 mm, and conversion peaks at approximately 700 mm (Fig. 10.2A, black diamonds). Although the broad area classified as cropland likely includes some remnant semiarid grasslands and woodland savannahs, we suggest the long-term potential for cheetah linkage habitat is low for these areas where crop conversion potential is high (Fig. 10.2B, black line). Human demographic trends throughout Africa suggest that it will be very difficult for conservation efforts to alter land conversion patterns to cropland at a meaningful scale. Therefore, high potential crop conversion areas are probably not ideal candidates for conservation investment, despite the fact that these areas represent historically productive cheetah habitat.

However, vast areas exist that have lower risk of crop conversion because they are too arid. Livestock production—commercial, pastoral, and/or subsistence—is the primary land-use in such areas (Fig. 10.2A, gray squares). In those

areas, competition from dominant predators is low, and human conflict and wild prey availability will continue to be the driving biotic forces. Such areas may be good candidates for successful implementation of landscape connectivity strategies for cheetahs (Fig. 10.2B). Nomadic, subsistence livestock production is one of the principal human activities in many such marginal and arid areas throughout much of Africa (Fig. 10.3) however, relatively little conservation investment currently addresses conflicts involving such agropastoralists and cheetahs.

We suggest that with appropriate conservation investment (e.g., addressing human conflict and ensuring sufficient prey), these vast, marginal areas could represent important linkage habitats. Additional analyses that incorporate more detailed sociological, prey, and land-use data would represent a reasonable next step. As discussed earlier in this chapter, diet and energetic studies suggest that the cheetah may be able to subsist on small prey in such areas, and could possibly utilize these areas for dispersal as functional linkages.

IMPLICATIONS FOR FUTURE CHEETAH CONSERVATION STRATEGIES

As the underlying drivers of habitat fragmentation and habitat destruction in Africa increase over the next century, the risk to cheetah populations will also increase. Conservation investment in linkage areas could facilitate animal movement, which is especially critical for small isolated populations that are below recommended viability targets. Although natural dispersal is the preferred process in which populations should be linked, in some cases conservation translocations may be required (Chapter 20).

Implementing strategies that minimize human-wildlife conflict (Chapters 13 and 15) remains cornerstone to connecting populations. Around the world, some progress is being made

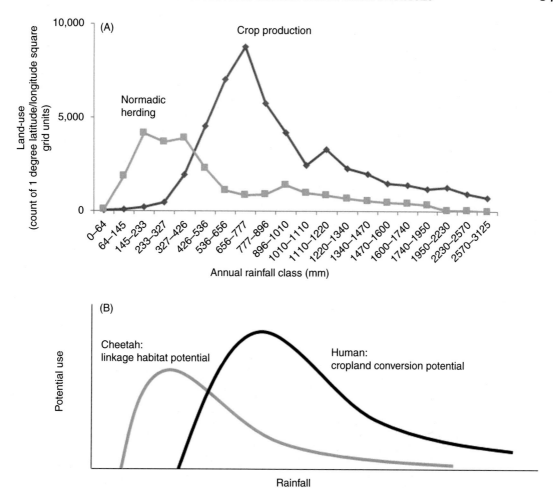

FIGURE 10.2 (A) Relative land-use area for crop production *(black diamonds)* and nomadic herding *(gray squares)* as a function of annual rainfall (mm). Land areas utilized for crop production peaks at 656–777 mm of rainfall, corresponding to semiarid grasslands and savannahs. Cropping use declines with higher rainfall because increasingly wetter areas have woodland and forest habitats that are less suitable for agriculture. Cheetahs historically used areas that have been converted to crops, but have been largely extirpated from such highly productive areas. (B) Conceptual model based on histograms in (A). The model illustrates relative suitability for human cropland conversion and potential suitability for cheetah linkage habitat as a function of rainfall. Cheetahs can also utilize more arid areas (i.e., less than ~500 mm rainfall per year) for movement that are generally unsuitable and at low risk for crop conversion. Nomadic herding *(gray squares)* is the primary land-uses in these arid areas. Land-use data from FAO/UNEP land-use systems (www.fao.org) and rainfall data from University of East Anglia Climatic Research Unit (http://harvestchoice.org).

both conceptually and in practice. Alternative models that allow people and predators to live together—that is, coexistence models—have started to demonstrate success (Phalan et al., 2011). Carnivores in Europe and North America have experienced recent population recovery and distribution expansion in human-dominated landscapes (Chapron et al., 2014). Carnivore range expansion after over a century of population declines has been demonstrated in North America,

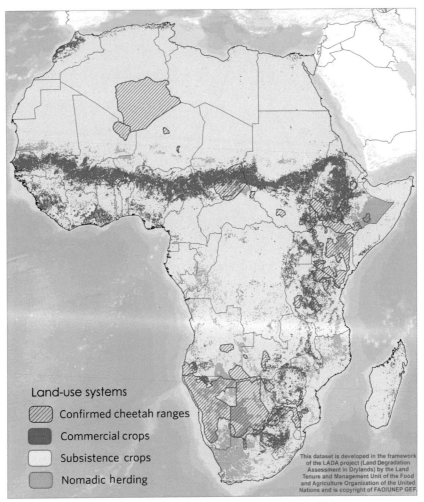

FIGURE 10.3 **Map showing confirmed resident cheetah ranges on generalized land-uses of commercial and subsistence cropland and nomadic herding across the African continent.** We suggest that marginal habitat areas where the primary land-use is nomadic herding and livestock raising could serve as linkage areas within and between extant populations if human tolerance and prey base are both adequate. Land-use data from FAO/UNEP land-use systems (www.fao.org) and cheetah ranges based on Durant et al. (2017).

with species including the grizzly bear (*Ursus arctos*), gray wolf (*Canis lupus*), mountain lion (*Puma concolor*), fisher (*Martes pennanti*), and coyote (*Canis latrans*) (Allen et al., 2015; Gompper, 2002; LaRue et al., 2012; Oakleaf et al., 2006; Pyare et al., 2004; Swenson et al., 1995).

In many ways, the cheetah's natural history and biology provide many potential advantages

for its conservation. Cheetahs do not kill people nor do they typically prey on adult cattle, and effective calf and smallstock management options have already been demonstrated to be successful (Marker and Dickman, 2004). Furthermore, the cheetah's elusive nature and ability to live in marginal, low productivity, arid areas may lead to underestimation of distribution. Most distribution maps are based on a combination of occurrence data, survey results, and expert opinion. Two recent studies support the assertion of range underestimate. Andresen et al. (2014) found that detection was 16% higher using site occupancy models and replicated spoor and camera-trap surveys than with more traditional presence–absence estimates, and Mateo-Sánchez et al. (2015) found that based on genetic data populations of brown bear (*Ursus arctos*) in Spain were more connected than habitat models would predict.

However, even remote areas are increasingly under pressure from new human activities throughout Africa, such as oil, gas, and mineral development and associated road networks. The coming decades will bring new and unexpected types of human activities, making additional research and conservation investment even more urgently needed outside of traditional focal areas for cheetahs. Perhaps the most urgent conservation need for the cheetah continues to be the prevention of large-scale habitat conversion and infrastructure development in remaining intact areas associated with core population strongholds, especially as population pressure and technological ability of humans to exploit new areas increase. Remote, undocumented extant populations could be providing important demographic buffers and population connectivity for the known cheetah populations. New human development could certainly harm this fragile balance in unexpected and rapid ways.

Wolves were erroneously thought to require wilderness because they had been extirpated by human persecution from everywhere in the lower 48 United States except wilderness areas (Mech, 1995). Over a decade later, a study of wolf recolonization in the northern Rocky Mountains of the United States demonstrated that wolf habitat is essentially any place where there is both sufficient prey and human tolerance (Oakleaf et al., 2006). The evidence presented in this chapter suggests that cheetah habitat can be similarly defined. Carnivores around the world have demonstrated their resilience and ability to bounce back after decades of decline when there is a confluence of human tolerance and prey density. Although this concept is simple, successful implementation remains complex.

Understanding prey (including preference and minimum requirements in both core stronghold areas and in relation to dispersal and movement), human tolerance and conflict, as well as emerging human land-use with changing demographics are key areas in need of further research. Establishing functional linkages between existing strongholds could reduce extinction risk by providing new source populations, facilitating habitat refugia, and enabling connectivity between core population areas. Like other large carnivores around the world, cheetahs have the potential to expand and thrive in new areas—and their survival will depend on human behavior wherever they inevitably come into contact with humans.

References

Allen, M.L., Evans, B.E., Gunther, M.S., 2015. A potential range expansion of the coastal fisher (*Pekania pennanti*) population in California. Calif. Fish Game 101, 280–285.

Andresen, L., Everatt, K.T., Somers, M.J., 2014. Use of site occupancy models for targeted monitoring of the cheetah: cheetah occupancy and detectability. J. Zool. 292, 212–220.

Battin, J., 2004. When good animals love bad habitats: ecological traps and the conservation of animal populations. Conserv. Biol. 18, 1482–1491.

Bennett, A.F., 1999. Linkages in the Landscape: The Role of Corridors and Connectivity in Wildlife Conservation. ICUN, Gland, Switzerland.

Bissett, C., Bernard, R.T.F., 2006. Habitat selection and feeding ecology of the cheetah (*Acinonyx jubatus*) in thicket vegetation: is the cheetah a savanna specialist? J. Zool. 271, 310–317.

Carbone, C., Gittleman, J.L., 2002. A common rule for the scaling of carnivore density. Science 295, 2273–2276.

Carbone, C., Teacher, A., Rowcliffe, J.M., 2007. The costs of carnivory. PLoS Biol. 5, e22.

Caro, T.M., 1994. Cheetahs of the Serengeti Plains: Group Living in an Asocial Species. University of Chicago Press, Chicago, IL.

Channell, R., Lomolino, M.V., 2000. Trajectories to extinction: spatial dynamics of the contraction of geographical ranges. J. Biogeogr. 27, 169–179.

Chapron, G., et al., 2014. Recovery of large carnivores in Europe's modern human-dominated landscapes. Science 346 (6216), 1517–1519.

Chetkiewicz, C.-L.B., St. Clair, C.C., Boyce, M.S., 2006. Corridors for conservation: integrating pattern and process. Annu. Rev. Ecol. Evol. Syst. 37, 317–342.

Creel, S., 2001. Four factors modifying the effect of competition on carnivore population dynamics as illustrated by African wild dogs. Conserv. Biol. 15, 271–274.

Crooks, K.R., 2002. Relative sensitivities of mammalian carnivores to habitat fragmentation. Conserv. Biol. 16, 488–502.

Crooks, K.R., Burdett, C.L., Theobald, D.M., Rondinini, C., Boitani, L., 2011. Global patterns of fragmentation and connectivity of mammalian carnivore habitat. Philos. Trans. R. Soc. B Biol. Sci. 366, 2642–2651.

Crooks, K.R., Sanjayan, M., (Eds.), 2006. Connectivity Conservation. Cambridge University Press, Cambridge, United Kingdom.

Dickman, A., 2010. Complexities of conflict: the importance of considering social factors for effectively resolving human–wildlife conflict. Anim. Conserv. 13, 458–466.

Durant, S.M., 1998. Competition refuges and coexistence: an example from Serengeti carnivores. J. Anim. Ecol. 67, 370–386.

Durant, S.M., 2000. Living with the enemy: avoidance of hyenas and lions by cheetahs in the Serengeti. Behav. Ecol. 11, 624–632.

Durant, S.M., Mitchell, N., Groom, R., Pettorelli, N., Ipavec, A., Jacobson, A., Woodroffe, R., Bohm, M., Hunter, L., Becker, M., Broekuis, F., Bashir, S., Andresen, L., Aschenborn, O., Beddiaf, M., Belbachir, F., Belbachir-Bazi, A., Berbash, A., Brandao de Matos Machado, I., Breitenmoser, C., Chege, M., Cilliers, D., Davies-Mostert, H., Dickman, A., Fabiano, E., Farhadinia, M., Funston, P., Henschel, P., Horgan, J., de Iongh, H., Jowkar, H., Klein, R., Lindsey, P., Marker, L., Marnewick, K., Melzheimer, J., Merkle, J., Msoka, J., Msuha, M., O'Neill, H., Parker, M., Purchase, G., Saidu, Y., Samaila, S., Samna, A., Schmidt-Küntzel, A., Selebatso, E., Sogbohossou, E., Soultan, A., Stone, E., van der Meer, E., van Vuuren, R., Wykstra, M., Young-Overton, K., 2017. The global decline of cheetah and what it means for conservation. Proc. Natl. Acad. Sci. USA 114, 528–533.

Farhadinia, M.S., Akbari, H., Eslami, M., Adibi, M.A., 2016. A review of ecology and conservation status of Asiatic cheetah in Iran. Cat News 10, 18–26.

Frankham, R., Ballou, J.D., Ralls, K., Eldridge, M.D.B, Dudash, M.R., Fenster, C.B., Lacy, R.C., Sunnucks, P., Genetic Management of Fragmented Animal and Plant Populations, 2017, Oxford University Press.

Frankham, R., Bradshaw, C.J.A., Brook, B.W., 2014. Genetics in conservation management: revised recommendations for the 50/500 rules, Red List criteria and population viability analyses. Biol. Conserv. 170, 56–63.

Fuller, T.K., Sievert, P.R., 2001. Carnivore demography and the consequences of changes in prey availability. In: Gittleman, J.L. et al., (Ed.), Carnivore Conservation. Cambridge University Press, Cambridge, pp. 163–168.

García-Dorado, A., 2015. On the consequences of ignoring purging on genetic recommendations for minimum viable population rules. Heredity 115, 185.

Gittleman, J.L., 2001. Carnivore Conservation. Cambridge University Press, Cambridge, United Kingdom.

Gompper, M.E., 2002. Top carnivores in the suburbs? Ecological and conservation issues raised by colonization of north-eastern North America by coyotes. BioScience 52, 185–190.

Hale, R., Treml, E.A., Swearer, S.E., 2015. Evaluating the metapopulation consequences of ecological traps. Proc. R. Soc. B Biol. Sci. 282, 20142930.

Hayward, M.W., Hofmeyr, M., O'Brien, J., Kerley, G.I.H., 2006. Prey preferences of the cheetah (*Acinonyx jubatus*) (Felidae: Carnivora): morphological limitations or the need to capture rapidly consumable prey before kleptoparasites arrive? J. Zool. 270, 615–627.

Hayward, M.W., Kerley, G.I.H., 2008. Prey preferences and dietary overlap amongst Africa's large predators. South Afr. J. Wildl. Res. 38, 93–108.

Homewood, K., Lambin, E.F., Coast, E., Kariuki, A., Kikula, I., Kivelia, J., Said, M., Serneels, S., Thompson, M., 2001. Long-term changes in Serengeti-Mara wildebeest and land cover: pastoralism, population, or policies? Proc. Natl. Acad. Sci. 98, 12544–12549.

Kelly, M.J., Laurenson, M.K., FitzGibbon, C.D., Collins, D.A., Durant, S.M., Frame, G.W., Bertram, B.C., Caro, T.M., 1998. Demography of the Serengeti cheetah (*Acinonyx jubatus*) population: the first 25 years. J. Zool. 244, 473–488.

LaRue, M.A., Nielsen, C.K., Dowling, M., Miller, K., Wilson, B., Shaw, H., Anderson, C.R., 2012. Cougars are recolonizing the midwest: analysis of cougar confirmations during 1990–2008. J. Wildl. Manage. 76, 1364–1369.

Laurenson, M.K., 1994. High juvenile mortality in cheetahs (*Acinonyx jubatus*) and its consequences for maternal care. J. Zool. 234, 387–408.

Linnell, J.D., Swenson, J.E., Anderson, R., 2001. Predators and people: conservation of large carnivores is possible at high human densities if management policy is favourable. Anim. Conserv. 4, 345–349.

Marker, L., Dickman, A., 2004. Human aspects of cheetah conservation: lessons learned from the Namibian farmlands. Hum. Dimens. Wildl. 9, 297–305.

Marker, L.L., Dickman, A.J., Mills, M.G.L., Jeo, R.M., Macdonald, D.W., 2008. Spatial ecology of cheetahs on north-central Namibian farmlands. J. Zool. 274, 226–238.

Marker, L., Dickman, A., Mills, M.G., Macdonald, D., 2003a. Aspects of the management of cheetahs, *Acinonyx jubatus jubatus*, trapped on Namibian farmlands. Biol. Conserv. 114, 401–412.

Marker, L.L., Mills, M.G.L., Macdonald, D.W., 2003b. Factors influencing perceptions of conflict and tolerance toward cheetahs on Namibian farmlands. Conserv. Biol. 17, 1290–1298.

Mateo-Sánchez, M.C., Balkenhol, N., Cushman, S., Pérez, T., Domínguez, A., Saura, S., 2015. Estimating effective landscape distances and movement corridors: comparison of habitat and genetic data. Ecosphere 6, 1–16.

Mech, L.D., 1995. The challenge and opportunity of recovering wolf populations. Conserv. Biol. 9, 270–278.

Mills, L.S., Allendorf, F.W., 1996. The one-migrant-per-generation rule in conservation and management. Conserv. Biol. 10, 1509–1518.

Mills, M.G., Gorman, M.L., 1997. Factors affecting the density and distribution of wild dogs in the Kruger National Park. Conserv. Biol. 11, 1397–1406.

Mitchell, M.S., Hebblewhite, M., 2012. Carnivore habitat ecology: integrating theory and application. In: Boitani, L., Powell, R.A. (Eds.), Carnivore Ecology and Conservation: A Handbook of Techniques. Oxford University Press, London, UK, pp. 218–255.

Nghikembua, M., Harris, J., Tregenza, T., Marker, L., 2016. Spatial and temporal habitat use by GPS collared male cheetahs in modified bushland habitat. Open J. For. 06, 269–280.

Nowell, K., Jackson, P., 1996. Wild Cats: Status Survey and Conservation Action Plan. IUCN, Gland, Switzerland.

Oakleaf, J.K., Murray, D.L., Oakleaf, J.R., Bangs, E.E., Mack, C.M., Smith, D.W., Fontaine, J.A., Jimenez, M.D., Meier, T.J., Niemeyer, C.C., 2006. Habitat selection by recolonizing wolves in the northern Rocky Mountains of the United States. J. Wildl. Manage. 70, 554–563.

Ogutu, J.O., Bhola, N., Reid, R., 2005. The effects of pastoralism and protection on the density and distribution of carnivores and their prey in the Mara ecosystem of Kenya. J. Zool. 265, 281–293.

Phalan, B., Onial, M., Balmford, A., Green, R.E., 2011. Reconciling food production and biodiversity conservation: land sharing and land sparing compared. Science 333, 1289–1291.

Pyare, S., Cain, S., Moody, D., Schwartz, C., Berger, J., 2004. Carnivore re-colonisation: reality, possibility and a non-equilibrium century for grizzly bears in the southern Yellowstone ecosystem. Anim. Conserv. 7, 71–77.

Ray, J.C., Hunter, L., Zigouris, J., 2005. Setting Conservation and Research Priorities for Larger African Carnivores. Wildlife Conservation Society, Bronx, NY.

Ripple, W.J., Estes, J.A., Beschta, R.L., Wilmers, C.C., Ritchie, E.G., Hebblewhite, M., Berger, J., Elmhagen, B., Letnic, M., Nelson, M.P., et al., 2014. Status and ecological effects of the world's largest carnivores. Science 343, 1241484.

Schipper, J., Chanson, J.S., Chiozza, F., Cox, N.A., Hoffmann, M., Katariya, V., Lamoreux, J., Rodrigues, A.S., Stuart, S.N., Temple, H.J., et al., 2008. The status of the world's land and marine mammals: diversity, threat, and knowledge. Science 322, 225–230.

Soule, M.E., Wilcox, B.A., 1980. Conservation Biology: An Evolutionary-Ecological Perspective. Sinauer Associates, Sunderland, MA.

Swenson, J.E., Wabakken, P., Sandegren, F., Bjärvall, A., Franzén, R., Söderberg, A., 1995. The near extinction and recovery of brown bears in Scandinavia in relation to the bear management policies of Norway and Sweden. Wildl. Biol. 1, 11–25.

Traill, L.W., Brook, B.W., Frankham, R.R., Bradshaw, C.J.A., 2010. Pragmatic population viability targets in a rapidly changing world. Biol. Conserv. 143, 28–34.

United Nations, Department of Economic and Social Affairs, Population Division, 2015. World Population Prospects: The 2015 Revision, Methodology of the United Nations Population Estimates and Projections.

Watson, F.G., Becker, M.S., Milanzi, J., Nyirenda, M., 2015. Human encroachment into protected area networks in Zambia: implications for large carnivore conservation. Reg. Environ. Change 15, 415–429.

Williamson, D., Williamson, J., 1984. Botswana's fences and the depletion of Kalahari wildlife. Oryx 18, 218–222.

Winterbach, H.E.K., Winterbach, C.W., Boast, L.K., Klein, R., Somers, M.J., 2015. Relative availability of natural prey versus livestock predicts landscape suitability for cheetahs *Acinonyx jubatus* in Botswana. PeerJ 3, e1033.

Winterbach, H.E.K., Winterbach, C.W., Somers, M.J., 2014. Landscape suitability in Botswana for the conservation of its six large African carnivores. PLoS One 9, e100202.

Winterbach, H.E.K., Winterbach, C.W., Somers, M.J., Hayward, M.W., 2013. Key factors and related principles in the conservation of large African carnivores: Factors and principles in carnivore conservation. Mammal Rev. 43, 89–110.

Woodroffe, R., 2000. Predators and people: using human densities to interpret declines of large carnivores. Anim. Conserv. 3, 165–173.

Woodroffe, R., Ginsberg, J.R., 1997. Past and future causes of wild dogs' population decline. In: Woodroofe, R., Ginsberg, J., Macdonald, D. (Eds.), The African Wild Dog: Status Survey and Conservation Action Plan. IUCN, Gland, Switzerland.

Woodroffe, R., Thirgood, S., Rabinowitz, A., 2005. People and Wildlife, Conflict or Co-existence? Cambridge University Press, Cambridge.

The Status of Key Prey Species and the Consequences of Prey Loss for Cheetah Conservation in North and West Africa

Laurie Marker, Thomas Rabeil**, Pierre Comizzoli†,
Hayley Clements‡,§, Matti T. Nghikembua*,
Matt W. Hayward‡,¶,‡‡, Craig J. Tambling‡,††*

*Cheetah Conservation Fund, Otjiwarongo, Namibia
**Sahara Conservation Fund, Bussy St. Georges, France
†Smithsonian Conservation Biology Institute, Washington, DC, United States
‡Nelson Mandela University, Port Elizabeth, South Africa
§Monash University, Melbourne, VIC, Australia
¶Bangor University, Bangor, United Kingdom
††University of Fort Hare, Alice, South Africa
‡‡University of Newcastle, Callaghan, NSW, Australia

INTRODUCTION

Conservation of cheetahs (*Acinonyx jubatus*) throughout their remaining range can only be accomplished if the conservation of their prey is ensured (Wolf and Ripple, 2016). Today, many of the world's wildlife species are threatened by habitat loss (Chapter 10), persecution, and unsustainable harvesting, with global declines in the numbers of large carnivores and their prey accelerating in the past 40 years (Ripple et al., 2014, 2015). For the cheetah, the declining abundance of prey species threatens its survival throughout much of its range as reductions in prey numbers lower predator carrying capacity (Hayward et al., 2007). In areas with prey reductions predators may be more likely to kill domestic species, in turn driving human-wildlife conflict and potential predator persecution (Wolf and Ripple, 2016; Chapter 13). Global drivers of ungulate extirpation include unsustainable hunting and poaching for meat (bushmeat), use of poison by herders (targeting competing ungulates), and growing human populations that lead to habitat loss and competition with livestock for grazing (Wolf and Ripple, 2016).

The global decline of ungulate species is particularly relevant for cheetah survival, considering their morphological adaptations (Chapter 7) to capture prey animals which pose a low risk of injury (Hayward et al., 2006). This specialization has resulted in cheetahs having a narrow preferred prey weight range (14–40 kg), and avoiding ungulates weighing less than 14 kg or more than 135 kg (Clements et al., 2014). The majority (78%) of recorded preferred cheetah prey are listed as Least Concern by the International Union for Conservation of Nature (IUCN); a small percentage (~22%) as Near Threatened, and none as Vulnerable, Endangered, or Critically Endangered (Wolf and Ripple, 2016). However, this seemingly positive outlook draws on data primarily collected in southern and eastern Africa; regions that support adequately managed protected area systems. In contrast, wildlife declines are more severe in North and West Africa (Craigie et al., 2010; Durant et al., 2014), raising the question of whether the prey species of cheetah outside of southern and eastern Africa are faring as well.

In this chapter, we assess the conservation status of cheetah prey, primarily focusing on North and West Africa where remaining ungulate prey species are threatened or endangered. Key threats to cheetah prey populations are discussed.

STATUS OF THE NORTHWEST AFRICAN CHEETAH

The conservation status of the northwest African cheetah (*A. j. hecki*), also called the Saharan cheetah, is poorly documented (Caro, 2013; Claro

et al., 2006; Newby and Grettenberger, 1986; Wacher et al., 2005). Once broadly distributed across the Sahelo-Saharan region, the northwest African cheetah now occurs in only 4.7% of its former range (Fig. 11.1; Chapter 4) and is categorized as Critically Endangered by IUCN (Belbachir, 2008; Durant et al., 2014). It is now reported in only four or five countries: Algeria, Benin, Burkina Faso, and Niger, and possibly Mali (Durant et al., 2017; Chapter 4). The region where the cheetah is found is referred to as Sahelo-Sahara; and the local habitats are classified into Sahelo-Sudanian and Saharan biomes. Over the last 20 years, sightings, including pictures from camera traps, have been reported in only three areas (Fig. 11.1—note, sighting dates precede publication dates by extended periods of time in most occasions):

- the southern Algerian Sahara in the Ahaggar and the Tassili National Parks (Belbachir et al., 2015; Saharan biome)
- the Termit Massif in Niger (Claro et al., 2006; Rabeil et al., 2008; Sillero-Zubiri et al., 2015; Saharan biome)
- the W–Arly–Pendjari (WAP) complex in Niger, Benin and Burkina Faso (Berzins et al., 2007; Caro, 2013; Claro et al., 2006; Sahelo-Sudanian biome).

DIET OF THE NORTHWEST AFRICAN CHEETAH

The cheetah is an opportunistic predator with specific prey preferences (Hayward et al., 2006; Marker et al., 2003; Chapter 8); however, very little is known about the northwest African cheetah's feeding ecology. Extrapolating the data from southern and eastern Africa and assuming that the northwest African cheetah would show similar prey weight preferences, eight ungulate species occurring in North and West Africa are predicted to be "preferred" prey of the northwest African cheetah [i.e., fall within the 14–40 kg (three-quarter adult female weight) range identified as "preferred"; Clements et al., 2014; Table 11.1]. An additional six ungulate species

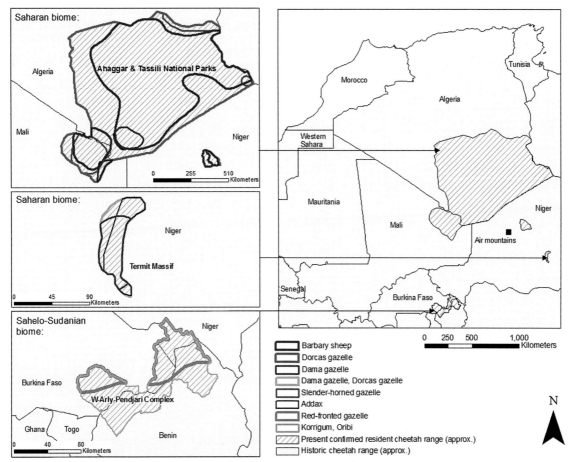

FIGURE 11.1 **Map showing the historic and current distribution of the northwest African cheetah.** The distribution of the cheetah's three main prey species (red fronted gazelle, oribi, and dorcas gazelle) and other potential key prey species that are vulnerable, endangered, or critically endangered are shown. Note: The potential prey species listed as least concern and near threatened in Table 11.1 also occur in the W–Arly–Pendjari complex. Data source: The IUCN (International Union for Conservation of Nature) red list of threatened species, http://www.iucnredlist.org/.

found in the area are within the accessible prey weight range of the northwest African cheetah (i.e., "accessible" prey; 14–135 kg as defined by Clements et al., 2014; Table 11.1).

Indeed, ungulate prey species falling within the preferred prey weight range of cheetahs have been identified as cheetah prey in diet studies of the northwest African cheetah (see below for details). Furthermore, prey of suboptimal size (e.g., hares, rodents, birds) are often detected in northwest African cheetah diets, representing a possible alternative food source in times

of ungulate scarcity. However, it is unknown whether this alternate food source would be sufficient for cheetah survival, should ungulate populations further decline, given that even in large quantities it would not meet the metabolic requirements of large felids like the cheetah (Carbone et al., 2007). The prevalence of suboptimal size prey in the northwest African cheetah's diet may, therefore, rather suggest a cheetah population about to decline toward extinction.

In the Sahelo-Sudanian biome, small and medium sized ungulates, such as red-fronted

TABLE 11.1   Potential Cheetah Ungulate Prey Species (Under 135 kg Including Accessible and Preferred Size Classes) in the Saharan and Sahelo-Sudanian Biomes. The 3/4 adult female body mass, cheetah preference class, and International Union for the Conservation of Nature (IUCN) status are shown. All populations are in decline in North and West Africa

	3/4 adult female bodymass (kg)[a]	Cheetah preference class[b]	IUCN status[c]
SAHELO-SUDANIAN BIOME			
Bushbuck (*Tragelaphus scriptus*)	24–32	Preferred	LC
Duiker, common (*Sylvicapra grimmia*)	8–20	Preferred	LC
Gazelle, red-fronted (*Gazella rufifrons*)	19–23	Preferred	V
Hartebeest (*Alcelaphus buselaphus major*)	90+	Accessible	NT
Kob, buffons (*Kobus kob kob*)	34	Preferred	LC
Korrigum (*Damaliscus lunatus korrigum*)	92–102	Accessible	V
Oribi (*Ourebia ourebi*)	6–15	Preferred	LC
Reedbuck, bohor (*Redunca redunca*)	26–34	Preferred	LC
Warthog, common (*Phacochoerus africanus*)	56	Accessible	LC
SAHARAN BIOME			
Addax (*Addax nasomaculatus*)	45–68	Accessible	CE
Barbary sheep (*Ammotragus lervia*)	23–109	Accessible	V
Gazelle, dama (*Nanger dama*)	30–56	Accessible	CE
Gazelle, dorcas (*Gazella Dorcas*)	11–15	Preferred	V
Gazelle, slender-horned (*Gazella leptoceros*)	15–23	Preferred	E

[a]*Body mass data from Castelló, J.R. (2016), with the exception of warthog data from Bodendorfer et al. (2006).*
[b]*Preferred: 14–40 kg, avoided: <14 kg and >135 kg, accessible: 40–135 kg (Clements et al., 2014).*
[c]*IUCN red list; CE, Critically Endangered; E, Endangered; LC, Least Concern; NT, Near Threatened; V, Vulnerable—http://www.iucnredlist.org/.*

gazelle (*Eudorcas rufifrons*) and oribi (*Ourebia ourebi*) are preyed upon by cheetahs (Rabeil 2003). Cape hare (*Lepus capensis*) and helmeted guineafowl (*Numida meleagris*) are abundant in W Park, and are also common in the cheetah's diet (Rabeil, unpublished data).

In the Saharan biome, predation of Dorcas gazelles (*Gazella dorcas*) by cheetahs has been reported in the Termit Massif (Claro and Sissler, 2003), the Aïr Mountains in Niger (Poilecot, 1996), and in the Ahaggar Massif in Algeria (Belbachir et al., 2015). Within Termit and Tin-Toumma National Nature Reserve in Niger cheetah's diet also includes Cape hare and small rodents, such as jerboas (*Jaculus orientalis*) and gerbils (*Dipodillus campestris*) (Rabeil,

unpublished data), similar to cheetahs in Algeria, which have been found to prey on Cape hare (Wacher et al., 2005; Chapter 8). In addition, cheetahs in the Saharan biome are likely to supplement their diet with birds [e.g., helmeted guinea fowl, Nubian bustard (*Neotis nuba*) and Arabian bustard (*Ardeotis arabs*)] (Claro et al., 2006; Sillero-Zubiri et al., 2015).

Depredation of smallstock (goats and sheep) by northwest African cheetahs is known to occur in and around WAP complex (Garba and Di Silvestre, 2008; Sogbohossou et al., 2011). In the Saharan biome, depredation of smallstock has also been reported by the local communities, in addition to reports and evidence of northwest African cheetahs killing young camels in the

Termit Massif (Matchano, 2011) and the Ahaggar Massif (Wacher et al., 2005).

STATUS OF THE NORTHWEST AFRICAN CHEETAH'S MAIN PREY SPECIES

Most large mammals in the Sahelo-Saharan region, including ungulates within the cheetah's accessible prey weight range (14–135 kg; Table 11.1), have suffered severe population declines and geographic range contractions of over 90% in the last 30 years (Durant et al., 2014; Walther, 2016). Moreover, most of the large mammals of West Africa are primarily found in protected areas (Craigie et al., 2010). Of the preferred and accessible ungulate prey species in the region, 50% are listed by the IUCN as Vulnerable, Endangered, or Critically Endangered (Table 11.1). Two additional species, the scimitar horned oryx (*Oryx dammah)* and bubal hartebeest (*Alcelaphus buselaphus buselaphus*), which could have been cheetah prey in the past, are now extinct in the wild (Durant et al., 2014).

As the three main ungulate species that northwest African cheetahs have been observed to prey upon are the red-fronted gazelle, oribi, and Dorcas gazelle, further information about their status is provided later. The limited amount of data available regarding cheetah diet precludes from knowing the extent to which less abundant species are preyed upon. Predation on the Critically Endangered Dama gazelle (*Nanger dama)*—found in the Termit Massif and the Aïr Mountains (extinct in Algeria)—has not been documented nor reported; likely due to their small numbers (less than 250 remaining of which each remaining subpopulation has <50 mature individuals, IUCN SSC Antelope Specialist Group, 2016b). Predation on the Auodad or Barbary sheep (*Ammotragus lervia*) is likely low, but underreported due to the inaccessibility of the mountainous habitat that Barbary sheep inhabit.

Red-fronted gazelle: Being less water dependent than other sympatric ungulates (Rabeil, 2003; Clément et al., 2015), the red-fronted gazelle occurs in a broad range of habitats, including both the wooded savannas and grasslands of the Saharan and Sahelo-Sudanian biome (Fig. 11.1). The red-fronted gazelle is listed as Vulnerable by the IUCN (IUCN SSC Antelope Specialist Group, 2008b). However, its conservation status is poorly documented and several populations have been declining since the last assessment in 2008 (Clément et al., 2015). Despite the rising concern for the future of the red-fronted gazelle, studies aiming to address these declines are limited. The red-fronted gazelle is less abundant than other ungulates with which it coexists (i.e., oribi) with density estimates between 0.01 and 0.2 individual/km^2 in W Park, Niger (Rabeil, 2003).

Oribi: The oribi is found mainly in and bordering protected areas in the Sahelo-Sudanian biome (Fig. 11.1). Although listed as Least Concern by the IUCN, populations are declining across most of Africa (IUCN SSC Antelope Specialist Group, 2016a). Oribi inhabit wooded savanna and open grasslands, with a preference for tall grasses where they can hide from predators (Mduma and Sinclair, 1994). They avoid densely vegetated habitat, such as riparian or gallery forest. In W Park, Niger, although the oribi are more water-dependent than red-fronted gazelle, their distribution is very similar, with densities of oribi (0.1–1 individual/ km^2) considerably higher than red-fronted gazelle (Rabeil, 2003).

Dorcas gazelle: Dorcas gazelles are adaptable, inhabiting a wide range of habitats from grasslands to the high plateau of the Ahaggar Mountains in Algeria. Listed as Vulnerable by the IUCN (IUCN SSC Antelope Specialist Group, 2008a), Dorcas gazelles were formerly found in large numbers from the Atlantic Coast to the Red Sea. Globally, the numbers of Dorcas gazelles are declining primarily as a consequence of overhunting and poaching. Yet of the predicted, preferred, and accessible species of the Saharan region, only the Dorcas gazelle is sufficiently widespread and abundant to likely represent a key prey species for cheetah. Densities of the Dorcas gazelle vary considerably across the Saharan countries (Beudels-Jamar et al., 2005). The highest densities of 6–8 gazelles/km² are recorded in the grasslands of the Ouadi Rimé—Ouadi Achim Game Reserve in Chad (Wacher and Newby, 2012). In areas where northwest African cheetahs and Dorcas gazelle coexist, densities are lower and declining rapidly, with only 1–3 gazelles/km² in the Termit Massif (Rabeil, 2015; Wacher and Newby, 2012), and even fewer in the Ahaggar Massif (Wacher et al., 2005; Fig. 11.1) and in the Aïr Mountains (Rabeil, 2014) where densities range from 0.05 to 0.5 gazelles/km².

PRIMARY THREATS TO THE NORTHWEST AFRICAN CHEETAH'S KEY PREY SPECIES

Although exploitation (hunting and poaching) has been the principal cause of the decline and disappearance of desert ungulates throughout their range, new threats are rapidly growing in significance in the Sahelo-Saharan region, including land use change (agricultural encroachment, habitat degradation, and resource extraction) and climate change (Beudels-Jamar et al., 2005; Newby et al., 2016). Exploitation and land use changes are expected to be further exacerbated by continued human population growth.

Hunting and Poaching

In West and Central Africa, bushmeat hunting is a survival strategy for large numbers of people, sometimes comprising the majority of animal protein consumed, and contributing significantly to household incomes (Brashares et al., 2004; Wilkie and Carpenter, 1999). Traditionally, hunting in the Sahelo-Saharan region

occurred on foot, or with horses, and dogs, using primitive weapons and traps (Newby et al., 2016). In recent decades however, increased access to firearms and motorized vehicles, together with an increase in road infrastructure, has enabled people to (1) access remote areas, (2) harvest wildlife more productively and efficiently, and (3) reach new local and global markets for bushmeat products (Child et al., 2012). Additionally, as urban centers grow, the demand for bushmeat in these urban centers is anticipated to increase, perpetuating the trade (Fa, 2016). Bushmeat hunting in the Sahelo-Saharan region has also been attributed to the presence of mining, military, and administrative personnel (Belbachir, 2006; Beudels et al., 2006). The additional offtake perpetrated by poorly regulated hunting tourism is well documented for Niger and Mali (Beudels et al., 2006). Given these increased levels of exploitation, hunting and poaching are key drivers of ungulate population declines in the region (Craigie et al., 2010; Newby et al., 2016). This trend has been observed in potentially important prey species for the northwest African cheetah, such as Dorcas, slender horned, and Dama gazelle species and addax (Belbachir, 2006; Beudels-Jamar et al., 2005). These regional declines in prey availability could have consequences for human food security, and the viability of wildlife populations (Lindsey et al., 2011a,b).

Protected areas remain in the Saharan region, into which some of the Saharan antelope species have been reintroduced (Beudels et al., 2006). However, the protected area networks lack the resources to effectively protect wildlife from poaching, with underfunding being a major challenge (Craigie et al., 2010; Ripple et al., 2015). As the numbers of wild ungulates decrease, poachers have been reported to penetrate further into protected areas to supply the demand for bushmeat. Bushmeat hunters and predators effectively compete for a finite resource (Rogan et al., 2017), and the continuing decline of ungulate species through poaching may severely limit conservation efforts aimed at restoring northwest African cheetah numbers (Beudels-Jamar et al., 2005).

Land Use Changes

Many of the savanna ecosystems in Africa are under pressure from land-use changes due to the world's growing human population, including an increase in resource prospecting, mining, and pastoralism. These threats extend into the Sahelo-Saharan region and have been identified as key threats to biodiversity conservation (Brito et al., 2014).

Prospecting and Mining: North and West Africa are rich in deposits of gold, copper, iron ore, zinc, and other heavy minerals (https://www.projectsiq.co.za/mining-in-north-africa.htm), as well as coal, oil, and natural gas. Many areas of suitable cheetah habitat in North and West Africa overlap with the distribution of these resources. As a result, resource prospecting and mining, included within the protected areas, poses a key threat to the biodiversity in the Sahelo-Saharan region, as it does in other regions in Africa (Brito et al., 2014; Poulsen et al., 2017).

The development of the mining industry has resulted in major road networks penetrating the desert areas, facilitating access to areas where wildlife occurs, as well as degrading natural habitat and increasing the risk of wildlife mortalities from vehicle collisions. The mining industry has brought modern technology and improved communications to remote locations, but such activities provide limited opportunities for local employment (Duncan et al., 2014) and local community demands on natural resources remain. Illegal bushmeat hunting surrounding and emanating from these mining operations has been on the increase (Duncan et al., 2014; Newby et al., 2016). Intensification of petroleum exploration and exploitation in countries, such as Mauritania, Niger, and Chad, have recently impacted those

last strong holds of the Sahelo-Saharan ungulates, not so much by the activity itself, but from the disturbance caused by the development of chaotic road networks and the hunting carried out by the military in charge of protecting oil workers from terrorism and banditry (Newby et al., 2016; Rabeil, 2016). Thus, without the support of strong governments to regulate exploitation and natural resource extraction, ungulates in the region are at risk of being decimated (Ripple et al., 2015).

Pastoralism: The Sahelo-Saharan region exhibits wide environmental heterogeneity (Brito et al., 2016) and wildlife needs to be adapted to hyperarid conditions, fierce winds, intense heat, and wide temperature variations. As a result of the harsh conditions, the Sahara historically lacked human presence, with the exception of scattered groups of nomadic people and animals wherever vegetation or reliable water sources occurred (Nicolson et al., 1998). Human ranges and domestic stock numbers have increased through the drilling of deep water wells, reducing people's nomadic nature and resulting in intensive overgrazing of fragile ecosystems (Beudels et al., 2006). Cattle are limited in the Saharan region and mostly found at the southern borders of the Sahel, but sheep, goats, and camels are the mainstays of pastoralists within the region. In the desert itself, sedentary human occupation is confined to the oases, where cultivation is no longer limited by irrigation permits, thereby increasing the pressure on the environment. In this region where rainfall is spatially and temporally variable, high levels of competition between domestic and wild ungulates for grazing are inevitable. In addition, the removal of woody species of plants for fire wood and livestock fodder places further pressure on the region's wild ungulate species which, already at their heat tolerance limit, require the plants for shade (Beudels et al., 2006).

In the less arid Sahelo-Sudanian biome, competition between livestock and wildlife for grazing also occurs as pastoralists are increasingly encroaching protected areas. In the WAP complex, cattle numbers have increased from less than 6,000 to more than 50,000 between 2004 and 2012 (Bouché, 2012), undoubtedly increasing the pressure on local prey communities. This rapid increase in livestock numbers has exacerbated the conflict between resident large predators [predominantly lion (*Panthera leo*)] and pastoralists due to the depredation of livestock (Henschel et al., 2016). Retaliatory killings of lions and cheetahs have been reported in the complex (Henschel et al., 2016).

Climate Change

There is increasing evidence that climate change will strongly affect the African continent, particularly in the drier regions, such as the Sahelo-Saharan region (Adger et al., 2007; Kurukulasuriya et al., 2006; Chapter 12). While climate change scenarios for West Africa point to increased temperatures in the Sahel, there is disagreement regarding changes in precipitation. Predictions of rainfall changes in the Sahel and Sahara, based on 21 global model outputs were inconclusive (Christensen et al., 2007), limiting the ability to predict wildlife responses. If global warming increases, the already existing characteristic of extreme environmental heterogeneity in the Sahelo-Saharan region (Brito et al., 2016) and the ungulates' need for large ranges and movement between patchy resources will be further exacerbated.

Climate projections from three different Atmosphere-Ocean Global Circulation Models suggest that the potential ranges of the 73 extant African antelope species will decline by 44% on average by 2080 (Payne and Bro-Jørgensen, 2016a,b). Climate-induced declines in species richness are predicted to occur in the Sahelo-Saharan region, including range shifts and declines for the addax and Dama gazelle (Payne and Bro-Jørgensen, 2016a,b). A greater frequency of catastrophic droughts in the Saharan region in the second half of the 20th century has already been implicated in the declines of many antelope species (Beudels et al., 2006). In this region, the desert has been reported to be advancing at an alarming rate (by as much as 5–6 km/year; Hassabella and Nimir, 1991), reducing the available range for these species. Furthermore, the desert in the Sahelian zone fluctuates from year to year, tracking the interannual variability in rainfall (Nicolson et al., 1998). These range contractions and fluctuation will undoubtedly put increased pressure on populations of the cheetah's key prey species.

Human Population Growth and Associated Pressures

The human population in Africa, currently 1.1 billion, will reach 1.9 billion by 2050 (United Nations, 2015). In North Africa alone the human population has more than doubled from 107 million in 1980 to 232 million in 2017 (http://www.worldometers.info/world-population/northern-africa-population/). The exponential growth of the human population in conjunction with widespread poverty is likely to result in increased habitat degradation and increased reliance on bushmeat for survival needs in North Africa.

With a rising human population, livestock numbers are also likely to increase. Livestock numbers have tripled globally between 1980 and 2002, increasing by 25 million annually (Ripple et al., 2014). In many areas of Africa, including the Sahelo-Saharan region, livestock are kept not only for consumption but also as status symbols or as evidence of wealth (Ripple et al., 2015). The result of this phenomenon is higher livestock numbers than the environment can sustainably support, resulting in a higher intensity of competition with indigenous ungulates, and leading to a measurable impact on ecosystem functioning at the landscape scale (Taylor et al., 2016). These factors will limit the resource base on which conservation actions depend, to recover the northwest African cheetah population.

CONCLUSIONS

The future of the cheetah in the Sahelo-Saharan region is largely dependent on the availability of accessible ungulate prey that fall between 14 and 135 kg. The majority of these prey species are endangered in the Saharan region, and experiencing declines in the Sahelo-Sudanian region. Illegal bushmeat hunting and disturbance caused by mining activities are the biggest threat to the cheetah's prey and, in light of the growing human population, are likely to be exacerbated in the future. To protect ungulate populations from overexploitation, national planning in natural resource management and land-use in and around protected areas is necessary, including wildlife law enforcement, corridors for wildlife movement, incentives for local communities to conserve wildlife, as well as better availability of alternative protein and income sources (Lindsey et al., 2013). Debate remains regarding the effectiveness of law enforcement versus incentive programs in protecting wildlife from exploitation, particularly given the resource-constraints of African countries to carry out effective enforcement (Watson et al., 2014). Community-based conservation and the promotion of tourism can play an important role in contributing positively toward biodiversity conservation and economic development, as well as preventing wildlife overexploitation (Chapters 16 and 17). For the effective protection of wildlife, increased conservation funding will be necessary to protect ungulate species within protected areas and ideally expand the current protected area network (Waldron et al., 2013). Prey populations can be recovered if managed (Wolf and Ripple, 2016). However, increased legislative support and enforcement, as well as broad-scale public education, regarding the importance of wildlife and the causes of its decline, are likely to be needed to deter the threats throughout the Saharan region of Africa. Through identification of the conservation actions that will be most appropriate for each of the cheetah's key prey

species, strategies to conserve these ungulate species could lead to increased cheetah densities in the region. Given the significant and continuing declines in the range and abundance of the northwest African cheetah, immediate action is necessary to conserve this iconic species before it is too late.

References

Adger, N., Agrawala, S., Mirza, M.M.Q., Conde, C., O'Brien, K., Pulhin, J., Pulwarty, R., Smit, B., Takahashi, T., 2007. Assessment of adaptation practices, options, constraints and capacity. In: Parry, M.L., Canziani, O.F., Palutikof, J.P., van der Linden, P.J., Hanson, C.E. (Eds.), Climate Change 2007: impacts, adaptation and vulnerability. Contribution of Working Group II to the Fourth Assessment Report of the Intergovernmental Panel on Climate Change. Cambridge University Press, Cambridge, UK, pp. 717–743.

Belbachir, F., 2006. Human attitudes and conservation of Sahelo-Saharan antelopes and cheetah: Alegeria in context. Proceedings of the Seventh Annual SSIG Meeting, Douz, Tunisia, May 8th–11th, 2006.

Belbachir, F., 2008. *Acinonyx jubatus* ssp. *hecki*. The IUCN Red List of Threatened Species 2008: e.T221A13035738. Available from: http://dx.doi.org/10.2305/IUCN.UK.2008. RLTS.T221A13035738.en.

Belbachir, F., Pettorelli, N., Wacher, T., Belbachir-Bazi, A., Durant, S.M., 2015. Monitoring rarity: the critically endangered Saharan cheetah as a flagship species for a threatened ecosystem. PloS One 10 (1), e0115136.

Berzins, R., Claro, F., Akpona, A.H., Alfa Gambari Imorou, S., 2007. Conservation du guépard et développement durable dans les aires protégées du nord Bénin—Mission d'enquête auprès des villageois et des agents d'aires protégées (16/12/2005–26/02/2006). Société Zoologique de Paris, Paris, France.

Beudels, R.C., Devillers, P., Lafontaine, R-M., Devillers-Terschuren, J., Beudels M-O. (Eds.), 2006. Sahelo-Saharan Antelopes. Status and Perspectives. Report on the conservation status of the six Sahelo-Saharan Antelopes. CMS SSA Concerted Action. second ed. CMS Technical Series Publication No. 10, 2005. UNEP/CMS Secretariat, Bonn, Germany.

Beudels-Jamar, R.C., Devillers, P., Lafontaine, R.M., Devillers-Terschueren, J., Beudels, M-O., 2005. Les Antilopes Sahélo-Sahariennes. Statut et Perspectives. Rapport sur l'état de conservation des six Antilopes Sahélo-Sahariennes. Action Concertée CMS ASS. 2ème édition. CMS Technical Series Publication No. 10. UNEP/CMS Secretariat, Bonn, Allemagne.

Bodendorfer, T., Hoppe-Dominik, B., Fischer, F., Linsenmair, K.E., 2006. Prey of the leopard (*Panthera pardus*) and the lion (*Panthera leo*) in the Comoe and Marahoue National Parks, Cote d'Ivoire, West Africa. Mammalia, 70, 231–246.

Bouché, P., 2012. InventaireAériendel'écosystèmeW-Arly-Pendjari, Mai-Juin 2012. Ouagadougou:CITES-MIKE, WAP/UNOPS.

Brashares, J.S., Arcese, P., Sam, M.K., Coppolillo, P.B., Sinclair, A.R., Balmford, A., 2004. Bushmeat hunting, wildlife declines, and fish supply in West Africa. Science 306 (5699), 1180–1183.

Brito, J.C., Godinho, R., Martínez-Freiría, F., Pleguezuelos, J.M., Rebelo, H., Santos, X., Vale, C.G., Velo-Antón, G., Boratyński, Z., Carvalho, S.B., Ferreira, S., Gonçalves, D.V., Silva, T.L., Tarroso, P., Campos, J.C., Leite, J.V., Nogueira, J., Álvares, F., Sillero, N., Sow, A.S., Fahd, S., Crochet, P.-A., Carranza, S., 2014. Unravelling biodiversity, evolution and threats to conservation in the Sahara-Sahel. Biol. Rev. 89, 215–231.

Brito, J.C., Tarroso, P., Vale, C.G., Martínez-Freiría, F., Boratyński, Z., Campos, J.C., Ferreira, S., Godinho, R., Gonçalves, D.V., Leite, J.V., Lima, V.O., Pereira, P., Santos, X., da Silva, M.J.F., Silva, T.L., Velo-Antón, G., Veríssimo, J., Crochet, P.-A., Pleguezuelos, J.M., Carvalho, S.B., 2016. Conservation Biogeography of the Sahara-Sahel: additional protected areas are needed to secure unique biodiversity. Divers. Distrib. 22, 371–384.

Carbone, C., Teacher, A., Rowcliffe, J.M., 2007. The costs of carnivory. PLoS Biol 5 (2), e22.

Caro, T.M., 2013. Genus *Acinonyx* Cheetah. In: Kingdon, J., Hoffmann, M. (Eds.), Mammals of Africa. Volume V. Carnivores, Pangolins, Equids and Rhinoceroses,. Bloomsbury Publishing, London, UK.

Castelló, J.R., 2016. Bovids of the World: Antelopes, Gazelles, Cattle, Goats, Sheep, and Relatives. Princeton University Press, Princeton, NJ, USA.

Child, B.A., Musengezi, J., Parent, G.D., Child, G.F.T., 2012. The economics and institutional economics of wildlife on private land in Africa. Pastoralism: Res. Policy Prac. 2, 1–32.

Christensen, J.H., Hewitson, B., Busuioc, A., Chen, A., Gao, X., Held, I., Jones, R., Kolli, R.K., Kwon, W.-T., Laprise, R., Magana Rueda, V., Mearns, L., Menéndez, C.G., Raisanen, J., Rinke, A., Sarr, A., Whetton, P., 2007. Regional climate projections. In: Solomon, S., Qin, D., Manning, M., Chen, Z., Marquis, M., Averyt, K.B., Tignor, M., Miller, H.L. (Eds.), Climate Change 2007: the physical science basis. Contribution of Working Group I to the Fourth Assessment Report of the Intergovernmental Panel on Climate Change. Cambridge University Press, Cambridge, UK, pp. 847–940.

Claro, F., Leriche, H., Van Syckle, S., Rabeil, T., Hergueta, S., Fournier, A., Alou, M., 2006. Survey of the cheetah in W National Park and Tamou fauna reserve, Niger. Cat News 45, 4–7.

Claro, F., Sissler, C., 2003. Saharan cheetahs in the Termit region of Niger. Cat News 38, 23–24.

Clément, C., Ceulers, V., Medina, R., Welle, M., Rabeil, T., Jullien, E., 2015. The Red-fronted gazelle (*Eudorcas rufifrons*) in the Boundou Community Nature Reserve. Senegal. Gnuslett. 32 (2), 9–10.

Clements, H.S., Tambling, C.J., Hayward, M.W., Kerley, G.I.H., 2014. An objective approach to determining the weight ranges of prey preferred by and accessible to the five large African carnivores. PLoS One 9, e101054.

Craigie, I.D., Baillie, J.E.M., Balmford, A., Carbone, C., Collen, B., Green, R.E., Hutton, J.M., 2010. Large mammal population declines in Africa's protected areas. Biol. Conserv. 143, 2221–2228.

Duncan, C., Kretz, D., Wegmann, M., Rabeil, T., Pettorelli, N., 2014. Oil in the Sahara: mapping anthropogenic threats to Saharan biodiversity from space. Philos. Trans. R. Soc. 369 (1643), 20130191.

Durant, S.M., Mitchell, N., Groom, R., Pettorelli, N., Ipavec, A., Jacobson, A.P., Woodroffe, R., Böhm, M., Hunter, L.T.B., Becker, M.S., Broekhuis, F., Bashir, S., Andresen, L., Aschenborn, O., Beddiaf, M., Belbachir, F., Belbachir-Bazi, A., Berbash, A., Brandao de Matos Machado, I., Breitenmoser, C., Chege, M., Cilliers, D., Davies-Mostert, H., Dickman, A.J., Ezekiel, F., Farhadinia, M.S., Funston, P., Henschel, P., Horgan, J., de Iongh, H.H., Jowkar, H., Klein, R., Lindsey, P.A., Marker, L., Marnewick, K., Melzheimer, J., Merkle, J., M'soka, J., Msuha, M., O'Neill, H., Parker, M., Purchase, G., Sahailou, S., Saidu, Y., Samna, A., Schmidt-Küntzel, A., Selebatso, E., Sogbohossou, E.A., Soultan, A., Stone, E., van der Meer, E., van Vuuren, R., Wykstra, M., Young-Overton, K., 2017. The global decline of cheetah *Acinonyx jubatus* and what it means for conservation. Proc. Natl. Acad. Sci. 114, 528–533.

Durant, S.M., Wacher, T., Bashir, S., Woodroffe, R., DeOrnellas, P., Ransom, C., Newby, J., Abaigar, T., Abdelgader, M., el Alqamy, J., Baillie, J., Beddiaf, M., Belbachir, F., Belbachir-Bazi, A., Berbash, A.A., Bemadjim, N.E., Beudels-Jamar, R., Boitani, I., Breitenmoser, C., Cano, M., Chardonnet, P., Collen, B., Cornforht, W.A., Cuzin, F., Gerngross, P., Haddane, B., Hadjeloum, M., Jacobson, A., Jebali, A., Lamarque, F., Mallon, D., Minkowski, K., Monfort, S., Ndoassal, B., Niagate, B., Purchase, G., Samaila, A.K., Sillero-Zubiri, C., Soultan, A.E., Price, M.R.S., Pettorelli, N., 2014. Fiddling in biodiversity hotspots while deserts burn? Collapase of the Sahara's megafauna. Div. Distrib. 20, 114–122.

Fa, J.E., 2016. Importance of antelope bushmeat consumption in Africa wet and moist tropical forests. In: Bro-Jorgensen, J., Mallon, D. (Eds.), Antelope Conservation: From Diagnosis to Action. John Wiley & Sons, Oxford, UK, pp. 78–91.

Garba, H.M., Di Silvestre, I., 2008. Conflicts between large carnivores and domestic livestock in the peripheral zone of the W transboundary Park in Niger. In: Croes, B., Buij, R., de Iongh, H.H., Bauer, H., (Eds.), Conservation of large carnivores in West, Central, Africa, Proceedings of an International Seminar, Maroua, Cameroon, 15–16 November, 2006, Institute of Environmental Sciences, Leiden, The Netherlands.

Hassabella, E.R.O., Nimir, N.B., 1991. Towards a national conservation policy in the Sudan. In: D. Ernst, (Ed.), Proceedings, Seminar on wildlife conservation and management in the Sudan, Khartoum, March 16–21, 110, 1985. Wildlife Conservation Forces and German Agency for Technical Co-operation (GTZ). Hamburg, Günter Stubbemann. pp. 137–151.

Hayward, M.W., Hofmeyr, M., O'Brien, J., Kerley, G.I.H., 2006. Prey preferences of the cheetah (*Acinonyx jubatus*) (Felidae: Carnivora): morphological limitations or the need to capture rapidly consumable prey before klepto-parasites arrive? J. Zool. 270, 615–627.

Hayward, M.W., O'Brien, J., Kerley, G.I.H., 2007. Carrying capacity of large African predators: predictions and tests. Biol. Conserv. 139, 219–229.

Henschel, P., Petracca, L.S., Hunter, L.T.B., Kiki, M., Sewadé, C., Tehou, A., Robinson, H.S., 2016. Determinants of distribution patterns and management needs in a critically endangered lion *Panthera leo* population. Front. Ecol. Evol. 4, e110.

IUCN SSC Antelope Specialist Group. 2008. *Gazella dorcas*. The IUCN Red List of Threatened Species 2008: e.T8969A12941858. Available from: http://dx.doi.org/10.2305/IUCN.UK.2008.RLTS.T8969A12941858.en.

IUCN SSC Antelope Specialist Group. 2008. *Eudorcas rufifrons*. The IUCN Red List of Threatened Species 2008: e.T8973A12943749. Available from: http://dx.doi.org/10.2305/IUCN.UK.2008.RLTS.T8973A12943749.en.

IUCN SSC Antelope Specialist Group. 2016. *Ourebia ourebi*. The IUCN Red List of Threatened Species 2016: e.T15730A50192202. Available from: http://dx.doi.org/10.2305/IUCN.UK.2016-1.RLTS.T15730A50192202.en.

IUCN SSC Antelope Specialist Group. 2016. *Nanger dama*. The IUCN Red List of Threatened Species 2016: e.T8968A50186128. Available from: http://dx.doi.org/10.2305/IUCN.UK.2016-2.RLTS.T8968A50186128.en.

Kurukulasuriya, P., Mendelsohn, R., Hassan, R., Benhin, J., Deressa, T., Diop, M., Eid, H.M., Fosu, K.Y., Gbetibouo, G., Jain, S., Mahamadou, A., Mano, R., Kabubo-Mariara, J., El Marsafawy, S., Molua, E., Ouda, S., Ouedraogo, M., Sene, I., Maddison, D., Seo, S.N., Dinar, A., 2006. Will African agriculture survive climate change? World Bank Econ. Rev. 20, 367–388.

Lindsey, P.A., Barnes, J., Nyirenda, V., Pumfrett, B., Tambling, C.J., Taylor, W.A., Rolfes, MtS., 2013. The Zambian Wildlife Ranching Industry: scale, associated benefits, and limitations affecting its development. PLoS One 8, e81761.

Lindsey, P., Romanach, S., Tambling, C.J., Chartier, K., 2011a. Ecological and financial impacts of the bushmeat trade in Zimbabwe. Oryx 45, 96–111.

Lindsey, P.A., Tambling, C.J., Brummer, R., Davies-Mostert, H.T., Hayward, M.W., Marnewick, K., Parker, D., 2011b. Minimum prey and area requirements of cheetahs: implications for reintroductions and management of the species as a managed metapopulation. Oryx 45, 587–599.

Marker, L.L., Muntifering, J.R., Dickman, A.J., Mills, M.G.L., Macdonald, D.W., 2003. Quantifying prey preferences of free-ranging Namibian cheetahs. S. Afr. J. Wildl. Res. 33, 43–53.

Matchano, A., 2011. Mise en ńuvre d'une stratégie dans le cadre du conflit homme/carnivore dans le massif de Termit et ses environs. Carnivore Project, Sahara Conservation Fund, Niamey, 32 p.

Mduma, S.A.R., Sinclair, A.R.E., 1994. The function of habitat selection by oribi in Serengeti. Tanzan. Afr. J. Ecol. 32, 16–29.

Newby, J.E., Grettenberger, J.F., 1986. The human dimension in natural resource conservation: a Sahelian example from Niger. Environ. Conserv. 13 (03), 249–256.

Newby, J., Wacher, T., Durant, S.M., Petorelli, N., Gilbert, T., 2016. Desert antelopes on the brink: how resilient is the Sahelo-Saharan ecosystem. Antelope Conservation: From Diagnosis to Conservation. Wiley Blackwell, Chichester, pp. 253–279.

Nicolson, S.E., Tucker, C.J., Ba, M.B., 1998. Desertification, drought, and surface vegetation: an example from the West African Sahel. Bull. Am Meteorol. Soc. 79, 815–829.

Payne, B.L, Bro-Jørgensen, J., 2016a. Disproportionate climate-induced range loss forecast for the most threatened African antelopes. Curr. Biol. 26, 1200–1205.

Payne, B.L., Bro-Jørgensen, J., 2016b. A framework for prioritizing conservation translocations to mimic natural ecological processes under climate change: A case study with African antelopes. Biol Conserv. 201, 230–236.

Poilecot, P., 1996. La faune de la Réserve Naturelle Nationale de l'Aïr et du Ténéré. MH/E, WWF & UICN. Sous la direction de F. Giazzi. La Réserve Naturelle Nationale de l'Aïr et du Ténéré (Niger). UICN, Gland. Réintroduction, gestion et aménagement. Eschborn, GTZ, 181–265.

Poulsen, J.R., Koerner, S.E., Moore, S., Medjibe, V.P., Blake, S., Clark, C.J., Akou, M.E., Fay, M., Meier, A., Okouyi, J., Rosin, C., White, L.J.T., 2017. Poaching empties critical Central African wilderness of forest elephants. Curr. Biol. 27, R134–R135.

Rabeil, T., 2003. Distribution potentielle des grands mammifères dans le Parc du W au Niger (Doctoral dissertation, Université Paris-Diderot-Paris VII).

Rabeil, T., 2014. Evaluation de l'état de conservation de la faune, des habitats et des menaces dans la Réserve Naturelle Nationale de l'Aïr et du Ténéré (RNNAT), Niger. UNESCO World Heritage Center, technical assistance,

Ministère de l'environnement, de la salubrité urbaine et du Développement durable / Sahara Conservation Fund, Niger, 43 p.

Rabeil, T., 2015. Mission report in the Termit massif. June 2015, Niger Fauna Corridor Project/Sahara Conservation Fund, Niamey, 22 p.

Rabeil, T., 2016. Action plan for the worlds' last wild population of addax antelope. Gnusletter 33 (1), 21–24.

Rabeil, T., Newby, J., Harouna, A., 2008. Conservation of the Termit and Tin Touma (Niger) Annual report for 2007 of the Sahara Conservation Fund (SCF).

Ripple, W.J., Estes, J.A., Beschta, R.L., Wilmers, C.C., Ritchie, E.G., Hebblewhite, M., Berger, J., Elmhagen, B., Letnic, M., Nelson, M.P., Schmitz, O.J., Smith, D.W., Wallach, A.D., Wirsing, A.J., 2014. Status and ecological effects of the world's largest carnivores. Science 343 (6167), 1241484.

Ripple, W.J., Newsome, T.M., Wolf, C., Dirzo, R., Everatt, K.T., Galetti, M., Hayward, M.W., Kerley, G.I.H., Levi, T., Lindsey, P.A., Macdonald, D.W., Malhi, Y., Painter, L.E., Sandom, C.J., Terborgh, J., Van Valkenburgh, B., 2015. Collapse of the world's largest herbivores. Sci. Adv. 1, e1400103.

Rogan, M.S., Lindsey, P.A., Tambling, C.J., Golabek, K.A., Chase, M., Collins, K., McNutt, J.W., 2017. Illegal bushmeat hunting in the Okavango Delta, Botswana, competes with large carnivores, jeopardises ungulate populations, and threatens a global tourism hotspot. Biol. Conserv. 210, 233–242.

Sillero-Zubiri, C., Rostro-García, S., Burruss, D., Matchano, A., Harouna, A., Rabeil, T., 2015. Saharan cheetah *Acinonyx jubatus hecki*, a ghostly dweller on Niger's Termit massif. Oryx 49, 591–594.

Sogbohossou, E.A., de Iongh, H.H., Sinsin, B., de Snoo, G.R., Funston, P.J., 2011. Human–carnivore conflict around Pendjari Biosphere Reserve, northern Benin. Oryx 45, 569–578.

Taylor, A., Lindsey, P.A., Davies-Mostert, H.T., 2016. An Assessment of the Economic, Social and Conservation Value of the Wildlife Ranching Industry and its Potential to Support the Green Economy in South Africa. Endangered Wildlife Trust, Johannesburg, South Africa.

United Nations DoEaSA, Population Division, 2015. World Population Prospects: The 2015 Revision, Methodology of the United Nations Population Estimates and Projections, Working Paper No. ESA/P/WP.242.

Wacher, T., De Smet, K., Belbachir, F., Belbachir-Bazi, A., Fellous, A., Belghoul, M., Marker, L., 2005. Sahelo-Saharan Interest Group Wildlife Surveys, Central Ahaggar Mountains.

Wacher, T., Newby, J., 2012. Summary of results and achievements of the Pilot Phase of the Pan Sahara Wildlife Survey 2009–2012. SCF PSWS Technical Report 12, November 2012. Sahara Conservation Fund.

Waldron, A., Mooers, A.O., Miller, D.C., Nibbelink, N., Redding, D., Kuhn, T.S., Roberts, J.T., Gittleman, J.L., 2013. Targeting global conservation funding to limit immediate biodiversity declines. Proc. Natl. Acad. Sci. 110, 12144–12148.

Walther, B.A., 2016. A review of recent ecological changes in the Sahel with particular reference to land-use changes, plants, bird and mammals. Afr. J. Ecol. 54, 268–280.

Watson, J.E., Dudley, N., Segan, D.B., Hockings, M., 2014. The performance and potential of protected areas. Nature 515 (7525), 67–73.

Wilkie, D., Carpenter, J., 1999. Bushmeat hunting in the Congo Basin: an assessment of impacts and options for mitigation. Biodiv. Conserv. 8, 927–955.

Wolf, C., Ripple, W.J., 2016. Prey depletion as a threat to the worlds large carnivores. R. Soc. Open Sci. 3, 160252.

The Impact of Climate Change on the Conservation and Survival of the Cheetah

Matti T. Nghikembua*, Flavio Lehner**, Wilbur Ottichilo[†,‡,§], Laurie Marker*, Steven C. Amstrup[¶,††]

*Cheetah Conservation Fund, Otjiwarongo, Namibia
**National Center for Atmospheric Research, Boulder, CO, United States
[†]Republic of Kenya, Parliament, Nairobi, Kenya
[‡]Parliament Conservation Caucus, Nairobi, Kenya
[§]Environment Committee, Nairobi, Kenya
[¶]Polar Bears International, Bozeman, MT, United States
[††]University of Wyoming, Laramie, WY, United States

INTRODUCTION

Global Warming and the Cheetah

Current global warming, which multiplies all historic threats to cheetah persistence (Durant et al., 2015; Marker, 2002; Chapters 10, 11, 13, and 14), is caused by human interference with earth's energy balance (Schneider, 1989). Incoming shortwave radiation from the sun ultimately must be balanced by the outgoing long-wave radiation emitted back into space from the earth and its atmosphere (Lutgens and Tarbuck, 2004). A climate forcing is a factor that perturbs this balance (Hansen and Sato, 2004). For example,

aerosols (small particles that become suspended in the atmosphere) released by volcanoes or human industrial activities can provide a <1–2 years negative (cooling) forcing, by reflecting the sun's energy back into space before it reaches earth's surface. Increased concentrations of CO_2, methane, and other greenhouse gases (GHGs) provide a positive (warming) forcing, prolonging retention of the sun's energy before it escapes back into space. Unlike aerosols, GHGs persist in the atmosphere for decades to centuries.

Natural climate fluctuations (e.g., El Niño, La Niña) are independent of anthropogenic climate forcings, and make some periods cooler and

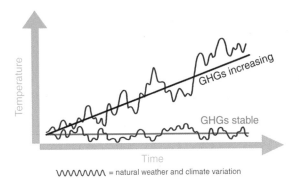

FIGURE 12.1 **General relationship between temperature and time, with and without persistent greenhouse
gas (GHG) forcing.** With GHG levels stable, the average of
natural climate fluctuations is a level line *(green)*. But with
rising GHG levels, average temperatures must rise *(red)*. Despite the chaotic and unpredictable nature of natural variations *(purple lines)*, it always will be warmer than it would
have been without the higher GHG concentration. *Source:
Designed by S.C. Amstrup, graphic by Peppermint Narwhal.*

Climate Change: Temperature and Precipitation

Although global warming is indeed global,
different regions will warm at different rates, and
they may experience differing climate change
symptoms at different times. Within the mostly
arid remaining range of cheetahs (portions of
southern, central, and northern Africa, and Iran;
Chapters 4 and 5), the biggest climate change
concerns are related to increasing temperatures
and changing precipitation patterns. Across the
cheetah's current range, average temperatures
have increased between 1 and 2°C, since the early
part of the 20th century (Fig. 12.2). Following current GHG emissions [Representative Concentration Pathway 8.5 (RCP 8.5); Wayne, 2013] another
~3°C warming is likely by the end of this century.
However, aggressive GHG mitigation (RCP 2.6)
could limit additional average temperature rise
to between 1 and 2°C (Fig. 12.2).

Precipitation changes are more difficult to
predict. Globally, more precipitation is assured
because warmer air holds more water vapor,
but the distribution of precipitation is spatially
heterogeneous. Observations of recent decades
show that some historically wet areas have become wetter, and dry areas have become drier
as the world has warmed (Greve et al., 2014).
This pattern is projected to continue with further warming (Durack et al., 2012; Held and
Soden, 2006). In addition to altered average
moisture distribution, frequencies of droughts,
floods, and generally extreme weather are expected to increase (National Academies of Sciences, Engineering, and Medicine, 2016).

others warmer than average (Deser et al., 2012).
In the natural state where GHG concentrations
are stable, the long-term average or "baseline"
around which these oscillations occur can be
represented as a horizontal line. With chronical-
ly increasing GHG forcing, natural fluctuations
in the climate system will continue, but over a
higher and rising baseline (Fig. 12.1).

Existing GHG emission patterns have mul-
ticentennial and even millennial ramifications.
Avoiding centennial climate change milestones,
however, is critical to preserve cheetahs and their
habitats in those longer terms. Therefore, we focus on climate change thresholds that, without
GHG mitigation, are likely to be exceeded by
the latter 21st century. As global warming is primarily caused by human activities in the indus-
trialized world, cheetah conservation depends
on societies, distant from cheetahs, adopting
sustainable economies based on renewable energy sources. Here, we describe climate change
symptoms of warming that have the potential
to determine the cheetah's future, and we summarize on-the-ground conservation actions that
will maximize the potential for cheetahs to persist under a changing global climate.

IMPACTS OF INCREASING TEMPERATURES AND ALTERED PRECIPITATION

Water Availability

In response to increased temperatures and
changes in the hydrologic cycle, surface water
availability in most of the cheetah's range will

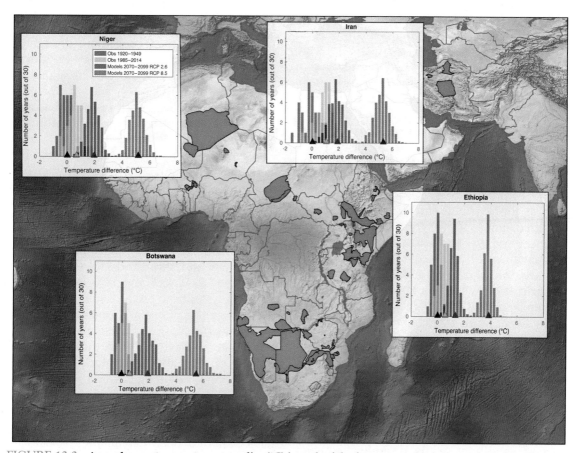

FIGURE 12.2   **Annual mean temperature anomalies (°C) in each of the four geographic regions where cheetahs occur, as represented by data from Niger (west Africa), Ethiopia (East Africa), Botswana (southern Africa), and Iran.** For each region, three 30-year periods (historical: 1920–1949 and 1985–2014; and two projections for: 2070–2099), are represented. The number of years for which the annual mean temperature differs from the mean temperature recorded in 1920–1949, is shown. Triangles represent the average for each of the 30 year periods. The Representative Concentration Pathway (RCP) 8.5, or "business as usual," projection assumes GHG emissions will continue unabated at the present-day rate. The RCP 2.6 projection assumes dramatic reductions of GHG emissions over the course of this century, with a chance of keeping global mean temperature rise below 2°C. Data for historical and current periods are based on gridded temperature observations from the Berkeley Earth Surface Temperature dataset (Rohde et al., 2013). The two different projections for the late 21st century, 2070–2099, are based on the multimodel mean of 25 climate models of the 5th Coupled Model Intercomparison Project (CMIP5; Taylor et al., 2012). *Source: Cheetah distribution data from Durant, et al., 2017. Graphic by S.C. Amstrup and D.C. Douglas*

become more limited. Increased drought frequency is expected to accompany declines in average water availability. For example, Iran, home of the most threatened population of cheetahs (Chapter 5), where the historic mean annual precipitation base is ~250 mm, has seen dramatic declines in moisture over much of the country (Amiri and Eslamian, 2010). Lake Urmia, historically the largest lake in the Middle East, and a critical water source for Iran's agriculture, has lost 88% of its surface area since the 1970s (Mirchi et al., 2015). In southern Africa, at the opposite end of the cheetah distribution, unusually dry spells became more frequent and wet spells less so after the 1960s (Garanganga, 2011). The 1981–1983 drought in southern Africa directly and indirectly impacted agriculture, wildlife, and human livelihoods (Garanganga, 2011; Hillman and Hillman, 1977; Klein, 2006; Marker-Kraus et al., 1996; Verlinden, 1997). As temperatures increased over the last 60 years, frequencies of intense rains and widespread drought increased in Namibia (Turpie et al., 2010), and estimated water balance declined at 1/3 of water monitoring stations. Heat waves in the southern part of Africa are projected to increase 3.5-fold over current conditions while drought frequency increases (Lyon, 2009).

Lyon and DeWitt (2012) reported that a clear decline in East African precipitation beginning in the 1980s accelerated after 1999. Average precipitation in the Sahel and Greater Horn of Africa decreased when comparing the mean of 1990–2009 to that of 1970–1989 (Williams et al., 2012). The higher drought frequency and more pulsatile precipitation observed in recent decades across southern and eastern Africa are projected to increase (Funk et al., 2012; Garanganga, 2011; Williams and Funk, 2011). Projections of reduced rainfall and increased frequency of prolonged droughts where the climate historically has been arid to semiarid (Omari, 2010) can only exacerbate recently observed problems.

Altered Habitat

Some woody plants are more drought-tolerant and efficient than grasses in utilizing the elevated atmospheric CO_2 (Higgins and Scheiter, 2012; MET, 2008), and woody plants are already invading much of the open savannah habitat that cheetahs prefer (Muntifering et al., 2006). This "bush encroachment", occurring across varying land use practices, has been attributed to warmer and drier conditions and increased atmospheric CO_2 concentrations (Wigley et al., 2009, 2010). Midgley et al. (2005) projected that with current GHG emission trends, Namibian savanna habitats will decline 22%–67% by the latter part of the century, while desert habitats and shrublands were projected to increase by 27%–45% and 14%–18%. Although harvesting shrubs to produce sources of alternative energy (e.g., Cheetah Conservation Fund's Bushblok program) and other habitat restoration programs have emerged, they are unlikely to keep up with the accelerated habitat alterations of the future (Higgins and Scheiter, 2012; MET, 2008). Bush encroachment reduces visibility, which may affect the hunting efficiency of cheetahs, and lowers densities of the cheetah's preferred open habitat prey species like eland (*Tragelaphus oryx*), oryx (*Oryx gazella*), red hartebeest (*Alcelaphus buselaphus*), and warthog (*Phacochoerus africanus*) (Marker, 2002; Muntifering et al., 2006).

Parasites and Disease

As livestock and wildlife both aggregate at water points, especially during the dry seasons, disease transfer between livestock and wildlife is likely to increase as temperature rises and water becomes more scarce. In addition, the projected higher frequency of alternating drought and flood periods may create favorable conditions for arthropod disease vectors, such as ticks and fleas (Roach, 2008; Sejian et al., 2016). Olwoch et al. (2008) reported a range shift and expansion in some southern African countries of the tick *Rhipicephalus appendiculatus* in response to the temperature and rainfall changes. This tick is a vector for Theileriosis (East Coast fever), a disease causing livestock industry losses of US $168 million annually (Norval et al., 1992; Olwoch et al., 2008). Other serious tick-borne diseases include African swine fever, transmitted by the tick *Ornithodorus moubata* and mostly carried by warthogs. Heartwater, a tick-borne rickettsial disease of ruminants, including eland and springbok (*Amphipneustes marsupialis*), is caused by the tick *Amblyomma hebraeum*.

Despite the anticipated drying of most cheetah habitats, canine distemper virus (CDV) could become an increased problem. Outbreaks of feline infections with CDV in the Serengeti ecosystem were linked to climate change in 1994 and 2001, with losses of up to 1/3 of lions (*Panthera leo*) in the area (Munson et al., 2005; Roach, 2008; Roelker-Parker et al., 1996). Prevalence of CDV antibodies among the free-ranging Namibian cheetah population is confirmed (Munson et al., 2005), although low population densities, large territories, and dry environments apparently keep infection rates low (Marker, 2002).

Alternating droughts and heavy rains also drive pathogenic outbreaks of anthrax (Hampson et al., 2011; Lelenguyah, 2012), a disease to which cheetahs are highly susceptible (Good et al., 2008; Jäger et al., 1990; Lembo et al., 2011; Switzer et al., 2016; Turnbull et al., 2004). In southern and eastern Africa, anthrax is well documented in cheetah prey species (Hampson et al., 2011), and deaths from anthrax have been documented in wild and captive cheetahs in Namibia and Botswana (Good et al., 2008; Jäger et al., 1990; Switzer et al., 2016). Impacts of anthrax and other diseases on survival of free-ranging cheetahs in a hotter drier and more unpredictable environment are as of yet largely unknown, but certainly cause for concern.

Conflicts with Humans

Escalating conflicts between humans and cheetahs are virtually assured in a warming world (Marker et al., 2008), even if human numbers were to remain stable. At the height of the 1980s drought, for example, approximately 890 Namibian cheetahs were captured or killed annually due to escalated conflict with livestock farmers (Marker-Kraus et al., 1996). Growing human populations exacerbate such conflicts. In southern African cheetah range countries, the human population of 144.49 million is projected to increase by a factor of 1.8 by 2050 and triple by century's end. Similar increases in human numbers are expected across most of the cheetah's range (The World Bank, 2016). At the same time, both predators and natural prey will need to search larger landscapes for food and water, while human food and water stresses increase (Asseng et al., 2014; Liu et al., 2016).

With rapid human population growth, increases in poaching, deforestation, and other land conversions will magnify declines in habitat (Chapter 10) and wildlife security (Chapter 11) resulting from a warming climate. As wildlife have greater ability to adapt to limited moisture distribution than livestock (Marker-Kraus et al., 1996), game farming has increased in drought periods and may further gain in popularity as drought frequencies rise. Often accompanied by 2.5-m-high fences, game farms can restrict normal wildlife movements restricting cheetah access to habitat and prey (Marker-Kraus et al., 1996; Marker et al., 2003; Schumann et al., 2008), at the same time landowner disputes over wildlife ownership are likely to increase (Marker-Kraus et al., 1996).

DISCUSSION

Effects of observed 20th century warming across the cheetah's range are harbingers of what is to come in the absence of aggressive

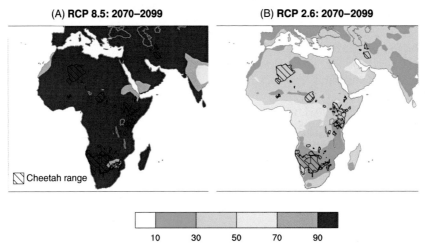

FIGURE 12.3   **Probability that a given summer across the current cheetah range, during 2070–2099, will be warmer than the warmest on record during the historical period of 1920–2014.** The maps depict the much lower probability of unprecedented hot summers if societies adopt aggressive GHG mitigation (B vs. A). Following business as usual emissions [RCP 8.5; panel (A)] will lead to almost every future summer likely being hotter than any we previously have experienced, while with aggressive emissions reduction [RCP 2.6; panel (B)] the probability that future summers will produce record heat is predicted to be much less. Data are the same as used in Fig. 12.2.

GHG mitigation. Unabated emissions (RCP 8.5) will lead to annual mean temperatures 3–6°C higher than preindustrial conditions by the end of this century. Average summer temperatures at that time will exceed the hottest summers ever recorded across most of the cheetah's current range (Fig. 12.3; Lehner et al., 2016). Even with less certainty about future precipitation, the virtually certain increase in temperature will increase evapotranspiration and put additional stress on water resources. The complications that already have accompanied observed warming will be magnified while competition between animals and people increases.

Conservation Action

If cheetahs are to survive, stakeholders must engage conservation at local, national, regional, and global levels. Development of integrated livestock and wildlife management plans, like the conservancy model used in Namibia

(Chapter 17) would empower communities to manage local natural resources to benefit directly from wildlife through trophy hunting, ecotourism, and collection of veld products. On a national and regional scale, efforts to save cheetahs and their prey must be embedded in each country's regional development plans. Adaptation strategies that reduce livestock losses and rangeland degradation, and promote water conservation, will be necessary. On a global scale, efforts to minimize GHG emissions must become every nation's priority. Natural variations in the climate system impair our ability to exactly predict the local and regional pace and severity of climate change effects. Unabated GHG-forced global warming, however, guarantees that we ultimately will exceed thresholds important to cheetah survival. The ever-increasing human footprint on the landscape, combined with unabated "warming-driven" habitat alteration, likely will spell the demise of cheetahs

CONCLUSIONS

We have examined ongoing changes that will profoundly affect the future welfare of cheetahs and the environments of which they are a part. Temperature increases driven by rising GHG concentrations could introduce extreme weather patterns including higher frequencies of drought and in some places more pulsatile rains. The past has shown that cheetahs are directly and indirectly vulnerable to such climate changes. As climate patterns change, a multidisciplinary and adaptive approach, built upon past management successes, will be necessary. Even as world societies work to curb global temperatures, conservation efforts must continue on the ground.

The urgency of action cannot be overstated. Procrastination now will assure catastrophe later. Unlike poaching, overgrazing, or deforestation, the impacts of continuing to pollute our atmosphere with ever-higher concentrations of GHGs will be felt for centuries or even millennia (Matthews and Zickfeld, 2012). If society waits to halt GHG rise, the resulting human nutritional challenges and refugee crises are likely to be greater. The need to address these human needs is virtually certain to shift attention away from conservation of cheetahs and other wildlife. The good news is that most future climate uncertainty is in our hands. We cannot control the natural variation in the climate system but we can control the slope of the rising baseline (Fig. 12.1). At the 2015 Paris Climate Talks in December 2015, nations set a goal of keeping global mean temperature rise below 2°C above preindustrial times (or approximately 1°C above present). The RCP 2.6 scenario approximates the Paris goals, and if world societies can get on that or a similar path, temperatures across the cheetah's range could stabilize at levels much more similar to those experienced in recent history (Figs. 12.2 and 12.3).

The future, therefore, is largely up to us. We can stay on the current path and let our climate continue to change in ways that will negatively impact the lives of all future generations. We can choose a more gradual slope, or ideally, we can choose a path, like that outlined in Paris, that stops the rise in GHG concentrations, and brings us to a new level baseline. Taking this path could avoid the worst that future global warming has to offer, and engenders hope for the long-term persistence of cheetahs.

References

Amiri, M.J., Eslamian, S.S., 2010. Investigation of climate change in Iran. J. Environ. Sci. Technol. 3 (4), 208–2016.

Asseng, S., Ewert, F., Martre, P., Rötter, R.P., Lobell, D.B., Cammarano, D., Kimball, B.A., Ottman, M.J., Wall, G.W., White, J.W., Reynolds, M.P., Alderman, P.D., Prasad, P.V.V., Aggarwal, P.K., Anothai, J., Basso, B., Biernath, C., Challinor, A.J., De Sanctis, G., Doltra, J., Fereres, E., Garcia-Vila, M., Gayler, S., Hoogenboom, G., Hunt, L.A., Izaurralde, R.C., Jabloun, M., Jones, C.D., Kersebaum, K.C., Koehler, A.-K., Müller, C., Naresh Kumar, S., Nendel, C., O'Leary, G., Olesen, J.E., Palosuo, T., Priesack, E., EyshiRezaei, E., Ruane, A.C., Semenov, M.A., Shcherbak, I., Stöckle, C., Stratonovitch, P., Streck, T., Supit, I., Tao, F., Thorburn, P.J., Waha, K., Wang, E., Wallach, D., Wolf, J., Zhao, Z., Zhu, Y., 2014. Rising temperatures reduce global wheat production. Nat. Clim. Change 5, 143–147.

Deser, C., Knutti, R., Solomon, S., Phillips, A.S., 2012. Communication of the role of natural variability in future North American climate. Nat. Clim. Change 2, 775–779.

Durack, P.J., Wijffels, S.E., Matear, R.J., 2012. Ocean salinities reveal strong global water cycle intensification during 1950–2000. Science 336, 455–458.

Durant, S., Mitchell, N., Ipavec, A., Groom, R., 2015. *Acinonyx jubatus*. The IUCN Red List of Threatened Species 2015: e.T219A50649567. Available from: http://dx.doi.org/10.2305/IUCN.UK.2015-4.RLTS.T219A50649567.en.

Durant, S.M., Mitchell, N., Groom, R., Pettorelli, N., Ipavec, A., Jacobson, A., Woodroffe, R., Bohm, M., Hunter, L., Becker, M., Broekuis, F., Bashir, S., Andresen, L., Aschenborn, O., Beddiaf, M., Belbachir, F., Belbachir-Bazi, A., Berbash, A., Brandao de Matos Machado, I., Breitenmoser, C., Chege, M., Cilliers, D., Davies-Mostert, H., Dickman, A., Fabiano, E., Farhadinia, M., Funston, P., Henschel, P., Horgan, J., de Iongh, H., Jowkar, H., Klein, R., Lindsey, P., Marker, L., Marnewick, K., Melzheimer, J., Merkle, J., Msoka, J., Msuha, M., O'Neill, H., Parker, M., Purchase, G., Saidu, Y., Samaila, S., Samna, A., Schmidt-Küntzel, A., Selebatso, E., Sogbohossou, E., Soultan, A., Stone, E., van der Meer, E., van Vuuren, R., Wykstra, M., Young-Overton, K., 2017. The global decline of cheetah and what it means for conservation. Proc. Natl. Acad. Sci. USA 114, 528–533.

Funk, C., Michaelsen, J., Marshall, M., 2012. Mapping recent decadal climate variations in precipitation and temperature across eastern Africa and the Sahel. In: Wardlow, B.D., Anderson, M.C., Verdin, J.P. (Eds.), Remote Sensing of Drought: Innovative Monitoring Approaches. CRC Press, Boca Raton, FL, USA, pp. 331–358.

Garanganga, B.J., 2011, Drought risk Management in Southern Africa, Africa-Asia Drought Risk Management Peer Assistance Project. Available from: www.web.undp. org/drylands/docs/drought/AADAF1/1.7.

Good, K.M., Houser, A., Arntzen, L., Turnbull, P.C., 2008. Naturally acquired anthrax antibodies in a cheetah (*Acinonyx jubatus*) in Botswana. J. Wildl. Dis. 44 (3), 721–723.

Greve, P., Orlowsky, P., Mueller, B., Sheffield, J., Reichstein, M., Seneviratne, S.I., 2014. Global assessment of trends in wetting and drying over land. Nat. Geosci. 7, 716–721.

Hampson, K., Lembo, T., Bessel, P., Auty, H., Packer, C., Halliday, J., Beesley, C.A., Fyumagwa, R., Hoare, R., Ernest, E., Mentzel, C., Metzger, K.L., Mlengeya, T., Stamey, K., Roberts, K., Wilkins, P.P., Cleaveland, S., 2011. Predictability of anthrax infection in the Serengeti, Tanzania. J. Appl. Ecol. 48, 1333–1344.

Hansen, J., Sato, M., 2004. Greenhouse gas growth rates. Proc. Natl. Acad. Sci. 101 (46), 16109–16114.

Held, I.M., Soden, B.J., 2006. Robust responses of the hydrological cycle to global warming. J. Clim. 19, 5686–5699.

Higgins, S., Scheiter, S., 2012. Atmospheric CO_2 forces abrupt vegetation shifts locally, but not globally. Nature 488, 209–212.

Hillman, J.C., Hillman, K.K., 1977. Mortality of wildlife in Nairobi National Park, during the drought of 1973–1974. Afr. J. Ecol. 15, 1–18.

Huber, D.G., Gulledge, J., 2011. Extreme Weather and Climate Change: Understanding the Link, Managing the Risk. Pew Center on Global Climate Change, Arlington, VSA, Available from: www.pewclimate.org/publications/ extreme-weather-and-climatechange.

Jäger, H.G., Booker, H.H., Hübschle, O.J., 1990. Anthrax in cheetahs (*Acinonyx jubatus*) in Namibia. J. Wildl. Dis. 26 (3), 423–424.

Klein, R., 2006. Status report for the Cheetah in Botswana. Cat News Spec. Issue 3, 13–21.

Lehner, F., Deser, C., Sanderson, B.M., 2016. Future risk of record-breaking summer temperatures and its mitigation. Clim. Change, 3–13.

Lelenguyah, G.L., 2012. Drought, diseases and Grevy's zebra (*Equus grevyi*) mortality—the Samburu people perspective. Afr. J. Ecol. 50, 371–376.

Lembo, T., Hamson, K., Auty, T., Beesley, C.A., Bessel, P., Packer, C., Halliday, J., Fyumagwa, R., Hoare, R., Ernest, E., Mentzel, C., Mlengcya, T., Stamey, K., Wilkins, P.P., Cleveland, S., 2011. Serologic surveillance of Anthrax in the Serengeti ecosystem, Tanzania, 1996–2009. Emerg. Infect. Dis. 17 (3), 387–394.

Liu, B., Asseng, S., Müller, C., Ewert, F., Elliott, J., Lobell, D.B., Martre, P., Ruane, A.C., Wallach, D., Jones, J.W., Rosenzweig, C., Aggarwal, P.K., Alderman, P.D., Anothai, J., Basso, B., Biernath, C., Cammarano, D., Challinor, A., Deryng, D., De Sanctis, G., Doltra, J., Fereres, E., Folberth, C., Garcia-Vila, M., Gayler, S., Hoogenboom, G., Hunt, L.A., Izaurralde, R.C., Jabloun, M., Jones, C.D., Kersebaum, K.C., Kimball, B.A., Koehler, A.K., Kumar, S.N., Nendel, C., O'Leary, G.J., Olesen, J.E., Ottman, M.J., Palosuo, T., Vara Prasad, P.V., Priesack, E., Pugh, T.A.M., Reynolds, M., Rezaei, E.E., Rötter, R.P., Schmid, E., Semenov, M.A., Shcherbak, I., Stehfest, E., Stöckle, C.O., Stratonovitch, P., Streck, T., Supit, I., Tao, F., Thorburn, P., Waha, K., Wall, G.W., Wang, E., White, J.W., Wolf, J., Zhao, Z., Zhu, Y., 2016. Similar estimates of temperature impacts on global wheat yield by three independent methods. Nat. Clim. Change 6, 1130–1136.

Lutgens, F.K., Tarbuck, E.J., 2004. The Atmosphere. Pearson Prentice Hall, Upper Saddle River, New Jersey, USA, p. 508.

Lyon, B., 2009. Southern Africa summer drought and heat waves: observations and coupled model behavior. J. Clim. 22 (22), 6033–6046.

Lyon, B., DeWitt, D.G., 2012. A recent and abrupt decline in the East African long rains. Geophys. Res. Lett. 39 (2), L02702.

Marker, L.L., 2002. Aspects of Cheetah (*Acinonyx jubatus*) Biology, Ecology and Conservation Strategies on Namibian Farmlands. PhD thesis, University of Oxford, United kingdom.

Marker, L.L., Dickman, A.J., Mills, M.G.L., Jeo, R.M., Mackdonald, D.W., 2008. Spatial ecology of cheetahs on north-central Namibian farmlands. J. Zool. 274, 226–238.

Marker, L., Mills, M.G.L., Macdonald, D.W., 2003. Factors influencing perceptions and tolerance toward cheetahs (*Acinonyx jubatus*) on Namibian farmlands. Conserv. Biol. 17, 1–9.

Marker-Kraus, L., Kraus, D., Barnet, D., Hurlbut, S., 1996. Cheetah Survival on Namibian Farmlands. Cheetah Conservation Fund, Windhoek.

Matthews, H.D., Zickfeld, K., 2012. Climate response to zeroed emissions of greenhouse gases and aerosols. Nat. Clim. Change 2, 338–341.

Midgley, G., Hughes, G., Thuiller, W., Drew, G., Foden, W., 2005. Assessment of Potential Climate Change Impacts on Namibia's Floristic Diversity, Ecosystem Structure and Function. Namibian National Biodiversity Programme, Directorate of Environmental Affairs, Windhoek.

Ministry of Environment and Tourism (MET), 2008. Climate change vulnerability & adaptation assessment. Namibia Final Report.

Mirchi, A., Madani, K., Aghakouchak, A., 2015. Lake Urmia: How Iran's Most Famous Lake is Disappearing. the

Guardian, 23 January Available from: www.theguardian.com/world/iran-blog/2015/jan/23/iran-lake-urmia-drying-up-new-research-scientists-urge-action.

Munson, L., Terio, K.A., Worley, M., Jago, M., Bagot-Smith, A., Marker, L., 2005. Extrinsic factors significantly affect patterns of disease in free-ranging and captive cheetah (*Acinonyx jubatus*) population. J. Wildl. Dis. 41 (3), 542–548.

Muntifering, J.R., Dickman, A.J., Perlow, L.M., Hruska, T., Ryan, P.G., Marker, L.L., Jeo, R.M., 2006. Managing the matrix for large carnivores: a novel approach and perspective from cheetah (*Acinonyx jubatus*) habitat suitability modelling. Animal Conserv. 9, 103–112.

National Academies of Sciences, Engineering, and Medicine, 2016. Attribution of Extreme Weather Events in the Context of Climate Change. The National Academies Press, Washington, DC.

Norval, R.A.I., Perry, B.D., Young, A.S., 1992. The Epidemiology of Theileriosis in Africa. Academic Press, London.

Olwoch, J.M., Reyers, B., Engelbrecht, F.A., Erasmus, B.F.N., 2008. Climate change and the tick-borne disease, Theileriosis (East Coast fever) in sub-Saharan African Africa. J. Arid Environ. 72 (2008), 108–120.

Omari, K., 2010. Climate Change Vulnerability and Adaptation Preparedness in Southern Africa—A Case Study of Botswana. Available from: www.boell.org/sites/default/files/downloads/HBF-web_Bots.pdf.

Roach, J., 2008. Major Lion Die-Offs Linked to Climate Change. National Geographic News, 25th June Available from: http://news.nationalgeographic.com/news/2008/06/080625-warming-lions.html.

Roelker-Parker, M.E., Munson, L., Packer, C., Kock, R.A., Cleaveland, S., Carpenter, M., O'brien, S.J., Pospichil, A., Hoffman-Lehmann, R., Lutz, H., Mwamengele, G.L.M., Mgasa, M.N., Machange, G.A., Summers, B.A., Appel, M.J.G., 1996. A canine distemper virus epidemic in Serengeti lions (*Panthera leo*). Nature 379, 441–445.

Rohde, R., Muller, R.A., Jacobsen, R., Muller, E., Perlmutter, S., Rosenfeld, A., Wurtele, J., Groom, D., Wickham, C., 2013. A new estimate of the average Earth surface land temperature spanning 1753 to 2011. Geoinf. Geostat. 1, 1.

Schneider, S.H., 1989. The greenhouse effect: science and policy. Science 243, 771–781.

Schumann, M., Watson, L.H., Schumann, B.D., 2008. Attitudes of Namibian commercial farmers toward large carnivores: the influence of conservancy membership. S. Afr. J. Wildl. Res. 38 (2), 123–132.

Sejian, V.,Gaughan, J.B.,Raghavendra, B., Naqvi, S.M.K., 2016. Impact of Climate Change on Livestock Productivity. Feedipedia no. 26, February 2016 Available from: http://www.feedipedia.org/sites/default/files/public/BH_026_climate_change_livestock.pdf.

Switzer, A., Munson, L., Beesley, C., Wilkins, P., Blackburn, J.K., Marker, L., 2016. Namibian farmland *cheetahs* (*Acinonyx jubatus*) demonstrate seronegativity for antibodies against *Bacillus anthracis*. Afr. J. Wildl. Res. 46, 139–143.

Taylor, K.E., Stouffer, R.J., Meehl, G.A., 2012. An overview of CMIP5 and the experiment design. Bull. Am. Meteorol. Soc. 93, 485–498.

The World Bank, 2016. World Bank Open Data. Available from: http://data.worldbank.org/indicator/SP.POP.

Turnbull, P.C., Tindall, B.W., Coetzee, J.D., Conradie, C.M., Bull, R.L., Lindeque, P.M., Huebschle, O.J., 2004. Vaccine-induced protection against anthrax in cheetah (*Acinonyx jubatus*) and black rhinoceros (*Diceros bicornis*). Vaccine 3 (22), 25–26.

Turpie, J., Midgley, G., Brown, C., Barnes, J., Pallet, J., Desmet, P., Tarr, J., Tarr, P., 2010. Climate Change Vulnerability and Adaptation Assessment for Namibia's Biodiversity and Protected Area System. Final Report. *Strengthening the Protected Area Network (SPAN) Project.* Ministry of Environment and Tourism, Windhoek.

Verlinden, A., 1997. Human settlements and wildlife distribution in the southern Kalahari of Botswana. Biol. Conserv. 82, 120–136.

Wayne, G.P., 2013. The Beginners Guide to Representative Concentration Pathways. Skeptical Science. Available from: http://www.skepticalscience.com/rcp.php.

Wigley, B.J., Bond, W.J., Hoffman, M.T., 2009. Bush encroachment under three contrasting land-use practices in a mesic South African savanna. Afr. J. Ecol. 47 (Suppl. 1), S62–S70.

Wigley, B.J., Bond, W.J., Hoffman, M.T., 2010. Thicket expansion in a South African savanna under divergent land use: local versus global drivers? Glob. Change Biol. 16 (3), 964–976.

Williams, A.P., Funk, C., 2011. A westward extension of the warm pool leads to a westward extension of the Walker circulation, drying eastern Africa. Clim. Dyn. 37 (11–12), 2417–2435.

Williams, A.P., Funk, C., Michaelsen, J., Rauscher, S.A., Robertson, I., Wils, T.H.G., Koprowski, M., Eshetu, Z., Loader, N.J., 2012. Recent summer precipitation trends in the Greater Horn of Africa and the emerging role of Indian Ocean Sea surface temperature. Clim. Dyn. 39 (9–10), 2307–2328.

The Costs and Causes of Human-Cheetah Conflict on Livestock and Game Farms

Amy Dickman*, Niki A. Rust**,‡‡, Lorraine K. Boast†, Mary Wykstra‡, Louisa Richmond-Coggan§, Rebecca Klein†, Moses Selebatso¶, Maurus Msuha††, Laurie Marker§

*University of Oxford, Tubney, Abingdon, United Kingdom
**WWF-UK, Woking, United Kingdom
†Cheetah Conservation Botswana, Gaborone, Botswana
‡Action for Cheetahs in Kenya Project, Nairobi, Kenya
§Cheetah Conservation Fund, Otjiwarongo, Namibia
¶Kalahari Research and Conservation, Gaborone, Botswana
††Tanzania Wildlife Research Institute, Arusha, Tanzania
‡‡University of Kent, Canterbury, United Kingdom

INTRODUCTION

Human-wildlife conflict can be defined as "the situation that arises when behavior of a nonpest, wild animal species poses a direct and recurring threat to the livelihood or safety of a person or a community and, in response, persecution of the species ensues" (Inskip and Zimmerman, 2009). Cheetahs have undergone striking declines in distribution and numbers over the past century, with only around 7100 adults and adolescents thought to remain as of 2016 (Durant et al., 2017). Along with habitat and prey loss (Chapters 10 and 11) conflict with humans is one of the primary reasons for the decline (IUCN, 2007a; RWCP & IUCN/SSC, 2015).

Indeed, cheetahs are one of nine felid species for which conflict with humans is deemed a great conservation threat (Inskip and Zimmerman, 2009). One of the reasons that cheetahs are particularly susceptible to conflict is because, at a range wide scale, the majority (77%) of known cheetah range is on unprotected land (Durant et al., 2017). Furthermore, cheetahs can be extremely wide-ranging (Chapter 8), for example, in Namibia the mean overall home range was 1651 km^2, which could potentially overlap 21 farms (Marker et al., 2008). These vast home ranges increase the likelihood of encountering humans who are often intolerant to the cheetahs' presence thus increasing their susceptibility to conflict.

Cheetahs predominantly enter into conflict with humans due to their potential to predate on sheep and goats (Fig. 13.1; Selebatso et al., 2008) although they can also kill cattle with a preference for young animals (Inskip and Zimmerman, 2009). Conflict also occurs on game farms, particularly if cheetahs kill exotic, expensive game species, potentially leading to high rates of large carnivore removal (Schumann et al., 2006). In an analysis of all felid species, cheetahs fell into the "moderate" conflict category, which placed them higher than 27 other felid species, but the lowest of all big cat species (Inskip and Zimmerman, 2009). In Tanzania, far more

respondents reported attacks on livestock from lions (*Panthera leo*), spotted hyenas (*Crocuta crocuta*), leopards (*Panthera pardus*), or African wild dogs (*Lycaon pictus*) than from cheetahs (Dickman, 2009; Maddox, 2003). None of the 703 reported depredation events outside Serengeti National Park and none of the 106 investigated depredation events in Tanzania's Ruaha landscape were attributed to cheetahs (Dickman, 2009; Holmern et al., 2007). In Botswana, cheetahs have also been reported to kill far fewer livestock than other carnivores (Central Statistics Office, 2013; Gusset et al., 2009; Table 13.1). As such, while cheetahs certainly do cause conflict with people, they are often not the most important conflict-causing carnivore in the landscape, at least if other larger species are present.

In this chapter, we examine both sides of human-cheetah conflict: the costs imposed upon humans (e.g., through stock loss) and the costs imposed upon cheetahs (e.g., by being killed). We then examine the factors that influence the magnitude of damage caused by cheetahs and by humans. Lastly, we explain technical approaches used to reduce human-cheetah conflict and highlight necessary approaches for easing human-cheetah conflict in the future.

COSTS OF CONFLICT TO HUMANS

Cheetah presence can incur significant direct and indirect costs to humans in both financial and nonfinancial ways.

Direct Financial Costs

Depredation upon livestock and farmed game by cheetahs can cause substantial financial costs to households (Table 13.2), especially when expensive species or breeds are predated upon (Marnewick et al., 2007), or in areas that constitute particular "depredation hotspots" (Abade et al., 2014). Conflict due to the depredation of farmed game is of particular concern in South

FIGURE 13.1   **Cheetah preying on a goat in Namibia.**
Source: Laurie Marker, Cheetah Conservation Fund.

TABLE 13.1 Recorded Livestock and Farmed Game Losses to Cheetahs and Other Predators in Three African Countries as Reported in Interviews With Farmers and National Problem Animal Control Incidents

Country	Type of loss	Recorded losses
Botswana	Farmed game	93% of farmers reported losing farmed game to predators in previous 12 months ($N = 44$), equivalent to 3% of a farm's game herd; 24% of which was killed by cheetahs (Boast, 2014)
	Livestock	35% of farmers in Ghanzi district lost livestock to cheetahs annually ($N = 123$) (Selebatso et al., 2008)
		1% of livestock were reportedly killed by cheetahs in Northern Botswana compared to 77% killed by jackals (Gusset et al., 2009)
		4% of 7029 national problem animal control incidents caused by mammalian predators in Botswana were due to cheetahs, compared with 40% due to leopards, 27% to lions, and 24% to African wild dogs, and the remaining 5% to caracal, jackal, and hyenas (Central Statistics Office, 2013)
		Farmers reported higher levels of livestock losses to leopards (30%) and jackals (28%) than cheetahs (14%) (Klein, 2013)
Namibia	Cattle	13% of reported cattle losses were due to predators—of that, 29% were reported to have been lost to cheetahs (Marker, 2002)
	Sheep and goats	35% of recorded smallstock losses were due to predators—of that, 3% were reported to have been lost to cheetahs (Marker, 2002)
	Livestock	22% of farmers reported cheetah problems compared with 36% for leopards, 48% for caracals, and 68% for jackals ($N = 241$) (Marker, 2002)
Tanzania	Livestock	11%–13% of respondents had suffered stock lost to cheetahs in the past 12 months ($N = 179$) (Maddox, 2003)

Africa and Namibia. These countries have the most developed game farming industries, which encompass an area of almost 500,000 km^2 and largely overlap the countries' cheetah distribution (Cousins et al., 2008; Lindsey et al., 2013b; Taylor et al., 2016). Game farming can generate significant income through photographic tourism, trophy hunting, meat production, and the sale of live antelope. But game species are the cheetah's natural prey and can be difficult/expensive to protect from predation. Cheetahs are blamed for a quarter of depredation events on game farms in Botswana (Boast, 2014) and for high rates of predation on blesbok (*Damaliscus pygargus*) and springbok (*Antidorcas marsupialis*) (Boast et al., 2016; CCF unpublished data). Therefore, game farmers often dislike the presence of cheetahs, given their potentially negative impact on revenue (Marnewick et al., 2007).

Game farmers in Botswana reported financial losses 7 times greater than livestock farmers (Boast, 2014), although losses were likely overestimated due to difficulties in accurately monitoring game numbers (Boast et al. 2016). Damage caused by cheetahs is often exacerbated by the fact that, in most range countries, cheetahs cannot be trophy hunted so are often seen as "worthless problem" animals (Marnewick et al., 2007). However, these relationships are not always clear-cut: surveys in South Africa, Botswana, Namibia, and Zimbabwe suggest that game farmers are more tolerant of carnivores than livestock farmers are, but where conflict occurs financial losses and persecution of predators can be more intense (Lindsey et al., 2013a; Thorn et al., 2012).

Cheetah depredation on livestock, often due to inadequate livestock protection and pasture

TABLE 13.2   Financial Losses to Cheetahs and Other Predators in Five African Countries as Reported in Interviews With Farmers or From Compensation Payouts

Country	Type of loss	Recorded financial losses
Botswana	Farmed game	Median loss to carnivores: US $6,536 per farm per year (range US $0–235,325) (N = 53) (Boast, 2014)
	Livestock farms: commercial farms	Median loss to carnivores: US $3399 per farm per year (range US $0–32,684) (N = 55) (Boast, 2014)
	Livestock farms: 75% communal, 25% commercial	Mean loss to carnivores: US $149 per farm per year (range US $0–4390) (N = 226) (Klein, 2013)
Kenya	Livestock: commercial farm	Estimated cost per 5700 km^2 of farmland per year to support 1 cheetah: US $110 (Frank et al., 2005; Woodroffe et al., 2005): One spotted hyena US $35 One African wild dogs US $15 One leopard US $211 One lion US $360
	Olgulului Group Ranch (livestock)	2008–12: Compensation payouts for cheetah = US $23,255 for 696 cheetah depredations on 845 livestock (Okello et al., 2014)
	Mbirikani Group Ranch (livestock)	2008–12: Compensation payouts for cheetah = US $25,461 for 437 cheetah depredations on 566 livestock (Okello et al., 2014)
Namibia	Livestock: communal farms	Mean loss to carnivores: US $3461 per farm per year (N = 147) (Rust and Marker, 2013b)
South Africa	Farmed game	Carnivore depredation accounted for 0.2%–0.3% of a farm's net annual operating profit (Thorn et al., 2012)
Zimbabwe	Livestock: communal farms	Mean loss to carnivores: US $13 per household per year (12% of household's net annual income; N = 98) (Butler, 2000)

management, can occur frequently and may cause significant conflict, especially in certain hotspots, such as where natural prey populations are scarce (Chapter 11). Reviews of studies from Botswana, Namibia, and Tanzania revealed that 11%–35% of farmers reported losing livestock to cheetahs annually (Table 13.1). On farmland, where larger carnivores, such as lions and spotted hyenas have largely been extirpated, the relative importance of cheetahs as livestock predators can be high. For example, just under a third (29%) of recorded cattle depredations in Namibia were attributed to cheetahs (Marker, 2002).

There are few evaluations of the actual financial costs of living with cheetahs (vs. the perceived or reported costs), but Woodroffe et al. (2005) estimated that the regional cost of supporting 1 cheetah across an area of 5700 km^2 of commercial farmland in Kenya was US $110 annually. This is more than the costs for spotted hyena or African wild dog, but less than for leopards or lions (Table 13.2). Even though this cost estimate seems relatively low, cheetahs can cause significant financial impact at the household level, especially in areas where the communities are very poor and rely heavily on a few livestock for their financial security.

Indirect Financial Costs

Carnivores can impose additional financial burdens on livestock and game farmers through lost potential revenue, such as that from wool, milk, or future offspring (Mertens and Promberger, 2001). Farmers also bear indirect costs by investing money in strategies to prevent depredation. These costs can be significant—for instance, one of the most effective ways of protecting farmed game from carnivores is through electric fencing, but a study in Namibia estimated installation costs at US $781 per kilometer, and maintenance costs, including vehicle costs to regularly check the fence, at US $952 per kilometer per year (Schumann et al., 2006). Using livestock guarding dogs can also incur substantial costs—specialized-breed livestock guarding dogs in Namibia and South Africa were reported to cost between US $469 and US $2780 to purchase and maintain the dog for the first year of its life (Chapter 15). Even lethal carnivore control measures, such as gin traps and hunting, can have significant costs (often higher than some nonlethal measures) (McManus et al., 2014).

Cheetah presence alone, even when cheetahs do not kill farmed animals, can cause problems. In some cases cheetahs have been reported to cause cattle to panic and break fences, incurring maintenance costs for the farmer, as well as the risk that livestock will be lost or injured (Marker, unpublished data.). Prey, including livestock, may be more vigilant and stressed when carnivores are present, reducing feeding time and slowing weight gain (Howery and DeLiberto, 2004), which can result in significant costs. Some farmers have suggested that enclosing calves at night to protect them from predators, rather than letting them roam freely, may reduce weight gain (Rothman, unpublished data), while increasing disease transmission (e.g., increased parasitic pressure) and therefore veterinary costs (Rust, unpublished data). However, according to a Namibian farmer, if proper management strategies are implemented, calves penned for the first 2 months of their lives do not have reduced weights by weaning age (8 months) (Marker, unpublished data).

In addition to direct and indirect costs to the individual livestock owner, carnivore depredation can incur significant financial costs to government departments or conservation organizations through human-wildlife conflict mitigation efforts, compensation payments for predated stock, or problem animal control. In the Amboseli area of Kenya, US $23,255 was paid in compensation for cheetah depredations between 2008 and 2012 on the Olgulului Group Ranch (Okello et al., 2014). However, it is to be noted that depredation reports tend to be exaggerated when compensation is involved (Selebatso et al., 2008).

Nonfinancial Costs

Carnivores can also impose direct nonfinancial costs. For example, in traditional pastoralist societies, the ownership of livestock confers social and cultural value, which is at least as important as the financial value of the animals owned (Dickman, 2009). Loss of stock is associated with a loss of social standing and individuals can be ostracized from their communities if they experience high levels of livestock loss. In some rural Tanzanian communities, the loss of livestock is associated with witchcraft, and in an extreme example if someone (especially an older female widower) loses all her livestock, then she is at risk of being murdered as a witch (Dickman, unpublished data). Pastoralist societies are prevalent in some of the most important remaining cheetah range countries, making these traditions relevant to cheetah survival.

Another often-overlooked cost of human–carnivore coexistence is the fear factor induced by living alongside potentially dangerous species. Such fear of carnivores can have an important effect on people; so much so that, in Japan it is said that just seeing a large carnivore can impose "spiritual damage" (Knight, 2000). Furthermore, in some African countries, the presence of carnivores and other dangerous wildlife impedes everyday activities like the ability of children to walk to school (Nyamwaro et al., 2006). Even though there has never been a confirmed attack on humans by a wild cheetah, cheetahs are commonly confused with leopards (e.g., only a quarter of Tanzanian villagers surveyed correctly identified the cheetah from photographs; Dickman, 2009) and are often thought to pose a risk to human life. Around Tanzania's Ruaha National Park, 7% of villagers who disliked cheetahs said that it was because they posed a threat to both humans and livestock (Dickman, 2009).

There are also opportunity costs from living alongside large carnivores. Children are often used as herders to protect stock against attacks from cheetahs and other carnivores (Fig. 13.2), which can lead to missing school and limits future opportunities (Dickman, 2009). Although hard to quantify, these impacts influence people's perceptions and should not be discounted in terms of their significance.

FIGURE 13.2   **Herders can help reduce livestock depredation.** Source: Niki Rust.

FACTORS AFFECTING THE MAGNITUDE OF COSTS TO HUMANS

Factors increasing the risk of depredation on livestock and farmed game include intrinsic factors (i.e., linked to the cheetahs themselves) and factors related to the environment.

Intrinsic Factors

The age, sex, and condition of cheetahs can influence the likelihood of them becoming embroiled in conflict. A Namibian study found that cheetahs trapped as a perceived threat on game farms were more likely to be adults (either young or prime adults), while those on livestock farms were more likely to be younger animals, such as dispersing subadults (Marker et al., 2003). Females were more likely to be trapped on game farms (42% of cheetahs caught) than livestock farms (26%; Marker et al., 2003). However, it should be noted that most of these cheetahs were captured as a precautionary measure against potential depredation and there was no evidence for cheetahs of a particular age or sex being more likely to attack livestock (Marker et al., 2003).

Injury or other poor physical conditions are often associated with a propensity for carnivores to prey on livestock instead of wild game (Rabinowitz, 1986), and this relationship has been observed for cheetahs. In Namibia, of 198 cheetahs captured on farmland, only 6 (3%) incidents had circumstantial evidence that the cheetah had killed livestock (Marker et al., 2003). Of the 6 cheetahs, 4 (67%) had a physical impediment that could hinder their hunting ability —far more than the 11% with similar impediments in the sample population (Marker et al., 2003).

Environmental Factors

Levels of conflict can also be affected by environmental factors, such as the climate and seasonality. In general cheetah depredation rates on livestock are higher during the dry season or periods of drought due to a lack of natural prey availability (Dickman, 2009; Maddox, 2003; Okello et al., 2014). There have also been reports of increased levels of cheetah depredation during the calving season, especially among cattle who are first-time mothers (Marker-Kraus et al., 1996).

Wild prey depletion caused by bushmeat hunting has also been linked to increased human–felid conflict (Inskip and Zimmerman, 2009), but there is relatively little data on this, specifically for cheetahs. However, Rust and Marker (2013b) found no link between the frequencies of wild prey sightings and reported livestock depredation by all carnivores (including cheetahs) between communal and resettled land in Namibia. In Salama, Kenya, cheetah conflicts actually decreased as wild prey numbers did (Wykstra, unpublished data). Differences in conflict trends with reduced wild prey numbers may be explained by a difference between low naturally occurring prey numbers, which would support lower cheetah populations and thus reduced conflict, while

recent prey number reduction is more likely to lead to increased conflict as existing cheetahs turn to livestock as an alternative food source.

Schiess-Meier et al. (2007) found that livestock depredation from all carnivores including cheetahs increased with proximity to a prey-rich protected area in Botswana but this relationship was not observed in Tanzania, although proximity to Ruaha National Park was associated with greater human–lion conflict (Dickman, 2009). Conflict mitigation strategies to reduce carnivore depredation even in risky environments, such as farms abutting protected areas, are reviewed briefly in section, "Human Factors Affecting the Magnitude of Costs to Cheetahs."

COSTS OF CONFLICT TO CHEETAHS

Conflict-related killings outside of protected areas often (although not always) outnumber other forms of cheetah mortality (Table 13.3).

TABLE 13.3   Anthropogenic Mortality of Cheetahs in Four African Countries

Country	Cause of death	Methodology	Level of cheetah mortality
Botswana	Conflict on commercial livestock and game farms	Interviews	Cheetahs killed on 20% of farms on which they occurred ($N = 64$) in the 12 months previous to the survey (Boast, 2014)
Kenya	Conflict with pastoralists	Investigated deaths	Intentional killing accounted for 11% of deaths ($N = 19$) (Wykstra, unpublished data)
Namibia	Conflict on livestock and game farms	Interviews	1980–93: Game farmers killed 26.1 cheetahs per farm ($N = 49$) Livestock farmers killed 12.6 cheetahs per farm ($N = 108$) (Marker et al., 2003)
		Investigated deaths	Intentional killing accounted for 63% of adult deaths ($N = 63$), 25 of these cheetahs were shot plus an additional 5 cubs were presumed to have died due to these killings (Marker et al., 2003)
		CITES permits	1980–91: 6293 Cheetahs killed by farmers (average 572.1 per year) (CITES, 1992)
	Trophy hunting	CITES permits	1980–91: 190 Cheetahs (15.8 per year) (CITES, 1992) 1990–2012: 1720 Cheetahs (74.8 per year) (Nowell, 2014)
		Investigated deaths	10% of investigated deaths ($N = 63$) (Marker et al., 2003)
Tanzania	Conflict with pastoralists	Interviews	Mean number of cheetahs killed = $0.07 \pm$ SD 0.03 per person in lifetime ($N = 179$) (Maddox, 2003)

Farmers in Namibia reported killing 6293 cheetahs in defense of their livestock between 1980 and 1991, averaging 572 annually (Table 13.3) compared with just 190 cheetahs killed by trophy hunting during this time (CITES, 1992). In contrast, in the Salama area of Kenya, intentional killing of cheetahs accounted for only 11% of all recorded mortality sources, markedly less than highway mortality, which accounted for 50% of recorded deaths (Wykstra, unpublished data).

In Botswana, 20% of commercial farmers reported having killed a cheetah during the year previous to the survey (Boast, 2014). Yet, in East Africa, killing of cheetahs due to conflict is reportedly low, and it is unclear how much of a direct threat intentional killing poses to cheetahs across some of their most important range (TAWIRI, 2007). In Tanzania's Ruaha landscape, even though many people viewed cheetahs and other large carnivores as problematic, only 2% of villagers said that they used lethal methods to control carnivores (Dickman, 2009). While this could result from fear of admitting illegal killing to outsiders, investigations of over 80 carnivore killing events around Ruaha revealed that none of the carnivores killed were cheetahs (Dickman, unpublished data). Interestingly, in northern Tanzania, reported rates of carnivore killings were much higher than in the south, with 60% of people saying they had killed a carnivore, although it was still relatively uncommon for people to report killing cheetahs (Maddox, 2003).

Besides being killed, there can be subtler costs of conflict on cheetahs as well. For example, the erection of electrified predator-proof perimeter fencing on game farms or intensive wildlife farming camps (small holding pens where expensive game species are kept) results in significant habitat loss and fragmentation (Schumann et al. 2006; Chapter 10). It is estimated that 10% of the area on South African game farms have now been fenced into intensive wildlife farming camps. Any carnivore found within the camps is removed using either poison, shooting, or capture (Taylor et al., 2016).

HUMAN FACTORS AFFECTING THE MAGNITUDE OF COSTS TO CHEETAHS

The determination of people to kill or exclude cheetahs is partly driven by how significant a problem the species are perceived to be, and by the person's willingness and ability to act on those perceptions. A complex array of social, economic, psychological, political, and cultural factors affect these drivers, some of which will be reviewed later.

Perceptions of the Magnitude of Damage Caused

There is considerable evidence that people are poor judges of the "actual" damage caused by carnivores. In Ruaha, over 80% of surveyed farmers viewed cheetahs as problematic (with over 50% saying they were a big problem), despite very little evidence for significant financial impacts (Dickman, 2009). Similarly, Marker et al. (2003) found that, while the motivation for trapping 91% ($N = 198$) of cheetahs in Namibia was a perceived threat to livestock or farmed game, there was only reasonable evidence for them causing depredation in 3% of cases. Some of the exaggerated perceptions of cheetahs as problem animals may be due to a lack of differentiation between cheetahs and leopards or because people inherently weight negative experiences more than positive ones (Kansky and Knight, 2014). That said, people are more likely to act on perceived rather than real risks, so it is important that conservationists address both perceived and actual risks when mitigating conflict.

People can also be more hostile toward carnivores if they feel they are "wasteful" of their food. Research in Botswana (Kent, 2011) showed that farmers are less tolerant of cheetahs than leopards because leopards will return to and consume the whole carcass, whereas cheetahs will leave after the initial feeding, probably to decrease the risk of being detected by larger carnivores (Hunter et al., 2007).

Views toward cheetahs may also be affected by experiences with and attitudes toward other species. For example, some individuals are likely to collectively view all carnivores in a similar way, and assume that they must all cause damage (Rust and Taylor, 2016), so-called "contagious conflict" (Dickman et al., 2014).

Impacts and Context of Damage Caused

The context of any depredation is crucial, especially in terms of the vulnerability of the household concerned. People who heavily depend on livestock, for economic and/or cultural reasons, are likely to be more sensitive to depredation than others, so pastoralist societies can be particularly antagonistic toward large carnivores (Kansky et al., 2014) and engage in high levels of carnivore killing (Dickman et al., 2014; Hazzah et al., 2017). Commercial farmers in Botswana whose income was largely derived from livestock were also less tolerant of cheetahs than farmers whose income relied upon a range of sources (Klein, 2013).

When household-level economic security is reduced (e.g., by stock die-offs during droughts), the perceived risks of carnivore presence are heightened as people are more vulnerable to any additional costs imposed from depredation (Dickman, 2009). Poverty-stricken households (such as those across much of the remaining cheetah range) are at particular risk of "compounding vulnerability"; they often own very few animals, are unable to invest in effective methods for protecting their assets (such as strong kraals or livestock guarding dogs), and are also the least able to recover from any damage suffered (Naughton-Treves, 1997). Given the potentially catastrophic impacts of a depredation event, individuals may turn to low-cost lethal measures, such as poison and snares to reduce or prevent future depredation, regardless of present depredation levels. Also, depre-

dation should be viewed in the context of past experiences: large numbers or high frequencies of past depredation events are likely to increase hostility.

Relative Value of Live and Dead Cheetahs

One of the most interesting results to emerge from the long-term studies in Namibia was that, even among farmers who did not view cheetahs as problematic, a majority (59%) still removed them (Marker, 2002). This suggests that, to these farmers the value of live cheetah presence is outweighed by the value of having a landscape free from cheetahs or by the financial value of dead cheetahs (e.g., by selling their skins) (Rust, 2016). Ultimately, people are more willing to coexist with potentially dangerous wildlife if they receive direct, tangible benefits, which they perceive to outweigh the costs imposed. Namibian conservancies, where members receive revenue from utilizing wildlife (Chapter 17), provide a good example of this, as members were shown to be more willing to tolerate the presence of large carnivores (including the risks of depredation) (Schumann et al., 2008).

Social, Cultural, and Political Factors

Whether someone kills or excludes carnivores is often grounded in culture as much as in economic realities. People are less tolerant of risks if they incite deep-seated feelings of fear or dread (Sjoberg et al., 2004), which tend to be elicited by carnivore attacks (Berg, 2001). However, people are more likely to tolerate risky situations if they are undertaken voluntarily rather than externally imposed; all too often, large carnivore presence is perceived as imposed by disliked and remote external groups (Dickman, 2010). Local communities may also feel that the needs of wildlife and/or tourism are being prioritized over their own needs, which often inflames anger and carnivore killing (Dickman, 2009;

Goldman et al., 2013). This is exacerbated if their own views seem less influential than distant elites', leading to a sense of powerlessness (Dickman, 2010; Nie, 2004).

At the scale of the wider political landscape, there seems to be an increasing disconnect between the opinions of those who live alongside large carnivores and those who influence international decisions on wildlife management. Pressure to ban trophy hunting was fueled by the heavily publicized case of "Cecil the lion" in 2015, despite many African countries viewing trophy hunting as an important form of wildlife management. If successful, the current international pressure to stop trophy hunting (especially of big cats) could have unintended negative consequences. Trophy hunting protects large areas of wildlife habitat and is thought to promote tolerance of carnivores, so banning trophy hunting could have far-reaching negative consequences for biodiversity (Di Minin et al., 2016). For example, modeling suggests that removing trophy hunting would mean that only 16% of Namibian conservancies would be profitable, with "substantial reductions in overall benefit generation and incentives for wildlife conservation throughout Namibia" (Naidoo et al., 2016).

Political issues can also have long-standing impacts on human–carnivore coexistence. For example, instances of carnivore conflict in Namibia have been linked to knock-on effects of the previous apartheid regime on worker employment and livestock management, that influenced carnivore management practices (Rust et al., 2016). The power differentials and inequality between white commercial farmers and indigenous workers was linked to poaching of livestock and farmed game by workers for food and money, while workers blamed carnivores for this loss. Farmers therefore thought that they had a worse problem with carnivores than in reality.

Although studies have shown a clear link between attitudes and actions, in that people who have more negative views are more likely to kill carnivores (Hazzah et al., 2017), social norms can mediate the chance of such killing occurring. For example, peer pressure among farmers can lead to increased or decreased carnivore removal regardless of personal attitudes and this peer pressure can cause substantial interpersonal conflict (Boast and Marker, personal observation). Pressure to keep cheetahs is more likely in areas where they yield financial benefits (CCF, unpublished data). In traditional pastoralist societies, carnivore killing can be an important rite of passage to adulthood, as well as representing a form of community protection, so personal attitudes toward carnivores become less important than the cultural expectation of killing them (Dickman, personal observation).

It is clear that human attitudes and behaviors toward carnivores are shaped by a complex myriad of factors, including direct experience, perceived behavioral control, anecdotal information, ideologies, values, cultural norms and expectations, power, fear, and many others (Dickman, 2012; Rust, 2015). Although these factors have not all been studied in relation to cheetahs, it seems likely that human-cheetah conflict will be similarly complex and multifaceted. Therefore, long-term conflict mitigation will require not only reducing costs of human-cheetah coexistence, but also influencing underlying beliefs and values, which will give people a reason to tolerate cheetah presence. This requires detailed understanding of the specific conflict situation (which is likely to be highly dynamic) and long-term engagement with the human communities involved (Chapter 35 for social science methodologies).

REDUCING HUMAN-CHEETAH CONFLICT

In the short term, the most urgent priorities for reducing human-cheetah conflict are to limit costs (direct, indirect, financial, and nonfinancial), especially to livestock or game producers and to provide or improve benefits (e.g., incentives) associated with coexisting with cheetahs (Chapter 16). Long-term mitigation may be more challenging as this is likely to involve reducing the underlying drivers of conflict, some of which may be extremely difficult to change (Rust 2015).

Reducing the Economic Costs of Cheetah Presence by Reducing Depredation

Numerous techniques have been tested and implemented in order to reduce depredation on both livestock and farmed game. Most of the strategies focus on better livestock guarding and game management. While some farmers use lethal control as their first option, nonlethal methods were more cost-effective in South Africa (McManus et al., 2014). We provide a brief overview of some of the key nonlethal techniques here.

On livestock farms, keeping guarding dogs with smallstock herds has proven to be one of the most effective methods for reducing cheetah depredation (Chapter 15). Other kinds of guardian animals, especially female donkeys with young ones, can also be effective livestock guardians (Marker-Kraus et al., 1996). Fitting livestock with protective collars can be another effective and cost-efficient way of preventing attacks (McManus et al., 2014). An additional possibility is changing from smallstock to cattle production if ecological conditions allow (Rust and Marker, 2013b). Other husbandry methods that have been suggested to work by farmers (but not scientifically tested yet) include using indigenous breeds of livestock that have stronger antipredator instincts, alongside keeping horns on livestock and using holistic rangeland management (Rust, 2016). In addition to directly protecting stock, it is important to maintain enough natural prey on farmland (Marker et al., 2010).

On game farms, electric fences are an effective but expensive form of predator exclusion. Installing "swing-gates" in game fencing is a low-cost technique that may prevent species, such as warthogs (*Phacochoerus aethiopicus*) and aardvarks (*Orycteropus afer*) from digging

under fences and thereby creating holes which allow the entry of carnivores into fenced game farms (Schumann et al., 2006). However, a long-term study on the effectiveness of swing gates showed that jackals learned to use the gates (Rust et al., 2015), and it is unclear whether cheetahs could learn to do this. It is also recommended that game farmers stock native game rather than exotic species, as native game is more wary of local predators and thus better-adapted to avoiding them (Marker-Kraus et al., 1996).

While costs can be reduced substantially by implementing better husbandry methods, some losses may still happen, and just the fear of incurring financial costs can drive negative attitudes toward large carnivores. The likelihood of substantial costs being imposed at a household level can be reduced if a compensation scheme is in place. Such compensation schemes can be financed externally or run as a locally managed insurance scheme. In Kenya, compensation was possibly linked to a reduction in cheetah killing, and has appeared valuable in reducing the killings of other carnivores, such as lions (Maclennan et al., 2009). However, compensation is fraught with issues (Dickman et al., 2011; Nyhus et al., 2003, 2005), such as over-reporting, difficulty in ensuring rapid and accurate verification of kills, lack of access to the system in remote areas, not accounting for the full market value of animals (e.g., pregnant animals, or milk production), or overpaying for animals if market value is not matched in real time (e.g., in the case of drought), and dependency on external funding. In Botswana, a government compensation program exists for livestock losses to most large carnivore species including cheetahs, but since it rarely pays full market value for losses it is not successful in improving perceptions toward carnivores (Klein, 2013). There is also a risk of "moral hazard," where compensation reduces the incentive to prevent damage occurrence, and it can increase the perception that carnivores are a problem whose costs should be borne by an external entity (such as a government or conservation organisation) (Dickman et al., 2011; Nyhus et al., 2003). Alternatively, livestock could be replaced "like for like," from a livestock breeding center, but even this is unlikely to address all the issues related to compensation, especially as many farmers have their own specific breeds of livestock. Some of the issues associated with compensation schemes can theoretically be avoided by community insurance schemes, as there is more timely verification and oversight, but they still require some degree of external investment, and there is often relatively poor buy-in as individuals are often unwilling to pay up-front even if they are likely to suffer livestock losses later (Dickman et al., 2011).

Improving the Benefits Associated With Cheetah Presence

Regardless of how successful damage reduction might be, conflict is unlikely to be effectively resolved unless the local stakeholders receive (and recognize) sufficient benefits to outweigh any remaining costs associated with coexisting with cheetahs (including perceived threats or costs). Having ownership of, or tenure rights to, wildlife resources and/or land is a key requirement for people receiving meaningful individual benefits (Romanach et al., 2007). This model has worked relatively well for cheetah conservation in Namibia, and attitudes to cheetahs are more positive on farms where farmers can directly benefit from wildlife-related tourism or trophy hunting (Rust and Marker, 2013a). Generating sufficient household-level benefits is more challenging where the government rather than the landowner owns and manages wildlife. Direct financial incentives to protect wildlife include "payments for presence" (e.g., Zabel and Engel, 2010; Zabel and Holm-Muller, 2008) where landowners are rewarded monetarily for some stipulated target, such as the number of cheetahs persisting in the area over a given time frame. Indirect financial incentives include certification schemes, such as a beef certification,

which has been proposed in Namibia (associated with a premium price for the beef) if farmers adhere to wildlife-friendly management techniques (Marker et al., 2010; Chapter 16). While such economic incentives may be less useful for individuals who are already financially secure (MacMillan and Phillip, 2010), many people living in the cheetah's range are living below international poverty lines (Chapter 16).

Alternatives to cash payments include the provision of vital community resources, such as schoolbooks, human medical supplies, and veterinary medicines. In Tanzania's Ruaha landscape, communities earn points for camera trap pictures of cheetahs and other wildlife, which they exchange for these community benefits (Ruaha Carnivore Project, 2015). However, nearly all payment- and resource-based initiatives will likely have issues sustaining program funding (Dickman et al., 2011), and if funding stops, then affected individuals are likely to be even less tolerant of predators than they were before the benefit system was initiated.

Noneconomic benefits associated with cheetahs and other wildlife include their ecological and existence values (Vucetich et al., 2015). Reinforcing these intrinsic values is an important strand of a long-term conservation strategy. However, it is unlikely that many people would decide that the intrinsic value of cheetahs outweighs their own needs, especially if the individual is already poverty-stricken and vulnerable. Cost reduction and benefit provision strategies can be expensive. Therefore awareness should be raised among funders that it is imperative that the developed world invests at a larger and committed long-term scale in large carnivores and the people who coexist with them, if we want to ensure their continued conservation.

Addressing the Underlying Drivers of Conflict

While providing economic benefits and education may improve tolerance among individuals who may already have a positive attitude toward cheetahs, it might not work for those individuals who are least tolerant and most likely to kill cheetahs, or for whom money is not a significant motivating factor (Rust, unpubublished data). Changing the behavior of staunch carnivore opponents requires addressing the underlying drivers of conflict. This is not an easy task, as it often requires tackling complex societal issues, such as power, ideology, and corruption (Rust and Taylor 2016).

A Namibian study looking at stakeholder acceptance of carnivore conflict management strategies showed that while most people preferred nonlethal methods, a third of participants (mostly smallstock farmers) preferred lethal methods (Rust, 2016). Offering money, education, or technical fixes is unlikely to make a difference to this latter group. They may instead require a different tactic, for example, well-regulated trophy hunting, as well as more indirect approaches to mitigate conflict, such as civic mediation or conflict transformation (Madden and McQuinn, 2014). Improving the attitudes and behaviors of these individuals is likely to be extremely challenging, as these are often driven by deep-seated ideologies that are resistant to change, especially in response to proponents of carnivores. As such, collaborations with social scientists may be useful when trying to understand and change the behavior of these individuals. Use of pressure or value confrontation from respected peers, especially if from the same general background, may also help (Rust and Taylor 2016; Chapter 35).

DEVELOPING STRATEGIES FOR LONG-TERM HUMAN–CHEETAH COEXISTENCE

Regardless of the success of any of the aforementioned strategies, effective mitigation of human-cheetah conflicts, and fostering a broader view for cheetah conservation, requires a shift from short-term problem solving to long-term conservation, land-use planning and governance. Cheetah conservation strategies have now been developed for regions across Africa (IUCN/SSC 2007a,b; IUCN/SSC 2012, RWCP & IUCN/SSC 2015), thereby providing a structure under which national action plans have and can be developed for individual cheetah range countries. The Regional Conservation Strategies developed multiple objectives, one of which was to "Develop and implement strategies to promote coexistence of cheetah with people and domestic animals." The objective's targets include: reducing indiscriminate hunting and illegal off-take of wild ungulates; developing sustainable tools to reduce cheetah impacts on livestock; initiating programs for local people to derive sustainable economic benefits from cheetahs and their prey; and developing awareness-creation programs relevant to cheetah conservation. These targets are being addressed in various ways across the cheetah range countries, but there is an urgent need for effective methods to be expanded and implemented at a far greater scale to surmount the magnitude of conservation threats. Given that conservation interventions are highly dependent on local participation and acceptance, it is imperative that these range-wide strategies also integrate local knowledge and priorities at both household and community levels.

Cheetahs will undoubtedly face daunting challenges over the next decades. Africa's human population is predicted to increase from around 1.2 billion in 2015 to 4.1 billion by 2100 (United Nations, 2015). This will have huge impacts on cheetah habitat in terms of habitat degradation, fragmentation (Chapter 10), and wild prey loss (Chapter 11). Increasing human demands for food may well make the presence of carnivores on farmland even less tolerable for future landowners.

The likely future threats facing cheetahs are particularly concerning given the current vulnerable status of the species (Durant et al. 2017), and the fact that cheetahs are one of the highest-priority felid species in terms of their need for conservation action (Dickman et al., 2015). If cheetah populations are to be maintained in such a human-dominated landscape as is likely to be the case as Africa develops, it is imperative that Africans themselves are as fully engaged and empowered as possible in managing long-term human–wildlife coexistence. It is crucial that farmers directly affected by conflict be informed of the best strategies for successfully maintaining livestock and farmed game alongside wildlife, through knowledge and capacity building (Chapter 18). Given the extensive ranging patterns of cheetahs, successful approaches from the level of the individual landowner or community need to be spatially scaled up to a wider, more sustainable landscape-level approach.

Governance mechanisms, both locally and nationally, will play a critical role in determining the future of human–wildlife coexistence in Africa. Politicians need to be educated about the severity of the threat to cheetahs and other carnivores, and encouraged to develop conservation-minded policies around issues, such as sustainable wildlife use, tourism, poverty alleviation, land-use planning and human-wildlife conflict mitigation. Policies specifically aimed at reducing human-wildlife conflict have already been developed for Botswana and Namibia, and these countries' national cheetah action plans can also play an important role in helping guide future governmental and nongovernmental actions (Marker and Boast, 2015). Despite the clear threats posed by human-cheetah conflict, the development and implementation of well-considered strategies could have significant benefits not only in terms of cheetah conservation, but also in terms of improving the long-term economic security of the local communities on whose land so many cheetahs depend. Ultimately, action is needed at all levels, from individuals to communities and governments, in order to encourage human–cheetah coexistence through the 21st century and beyond.

References

Abade, L., Macdonald, D.W., Dickman, A.J., 2014. Assessing the relative importance of landscape and husbandry factors in determining large carnivore depredation risk in Tanzania's Ruaha landscape. Biol. Conser. 180, 241–248.

Berg, K.A., 2001. Historical Attitudes and images and the implications on carnivore survival. Endanger. Species Update 18 (4), 186–189.

Boast, L., 2014. Exploring the Causes of and Mitigation Options for Human-Predator Conflict on Game Ranches in Botswana. PhD thesis, University of Cape Town, South Africa.

Boast, L., Houser, A.-M., Horgan, J., Reeves, H., Phale, P., Klein, R., 2016. Prey preferences of free-ranging cheetahs in farmland: scat analysis versus farmers' perceptions. Afr. J. Ecol. 54, 424–433.

Butler, J.R.A., 2000. The economic costs of wildlife predation on livestock in Gokwe communal land Zimbabwe. Afr. J. Ecol. 38, 23–30.

Central Statistics Office, 2013. Botswana Environment Statistics 2012, Statistics Botswana, Gabarone, Botswana.

CITES, 1992. Quotas for trade in specimens of cheetah. Eighth meeting of the Convention of International Trade in Endangered Species of Wild Fauna and Flora, pp. 1–5.

Cousins, J.A., Sadler, J.P., Evans, J., 2008. Exploring the role of private wildlife ranching as a conservation tool in South Africa: stakeholder perspectives. Ecol. Soc. 13 (2), 43.

Department of Economic and Social Affairs, Population Division, United Nations, 2015. World Population Prospects: The 2015 Revision, Methodology of the United Nations Population Estimates and Projections, Working Paper No. ESA/P/WP.242.

Di Minin, E., Leader-Williams, N., Bradshaw, C.J., 2016. Banning trophy hunting will exacerbate biodiversity loss. Trends Ecol. Evol. 31 (2), 99–102.

Dickman, A.J., 2009. Key determinants of conflict between people and wildlife, particularly large carnivores, around Ruaha National Park, Tanzania. PhD thesis, University College London, United Kingdom.

Dickman, A., 2010. Complexities of conflict: the importance of considering social factors for effectively resolving human–wildlife conflict. Anim. Conserv. 13, 458–466.

Dickman, A.J., 2012. From cheetahs to chimpanzees: a comparative review of the drivers of human-carnivore conflict and human-primate conflict. Folia Primatol. 83, 377–387.

Dickman, A., Hazzah, L., Carbone, C., Durant, S., 2014. Carnivores, culture and 'contagious conflict': multiple factors influence perceived problems with carnivores in Tanzania's Ruaha landscape. Biol. Conser. 178, 19–27.

Dickman, A.J., Hinks, A.E., Macdonald, E.A., Burnham, D., Macdonald, D.W., 2015. Global priorities for felid conservation. Conser. Biol. 29 (3), 854–864.

Dickman, A.J., Macdonald, E.A., Macdonald, D.W., 2011. A review of financial instruments to pay for predator conservation and encourage human–carnivore coexistence. Proc. Natl. Acad. Sci. USA 108 (34), 13937–13944.

Durant, S., Mitchell, N., Groom, R., Pettorelli, N., Ipavec, A., Jacobson, A., Woodroffe, R., Bohm, M., Hunter, L., Becker, M., Broekhuis, F., Bashir, S., Andresen, L., Aschenborn, O., Beddiaf, M., Belbachir, F., Belchabir-Bazi, A., Berbash, A., Brandao de Matos Machado, I., Breitenmoser, C., Chege, M., Cillers, D., Davies-Mostert, H., Dickman, A.J., Ezekiel, F., Farhardinia, M.S., Funston, P., Henschel, P., Horgan, J., de Iongh, H., Jowkar, H., Klein, R., Lindsey, P., Marker, L., Marnewick, K., Melzheimer, J., Merkle, J., M'soka, J., Msuha, M., O'Neill, H., Parker, M., Purchase, G., Sahailou, S., Saidu, Y., Samna, A., Schmidt-Küntzel, A., Selebatso, E., Sogbohossou, E.A., Soultan, A., Stone, E., van der Meer, E., van Vuuren, R., Wykstra, M., Young-Overton, K., 2017. Disappearing spots, the global decline of cheetah and what it means for conservation. Proc. Natl. Acad. Sci. USA 114 (3), 528–533.

Frank, L.G., Woodroffe, R., Ogada, M.O., 2005. People and predators in Laikipia district, Kenya. In: Woodroffe, R., Thirgood, S., Rabinowitz, A. (Eds.), People and Wildlife: Conflict or Coexistence. Cambridge University Press, Cambridge, United Kingdom, pp. 286–304.

Goldman, M.J., de Pinho, J.R., Perry, J., 2013. Beyond ritual and economics: Maasai lion hunting and conservation politics. Oryx 47 (04), 490–500.

Gusset, M., Swarner, M., Mponwane, L., Keletile, K., McNutt, J., 2009. Human–wildlife conflict in northern Botswana: livestock predation by Endangered African wild dog Lycaon pictus and other carnivores. Oryx 43 (01), 67–72.

Hazzah, L., Bath, A., Dolrenry, S., Dickman, A., Frank, L., 2017. From attitudes to actions: predictors of lion killing by Maasai warriors. PloS One 12 (1), e0170796.

Holmern, T., Nyahongo, J., Roskaft, E., 2007. Livestock loss caused by predators outside the Serengeti National Park. Tanzan. Biol. Conserv. 135, 518–526.

Howery, L.D., DeLiberto, T.J., 2004. Indirect effects of carnivores on livestock foraging behavior and production. Sheep and Goat Research Journal 19, 53–57.

Hunter, J.S., Durant, S., Caro, T.M., 2007. To flee or not to flee: predator avoidance by cheetahs at kills. Behav. Ecol Sociobiol. 61 (7), 1033–1042.

Inskip, C., Zimmerman, A., 2009. Human-felid conflict: a review of patterns and priorities worldwide. Oryx 43 (1), 18–34.

IUCN/SSC, 2007a. Regional Conservation Strategy for the Cheetah and African Wild Dog in Eastern Africa. IUCN/SSC, Gland, Switzerland.

IUCN/SSC, 2007b. Regional Conservation Strategy for the Cheetah and African Wild Dog in Southern Africa. IUCN/SSC, Gland, Switzerland.

IUCN/SSC, 2012. Regional Conservation Strategy for the Cheetah and African Wild Dog in Western, Central, and Northern Africa. IUCN/SSC, Gland, Switzerland.

Kansky, R., Kidd, M., Knight, A.T., 2014. Meta-analysis of attitudes toward damage-causing mammalian wildlife. Conserv. Biol. 28 (4), 924–938.

Kansky, R., Knight, A.T., 2014. Key factors driving attitudes towards large mammals in conflict with humans. Biol. Conserv. 179, 93–105.

Kent, V.T., 2011. The status and conservation potential of carnivores in semi-arid rangelands, Botswana the Ghanzi Farmlands: A case study. PhD thesis, Durham University, United Kingdom.

Klein, R., 2013. An Assessment of Human Carnivore Conflict in the Kalahari Region of Botswana. MSc thesis, Rhodes University, South Africa.

Knight, J., 2000. Culling demons: the problem of bears in Japan. In: Knight, J. (Ed.), Natural Enemies: People-Wildlife Conflicts in Anthropological Perspective. Routledge, London, U.K.

Lindsey, P.A., Havemann, C.P., Lines, R., Palazy, L., Price, A.E., Retief, T.A., Rhebergen, T., Van der Waal, C., 2013a. Determinants of persistence and tolerance of carnivores on Namibian ranches: implications for conservation on southern African private lands. PLoS One 8 (1), e52458.

Lindsey, P.A., Havemann, C.P., Lines, R.M., Price, A.E., Retief, T.A., Rhebergen, T., Van Der Waal, C., Romañach, S.S., 2013b. Benefits of wildlife-based land uses on private lands in Namibia and limitations affecting their development. Oryx 47 (1), 41–53.

Maclennan, S.D., Groom, R.J., Macdonald, D.W., Frank, L.G., 2009. Evaluation of a compensation scheme to bring about pastoralist tolerance of lions. Biol. Conserv. 142 (11), 2419–2427.

MacMillan, D.C., Phillip, S., 2010. Can economic incentives resolve conservation conflict: the case of wild deer management and habitat conservation in the Scottish highlands. Hum. Ecol 38 (4), 485–493.

Madden, F., McQuinn, B., 2014. Conservation's blind spot: the case for conflict transformation in wildlife conservation. Biolo. Conserv, 178, 97–106.

Maddox, T., 2003. The Ecology of Cheetahs and Other Large Carnivores in A Pastoralist-Dominated Buffer Zone. PhD thesis, University College London, United Kingdom.

Marker, L., 2002. Aspects of Cheetah (Acinonyx jubatus) Biology, Ecology and Conservation Strategies on Namibian Farmlands. PhD thesis, University of Oxford, United Kingdom.

Marker, L.L., Boast, L.K., 2015. Human-Wildlife conflict 10 years later: lessons learned and their application to cheetah conservation. Hum. Dimens. Wildl. 20, 302–309.

Marker, L.L., Dickman, A.J., Mills, M.G.L., Jeo, R.M., Macdonald, D.W., 2008. Spatial ecology of cheetahs (Acinonyx jubatus) on north-central Namibian farmlands. J. Zool. 274, 226–238.

Marker, L.L., Dickman, A.J., Mills, M.G.L., Macdonald, D.W., 2003. Aspects of the management of cheetahs, *Acinonyx jubatus jubatus*, trapped on Namibian farmlands. Biol. Conserv. 114 (3), 401–412.

Marker, L., Dickman, A.J., Mills, M.G.L., Macdonald, D.W., 2010. Cheetahs and ranchers in Namibia: a case study. In: Macdonald, D.W., Loveridge, A.J. (Eds.), Biology and Conservation of Wild Felids. Oxford University Press, Oxford, pp. 353–372.

Marker-Kraus, L., Kraus, D., Barnett, D., Hurlbut, S., 1996. Cheetah survival on Namibian farmlands. Cheetah Conservation Fund, Windhoek.

Marnewick, K., Beckhelling, A., Cilliers, D., Lane, E., Mills, G., Herring, K., Caldwell, P., Hall, R., Meintjes, S., 2007. The status of the cheetah in South Africa. Cat News Special Issue 3, 27–31.

McManus, J., Dickman, A., Gaynor, D., Smuts, B., Macdonald, D., 2014. Dead or alive? Comparing costs and benefits of lethal and non-lethal human–wildlife conflict mitigation on livestock farms. Oryx 49 (4), 687–695.

Mertens, A., Promberger, C., 2001. Economic aspects of large carnivore-livestock conflicts in Romania. Ursus 12, 173–180.

Naidoo, R., Weaver, L.C., Diggle, R.W., Matongo, G., Stuart-Hill, G., Thouless, C., 2016. Complementary benefits of tourism and hunting to communal conservancies in Namibia. Conserv. Biol. 30 (3), 628–638.

Naughton-Treves, L., 1997. Farming the forest edge: vulnerable places and people around Kibale National Park. Uganda Geogr. Rev. 87, 27–46.

Nie, M., 2004. State wildlife governance and wildlife conservation. In: Fascione, N., Delach, A., Smith, M.E. (Eds.), People and Predators: From Conflict to Coexistence. Island Press, Washington D.C, pp. 197–218.

Nowell, K., 2014. An assessment of the conservation impacts of legal and illegal trade in cheetahs Acinonyx jubatus. SC65 Doc. 39 Rev. 2, 3. Available from: https://cites.org/sites/default/files/eng/com/sc/65/E-SC65-39.pdf.

Nyamwaro, S.O., Murilla, G.A., Mochabo, M.O.K., Wanjala, K.B., 2006. Conflict minimising strategies on natural resource management and use: the case for managing and coping with conflicts between wildlife and agro-pastoral production resources in Transmara district, Kenya. Policy Research Conference on Pastoralism and Poverty Reduction in East Africa, Nairobi.

Nyhus, P., Fischer, F., Madden, F., Osofsky, S., 2003. Taking the bite out of wildlife damage: the challenge of wildlife compensation schemes. Conserv. Pract. 4 (2), 37–40.

Nyhus, P., Osofsky, S.A., Ferraro, P., Madden, F., Fischer, F., 2005. Bearing the costs of human-wildlife conflict: the challenges of compensation schemes. In: Woodroffe, R., Thirgood, S., Rabinowitz, A. (Eds.), People and Wildlife: Conflict or Coexistence? Cambridge University Press, Cambridge, United Kingdom, pp. 107–121.

Okello, M.M., Bonham, R., Hill, T., 2014. The pattern and cost of carnivore predation on livestock in Maasai homesteads of Amboseli ecosystem, Kenya: Insights from a carnivore compensation programme. Int. J. Biodivers. Conserv. 6 (7), 502–521.

Rabinowitz, A.R., 1986. Jaguar predation on domestic livestock in Belize. Wildl. Soc. Bull. (1973–2006) 14 (2), 170–174.

Romanach, S.S., Lindsey, P.A., Woodroffe, R., 2007. Determinants of attitudes towards predators in central Kenya and suggestions for increasing tolerance in livestock dominated landscapes. Oryx 41, 185–195.

Ruaha Carnivore Project, 2015. Annual Report 2015, Ruaha Carnivore Project, Wildlife Conservation Research Unit, University of Oxford, Oxford, United Kingdom.

Rust, N., 2015. Understanding the Human Dimensions of Coexistence Between Carnivores and People: A Case Study in Namibia. PhD thesis, University of Kent, United Kingdom.

Rust, N.A., 2016. Can stakeholders agree on how to reduce human–carnivore conflict on Namibian livestock farms? A novel Q-methodology and Delphi exercise. Oryx 51 (2), 339–346.

Rust, N.A., Humle, T., Tzanpoulos, J., Macmillan, D., 2016. Why has human–carnivore conflict not been resolved in Namibia? Soc. Nat. Resour. 29 (9): 1079–1094).

Rust, N.A., Marker, L.L., 2013a. Attitudes toward predators and conservancies among Namibian farmers. Hum. Dimens. Wildl. 18 (6), 463–468.

Rust, N.A., Marker, L.L., 2013b. Costs of carnivore coexistence on communal and resettled land in Namibia. Environ. Conserv. 41 (1), 45–53.

Rust, N.A., Nghikembua, M.T., Kasser, J.J., Marker, L.L., 2015. Environmental factors affect swing gates as a barrier to large carnivores entering game farms. Afr. J. Ecol. 53 (3), 339–345.

Rust, N.A., Taylor, N., 2016. Carnivores, colonization, and conflict: a qualitative case study on the intersectional persecution of predators and people in Namibia. Anthrozoös 29 (4), 653–667.

RWCP & IUCN/SSC, 2015. Regional conservation strategy for the cheetah and African wild dog in southern Africa. Revised and Updated, August 2015. IUCN/SSC, Gland, Switzerland.

Schiess-Meier, M., Ramsauer, S., Gabanapelo, T., König, B., 2007. Livestock predation—insights from problem animal control registers in Botswana. Hum. Dimens. Wildl 71, 1267–1274.

Schumann, M., Schumann, B., Dickman, A., Watson, L.H., 2006. Assessing the use of swing-gates in game fences as a potential non-lethal predator exclusion technique. S. Afr. J. Wildl. Res. 36 (2), 173–181.

Schumann, M., Watson, L.H., Schumann, B., 2008. Attitudes of Namibian commercial farmers towards large carnivores: The influence of conservancy membership. S. Afr. J. Wildl. Res. 38, 123–132.

Selebatso, M., Moe, S., Swenson, J., 2008. Do farmers support cheetah *Acinonyx jubatus* conservation in Botswana despite livestock depredation? Oryx 42, 430–436.

Sjoberg, L., Moen, B.-E., Rundmo, T., 2004. Explaining risk perception: an evaluation of the psychometric paradigm in risk perception research. Rotunde 84, 1–39.

TAWIRI, 2007. Proceedings of the First Tanzania Cheetah Conservation Action Plan Workshop, TAWIRI, Arusha, Tanzania.

Taylor, A., Lindsey, P., Davies-Mostert, H., 2016. An Assessment of the Economic, Social and Conservation Value of the Wildlife Ranching Industry and its Potential to Support the Green Economy in South Africa. The Endangered Wildlife Trust, Johannesburg, South Africa.

Thorn, M., Green, M., Dalerum, F., Bateman, P.W., Scott, D.M., 2012. What drives human–carnivore conflict in the North West Province of South Africa? Biol. Conserv. 150 (1), 23–32.

Vucetich, J.A., Bruskotter, J.T., Nelson, M.P., 2015. Evaluating whether nature's intrinsic value is an axiom of or anathema to conservation. Conserv. Biol. 29 (2), 321–332.

Woodroffe, R., Lindsey, P., Romanach, S., Stein, A., Symon, M.K. ole Ranah, 2005. Livestock predation by endangered African wild dogs (*Lycaon pictus*) in northern Kenya. Biol. Conserv. 124, 225–234.

Zabel, A., Engel, S., 2010. Performance payments: a new strategy to conserve large carnivores in the tropics? Ecol. Econ. 70 (2), 405–412.

Zabel, A., Holm-Muller, K., 2008. Conservation performance payments for carnivore conservation in Sweden. Conserv. Biol. 22, 247–251.

14

Pets and Pelts: Understanding and Combating Poaching and Trafficking in Cheetahs

Patricia Tricorache*, Kristin Nowell**, Günther Wirth†,
Nicholas Mitchell‡,§, Lorraine K. Boast¶, Laurie Marker††

*Cheetah Conservation Fund, Islamorada, FL, United States
**Cat Action Treasury and World Conservation Union (IUCN) Red List Programme, Cape Neddick, ME, United States
†Independent Researcher, Hargeisa, Somaliland
‡Zoological Society of London, London, United Kingdom
§Wildlife Conservation Society, New York, NY, United States
¶Cheetah Conservation Botswana, Gaborone, Botswana
††Cheetah Conservation Fund, Otjiwarongo, Namibia

INTRODUCTION

Illegal wildlife trade is the sale, purchase, or exchange of wildlife species, which is prohibited by law and, therefore, by definition, operates outside government regulations (Nellemann et al., 2014). It has a substantial impact on both global biodiversity and the economic and structural growth of developing countries (Haken, 2011). It is estimated to be the fourth most valuable global illegal activity after narcotics, counterfeit products, and human trafficking (Haken, 2011), and is valued at US $50–150 billion per year (UNEP, 2014). Unfortunately, the rates of detection, arrest, and conviction for wildlife crime are low (Akella and Cannon, 2004), and illegal wildlife trade has been associated with the decline of large carnivore species, including tigers (*Panthera tigris*) (Goodrich et al., 2015) and most recently cheetahs (*Acinonyx jubatus*) (Nowell, 2014).

Historically, cheetahs were kept by the aristocracy as pets or as coursing companions to hunt antelopes for sport (Chapter 2). This demand for hunting pets continued in modern times, and is believed to have significantly contributed to the near disappearance of cheetahs in Asia by the mid-1980s (Divyabhanusinh, 1995; Chapter 4).

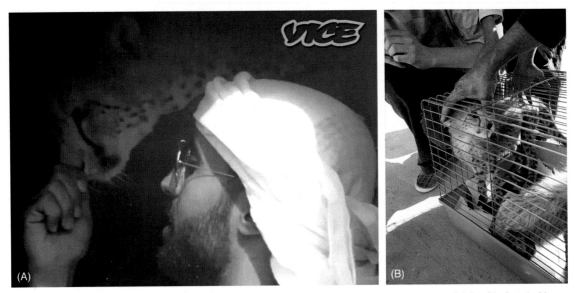

FIGURE 14.1   (A) Pet cheetah in car in Kuwait. (B) Cheetah cubs at School Pet Day, International School in Sana'a, Yemen. *Source: Part A, VICE Media LLC; part B, David Stanton.*

Wild African cheetahs were commonly exported from Africa for zoo exhibits; available data indicate that 677 cheetahs were imported to supply zoos, mostly in Europe and North America, between 1955 and 1975 (Marker-Kraus, 1829; Chapter 22). Cheetahs were also traded internationally for their skins, which were popular as clothing and ornaments in Europe and North America, where an estimated 3000–5000 skins were imported annually in the 1960s (Nowell and Jackson, 1996).

Although knowledge of wildlife trade prior to the 1970s is sparse, the trade is considered to have been substantial. In response to advocacy by conservation groups, in 1971, the International Fur Trade Federation recommended to its members to cease trade in skins from many endangered big cat species (Nowell and Jackson, 1996). In 1975, all international trades in cheetahs were formally addressed by the Convention on International Trade in Endangered Species of Wild Fauna and Flora (CITES). This international agreement regulates the trade of species whose collection from the wild would endanger their survival. Cheetahs are listed as Appendix I species, meaning Parties signatory to CITES cannot trade cheetahs internationally for commercial purposes. Botswana, Zimbabwe, and Namibia since 1992 have had limited noncommercial annual export quotas of 5, 50, and 150 wild cheetahs, respectively, for live specimens or hunting trophies (CITES, 1992). In addition, captive-bred cheetahs may be traded for both commercial and noncommercial purposes (Chapter 21).

Although international legal trade in wild cheetahs (with the aforementioned exceptions) ceased for participating Parties in 1975, cheetahs are still taken from the wild to be exploited as pets (Fig. 14.1A–B) or tourist attractions, entered into illegal captive breeding operations, killed in illegal trophy hunts, or their body parts sold as ornaments, traditional medicines, and clothing (Nowell, 2014). Illegal trade, in conjunction with other threats, such as habitat loss (Chapter 10) and human-cheetah conflict (Chapter 13), is considered a threat to the survival of the remaining populations throughout their range (IUCN/SSC, 2007a,b, 2012; RWCP & IUCN/SSC, 2015).

Combating illegal trade in cheetahs was first tabled for discussion at the CITES 16th Conference of the Parties in 2013, and designated a priority at their 27th Animal Committee Meeting in 2014 (CITES, 2014; Chapter 21). In 2016, the CITES 66th Standing Committee Meeting adopted a number of recommendations to be taken by the Parties to improve their ability to monitor illegal cheetah trade and enforce trade restrictions. These included improving communications and collaboration in the areas of enforcement and disposal of confiscated animals, as well as developing public awareness campaigns to reduce demand. Subsequently, two decisions were adopted at the CITES 17th Conference of the Parties (CoP17): to commission the development of a cheetah trade resource kit for law enforcement agencies (recommended best practices, procedures, and protocols, especially concerning live cheetahs), and to assess the feasibility of creating a web-based Cheetah Forum for all stakeholders to share information about cheetah (CITES, 2016a). A third decision of the CITES Secretariat to engage online services to address e-commerce of live cheetahs and raise awareness was adopted as part of the broader Combatting Wildlife Cybercrime decisions, which apply to all CITES species (CITES, 2016b).

Due to the clandestine nature of poaching and illegal trade, it is difficult to establish the exact numbers of cheetahs lost to illegal trade, or to precisely establish the main sources and supply routes. This chapter provides an overview of the illegal trade in cheetahs. We will discuss the drivers of the demand for cheetahs, and then look at how the demand is supplied from range countries, focusing on the regions which have been identified as the major areas for illegal trade, that is, the Horn of Africa and the Gulf Cooperation Council States [GCC; e.g., Kuwait, Oman, Qatar, Saudi Arabia, United Arab Emirates (UAE)] (Nowell, 2014). Live animal trade within southern Africa destined for the South African captive cheetah export industry is a secondary concern that we will also discuss.

The information summarized in this chapter is based on data collected between 2005 and 2015 from:

1. the CITES trade database (UNEP/WCMC, 2016);
2. a commissioned report on the trade and its impacts on wild cheetah populations (Nowell, 2014) which arose from the recent CITES initiative to address illegal cheetah trade;
3. a subsequent survey of all CITES Parties about illegal cheetah trade (CITES, 2016c)—the survey was completed by 33 countries: 16 cheetah-range countries (48%) and 17 nonrange countries from Europe, North America, and Oceania, including 6 from the Arabian Peninsula (18%); and
4. records compiled by Cheetah Conservation Fund (CCF) of all reports of illegal cheetah trade based on direct observations, information gathered from conservationists, government officials, and members of the public, as well as publicly available information on media articles and the Internet.

CCF makes every effort to carefully establish the veracity of these reports, and to cross-reference all sources in its illegal trade database to remove any duplications. It is important to note that the data contained in this chapter are limited to the extent to which information is available.

ILLEGAL TRADE OVERVIEW

Until 2016, Parties were not required to report annually on illegal trade to CITES (CITES, 2016d). As a result, data in the CITES trade database cannot be viewed as comprehensive, and from 2005 to 2015, only 27 live cheetahs were reported as being confiscated or seized (Source Code I) in the database (UNEP/WCMC, 2016). Moreover, data are not collected on domestic seizures because the Convention

TABLE 14.1   The Number of Cases and Number of Cheetahs Poached or Trafficked in Cheetah-Range Countries Recorded by Cheetah Conservation Fund Between November 2005 and December 2015[a]

Region	Country	Live animals			Parts and products[b]		
		Cases	No. of cheetahs		Cases	No. of cheetah units	
			Observed[c]	Confiscated		Observed[c]	Confiscated
Asia	Iran	2	1	1	0	0	0
Eastern Africa incl. Horn of Africa (HOA)	Djibouti (HOA)	2	0	6	0	0	0
	Ethiopia (HOA)	25	87	16	0	0	0
	Kenya (HOA)	4	0	10	27	0	44
	Somalia[d] (HOA)	7	54	26	0	0	0
	Somaliland[e]	65	192	96	1	1	0
	South Sudan (HOA)	0	0	0	2	0	3
	Tanzania	2	0	4	8	0	9
	Uganda (HOA)	0	0	0	2	0	2
	Total	105	333	158	40	1	58
Southern Africa	Angola	0	0	0	3	7	0
	Botswana	5	0	18	4	0	4
	Namibia	4	0	20	4	0	5
	South Africa	4	2	3	3	0	5
	Zambia	0	0	0	1	0	1
	Zimbabwe	2	0	4	0	0	0
	Total	15	2	45	15	7	15
Central and West Africa	Benin	0	0	0	1	1	0
Total		122	336	204	56	9	73

[a]Data were compiled from the CITES trade database, Internet searches, personal correspondence with informants, and CITES (2016c).
[b]Only reports involving cheetah parts/products that are equivalent to a minimum of at least one animal are included (e.g., whole skin, whole skeleton, one skull).
[c]Observed incidents were those where cheetahs or cheetah parts and products were not confiscated
[d]Excluding the autonomous region of Somaliland.
[e]Autonomous region of Somalia.

covers only international wildlife trade. As a result, the recording of illegal trade in cheetahs is incomplete and vastly underestimated.

Reports collected by CCF between 2005 and 2015 identified 280 cases of trafficking (i.e., trade and/or transportation) in live cheetahs or their skins and other body parts, involving a minimum of 1108 individual cheetahs (Tables 14.1 and 14.2); the majority (89% of the total cheetah numbers shown in the two tables) involved live animals, and these mainly consisted of young cubs (Nowell, 2014). Most of the illegal trade recorded herein was not detected or intercepted by law enforcement authorities (672 cheetahs, including live cheetahs and body parts, 61%). Only the remaining 39% (436 cheetahs) were confiscated, which makes up the only data for some countries. The low percentage of confiscations

TABLE 14.2   The Number of Cases and Number of Cheetahs Trafficked Outside Cheetah-Range Countries Recorded by Cheetah Conservation Fund Between November 2005 and December 2015[a]

Region	Country	Live animals			Parts and products[b]		
		Cases	No. of cheetahs		Cases	No. of cheetah units	
			Observed[c]	Confiscated		Observed[c]	Confiscated
Americas	Mexico	1	0	4	0	0	0
	USA	0	0	0	13	0	22
	Total	*1*	*0*	*4*	*13*	*0*	*22*
Arabian Peninsula incl. Gulf Cooperation Council (GCC)	Iraq	1	2	0	0	0	0
	Kuwait (GCC)	7	11	7	1	0	3
	Oman (GCC)	1	0	7	0	0	0
	Qatar (GCC)	3	2	1	0	0	0
	Saudi Arabia (GCC)	11	107	37	0	0	0
	UAE (GCC)	30	77	52	3	0	3
	Yemen	8	120	5	0	0	0
	Total	*61*	*319*	*109*	*4*	*0*	*6*
Asia	Afghanistan	0	0	0	4	4	0
	India	0	0	0	2	0	1
	Singapore	0	0	0	1	0	Unknown
	Total	*0*	*0*	*0*	*7*	*4*	*1*
Africa	Cameroon	1	2	0	0	0	0
	Egypt	1	2	0	0	0	0
	Morocco	1	0	4	0	0	0
	Sudan	0	0	0	1	0	1
	Total	*3*	*4*	*4*	*1*	*0*	*1*
Europe	France	0	0	0	1	0	Unknown
	Greece	1	0	1	0	0	0
	The Netherlands	0	0	0	1	0	1
	Portugal	0	0	0	1	0	1
	Spain	0	0	0	2	0	2
	Switzerland	0	0	0	1	0	3
	United Kingdom	1	0	1	4	0	4
	Total	*2*	*0*	*2*	*10*	*0*	*11*
Total		*67*	*323*	*119*	*35*	*4*	*40*

[a]Data were compiled from the CITES trade database, Internet searches, personal correspondence with informants, and CITES (2016c).
[b]Only reports involving cheetah parts/products that are equivalent to a minimum number of animals are included (e.g., whole skin, whole skeleton, one skull).
[c]Observed incidents were those where cheetahs or cheetah parts and products were not confiscated.

by enforcement authorities demonstrates the need for greater monitoring efforts.

Table 14.1 summarizes illegal trade cases recorded within cheetah-range countries, 81% (145 cases involving a minimum of 550 cheetahs) of which were in eastern Africa. Within eastern Africa, the greatest numbers of illegally traded live cheetahs (288 cheetahs, 59% of the regional live total) were recorded in Somaliland, an autonomous region in northeast Somalia. As a self-declared state, Somaliland does not consider itself to be part of the Federal Republic of Somalia nor is it a member of the United Nations, the African Union or a Party to CITES. Following Somaliland were Ethiopia (103 live cheetahs, 21%) and the remainder of Somalia (80 live cheetahs, 16%). Kenya was the country with the greatest number of illegally trafficked cheetah parts reported in the eastern Africa region (44 parts; 75%), and 54% in all cheetah-range countries. The majority of live animals detected in illegal trade in southern Africa were in Namibia (20 cheetahs, 43%) and Botswana (18 cheetahs, 38%). In eastern and southern Africa more illegal trade was detected in live animals than parts and skins. Data are sparse for other cheetah-range countries: 1 cheetah skin was reported in Benin (Table 14.1). In addition, 8 live animals and 1 cheetah skin were recorded as having been illegally trafficked in four former-cheetah range African countries (Table 14.2).

Although there is some demand for pet cheetahs in cheetah-range countries (as reported by Kenya, Somalia, and South Africa in the CITES survey on illegal cheetah trade; CITES, 2016c), all available information points to a major export trade from Ethiopia, Somalia, and Kenya in the Horn of Africa to the Gulf States, with Yemen serving as the transit point (Nowell, 2014). Outside cheetah-range countries, 97% of the live cheetahs recorded as being illegally traded occurred in the Arabian Peninsula (GCC States, Yemen and Iraq: 428 out of 442 animals; Table 14.2). Within this region, the countries with the greatest numbers of known illegally traded

live cheetahs were Saudi Arabia (144 cheetahs, 34%), the UAE (129 cheetahs, 30%), and Yemen (125 cheetahs, 29%), nearly all were observed (75%) rather than confiscated.

Although there is possibly some overlap between the recorded cases in the cheetah-range and nonrange countries (e.g., a cheetah observed in illegal trade in the Horn of Africa may have later been confiscated in the Arabian Peninsula), many more cases are likely to remain undetected.

THE DEMAND: DRIVERS AND REGULATIONS

Live Cheetahs

Unlike other big cats, cheetahs are relatively docile and do not present a threat to the life of adult humans, making them a highly prized pet. Images of people in the Arabian Peninsula posing with their pet cheetah are often circulated on social media (Nowell, 2014). Owning cheetahs and other exotic pets in this region is thought to convey social and economic status, in emulation of individuals in positions of power or leadership (Mohamed, 2016). Other motivating factors are the intent to rescue sick-looking cubs kept by dealers (as revealed by direct interviews with pet cheetah owners and veterinarians), or to protect animals from the threats they face in their natural environment (as revealed by comments on social media). These purchases, although well intentioned, keep the demand up and the smuggling of cheetahs profitable.

Cheetahs once roamed areas of the Arabian Peninsula (Chapter 4), and were commonly kept as pets or hunting companions (Chapter 2). This centuries-old tradition, along with a solid economy throughout the region, has supported a high demand for cheetahs and other exotic animals among the region's affluent population. A live cheetah can be sold for up to US $15,000, substantially more than the average asking price of US $200–300 received by dealers in the Horn of Africa.

Cheetah owners tend to have insufficient knowledge about cheetah care and cheetahs are often kept in inappropriate conditions (e.g., small indoor rooms without suitable exercise). Poor diet, in particular, has been linked to a host of health problems including ataxia and hind limb paralysis (Kaiser et al., 2014). As a result, cheetah survivorship is low; for example, a veterinarian in Jeddah, Saudi Arabia, stated that approximately 100 pet cheetah cubs had died in Jeddah and Riyadh over a 6-month period in 2007 (Bahrain Tribune, 2007).

All of the GCC States are Parties to CITES, all prohibit the import of CITES-listed species or predators in general, and all have stated that captive cheetah imports are only permitted to licensed facilities, such as zoos (CITES, 2016c). The low numbers of cheetahs legally imported into the region or born within its borders do not tally with the apparently high numbers of cheetahs in private hands. It can be difficult to breed cheetahs in captivity, and therefore, it is unlikely that private facilities or individuals are successfully breeding cheetahs to fulfill the demand. The International Cheetah Studbook (ICSB) comprises approximately 250 cheetah-holding facilities worldwide, including 6 breeding facilities in 2 Gulf States (UAE and Qatar). Between 2005 and 2015, these 6 facilities produced 161 viable cubs (217 cubs minus 56 deaths <6 months); all but 3 were placed in facilities reporting to the ICSB. It is therefore likely that the majority of pets in the GCC were taken from the wild. Although some cheetah owners undoubtedly operate under the assumption that their animals are captive bred, others are well aware that their animals are illegally sourced (Nowell, 2014).

Some GCC countries highlighted their difficulties in policing online trade in the CITES cheetah survey (CITES, 2016c). An investigation initiated by CCF in September 2015 has shown that online trade, particularly on social media (Instagram in particular), is a major facilitator for illegal wildlife trade in the region. The investigation has identified 369 user accounts advertising live cheetahs for sale online. Often, these dealers also offer other endangered species, including tigers, great apes, reptiles, and birds. Most of these dealers are based in the Arabian Peninsula: 135 in Saudi Arabia, 119 in Kuwait, 77 in the UAE, 34 in Qatar, and 1 in Bahrain. Cheetah sellers were also found in Yemen (1) and Pakistan (2). Ninety-three of the dealers were identified as frequent sellers. Prices asked online range between US $5,000 and 15,000 for individual cheetahs. An ongoing analysis of advertisements by these sellers has recorded at least 1000 cheetahs offered for sale over a 4-year period (2012–16) (CCF, unpublished data).

Recently there has been a welcome trend toward greater restrictions and policing of exotic animal ownership in the GCC countries. In 2011, after growing concerns for human safety, Ajman became the first of the seven emirates in the UAE to ban the breeding and keeping of dangerous animals in private homes (UAE Ministry of Climate Change and Environment, 2016). In December 2014, the Sharjah emirate introduced a similar regulation banning the breeding and private ownership of dangerous predators in residential areas (WAM, 2014). A 1-month amnesty following the implementation of this law resulted in hundreds of animals of various species being handed over to zoos and sanctuaries, although many more were reported to have been moved to neighboring Emirates (Ali, 2014). In June 2016, the Sharjah Environment and Protected Areas Authority announced the construction of a new center to house animals handed over or confiscated as a result of the ban (EPAA, 2016). In December 2015, the Kuwait's National Assembly unanimously passed the Animal Rights Bill that penalizes the illegal ownership and sale of predators (KUNA, 2015). Subsequently, in late December 2016, the UAE enacted Federal Law No. 22/2016, which regulates the trade and private possession of exotic animals, including cheetah, nationwide, and carries jail terms

FIGURE 14.2   (A) Confiscated cheetah skin at the Ministry of Wildlife Conservation and Tourism in Juba, South Sudan. (B) 3 cheetah cubs confiscated in Berbera, Somaliland, in 2012. The cub at the bottom had just died when this image was taken. *Source: Part A, Nick Mitchell; part B, Günther Wirth.*

and/or penalties from 10,000 to 700,000 Emirati Dirhams (US $2,700–190,000) (Mohamed, 2017).

CCF has also received email enquiries from private individuals in eastern Europe and Asia asking about the acquisition or care of cheetahs, most likely sourced through illegal trade. These enquiries are answered with information about international laws and conservation issues, in an attempt to discourage potential buyers.

In Africa, live cheetahs are in demand for many of the same reasons as described earlier for the Arabian Peninsula, although the volume is probably lower. Cheetah pets have been observed at the homes of high-ranking officials in the Horn of Africa region (CCF, unpublished data). In addition, they are in demand as tourist attractions, particularly in South Africa (Marnewick, 2012).

Cheetah Parts

Demand for cheetah pelts and other body parts is rarely specifically aimed at the species (Fig. 14.2A) and is instead generalized for spotted or big cat products (such as leopard *Panthera* *pardus* or lion *Panthera leo*). Big cat skins, including cheetah, are in demand as whole skins for decorative or ceremonial purposes or in pieces for garments and accessories (e.g., shoes, bags, and garment trim). In Sudan traditional men's shoes made of spotted cat fur (*markoob*) are highly prized. Big cat skin or parts are also used in traditional medicine (*magie médicale*, as it is known in West Africa, or *muti* in southern Africa). Their use encompasses a range of practices from medical (seeking to cure a physical ailment) to spiritual and mystical (Nowell, 2014; Chapter 2). In Benin, where cheetahs are rare, Sogbohossou (2006) was able to document one use for a cheetah part: its anus will purportedly attract the man or woman of the user's desire. Although some people may have had positive experiences using big cat parts, real or fake, as spiritual or consumptive medicine, this is not only illegal but also potentially dangerous to the consumer if it replaces appropriate medical care.

A complicating factor in understanding the scope of the trade in cheetah skins is the prevalence of fakes. Fake skins (skins of common domestic animals painted with black spots) have been observed in West Africa and Sudan. Some of these fake examples have been mistakenly reported by well-intentioned observers as actual cheetah products (Nowell, 2014); while not representing loss of animals from the wild, they do indicate enterprising attempts to meet demand.

THE SUPPLY: TRAFFICKING WITHIN AND OUT OF AFRICA

The Horn of Africa

The Horn of Africa is the region where illegal trade is likely having the greatest impact on wild cheetah populations, due to both the sheer magnitude of the trade and the threatened status of cheetah populations in the region (Nowell, 2014). Not much is known about the suppliers or the specific source populations because trade is usually intercepted further downstream: in transit or at the final destination. However, all available information indicates that cheetah cubs are largely taken from ethnic Somali regions in parts of Ethiopia, possibly Kenya, and Somalia itself. Most cheetah smuggling cases across the region are organized or supported by groups of individuals from within a clan (interrelated family groups). In addition, if cheetahs are perceived as a threat to their livelihoods (Chapter 13), herders may catch and sell them to passing traders, further perpetuating the illegal trade market.

A haven for piracy, the ca. 1100-km northern coast of Somalia is the way out of Africa for poached live cheetah cubs. From here, they are transported by boat to the coast of Yemen and on to the GCC (Nowell, 2014). A 2006 survey on illegal wildlife trade in Somalia (Amir, 2006) and recent interviews conducted in Somaliland and Kuwait (CCF, unpublished data) also indicate that animals and animal products are flown by private aircraft into the Arabian Peninsula from private airstrips in Somalia.

Of particular concern is the autonomous region of Somaliland. Somaliland covers over 50% of the northern Somali coastline facing Yemen, and shares a border of ca. 500 miles with Ethiopia to the southwest. Sixty-two percent of live cheetah illegal trade cases in eastern Africa were recorded in Somaliland (Table 14.1). With the fourth lowest gross domestic product per capita in the world (World Bank, 2014), rampant poverty is compounded by easy access to wealthy consumer markets and a lack of understanding of the conservation plight of wildlife, such as the cheetah. Some cheetahs are confiscated by officials, but instances of cubs being sold back to the smugglers have been encountered (CCF, unpublished data). It should be noted that monitoring networks are better in Somaliland than in the rest of eastern Africa, and therefore, the territory's relative importance as the main cheetah trafficking area may be partially due to underestimation in other countries. Even so, the actual number of live cheetahs smuggled through Somaliland could be much higher than the recorded 288 live cheetahs between 2005 and 2015; informants in Somaliland estimate that up to 300 cheetah cubs may be exported every year.

The asking prices for cheetah cubs in Somaliland vary greatly, from as little as US $80 for an unhealthy cub to US $1000 for a healthy cub; however, most traders ask for US $200–300 per cub. Cubs are often inadequately cared for (Fig. 14.2B). In the 33 Somaliland cases involving 142 cheetahs with known outcomes (i.e., where the cheetah was reported as being alive at the time of confiscation or where it is known whether the cheetah lived or died in observed cases), the survival rates were 17% compared to >33% across all regions. Somaliland has had some success with confiscations in recent years, confiscating 96 cheetahs (33%) of the recorded 288 live cheetahs trafficked between 2005 and 2015 (Table 14.1). However, of these 96 confiscated cubs, only 21 were transferred to safety, while the rest were reported by officials as dead or missing. In general, limitations in implementing and policing government policy, largely due to weak government institutions and lack of resources in the territory, are allowing the chee-tah trade to continue.

Southern Africa

Namibia, Botswana, and Zimbabwe are the only countries from which wild cheetahs may be legally exported. From 2002 to 2011, their wild cheetah exports averaged 153 per year, mainly from hunting trophies from Namibia (Nowell, 2014). This trade is considered overall to be well regulated and sustainable, and is not thought to have had a negative impact on chee-tah populations (Nowell, 2014).

The majority of legal trade in captive-bred cheetahs also originates in southern Africa (Marker, 2015). South Africa is the world's larg-est exporter of captive-bred cheetahs, and the only country to have registered commercial cheetah breeding facilities (two) with the CITES Secretariat (as is required for the commercial breeding of Appendix I species: Nowell, 2014). Between 2005 and 2015, CITES export permits were issued for 1065 live cheetahs worldwide. Of these, 780 (747 captive bred, 13 ranched, and 20 wild-born) were issued in South Africa (CCF unpublished analysis of CITES trade data, ad-justed for duplicates and reexports). The legal trade creates a potential mechanism by which wild-caught cheetahs could be miscoded as captive bred and illegally exported, and there is anecdotal evidence that this is taking place (Nowell, 2014). A CITES Inter-Sessional Work-ing Group investigated the causes and effects of such miscoding, an issue that has been ta-bled at various Convention meetings under the title "Implementation of the Convention relat-ing to captive-bred and ranched specimens" (CITES, 2016e).

According to the Model Law by CITES (CITES, 2016f) (parts of which have been ad-opted into South African law) for an Appen-dix I captive-bred, live animal to be exported, it must be "individually and permanently marked in a manner so as to render alteration

or modification by unauthorized persons as dif-ficult as possible" (Department of Environmen-tal Affairs, 2010). Microchips are currently used for this purpose, but there are concerns over the legitimacy of this method, as microchips could also be inserted into wild-caught cheetahs (Nowell, 2014). To improve controls on captive-bred cheetahs, South Africa is currently making regulatory amendments including the require-ment to supply parental DNA when a cheetah is exported, thereby, proving its captive-bred sta-tus (CITES, 2016c).

A draft resolution tabled at CoP17 entails a re-view of significant trade in specimens declared as produced in captivity, which could reveal unusually high export volumes of cheetahs by some countries. In addition, requiring that cap-tive breeding facilities report to a regional or in-ternational cheetah studbook, which is currently not mandatory, would improve trade controls while also improving strategies to maintain ge-netic diversity in captive populations. A com-parison between the CITES trade data and the international studbook data for South Africa for the period 2005–14 showed only 28% of the cheetahs issued export permits by CITES were registered in the studbook; the 72% difference indicates that a vast number of cheetahs are be-ing traded between facilities not reporting to the international cheetah studbook, which could be facilitating illegal trade (CCF, unpublished data).

In addition, incidences of cheetahs be-ing illegally caught and transported between Botswana, Namibia, and South Africa have been recorded (Marnewick et al., 2007) and investigat-ed (CITES, 2016c). Twenty-four cases of illegal trade between these countries (43 live cheetahs and 14 skins/skeletons) were recorded between 2005 and 2015 (Table 14.1). However, the num-bers could be much higher with some conser-vationists estimating 50–60 cheetahs being re-moved annually for illegal trade from Botswana alone (Cilliers, NCMP, personal communica-tion; Houser, CCB, personal communication;

Klein, 2007). The trade is thought to primarily exist to supplement existing captive breeding operations in South Africa (Marnewick, 2012). In addition, cheetahs may be captured to feed the demand for tourist attractions, trophy hunting, body parts, or the greater illegal trade of cheetahs beyond their natural range. In Namibia, farmers who resort to killing captured cheetahs have been reported to sell them to the Chinese medicinal market as a substitute for tiger bones. This practice could be on the rise.

REDUCING THE SUPPLY OF CHEETAHS FOR ILLEGAL TRADE

Poverty, corruption, weak legislation, and inadequate enforcement are considered as some of the main drivers of illegal wildlife trafficking in supply countries (UNEP, 2014). Reducing the supply of cheetahs for the illegal trade market may require the development of alternative forms of income (Chapter 16) while also increasing law enforcement, prosecution, and penalties to deter individuals from capturing and selling cheetahs. Penalties for being convicted of poaching, trafficking, or illegal possession of cheetah specimens vary across the cheetah-range states. The reported maximum sentence ranges from 6 months, in Somalia, to life imprisonment in Kenya, and the reported maximum fines range from US $839 in Chad to US $752,431 in South Africa (data available from 13 countries) (CITES, 2016c). However, data indicates that most illegal trade is undetected, let alone investigated, prosecuted, and penalized. Lack of resources and funding continue to be a major issue in policing wildlife trafficking in cheetah-range countries (CITES, 2016c). To act as a sufficient deterrent, penalties should reflect international recommendations to consider wildlife crime as a serious transnational crime on par with drugs and human trafficking (TRAFFIC, 2013) and monetary fines should be above the retail value of the animal on the international market.

With illegal cheetah trafficking now being recognized as a significant issue under CITES, the species should receive greater attention from African regional and subregional bodies and collaborative platforms established to focus on transnational wildlife crime. These platforms include the Nairobi-based Lusaka Agreement Task Force, the Wildlife Enforcement Network being established in the Horn of Africa, the Law Enforcement Action Plans in Southern and Central Africa (Nowell, 2014), and possibly the Trade in Wildlife Information Exchange in Central Africa (TRAFFIC, 2015). Programs driven by non-governmental organizations in cheetah-range countries are seen as key to help improve the capacity of governments to control domestic and transboundary poaching and trafficking in wildlife. In the Horn of Africa, organizations such as the Range Wide Conservation Program for Cheetahs and African Wild Dog (a project of the Wildlife Conservation Society and the Zoological Society of London), the International Fund for Animal Welfare (IFAW), TRAFFIC, and the African Wildlife Foundation work with relevant government agencies and international institutions to, for instance, train personnel and strengthen environmental legislation. In Ethiopia, the Born Free Foundation launched the Border Point Project in 2015 to facilitate intergovernmental dialogue between Ethiopia and its neighbors, through cooperation between the Ethiopian Defense Forces, Customs, and Federal and Regional Police. However, lack of capacity and funding continue to be a major issue in policing wildlife trade in cheetah-range countries (CITES, 2016c), thus deterring their ability to enforce international treaties and conventions.

REDUCING DEMAND

To combat the illegal trade in cheetahs, it is essential to understand who the consumers are and what drives their attitudes and motivations. Targeted public awareness campaigns should be designed to modify behaviors by addressing the relevant drivers of demand in each region, such as medicinal or ceremonial use, or an individual's need to highlight social status and personal success. In the case of pets, the keeping of dangerous animals is increasingly being prohibited. However, in instances where they are still allowed, it is important to make potential buyers aware that cheetahs do not do well in the same settings as domestic animals and will probably perish despite well-meaning intentions. Thus, developing appropriate educational messages that wild animals should not be pets should be the priority. Demand reduction should also be addressed through social media with campaigns that publicize the conservation impacts of cheetah poaching and trafficking, and law enforcement outcomes (CITES, 2016g). And greater efforts should be made to involve religious authorities in education efforts, as the public often turns to these figures for guidance. For example, Sudan reported asking religious figures to discourage the wearing of shoes made from endangered species (CITES, 2016c). And there have been questions on Internet sites whether the eating of meat from an animal killed by one's hunting cheetah is in conformity with Islam (Nowell, 2014). Substitution can be an effective tool for big cat skins, such as the provision of quality fake leopard furs to a religious community in South Africa (AFP, 2014). In addition, international opinion makers, media in particular, must continue to increase public engagement. Similarly, governments and community leaders in consumer countries must actively discourage and counteract the perception that a big cat pet is a symbol of prestige. The public must understand not only that it is illegal to purchase a cheetah cub, but also that it was likely taken from the wild and has major impacts on the chances of survival of the species.

Several efforts are currently underway in the Gulf States to address illegal wildlife trade and exotic pet ownership. For example, the campaign "Belong to the Wild," aimed at 10–11-year-old learners, is being conducted by IFAW in collaboration with authorities in the UAE, Kuwait, Bahrain, and Lebanon. Children are targeted as anecdotal evidence suggests that the motivation for an exotic pet often comes from them, and as such they can be influential in communicating environmental messages to their parents. Approximately 100,000 learners have learned about conservation and the human costs (safety and zoonotic diseases) associated with ownership of exotic pets (Mohamed, 2016). However, more work needs to be done in the region to specifically research and target the motivations for exotic pet ownership.

Regarding cheetahs that are already in captivity, both as pets and in registered facilities, CCF is working to improve their care by training Emirati cheetah-holding facilities and veterinarians. The organization has also trained veterinarians and biologists in the collection and viable storage of reproductive samples and initiated the first Genome Resource Bank for cheetah in the country. Captive specimens in the UAE are of great genetic value as they are believed to originate mostly in areas of Ethiopia and Kenya adjacent to Somalia, where cheetah populations are small. Building a genetic database can assist with determining the origin of confiscated cheetahs not only to ensure that they can be included in legitimate captive breeding programs, but, importantly, to support enforcement investigations. Therefore, a collaborative effort to collect genetic samples from confiscated cheetahs in Somaliland, Ethiopia, Yemen, and the UAE has begun to build a database that will assist with identifying the origin of confiscated cheetahs.

The international community through global programs, such as the Coalition Against Wildlife Trafficking, the 2014 London Conference on the Illegal Wildlife Trade (UK Government, 2014), and the International Consortium on Combating Wildlife Crime (ICCWC)—a collaborative effort of CITES, INTERPOL, UNODC, World Bank, and the WCO—has recognized the need to deploy a broader spectrum of government resources to counter illegal wildlife trafficking. The continued goal should be to ensure that, in addition to iconic species, such as elephants *elephantidae spp.*, rhinoceros *rhinocerotidae spp.*, and tigers, cheetahs are consistently included in all initiatives, policies, legislation, and enforcement actions, relevant to wildlife trafficking.

CONCLUSIONS

Due to the very nature of illegal trade, collecting detailed information on the extent of poaching and trafficking in cheetahs is difficult. Yet available data suggest that the magnitude of the trade is likely to be substantially impacting wild cheetah populations, particularly those in Ethiopia and northern Kenya; an area with less than 300 cheetahs (Durant et al., 2017). The recorded loss of at least 50 cheetah cubs from the Horn of Africa into the Arabian Peninsula every year (with online research by CCF and informants in the region indicating that the actual numbers could be significantly higher) has the potential to decimate cheetahs in this region where populations are already fragmented (Chapter 10).

Increased law enforcement and stricter repercussions for convicted traffickers are required in the primary source, transit, and demand regions to tackle the trade. In particular, countries with the greatest demand are urged to support and assist cheetah-range states with cheetah conservation and prevention of poaching and trafficking.

In addition, international cooperation to detect and police illegal wildlife trafficking, including the effective monitoring of transboundary activities, must be improved by encouraging the utilization of existing resources. In particular, law enforcement personal should utilize INTERPOL's policing capabilities, including use of the secured information network, deployment of Investigative Support Teams, and other operational support. Wildlife Incident Support Teams (led by INTERPOL in collaboration with ICCWC partners) could be requested to form a multiagency team. International antipiracy crews already patrolling the Indian Ocean, particularly the Gulf of Aden, a well-known route for transporting illegal wildlife and wildlife products between East Africa and Yemen, should be able to exchange information with relevant stakeholders.

It is generally accepted that enforcement alone will not stop poaching and trafficking, but that interventions must also focus on the factors that are driving illegal wildlife trafficking (Challender and MacMillan, 2014), along with community-driven approaches in range countries (IUCN SULi, IIED, CEED, Austrian Ministry of Environment and TRAFFIC, 2015). These should include an evidence-based understanding of the motivations and patterns of consumer behaviors in demand countries (TRAFFIC, 2016).

Whether as pets, shoes, rugs, or medicine, cheetahs throughout their range are being poached, captured, and traded for status, healing, or fashion. The supply is largely driven both by the low risk and high profit of the trade, along with poverty and the lack of alternative livelihoods in rural Africa. Simple ignorance regarding the origin of cheetahs available for sale and the impact of the poaching and trafficking in such a wide-ranging and thinly distributed species is also a major factor contributing to the continuation of the illegal trade in cheetahs. Focused education outreach and strong, international, cooperation to enforce international treaties and conventions to better police CITES provisions will be crucial to halting the poaching and trafficking of cheetahs and other endangered wildlife.

References

AFP, 2014. Zulu false dawn: Shembe faithful swap leopard skin for faux fur. Agence France Presse, The Guardian 19th February. Available from: https://www.theguardian.com/world/2014/feb/19/zulu-shembe-leopard-skin-south-africa.

Akella, A., Cannon, J., 2004. Strengthening the Weakest Links: Strategies for Improving the Enforcement of

Environmental Laws Globally. Conservation International, Washington, DC.

Ali, A., 2014. Deadline for exotic pet-owners in Sharjah ends on Tuesday, Sharjah: Gulf News, 22nd December. Available from: http://gulfnews.com/news/uae/crime/deadline-for-exotic-pet-owners-in-sharjah-ends-on-tuesday-1.1430428.

Amir, O.G., 2006, Wildlife trade in Somalia. A report to the IUCN/SSC/Antelope Specialist Group—North-east African Subgroup. Darmstadt, Germany.

Bahrain Tribune, 2007. 100 Smuggled Cheetahs died in Saudi in six months, 16th September. Available from: http://article.wn.com/view/2007/09/16/Wild_practice/#/fullarticle.

Challender, D.W., MacMillan, D.C., 2014. Poaching is more than an enforcement problem. Conserv. Lett. 7 (5), 484–494.

CITES, 1992. Amendments to appendices I and II of the convention adopted by the Conference of the Parties at its eighth meeting in Kyoto (Japan), 2–13 March. CITES, Geneva, Switzerland.

CITES, 2014. Illegal Trade in Cheetahs (Acinonyx jubatus) (Decision 16.72) (Agenda item 18). AC27 WG3 Doc 1. CITES, Geneva, Switzerland.

CITES, 2016a. Species-specific matters: illegal trade in Cheetahs Acinonyx jubatus. COP17 Doc. 49. In: Seventeenth meeting of the Conference of the Parties Johannesburg (South Africa), 24 September–5 October.

CITES, 2016b. Combatting Wildlife Cybercrime. COP17 Doc. 36. In: Seventeenth meeting of the Conference of the Parties Johannesburg (South Africa), 24 September–5 October.

CITES, 2016c. Questionnaire results on combating illegal trade in cheetahs, SC66 Doc. 32.5 Annex. CITES, Geneva, Switzerland.

CITES, 2016d. New annual illegal trade report. Notification 2016/007. CITES, Geneva, Switzerland.

CITES, 2016e. Summary record of the eighth session of committee II, CITES CoP17 Com. II, Rec. 8 (Rev. 1). CITES, Geneva, Switzerland, pp. 2–3.

CITES, 2016f. Model Law on International Trade in Wild Fauna and Flora. CITES, Geneva, Switzerland.

CITES, 2016g, Illegal trade in cheetahs (Acinonyx jubatus) report of the working group—SC66 Doc. 32.5. CITES, Geneva, Switzerland.

Department of Environmental Affairs, 2010. Convention on International Trade in Endangered Species of Wild Fauna and Flora (CITES) Regulations. Government Gazette, Pretoria, South Africa.

Divyabhanusinh, C., 1995. End of a Trail: The Cheetah in India. Banyan Books, New Delhi, India.

Durant, S., Mitchell, N., Groom, R., Pettorelli, N., Ipavec, A., Jacobson, A.P., Woodroffe, R., Böhm, M., Hunter, L.T.B., Becker, M.S., Broekhuis, F., Bashir, S., Andresen, L., Aschenborn, O., Beddiaf, M., Belbachir, F., Belbachir-Bazi, A., Berbash, A., Brandao de Matos Machado, I., Breitenmoser, C., Chege, M., Cilliers, D., Davies-Mostert, H., Dickman, A.J., Ezekiel, F., Farhadinia, M.S., Funston, P., Henschel, P., Horgan, J., de Iongh, H.H., Jowkar, H., Klein, R., Lindsey, P.A., Marker, L., Marnewick, K., Melzheimer, J., Merkle, J., M'soka, J., Msuha, M., O'Neill, H., Parker, M., Purchase, G., Sahailou, S., Saidu, Y., Samna, A., Schmidt-Küntzel, A., Selebatso, E., Sogbohossou, E.A., Soultan, A., Stone, E., van der Meer, E., van Vuuren, R., Wykstra, M., Young-Overton, K., 2017. The global decline of cheetah Acinonyx jubatus and what it means for conservation. Proc. Natl. Acad. Sci. USA 114 (3), 528–533.

EPAA, 2016. Hearing the Call of the Wild. Environment and Public Areas Authority, Sharjah, June. Available from: http://www.epaashj.ae/news/hearing-call-of-the-wild/.

Goodrich, J., Lynam, A., Miquelle, D., Wibisono, H., Kawanishi, K., Pattanavibool, A., Htun, S., Tempa, T., Karki, J., Jhala, Y., Karanth, U., 2015. Panthera tigris. The IUCN Red List of Threatened Species 2015, e.T15955A50659951. Available from: http://dx.doi.org/10.2305/IUCN.UK.2015-2.RLTS.T15955A50659951.en.

Haken, J., 2011. Transnational Crime in the Developing World. Global Financial Integrity, Washington, DC.

IUCN SULi, IIED, CEED, Austrian Ministry of Environment and TRAFFIC, 2015. Beyond enforcement: communities, governance, incentives and sustainable use in combating wildlife crime. In: Symposium Report, 26–28 February. Glenburn Lodge, Muldersdrift, South Africa.

IUCN/SSC, 2007a. Regional Conservation Strategy for the Cheetah and African Wild Dog in Eastern Africa. IUCN/SSC, Gland, Switzerland.

IUCN/SSC, 2007b. Regional Conservation Strategy for the Cheetah and African Wild Dog in Southern Africa. IUCN/SSC, Gland, Switzerland.

IUCN/SSC, 2012. Regional Conservation Strategy for the Cheetah and African Wild Dog in Western, Central and Northern Africa. IUCN/SSC, Gland, Switzerland.

Kaiser, C., Wernery, U., Kinne, J., Marker, L., Liesegang, A., 2014. The role of copper and vitamin A deficiencies leading to neurological signs in captive cheetahs (Acinonyx jubatus) and lions (Panthera leo) in the United Arab Emirates. Food Nutr. Sci. 5, 1978–1990.

Klein, R., 2007. Status report for the cheetah in Botswana. Cat News Special Issue 3, 14–21.

KUNA, 2015. National Assembly Approves Animal Rights Bill. Kuwait News Agency, Kuwait.

Marker-Kraus, L., 1829. History of the Cheetah (Acinonyx jubatus) in zoos. Int. Zoo Yearbook 35, 27–43.

Marker, L.L., 2015. 2014 International Cheetah (Acinonyx jubatus) Studbook. Cheetah Conservation Fund, Otjiwarongo, Namibia.

Marnewick, K., 2012. Captive Breeding, Keeping and Trade of Cheetahs in South Africa. Endangered Wildlife Trust, South Africa.

Marnewick, K., Beckhelling, A., Cilliers, D., Lane, E., Mills, G., Herring, K., Caldwell, P., Hall, R., Meintjes, S., 2007.

The status of the cheetah in South Africa. Cat News Special Issue 3, 22–31.

Mohamed, E.A., 2016. Tiger loose in Doha: exotic pet ownership, trade must stop in Middle East. International Fund for Wildlife, 11th March. Available from: http://www.ifaw.org/united-states/news/tiger-loose-doha-exotic-pet-ownership-trade-must-stop-middle-east.

Mohamed, E.A., 2017. Victory: UAE Bans Personal Ownership of Wild Animals. International Fund for Animal Welfare, 5th January. Available from: http://www.ifaw.org/international/news/victory-uae-bans-personal-ownership-wild-animals.

Nellemann, C., Henriksen, R., Raxter, P., Ash, N., Mrema, E. (Eds.), 2014. The Environmental Crime Crisis—Threats to Sustainable Development from Illegal Exploitation and Trade in Wildlife and Forest Resources. A UNEP Rapid Response Assessment. United Nations Environment Programme, Nairobi, Kenya and GRID-Arendal. Available from: http://hdl.handle.net/20.500.11822/9120.

Nowell, K., 2014. An assessment of the conservation impacts of legal and illegal trade in cheetahs *Acinonyx jubatus*. In: IUCN SSC Cat Specialist Group report prepared for the CITES Secretariat, 65th meeting of the CITES Standing Committee, Geneva, 7–11 July. CITES SC65 Doc. 39. Available from: http://cites.org/sites/default/files/eng/com/sc/65/E-SC65-39.pdf.

Nowell, K., Jackson, P., 1996. Wild Cats: Status Survey and Conservation Action Plan. Burlington Press, Cambridge, UK.

RWCP & IUCN/SSC. 2015. Regional conservation strategy for the cheetah and African wild dog in southern Africa; revised and updated, August 2015. RWCP & IUCN/SSC.

Sogbohossou, E.A., 2006. Conservation des grands carnivores en Afrique de l'Ouest: Perception par les populations et commerce des sous-produits. Technical Report.

Wildlife Conservation Society Small Grant for Africa. Calavi, Bénin.

TRAFFIC, 2013. Two year jail sentence in Malaysia's biggest tiger part smuggling case, TRAFFIC, 7th February. Available from: http://www.traffic.org/.

TRAFFIC, 2015, Groundwork laid for intelligence information exchange in Central Africa. TRAFFIC, 21st December. Available from: http://www.traffic.org/.

TRAFFIC, 2016. Changing Consumer Choice Advice a Few Clicks Away, TRAFFIC, 25th July Available from: http://www.traffic.org/.

UAE Ministry of Climate Change and Environment, 2016. Illegal Trade of Cheetahs in the UAE. UAE Ministry of Climate Change and Environment, Abu Dhabi.

UK Government, 2014. Declaration on Illegal Wildlife Trade. In: London Conference on the Illegal Wildlife Trade. London, UK. Available from: https://www.gov.uk/government/uploads/system/uploads/attachment_data/file/281289/london-wildlife-conference-declaration-140213.pdf.

UNEP, 2014. UNEP Year Book 2014: Emerging Issues in Our Global Environment. United Nations Environment Programme, Nairobi, Kenya.

UNEP/WCMC, 2016. CITES Trade Statistics Derived from the CITES Trade Database. UNEP World Conservation Monitoring Centre, Cambridge, UK.

WAM, 2014. EPAA: Tuesday is the deadline for delivery of dangerous predators in implementation of Sharjah Ruler's Order. Emirates News Agency, 22nd December. Available from: http://www.wam.ae/en/print/1395274162497.

World Bank, 2014. New World Bank GDP and poverty estimates for Somaliland. Hargeisha, Somaliland, 29th January. Available from: http://www.worldbank.org/en/news/press-release/2014/01/29/new-world-bank-gdp-and-poverty-estimates-for-somaliland.

CONSERVATION SOLUTIONS

Use of Livestock Guarding Dogs to Reduce Human-Cheetah Conflict

Amy Dickman, Gail Potgieter**, Jane Horgan[†],
Kelly Stoner[‡], Rebecca Klein[§], Jeannine McManus[¶],
Laurie Marker[††]*

*University of Oxford, Tubney, Abingdon, United Kingdom
**Tau Consultants, Maun, Botswana
[†]Cheetah Conservation Botswana, Maun, Botswana
[‡]Ruaha Carnivore Project, Iringa, Tanzania
[§]Cheetah Conservation Botswana, Gaborone, Botswana
[¶]Landmark Foundation Leopard Project, Riversdale, South Africa
[††]Cheetah Conservation Fund, Otjiwarongo, Namibia

INTRODUCTION

Human-cheetah (*Acinonyx jubatus*) conflict can impose significant direct and indirect costs on landowners through attacks on livestock and farmed game (Chapter 13). This often leads to retaliatory or preventative killing of cheetahs, posing one of the greatest conservation threats to the species (Durant et al., 2015). There is a pressing need to reduce this conflict by preventing damage caused by cheetahs and therefore livestock farmers' antagonism toward the species. In Africa, cheetahs are known to persist in only 9% of their historic range; the majority of which (77%) is outside formally protected areas (Durant et al., 2017). Improving human–cheetah coexistence on unprotected land is thus important for future cheetah conservation (IUCN, 2007, 2012; RWCP and IUCN 2015). Numerous strategies have been employed to try to reduce human-cheetah conflict. One of the most well-known mitigation methods is the use of guardian animals, particularly livestock guarding dogs (LGDs) (Marker et al., 2010).

Cheetahs: Biology and Conservation
http://dx.doi.org/10.1016/B978-0-12-804088-1.00015-0

LGDs, sometimes called livestock protection dogs, spend all their time with livestock to actively deter predators (Rigg, 2001). LGDs usually deter predators by barking at them and by alerting the herd and herder (if present) to the predator's presence, thus directly preventing attack. LGDs may also chase and occasionally kill predators (Potgieter et al., 2016; Urbigkit and Urbigkit, 2010), but primarily reduce predator presence in the area through barking and territorial markings (Gehring et al., 2010; Hansen et al., 2002; Potgieter et al., 2016; Urbigkit and Urbigkit, 2010). Guarding dogs are bred and selected based on their tendency to treat livestock as conspecifics; this is in contrast to herding dogs, which exhibit predatory behavior toward livestock (e.g., stalk/chase) (Coppinger and Coppinger, 2001).

In much of rural Africa, the larger predators, such as lions (*Panthera leo*) and spotted hyena (*Crocuta crocuta*) have been removed from farming areas. As a result LGDs are mostly used to protect smallstock (i.e., sheep and goats) against cheetah, leopard (*Panthera pardus*), African wild dog (*Lycaon pictus*), black-backed jackal (*Canis mesomelas*), caracal (*Caracal caracal*), and Chacma baboon (*Papio ursinus*) (Horgan, 2015; McManus et al., 2015; Potgieter et al., 2016). Trials using LGDs to protect cattle and game species, such as springbok (*Antidorcas marsupialis*) and nyala (*Tragelaphus angasii*), have also been undertaken (Cheetah Outreach, 2016), and in some areas LGDs are used to protect livestock against the full guild of carnivores (Dickman, unpublished data).

In this chapter, we review the history of guarding dogs and their use on cheetah rangeland, particularly their effectiveness as a conflict mitigation method. The issues associated with their use, and the future prospects of using this method to reduce human-cheetah conflict are discussed. We focus on LGDs that have been used to protect relatively intensively managed smallstock (i.e., animals are placed in enclosures at night and herders sometimes accompany smallstock in the day) against cheetahs, mainly in Namibia and Botswana where most of the LGD research has been conducted.

HISTORY OF LIVESTOCK GUARDING DOGS AND THEIR USE ON CHEETAH RANGELAND

LGDs are thought to have originated in and around Mesopotamia (Rigg, 2001). Domestic dogs and sheep were discovered together in archaeological sites in present day Iraq and Iran that date from 3585 years BCE (Gehring et al., 2010; Olsen, 1985). As one of the earliest types of working dogs, LGDs have been found in nearly every continent, and many traditional breeds of guarding dogs have been selectively bred to cope with the varying climatic conditions and to protect stock against predator species found in each region. Two of the most well-known breeds used in cheetah conservation are the Anatolian Shepherd and Kangal dogs, both originating from Turkey (Rigg, 2001; for a list of traditional LGD breeds). The first studies on LGD effectiveness were conducted in North America, focusing on the ability of purebred LGDs to guard sheep against coyotes (*Canis latrans*) and other North American predators (Andelt, 1992; Coppinger et al., 1983). Around the same time, others found that nonpurebred native American dogs were also effective LGDs (Black and Green, 1985). More recently, the use of purebred LGDs has been investigated on sheep farms in Australia, where they have reduced livestock depredation from dingoes (*Canis lupus dingo*) and red foxes (*Vulpes vulpes*) (van Bommel and Johnson, 2012).

There is little published information on the use of LGDs in sub-Saharan Africa before the 1990s, although they are reported to have been used by African farmers (Marker-Kraus et al., 1996). It was not until 1994 that 10 purebred Anatolian Shepherds were imported into Namibia by the Cheetah Conservation Fund (CCF) in collaboration with the Hampshire College Livestock Guarding Dog Project. This was the first time a cheetah conservation organization directly used LGDs and the project aimed to test LGDs' ability to reduce cheetah attacks on livestock, and consequently mitigate human-cheetah conflict (Marker et al., 2005a). Between 1994 and 2015, CCF placed 490 LGDs with farmers in Namibia. These dogs were primarily purebred Anatolian Shepherds or Kangal dogs, but also included one Anatolian/Rhodesian Ridgeback litter and several three-quarter Anatolian/one-quarter mongrel litters (Marker et al., 2005c). Cheetah Outreach expanded the use of purebred Anatolian Shepherds into South Africa in 2005 and in 2014 the project began placing Lesotho Highland dogs (an indigenous African breed) with smallstock farmers. Other cheetah conservation projects with LGD programs include Cheetah Conservation Botswana, which promotes the use of indigenous Tswana dogs as an alternative to purebred LGDs (Klein, 2007; Fig. 15.1), and the Ruaha Carnivore Project in Tanzania, which in collaboration with CCF, has used Anatolian Shepherds (Fig. 15.2) and plans to cross Anatolian Shepherds with dogs

FIGURE 15.2 **An Anatolian Shepherd dog placed in Tanzania to help Massai guard the herd.** Children are often used to care for the family livestock. *Source: Amy Dickman.*

FIGURE 15.1 **A Tswana dog with its goat herd.** The dog was the winner of the Cheetah Conservation Botswana's competition in 2008 for the best livestock guarding dog (LGD) in Botswana. *Source: Cheetah Conservation Botswana.*

from the local Sukuma tribe in the future (Dickman, unpublished data).

CHARACTERISTICS OF A SUCCESSFUL LIVESTOCK GUARDING DOG

According to Coppinger and Coppinger (1980), LGDs require three key characteristics: attentiveness, trustworthiness, and protectiveness to be successful guardians. Attentiveness means that the LGD should bond and remain with the herd, appear to be part of the herd, and display vigilant and investigative behavior toward livestock. Trustworthiness means that the

FIGURE 15.3 **Anatolian Shepherd dog protecting its stock in Namibia** *Source: Laurie Marker.*

dog should display calm, submissive behavior toward the herd, and should not display undesirable behaviors, such as playing roughly with livestock (i.e., chasing and biting) and chasing wildlife (Marker et al., 2005a). "Protectiveness" means that the dog should show protective behavior toward livestock and have the capacity to guard effectively (Fig. 15.3). Submissive behavior in an LGD is also desirable, as this will ease bonding with the herd, simplify training, and reduce behavioral problems (Lorenz and Coppinger, 1986). Last, LGDs should be approachable by trusted humans to allow for routine medical interventions (e.g., vaccinations). LGDs tend to be particularly active between dusk and dawn, which is beneficial, as predators are likely to be active during this time (Stannard, 2006a). These characteristics of LGDs should not require intensive training but should be largely innate (Andelt, 2004).

LIVESTOCK GUARDING DOGS AS A CONFLICT MITIGATION MEASURE

Ideally, a successful human-wildlife conflict mitigation measure should reduce (and/or be perceived to reduce) the impact of wildlife damage with persistent efficacy, be selective toward problematic individuals rather than entire species, be relatively cost effective, practical to implement and maintain, impact minimally on the environment, and should be a socially acceptable method to local people (McManus et al., 2015; Shivik, 2004, 2006). Ultimately it should improve local attitudes toward conflict-causing wildlife, and, especially for endangered species, it should reduce retaliatory or preventative killing of wildlife (Shivik, 2004, 2006). Here, we review evidence pertaining to the efficacy of guarding dogs in reducing human-cheetah conflict in terms of these key factors. It should be noted that all the studies discussed herein define an "effective" LGD slightly differently; however, these definitions are all based on the dog showing suitable LGD behavior, a reduction in livestock losses, farmer satisfaction with the LGD, and a reduction in carnivore killing as reported by farmers (Marker et al., 2005a; McManus et al., 2015; Horgan, 2015; Potgieter, 2011).

Reduced Impact of Predation on Smallstock

In a 1994–2001 survey in Namibia, 73% of farmers (N = 117) reported a "large decline" in smallstock losses to predators after receiving a guarding dog from the CCF program (categorized as a decrease from ≥10 losses/year to ≤4 losses/year, or a decrease from 5–9 losses/year to no losses) (Marker et al., 2005a). In a 2009–10 follow-up survey, Potgieter et al. (2016) found that 85% of farmers (N = 73) reported losing livestock to predators in the year before receiving a guarding dog compared to 35% of farmers in the most recent year with the dog. Similarly, in Botswana 95% of farmers (N = 75) who had experienced livestock losses prior to getting an LGD, reported a reduction in their losses after acquiring an LGD, with a mean reduction in losses of 85% (Horgan, 2015). In both countries, a few farmers reported no change in livestock losses after getting an LGD (5% of farmers in Namibia and 1% in Botswana) or an increase in livestock depredation (3% of farmers in Namibia and 4% of farmers in Botswana) (Horgan, 2015; Potgieter et al., 2013). Farmers are prone to over-estimate losses to predators (Chapter 13), as well as to overestimate the impact of their dogs, (see next section) and although the reported changes were not independently verified, it is clear that LGDs at least reduce perceived livestock depredation. Indeed, farmer satisfaction with their dogs appeared to be more closely correlated to their witnessing the dogs' attentive and trust-worthy behaviors with the livestock rather than the actual numbers of livestock lost during the year of the survey (Potgieter et al., 2013).

Cost Effectiveness and Farmer Satisfaction

The major costs associated with using LGDs include initial purchase costs (range US $14–553; Table 15.1), as well as regular maintenance costs (i.e., food and routine veterinary care including vaccinations, range US $138–432 for the first year, Table 15.1). There are additional indirect costs associated with emergency medical care and the time, and resources used to purchase and place the LGD, including regular site visits

for those dogs that are placed by conservation organizations. These factors can greatly increase the costs of LGD programs, for example, Rust et al. (2013) documented the first year costs to be US $2780 per dog when including the costs of monthly site visits (Table 15.1). These additional indirect costs vary depending on LGD programs, and monitoring is rarely accounted for in cost–benefit analyses.

The costs associated with using LGDs can be exacerbated if the dog is large (as it requires greater food requirements, both in terms of quality and quantity), if it is a rare breed (as the purchase costs will be higher) or if it is not well adapted to the local environment (which could result in higher veterinary costs) (Horgan, 2015). Due to these factors, large, purebred LGD breeds tend to be substantially more expensive than indigenous African local breeds of LGDs, in both purchase and maintenance costs (Table 15.1). Some organizations in South Africa, Namibia, and Tanzania have attempted to mitigate this problem by providing farmers with pure-bred LGDs free or at a subsidized rate, and/or with free or subsidized food and veterinary care (Dickman, unpublished data; McManus

TABLE 15.1 Comparison of Costs Associated With Different LGD Breeds

LGD breed	Country	Purchase price (US $)	Maintenance (US $/year)[a]	References
Purebred Anatolian	South Africa	2780[b]		Rust et al. (2013)
	Botswana	79	145	Horgan (2015)
	Namibia	84	385[a]	Potgieter (2011)
	South Africa	553	432	McManus et al. (2015)
Anatolian cross mixbreed[c]	Namibia	84	204	Potgieter (2011)
Tswana crossbreed[d]	Botswana	51	292	Horgan (2015)
Tswana	Botswana	14	138	Horgan (2015)

All figures represent mean costs converted to US $ at the time of the respective studies. Maintenance costs include food and routine veterinary care.
[a]The method used to calculate maintenance costs varied. For example, costs were calculated by Potgieter (2011) according to what the farmer should spend on food to maintain the dog's body condition. While Horgan (2015) recorded the farmers' estimates of their purchase and maintenance costs for their LGDs.
[b]Purchase and maintenance costs are combined and include the costs of regular site visits.
[c]Three-quarter Anatolian Shepherd, one-quarter mixed breed local dog.
[d]Tswana dogs crossed with purebred dogs not traditionally used as LGDs (Bulldog, Ridgeback, Rottweiler, German Shepherd, and Greyhound).

et al., 2015; Potgieter et al., 2016). However, the maintenance costs are still high enough to be prohibitive for low-income households (Potgieter, 2011). This, unfortunately, can lead to LGDs being malnourished or inadequately cared for (Landry et al., 2005; Potgieter, 2011). Local dogs may therefore be a more suitable alternative for rural farmers who cannot afford the purchase and maintenance costs of purebred LGDs.

Although the costs associated with keeping an LGD can be high for some rural African farmers, the financial benefits experienced through livestock that is protected against depredation can overshadow costs within the first few years of an LGD's life. In Namibia, 47% of the surveyed farmers in CCF's LGD program ($N = 64$) reported that their dogs saved enough livestock (a minimum of seven goats or nine sheep) to be considered economically beneficial in 1 year, had they have covered all the maintenance expenses themselves (Potgieter, 2011). With cheaper three-quarter Anatolians, 60% of the farmers would have saved enough livestock (a minimum of four goats or sheep annually) for their dogs to be economically beneficial. In Botswana, the average cost of purchasing and maintaining a Tswana LGD without subsidization was so low that having the LGD was economically beneficial if the dog saved only one head of livestock each year (either a sheep or goat) (Horgan, 2015). In both Namibia and Botswana, farmer satisfaction with their LGDs was higher than would be expected from economic cost–benefit analyses, indicating that intangible benefits play a role in farmer perceptions of LGDs (Horgan, 2015; Potgieter et al., 2013).

Improvement of Farmer Attitudes Toward Predators

In addition to physically protecting livestock from attacks by cheetahs and other predators, LGDs provide the additional benefit of improving pastoralist attitudes toward predators. For example, 79% of South African farmers who participated in Cheetah Outreach's LGD program ($N = 94$) reported that their tolerance toward cheetahs had greatly increased, despite a perceived increase in local cheetah populations (Rust et al., 2013). The presence of LGDs can improve attitudes toward predators even in cases when the dogs have no measurable impact on depredation rates (Horgan, 2015).

Reduction of Number of Cheetahs Killed by Farmers

In the survey of farmers in Botswana, 51% ($N = 94$) claimed that they would be less likely to use lethal control against predators now that they owned an LGD (Horgan, 2015). In Namibia, of 67 responding farmers who owned LGDs, none of them reportedly killed cheetahs in the year of the survey, whereas 4 farmers (6%) reportedly killed cheetahs in the year before they received their dogs (Potgieter et al., 2016). It is encouraging that the farmers' reported use of lethal control decreased as a result of owning an LGD. However, there are no independently collected data on the rate of predator removal before and after LGDs were introduced. Although it is often assumed that fewer livestock losses and improved attitudes toward predators leads to less lethal control of predators, the link between attitude and behavior is tenuous and nonlinear (Dickman, 2010). For example, Horgan (2015) found that 21% of LGD owners ($N = 108$) said they would continue to kill predators regardless of whether or not they were experiencing livestock losses.

On occasion LGDs will kill predators; however, this behavior has not been thoroughly reported or investigated. This primarily affects medium-sized, common predators, such as black-backed jackal and caracal (Potgieter et al., 2016). This is viewed as an undesirable behavior by some conservationists and could potentially put the dog at risk of injury. However, an LGD killing a predator in defense of its herd has potentially less negative side effects than other more indiscriminate forms of lethal control (e.g., trapping or poisoning), as the dog is more likely to kill the specific predator responsible for depredation (Gehring et al., 2010; Potgieter et al., 2016). This remains to be empirically tested.

PREVALENCE AND POTENTIAL SOLUTIONS FOR LIVESTOCK GUARDING DOG BEHAVIORAL PROBLEMS

The original studies on how LGDs work (Coppinger et al., 1983) have guided breeding and training protocols for LGD use in cheetah range countries. Although most farmers are satisfied with the performance of their guarding dogs, behavioral problems are widespread in LGDs. In Botswana, 62% of 176 LGDs had behavioral problems, although a third of those (30%) occurred infrequently (Horgan, 2015). In this study the most common problems were chasing and injuring livestock (exhibited by 56% of LGDs), leaving the livestock (55%), and chasing wildlife (45%). Anatolian Shepherds (n = 22) were more likely than other LGDs in this study to display behavioral problems, in particular abandoning the livestock (Horgan, 2015). Farmers in Namibia reported that 38% of the Anatolian Shepherds displayed behavioral problems (N = 164), the most common of which were staying at home rather than going out with livestock (18% of dogs), chasing wildlife (15%), and biting livestock (9%) (Potgieter et al., 2013).

Potgieter (2011) investigated potential reasons for behavioral problems in purebred and three-quarter Anatolians in the CCF program. The issue of dogs staying at home, rather than accompanying livestock, was linked to a lack of farmer care for the dog among subsistence farmers. Farmer care was calculated using the type of diet provided to the dog and the level of farmer involvement with the dog. Eighteen percent of farmers surveyed in Namibia (N = 83)

reported their LGD killed and/or chased wildlife. Dogs chasing and/or killing wildlife in this program was not linked to farmer care, the presence of a herder, or changes in herders during the dogs' lifetimes (Potgieter, 2011). LGDs in Botswana were more likely to chase wildlife when accompanied by herders (Horgan, 2015), suggesting that herders may have encouraged the LGDs to hunt wildlife species. LGDs in Botswana and Namibia rarely attacked humans (reported by 0.5 and 2% of farmers, respectively; Horgan, 2015; Potgieter, 2011).

To improve the effectiveness of LGDs and prevent these behaviors from developing and persisting, owners and herders should be trained to monitor and curb unwanted behaviors in the LGD. In Namibia, 61% of behavioral problems were effectively resolved using corrective training methods (Marker et al., 2005a). Many cheetah organizations implement training sessions for current and future LGD owners, and it is important that regular contact be maintained between the organization and the owners to assist with resolving behavioral problems. Cheetah Outreach in South Africa visit their LGDs on a monthly basis (Rust et al., 2013; Stannard, 2006b), in Tanzania's Ruaha landscape they are visited weekly, whereas CCF in Namibia and CCB in Botswana visit their dogs at 3 months, 6 months, and 1 year after placement, and annually thereafter (Potgieter, 2011).

FACTORS AFFECTING THE PERFORMANCE OF LIVESTOCK GUARDING DOGS

Breed and Lineage

Darwin (1909) described raising an LGD as "take any local puppy and raise it properly with stock and you have a decent livestock guarding dog." Black and Green (1985) developed a similar training theory after observing local Navajos dogs used as stock guardians. Some breeds of dogs have been selectively bred as livestock guardians and initial research suggested that LGDs bred from a lineage of working LGDs were more likely to show the necessary characteristics associated with being a good guardian (Coppinger and Coppinger, 2001).

Of all of the traditional LGD breeds, the Anatolian Shepherd is the one, most often associated with cheetah conservation. The Namibian LGD program initially used purebred Anatolian Shepherds as LGDs, as they had short hair, were adapted to high and fluctuating temperatures, and worked independently over vast areas (Marker, 2002). Over time, CCF bred three-quarter Anatolians ($n = 21$) (one-quarter mongrel) and found them to be no less effective at guarding livestock than purebred Anatolians ($n = 73$) (CCF, unpublished data). Local mixed breed dogs have also been found to show admirable guarding qualities, and local Tswana dogs in Botswana ($n = 126$) were reportedly more effective at reducing livestock depredation than Anatolian Shepherds ($n = 22$). These studies also found that LGDs that were obtained from litters with nonworking parents were equally effective as livestock-guardians as dogs with working parents (CCF, unpublished data; Horgan, 2015). Therefore, although using dogs from a working breed and lineage may enhance an LGD's performance, it is not an essential factor for success. Indeed, in the 1970s early research on LGDs was conducted on a variety of European dog breeds from working bloodlines. The results from this study showed that neither the breed nor the genetics of the dog were as critical to raising a successful LGD as the environment in which the puppy was raised (Coppinger and Coppinger, 2001).

Due to differences in measurements of LGD effectiveness, it is not possible to compare breeds and LGD placement programs across the studies mentioned here. A metaanalysis using raw data from all African LGD studies would be a valuable endeavor to allow for a direct comparison between breeds and their efficacy under local conditions.

Puppy Selection and Training

The process of puppy selection and training can influence LGD effectiveness. Although breed and lineage may not be crucial to LGD success, selecting a strong, healthy puppy can reduce the likelihood of health problems (Lorenz and Coppinger, 1986). Puppies should be kept with livestock continuously, while interactions with people should be limited, to ensure the LGD bonds with livestock more than people (Black and Green, 1985; Coppinger et al.,1985; Lorenz and Coppinger, 1986). However, the puppy should occasionally be handled in the presence of livestock such that it will allow veterinary care as an adult. Adult LGDs can help and train puppies through leading by example, and trained herders or other farm workers can reduce undesirable puppy behavior (e.g., rough play with livestock) with timely correction (Horgan, 2015). Puppies reportedly deter predators from as young as 6 months of age (Potgieter, 2011); however, the training process generally lasts until a dog reaches maturity (12–30 months) (Coppinger et al., 1988; Green and Woodruff, 1990; Rigg, 2001). Although initial training is not intensive, LGDs should ideally be monitored continuously to correct behavioral problems (e.g., chasing wildlife), which can develop at any age (Coppinger and Coppinger, 2001; Green and Woodruff, 1990; Potgieter, 2011).

Age at Placement

One of the most important factors determining the success of LGDs is how well they bond with the livestock. It has been recommended that dogs should be placed with their herd from a young age to develop social bonding between the livestock and the dog (Coppinger and Coppinger, 2001). Between 2 and 16 weeks of age is the critical period of "social development" (Scott and Marston, 1950) or "imprinting," which is the time when the puppy has the greatest capacity to learn social skills. This is also a time of major development in the puppy's brain and exposing them to the sights, smells, and sounds of what the LGD will protect helps influence their cognitive development (Coppinger and Coppinger, 2001). In Namibia and Botswana, 91% ($N = 164$) and 93% ($N = 107$) of puppies were placed on a farm by 12 weeks of age, respectively, and all of the dogs had either guarded livestock before placement or were raised on a livestock farm (but not necessarily in the corral) (Horgan, 2015; Marker et al. 2005a). Questionnaire surveys showed no effect of the age at placement on subsequently assigned scores of attentiveness, protectiveness, or trustworthiness of the LGD, or the satisfaction of their owners with the LGD (Horgan, 2015; Marker et al., 2005a). However, association with livestock from a young age remains an important consideration even if the LGD does not grow up with the specific livestock herd it will guard in the future.

Sex and Neutering

No difference was found in the effectiveness of male or female LGDs in Botswana (Horgan, 2015) or in Namibia (Marker et al., 2005a). Neutering of LGDs is believed to reduce problematic behaviors, such as roaming away from livestock when dogs are in estrus (Marker, unpublished data). Neutered dogs studied in Botswana ($N = 176$) were significantly more effective at guarding livestock than intact dogs, although there was no reported difference in terms of behavioral problems (Horgan, 2015).

Age and Size of the Livestock Guarding Dog

In Botswana and Namibia, the range of LGD age assessed was 1.5–10 years and 0.5–10 years, respectively (Horgan, 2015; Potgieter, 2011). Both studies found that the age of the LGD did not affect its ability to deter predators. Younger dogs, however, were more likely to display problem behavior by biting livestock (Potgieter, 2011) or chasing wildlife (Horgan, 2015). LGDs

in Botswana weighed from 12 to 45 kg (mean 17 kg ± SD 7.9), but size was not related to measures of effectiveness (Horgan, 2015).

Diet and Health Care

The level of farmer care provided to LGDs in the form of diet and general health care is an important contributor to LGDs effectiveness. In Namibia, less care (diet and farmer involvement with the dog) on subsistence farms increased the tendency for dogs to stay at home (Potgieter et al., 2013). In Botswana, the level of care (diet and veterinary care) provided was positively correlated with LGD effectiveness (Horgan, 2015). However, it is interesting to note that Tswana and crossbreed LGDs (i.e., local breeds) received less care than purebred Anatolian Shepherds (i.e., imported breeds), and yet were reported to to suffer from less diseases than purebred dogs, possibly due to a greater level of natural immunity to local diseases (Horgan, 2015). Typical LGD ailments reported in Botswana include parasites, physical injuries, and unidentified diseases (Horgan, 2015), while squamous cell carcinoma (tongue cancer) has been identified in LGDs in Namibia (Lester et al., 2008). Sourcing a quality diet for an LGD can be a challenge on remote farms, and may be especially difficult to afford for subsistence farmers. However, a combination of maize meal, milk, meat, and tinned or pellet dog food is generally considered the most nutritious and accessible diet for LGDs in Africa (Horgan, 2015; Marker et al., 2005a).

Presence of Herders

The presence of a herder is often thought to increase a dog's effectiveness (Hansen and Smith, 1999; Linnell et al., 1996) and reduce the risk of predator attacks on the dog itself (Bangs et al., 2006). However, paying a herder can increase the cost to the farmer and may not always be cost-effective (Shivik, 2004). In Namibia, the presence of a herder had no obvious impact on

LGD effectiveness (Marker et al., 2005a; Potgieter, 2011). In Botswana, LGDs were less effective if accompanied by a herder, as they were more likely to chase wildlife (Horgan, 2015). Herder training could thus improve the effectiveness of herder–LGD teams protecting livestock.

Herd Size

Horgan (2015) found that LGD effectiveness was negatively correlated to the number of livestock that it was guarding, so that the fewer livestock the dog was responsible for, the more effectively it could protect them from predators. As a result of this, farmers with large herds may see a decline in the effectiveness of LGDs, unless the herds are split or additional LGDs are sourced to work alongside each other.

Farm Type

While the type of farm on which a dog is placed is not directly related to the traits of the dog itself, many of the studies examining LGDs in Africa also tested for the effect of farm type (commercial versus subsistence) on LGD effectiveness (Horgan, 2015; Marker et al., 2005a; Potgieter et al., 2013). In Botswana, LGDs were reported to be more effective on subsistence ($n = 78$) than commercial ($n = 30$) farms (Horgan, 2015). In Namibia, the type of farm did not impact the dog's effectiveness or protectiveness score (Marker et al., 2005a; Potgieter et al., 2013). However, on subsistence farms ($n = 35$) care provided to the dog declined with age of the LGD, whereas care remained constant on commercial farms ($n = 111$) (Potgieter et al., 2013).

WORKING LIFE SPAN AND CAUSES OF MORTALITY FOR LIVESTOCK GUARDING DOGS

A key factor in determining the overall success of LGDs is the working life span of the dog. The longer a dog works, the more cost-efficient it is (Lorenz et al., 1986; Potgieter, 2011). However, 43%–48% of LGDs are removed from the herd (nonlethal removal) or die (lethal removal) before reaching retirement age (around 10 years for large-breed LGDs) (Lorenz et al., 1986; Marker et al., 2005b; Rust et al., 2013). In South Africa, the average life span of LGDs was just 2.3 years (maximum 6 years) (Rust et al., 2013), compared to 4.2 years (maximum 10 years) in Namibia (Marker et al., 2005b). Nonlethal removals usually occurred in response to behavioral problems resulting in the dogs being removed from working situations, either by the owner or by the conservation organization that placed them. Nonlethal removals accounted for 4% of all removals in Botswana (Horgan, 2015), 22% in South Africa (Rust et al., 2013), and 29% in Namibia (Marker et al., 2005b).

Lethal removals were either intentional (e.g., culling by the owner) or accidental (e.g., hit by car). Owners culled dogs that displayed aggressive or predatory behavior toward livestock (Horgan, 2015; Lorenz et al., 1986; Marker et al., 2005b). Accidental deaths occurred for a variety of reasons; snakebite was the most commonly reported cause. LGDs have also reportedly been injured and sometimes killed by predators such as baboons, leopards, or other dogs (Horgan, 2015; Marker et al., 2005b; Rust et al., 2013).

Less common causes of death included disease (Horgan, 2015; Marker et al., 2005b), culling by neighboring farmers who did not realize the dog was allowed to be with the herd (Marker et al., 2005b), farm workers intentionally poisoning a dog (Rust et al., 2013), and one case in Tanzania where a dog was speared during an intertribal conflict (Dickman, unpublished data).

Puppies (less than 1-year-old) and juveniles (from 1 year to sexual maturity) are more vulnerable to accidental death than adult dogs (Marker et al., 2005b). Therefore, special care and attention should be paid to young LGDs to ensure their survival to adulthood.

FUTURE PROSPECTS REGARDING THE USE OF LIVESTOCK GUARDING DOGS TO REDUCE HUMAN-CHEETAH CONFLICT

The use of LGDs is an effective and well-tested method to address conflict mitigation in southern Africa. Farmers report that LGDs considerably reduce predator attacks on livestock and surveys show that farmers' attitudes toward predators improved after acquiring an LGD. Farmers also reported they conducted less retaliatory or preventative killing of cheetahs and other predators after acquiring an LGD (Horgan, 2015; Potgieter, 2011). However, this was not independently verified and attempts to verify the impact of LGDs on cheetah populations should be done in the future.

Although LGDs are promoted as a conservation tool in cheetah range countries in southern Africa, there has been relatively little use of LGDs elsewhere in the species' range. They have recently been introduced in East Africa, a region that is critically important for cheetah persistence. In this region there is some doubt about whether subsistence farmers can afford to feed and care for purebred LGDs, even if the initial acquisition costs were subsidized. There are also concerns about how the dogs would be cared for in communities where dogs are not traditionally viewed as valuable. In addition, LGDs in East Africa would be exposed to diseases (e.g., trypanosomiasis), which are common in East Africa but not present in the countries where LGDs are currently used. The only way to examine the real magnitude of these concerns is to trial the use of LGDs on a broader scale. So far, only one trial of specialized guarding dogs in East Africa has been initiated, in Tanzania's Ruaha landscape (Dickman, unpublished data). The trial revealed that although the communities are surprisingly positive about caring for the dogs, the amount of food required to feed large dogs is often prohibitive, especially in pastoral households that regularly move long distances

with their stock. The use of smaller local breeds of dogs is a possibility; however, smaller dogs may be less effective against lions, which is a common threat in that area.

Despite their effectiveness for reducing livestock losses in cheetah range countries, LGDs are unlikely to be a "silver bullet" for resolving human–carnivore conflict across Africa. Holistic livestock protection, for example, using predator-proof enclosures at night and herders during the day, which includes LGDs, would increase protection of livestock and may be especially beneficial in areas with large predators, such as lions and spotted hyenas (Manoa and Mwaura, 2016).

LGDs have primarily been used for guarding smallstock, but cattle depredation causes proportionally greater human–carnivore conflict, especially with the larger carnivores. Although LGDs have shown some success in limited trials with cattle in South Africa (Cheetah Outreach, 2016), this technique needs to be explored more extensively. Additionally, and potentially most importantly, human–carnivore conflict is often not directly related to livestock losses, and addressing human attitudes and perceptions toward predators is a long-term, multifaceted endeavor (Chapter 13).

LGDs have been used to facilitate human–carnivore coexistence for centuries in many parts of the world. Africa brings a suite of new challenges and opportunities for using LGDs to mitigate human–carnivore conflict. With careful planning and adaptation to local situations, the use of LGDs can be expanded in cheetah range countries to benefit both people and predators.

References

Andelt, W.F., 1992. Effectiveness of livestock guarding dogs for reducing predation on domestic sheep. Wildl. Soc. Bull. 20, 55–62.

Andelt, W.F., 2004. Use of guarding animals to reduce predation on livestock. Sheep Goat Res. J. 19, 72–75.

Bangs, E., Jimenez, M., Niemeyer, C., Fontaine, J., Collinge, M., Krsichke, R., Handegard, L., Shivik, J., Sime, C.,

Nadeau, S., Mack, C., Smith, D.W., Asher, V., Stone, S., 2006. Non-lethal and lethal tools to manage wolf-livestock conflict in the northwestern United States. In: Timm, R.M., O'Brien, J.M. (Eds.), Proceedings of the 22nd Vertebrate Pest Conference. University of California, Davis, CA, (pp. 7–16).

Black, H.L., Green, J.S., 1985. Navajo use of mixed-breed dogs for management of predators. J. Range. Manage. 38, 11–15.

Cheetah Outreach, 2016. Livestock Guarding Dog Program, South Africa. Available from: http://www.cheetah.co.za/an_project.html.

Coppinger, R., Coppinger, L., 1980. Livestock guarding dogs: an old world solution to an age-old problem. Country J. 7 (4), 68–77.

Coppinger, R., Coppinger, L., 2001. Behavior and evolution. Scribner, New York, NY.

Coppinger, R., Coppinger L., Langeloh G., Gettler L., Lorenz J., 1988. A decade of use of livestock guarding dogs. In: Crabb, A.C., Marsh, R.E., (eds.), Proceedings of the 13th Vertebrate Pest Conference. University of California, Davis, pp. 209–214.

Coppinger, R., Lorenz, J., Glendinning, J., Pinardi, P., 1983. Attentiveness of guarding dogs for reducing predation on domestic sheep. J. Range. Manage. 36, 275–279.

Coppinger, R., Smith, C.K., Miller, L., 1985. Observations on why mongrels may make effective livestock protecting dogs. J. Range Manag. 38 (6), 560–561.

Darwin, C., 1909. The Voyage of the Beagle. P.F. Collier and Son, New York, p. 163.

Dickman, A.J., 2010. Complexities of conflict: the importance of considering social factors for effectively resolving human–wildlife conflict. Anim. Conserv. 13, 458–466.

Durant, S.M., Mitchell, N., Groom, R., Pettorelli, N., Ipavec, A., Jacobson, A., Woodroffe, R., Bohm, M., Hunter, L., Becker, M., Broekhuis, F., Bashir, S., Andresen, L., Aschenborn, O., Beddiaf, M., Belbachir, F., Belbachir-Bazi, A., Berbash, A., Brandao de Matos Machado, I., Breitenmoser, C., Chege, M., Cilliers, D., Davies-Mostert, H., Dickman, A., Fabiano, E., Farhadinia, M.S., Funston, P., Henschel, P., Horgan, J.E., de Iongh, H.H., Jowkar, H., Klein, R., Lindsey, P., Marker, L.L., Marnewick, K., Melzheimer, J., Merkle, J., M'Soka, J., Msuha, M., O'Neill, H., Parker, M., Purchase, G., Saidu, Y., Samaila, S., Samna, A., Schmidt-Küntzel, A., Selebatso, E., Sogbohossou, E., Soultan, A., Stone, E., van der Meer, E., van Vuuren, R., Wykstra, M., Young-Overton, K., 2017. The global decline of cheetah *Acinonyx jubatus* and what it means for conservation. Proc. Natl. Acad. Sci. USA 114, 528–531.

Durant, S., Mitchell, N., Ipavec, A., Groom, R., 2015. *Acinonyx jubatus*. The IUCN Red List of Threatened Species 2015. Available from: http://dx.doi.org/10.2305/IUCN.UK.2015-4.RLTS.T219A50649567.en.

Gehring, T.M., VerCauteren, K.C., Landry, J., 2010. Livestock protection dogs in the 21st century: is an ancient tool relevant to modern conservation challenges? BioScience 60, 299–308.

Green, J.S., Woodruff, R.A., 1990. ADC Guarding Dog Program update: a focus on managing dogs. In: Davis, L.R., Marsh, R.E. (Eds.), Proceedings of the 14th Vertebrate Pest Conference. University of California, Davis, CA, pp. 233–236.

Hansen, I., Smith, M.E., 1999. Livestock guarding dogs in Norway part II: different working regimes. J. Range. Manage. 52, 312–316.

Hansen, I., Staaland, T., Ringsø, A., 2002. Patrolling with livestock guard dogs: a potential method to reduce predation on sheep. Acta Agric. Scand. A 52, 43–48.

Horgan, J.E., 2015. Testing the Effectiveness and Cost-efficiency of Livestock Guarding Dogs in Botswana. MSc thesis, Rhodes University, South Africa.

IUCN/SSC, 2007. Regional Conservation Strategy for the Cheetah and African Wild Dog in Eastern Africa. IUCN/SSC, Gland, Switzerland.

IUCN/SSC, 2012. Regional conservation strategy for the cheetah and African wild dog in Western, Central and Northern Africa. IUCN/SSC, Gland, Switzerland.

Klein, R., 2007. Status report for the cheetah in Botswana. Cat News (Special Issue 3), 14–21.

Landry, J.M., Burri, A., Torriani, D., Angst, C., 2005. Livestock guarding dogs: a new experience for Switzerland. Carnivore Damage Prevention News 8, 40–48.

Lester, E.M., Hartmann, A.M., Schmidt-Küntzel, A., 2008. Lingual Squamous Cell Carcinoma in Working Dogs in Namibia; Investigation Into the Incidence, Predisposing Factors, Causes And Treatment Options; Preliminary Results of a Pilot Study. Veterinary Association of Namibia Congress, Namibia, pp. 1–9.

Linnell, J.D.C., Smith, M.E., Odden, J., Kaczensky, P., Swenson, J.E., 1996. Carnivores and sheep farming in Norway. 4. Strategies for the reduction of carnivore-livestock conflicts: a review. NINA Oppdragsmelding.

Lorenz, J.R., Coppinger, L., 1986. Raising and Training a Livestock Guarding Dog (Extension Circular No. 1238). Oregon State University Extension Service, USA.

Lorenz, J.R., Coppinger, R.P., Sutherland, M.R., 1986. Causes and economic effects of mortality in livestock guarding dogs. J. Range. Manage. 39, 293–295.

Manoa, D.O., Mwaura, F., 2016. Predator-proof bomas as a tool in mitigating human-predator conflict in Loitokitok sub-county Amboseli region of Kenya. Nat. Resour. 07, 28–39.

Marker, L.L., 2002. Aspects Of Cheetah (*Acinonyx jubatus*) Biology, Ecology And Conservation Strategies on Namibian Farmlands. PhD thesis, University of Oxford, United Kingdom.

Marker, L.L., Dickman, A.J., Macdonald, D.W., 2005a. Perceived effectiveness of livestock guarding dogs placed on Namibian farms. Range. Ecol. Manage. 58, 329–336.

Marker, L.L., Dickman, A.J., Macdonald, D.W., 2005b. Survivorship and causes of mortality for livestock guarding dogs on Namibian Rangeland. Range. Ecol. Manage. 58, 337–343.

Marker, L.L., Dickman, A.J., Mills, M.G.L., Macdonald, D.W., 2010. Cheetahs and ranchers in Namibia: a case study. In: Macdonald, D.W., Loveridge, A.J. (Eds.), Biology and Conservation of Wild Felids. Oxford University Press, Oxford, United Kingdom, pp. 353–372.

Marker, L.L., Dickman, A.J., Schumann, M., 2005c. Using livestock guarding dogs as a conflict resolution strategy on Namibian farms. Carnivore Damage Prevention News 8, 28–32.

Marker-Kraus, L., Kraus, D., Barnett, D., Hurlbut, S., 1996. Cheetah Survival on Namibian Farmlands. Cheetah Conservation Fund, Windhoek, Namibia.

McManus, J.S., Dickman, A.J., Gaynor, D., Smuts, B.H., Macdonald, D.W., 2015. Dead or alive? Comparing costs and benefits of lethal and non-lethal human–wildlife conflict mitigation on livestock farms. Oryx 49 (4), 687–695.

Olsen, J.W., 1985. Prehistoric dogs in mainland East Asia. In: Olsen, S.J. (Ed.), Origins of the Domestic Dog: The Fossil Record. University of Arizona Press, USA.

Potgieter, G.C., 2011. The Effectiveness of Livestock Guarding Dogs for Livestock Production and Conservation in Namibia. MSc thesis, Nelson Mandela Metropolitan University, South Africa.

Potgieter, G.C., Kerley, G.I.H., Marker, L.L., 2016. More bark than bite? The role of livestock guarding dogs in predator control on Namibian farmlands. Oryx 50 (3), 514–522.

Potgieter, G.C., Marker, L.L., Avenant, N.L., Kerley, G.I.H., 2013. Why Namibian farmers are satisfied with the performance of their livestock guarding dogs. Hum. Dimens. Wildl. 18, 403–415.

Rigg, R., 2001. Livestock Guarding Dogs: Their Current Use World Wide. IUCN/SSC Canid Specialist Group, Occasional Paper No. 1.

Rust, N.A., Whitehouse-Tedd, K.M., MacMillan, D.C., 2013. Perceived efficacy of livestock guarding dogs in South Africa: implications for cheetah conservation. Wildl. Soc. Bull. 37, 690–697.

RWCP and IUCN/SSC, 2015. Regional Conservation Strategy for the Cheetah and African Wild Dog in Southern Africa. RWCP and IUCN/SSC (revised and updated, August 2015).

Scott, J.P., Marston, M.V., 1950. Critical periods affecting the development of normal and mal-adjustive social behavior of puppies. Pedagog. Semin. J. Genet. Psychol. 77 (1), 25–60.

Shivik, J.A., 2004. Non-lethal alternatives for predation management. Sheep Goat Res. J. 19, 64–71.

Shivik, J.A., 2006. Tools for the edge: what's new for conserving carnivores. BioScience 56, 253–259.

Stannard, C., 2006a. Livestock protection dogs: resolving human-wildlife conflict. In: Daly, B., Davies-Mostert, H., Davies-Mostert, W., Evans, S., Friedman, Y., King, N., Snow, T., Stadler, H. (Eds.), Prevention Is the Cure: Proceedings of a Workshop on Holistic Management of Human-Wildlife Conflict in the Agricultural Sector of South Africa. Endangered Wildlife Trust, Johannesburg, South Africa, pp. 43–46.

Stannard, C., 2006b. Breeding Program for Livestock Guarding Dogs. The Grootfontein Agricultural Development Institute, Middelburg, South Africa.

Urbigkit, C., Urbigkit, J., 2010. A review: the use of livestock protection dogs in association with large carnivores in the Rocky Mountains. Sheep Goat Res. J. 25, 1–8.

Van Bommel, L., Johnson, C.N., 2012. Good dog! Using livestock guardian dogs to protect livestock from predators in Australia's extensive grazing systems. Wildl. Res. 39, 220–229.

Improved and Alternative Livelihoods: Links Between Poverty Alleviation, Biodiversity, and Cheetah Conservation

Mary Wykstra*, Guy Combes**, Nick Oguge[†],
Rebecca Klein[‡], Lorraine K. Boast[‡], Alfons W. Mosimane[§],
Laurie Marker[¶]

*Action for Cheetahs in Kenya Project, Nairobi, Kenya
**Guy Combes Studio, Antioch, CA, United States
[†]University of Nairobi, Nairobi, Kenya
[‡]Cheetah Conservation Botswana, Gaborone, Botswana
[§]University of Namibia, Neudamm, Namibia
[¶]Cheetah Conservation Fund, Otjiwarongo, Namibia

INTRODUCTION

A livelihood is a means of securing the necessities of life: food, water, and shelter (Chambers and Conway, 1991). Livelihoods in Africa are often traditional or cultural practices whereby people often raise livestock and/or harvest natural resources (e.g., wild animals and plants harvested for food or medicines, building materials, and energy sources; Roe, 2010) for subsistence needs and income (herein referred to as traditional livelihoods). In rural sub-Saharan Africa the majority of people support themselves and their families through these traditional livelihoods, primarily working as subsistence or small-scale farmers. Fourty percent of these rural residents live below the poverty line (less than US $1.25 per day; UN, 2015). They share the land with wildlife species, such as the cheetah (Table 16.1), but competition and conflict over resources can pose direct and indirect threats to both people and wildlife (Chapter 13).

Cheetah survival is particularly threatened from unsustainable land uses and conflict with humans because 77% of the land supporting resident cheetah populations is located outside of protected areas (Chapters 4 and 39). Indeed, one of the primary threats to the cheetah's survival is habitat loss and fragmentation caused by the growing human population (Chapter 10). Habitat loss impacts cheetah populations directly, as well as indirectly as it results in a decline in biodiversity and prey loss (Chapter 11), which in turn compounds human-cheetah conflict (Chapter 13). Improved management of human-dominated landscapes, is critical to cheetah survival and conservation tools must look beyond the biology of the cheetah and deeper into the relationship between human poverty alleviation and biodiversity conservation throughout the cheetah's range.

Poverty alleviation aims to improve the quality of life for people currently living in poverty. To permanently lift people out of poverty, both economic and humanitarian measures, use two primary links between human livelihoods and biodiversity conservation: (1) biodiversity as a means of providing income or input into livelihoods, or (2) biodiversity as an insurance to buffer livelihoods in future times of need (Roe, 2010). Biodiversity loss, therefore negatively impacts human well being in addition to the environment. For example, overgrazing and bush encroachment negatively affect the carrying capacity for cheetahs and their prey, but also seriously threaten livestock grazing areas resulting in decreased profits from livestock farming (Marker et al., 2010). In the past, most documented studies on the link between poverty alleviation and biodiversity conservation have focused on forest products or related resource use. However, there is an increasing emphasis on rangeland ecosystems (Ives et al., 2005), particularly pertaining to human–predator–prey interactions.

Programs in areas of high biodiversity, which use natural resource management to link conservation with socioeconomic interests on both local and international scales, are known as Integrated Conservation and Development Programs (ICDPs) (Silva and Khatiwada, 2014). ICDPs were introduced to the developing world in the mid-1960s (Silva and Khatiwada, 2014), and formally defined in the 1990s with an aim to improve or provide alternative income sources to poor communities (Davies et al., 2013). In 2011, it was estimated that the world spent

TABLE 16.1   Cheetah Range, Nationally Protected Areas, and Poverty Index in Four Cheetah Range Countries With Improved and Alternative Livelihood Programs Targeting Cheetah Conservation

Country	Total land area (ha)	Protected area[a] (percentage of total land area) (%)	Cheetah range[b] (percentage of total land area) (%)	Poverty index[c] (year of data) (%)
Botswana	60,037,000	29.20	95	19.3 (2010)
Kenya	58,265,000	12.40	48	45.9 (2005)
Namibia	82,329,000	37.90	60	28.7 (2009)
South Africa	121,309,000	8.90	17	53.8 (2010)

[a] Protected area refers to a land mass of 1000 ha or more under total or partial protection as designated by national authorities.
[b] Percentage of land mass inhabited by cheetahs uses the resident, possible, and connected land as identified in regional strategies and used by IUCN for population estimates published in 2015 (Durant et al., 2015).
[c] The poverty index is the percentage of the population under international poverty lines, also known as the headcount ratio (Kinyua et al., 2000; World Bank 2015).

US $126 billion in annual aid addressing global poverty and an additional US $8–12 billion on issues of biodiversity loss (Roe et al., 2011) without significant success. The combined aims are highlighted by the United Nations' "Sustainable Development Goals" (Fig. 16.1), which include goals to alleviate human poverty (#1), reduce marine (#14), and terrestrial (#15) biodiversity loss and address climate change (#13) (UN, 2015). Integrated programs that address biodiversity loss and poverty alleviation can impact cheetah conservation by reducing threats to the species from habitat degradation, biodiversity loss, and human-wildlife conflict. Similarly, the "One-Health" vision promoted by the World Health Organization, links biodiversity conservation, environmental stability, and human health (Mazet et al., 2009), and is providing

support for projects like cheetah programs that incorporate human livelihoods with species-specific research (Okello, 2015).

Improving the economic situation (i.e., livelihoods) of individuals is likely to increase tolerance toward cheetahs and facilitate cheetah–human coexistence. Economic costs of sharing the land with cheetahs are either direct through depredation of livestock or indirect through additional costs associated with protecting livestock from predators (Chapter 13). These costs can have a significant impact on household livelihoods, and are one of the drivers of human-cheetah conflict, although they can sometimes be perceived as worse than they are in reality. Improving livelihoods of individuals living with cheetahs is achieved when income is directly increased or losses of income (e.g., due to

FIGURE 16.1   **Sustainable development goals provide leadership and catalyze actions under the United Nations Division for Sustainable Development (DSD).** The 17 Sustainable Development Goals (SDGs) promote and coordinate implementation of internationally agreed development goals adopted by Heads of State and Governments in September 2015. *Source: Printed with permission from UN SDGs.*

predation) are reduced, enabling individuals to better support their families. Additionally, by diversifying livelihoods outside of the cultural or traditional sources of income, rural citizens can achieve more economical viability and reduce their reliance on livestock farming alone (herein these programs are referred to as alternative livelihood projects; Write et al., 2015). Programs that provide financial income from the sustainable use of natural resources are likely to promote coexistence beyond that achieved by the financial incentive alone, as individuals may be less likely to view wildlife species as a threat to their existence if they are able to profit directly from them.

Targeted poverty alleviation in connection with biodiversity conservation is therefore a significant aspect of cheetah conservation strategies. Many cheetah conservation organizations have instigated or promoted programs that have the aim of reducing poverty by improving traditional livelihoods (primarily livestock farming) or by promoting alternative income streams. These programs, many of which are still in the implementation phase, can be applied across the matrix of private and communally owned land upon which cheetahs occur.

This chapter describes examples of improved and alternative livelihood programs that impact cheetah conservation and discusses the development of such programs. There is only a small amount of literature documenting cheetah-conservation related community projects and their successes. Therefore, the majority of information in this chapter comes from internal reports and personal communication.

IMPROVED LIVELIHOODS

Many of the cheetah-conservation organizations in Africa, including Action for Cheetahs in Kenya (ACK), Cheetah Conservation Botswana (CCB), Cheetah Conservation Fund (CCF) in Namibia, and Cheetah Outreach (CO) in South

Africa provide opportunities for rural farmers to improve their livelihood while educating individuals about cheetahs and ecosystem conservation.

Livestock Management

Healthy livestock herds in appropriate numbers for the environmental conditions are less likely to be targeted by predators (Marker and Dickman, 2004; Ogada et al., 2003) and carry greater overall economic value. Integrated programs developed by cheetah conservation organizations have aimed at incorporating indigenous knowledge and culture with modern concepts that improve livestock care, increase productivity, and reduce losses to predators. Materials and advice on improved conflict mitigation and livestock husbandry are disseminated to rural communities through educational programs and are successfully raising awareness, leading to positive behavior change (Chapters 13 and 18). Evaluation surveys of community outreach activities in the Ghanzi district of Botswana, in 2015 and 2016, indicated that 83% of participants in farmer training workshops ($n = 108$) made direct improvements to livestock management and of those, 89% had seen a decrease in livestock losses (CCB, unpublished data). Similarly, 94% of conservancy members surveyed in Namibia after attending training workshops ($n = 85$) reported that the health of their livestock had improved, and 70% of respondents ($n = 63$) reported that they were using the provided resources to implement best practices in livestock husbandry (CCF, unpublished data).

In addition to the general programs, cheetah organizations run specific programs targeting livestock health and protection from predators. A "Build a Better Boma" campaign conducted by ACK provides information (posters) and physical assistance to improve livestock enclosures (also known as bomas) to reduce the opportunity for predator attacks on livestock in the enclosure. From January 2013 to December 2016, 1000 posters were distributed and field staff assisted in building over 300 bomas, with 98% of recipients reporting to have had zero predator attacks in the boma since receiving assistance (ACK, unpublished data). Similarly, to reduce depredation, CCF, CO, and CCB have all implemented livestock guarding dog programs, collectively placing nearly 1000 livestock guarding dogs on farms and assisting with veterinary care (Chapter 15).

To improve livestock health, in 2004, ACK restored four community-managed dips (cement ponds which are filled with diluted tick repellent to prevent tick infestation of livestock and dogs), which served about 400 households, each with 3–65 head of livestock. Tick infestation causes livestock death and lack of production accounting for a loss of over US $364 million annually in eastern Africa (Ragwa, 2012). As part of the cattle dip restoration project, ACK conducted seminars on livestock health and dip management to improve community capacity in livestock management. Out of 247 interviews conducted at the dips between 2007 and 2009, 49% of the respondents reported attending the dips weekly (ACK, unpublished data). All of those attending with regularity reported increased milk production and livestock weight, and decreased veterinary costs related to tick borne diseases. After ACK's support ended in 2007, the community managers continued the project as a business and, as of 2016, three of the four dips continue thriving as a source of community income while improving livestock management for small-scale farmers.

Programs have observed that improved husbandry management increases income, by reducing losses and improving herd health and livestock productivity (i.e., milk/calf production), and simultaneously can improve tolerance for the presence of predators through increased awareness. However, the economic impact of integrated management and reduced depredation for local communities was not evaluated at a household level and remains an important area for further study.

Ecocertification of Meat and Dairy Products

Ecocertification programs follow recent trends of environmentally conscious consumers becoming aware of food origins and the ecological impacts of food production (Aquino and Falk, 2001). Businesses use a marketable label assuring consumers that their products comply with environmental and social standards. This empowers consumers, sometimes in distant markets, to make environmentally responsible purchases; giving them confidence in the accredited environmental impact of the product (Stein, personal communication). For example, "Predator-friendly" certification of meat and dairy products increases consumer confidence that predator species were not harmed during production; this concept is applied by the "Dingo for Biodiversity Project" in Australia (DBP, 2016) and "Wolf-friendly" beef in the USA (AWI, 2017; WFEN, 2015). Certified farmers commit to using nonlethal conflict mitigation tools; for example, guarding animals or deterrent systems (lights, electric fences, sound boxes)—to reduce predation on livestock. Ecocertification aims to give the farmers a premium for their product (Lewis and Alpert, 1997). Because consumers often expect environmentally friendly products to be competitive on price, the premium is not always reflected as a per unit price increase. Instead the benefits come in indirect ways, such as increased demand through access to new markets or consumer loyalty over the long term (Stein, personal communication). Although the predator-friendly concept does not always generate a premium, many of the dedicated members participate because of their philosophical and ethical commitments to a healthy environment (Early, 2012).

CCF's "Cheetah Country Beef" business plan predicts an increase in profits for ecocertified farmers (Bell, 2006; Marker et al., 2010) through price premiums and other associated benefits. Farmers must abide by guidelines for predator-friendly farming practices, including nonlethal cheetah conflict mitigation to maintain third party certification. Developing a market and the associated premiums for producers can be time-consuming, costly, and technically challenging (Early, 2012; Treves and Jones, 2010) and as of yet predator-friendly meat products originating from within the cheetah's range are not being sold on a commercial scale.

Predator-friendly labeled cheese, ice cream, soap, and fudge made from goat milk are sold in Namibia at the CCF facility and through national outlets. The dairy products are produced on CCF's model farm as part of an initiative to develop alternative income sources. CCF's *Dancing Goat Creamery* label is branded with the Wildlife Friendly Enterprise Network (WFEN) trademark. WFEN is a "global community dedicated to the development and marketing of products that conserve threatened wildlife while contributing to the economic vitality of rural communities" (WFEN, 2015). The Wildlife Friendly label is a certified ecolabel that denotes enterprises meeting the highest standards of integrated farming practices. The goat creamery has generated employment for five women in the local community. The United Nations (UN) documented that on average women use 90% of their income to benefit family and community (GACC, 2012), thus promotion of women into income positions aids in poverty alleviation. Since women are actively involved in livelihood decision-making (Roberts, 1996), assuring the link between economic income and cheetah conservation is essential.

ALTERNATIVE LIVELIHOODS

In the context of cheetah conservation, alternative livelihood programs generally provide an additional source of income outside of the realm of livestock farming. Natural resource management can be a valuable aspect of alternative livelihood projects as it encourages people to see an economic value in the biodiversity around them. Integrated wildlife farming with livestock agriculture can increase economic and environmental stability. Other alternative livelihood programs, which benefit cheetah conservation, include the production of fuel logs, cooking stoves and the promotion, and sale of handcraft products.

Wildlife Farming, Tourism, and Hunting

Wildlife farming, photographic-tourism, and game hunting can provide financial returns on public and private land, especially in areas where wildlife numbers or landscape features are sufficient to attract clients. In South Africa, tourists spend US $90 billion in combined domestic and international tourism spread across several provinces (SAT, 2015). Similarly, in Kenya, the tourism industry supplies over 700,000 jobs (direct, indirect, and induced) and contributes US $6.4 billion to Kenya's Gross Domestic Product (Turner, 2015). However, revenues from national tourism do not always reach the population or communities who are coexisting with wildlife, and who may have had livestock losses due to predation (Roe, 2010). Additionally, negative effects of land designation for tourism may result in eviction of inhabitants from land with natural resources and exclude historical land use options like grazing or water and firewood collection (Norton-Griffiths and Southey, 1995). These drawbacks have historically exacerbated poverty or have been viewed as a threat to human rights, especially pertaining to water access (Brockington, 2002).

Opportunities for communities to directly manage wildlife tourism and game hunting operations for improved income distribution, has increased through the development of conservancies (Chapter 17). These conservancies are either community or private owned and consist of farms or tribal lands utilizing a unified management plan. Larger landscape management reduces costs and makes wildlife farming (often in conjunction with livestock farming) a more viable option. Wild game species can be utilized for photographic tourism, hunting, or meat, potentially generating employment opportunities and increased profits for conservancy members. In 2013, an evaluation of income from Namibian conservancies revealed that 47% of the overall conservancy income was generated from game use (hunting, meat sales, and live sale), 48% from tourism (joint-venture, community conservancies, and craft sales), and 5% from plants and other interests (Denker, 2015). At the end of 2014, the 83 communal conservancies in Namibia were employing approximately 1700 full time employees, 400 part time employees, plus 2100 indigenous plant product harvesters, and 760 craft producers. The total cash income and inkind benefits to rural communities was estimated to be over US $6.9 million (Denker, 2015). However, the impact of tourism income on poverty alleviation and alternative livelihood generation depends heavily on local conditions and management policies, particularly in regards to revenue sharing (Archibald and Naughton, 2001).

Even when community based tourism or hunting projects have successfully increased employment opportunities and raised household income (for at least some community members), people living in the area do not always acknowledge such benefits due to negative perceptions of predators or the affiliated organizations (Silva and Khatiwada, 2014). Additional challenges include removing barriers to private sector investment in communal areas, developing revenue streams in areas with low tourism potential and improving the quality of community-run tourism enterprises (Denker, 2015).

Large carnivores are a particular attractant to both tourists and hunters. Tourists ranked leopard (*Panthera pardus*), lion (*Panthera leo*), and cheetah as the species that they most wanted to see, and expressed a willingness to pay extra fees for places known to have high probabilities of seeing big cats in the wild (Hazzah, 2006; Lindsey et al., 2007; Stein, 2008). Although carnivore hunting is a controversial conservation management option, it has the potential to generate substantial income and community employment (Muposhi et al., 2016), which could lead to improved tolerance toward carnivores. Trophy hunting of cheetahs is only permitted by international and national law in Namibia and Zimbabwe. It is carefully regulated by CITES and the respective governments, and is not thought to have had a negative impact on cheetah populations (Chapter 21). While the presence of cheetahs (e.g., through tourism and hunting) may contribute to the overall income in a country, the distribution of benefits to rural communities or the awareness of such benefits can be improved.

Alternatives to Charcoal—Fuel Logs and Cooking Stoves

Bush encroachment—the increased growth and invasion of woody bushes at the detriment of grasses—is a widespread problem across livestock farming areas in sub-Saharan Africa. It is primarily caused by overgrazing by livestock in arid ecosystems coupled with a reduction of large, wild browsers [i.e., elephant (*Loxodata* spp.), black rhinoceros (*Diceros bicornis*) and giraffe (*Giraffa* spp.)]. Bush encroachment reduces the productivity of livestock grazing areas, negatively affecting both human livelihoods and biodiversity (Marker et al., 2010). In Namibia, CCF launched a research project into the feasibility of harvesting and processing bush, and manufacturing and marketing ecofuel logs made from the compressed branches, under the trade name "Bushblok" (CCF, 2015). The ecofriendly fuel log has a similar energy value (4600 cal/g) as good-quality wood and is a clean burning, low emission home heating, and cooking energy fuel source that could replace charcoal. Feasibility studies determined that approximately, 10 metric tons of excess woody biomass were available per hectare from thickened bush.

CCF's Business plans estimate one production plant could process about 5000 tons of woody biomass per year, and could thereby, clear large tracts of land for human, livestock, and wildlife use. The "Bushblok" product has a dual certification with the Wildlife Friendly Environmental Network (WFEN, 2015), and the Forest Stewardship Council (FSC), the highest standard achieved in forest certification. As of 2016, the project employs 30 harvesters and factory workers, it is building incountry capacity, and has the potential to locally increase biodiversity and restore cheetah habitat.

A related project in Kenya is the Improved Cooking Stoves Program started in 2011 by Sustainable Community Development Services and the Global Alliance for Clean Cookstoves (GACC) under UN guidance/financing. In Kenya, over 85% of the population relies on traditional fuels, such as wood, charcoal, dung, and agricultural residues for cooking and heating (KME, 2011). Despite alternative fuel sources, firewood remains the predominant fuel for cooking. The nonsustainable harvesting of plant species strongly contributes to land degradation, accelerating the loss of cheetah habitat. Traditional use of biomass as fuel is often inefficient, with an energy conversion efficiency of only 2%–20%, and can result in indoor air pollution that adversely affects human health (UN, 2012). Clean cook stoves evolved dramatically over the past decade, and modern bioenergy technologies offer energy conversion efficiencies of up to 90% (Haines et al., 2007), lowering the quantity of wood harvested and reducing harmful emissions (GACC, 2012; UN, 2012). In Kenya, the Improved Cooking Stoves Program has distributed or installed more than 10,000 stoves between 2011 and 2014, thus reducing firewood consumption by approximately 60%. Potential production is estimated at 11,000 additional stoves annually with a profit for employees comparable to wages in rural areas. Since 2011, 13 women's groups (200 people) were trained and participated in business management training to make stoves

in the Rural Stoves West Kenya Project, creating employment opportunities and financial independence (HIVOS, 2014). As a result, the women have gained status, self-confidence, and financial independence (Holm, 2005). Many cheetah range areas could greatly benefit from expansion of the Clean Stove Program. ACK is exploring the program specifically at sites in the Ngare Mara region of the Samburu district in northern Kenya and the Salama region of Kenya's Makueni district as a means of habitat conservation and alternative financial incentive.

Local Artisans, Traditional Handcrafts, and Conservation

The cheetah has been depicted in art throughout the millennia (Chapter 2). Still today, local art in cheetah range countries is expressive in character. The traditional art of beading and craft making is common throughout much of Africa and the cheetah can be seen in traditional batiks, carvings, beadwork, and embroidery. Traditional items, such as jewelry, decorative gourds, and spears are highly sought after as souvenirs and decorations, providing potential to generate substantial sales through the craft markets. For example, in Namibia, 2% of income on surveyed commercial and communal conservancies is derived from craft sales (Denker, 2015). Several cheetah projects (e.g., ACK, CCB, CCF, and CO) work with the art and craft industry to market hand-crafted items and assist communities in designing cheetah specific items. For example, CCF supports artisans making crafts using traditional makalani nuts, leather, embroidery, and beadwork and promotes these products under the Wildlife Friendly Cheetah Country Craft brand, to local and international marketplaces.

A similar project is ACK's COOL (Cheetah, Owl, Otter, and Lion) Craft project (Fig. 16.2A). Initiated in 2010, the project annually purchases a minimum of US $8000 of species themed handcrafts from nine community groups. The COOL project promotes quality workmanship

FIGURE 16.2 **Cheetah branded products show the level of community support, as well as the link to the cheetah in their designs and marketing of improved and alternative livelihood items.** The branding appeals to the buyer and represents the value of the product in conservation. Artisans and farmers are proud to see their products in the branded line (A) Action for Cheetahs in Kenya (ACK) branded products are called COOL Crafts in partnership with other species conservation (i.e., Cheetah, Otter, Owl, and Lion). (B) Cheetah Conservation Fund (CCF) branded products hold Forest Stewardship Council and Wildlife Friendly certifications. *Source: Part A, ACK; part B, CCF.*

marketing traditional, and new designs that utilize recycled and renewable resources, to an overseas market. Due to internal market saturation and an influx of cheap, mass-produced items from overseas factories resembling locally made crafts, traditional artisan skills are often viewed as unprofitable (Wykstra, personal observation). As a result, young people are increasingly drawn into illegal activities, such as prostitution, poaching, unauthorized sand harvesting, and charcoal burning (cutting trees to produce charcoal). ACK provides seminars on business management and environmental friendly livelihoods and responsibility, along with assisting the community groups to find lucrative local markets. By increasing the profitability of craft making and emphasizing the link between sustainable natural resource management and livelihoods, ACK is enticing younger generations to embrace the art of handcrafts and hopefully also to value wildlife like the cheetah. ACK's COOL project sales increased from US $12,040 in 2015 to $17,005 in 2016, thus building the foundation for project expansion.

The Painted Dog Project in Zimbabwe addresses issues of youth, poaching, and predator conservation by producing "snare art" made from poachers' snares under the *Iganyana Arts* brand (PDC, 2016). Similar snare art is marketed by the "Catching Hope Re-Purposed Poaching Snare Program" run by the Henry Vilas Zoo in partnership with the IUCN Saola (antelope) Working Group based in Laos and Vietnam (Flynn, personal communication). Both projects turn a stark symbol of wildlife destruction into something that benefits conservation and offers an alternative source of income (to poaching among others). An antipoaching measure is achieved through the use of removed wire snares, and the snare art brings immediate financial benefit to communities that are provided with materials, equipment, a place to work, and training. Through such projects, tourists, retail purchasers, and local residents learn about the target species and their fight for survival.

Involvement in artisan groups establishes relationships between conservation organizations and the community, particularly nonfarming groups. Sales of community-based crafts provide an added livelihood to household income, thus reducing the dependency on the competitive farming market and promoting biodiversity conservation. Conservation organizations can help build the capacity to produce quality products in sufficient quantity and link them to sales outlets. In the past, gift shops required high-turnover value for the products sold, but increasingly, individual zoo and aquarium souvenir shops promote conservation projects and their community-based one-of-a-kind craft products despite lower profits. Additionally, many handcraft programs use online retail sites and zoo-related conferences to increase sales.

No studies of household impact from community craft sales have been conducted specifically on crafts produced in cheetah study sites. However, the Snow Leopard Enterprise Program, which works with women to produce and sell Wildlife Friendly (WFEN) certified handwoven crafts (specifically representing the relationship between snow leopards and livestock) documents a 40% increase in household income to families participating in the program, along with decreased livestock loss to depredation (Jackson and Lama, 2016; SLT, 2016).

Product Branding to Maximize Conservation Impact

The conservation impact of many of these alternative livelihood programs is not just through poverty alleviation but also through awareness raising of predator conservation to both communities and product consumers. Telling a story about the product will make the consumer aware of the bigger issues and inspired by their purchases and the companies from whom they buy (Stein, personal communication). Product branding is important in relaying that message and ensuring the buyer is fully aware of the conservation value of their purchase in order to maximize the return to the project sponsors, beneficiary households, and communities. Bushblok and Dancing Goat Creamery, for example, are well thought out businesses that carry their own creative logo and ecocertification labels while also having the CCF logo to increase the link between the product and cheetah conservation (Fig. 16.2B). Bushblok in particular has captured the imagination of major grocery retailers in the United States (Stein, personal communication), with the potential to introduce sales there in the future. ACK's COOL Crafts project (Fig. 16.2A) is also an example of a branded program where promoting a sound business and community empowerment plan incorporates livelihood improvement actions into cheetah conservation in Kenya.

DEVELOPMENT OF LIVELIHOOD PROGRAMS

The launch of a new livelihood program can be a challenging task because community members are often hesitant to reconsider old methods or try new practices, especially when old practices are deep-seated in local cultures (Dickman, 2008). Conservation scientists tend to focus on the biological indicators of success and often fail to set up appropriate social or economic indicators. This oversight contributes to the public perception that conservationists care more about animals than they do about the people who use the same land as wildlife (Goldman et al., 2013), and can lead to a degree of mistrust toward initiatives proposed by conservationists.

Conservation Measures Partnership (CMP) was developed in 2004 by a group of prominent conservation NGO's as a means to design, manage, and measure impacts of conservation actions using an "Open Standards" approach in stakeholder participation (CMP, 2012). CMP used successful models in conservation (including threat reduction assessment of biological conservation success; Margoluis and Salafsky, 2001), public health, family planning, international development, social services, education, and business to determine common concepts and approaches. The basic steps in CMP are widely used in project-based programs to achieve clearly defined goals in program design, implementation, monitoring, and evaluation that are hinged on a policy science approach that stems from philosophies of John Dewey (1910) and Harold Laswell (1948) using a social decision making process.

Program Implementation

Elements of the decision making process necessary during the implementation of livelihood projects are intelligence, promotion, and prescription (Table 16.2). Gathering information (*intelligence*) to address the status and threats to cheetahs lays the foundation for species conservation, and is a starting point in identifying a meaningful project within the community. Intelligence also includes identifying meaningful participation by all stakeholders—from administrators (government, NGOs, conservationists, etc.) to land owners to herders—that is crucial in considering which livelihood projects will serve shared common interests that benefit the species and the community (Jackson and Lama, 2016). For a project to be successful, the stakeholders must understand and promote a clear link between a conservation target (e.g., the cheetah), and the economic benefits of the projects implemented by the conservation organization.

It is also important that during the *promotional* element of the decision-making process, stakeholders are given the opportunity to discuss livelihood threats and related solutions (Table 16.2). Stakeholder meetings provide opportunities to encourage support and collaboration for the introduction and development of an alternative or improved livelihood program. For programs to be effective key human values of enlightenment—wealth, power, respect, well-being, skill, affection, and rectitude—should be considered (Lasswell, 1971). During the *prescription* element a livelihood program is selected, the project plan is formalized and the planning document is developed and approved by stakeholders, including clear measures for evaluation (indicators for success) in both the short and long term. Conservation scientists tend to focus on the biological indicators and often fail to set up appropriate social or economic indicators (Child et al., 2012), even when biodiversity and human livelihoods are so critically linked. These indicators should be developed and endorsed through an equitable participatory planning process with community members to reduce potential future

TABLE 16.2   Elements of the Decision Process Used During the Development, Implementation, and Appraisal of Alternative and Improved Livelihood Projects

Elements	Definitions	Actions
IMPLEMENTATION		
Intelligence	Research dependable information relevant to the project area	• Identify key stakeholders. • Understand community needs and current livelihood dependencies. • Understand relevant cultural aspects of community including the value of biodiversity and the cheetah's role in the ecosystem. • Analyze previous livelihood projects and outcomes.
Promotion	Discussion of problems and potential livelihood projects with relevant stakeholders	• Workshops involving international experts, community members, wildlife authorities, and other stakeholders. • Open debate of problems and potential livelihood projects. • Clarify expectations of all stakeholders. • Develop a media plan.
Prescription	Formalize the project plan and develop the planning document	• Develop management document for project; it should include clear actions and timeline. • Obtain support and endorsements for planning document from various stakeholders.
Invocation	Implementation of the project	• Actions produce results that are timely and dependable.
MONITORING AND EVALUATION		
Appraisal	Assessment of the project	• Conduct timely, regular assessment of the project's actions and results. Appraisal should be independently conducted, ongoing, and the findings acted upon to develop and improve the project. • Review intelligence, promotion and prescription based on findings.
Termination	Ending the project or moving to a new phase	• Repeal or large-scale adjustment of prescription.

Adapted from: Clark, 2002.

disappointments if community members' expectations are not met. ACK conducted community meetings and stakeholder forums providing opportunities for sharing ideas that led to the cattle dip project, for example, whereby the community put forward several projects with which ACK could assist. A common goal was evaluated—*promoting a healthy ecosystem through tick control*—(*promotional element*) and agreements were made between ACK and the community as to the roles each would play in construction, management, data collection, and data distribution within the community (*prescription element*).

Upon agreement, the program can be enacted (*invocated*) within the agreed timelines and conditions (Table 16.2). The frequency of meetings and reviews during invocation depend on the guidelines set by the project team.

Monitoring and Evaluation

Conservation-driven poverty alleviation projects often fail somewhere between invocation and *appraisal*, as disputes arise when common interests are not adequately recognized or benefits to community members are not acknowledged quickly enough (Clark, 2002). Appraisal of the cattle dip project included interviews during and after the implementation of the project. Out of the 374 conflict investigations conducted in the Salama area between 2006 and 2014, 82% of the interviews reported that one of the valuable aspects of the cheetah is the implementation of the cattle dips, and associated training seminars in the community (ACK, unpublished data) showing that the community understood the link between the project and the survival of the cheetah in this region. Additionally, 100% of the members who regularly attended (once per week) the dipping sessions reported to have zero incidents of tick related disease mortality in their herd exceeding the expectation of the community members.

Cheetah organizations are burdened with the problematic issue of human–predator conflict (Chapter 13). As a consequence, community members' receptiveness to conservation organizations and livelihood programs can be reduced, especially during project promotion. Livestock losses to predators can reduce household income, potentially to levels above that which can be offset by traditional livelihood programs. Donor investment along with technical support for driving a new project's start-up is, therefore vital during the first difficult stages until projects become established and self-sustainable. Both government and donor support enabled the implementation of the cattle dip project, which intended to reduce donor dependence to zero by the end of the 3-year plan. Project management seminars during the 3-year program enabled the dip management committees to sustain funding through product sales—that is, the cost of dipping charged to the clients (members). The three dips that are still functioning have implemented product sale increases based on the business skills learned in the seminars, thus enabling them to maintain their business to date, 10 years after implementation.

Programs should conduct rigorous monitoring and regular reporting to community members and all relevant stakeholders, focusing on the indicators of success defined in the planning stages. The use of branding, strong conservation messaging, and continuous outreach programs must clearly promote the link between the environment and community livelihoods. Signage at each cattle dip clearly states the partnership of ACK and the dip management in establishing the facility, and announcement postings include the ACK logo as an endorsement of policy changes.

No studies of the economic impact on households from individual alternative or improved livelihood programs have been conducted in cheetah study sites. However, CCF measured the joint economic output of its programs (including the *International Research and Education Centre*, the *Bushblok Project*, *Dancing Goat Dairy*, and *Cheetah Country Crafts*) to Namibia due both to on-site spending and to off-site spending throughout Namibia by visitors coming to CCF. The principal finding revealed that the overall economic impact of CCF on Namibia in 2014 was approximately US $10 million (Humphreys, 2016).

CONCLUSIONS

Cheetah conservation related improved and alternative livelihood programs, such as those discussed in this chapter, have the potential to be viable solutions that can boost income levels for community members and help build positive attitudes toward cheetah conservation. However, cheetah conservationists are now challenged with scaling up their economic models to reach a greater number of communities. Such expansion will require the empowerment of land-owners and communities to manage and plan for sustainable natural resource conservation.

Conservation efforts associated with community development strive to achieve strong, visible relationships between their achievements in biodiversity conservation and poverty alleviation. Although difficult to quantify, the economic value of cheetah conservation efforts through improved and alternative livelihoods needs to be made available to policy makers, donors, and the community. Strengthening the promotion, monitoring and evaluation of existing and all future projects will therefore be the key in embedding cheetah conservation action into the homes of the people who live with the cheetah.

References

World Bank, 2015. Terrestrial Protected Areas, World Bank. Available from: http://www.worldbank.org/.

Aquino, H.L., Falk, C.L., 2001. A case study in the marketing of "Wolf-Friendly" beef. Rev. Agric. Econ. 23, 524–537.

Archibald, K., Naughton, L., 2001. Tourism revenue-sharing around national parks in Western Uganda: early efforts to identify and reward local communities. Found. Environ. Conserv. 28, 135–149.

AWI, 2013. New Certifications Recognize Farmers and Ranchers who Make Peace With Predators and Other Wildlife, Animal Welfare Network. AWI Quarterly, Spring. Available from: https://awionline.org/awi-quarterly/2013-spring/new-certifications-recognize-farmers-and-ranchers-who-make-peace-predators.

Bell, D., 2006. Cheetah Country Beef Business Plan. MeatCo, Windhoek, Namibia.

Brockington, D., 2002. Fortress Conservation: The Preservation of the Mkomazi Game Reserve, Tanzania. Indiana U Press, North America.

CCF, 2015, Cheetah Conservation Fund, Available from: http://www.bushblok.com/about.

Chambers, R., Conway, G., 1991. Sustainable Rural Livelihoods: Practical Concepts for the 21st Century. Institute of Development Studies Discussion paper 296.

Child, B., Musengezi, J., Parent, G., Child, G., 2012. The economics and institutional economics of wildlife on private land in Africa. Pastoralism 2, 32.

Clark, T.W., 2002. The policy process: A Practical Guide for Natural Resource Professionals. Yale University Press, New Haven, CT, USA.

CMP, 2012. Conservation Measures Partnership. Available from: www.conservationmeasures.org.

Davies, T.E., Fazey, I.R.A., Cresswell, W., Pettorelli, N., 2013. Missing the trees for the wood: why we are failing to see success in pro-poor conservation. Anim. Conserv. 17, 303–312.

DBP, 2016. Dingo for Biodiversity Project—Predator Friendly Network. Available from: http://www.dingobiodiversity.com/predator-friendly.html.

Denker, H., 2015. Impact of hunting bans on communal conservancies Namibia. Internal Report., NASCO, Windhoek, Namibia.

Dickman, A.J., 2008. Key Determinants of Conflict Between People and Wildlife, Particularly Large Carnivores, Around Ruaha National Park, Tanzania. PhD thesis, University College London, United Kingdom.

Durant, S., Mitchell, N., Ipavec A., Groom, R., 2015. Acinonyx jubatus. The IUCN Red List of Threatened Species 2015. e.T219A50649567. Available from: http://dx.doi.org/10.2305/IUCN.UK.2015-4.RLTS.T219A50649567.en.

Early, M., 2012. An Enquiry into Eco-labeling: The Promise of Predator Friendly Certification (PFC). MSc thesis, University of Montana, United States.

GACC, 2012. Results Report: Sharing Progress on the Path to Adoption of Clean Cooking Solutions, Global Alliance for Clean Cookstoves. Available from: http://cleancookstoves.org/resources_files/results-report-2012.pdf.

Goldman, M.J., de Pinho, J.R., Perry, J., 2013. Beyond ritual and economics: Maasai lion hunting and conservation politics. Oryx 47, 490–500.

Haines, A., Smith, K.R., Anderson, D., Epstein, P.R., McMichael, A.J., Roberts, I., Wilkinson, P., Woodcock, J., Woods, J., 2007. Policies for accelerating access to clean energy, improving health, advancing development, and mitigating climate change. Lancet 370, 1264–1281.

Hazzah, L.N., 2006. Living Among Lions (Panthera leo): Coexistence or Killing? Community Attitudes Towards Conservation Initiatives and the Motivations Behind Lion Killing in Kenyan Maasailand. MSc thesis, University of Wisconsin-Madison, United States.

HIVOS, 2014. The Improved Cook Stoves for Households and Institutions Project. Available from: https://east-africa.hivos.org/news/womens-participation-and-empowerment-renewable-energy-projects.

Holm, D., 2005. Renewable Energy Future for the Developing World. International solar Energy Society, Frieburg, Germay.

Humphreys, F.T., 2016. The Economic Impact of the Cheetah Conservation Fund on the Nation of Namibia (Redone for 2014). Cheetah Conservation Fund, Internal Report. Otjiwarongo, Namibia.

Ives, A.R., Cardinale, B.J., Snyder, W.E., 2005. A synthesis of subdisciplines: predator–prey interactions, and biodiversity and ecosystem functioning. Ecol. Lett. 8, 102–116.

Jackson, R.M., Lama, W.B., 2016. The role of mountain communities in snow leopard conservation. In: McCarthy, T., Mallon, D. (Eds.), Snow Leopards. Elsevier Inc., London, pp. 139–148.

Kinyua, P.I.D., Cornelis van Kooten, G., Bulte, E.H., 2000. African wildlife policy: protecting wildlife herbivores on private game ranches. Eur. Rev. Agric. Econ. 27, 227–244.

KME, 2011. Official Website, Government of Kenya, Kenya Ministry of Energy, Republic of Kenya. Available from: www.energy.go.ke.

Lasswell, H.D., 1971. A Pre-View of the Policy Sciences. American Elsevier, New York.

Lewis, D., Alpert, P., 1997. Trophy hunting and wildlife conservation in Zambia. Conserv. Biol. 11, 59–68.

Lindsey, P.A., Alexander, R., Mills, M.G.L., Romañach, S., Woodroffe, R., 2007. Wildlife viewing preferences of visitors to protected areas in South Africa: implications for the role of ecotourism in conservation. J. Ecotour. 6, 19–33.

Margoluis, R., Salafsky, N., 2001. Is our project succeeding? A guide to Threat Reduction Assessment for conservation. Biodiversity Support Program, Washington, DC.

Marker, L., Dickman, A., 2004. Human aspects of cheetah conservation: lessons learned from the Namibian farmlands. Hum. Dimens. Wildl. 9, 297–305.

Marker, L., Dickman, A.J., Mills, G., Macdonald, D., 2010. Cheetahs and Ranches in Namibia: A Case Study. In: Macdonald, D.W., Loveridge, J. (Ed.), Biology and Conservation of Wild Felids. Oxford University Press, London, pp. 353–373 (Chapter 15).

Mazet, J., Clifford, D., Coppololo, P., Deolalikar, A., Erickson, J., Kazwala, R., 2009. A "One Health" approach to address emerging zoonoses: the HALO project in Tanzania. PLoS Med. 6, 1–6.

Muposhi, V., Gandiwa, E., Bartels, P., Makuza, S.M., 2016. Trophy hunting, conservation, and rural development in Zimbabwe: issues, options, and implication. Int. J. Biodivers., 16, 2016.

Norton-Griffiths, M., Southey, C., 1995. The opportunity costs of biodiversity conservation in Kenya. Ecol. Econ. 12, 125–139.

Ogada, M.O., Woodroffe, R., Oguge, N.O., Frank, L.G., 2003. Limiting depredation by African carnivores: the role of livestock husbandry. Conserv. Biol. 17, 1–10.

Okello, A., 2015. The control of neglected Zoonotic diseases: from advocacy to action. World Health Organization, Geneva, Switzerland.

PDC, 2016. Painted Dog Conservation—Arts Centre. Available from: http://www.painteddog.org/education-programs/arts-center/.

Ragwa, I.M., 2012. The Factors Influencing the Collapse of Community Cattle Dips In Kenya. MA thesis, University of Nairobi, Kenya.

Roberts, B.D., 1996. Livestock production, age, and gender among the Keiyo of Kenya. Hum. Ecol. 24, 215–230.

Roe, D.E., 2010. Linking biodiversity conservation and poverty reduction: a state of knowledge review. In: Roe, D. (Ed.), CBD Technical Series. Convention on Biological Diversity, Montreal, Canada.

Roe, D., Thomas, C.D., Smith, J., Walpole, M., Elliott, J., 2011. Biodiversity and Poverty—Ten Frequently Asked Questions—Ten Policy Implications. Gatekeeper, IIED, London.

SAT, 2015. Tourism Performance Highlights 2015 South African Tourism. Available from: http://www.south-africa.net/gl/en.

Silva, J.A., Khatiwada, L.K., 2014. Transforming conservation into cash? Nature tourism in Southern Africa. Afr. Today 61, 17–45.

SLT, 2016. Snow Leopard Enterprise, Snow Leopard Trust. Available from: http://www.snowleopard.org/learn/community-based-conservation/snow-leopard-enterprises.

Stein, A.B., 2008. Ecology and Conservation of the Leopard (*Panthera pardus*) in Northcentral Namibia. PhD University of Massachusetts, United States.

Treves, A., Jones, S., 2010. Strategic tradeoffs for wildlife-friendly eco-labels. Front. Ecol. Environ. 8, 491–498.

Turner, R., 2015. Travel and Tourism Economic Impact Kenya. World Travel & Tourism Council, London.

UN, 2012. Global alliance for Clean Cookstoves: Enabling Markets Worldwide 2011–2012. United Nations, Washinton, DC.

UN, 2015. The Millennium Development Goals Report 2015. United Nations, New York.

WFEN, 2015. Wildlife Friendly Enterprise Network. Available from: http://wildlifefriendly.org/.

Write, J.H., Hill, N.A.O., Roe, D., Marcus, R.J., Kumpel, N.F., Day, M., Booker, F., Gulland, E.J.M., 2015. Reframing the concept of alternative livelihoods. Conserv. Biol. 30 (1), 30.

Coordination of Large Landscapes for Cheetah Conservation

Larkin A. Powell, Reinold Kharuxab**,
Laurie Marker†, Matti T. Nghikembua†,
Sarah Omusula‡, Robin S. Reid§, Andrei Snyman*,¶,
Chris Weaver††, Mary Wykstra‡*

*University of Nebraska-Lincoln, Lincoln, NE, United States
**Namibia University of Science and Technology, Windhoek, Namibia
†Cheetah Conservation Fund, Otjiwarongo, Namibia
‡Action for Cheetahs in Kenya Project, Nairobi, Kenya
§Colorado State University, Fort Collins, CO, United States
¶Northern Tuli Predator Project and Research Mashatu,
Mashatu Game Reserve, Botswana
††World Wildlife Fund, Windhoek, Namibia

CONSERVANCIES AS LARGE, MANAGED, COORDINATED LANDSCAPES

The global range of the cheetah (*Acinonyx jubatus*) is now limited to approximately 16 countries in sub-Saharan Africa and Iran (Durant et al., 2015). Cheetahs exist on freehold land (title owned by private owner), communal land (granted to a community—often indigenous peoples, but government holds title), tribal land (set aside for an ethnic group, government usually holds title), national protected land (e.g., National Parks), or other state-owned land. Some countries have protected a large proportion of their area as national parks or other types of protected land that may give refuge to cheetahs (Table 17.1). However, approximately 75% of the cheetah's current global range falls on lands controlled by private individuals or communal/tribal groups (IUCN/SSC, 2007). Each cheetah alive today faces distinctive, daily risks that are affected by the humans that use, manage, or otherwise affect decisions on the land.

TABLE 17.1   Land Ownership Trends in Countries That Comprise the Majority of the Global Range of the Cheetah (*Acinonyx jubatus*) in Africa and Asia

Country	Land controlling entities	Percent (%) private ownership	Percent (%) nationally protected lands
AFRICA			
Algeria	State	5	5
Angola	State and tribal	<5	10
Botswana	Tribal and state	5	31
Central African Republic	State and tribal	<1	15
Chad	State	0	9
Ethiopia	State	10	19
Kenya	Tribal and private	20	12
Mali[a]	State	0	2
Mauritania[a]	State	—[b]	<1
Namibia	Tribal and private	44	5
Niger	Tribal	—[b]	7
South Africa	Private and state	79	6
Sudan	State	—[b]	—[b]
Tanzania	Tribal	11	39
Uganda	Private and state	>50	32
Zambia	State	0	40
Zimbabwe	Tribal and private	42	15
ASIA			
Afghanistan[c]	State and tribal	—[b]	<1
India[c]	Private and state	52	5
Iran	State and private	40	10
Pakistan[c]	Private and tribal	>55	9
Turkmenistan[c]	State	0	3
Uzbekistan[c]	State	0	3

[a]*Recent information suggests presence of cheetahs here is uncertain and considered extinct.*
[b]*Information unavailable.*
[c]*Recent information suggests cheetahs have been extirpated.*
Lands may be controlled by the state, tribal (here, including both communal and noncommunal) groups, or private (leasehold or freehold) individuals; the two groups that control the most land are listed to characterize each country. Countries with 0% private ownership have state ownership of land with potential for leases or private use, but not freehold ownership.
Data source: Cahill and McMahon, 2010; U.S. Agency for International Development, 2016.

One fact is constant: cheetahs range widely and do not recognize property boundaries (Houser et al., 2009; Marker et al., 2008; Chapter 8). In Namibia, the size of a typical livestock farm is 70 km^2, but cheetah home ranges may exceed 1000 km^2 (Marker et al., 2008; Chapter 8). Conservation efforts for cheetahs will only be successful when the scale of the management effort matches the ecological scale (Prugh et al., 2008). Therefore, the future of the cheetah is spatially dependent on large landscapes that are managed in a coordinated fashion to reduce risk of mortality over time. In this chapter, we explore the benefits and complexities of providing such landscapes outside of nationally protected areas.

Strategies for protection on freehold or community/tribal lands are complex because conservation efforts for cheetahs and other wildlife in these areas are subject to volatility of economics. Landowners and users not only respond to threats to livestock profit, but also to potential profit opportunities from alternative uses of their land (Fig. 17.1). Can production landscapes

be managed in a coordinated fashion to benefit land users and cheetah conservation? What models exist to guide future efforts outside of state-protected areas? Here, we focus on cheetahs, but other species stand to gain from well-positioned strategies for coordinated conservation on lands outside of state-owned protected areas (Knight, 1999).

We are especially interested in groups of landowners or land-occupiers practicing forms of cooperative management based on a sustainable utilization strategy that promotes the conservation of natural resources and wildlife with the basic goal of sharing resources among all members (Shaw and Marker, 2011). Throughout the range of the cheetah, such collaborative efforts take many forms, but each serves as a model for conservation in which landowners or land users have pooled their resources for the purpose of the conservation of wildlife. Here, three types of shared management are presented: Namibian freehold and communal conservancies and Kenyan conservancies.

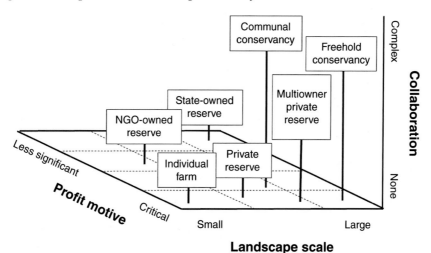

FIGURE 17.1 **Comparison of types of state-owned, freehold, and community lands used to reach conservation goals based on three factors: the landscape scale provided, the need for profit, and the degree of collaboration involved in the structure of management.** The three axes describe gradients of each characteristic. For example, state-owned reserves typically involve little collaboration among different landowners, have lower profit motives, and may provide large-scale landscapes. In contrast, individual farms provide a small scale for conservation initiatives, have high profit motive, and typically involve no collaboration, while a freehold conservancy provides for larger scale with high profit motive and complex arrangements to promote collaborations.

Namibian freehold conservancies: Livestock farmers (typically nonindigenous) with title to their land voluntarily join with neighbors to form freehold landowner groups, or conservancies. (We use the term "livestock farmers" in this chapter to refer to individuals who raise livestock for personal use or commercial sale. Local terminology differs in Namibia and Kenya, the case studies used in this chapter. Livestock farmers would be termed "pastoralists" in Kenya and many other regions of the world where "farmer" refers to those who raise plant crops.) Each landowner has utilization rights for wildlife that are conditional on height of fence and quotas prescribed by government on their land (Powell, 2010). The landowners, usually with mixed livestock and game farms, form agreements with neighbors about consumptive use of wildlife, ecotourism, and habitat and water management. Thus, these conservancies have provided a mechanism for neighbors to benefit from an integrated landscape.

There are currently 23 freehold conservancies in Namibia, covering about 4.7 million ha (6% of Namibia's land area) and supporting some 30,000 people (Shaw and Marker, 2011). Namibian freehold conservancies have around 5–58 farms, which range in size from 75,650 to 500,000 ha; size is generally limited socially, by the distances that neighbors are comfortable driving for meetings. Each conservancy is acknowledged by Namibia's Ministry of Environment and Tourism and constructs a constitution, which defines the relationship among its members and outlines its initial management plan. Namibian conservancy members list wildlife conservation and poaching protection as primary goals (Shaw and Marker, 2011), but also list social networking as a goal, which indicates the importance of communication and trust among members (Powell, 2010).

Namibian communal conservancies: On communal land in Namibia, indigenous residents acquire conditional rights over wildlife use and commercial tourism through the formation of a conservancy and its registration with the government through the Nature Conservation Amendment Act of 1996. As of 2014, there were 82 registered communal area conservancies that cover greater than 16 million ha (20% of Namibia's land area) and are home to one-fourth of rural Namibians (Namibian Association of CBNRM Support Organizations, 2014). A conservancy committee manages the activities of the community, including the development of zonation and management plans, the allocation of watering points, the reintroduction and monitoring of wildlife and other natural resources, and joint venture partnerships with tourism and hunting operators (Weaver and Skyer, 2003). A network of government and NGO support groups provide technical and financial assistance to the committee (Naidoo et al., 2016; Fig. 17.2). Lodges and camps, bases for nonconsumptive and consumptive tourism within the conservancy, are developed through joint venture agreements between conservancies and private sector investors. The income to communal conservancies in Namibia from game utilization, including trophy hunting, rose from US $1.3 million in 2006 to US $3.5 million in 2013. The benefits from nonhunting tourism rose from US $1.5 million in 2006 to US $3.0 million in 2013 (Naidoo et al., 2016).

Prior to passage of legislation that allowed Namibian communal community members to benefit from wildlife, many local people resented wildlife because of competition for livestock grazing and water, as well as livestock predation and damage to infrastructure. Now, wildlife is largely seen as a community asset because of the economic benefits that can be obtained (Weaver and Petersen, 2008). Income from wildlife ventures is returned to the committee and is used to pay salaries of game guards and other conservancy staff members. Community members also benefit from distribution of game meat, bursaries for education, and assistance with the development of small enterprises like crafts.

Kenyan conservancies: A conservancy in Kenya is a single parcel of land dedicated by an owner(s) (e.g., private rancher, or community) to

FIGURE 17.2 **Members of Ehirovipuka Conservancy in northwestern Namibia discuss the processes to be used by herders to graze cattle in rotation to improve range health.** *Source: Larkin Powell.*

be wildlife friendly while supporting livelihoods of local people (Reid et al., 2016). Early conservancies in Kenya were focused around state protected parks and reserves, which created a protected buffer around portions of the park to create dispersal zones and islands of lower mortality risk for predators and prey. Both community and private ranch conservancies are crucial migratory corridors and dispersal zones creating areas of protection for predators and prey.

Individual Kenyan conservancies fall under 13 regional forums governed by the Kenya Wildlife Conservancies Association (Kaelo, 2016). Regional forums (Jones et al., 2005) serve as an umbrella organization for multiple entities with their own individual boundaries, which may not be completely contiguous—a similarity with Namibian freehold conservancies. On community-owned lands in Kenya, conservancies are often run by elected committees of community members. An appointed manager makes management decisions on behalf of the community, and the managing board of the conservancy holds an annual meeting where the other members of the

community or supporting board can contribute ideas for the well-being of the conservancy.

Currently, 5 private and 120 community conservancies in Kenya protect over 5.6 million ha of land (10% of Kenya's land area; Kaelo, 2016). Conservancies are typically mixed-use areas with both wildlife and livestock. Frequently, portions of the conservancy are restricted from human activities or only used during certain seasons (e.g., dry-season grazing). The strictness of the conservation area and grazing plans depends on the managing committee; generally, a section of the conservancy is set aside for wildlife only and strict penalties are in place for those who break grazing rules.

BENEFITS OF COLLABORATIVE LANDSCAPES TO CHEETAH CONSERVATION

Conservation strategies for a species of concern should be aimed to increase resilience of the system (Powell, 2012). Species, such as cheetahs are at risk from ecological, political, social, and economic perturbations (Marker et al., 2008). As large, managed landscapes, conservancies provide many benefits for conservation of cheetahs (Marker et al., 2010).

Large, protected landscapes: Cheetahs need space; in fact, cheetahs occupy larger home ranges than their energy needs would dictate (Gittleman and Harvey, 1982). The mean of home ranges of males have been reported to range from 50 km^2 in the Serengeti (Caro, 1994) to over 1000 km^2 in Namibia (Marker et al., 2008; Chapter 8). Cheetahs also disperse widely (Marker et al., 2008). The annual home range of cheetahs goes far beyond the typical, single farm (Fig. 17.3), and cheetahs may even range beyond the boundaries of collaborative groups of farms (Marker et al., 2008). Similarly, cheetahs in the Masai Mara or Samburu National Reserves in Kenya often range beyond the protected area boundaries to community managed lands (Wykstra, personal observation). In addition, prey species move to find food during transitions between wet and dry seasons (Fynn and Bonyongo, 2011), and predators, such as cheetahs shift their own range as a result (Durant et al., 1988; Chapter 9). Thus, a single landowner's efforts to support cheetah conservation could be unsuccessful unless neighboring landowners have similar objectives for their management. To ensure conservation success, the scale of conservation efforts must match the scale of animal movements (Scott et al., 1999).

Collaborative landscapes provide such scale. Thus, conservancies have been the recipients of translocated animals, such as cheetahs (Marker, 2009) and black rhinoceros (*Diceros bicornis*). Such conservation actions are possible because

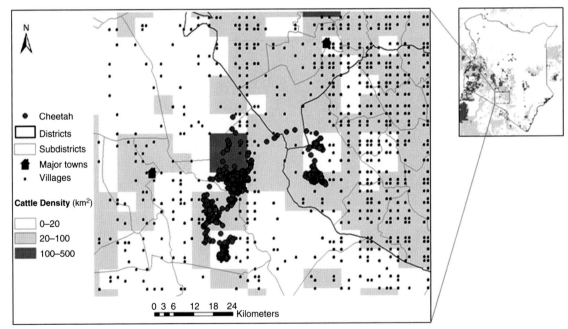

FIGURE 17.3 **Distribution of locations of a female radio-marked cheetah during 2009 in southwestern Kenya (country extent in inset map).** The area was a former conservancy with ongoing settlement by subsistence farmers. Cheetah locations are shown in context of villages and towns, contrasting cattle densities, and political boundaries of districts and subdistricts to emphasize the changing risks that the cheetah encounters as it moves in the landscape. The cheetah moved from East to West and then South, away from high cattle density areas.

conservancies are perceived to be relatively safe landscapes that have adequate area for species of concern.

Large-scale management may increase profits: Ecotourists respond favorably to large carnivores, such as the cheetah, which are often absent from nonprotected wildlife areas. Moreover conservation programs that have large spatial extents with buy-in from a broad community of landowners are key to the conservation success of ecotourism ventures (Lindsey et al., 2009). In Kenya, Western (1982) estimated that potential profits from cheetah-based ecotourism were projected to be greater if cheetahs had access to a wider landscape than just the focal national park. Responsible hunting of herbivores and bird species on freehold conservancies in Namibia may support conservation of wildlife if the local community is engaged at a large scale (Marker and Dickman, 2004).

In similar fashion, a larger group of livestock farmers on community and freehold lands enables unique, collective marketing programs for livestock, which increases profits and could promote carnivore conservation (Marker et al., 2010). The Northern Rangeland Trust (NRT) in Kenya works through a livestock trading and marketing program to give pastoralists economic security and incentives to conserve wildlife. NRT's program provides an alternative market that pays fair prices for livestock and purchases directly from the conservancies. In the future, the introduction of predator-friendly labeling of products could increase farm income (Marker et al., 2010). Farmers who are more financially secure may be less inclined to kill cheetahs and other predators (Stein et al., 2010).

The formation of a common-interest community among neighbors results in the joint management of parcels of land. A diverse set of economic ventures in collaborative landscapes is made possible when large-scale management allows structural heterogeneity of habitat to be established at multiple scales (Toombs et al., 2010). Coordinated management of large

landscapes may also allow spatial opportunity for more effective management responses to local variation in rainfall or other resources in a given year (Lindsey et al., 2009), which further supports diverse communities of wildlife and may protect rare species (Naidoo et al., 2011). Cheetahs and other predators benefit from large landscape management (Figs. 17.1 and 17.3), which allows a diverse complement of prey species to be supported. In turn, a more diverse predator community may attract more tourists (Lindsey et al., 2009).

Collaborations have potential to reduce human-wildlife conflict: Livestock farming is the main source of income and food for farmers throughout the range of the cheetah (Fig. 17.2; Marker-Kraus et al., 1996). Perceived risk to livestock is cited as the reason for many killings of cheetahs (Marker and Dickman, 2004; Chapter 13), which is the largest source of mortality for adult cheetahs in Namibian farmlands (Marker et al., 2003a). As an example cheetahs were perceived to be a significant source of predation and loss of income or food (Fig. 17.4) during a survey of 40 farmers conducted on the ≠Khoadi-//Hôas Conservancy in northwestern Namibia (Kharuxab, unpublished data). Such perceptions exist, even though predators are rarely seen by farmers (Brassine, 2014), and cheetahs are often wrongly identified as the cause of livestock deaths (Marker et al., 2003b).

Economic incentives are used globally to enhance conservation efforts (Powell, 2012), and farmers' access to alternative sources of income may reduce the retaliatory killing of cheetahs by farmers (Marker and Dickman, 2004). Farmers who participate in a collaborative community, such as a conservancy, can receive the distributed benefits from ecotourism, which may offset some potential economic loss and can lead to a higher level of acceptance of predators on the landscape (Marker et al., 2010). Conservation workers in Kenya have reported a variety of economic benefits—direct and indirect—that members of conservancies derive from participation

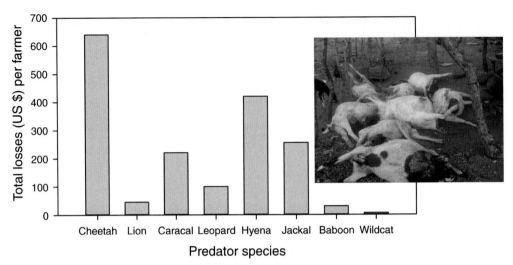

FIGURE 17.4   **Estimates for annual losses, per farmer, at a communal conservancy in Namibia in 2010.** US dollars assigned by farmers to cheetah, African lion (*Panthera leo*), caracal (*Caracal caracal*), leopard (*Panthera pardus*), hyena (spotted, *Crocuta crocuta*, and brown, *Hyaena brunnea*), black-backed jackal (*Canis mesomelas*), and wildcat (*Felis silvestris*). Losses are based on annual number of losses to predators identified by 40 farmers from the ≠Khoadi-//Hôas Conservancy in northwestern Namibia during in-person interviews. Estimated numbers of livestock lost were adjusted by the market value for cattle, goats, sheep, donkeys, horses, and chickens during 2010 to calculate total losses. Insert shows suspected cheetah predation on goats at a family's kraal in the Conservancy. *Source: Reinold Kharuxab.*

in a collaborative group: income from tourism, employment, development of clinic/dispensaries, solar power provision to schools, cattle treatments for pests, donations for livestock water tanks, construction or desilting of dams, and provision of scholarships for children. Farmers engaged in ecotourism activities in northwestern Namibia had more positive feelings toward cheetahs than farmers who were not engaged in ecotourism, despite considering cheetahs a significant source of predation (Powell, unpublished data).

Even in the absence of economic incentives, mortality of cheetahs and other predators may be reduced on lands managed by a community that has made a decision to establish objectives for wildlife. Only 3 (8%) of 40 farmers interviewed at ≠Khoadi-//Hôas Conservancy reported hunting predators after their livestock had been killed. Similarly, only 11% of farmers in conservancies located in Namibia's Greater Waterberg Landscape reported that they killed

predators following depredation, despite relatively large losses (Marker, unpublished data). A mixture of appreciation of the conservation program and fear of penalties if caught hunting predators could be responsible for the lack of retributions against them (Kharuxab, unpublished data).

Finally, improved animal husbandry can lessen human-cheetah conflict (Ogada et al., 2003; Chapter 13) by reducing losses of livestock. Many farmers on communal lands in Namibia and Botswana do not herd their livestock in the day (Kharuxab, unpublished data; Brassine, 2014), even though herding can reduce livestock losses (Ogada et al., 2003). The presence of landscape collaborations in Namibia is now allowing communal farmers to pool resources to support herders and livestock guarding dogs (Chapter 15) to guard livestock of many farmers at once. Herders also rotate livestock grazing areas to improve rangeland management, so livestock production should increase profit by better range management and reduced predator interactions (Fig. 17.2). This should result in the reduction of the farmers' desires to remove cheetahs from the landscape.

CHALLENGES TO THE USE OF COLLABORATIONS TO MANAGE LANDSCAPES

Collaborative groups may provide potential benefits for conservation of cheetahs, but challenges still remain. A member of a Namibian freehold conservancy, Harry Schneider-Waterberg, summed up most of the challenges: *"Conservancies are a hard road to go, because they involve people managing people"* (Powell, 2010). Some of the biggest challenges are:

The need for tangible benefits: The primary challenge to conservation on private or community lands is to provide an incentive to landowners and users to conserve wildlife (Powell, 2012) when conservation measures conflict with other economic ventures and investments, such as crop or livestock farming (Table 17.1; Fig. 17.1). A Namibian freehold conservancy leader, reflecting on the goals of their membership, stated: "In their eyes, the conservancy will only be valuable for them if the conservancy can increase their profit" (Powell, 2010). Similarly, Wunder (2000) stated that the success of conservation incentives depends on the choices available to participants—how does conservation compare with other activities that might generate income?

Namibian communal conservancies distribute many indirect benefits from wildlife to residents that include contributions to conservancy schools, food for elderly residents, support for livestock vaccination, diesel for water-points, and payments to offset damage from elephants (Jones and Weaver, 2009). However, on the ≠Khoadi-//Hôas and Ehirovipuka Conservancies, the perception of predators was more favorable by communal farmers who received direct monetary payments from conservancies and who received money directly from tourists than by farmers who only acknowledged receipt of indirect benefits from conservancy projects (Powell, unpublished data).

The need for support, coordination, and expertise: Wildlife management decisions are complex and have a high level of uncertainty, even when made by trained wildlife biologists (Powell, 2012). Although farmers and ranchers usually have lifelong experience with management of rangeland and domestic animals based on concepts of population growth and sustainable harvest, they are not typically trained in concepts or techniques of wildlife management or conservation biology (Powell, 2010). Thus, governments, agencies, NGOs, and universities have a critical role to work with landowners (Reid et al., 2009) to extend basic knowledge to achieve wildlife-related goals.

For example, communal farmers in Namibia's Greater Waterberg Landscape were enthusiastic to match their experience with new information provided during workshops on rangeland, livestock and predator management (Marker, unpublished data). And, work in Kenya provides evidence that external teams can work with communities to improve livestock herd quality and expand schemes to benefit financially from wildlife (Reid et al. 2009). The costs of support programs may be significant to the supporting institutions (Powell, 2012), and such work requires unique planning and continual engagement with the community. Conservation goals are typically only met when livelihood needs of communities are first met and acknowledged by the support team (Reid et al. 2009).

Coordination of large landscapes requires having useful information available to the group of decision-makers, and freehold conservancies in Namibia have struggled to obtain an inventory of the resources (e.g., wildlife stocks) that are within the groups of farmers to cooperatively manage these resources. Similar data is presented biannually on Namibian communal lands (Namibian Association of CBNRM Support Organizations, 2014). In northern Kenya, the NRT has helped to overcome this challenge by serving as facilitator to summarize and return information to the conservancies that collect the data on wildlife and livelihood.

Gaps in a cooperatively managed landscape: Successful conservation of cheetahs and their ungulate prey depends upon functional landscapes (Fynn and Bonyongo, 2011). Protected areas that are small and fragmented are difficult to manage, and the wildlife populations inhabiting such areas, are exposed to many threats that directly compromise their viability (Chapter 19). Edge effects are a major concern for populations of large predators living in small protected areas (Brashares et al., 2001). Wide-ranging predators, such as cheetahs, can experience high levels of mortality or emigration when protected areas they live in are too small in relation to their home ranges, which could lead to local extinctions (Snyman et al., 2015; Woodroffe and Ginsberg, 1998).

Collaborative networks of private or communal lands can fill gaps in the landscape that are not protected by public reserves (Fig. 17.1; Scott et al., 1993). A network of small protected areas connected by corridors can function as a larger body through which the viability and sustainability of wide-ranging, low density, large carnivores such as cheetahs can be maintained (Zeller et al., 2012). But, a network is only as strong as its weakest link, and gaps in protection may result in significant risk over time for local populations (Snyman, 2016). Gaps can be created with some landowners who do not agree to participate in collaborative efforts (Powell, 2012).

WAYS FORWARD

Collaborative management of landscapes has the potential to solve two of the largest conservation problems for cheetahs: (1) a lack of contiguous habitat at the scale of the cheetah's movements (Fig. 17.1), and (2) areas of high risk

in the landscape (Fig. 17.3). Facilitated discussions among land users have the potential to lower risk for cheetahs through support for livelihoods and increased knowledge. If strategies are developed to provide economic incentives through actions that also conserve cheetahs, the sustainability of conservation should be viable in the long term. However, coordination is inherently difficult and often requires the support and knowledge of external groups. The models for conservation that we provide here have one important common characteristic—they are locally developed as unique efforts for the location, the ecosystem, and the livelihoods of the people involved.

References

Brashares, J.S., Arcese, P., Sam, M.S., 2001. Human demography and reserve size predict wildlife extinction in West Africa. R Soc. Lond. 268, 2473–2478.

Brassine, E.I., 2014. The cheetahs of the Northern Tuli Game Reserve, Botswana: population estimates, monitoring techniques and human-predator conflict. MSc thesis, Rhodes University, South Africa.

Cahill, K., McMahon, R., 2010. Who owns the world: the surprising truth about every piece of land on the planet. Grand Central Publishing, New York, NY.

Caro, T.M., 1994. Cheetahs of the Serengeti plains. University of Chicago Press, Chicago, USA.

Durant, S.M., Caro, T.M., Collins, D.A., Alawi, R.M., FitzGibbon, C.D., 1988. Migration patterns of Thomson's gazelles and cheetahs on the Serengeti Plains. Afr. J. Ecol. 26, 257–268.

Durant, S., Mitchell, N., Ipavec, A., Groom, R., 2015. *Acinonyx jubatus*. The IUCN Red List of Threatened Species 2015: e.T219A50649567. Available from: http://dx.doi.org/10.2305/IUCN.UK.2015-4.RLTS.T219A50649567.en.

Fynn, R.W., Bonyongo, M.C., 2011. Functional conservation areas and the future of Africa's wildlife. Afr. J. Ecol. 49, 175–188.

Gittleman, J.L., Harvey, P.H., 1982. Carnivore home range size, metabolic needs and ecology. Behav. Ecol. Sociobiol. 10, 57–63.

Houser, A., Somers, M.J., Boast, L.K., 2009. Home range use of free-ranging cheetah on farm and conservation land in Botswana. S. Afr. J. Wildl. Res 39, 11–22.

IUCN/SSC, 2007. Regional Conservation Strategy for the Cheetah and African Wild Dog in Southern Africa. IUCN/SSC. Switzerland.

Jones, B.T., Stolton, S., Dudley, N., 2005. Private protected areas in East and southern Africa: contributing to biodiversity conservation and rural development. Parks 15, 67–76.

Jones, B., Weaver, C., 2009. CBNRM in Namibia: growth, trends, lessons and constraints. In: Suich, H., Child, B., Spenceley, A. (Eds.), Evolution, Innovation in Wildlife Conservation: Parks, Game Ranches to Transfrontier Conservation Areas. Earthscan, London, UK, pp. 223–242.

Kaelo, D., 2016. Kenya's wildlife conservancies. Pathways to Success Conference, Nanyuki, Kenya.

Knight, R.L., 1999. Private lands: the neglected geography. Conserv. Biol. 13, 223–224.

Lindsey, P.A., Alexander, R., Mills, M.G.L., Romanach, S., Woodroffe, R., 2009. Wildlife viewing preferences of visitors to protected areas in South Africa: implications for the role of ecotourism in conservation. J. Ecotour. 6, 19–33.

Marker, L., 2009. Conservation strategy for the long-term survival of the cheetah: Cheetah Conservation Fund Annual Report 2009. Cheetah Conservation Fund: Otjiwarongo, Namibia. NASCO.

Marker, L., Dickman, A., 2004. Human aspects of cheetah conservation: lessons learned from the Namibian farmlands. Hum Dimens. Wildl. 9, 297–305.

Marker, L.L., Dickman, A.J., Jeo, R.M., Mills, M.G.L., Macdonald, D.W., 2003a. Demography of the Namibian cheetah, Acinonyx jubatus jubatus. Biol. Conserv. 114, 413–425.

Marker, L.L., Dickman, A.J., Mills, M.G.L., Jeo, R.M., Macdonald, D.W., 2008. Spatial ecology of cheetahs (Acinonyx jubatus) on north-central Namibian farmlands. J. Zool. 274, 226–238.

Marker, L.L., Dickman, A.J., Mills, M.G.L., Macdonald, D.W., 2003b. Aspects of the management of cheetahs, Acinonyx jubatus jubatus, trapped on Namibian farmlands. Biol. Conserv. 114, 401–412.

Marker, L.L., Dickman, A.J., Mills, M.G.L., Macdonald, D.W., 2010. Cheetahs and Ranches in Namibia: a case study. In: Macdonald, D.W., Loveridge, J. (Eds.), Biology, Conservation of Wild Felids. Oxford University Press, Oxford, pp. 353–372.

Marker-Kraus, L., Kraus, D., Barnett, D., Hurlbut, S., 1996. Cheetah Survival on Namibian Farmlands. Cheetah Conservation Fund, Windhoek, Namibia.

Naidoo, R., Weaver, L.C., Diggle, R.W., Matongo, G., Stuart-Hill, G., Thouless, C., 2016. Complementary benefits of tourism and hunting to communal conservancies in Namibia. Conserv. Biol. 30, 628–638.

Naidoo, R., Weaver, L.C., Stuart-Hill, G., Tagg, J., 2011. Effect of biodiversity on economic benefits from communal lands in Namibia. J. Appl. Ecol. 48, 310–316.

Namibian Association of CBNRM Support Organizations, 2014. Summary of conservancies. Available from: www.nacso.org.na/SOC_profiles/conservancysummary.php.

Ogada, M.O., Woodroffe, R., Oguge, N.O., Frank, L.G., 2003. Limiting depredation by African carnivores: the role of livestock husbandry. Conserv. Biol. 17, 1521–1530.

Powell, L.A., 2010. Farming with Wildlife: Conservation and Ecotourism on Private Lands in Namibia. Lulu Press, Lincoln, Nebraska, USA.

Powell, L.A., 2012. Common–interest community agreements on private lands provide opportunity and scale for wildlife management. Anim. Biodiv. Conserv. 35, 295–306.

Prugh, L.R., Hodges, K.E., Sinclair, A.R., Brashares, J.S., 2008. Effect of habitat area and isolation on fragmented animal populations. Proc. Natl. Acad. Sci. 105, 20770–20775.

Reid, R.S., Kaelo, D., Galvin, K.A., Harmon, R., 2016. Pastoral Wildlife Conservancies in Kenya: A Bottom-up Revolution in Conservation, Balanced with Livelihoods? International Rangelands Congress Proceedings, July 16–22, Saskatoon, Saskatchewan, Canada.

Reid, R.S., Nkedianye, D., Said, M.Y., Kaelo, D., Neselle, M., Makui, O., Onetu, L., Kiruswa, S., Ole Kamuaro, N., Kristjanson, P., Ogutu, J., BurnSilver, S.B., Goldman, M.J., Boone, R.B., Galvin, K.A., Dickson, N.M., Clark, W.C., 2009. Evolution of models to support community and policy action with science: balancing pastoral livelihoods and wildlife conservation in savannas of East Africa. Proc. Natl. Acad. Sci. 113, 4579–4584.

Scott, J.M., Davis, F., Csuti, B., Noss, R., Butterfield, B., Groves, C., Anderson, H., Caicco, S., D'erchia, F., Edwards, Jr., T.C., Ulliman, J., Wright, R.G., 1993. Gap analysis: a geographic approach to protection of biological diversity. Wildl. Monogr. 123, 1–41.

Scott, J.M., Norse, E., Arita, H., Dobson, A., Estes, J., Foster, M., Gilbert, B., Jensen, D., Knight, R., Mattson, D., Soule, M., 1999. Considering scale in the identification, selection, and design of biological reserves. In: Soule, M., Terborgh, J. (Eds.), Continental Conservation: Scientific Foundations of Regional Reserve Networks. Island Press, Washington DC, pp. 19–38.

Shaw, D., Marker, L., 2011. The Conservancy Association of Namibia: an overview of freehold conservancies. CANAM, Windhoek, Namibia.

Snyman, A., 2016. Assessment of Resource Use and Landscape Risk for African Lions (Panthera leo) in Eastern Botswana. PhD thesis, University of Nebraska-Lincoln, USA.

Snyman, A., Jackson, C.R., Funston, P.J., 2015. The effects of alternative forms of hunting on the social organization of two small populations of lions Panthera leo in southern Africa. Oryx 49, 604–610.

Stein, A.B., Fuller, T.K., Damery, D.T., Sievert, L., Marker, L.L., 2010. Farm management and economic analyses of leopard conservation in north-central Namibia. Anim. Conserv. 13, 419–427.

Toombs, T.P., Derner, J.D., Augustine, D.J., Krueger, B., Gallagher, S., 2010. Managing for biodiversity and livestock:

a scale-dependent approach for promoting vegetation heterogeneity in western Great Plains grasslands. Rangelands 32, 10–15.

U.S. Agency for International Development, 2016. Country profiles: property rights and resource governance. Available from: www.usaidlandtenure.net/country-profiles.

Weaver, L.C., Petersen, T., 2008. Namibia communal area conservancies. Best Pract. Sustain. Hunt. 1, 48–52.

Weaver, L.C., Skyer, P., 2003. Conservancies: Integrating wildlife land-use options into the livelihood, development and conservation strategies of Namibian communities. The 5th World Parks Congress, September 8-17, Durban, South Africa.

Western, D., 1982. Amboseli National Park: enlisting landowners to conserve migratory wildlife. Ambio 11, 302–308.

Woodroffe, R., Ginsberg, J.R., 1998. Edge effects and the extinction of populations inside protected areas. Science 280, 2126–2128.

Wunder, S., 2000. Ecotourism and economic incentives: an empirical approach. Ecol. Econ. 32, 465–479.

Zeller, K., McGarigal, K., Whiteley, A., 2012. Estimating landscape resistance to movement: a review. Landscape Ecol. 27, 777–797.

Cheetah Conservation and Educational Programs

Courtney Hughes,**, Jane Horgan†, Rebecca Klein‡, Laurie Marker§*

*University of Alberta, Edmonton, AB, Canada
**Alberta Environment and Parks, Edmonton, AB, Canada
†Cheetah Conservation Botswana, Maun, Botswana
‡Cheetah Conservation Botswana, Gaborone, Botswana
§Cheetah Conservation Fund, Otjiwarongo, Namibia

INTRODUCTION

As a wide-ranging carnivore, cheetahs need large landscapes to thrive. However, the expansion of farmland and other anthropogenic land uses into prime cheetah habitat has resulted in population decline and human-cheetah conflict (Chapters 10 and 13). Other anthropogenic factors, such as expanding industrial activities, road networks, and the illegal trade and trafficking of cheetahs (Chapter 14), exacerbates population decline and ultimately jeopardizes long-term survival (Durant et al., 2015). Scientific research utilizing multi- and interdisciplinary methods to address these threats have been used by cheetah organizations across Namibia, Tanzania, South Africa, Botswana, Zimbabwe, Kenya, and Iran (Marker and Boast, 2015). Additionally, research and training is underway in

Algeria, Angola, Benin, Ethiopia, Mozambique, Niger, and Zambia (Marker and Boast, 2015). Over the past 14 years strategies including educational programs have been implemented to address conservation concerns, specifically to influence positive, sustained attitudes, and behaviors across the human population by raising awareness and understanding of the issues affecting cheetahs and the broader environment (Marker and Boast, 2015). Indeed, these programs are increasingly viewed as critical to achieve desired conservation outcomes, by fostering environmental knowledge and empathy toward cheetahs and encouraging coexistence with these cats (Marker and Boast, 2015).

Namibia's Cheetah Conservation Fund (CCF), South Africa's Cheetah Outreach (CO), Cheetah Conservation Botswana (CCB), the Iranian Cheetah Society (ICS), Cheetah Conservation

Cheetahs: Biology and Conservation
http://dx.doi.org/10.1016/B978-0-12-804088-1.00018-6

Project Zimbabwe (CCPZ), and Kenya's Action for Cheetahs (ACK), have reached hundreds of thousands of people through their cheetah education programs, often in partnership with other national, international, and governmental organizations. Broadly, goals and objectives of these organizations' programs are to foster a stewardship ethic for cheetah conservation and contribute to delivering on key policies, such as the United Nations Strategic Plan for Biodiversity 2011–20 and Aichi Targets (Convention on Biological Diversity, 2010). Cheetah education programs have been instrumental in helping achieve long-term conservation and coexistence goals (Jacobson et al., 2015; Monroe et al., 2007; Selebatso et al., 2008). This chapter provides considerations for conservationists on the philosophies, models, and methods for cheetah education programs, using practical examples from cheetah conservation organizations today. Considerations are also provided for designing, implementing, and evaluating future cheetah education programs.

EDUCATION IN CHEETAH CONSERVATION

Education for environmental or wildlife conservation is often captured under the banner of "environmental education" (EE). Broadly, EE is a learning process seeking to develop an informed citizenry by increasing knowledge and deepening people's understanding of specific problems, as well as developing decision-making, critical-thinking, and other practical skills, that encourage responsible and proactive behavior to help safeguard natural resources (Heimlich, 2010; Hungerford, 2009). Indeed, EE is an "interdisciplinary and complex field that offers a multitude of strategies for learning, dependent on [available] resources, time, space, curriculum, [and learner] characteristics" (Winther et al., 2010). Currently, many cheetah education programs blend the teaching of factual information with developing learners' empathy toward

cheetahs, or nature overall, a potentially effective combination to address conservation concerns (McPherson-Frantz and Mayer, 2014). For the purposes of cheetah conservation, EE programs can be categorized under formal, nonformal, and informal approaches that occur within rural or urban settings (Eshach, 2007). Programming considerations, among pedagogical and logistical ones, are important for conservation practitioners to reflect on to achieve success.

Theoretical insights and practical considerations outlined in this chapter largely refer to EE or educational theory, and draw on conservation psychology and other social sciences as relevant. More detailed information on social sciences in cheetah conservation is covered in Chapter 35.

Formal Approaches: School Programs for Children and Youth

Formal education refers to structured learning within an institutional framework, such as schools, colleges, universities, or other technical and professional environments (Eshach, 2007), where an established curriculum guides teaching and learning processes (Monroe et al., 2007). In terms of cheetah conservation, formal education is primarily conducted with children and youth in schools (Fig. 18.1). These programs help

FIGURE 18.1 **Formal environmental education (EE) in a school in Namibia.** *Source: Bobby Bradley, Cheetah Conservation Fund.*

integrate cheetah biology, ecology, and information regarding practical solutions to conservation threats into school-based learning (Monroe et al., 2007; Stern et al., 2014). The programs aim to positively influence future generations to conserve and coexist with cheetahs by providing engaging, often experiential, learning opportunities. Conducted as whole school assemblies, in class presentations and after school groups, hundreds of thousands of children have been engaged in formal cheetah educational programs. In Namibia alone, over 300,000 students have participated in educational programs in the past 20 years (Marker and Boast, 2015).

In formal systems the learner can be viewed as passive, where teacher-centered styles are used to impart information, or as active participants in their own knowledge construction. Pedagogical methods can include direct instruction and lectures, where information is prescriptively imparted through formatted lesson plans by the teacher or experiential and collaborative approaches, that is, learning by doing and peer-based problem-solving (Monroe et al., 2007). While direct instruction can be beneficial in teaching technical skills, experiential models are increasingly preferred as they engage and empower learners in tangible, real-world issues through meaningful, first-hand experiences (Stern et al., 2014). Combined with collaborative approaches, learners participate cooperatively through group projects to solve problems and colearn among peers (Barkley et al., 2005).

The educator's role in experiential, collaborative programming is to provide learning experiences, emphasizing on content and process, through, for example, inquiry-based instructional styles (Grasha, 2002; Stephenson, 2001). This offers opportunities to learn through problem-solving frameworks (Uzzell, 1999). Using reflection and critical-thinking, learners select and implement appropriate actions for cheetah conservation problems (Uzzell, 1999).

Formal cheetah conservation programs are often facilitated in schools by cheetah organization staff or schoolteachers. For example, CCF has reached over 50,000 school children since 2014 (CCF, unpublished data). A major benefit of teacher-based facilitation is the ability to reach more children over time, as cheetah education can be implemented by teachers in successive years. To do so, it is necessary to establish cooperative agreements with educational institutions or government departments so cheetah education can be formally integrated into existing systems. This can increase capacity for long-term program implementation and engage teachers to codevelop integrated curricular materials that both resonate with student needs and effectively link with broader student learning objectives. Establishing cooperative agreements provides opportunities for the professional development of teachers and in some cases can help secure funding and other resources (Monroe et al., 2007). Furthermore, cooperative agreements can incorporate evaluations of student learning through existing institutionalized assessments. Data can, thereby, be shared with cheetah organizations to determine the impact of their educational programs on cheetah conservation.

Many cheetah organizations have collaborated with national governments to develop curriculum-aligned resources and lesson plans for teachers. Resources, such as those developed by CCF, CO, and CCB, focus on improving children's natural science, literacy, and life skills by integrating cheetah biology and ecology into science, math, literacy, physical education, the arts, or other disciplines. These materials guide teachers in delivering conservation messages, and support knowledge and skill development of students, so they can take action for cheetah conservation. Value is added when an integrated approach is used. Establishing conservation-focused learning outcomes in existing curricula enables repetition of key messages, with children receiving environmental knowledge regularly throughout their schooling, as opposed to a one-off intervention. This repetition is likely to improve knowledge retention and facilitates positive behavioral change. An evaluation of CO's Animal Awareness for World and Regional Education (AAWARE) educational materials emphasized the importance of providing an adaptable program for different students and abilities, and fostering lasting relationships between schools and conservation organizations to ensure program receptivity (Glover, 2006). Also, targeting preschool and primary level children can be beneficial in shaping and developing beliefs and behaviors early on, in support of conservation efforts, as belief structures can be more easily influenced before children reach secondary school level (Kellert, 2002). CCF has expanded its education program to include early childhood education opportunities for underserved Namibian preschool children.

Nonformal Approaches: Bush Camps and Site Visits for Children and Youth

Nonformal programs refer to out-of-school learning opportunities, often developed and facilitated by cheetah organizations at venues of their choosing. Instructional styles can follow the aforementioned recommendations for formal approaches and apply to both nonformal programs in children and adult education. Nonformal programs, can also utilize established curricula specific to learning objectives and desired outcomes, though they typically use flexible instructional formats in order to meet learner needs (Eshach, 2007; Rogers, 2007). Educators' authenticity, knowledge, and passion are also important aspects to consider in programming, where the ultimate goal is to inspire lasting change (Stern et al., 2014). Examples of nonformal programs for children include bush camps, field excursions, or facility-based activities that engage students in hands-on learning. Organizations, such as CCF, CO, and CCB invite local students to visit their Field Research and Education Centres, where students and teachers participate in experiential learning facilitated by staff. Other activities include touring

demonstration farms and livestock guarding dog facilities (CCB, CCF), cheetah museum, and watching daily feedings of resident cheetahs (CCF, CO). Typically, cognitive, effective, and kinesthetic approaches are combined; these approaches are thought to increase the "fun factor" and help create positive associations between learning objectives and desired outcomes (Sandbrook et al., 2015; Tobias et al., 2014). Experiential programs that focus on developing an emotional connection can, for example, help forge a lasting stewardship ethic and inspire learners to take conservation action (Ballantyne et al., 2011; Stern et al., 2014). Indeed, the sensory experience of seeing a live cheetah being fed or running (Chapter 28), for example, during CCF and CO site visits, combined with learning factual information about cheetah conservation combines cognitive, emotional, and behavioral experiences that are likely to have lasting effects (Kellert, 2002). Experiences like this can be so powerful they have been the inspiration for CO's motto "See it, sense it, save it."

CCF's overnight educational program (for 1–3 nights) offers children, chaperoned by teachers, a unique adventure in a bush camp to participate in various learning activities, providing children with hands-on opportunities to learn about cheetahs through play-based, tactile and cooperative activities. Approximately, 200 children have annually participated in this overnight camp. CCB offers a similar program, located within the depths of the Kalahari bush, where youth are introduced to cheetah biology and ecology, as well as other place-based learning experiences (Fig. 18.2). CO also has an innovative nonformal program called Bus2Us initiative, which provides sponsored transport for under privileged school groups to visit their center and learn about cheetah conservation. Similarly, ACK utilizes conservation debates and tree planting programs to engage young learners in cheetah conservation dialogue and action.

Nonformal education programs have a greater opportunity than formal programs to integrate

FIGURE 18.2 **Children give a presentation on cheetahs and leopards at a weekend bush camp.** *Source: Cheetah Conservation Botswana.*

sports, art, poetry, music, or dance in their design and delivery (Jacobson et al., 2015). For example, CCB collaborates with the Botswana Predator Conservation Trust (BPCT) to deliver Coaching for Conservation, an after-school sports program where soccer is used as the delivery mechanism to teach youth about carnivore conservation. Over the past 2 years, 330 youth have participated in this program. Similarly, ICS uses drama to highlight the cheetah's plight in Iran via their theater show "Troubles of a cheetah," and CCF uses artwork and creative writing to promote knowledge and appreciation for cheetahs, as seen in their book *The Orphan Calf and the Magical Cheetah*. Using art as a learning platform can help identify what values and perceptions people have for cheetahs, or can be used as an evaluative strategy (Hughes, 2013). For example, ACK noted that in the Tsavo region of Kenya where there was a strong antipoaching campaign, most drawings by children in their education programs depicted spear-carrying Maasai warriors being arrested by Kenya Wildlife Service, while pictures from near the Maasai Mara National Reserve incorporated tourist vehicles viewing wildlife (Wykstra, personal communication). Enabling all ages to learn and express themselves through art, games, dance, or sport can be a powerful tool in helping drive home the conservation message and evaluate effects of programming.

Nonformal Approaches to Adult Education

Many organizations also offer adult education in a nonformal setting, involving learning opportunities that develop knowledge and skills or enhance expertise and abilities, using experiential, collaborative, or transformational processes. Of note, colearning opportunities can arise, where conservation practitioners and target audiences can share information on their perceptions and prior knowledge, insights can be gained, myths addressed, and solutions to problems codiscovered (Keen et al., 2005). Indeed, adult learning is best viewed as problem-based and engaging, where autonomous adults determine what and how they will learn, and if or how they will employ new knowledge or skills.

Common examples across cheetah organizations include farmer training, teacher, and conservation staff professional development, internships, and alternative livelihoods programs. Facilitating adult learning means ensuring an environment in which adults are supported in their pursuits, with recognition for their uniqueness, and flexibility in instruction to address their unique needs and abilities (McDonough, 2013). Educators should "encourage intellectual freedom, experimentation, and creativity" (McDonough, 2013). Commonly, adult education rests on assumptions where learning new information and changing behavior is based on personal and extrinsic motivations (Schwartz, 1968, 1977). Other theories that may also be relevant include the theory of reasoned action or planned behavior, value-belief-norm theory, or self-determination theory (Ajzen, 1985; Deci and Ryan, 1985; Fishbein and Ajzen, 1977; Stern, 2000; Vining and Ebreo, 2002).

Training programs should therefore span the breadth and depth of various disciplinary areas across natural and social sciences and humanities, and provide learners with opportunities to develop areas of interest alongside those necessary to their work.

The bulk of cheetah organizations' adult education programs, focus on farmer training. Organizations provide facility-based training or mobile workshops in outlying communities to improve awareness and understanding of cheetah biology and ecology, promote best practices in livestock and integrated land management, and develop or enhance practical skills (Barthwal and Mathur, 2012; Marker and Boast, 2015; Skupien et al., 2016). These programs aim to address the often inaccurate perceptions associated with cheetah-caused livestock depredation (Chapter 13) and reduce the killing of cheetahs. CCFs "Future Farmer of Africa" program alone has engaged over 3000 farmers in a week-long training program (Marker and Boast, 2015).

Fellowships, professional development, or staff training aims to build or enhance practical skills of adults working in conservation, wildlife, agriculture, or related fields. These learning opportunities include multidisciplinary curricula integrating theories, information, and techniques from biology, ecology, veterinary science, and EE. Teacher fellowships are used by CO, to build expertise, promote information exchange, and expose teachers to novel theories and approaches in educational instruction (Bodzin et al., 2010). Other cheetah organizations conduct teacher workshops to encourage conservation education. CCB alone has trained over 500 teachers, and found that 6-months postprofessional development, 86% of teachers utilized cheetah educational resources and integrated these into their curricula and instruction (CCB, unpublished data). In other professional development, CCF hosts a month-long conservation biology course for international and Namibian wildlife professionals, and has

trained over 300 people from 15 cheetah range countries (Marker and Boast, 2015). It also hosts ecoguiding courses for Namibian guides and tourism personnel, to ensure people are learning the most updated information and techniques possible. Staff training is also used by organizations, to expand learners' theoretic and practical knowledge and skills, and provide opportunities to critically examine new ways of thinking. Staff training is important, given the continued advances of conservation's body of knowledge, and helps ensure organizations can be innovative in their approaches toward achieving conservation success.

Internship programs are also used, providing learners with opportunities to work closely with researchers and experts in a field setting over longer periods of time. CCF interns learn administrative skills relevant to a nonprofit organization, captive cheetah husbandry, wild cheetah ecology and wildlife monitoring and census techniques, veterinary clinical skills, conservation genetics training, and processing and analyzing scientific data to help equip them as future conservationists. Additionally, exposure to different professional development opportunities can also spark interest and pursuit of graduate studies, which can advance knowledge and skills and prepare future leaders in cheetah conservation.

Other innovative activities include training programs that supports alternative livelihoods, linking economic development to human–cheetah coexistence (Chapter 16) (Table 18.1).

Informal Approaches

Informal education refers to learning outside of a formal or nonformal system, ranging from casual conversations (Eshach, 2007), to media-based communications (Dabbagh and Kitsantas, 2012). Informal models commonly adopt an information dissemination approach, referring to the often one-way transmission of factual, short key messages distilled from

TABLE 18.1 Examples of Environmental Education (EE) Programs Currently Used in Cheetah Conservation

Program type	
School-based programs or preschool day care	*Target Audience*: Children, adolescents *Goals*: Develop knowledge and skills about cheetahs, their environment and conservation. Provide safe, affordable, and accessible learning opportunities for preschool aged-children. Develop empathy for and national pride in cheetahs. *Delivery*: In schools. Assemblies or integrated lesson plans and curricular materials taught by conservation organization or school teaching staff. *Key Topics*: Predator's role in the ecosystem, cheetah behavior and ecology, broader environmental issues.
Bush camp programs	*Target Audience*: Children, adolescents *Goals*: Introduce participants to farmland and natural ecosystems, research and conservation. *Delivery*: Overnight bush camps, use of interactive, hands-on presentations and collaborative learning, team-building activities, drama, role-play, hiking, games, and sports at cheetah education centers. *Key Topics*: Predator species identification and biology, ecosystem dynamics, livestock and predator management, solutions to human-wildlife conflict, and broader environmental issues.
Livestock management training	*Target Audience*: Farmers, government staff *Goals*: Develop knowledge and improved responsibility for wildlife and the environment through active participation in mitigating and reducing conflict. *Delivery*: Workshops at an education facility or *in situ*, demonstration farms, hands-on course-based learning. *Key Topics*: Ecosystem dynamics, predator identification and biology, grassland and livestock health, best practices in livestock management.
Expert or teacher training	*Target Audience*: Community professionals, teachers *Goals*: Equip teachers with knowledge and skills to incorporate and teach EE into existing curricula. *Delivery*: Workshops aligned with specific needs, curricula. *Key Topics*: Cheetah ecology and biology, wildlife identification, broader environmental issues, and how to conduct integrated lesson planning and instruction.
Train the trainer courses	*Target Audience*: Government or NGO staff, scientists *Goals*: Promote capacity building and implementation of cheetah conservation policy and programs across local, national, and international communities. *Delivery*: Courses at cheetah education facilities. *Key topics*: Livestock management training, ecology and biological census techniques, data analysis, report writing and presentation skills, wildlife identification.
Ecotourism guide training	*Target Audience*: New or existing guides, community conservationists, students *Goals*: Equip guides to impart knowledge, act responsibly and encourage predator conservation, with a focus on interpreting information for tourists. *Delivery*: Courses and hands-on practicums in EE, public speaking, presentation design, trail etiquette, game counting, and antipoaching. *Key topics*: History, flora and fauna of area, antipoaching efforts, techniques to interpret, and share information.
Internship programs	*Target Audience*: Youths, adults *Goals*: Provide opportunities for individuals to learn alongside cheetah conservation professionals in the field, lab, and other settings in order to encourage the next generation of practitioners and researchers. *Delivery*: Job shadowing, face-to-face meetings. *Key Topics*: Cheetah ecology and biology, research techniques, and skill building.

(Continued)

3. CONSERVATION SOLUTIONS

TABLE 18.1   Examples of Environmental Education (EE) Programs Currently Used in Cheetah Conservation (*cont.*)

Program type	
Alternative livelihood training	*Target Audience*: Impoverished communities *Goals*: Reduce severity of cheetah conflict by building local capacity to diversify income streams and addresses poverty. *Delivery*: Single or multiday workshops in community meeting rooms or at education centers. *Key Topics*: Interconnections between natural resources and income generation, EE incorporating skill-based learning in traditional craft making, marketing and sales.
Specialized informational campaigns	*Target Audience*: General public *Goals*: Raise awareness and improve public knowledge on cheetah conservation, and encouraging action. *Delivery*: Information dissemination at community events, TV, radio, social media, websites, posters, and pamphlets. *Key Topics*: Predator conservation and best practices in livestock management, avoiding vehicle collisions with wildlife, dangers of exotic pet ownership, and stopping illegal wildlife trade.
Parliamentary Conservation Caucus	*Target Audience*: Members of Parliament (MPs) and other policy makers *Goals*: Inform members of parliament on issues relating to natural resource management to encourage informed voting choices at government level. *Delivery*: Scientific reports and regular information workshops involving key experts and stakeholders. *Key Topics*: Topics requested from the MPs themselves, usually in line with upcoming legislation requirements.

scientific findings, used stand-alone, or to complement other educational programs (Monroe et al., 2007). Conservation psychology, and more commonly social marketing and media communications theories and methodologies can be utilized to craft effective information dissemination messaging and delivery mechanisms.

The most widely used formats are print media, including pamphlets, brochures, posters, kids activity booklets, and "how to" guides, many of which are available for download on cheetah conservation websites, and are distributed during educational programs. Indeed, combining information dissemination with other education interventions can help reinforce key messages, elaborate on pertinent topics, and engage people in capacity-building.

Most cheetah organizations utilize websites and social media as information platforms. For instance, CCF utilizes a Facebook page as an information platform on illegal cheetah trafficking and to solicit public reporting. General public programs include public speaking events, interactive learning experiences, such as those carried out at zoos (Chapters 23 and 28), and informational booths at agricultural and other community events that provide pamphlets, stickers, or factsheets to visiting individuals (Fig. 18.3). Marketing and fundraising activities that include educational elements are also used to share information. For example, International Cheetah Day (December 4th) is a globally recognized day to celebrate cheetahs, and raise awareness on the plight of cheetahs across a broad public domain. TV and media also have the ability to reach local, national, and international audiences.

Other innovative approaches used in cheetah conservation include networking opportunities, cheetah-themed movie nights, puppet shows, blogs or chat forums, and emails. For example,

FIGURE 18.3 **Raising awareness of cheetah conservation and predator friendly livestock management at a community event in Botswana.** *Source: Cheetah Conservation Botswana.*

MEASURING THE IMPACT OF EDUCATION PROGRAMS ON CHEETAH CONSERVATION

Evaluating EE programming typically focuses on if and how learning has occurred, what strategies are most effective, and if attitudes or behaviors have changed as a result of educational intervention (Engels and Jacobson, 2007; Leeming et al., 1993; Ryan, 1991; Trewhella et al., 2005; Vaughan et al., 2003). Evaluation should also examine if skills were developed, if proactive behaviors are adopted and sustained, and ultimately what impacts the interventions had on conservation outcomes.

Evaluations can occur in schools, through teacher-directed efforts, such as tests or essays, or by cheetah conservation organizations (Blum, 2009; Espinosa and Jacobson, 2012; Mueller and Bentley, 2009). The evaluation process can be challenging and requires time and particular expertise to identify indicators, plan, collect, prepare, and analyze data, and write reports. Working with education specialists to improve rigor and methods of measuring success should be encouraged (Stern et al., 2014; Zelezny, 1999).

Formative (i.e., before or during) or summative (i.e., after, outcome-focused) methods can provide baseline information to identify areas of improvement, or individual gains in knowledge or skills (e.g., Bloom et al., 1971). Common formative and summative methods used by cheetah organizations include evaluations of farmer training workshops and children's programs (Skupien et al., 2016; Stern et al., 2014). These prepost tests can help identify immediate changes in awareness, knowledge, and attitudes. However, sustained attitude or behavior change is best monitored through long-term follow-ups, using control groups to compare results.

While difficult to assess, the impact of cheetah educational programs on raising awareness, improving knowledge, and fostering attitude and behavior change is thought to be substantial in proactively addressing the plight of the

CCF and CCB utilize "farmer networks" to facilitate peer-learning and liaising with other stakeholders, such as government. CCB uses the Spirit of the Kalahari film and theatrical production to encourage adoption of best practices in livestock management by conveying conservation messages in culturally appropriate ways. While ICS utilizes online and print platforms to promote safe, responsible vehicle operations to reduce vehicle-caused cheetah mortality (Chapter 5), and provides relevant wildlife themed books to school libraries.

Scientific reports are also a form of information dissemination, used to provide succinct research results to the scientific community and governments, which can influence policy change. Some of the most influential conservation education occurs when government officials and national and international policy-makers receive briefing reports on research, which in turn can be used in cooperative planning efforts. Various NGO's in Botswana, Kenya, Tanzania, Namibia, Zambia, and Zimbabwe, for example, provide objective information on sustainable natural resource management to parliamentarians, encouraging informed voting choices within government.

cheetah (Marker and Boast, 2015). For example, CCF's evaluations of the "Future Farmer of Africa" training program estimated 65% of livestock losses could be reduced by training farmers in basic management techniques (CCF, unpublished data). CCB also conducted evaluations of their farmer training using follow-up site visits and phone calls 6-months postworkshop. They found 83% of farmers improved their management practices to better protect livestock from predation, indicating a change in behavior (CCB, unpublished data).

Children's programs are also thought to have similar conservation impacts, as demonstrated by children who attended Botswana's "Coaching for Conservation" program. Students showed increased self-worth, increased empathy toward animals and greater knowledge of the environment after the coaching program, and long-term evaluations using control groups are underway (BPCT, unpublished data). Similar increases in awareness and knowledge were documented in Zimbabwe after showing children an educational DVD on cheetahs and climate change (Purchase et al., 2010). In addition to the importance of targeting the next generation, children's education programs can also have benefits to adults and Vaughan et al. (2003) found that under certain circumstances teaching young people creates an increase in knowledge in both their parents and the community.

CHALLENGES IN ENVIRONMENTAL EDUCATION PROGRAMMING

Difficulties in environmental educational programming largely include the sustained investments of time to affect long-term behavior change. For instance, there may be little motivation compelling people to change attitudes or behaviors simply by receiving one dose of information (De Young, 1993; Kanouse and Jacoby, 1988). Generally, recipients of conservation efforts can choose to ignore or disregard information without internalizing key messages, a particular problem in informal education due to the inability for ongoing dialogue with target audiences, particularly when information is not provided face-to-face. Language barriers and the need for translation, illiteracy, failure to provide socially or culturally-relevant messaging, political sensitivities that may prevent dissemination, or inability of target audiences to access particular formats (e.g., online) are also challenges to overcome. Moreover, recruiting participation by adults can be challenging and financial incentives are often required. As aforementioned, assessing the impact of education programs on behavior change for cheetah conservation is also challenging. That said, many of these pitfalls could be minimized by engaging social science and educational experts alongside other relevant stakeholders in designing educational programs, their materials, and in conducting their delivery (Kaplan, 2000; Schultz, 2011).

DESIGN AND IMPLEMENTATION OF CHEETAH EDUCATION PROGRAMS

Standard steps in designing and implementing a conservation education program are to: identify goals and objectives; clarify target audiences; determine scale and scope; identify and secure partners and resources; and, design specific activities. Measuring the success of educational programs and adapting the program as needed is also an essential step, as previously discussed. Conservation Education and Outreach by Jacobson et al. (2015) is an excellent resource that can guide organizations in their educational endeavors.

Goals and Objectives

The first step is to articulate the conservation problems at hand, and then determine what and how education can help achieve positive results (Jacobson et al., 2015) through identifying specific, measurable, achievable, relevant and timely objectives for the short- and long-term (Hughes, 2012; Jacobson et al., 2015). The goals and objectives are the guideposts for building a program, and will guide the scale, design, and delivery mechanisms, as well as becoming the basis for monitoring and evaluation.

Target Audiences and Partners

Understanding who the people are that will participate in cheetah education programs and then tailoring programs to suit cultural norms, including use of commonly spoken languages, dialects, and local knowledge or traditional folklore, will help improve reception of conservation messages. Target audiences typically include farmers, government, teachers, conservationists, children, or the broader public. Data on the target audience's current state of knowledge, skills, motivations, attitudes, and behaviors will help tailor the programming to the interests and needs, and help identify barriers and opportunities for success (Jacobson et al., 2015). Existing data or data collection can be used to find this information.

Involving a diversity of stakeholders to partner in education programs can help to address gaps in informational sources, identify potential financial opportunities, leverage existing resources, and build networks of support (e.g., Seng and Rushton, 2003). Often businesses, teachers, or ambassador farmers can bolster educational initiatives and government departments can play an important consultative role.

Scale and Scope

It is impossibile to reach every person in every cheetah range. As such, identifying where, when, and how a program will have the most impact (i.e., scope and scale) is important. Inventorying resources identifies what is required, including staff or volunteers, their capacity,

and skills, other expertise, funding, infrastructure, materials, and evaluation requirements. This can also assist in identifying feasibility of programming.

Design of Activities

Cheetah educators must decide what activities are most appropriate as related to the conservation problem, target audience, contextual variables, and available resources. Other factors to consider include the cultural, financial, and technological contexts under which programs operate. Successful education programs must consider the context of peoples' lives, and not just give facts and figures. Information shared must include a clear indication of what actions are desirable and what people can do to support cheetah conservation.

FUTURE DIRECTION

Education is the cornerstone of any conservation initiative. As human-cheetah conflict (Chapter 13) and illegal trade (Chapter 14) are two important threats to cheetah survival, education is considered critical to reversing the downward spiral of the species. EE helps stakeholders appreciate and understand the importance of cheetahs within the ecosystem, and the conservation actions people can adopt to help ensure the species' survival.

This chapter has provided some foundational considerations for cheetah education, including considerations for conservationists on the philosophies, models, and methods for cheetah education programs using practical examples provided by the aforementioned cheetah organizations. There is a broad body of knowledge available on the topic of EE and we suggest readers explore this further, to familiarize themselves with this content. Engaging educational experts alongside those in environmental sociology, psychology, social marketing, and other relevant fields will ultimately benefit cheetah conservation by building capacity and laying the foundations to secure the survival of this species for generations to come.

References

Ajzen, I., 1985. From intentions to actions: a theory of planned behavior. In: Kuhl, J., Beckman, J. (Eds.), Action Control: From Cognition to Behaviour. Springer, Heidelberg, Germany, pp. 11–39.

Ballantyne, R., Packer, J., Falk, J., 2011. Visitors learning for environmental sustainability: testing short and long-term impacts of wildlife tourism experiences using structural equation modelling. Tourism Manage. 32 (6), 1243–1252.

Barkley, E.F., Cross, K.P., Major, C.H., 2005. Collaborative Learning Techniques: A Handbook for College Faculty. Jossey-Bass, San Francisco, USA.

Barthwal, S.C., Mathur, V.B., 2012. Teachers' knowledge of and attitude toward wildlife and conservation: a case study from Ladakh, India. Mountain Res Dev. 32 (2), 169–175.

Bloom, B.S., Hastings, J.T., Madaus, G.F., 1971. Handbook on Formative and Summative Evaluation of Student Learning. McGraw Hill, New York, USA.

Blum, N., 2009. Teaching science or cultivating values? Conservation NGOs and environmental education in Costa Rica. Environ. Educ. Res. 15 (6), 715–729.

Bodzin, A., Klein, B.S., Weaver, S., 2010. The Inclusion of Environmental Education in Science Teacher Education. Springer Science & Business Media, The Netherlands.

Convention on Biological Diversity, 2010. Strategic Plan for Biodiversity Conservation 2011–2020. Aichi Biodiversity Targets.

Dabbagh, N., Kitsantas, A., 2012. Personal learning environments, social media, and self-regulated learning: a natural formula for connecting formal and informal learning. Internet High. Educ. 15 (1), 3–8.

De Young, R., 1993. Changing behavior and making it stick the conceptualization and management of conservation behavior. Environ. Behav. 25 (3), 485–505.

Deci, E.L., Ryan, R.M., 1985. Intrinsic Motivation and Self-determination in Human Behavior. Plenum Press, New York.

Durant, S., Mitchell, N., Ipavec, A., Groom, R., 2015. *Acinonyx jubatus*. The IUCN Red List of Threatened Species 2015: e.T219A50649567. Available from: http://dx.doi.org/10.2305/IUCN.UK.2015-4.RLTS.T219A50649567.en.

Engels, C.A., Jacobson, S.K., 2007. Evaluating long-term effects of the golden lion tamarin environmental education program in Brazil. J. Environ. Educ. 38 (3), 3–14.

Eshach, H., 2007. Bridging in-school and out-of-school learning: Formal, non-formal, and informal education. J. Sci. Educ. Technol. 16 (?), 171–190.

Espinosa, S., Jacobson, S.K., 2012. Human-wildlife conflict and environmental education: Evaluating a community program to protect the Andean bear in Ecuador. J. Environ. Educ. 43 (1), 55–65.

Fishbein, M., Ajzen, I., 1977. Belief, Attitude, Intention, and Behavior: An Introduction to Theory and Research. Addison-Wesley, Reading, Massachusetts, USA.

Glover, D., 2006. Environmental Education. The evaluation of Environmental Learning Support Material: A Case Study of the AAWARE Teacher's Guide. MSc thesis, Cape Town University, South Africa.

Grasha, A.F., 2002. Teaching With Style: A Practical Guide to Enhancing Learning by Understanding Teaching and Learning Styles. Alliance Publishers, Pittsburgh, PA.

Heimlich, J.E., 2010. Environmental education evaluation: Reinterpreting education as a strategy for meeting mission. Eval. Prog. Plan. 33 (2), 180–185.

Hughes, C., 2012. Environmental education for conservation: considerations to achieve success. Nat. Areas J. 32 (2), 218–219.

Hughes, C., 2013. Exploring children's perceptions of cheetahs through storytelling: implications for cheetah conservation. Appl. Environ. Educ. Commun. 12 (3), 173–186.

Hungerford, H.R., 2009. Environmental education (EE) for the 21st century: Where have we been? Where are we now? Where are we headed? J. Environ. Educ. 41 (1), 1–6.

Jacobson, S.K., McDuff, M.D., Monroe, M.C., 2015. Conservation Education and Outreach Techniques, second ed. Oxford University Press, Oxford, UK.

Kanouse, D.E., Jacoby, I., 1988. When does information change practitioners' behavior? Int. J. Technol. Assess. Health Care 4 (01), 27–33.

Kaplan, S., 2000. Human nature and environmentally responsible behavior. J. Soc. Issues 56 (3), 491–508.

Keen, M., Brown, V.A., Dyball, R., 2005. Social Learning in Environmental Management: Towards a Sustainable Future. Earthscan, London, UK.

Kellert, S.R., 2002. Experiencing nature: affective, cognitive, and evaluative development in children. In: Kahn, P.L.J., Kellert, S.R. (Eds.), Children and Nature: Psychological, Sociocultural, and Evolutionary Investigations. MIT Press, Cambridge, UK, pp. 117–151.

Leeming, F.C., Dwyer, W.O., Porter, B.E., Cobern, M.K., 1993. Outcome research in environmental education: a critical review. J. Environ. Educ. 24 (4), 8–21.

Marker, L.L., Boast, L.K., 2015. Human–wildlife conflict 10 years later: lessons learned and their application to cheetah conservation. Hum. Dimens. Wildl. 20 (4), 302–309.

McDonough, D., 2013. Similarities and differences between adult and child learners as participants in the natural learning process. Psychology 4 (3A), 345–348.

McPherson-Frantz, C.M., Mayer, F.S., 2014. The importance of connection to nature in assessing environmental education programs. Stud. Educ Eval. 41, 85–89.

Monroe, M.C., Andrews, E., Beidenweg, K., 2007. A framework for environmental education strategies. Appl. Environ. Educ. Commun. 6, 205–216.

Mueller, M.P., Bentley, M.L., 2009. Environmental and science education in developing nations: a Ghanaian approach to renewing and revitalizing the local community and ecosystems. J. Environ. Educ. 40 (4), 53–64.

Purchase, G., Kelly, D., Purchase, D., Nyoni, W., 2010. Cheetahs, wild dogs and climate change: (Pilot project). Stage 1. Proof of concept and increase in knowledge, Zimbabwe, Wildlife Conservation Society, British Council.

Rogers, A., 2007. Non-Formal Education: Flexible Schooling or Participatory Education? Comparative Educational Research Center. The University of Hong Kong, Hong Kong.

Ryan, C., 1991. The effect of a conservation program on schoolchildren's attitudes toward the environment. J. Environ. Educ. 22 (4), 30–35.

Sandbrook, C., Adams, W.M., Monteferri, B., 2015. Digital games and biodiversity conservation. Conserv. Lett. 8 (2), 118–124.

Schultz, P., 2011. Conservation means behavior. Conserv. Biol. 25 (6), 1080–1083.

Schwartz, S.H., 1968. Words, deeds and the perception of consequences and responsibility in action situations. J Pers. Soc. Psychol. 10 (3), 232.

Schwartz, S.H., 1977. Normative influences on altruism. In: Berkowitz, L. (Ed.), Advances in Experimental Social Psychology. Academic Press, New York, USA, pp. 221–279.

Selebatso, M., Moe, S.R., Swenson, J.E., 2008. Do farmers support cheetah *Acinonyx jubatus* conservation in Botswana despite livestock depredation? Oryx 42 (3), 430–436.

Seng, P., Rushton, S., 2003. Best Practices Workbook for Boating, Fishing, and Aquatic Resources Stewardship Education. Recreational Boating and Fishing Foundation, Alexandria, Virginia, USA.

Skupien, G.M., Andrews, K.M., Larson, L.R., 2016. Effects of conservation education programs on wildlife acceptance capacity for the American alligator. Hum. Dimens. Wildl. 21 (3), 264–279.

Stephenson, J. 2001. Teaching & Learning Online: Pedagogies for New Technologies. Kogan Page, United Kingdom.

Stern, P.C., 2000. Toward a Coherent theory of environmentally significant behavior. J. Soc. Issues 56 (3), 407–424.

Stern, M.J., Powell, R.B., Hill, D., 2014. Environmental education program evaluation in the new millennium: what do we measure and what have we learned? Environ. Educ. Res. 20 (5), 581–611.

Tobias, S., Fletcher, D., Wind, A., 2014. Game-based learning. In: Spector, J.M., Merrill, M.D., Elen, J., Bishop, M.J. (Eds.), Handbook of Research on Educational Communications and Technology. Springer, New York, USA, pp. 485–503.

Trewhella, W., Rodriguez-Clark, K., Corp, N., Entwistle, A., Garrett, S., Granek, E., Lengel, K., Raboude, M., Reason, P., Sewall, B., 2005. Environmental education as a component of multidisciplinary conservation programs: lessons from conservation initiatives for critically endangered fruit bats in the western Indian Ocean. Conserv. Biol. 19 (1), 75–85.

Uzzell, D., 1999. Education for environmental action in the community: new roles and relationships. Camb. J. Educ. 29 (3), 397–413.

Vaughan, C., Gack, J., Solorazano, H., Ray, R., 2003. The effect of environmental education on schoolchildren, their parents, and community members: a study of intergenerational and intercommunity learning. J. Environ. Educ. 34 (3), 12–21.

Vining, J., Ebreo, A., 2002. Emerging theoretical and methodological perspectives on conservation behaviour. In: Bechtel, R., Churchman, A. (Eds.), New Handbook of Environmental Psychology. Wiley, New York, USA, pp. 541–558.

Winther, A.A., Volk, T.L., Shrock, S.A., 2002. Teacher decision making in the 1st year of implementing an issues-based environmental education program: a qualitative study. J. Environ. Educ. 33 (3), 27–33.

Zelezny, L.C., 1999. Educational interventions that improve environmental behaviors: a meta-analysis. J. Environ. Educ. 31 (1), 5–14.

Protected Areas for Cheetah Conservation

Bogdan Cristescu*, Peter Lindsey**,†,‡, Olivia Maes§,
Charlene Bissett¶, Gus Mills††,‡‡, Laurie Marker*

*Cheetah Conservation Fund, Otjiwarongo, Namibia
**Panthera, New York, NY, United States
†University of Pretoria, Pretoria, South Africa
‡Wildlife Network, Palo Alto, CA, United States
§Cheetah Conservation Fund, Toddington, United Kingdom
¶South African National Parks, Kimberley, South Africa
††The Lewis Foundation, Johannesburg, South Africa
‡‡University of Oxford, Tubney, Abingdon, United Kingdom

INTRODUCTION

Setting aside land to preserve habitat for wildlife is a long-standing strategy for biological conservation. On lands where protection is enforced, wildlife species have the opportunity to find refuge from consumptive use or from persecution by humans. Yet only 14.8% of the world's land area is protected formally (UNEP-WCMC and WRI, 2014) and protected area (PA) effectiveness varies widely (Rodrigues et al., 2004). Africa has a relatively large PA system, with many of the biomes having 10%–15% of their surface area protected (Jenkins and Joppa, 2009). Several African countries have a much larger percentage of their land area under protection compared to the

global average (Lindsey et al., 2017). To increase PA effectiveness particularly for wide-ranging species, conservation efforts must focus not only within the PAs themselves, but also on lands surrounding them, thereby minimizing edge effects (Woodroffe and Ginsberg, 1998).

Cheetah (*Acinonyx jubatus*) conservation is an illustrative example of the challenges associated with safeguarding wide-ranging species that can come into conflict with people because of perceived or real impacts on livestock and farmed game animals (Inskip and Zimmermann, 2009; Chapter 13). While many of the world's free-ranging cheetahs are found outside PAs, some heavily persecuted cheetah populations that range outside PAs likely depend on the protection

TABLE 19.1 Percentages of Resident Cheetah Ranges and Population Sizes in Protected Areas (PAs) by Geographical Regions

Region	Resident range (km²)	Population size (adult and adolescents)	Resident range in PAs (km²)	Range in PAs (%)	Population size in PAs (adult and adolescents)	Populations in PAs (%)
Southern Africa	1,324,570	4,297	332,789	25.1	1,172	27.3
Eastern Africa	615,071	2,290	151,576	24.6	1,090	47.6
Western, central, and northern Africa	1,037,322	457	175,058	16.9	93	20.3
Total Africa	2,976,963	7,044	659,423	22.2	2,355	33.4
Total Asia	146,867	43	65,091	44.3	N/A	N/A
Total global	3,123,830	7,087	724,514	23.2	N/A	N/A

An expanded version of the table by population is provided in Chapter 39. For a spatial layout of populations, refer to Chapters 4 and 39.
Data from Durant et al., 2017.

within parks and reserves for their persistence. Of the world's estimated 7100 free-ranging adult and adolescent cheetahs, approximately 33% are found inside PAs (Durant et al., 2017; Table 19.1). In southern Africa (61% of the world's cheetah population), 27% of cheetahs are in PAs, whereas in eastern Africa (32% of the world's population), 48% are in PAs (Durant et al., 2017). In western, central, or northern Africa (6% of the world's population), 20% inhabit PAs, while in Asia (0.6% of the world's population), 44% of the cheetah's range is in PAs (Durant et al., 2017; Table 19.1).

FACTORS AFFECTING THE EFFECTIVENESS OF PAs FOR CHEETAH CONSERVATION

Cheetah growth rate is assumed to be higher inside versus outside PAs (Durant et al., 2017). Therefore, PAs can be strongholds for cheetah persistence and potentially serve as sources for cheetah recolonization of unprotected land where they are more likely to be extirpated or reduced in numbers

(Durant et al., 2017). However, the conservation potential of PAs depends on many factors, including their size, connectivity with other PAs and setting within the broader landscape, as well as management policy and law enforcement.

PA Sizes: The Larger the Better

Due to the species' high mobility and vast home ranges (Chapter 8), even in ideal cheetah habitat, a minimum area of 4000–8000 km² is needed for maximizing the chance of long-term cheetah population viability (Durant et al., 2007). This variation is largely driven by the effect of preferred prey density on cheetah density (Hayward et al., 2007a,b). Given comparable habitat conditions, cheetahs typically persist in higher numbers in large reserves. Therefore larger reserves will generally be preferable to conserve cheetahs and their habitats than smaller ones. Larger areas will also generally support a greater prey base (East, 1984); although this may vary with the productivity of the landscape and if appropriate management of the PA is conducted. Larger PAs can be exposed to lower edge effects, such as deleterious human influences along the reserve's borders, but the shape of the reserve (perimeter-to-area ratio) will also influence abundance and population viability of sensitive species (Noss, 1990). To date, many PAs in Africa are too small to support viable cheetah populations in the long term (Durant et al., 2017), and in such situations, populations might only persist based on immigration from neighboring unprotected lands or via a managed metapopulation (Lindsey et al., 2011).

Some PAs in southern and eastern Africa are much larger than the global average individual PA size (Lindsey et al., 2017). However, a cautionary note is that managing large PAs is difficult and expensive, and size does not always afford significant protection against some threats, such as illegal hunting for bushmeat, in cases where management is inadequate (Lindsey et al., 2017). Vast PAs are often more difficult to access, making it difficult to perform law enforcement operations. Where antipoaching efforts are weak or nonexistent, cheetah prey and cheetahs themselves are vulnerable to illegal take.

Interspecific Competition

Predation is an important cause of cheetah cub deaths (Mills and Mills, 2014). Although often ascribed to lions (*Panthera leo*) (Laurenson, 1994), Mills and Mills (2014) have shown that deaths caused by lions may be overestimated, with smaller predators also capable of killing cheetah cubs. Dominant predators, such as lions and spotted hyenas (*Crocuta crocuta*), have also been reported to have a nonlethal impact on cheetah ecology, potentially displacing them into suboptimal habitat with lower prey densities (Durant, 1998, 2000) and possibly influencing their habitat selection and movement (Broekhuis et al., 2013; Vanak et al., 2013). However, there is no evidence that cheetahs alter their predatory behavior in the presence of dominant predators (Hayward et al., 2006). Cheetahs have historically coexisted with lions and other

large predators, and recent analysis of long-term data from the Serengeti National Park, as well as current data from southern Africa, suggested little effect of lion presence on cheetah density (Swanson et al., 2014). The ability of cheetahs to survive in the presence of large predators may be jeopardized if the carrying capacity of the land decreases or densities of competing predators increase (e.g., some small PAs; Hayward et al., 2007a,b). Monitoring of the prey, cheetah, and other predator populations are recommended as part of routine operations inside PAs.

Connected Networks and Fencing

Many PAs are ecologically isolated from each other, due to the loss of suitable habitat outside of PAs (e.g., conversion to crop land) and persecution by humans. Isolation prevents the movement and dispersal of cheetahs, among other species, between PAs (Newmark, 2008). Emphasis should be placed on ensuring connectivity between PAs through delineation of ecological corridors that facilitate the mobility of individuals between resident cheetah populations (Chapter 10). These corridors need to be identified, for example, as recently done for cheetahs in Iran (Moqanaki and Cushman, 2017), and the ability of continued use ensured. The establishment of transfrontier conservation areas linked with corridors has the potential to promote transboundary cheetah populations and align management between PAs.

One important impediment to connectivity is the fencing of PAs, as well as fencing of land outside PAs (Løvschal et al., 2017). However, fencing can allow for the utilization of smaller fragments of land for the conservation of a wider range of species than would otherwise be possible (Lindsey et al., 2012). There has therefore been substantial debate in the scientific literature on the conservation impact of fences and how these might affect the ecology of wildlife (Creel et al., 2013; Hayward and Kerley, 2009; Packer et al., 2013). Fencing has the potential to reduce human-wildlife conflict, and limit human and livestock encroachment into PAs. However, fences are expensive and difficult to maintain, and thus an unrealistic prospect for PAs in many African countries at present (Hayward and Kerley, 2009). Poorly maintained fences are ineffective at limiting the movement of either people or animals, and can be vandalized by poachers to obtain snare wire (Kesch et al., 2015; Lindsey et al., 2012).

Fences can limit individual dispersal and population connectivity of wildlife species. The restriction of movement is a particularly significant problem during times of resource depletion, such as drought, and is further exacerbated due to the tendency of fenced wildlife populations in small PAs to become overabundant in the absence of suitable management (Lindsey et al., 2012). To prevent genetic or demographic problems associated with small isolated populations, fencing needs to be combined with metapopulation management approaches (unless the fenced area is exceptionally large). The Cheetah Metapopulation Project in South Africa represents a coordinated effort to resolve such issues, with considerable resources going into relocating cheetahs and managing prey (Lindsey et al., 2011).

Fencing should be limited to scenarios where alternative management options are not practical and where sufficient funding and expertise exist to allow for effective maintenance and active management. The need for fencing around PAs can be reduced by introducing improved livestock husbandry techniques in neighboring communal and private lands, to allow for wildlife coexistence with people and to decrease conflict (Chapter 13).

The Broader Landscape Context

PAs account for approximately 15% of land mass in sub-Saharan Africa (UNEP-WCMC and WRI, 2014). By and large, parks and reserves can therefore be seen as protective islands in a sea of utilitarian human land uses that often conflict with conservation goals. Matrices between PAs are critical for the effectiveness of the PA networks for conserving cheetahs and other wide-ranging species.

In particular in small PAs, predator home ranges might extend beyond the boundaries of the PA, exposing them to higher levels of anthropogenic mortality outside PAs (Balme et al., 2010; Chapter 13). If primary or sole means of livestock protection from predators is lethal management, ecological traps (sensu Schlaepfer et al., 2002) could develop. Source–sink dynamics can form, wherein the PA acts as a predator population source and the surrounding matrix acts as a sink. In such cases, economic incentives for species protection and educational programs in local communities surrounding PAs might increase tolerance to predators and decrease their persecution. Ideally, buffer zones should be established around reserves, wherein allowable human activities are those that are least likely to result in negative human–carnivore interactions. Livestock grazing in particular should be discouraged and kept to a minimum in these buffer areas and any livestock present should be managed effectively to reduce conflict (Chapters 13 and 15). In addition, mining and unregulated charcoal–related activities should be limited in the vicinity of PAs, as these activities have negative impacts on cheetah populations (Iranian Cheetah Society, 2016; Leowinata, 2014).

Management Policy and Law Enforcement: The Need for Greater Financing and Capacity

Many countries within the cheetah's range struggle with high-human population density, lack of funding and capacity building for conservation, corruption, and poverty. For cheetahs and other species, these issues affect the ability of PAs, and even more so the areas outside PAs, to be maintained as successful conservation areas (Lindsey et al., 2017). As a result, populations of cheetahs and their prey are likely to be significantly lower in PAs than they could potentially be.

Human encroachment into PAs is also reducing their effectiveness (Dobson and Lynes, 2008; Newmark, 2008). Human access inside protected land is difficult to control when PAs are vast, and in wet seasons when the road network in some PAs becomes extremely difficult to navigate. In parts of the continent, and notably in west, central, and east Africa, the incursion of PAs with livestock is increasingly becoming common (Lindsey et al., 2017). During drought periods, such incursions might be more frequent if PAs have permanent water sources that are absent on farmlands. This poses a number of threats to cheetahs, including direct persecution by people to reduce the perceived threat to their livestock and competition between livestock and wild prey for forage and water, which can ultimately result in reduced wildlife densities.

Bushmeat hunting is a major challenge to PA effectiveness, and consequently to cheetah conservation (Bruner et al., 2001; Lindsey et al., 2017; Chapter 11). It poses a challenge not only directly when snares set for small antelopes catch cheetahs, but also indirectly through the poaching of cheetah prey. (Lindsey et al., 2013). Cheetahs are also subject to targeted poaching for their skins or to obtain live cubs for trafficking in the illegal pet trade (Chapter 14).

There is a need for adequate financing and capacity with which effective management and law enforcement can be imparted to prevent illegal hunting and human and livestock encroachment (Lindsey et al., 2017; Watson et al., 2014). Unfortunately, there is declining governmental support for PA management, including funding cuts and staff reductions (Watson et al., 2014). This threatens PA effectiveness and can result in reduced or nonexistent protection to cheetahs and other wildlife within their boundaries. The resulting depletion of wildlife may cause PAs to be degazetted, as they start to act as a financial drain (both through direct and opportunity costs; Lindsey et al., 2017). However, some positive steps are being taken, such as those by the NGO African Parks, which partners with governments in the long-term management of PAs with the goals of conserving wildlife, restoring landscapes, and facilitating sustainable livelihoods for local communities. Ensuring that local communities have a stake in the management of and benefits from PAs is essential to foster local support for conservation efforts. PAs can provide substantial benefits for conservation and communities with regards to tourism and development programs (Scheyvens, 1999).

ROLES OF PAs BESIDES SPECIES PROTECTION

In addition to providing habitat for and direct protection of cheetahs and other wildlife within their boundaries, PAs also have a role to play in community development, awareness raising through tourism, and research opportunities to further cheetah conservation efforts.

Tourism and Development

PAs are an important employer for local communities (Scheyvens, 1999). Employment opportunities include maintenance of operations, tourism, law enforcement, and guiding. PAs also have the potential to act as hubs for the development of tourism, which could promote sustainable development in remote rural areas with few alternatives. Local businesses compatible with conservation, such as ecotourism operators, accommodation, and catering businesses can be encouraged by the presence of a PA. Tourism operations have the potential to help reduce human-wildlife conflict outside of PAs, although the assumption has rarely been tested (Dickman et al., 2011; Walpole and Thouless, 2005).

PAs play an important role in educating the next generation of land and wildlife stewards. Many PAs organize tours for school learners, engaging young people from disadvantaged backgrounds, as well as those from cities who would otherwise have little or no exposure to wildlife in natural settings. In addition, young people from developed nations visit and gain work experience in many African reserves and parks. This exposes them to the conservation challenges and their solutions, while helping conservation and management projects. Cheetahs and other charismatic large mammals are favorite viewing species, especially for foreign and first-time visitors to Africa (Lindsey et al., 2007). As such, having stable cheetah populations inside PAs can increase the economic revenue and hence conservation potential of the PAs.

Research Opportunities to Study Conservation Ecology

Cheetahs are understudied compared to many other large carnivores, especially in western, central, and northern Africa (Fig. 19.1). In southern Africa, a slightly greater number of studies have occurred outside compared to inside PAs, whereas in eastern Africa most studies have focused on parks and reserves (Fig. 19.1).

PAs can be used as reference systems or "natural laboratories," close to a natural state. They often have land management that allow natural processes to take place (Mills and Mills, 2017), whereas outside the protected land, crop cultivation and livestock farming modify the ecosystem extensively. Comparing cheetah ecology and behavior inside and outside PAs would advance our understanding of human-related influences on cheetahs. However, beyond recording human-caused mortality, anthropogenic impacts on cheetah populations have not been extensively studied. Behavior in other carnivore species has been shown to be affected by land protection status. For instance, both lion and spotted hyena space use and movement patterns are affected by human activities (Kolowski and Holekamp, 2008; Oriol-Cotterill et al., 2015).

In some cases, cheetah behavior within PAs may also be modified. For example, cheetahs in

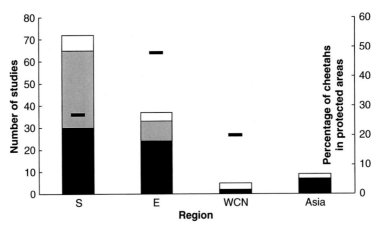

FIGURE 19.1 **Research effort (publication outputs; *n* = 99) on cheetah and PAs by geographical region, as assessed through a Thomson Reuters Web of Science keyword search by "cheetah*" and "protected area*" (February 20, 2017).** *Black shading* on the stacked bars indicates number of studies inside PAs, whereas *gray shading* indicates number of studies outside PAs. When study areas included both protected and unprotected land, the respective studies were counted in both land types. *White* indicates studies for which the location (protected vs. unprotected) was unclear. The *black horizontal lines* correspond to the secondary axis, which shows percentage of the cheetah populations residing inside PAs by region. Values for the secondary axis are from Durant et al. (2017), with no data available for Asia.
E, Eastern Africa; *S*, southern Africa; *WCN*, western, central, and northern Africa.

areas with photographic tourism are often habituated to vehicles.

UNCONVENTIONAL PAs: A ROLE FOR CONSERVANCIES AND BIOSPHERE RESERVES

Cheetahs can greatly benefit from wildlife-based land use, such as conservancies and biosphere reserves. Conservancies are lands set aside for wildlife conservation purposes (Chapter 17). They are self-governing and run by their members, which in turn benefit from wildlife conservation through income from tourism, hunting, and other joint ventures. The efficiency of conservancies for cheetah conservation has been demonstrated in Namibia and Kenya (Marker and Dickman, 2004; Chapter 17).

UNESCO biosphere reserves are areas that promote solutions reconciling the conservation of biodiversity with sustainable use of the ecosystem (UNESCO, 2017). These reserves are key areas for understanding and managing changes and interactions between social and ecological systems, including conflict prevention and management of biodiversity. Biosphere reserves are currently not common in cheetah range countries, but they have substantial potential to facilitate cheetah conservation.

PAs AND THE CHEETAH'S OUTLOOK

PAs are likely to become more important for cheetahs as human populations grow and urbanization and human use of land expand. However, some PAs of key importance for cheetahs are currently severely underfunded, putting at risk the future of the cheetah populations inhabiting them, for example, the tricountry WAP National Park complex in western Africa, Kafue National Park in Zambia, and the Angolan PAs. As cheetahs require vast areas, conservation efforts need to focus not only on the support of key PAs, but also on coexistence on the lands outside PAs and on connectivity between them.

Historically, many PAs were assigned because of unproductive land and not necessarily for their ecological functioning or biodiversity. The process of new PA establishment and expansion of existing PAs should prioritize acquisition of land with high value for biodiversity. Connectivity can be achieved through land use planning to ensure that wildlife-friendly land uses are practiced in key connectivity pathways (corridors); these corridors are also more likely to be successful in less productive land (Chapter 10).

To maximize PA effectiveness, information is necessary on cheetah conservation status in key landscapes, which then would allow decision making on how to best prioritize investment in cheetah conservation. When devising cheetah presence surveys in areas with unknown population statuses, PAs should generally receive priority due to the greater chance of persistence of cheetahs in such areas, as well as logistical and management framework associated with PAs. These include antipoaching units and road infrastructure used by rangers on patrols or by tourists. Information on cheetah distribution, ecological relationships, and behavior in PAs, as well as data on cheetah prey, will help conserve cheetahs if these are found in the area, or assist in their recovery if reintroduction programs are undertaken.

With the water regime, vegetation gradients, and spatial patterns of human land use predicted to shift with climate change (Chapter 12), prioritizing areas for cheetah conservation should incorporate these factors in the decision-making process. Systematic monitoring of PA effectiveness (Gaston et al., 2008; Naughton-Treves et al., 2005), including trends in populations of cheetah and their prey in established PAs, will be essential, affording opportunities for management while the patterns of decline are still reversible.

CONCLUSIONS

PAs have not been viewed as the panacea for cheetah conservation because the majority of the cheetah's range lies outside them (Durant et al., 2017). On unprotected lands, cheetah coexistence with humans is easier than for most other large carnivores because cheetahs are not dangerous to people and generally have a low proportion of livestock in their diet if there is a sustainable prey base (Marker et al., 2003). However, if the current trends in human population size, consumptive footprint, and habitat degradation continue, the last cheetahs to persist will probably do so inside PAs rather than on unprotected land, where they would likely be driven to extinction through prey loss and persecution. In the coming decades, the human population in Africa, and globally, is set to grow exponentially, with a projected associated increase in livestock numbers (Ripple et al., 2014; Chapter 11). As such, PAs will become increasingly important for all wildlife in Africa, as they stand the best chance of not being converted to alternative land uses as human populations grow.

Therefore, PAs need to be seen as strongholds for cheetahs and conservation efforts should preserve existing populations that inhabit PAs; reestablish cheetahs in parks and reserves where they occurred historically; expand the PA network, while considering movement connectivity; as well as work with land users, neighboring parks, and reserves to facilitate cheetah–human coexistence. For parks and reserves to fulfill their potential, major investment, a long-term vision, and strategic plan are often needed. The extent of conservation challenges suggests that we will be unable to save everything, and a triage approach is sometimes necessary for conservation prioritization.

PAs play an important role in the human aspect of cheetah conservation, through ecotourism, rural development, and education on conservation issues, thereby reaching those people who might otherwise have become antagonists

of cheetah conservation, such as rural farmers. Despite the fact that PAs occupy a relatively small proportion of cheetah range, they have distinct features from nonprotected land and functions that make them important for the conservation of cheetahs and other wildlife.

References

Balme, G.A., Slotow, R., Hunter, L.T.B., 2010. Edge effects and the impact of non-protected areas in carnivore conservation: leopards in the Phinda–Mkhuze Complex, South Africa. Anim. Conserv. 13, 315–323.

Broekhuis, F., Cozzi, G., Valeix, M., McNutt, J.W., Macdonald, D.W., 2013. Risk avoidance in sympatric large carnivores: reactive or predictive? J. Anim. Ecol. 82 (5), 1098–1105.

Bruner, A.G., Gullison, R.E., Rice, R.E., da Fonseca, G.A.B., 2001. Effectiveness of parks in protecting tropical biodiversity. Science 291, 125–128.

Creel, S., Becker, M.S., Durant, S.M., M'Soka, J., Matandiko, W., Dickman, A.J., Christianson, D., Droge, E., Mweetwa, T., Pettorelli, N., Rosenblatt, E., Schuette, P., Woodroffe, R., Bashir, S., Beudels-Jamar, R.C., Blake, S., Borner, M., Breitenmoser, C., Broekhuis, F., Cozzi, G., Davenport, T.R.B., Deutsch, J., Dollar, L., Dolrenry, S., Douglas-Hamilton, I., Fitzherbert, E., Foley, C., Hazzah, L., Henschel, P., Hilborn, R., Hopcraft, J.G.C., Ikanda, D., Jacobson, A., Joubert, B., Joubert, D., Kelly, M.S., Lichtenfeld, L., Mace, G.M., Milanzi, J., Mitchell, N., Msuha, M., Muir, R., Nyahongo, J., Pimm, S., Purchase, G., Schenck, C., Sillero-Zubiri, C., Sinclair, A.R.E., Songorwa, A.N., Stanley-Price, M., Tehou, C.A., Trout, C., Wall, J., Wittemyer, G., Zimmermann, A., 2013. Conserving large populations of lions—the argument for fences has holes. Ecol. Lett. 16, 1413.

Dickman, A.J., Macdonald, E.A., Macdonald, D.W., 2011. A review of financial instruments to pay for predator conservation and encourage human-carnivore coexistence. Proc. Natl. Acad. Sci. USA 108, 13937–13944.

Dobson, A.P., Lynes, L., 2008. How does poaching affect the size of national parks? Trends Res. Ecol. Evol. 23, 177–180.

Durant, S.M., 1998. Competition refuges and coexistence: an example from Serengeti carnivores. J. Anim. Ecol. 67, 370–386.

Durant, S.M., 2000. Living with the enemy: avoidance of hyenas and lions by cheetahs in the Serengeti. Behav. Ecol. 11, 624–632.

Durant, S.M., Bashir, S., Maddox, T., Laurenson, M.K., 2007. Relating long-term studies to conservation practice: the case of the Serengeti Cheetah Project. Conserv. Biol. 21, 602–611.

Durant, S.M., Mitchell, N., Groom, R., Pettorelli, N., Ipavec, A., Jacobson, A.P., Woodroffe, R., Böhm, M., Hunter, L.T.B., Becker, M.S., Broekhuis, F., Bashir, S., Andresen, L., Aschenborn, O., Beddiaf, M., Belbachir, F., Belbachir-Bazi, A., Berbash, A., de Matos Machado, I.B., Breitenmoser, C., Chege, M., Cilliers, D., Davies-Mostert, H., Dickman, A.J., Ezekiel, F., Farhadinia, M.S., Funston, P., Henschel, P., Horgan, J., de Iongh, H.H., Jowkar, H., Klein, R., Lindsey, P.A., Marker, L., Marnewick, K., Melzheimer, J., Merkle, J., M'soka, J., Msuha, M., O'Neill, H., Parker, M., Purchase, G., Sahailou, S., Saidu, Y., Samna, A., Schmidt-Küntzel, A., Selebatso, E., Sogbohossou, E.A., Soultan, A., Stone, E., van der Meer, E., van Vuuren, R., Wykstra, M., Young-Overton, K., 2017. The global decline of cheetah *Acinonyx jubatus* and what it means for conservation. Proc. Natl. Acad. Sci. USA 114, 528–533.

East, R., 1984. Rainfall, soil nutrient status and biomass of large African savanna mammals. Afr. J. Ecol. 22, 245–270.

Gaston, K.J., Jackson, S.F., Cantu-Salazar, L., Cruz-Pinon, G., 2008. The ecological performance of protected areas. Annu. Rev. Ecol. Evol. Syst. 39, 93–113.

Hayward, M.W., Hofmeyr, M., O'Brien, J., Kerley, G.I.H., 2006. Prey preferences of the cheetah *Acinonyx jubatus*: morphological limitations or the need to capture rapidly consumable prey before kleptoparasites arrive? J. Zool. 270, 615–627.

Hayward, M.W., Kerley, G.I.H., 2009. Fencing for conservation: restriction of evolutionary potential or a riposte to threatening processes? Biol. Conserv. 142, 1–13.

Hayward, M.W., Kerley, G.I.H., Adendorff, J., Moolman, L.C., O'Brien, J., Sholto-Douglas, A., et al., 2007a. The reintroduction of large carnivores to the Eastern Cape, South Africa: an assessment. Oryx 41 (2), 205–214.

Hayward, M.W., O'Brien, J., Kerley, G.I.H., 2007b. Carrying capacity of large African predators: predictions and tests. Biol. Conserv. 139, 219–229.

Inskip, C., Zimmermann, A., 2009. Human-felid conflict: a review of patterns and priorities worldwide. Oryx 43, 18–34.

Iranian Cheetah Society, 2016. Mining plan threatens critical cheetah habitat in central Iran. Iranian Cheetah Society Newsletter. Available from: http://www.wildlife.ir/en/2016/12/28/mining-plan-threatens-critical-cheetah-habitat-central-iran/.

Jenkins, C.N., Joppa, L., 2009. Expansion of the global terrestrial protected area system. Biol. Conserv. 142, 2166–2174.

Kesch, M.K., Bauer, D.T., Loveridge, A.J., 2015. Break on through to the other side: the effectiveness of game fencing to mitigate human–wildlife conflict. Afr. J. Wildlife Res. 45, 76–87.

Kolowski, J.M., Holekamp, K.E., 2008. Effects of an open refuse pit on space use patterns of spotted hyenas. Afr. J. Ecol. 46 (3), 341–349.

Laurenson, M.K., 1994. High juvenile mortality in cheetahs (*Acinonyx jubatus*) and its consequences for maternal care. J. Zool. 234 (3), 387–408.

Leowinata, B., 2014. Illegal charcoal trade engulfs cheetah habitat. Cat Watch. National Geographic. Available from: http://voices.nationalgeographic.com/2014/04/16/africas-illegal-charcoal-trade-engulfs-cheetah-habitat/.

Lindsey, P.A., Alexander, R., Mills, M.G.L., Romañach, S., Woodroffe, R., 2007. Wildlife viewing preferences of visitors to protected areas in South Africa: implications for the role of ecotourism in conservation. J. Ecotour. 6, 19–33.

Lindsey, P.A., Balme, G., Becker, M., Begg, C., Bento, C., Bocchino, C., Dickman, A., Diggle, R.W., Eves, H., Henschel, P., Lewis, D., Marnewick, K., Mattheus, J., McNutt, J.W., McRobb, R., Midlane, N., Milanzi, J., Morley, R., Murphree, M., Opyene, V., Phadima, J., Purchase, G., Rentsch, D., Roche, C., Shaw, J., Van der Westhuizen, H., Van Vliet, N., Zisadza-Gandiwa, P., 2013. The bushmeat trade in African savannas: impacts, drivers, and possible solutions. Biol. Conserv. 160, 80–96.

Lindsey, P., Masterson, C., Romanach, S., Beck, A., 2012. The ecological, financial and social issues associated with fencing as a conservation tool in southern Africa. In: Somers, M., Hayward, M. (Eds.), Fencing for Conservation—Restriction of Evolutionary Potential or a Riposte to Threatening Processes? Springer Publishers, New York, NY, pp. 215–234.

Lindsey, P.A., Petracca, L.S., Funston, P.J., Bauer, H., Dickman, A., Everatt, K., Flyman, M., Henschel, P., Hinks, A.E., Kasiki, S., Loveridge, A., Macdonald, D.W., Mandisodza, R., Mgoola, W., Miller, S.M., Nazerali, S., Siege, L., Uiseb, K., Hunter, L.T.B., 2017. The performance of African protected areas for lions and their prey. Biol. Conserv. 209, 137–149.

Lindsey, P.A., Tambling, C.J., Brummer, R., Davies-Mostert, H., Hayward, M., Marnewick, K., Parker, D., 2011. Minimum prey and area requirements of the Vulnerable cheetah *Acinonyx jubatus*: implications for reintroduction and management of the species in South Africa. Oryx 45, 587–599.

Løvschal, M., Bøcher, P.K., Pilgaard, J., Amoke, I., Odingo, A., Thuo, A., Svenning, J.-C., 2017. Fencing bodes a rapid collapse of the unique Greater Mara ecosystem. Sci. Rep. 7, 41450.

Marker, L.L., Dickman, A.J., 2004. Human aspects of cheetah conservation: lessons learned from the Namibian farmlands. Human Dimens. Wildlife 9 (4), 297–305.

Marker, L.L., Muntifering, J.R., Dickman, A.J., Mills, M.G.L., Macdonald, D.W., 2003. Quantifying prey preferences of free-ranging Namibian cheetah. S. Afr. J. Wildlife Res. 33, 45–53.

Mills, M.G.L., Mills, M.E.J., 2014. Cheetah cub survival revisited: a re-evaluation of the role of predation, especially by lions, and implications for conservation. J. Zool. 292 (2014), 136–141.

Mills, M.G.L., Mills, M.E.J., 2017. Kalahari Cheetahs: Adaptations to an Arid Region. Oxford University Press, Oxford.

Moqanaki, E.M., Cushman, S.A., 2017. All roads lead to Iran: predicting landscape connectivity of the last stronghold for the critically endangered Asiatic cheetah. Anim. Conserv. 20, 29–41.

Naughton-Treves, L., Holland, M.B., Brandon, K., 2005. The role of protected areas in conserving biodiversity and sustaining local livelihoods. Annu. Rev. Environ. Resour. 30, 219–252.

Newmark, W.D., 2008. Isolation of African protected areas. Front. Ecol. Environ. 6, 321–328.

Noss, R.F., 1990. Indicators for monitoring biodiversity: a hierarchical approach. Conserv. Biol. 4, 355–364.

Oriol-Cotterill, A., Macdonald, D.W., Valeix, M., Ekwanga, S., Frank, L.G., 2015. Spatiotemporal patterns of lion space use in a human-dominated landscape. Anim. Behav. 101, 27–39.

Packer, C., Loveridge, A., Canney, S., Caro, T., Garnett, S.T., Pfeifer, M., Zander, K.K., Swanson, A., MacNulty, D., Balme, G., Bauer, H., Begg, C.M., Begg, K.S., Bhalla, S., Bissett, C., Bodasing, T., Brink, H., Burger, A., Burton, A.C., Clegg, B., Dell, S., Delsink, A., Dickerson, T., Dloniak, S.M., Druce, D., Frank, L., Funston, P., Gichohi, N., Groom, R., Hanekom, C., Heath, B., Hunter, L., DeIongh, H.H., Joubert, C.J., Kasiki, S.M., Kissui, B., Knocker, W., Leathem, B., Lindsey, P.A., Maclennan, S.D., McNutt, J.W., Miller, S.M., Naylor, S., Nel, P., Ng'weno, C., Nicholls, K., Ogutu, J.O., Okot-Omoya, E., Patterson, B.D., Plumptre, A., Salerno, J., Skinner, K., Slotow, R., Sogbohossou, E.A., Stratford, K.J., Winterbach, C., Winterbach, H., Polasky, S., 2013. Conserving large carnivores: dollars and fence. Ecol. Lett. 16, 635–641.

Ripple, W.J., Estes, J.A., Beschta, R.L., Wilmers, C.C., Ritchie, E.G., Hebblewhite, M., Berger, J., Elmhagen, B., Letnic, M., Nelson, M.P., Schmitz, O.J., Smith, D.W., Wallach, A.D., Wirsing, A.J., 2014. Status and ecological effects of the world's largest carnivores. Science 343, 1241484.

Rodrigues, A.S.L., Andelman, S.J., Bakarr, M.I., Boitani, L., Brooks, T.M., Cowling, R.M., Fishpool, L.D.C., da Fonseca, G.A.B., Gaston, K.J., Hoffmann, M., Long, J.S., Marquet, P.A., Pilgrim, J.D., Pressey, R.L., Schipper, J., Sechrest, W., Stuart, S.N., Underhill, L.G., Waller, R.W., Watts, M.E.J., Yan, X., 2004. Effectiveness of the global protected area network in representing species diversity. Nature 428, 640–643.

Scheyvens, R., 1999. Case study ecotourism and the empowerment of local communities. Tour. Mgmt. 20, 245–249.

Schlaepfer, M.A., Runge, M.C., Sherman, P.W., 2002. Ecological and evolutionary traps. Trends Ecol. Evol. 17, 474–480.

Swanson, A., Caro, T., Davies-Mostert, H., Michael, G.L., Mills, M.G.L., Macdonald, D.W., Borner, M., Masenga, E., Packer, C., 2014. Cheetahs and wild dogs show contrasting patterns of suppression by lions. J. Anim. Ecol. 83, 1418–1427.

UNEP-WCMC, WRI, 2014. Terrestrial protected areas (% of total land area). United Nations Environmental Program—World Conservation Monitoring Centre and World Resources Institute. Available from: http://data.worldbank.org/indicator/ER.LND.PTLD.ZS?end=2014&start=2014&view=bar.

UNESCO, 2017. Biosphere reserves—learning sites for sustainable development. Available from: http://www.unesco.org/new/en/natural-sciences/environment/ecological-sciences/biosphere-reserves.

Vanak, A.T., Fortin, D., Thaker, M., Ogden, M., Owen, C., Greatwood, S., Slotow, R., 2013. Moving to stay in place: behavioral mechanisms for coexistence of African large carnivores. Ecology 94 (11), 2619–2631.

Walpole, M.J., Thouless, C.R., 2005. Increasing the value of wildlife through non-consumptive use? Deconstructing the myths of ecotourism and community-based tourism in the tropics. In: Woodroffe, R., Thirgood, S., Rabinowitz, A. (Eds.), People and Wildlife: Conflict or Coexistence? Cambridge University Press, Cambridge, pp. 122–139.

Watson, J.E.M., Dudley, N., Segan, D.B., Hockings, M., 2014. The performance and potential of protected areas. Nature 515, 67–73.

Woodroffe, R., Ginsberg, J.R., 1998. Edge effects and the extinction of populations inside protected areas. Science 280, 2126–2128.

CHAPTER

20

Cheetah Translocation and Reintroduction Programs: Past, Present, and Future

Lorraine K. Boast*, Elena V. Chelysheva**,
Vincent van der Merwe[†], Anne Schmidt-Küntzel[‡],
Eli H. Walker[‡], Deon Cilliers[§], Markus Gusset[¶],
Laurie Marker[‡]

*Cheetah Conservation Botswana, Gaborone, Botswana
**Mara-Meru Cheetah Project, Kenya Wildlife Service, Nairobi, Kenya
[†]Endangered Wildlife Trust, Modderfontein, Gauteng, South Africa
[‡]Cheetah Conservation Fund, Otjiwarongo, Namibia
[§]Cheetah Outreach Trust, Cape Town, South Africa
[¶]World Association of Zoos and Aquariums, Gland, Switzerland

INTRODUCTION

The deliberate movement of individuals of species from one site for release to another (i.e., translocation) is a conservation tool used with increasing frequency (Seddon et al., 2007). "Conservation translocations," as opposed to those conducted purely for commercial objectives or to reduce human-wildlife conflict, have the purpose to yield a "measurable conservation benefit at the levels of a population, species, or ecosystem" (IUCN SSC, 2013). In other words, the benefit should go beyond the translocated individual. Conservation translocations are conducted to maintain gene flow, for example, during metapopulation management, or as part of a reinforcement or reintroduction program, to restore animals to an area where they are threatened or no longer occur. Due to their prey requirements and potential for human-wildlife conflict, carnivores, especially large species, are considered harder to translocate than herbivores

(Wolf et al., 1996). However, there have been successful carnivore reintroduction programs, for example, gray wolves (*Canis lupus*), Eurasian lynx (*Lynx lynx*), and brown bears (*Ursus arctos*) (Hayward and Somers, 2009).

As cheetah (*Acinonyx jubatus*) populations continue to decline in numbers and range (Durant et al., 2017), the reintroduction of cheetahs into suitable areas of habitat has been suggested as a potential conservation measure. Cheetahs display characteristics necessary for successful translocation; for example, they can tolerate a wide variety of areas and consume a broad range of prey species (Caro, 1994). However, they naturally occur at low densities, have large home ranges, are susceptible to competition from larger carnivores, and their release may be viewed negatively by land users within or near release sites (Chapters 8, 9, and 13). This chapter provides an overview of the knowledge and past experience of cheetah translocations, in order to discuss the feasibility of reintroduction programs as a conservation measure for the species.

RATIONALE FOR THE TRANSLOCATION OF CHEETAHS

Reintroduction programs have the potential to increase the current distribution of cheetahs by reclaiming past distribution areas. Additionally, releasing cheetahs into small existing cheetah populations or reestablishing connectivity between fragmented cheetah populations has the potential to boost genetic diversity at a local scale (Johnson et al., 2010b), thereby minimizing the genetic and demographic problems associated with small populations (Chapter 6 and 10). As cheetahs disperse over long distances (Marker, 2002), recolonization of large areas and enhancement of gene flow between relatively distant populations might be achieved through the reintroduction of connector populations.

The reintroduction of cheetahs also has the potential to benefit other species. Large carnivores have cascading effects on lower trophic levels and are necessary for the maintenance of biodiversity and ecosystem functioning. For example, the recovery of Eurasian lynx and gray wolf populations has arguably restored ecological balance to areas where populations of these top predators had diminished (Ripple et al., 2014). The cheetah could also act as a flagship species for the reintroduction site, acting as an ambassador for its protection. Proponents of cheetah reintroductions also argue that, as humans, we have an ethical responsibility to reintroduce cheetahs, as it was human expansion that was responsible for their removal (Ranjitsinh and Jhala, 2010).

PAST CHEETAH TRANSLOCATIONS AND METAPOPULATION MANAGEMENT

The first translocated cheetahs were released into fenced and unfenced nationally protected areas during the 1960s and 1970s in Namibia and South Africa, to reintroduce or reinforce existing populations (Anderson, 1983; Du Preez, 1970). Subsequent cheetah releases into unfenced environments have been documented (Boast et al., 2016; Marker et al., 2008; Purchase and Vhurumuku, 2005; Weise et al., 2015). However, the vast majority of translocated cheetahs were released into fenced areas. Legislation passed in South Africa in the 1960s returned the right to utilize wildlife to landowners, paving the way for the development of private game reserves (McGranahan, 2008). In 1991, landowners began stocking private reserves with cheetahs for tourism purposes, and cheetah translocations intensified during the mid-1990s to mid-2000s (Chelysheva, 2011). These reserves are fenced with "predator-proof," often electrified fencing. Although no fence will retain all predators with 100% effectiveness, "predator-proof" fencing reduces the chance that predators will be able to leave the reserve. The number of cheetahs released in each reserve is usually small [2–8 cheetah in 74% of sites (n = 65) reviewed by Chelysheva (2011)]. In 2009, the decision was taken to manage the cheetahs in South African private game reserves as a metapopulation, with the aim of maintaining demographic and genetic viability (Lindsey and Davies-Mostert, 2009; see http://www.cheetahpopulation.org.za for further information). By 2016, cheetah reintroductions had been attempted at 72 fenced reserves in South Africa, and the metapopulation has increased naturally from 241 cheetahs in 41 reserves in 2011 to 325 individuals in 54 reserves (16 state owned and 38 private game reserves) in 2017 [Endangered Wildlife Trust (EWT), unpublished data]. The metapopulation program has the potential to support a viable population of cheetahs, but it is not sustainable without intensive management (Lindsey et al., 2009a).

SUCCESS OF PAST CHEETAH TRANSLOCATIONS

Evaluating the success of cheetah translocations is complicated. The outcomes of many incidences are unpublished and those that are published potentially suffer from positive publication bias (i.e., successes are more likely to be published than failures). Success is generally based on reproductive success, but programs often use different definitions of this term (Hayward et al., 2007a). A metaanalysis of documented cheetah translocations determined that at least 727 cheetahs were translocated into 64 sites in southern Africa between 1965 and 2010 (Chelysheva, 2011). Six of the 64 release sites were considered successful (Chelysheva, 2011) based on natural recruitment (i.e., births) exceeding adult mortality 3 years after reintroductions began (as defined in Hayward et al., 2007a). In many of the

other projects, the number of cheetahs released was small and long-term monitoring was not conducted. If such long-term monitoring had been implemented and documented, additional sites might have been deemed successful. Indeed, as of 2016, 71% of the 72 sites at which cheetah reintroductions have been attempted in South Africa have breeding populations of cheetahs, which are currently contributing to the South African metapopulation (EWT, unpublished data).

Four of the six sites, deemed successful by Chelysheva (2011), have persisting cheetah populations in 2016 (EWT, unpublished data). Three of these four sites are fenced reserves within the South African metapopulation and one is a free-ranging population in Zimbabwe's Matusadona National Park. However, after an initial population growth (Purchase and Vhurumuku, 2005) this free-ranging population has declined to three related individuals (van der Meer, 2016). Insufficient area of habitat (due to rising lake levels), increasing human-wildlife conflict on park borders (in response to a growing human population and an economic crisis in Zimbabwe), a lack of subsequent cheetah releases, and limited opportunity for natural colonization are likely to have contributed to its failure (van der Meer, 2016). It is considered inadvisable to conduct future translocations into the area and unfeasible to incorporate the population into a metapopulation plan (van der Meer, personal communication). Without this support, this free-ranging population is not viable in the long term (van der Meer, 2016).

However, since the Chelysheva (2011) metaanalysis, the successful translocation of rehabilitated orphaned cheetahs and cheetahs suspected of killing livestock into the unfenced NamibRand Nature Reserve in Namibia has been documented (Marker, et al., in preparation; Weise et al., 2015). Cheetahs were previously believed to be absent or transient in the reserve, and these releases have resulted in the establishment of a self-sustaining resident cheetah population in the larger pro-Namib ecosystem (Weise and Odendaal, personal communication).

POSTRELEASE MOVEMENTS OF CHEETAHS

The most critical period for the survival of translocated cheetahs is the first 3–4 months postrelease (Fontúrbel and Simonetti, 2011; Hunter, 1999; Weise et al., 2015). During this initial period, carnivores often make large exploratory and sometimes directional movements, often toward home (Linnell et al., 1997; Marker et al., 2008). For example, 5 out of 20 translocated cheetahs studied in Namibia roamed over 2000 km² during the first 3 months after release (Weise et al., 2015), and an individual in Botswana returned to its capture site after being translocated 170 km away (Boast et al., 2016).

Cheetahs tend to stop exploratory and directional movements within 2–3 months postrelease (Hunter, 1999); after 3 months, cheetahs translocated in Namibia showed no significant difference in home range sizes and daily movements compared with cheetahs released at the capture site (Marker et al., 2008). During this critical period in free-ranging environments, the large movements expose cheetahs to multiple threats, including crossing roads and farmland owned by multiple landowners, some of whom are likely to be intolerant of predators. As a result, high mortality rates of cheetahs released into free-ranging environments are often (although not always; Marker et al., 2008) recorded during this initial period (Boast et al., 2016; Weise et al., 2015) (Table 20.1). Human-mediated mortality is the primary cause of death for cheetahs translocated into free-ranging environments (Boast et al., 2016; Du Preez, 1970; Purchase and Vhurumuku, 2005; Weise et al., 2015).

TABLE 20.1   Survival Rates of Documented Cheetah Translocations

| Release site[b] | Source of cheetahs[c] | Percentage of adult animals surviving[a] | | Source |
		110 Days postrelease	1 Year postrelease	
Various *fenced reserves* in South Africa	Suspected damage causing	Not stated	85% (n = 92)	Marnewick et al. (2009)
Fenced reserve in Namibia	Orphaned	100% (n = 10)	60% (n = 10)	Marker et al. (in preparation)
Free-ranging reserve in Namibia	Orphaned	71% (n = 7)	71% (n = 7)	Marker et al. (in preparation)
Various *free-ranging* sites in Namibia	Suspected damage causing Orphaned	71% (n = 17) 40% (n = 5)	56% (n = 16)[d] 25% (n = 4)[d]	Weise et al. (2015)
Various *free-ranging* sites in Botswana	Suspected damage causing	33% (n = 11)	18% (n = 11)	Boast et al. (2016)
Free-ranging reserve in Botswana	Orphaned	100% (n = 3)	0% (n = 3)	Houser et al. (2011)

[a]110 days and 1 year postrelease survival were success rate criteria used by Fontúrbel and Simonetti (2011).
[b]Fenced refers to predator-proof fencing.
[c]Suspected damage-causing cheetahs are those perceived by landowners to be depredating on livestock, although evidence of depredation was not always present.
[d]Collar failed for 1 cheetah.

FACTORS ASSOCIATED WITH THE SUCCESSFUL TRANSLOCATION OF CHEETAHS

The principal factor associated with reproductive success in a carnivore translocation program is the suitability of the release site for the target species, and in the case of free-range releases, the suitability of the surrounding area. Important characteristics of the release site include habitat and prey availability, the potential for intra- and interspecific competition, and the cheetahs' ability to leave the site (Johnson et al., 2010a). Additional factors known to affect success are the individual cheetah's background, number, and grouping of cheetahs released, the method of release, and the availability of postrelease monitoring and care. The success of a reintroduction program will also require the support of people living near the release site. Reintroduced animals must not threaten resident wildlife, for example, through disease, genetic factors or competition, and the program must be cost effective. These factors will be discussed to offer insight into the feasibility of reintroduction as a conservation measure for cheetahs and to guide future programs.

Free-Range Versus Fenced Release Sites

In some areas, fenced release sites may be the only option available, for example, in densely populated and more developed regions of South Africa, Malawi, and Swaziland. Animals reintroduced into reserves with "predator-proof" fencing experience greater reproductive success than animals reintroduced into unfenced free-ranging environments (Chelysheva, 2011). Five of the six reintroduction sites considered as successful in Chelysheva's (2011) metaanalysis of cheetah translocations were reserves with "predator-proof" fencing (Bissett, 2004; Hayward et al., 2007a; Hofmeyr and van Dyk, 1998; Hunter, 1998a; Pettifer et al., 1982). In areas with "predator-proof" fencing, it is

difficult for cheetahs to leave the reserve; they are protected from external causes of mortality (e.g., human–predator conflict), and also generally receive greater follow-up and care than is possible for cheetahs released into free-ranging areas. However, despite cheetahs having a greater chance of survival in fenced environments (Table 20.1), these populations must be intensely managed to maintain genetic diversity and avoid exceeding the reserve's carrying capacity of carnivores. Not all cheetahs in fenced reserves are part of a metapopulation management plan, and even those that are included, are dependent on intensive management and are potentially vulnerable to changes in land ownership, land-use, and financial support. As a result, metapopulation management may not be the most appropriate long-term option across the cheetah's range. Where possible, reintroduction programs for cheetahs should focus on free-ranging areas, where cheetah dispersal could potentially facilitate natural colonization and connectivity with existing populations.

Characteristics of Release Sites: Ecological Factors

It is imperative that the threats responsible for the initial removal of cheetahs at potential release sites are addressed, or plans are in place to mitigate the threat (Hayward and Slotow, 2016). A habitat suitability study should be conducted at each site to ensure there is sufficient vegetation to support viable prey populations to sustain the reintroduced cheetah population in the long term. In fenced reserves sufficient prey should be available to sustain the reintroduced cheetahs and all other predators for at least 18 months before supplementation. The reintroduced cheetah population needs to be protected from anthropogenic threats, and the potential impact of unnaturally high inter- and intraspecific competition needs to be managed.

Due to the cheetah's large home ranges and tendency to occur at low densities (Chapter 8), release sites need to be part of a larger suitable landscape or else intensive metapopulation management becomes necessary. Using habitat modeling, Weise et al. (2015) found that the released cheetahs' movements would extend beyond the boundary of all protected areas in Namibia, regardless of the chosen release site. Nevertheless, cheetahs can survive in unprotected landscapes (e.g., 77% of the cheetah's current known range is on unprotected land: Durant et al., 2017). However, safeguarding areas of suitable habitat around the release site and providing wildlife corridors for natural cheetah dispersal and connectivity between populations will be necessary.

One of the key factors in determining reintroduction success in the metaanalysis of cheetah translocations by Chelysheva (2011) was the density of lions (*Panthera leo*) and spotted hyenas (*Crocuta crocuta*) in the release area. In addition to interspecific competition, intraspecific competition is also likely to have an impact on the survival of cheetahs released to reinforce existing populations. The presence of resident cheetahs can result in territorial fights, sometimes leading to the death of reintroduced individuals (Bissett, 2004; Hayward et al., 2007a; Marker et al., 2008). At the very least, competition is likely to result in reintroduced cheetahs needing to roam further to establish themselves in the area (Pettifer et al., 1982), potentially increasing their exposure to threats.

Source of Cheetahs for Reintroductions

Cheetah reintroductions are most likely to succeed if founder animals are healthy adults, sourced from similar environments (e.g., prey density, competitors) to those into which they will be released. If animals will be released in areas with other large carnivores, it is recommended that reintroduced cheetahs have experience with these competitors (Boast et al., 2016; Hayward et al., 2007b; Hunter, 1998a; Weise et al., 2015). A reintroduced cheetah's naivety

of large competitors is thought to contribute not only to the poorer survival of the individual but also poorer survival of cubs born to naive mothers (Marnewick et al., 2009). Ideally, cheetahs should be sourced from an area with these large predators (Hayward et al., 2007b); although exposure during soft release (when cheetahs are temporarily held in a holding-pen at the release site) is also thought to improve survival outcomes. For releases into fenced reserves, within a metapopulation management plan, the best source of cheetahs are those from within a viable existing metapopulation. For example, 5–10 cheetahs become available for reintroduction attempts per year from the South African metapopulation (EWT, unpublished data). These cheetahs are successful hunters and in the majority of cases have experience of large competitors (e.g., 73% of reserves in the South African metapopulation have lions). However, predominately, these cheetahs are highly habituated to the presence of game-drive vehicles, and habituated cheetahs make poor candidates for release into free-ranging environments where they are likely to encounter human presence.

For release into free-ranging environments, there is no ideal source of cheetahs. The removal of cheetahs from the wild, even from within healthy populations, should not be advocated due to the potential strain it could place on source populations. Past reintroduction programs have used cheetahs perceived by landowners to be predating on livestock (Marnewick et al., 2009; Purchase and Vhurumuku, 2005). These suspected damage-causing cheetahs are often threatened with lethal control if not removed (see Box 20.1 for a discussion on translocating cheetahs to reduce human-wildlife conflict). Although the risk of transferring conflict is thought to be low (Purchase and Vhurumuku, 2005) and these cheetahs are likely to have a fair chance of survival because they have experience of the wild; the primary message promoted by cheetah conservationists on farmland should be coexistence not removal.

BOX 20.1

TRANSLOCATING CHEETAHS TO REDUCE HUMAN-CHEETAH CONFLICT

Conservation organizations are often under pressure to translocate cheetahs that are believed to be responsible for livestock depredation to prevent them from being killed, and as such, translocation is often considered a humane method to mitigate human–carnivore conflict (Massei et al., 2010). Frequently; however, the demand for cheetah removal outweighs the availability of suitable reintroduction sites and resources. As a result, the majority of cheetahs captured due to perceived or actual depredation on livestock have been relocated within existing cheetah populations, without clear reintroduction aims. These translocations, although potentially promoting gene flow and survival of individuals, cannot be viewed as "conservation translocations" (IUCN SSC, 2013), as their benefit is largely restricted to individuals rather than populations.

In a bid to better utilize perceived conflict cheetahs, and as a trial conflict mitigation method, a compensation–relocation program was carried out in South Africa between 2000 and 2006. Cheetahs perceived to be predating on livestock were captured by landowners and relocated to private reserves and national parks (Lindsey et al., 2009a; Marnewick et al., 2009). The cheetahs captured and removed both in South Africa and as part of other translocation programs were generally perceived, but not known, to have killed livestock. As a result, although there is a chance some cheetahs will kill livestock at the release site (Boast et al., 2016), cheetah conflict with human populations neighboring free-ranging release sites does not necessarily increase (Purchase and Vhurumuku, 2005).

The translocation of those predators, suspected to be killing livestock, provides farmers with a perceived level of control over predation risk, as cheetahs can be removed if their presence can no longer be tolerated (Marnewick et al., 2009). Maintaining a degree of control often decreases the level of threat that cheetahs are perceived to pose to livestock, and as such is likely to increase tolerance toward cheetahs (Boast et al., 2016; Dickman, 2008). Although few studies published quantitative data on stock losses before and after predator translocations, those that did have shown conflicting and often inconclusive results (Linnell et al., 1997). Farmers generally perceive predator translocations to be ineffective at reducing stock losses (Boast et al., 2016), and the majority of farmers who resort to this method, request removal of other cheetahs within a short timeframe. For example, Weise et al. (2015) found that 64% of farmers ($n = 14$) requested the removal of additional cheetahs within 2 years of the first animal being removed. These repeated removals have the potential to create a population sink, which results in vacant territories and increased home ranges of resident cheetahs (Marker, 2002), potentially compromising the viability of the source population. Vacant territories, in turn, encourage immigration of new individuals, which as observed with the removal of pumas (Teichman et al., 2016), may increase the risk of human–predator conflict. Removing predators is counterproductive to encouraging landowners to coexist with large carnivores on their land, and the impact of repeated removals on wild populations was the primary reason the compensation–relocation program in South Africa was suspended (Marnewick, personal communication cited in Weise et al., 2015).

As a result, translocation conducted to reduce human–carnivore conflict is unlikely to be justified in areas, such as in southern and eastern Africa where cheetah populations rather than individuals are the management units (Boast et al., 2016; Fontúrbel and Simonetti, 2011; Linnell et al., 1997; Massei et al., 2010; Weise et al., 2015). However, for critically endangered populations, where every individual is crucial, human-cheetah conflict translocations are likely to remain a valuable tool.

An alternative is to use wild-born cheetahs that have been held in captivity (e.g., orphaned cheetahs or injured adults that have been rehabilitated). The instinct to hunt is innate and rehabilitated orphaned cheetahs show hunting, killing, and feeding behaviors similar to those of wild conspecifics (for details of the rewilding methodology, refer to Houser et al., 2011 and Marker et al., in preparation). Several releases of captive-raised, wild-born cheetahs have been documented (Adamson, 1969; Houser et al., 2011; Marker et al., in preparation; Weise et al., 2015). At the end of 2015, there were 160 wild-born captive cheetahs registered in the International Cheetah Studbook in Namibia alone (Marker, 2016); most were orphaned or placed in captivity due to human-wildlife conflict. Returning suitable individuals to fenced reserves (within a metapopulation) or to free-ranging environments would enable these cats to contribute to the gene pool, reduce the pressure on captive facilities, and allow the reintroduction of cheetahs into new populations without the risk of depleting existing wild cheetah populations (Marker et al., in preparation). The reintroduction of rehabilitated cheetahs requires intensive postrelease monitoring, possibly including supplementary feeding (Marker et al., in preparation; Marnewick et al., 2009). Rehabilitated cheetahs have also been reported to choose inappropriate prey resulting in potential injury, they lack experience identifying danger, and they are potentially more susceptible to infectious diseases due to limited previous exposure (Jule et al., 2008). As a result, they generally have a lower chance of survival compared with those born and raised in the wild (Chelysheva, 2011; Hayward et al., 2007a; Jule et al., 2008; Marnewick et al., 2009; Weise et al., 2015) (Table 20.1). Also, the extended captive care of rehabilitated cheetahs often results in their habituation to humans, which is associated with a greater potential for human–carnivore conflict and poorer survival in free-ranging environments (Bauer, 2005; Weise et al., 2015).

Captive-bred cheetahs, although another abundant source of cheetahs that can be rehabilitated for release into the wild (Ferguson, 1993; Pettifer, 1981), are not well suited for release as they have usually been exposed to intense human contact and have never witnessed any wild behavior from conspecifics. The release of these animals can be justified only as a last resort and only from reputable breeding programs using individuals with a known genetic background (Chapter 22).

Interactions with Resident Cheetahs: Disease and Genetics

Cheetahs from different regions are likely to have been exposed to different pathogens and parasites (Castro-Prieto et al., 2012), including those found in domestic felines and canines (Munson et al., 2004). To avoid cheetahs spreading or acquiring disease through translocation, it is important to know the disease status of the translocated individual(s) and recipient populations. Translocated individuals should be screened for infectious disease and should be free of obvious health or reproductive impairments.

Cheetahs are genetically very similar (Chapter 6), and animals pertaining to the same subspecies should be able to be translocated between populations. Even when introducing cheetahs into a small isolated population, the risk of inbreeding depression (if the population was not reinforced) is expected to outweigh the risk of outbreeding depression, potentially imposed by introducing cheetahs. However, in some cases, it will not be feasible to source cheetahs for reintroduction from the same subspecies. For instance, the remaining population of Asiatic cheetahs (*A. jubatus venaticus*) in Iran is too small to sustain any offtake for reintroduction into other areas of the subspecies' former range. If cheetahs are absent from the area that is to receive translocated animals, as is the case in India, the only consideration relates to the chance of survival of the translocated individuals in their new environment. Given the relatively short separation time between cheetah subspecies (e.g., 4,700–67,400 years ago between *A. jubatus jubatus* and *A. jubatus venaticus*) (Charruau et al., 2011), extreme differentiation leading to the inability to survive in the new environment is not likely to have arisen between the extant cheetah subspecies.

If a remnant cheetah population needed supplementation from another subspecies (i.e., genetic rescue), additional factors become relevant. The available prey populations, the number of resident animals, and the level of genetic differentiation between the source and target population must be assessed. The number of animals released, as well as the length of time they remain in the population must be carefully considered to avoid new individuals outcompeting the existing ones, leading to a shift from one subspecies to the other, rather than a genetic rescue. Two successful genetic rescues were executed in the puma (*Puma concolor*) and lion, where a puma population in Florida and a lion population in Hluhluwe-iMfolozi Park were reinforced with females originating from Texas and Etosha National Park, respectively. As a result, animal numbers increased, and inbreeding correlates declined significantly (Johnson et al., 2010b; Trinkel et al., 2008); genetic heterozygosity was measured in the Florida puma and was found to have doubled (Johnson et al., 2010b). It is important to note that hybrid animals might not benefit from the same protection status under IUCN as individuals of a pure subspecies do (Fitzpatrick et al., 2015; O'Brien and Mayr, 1991).

Optimal Number of Cheetahs for Release

To determine the optimal number of cheetahs for release to ensure the long-term persistence of a reintroduced population, a population viability analysis (PVA) needs to be conducted for the release site. A PVA uses different demographic parameters to estimate survival probabilities for a population while retaining a specified level of genetic diversity (Chapter 38). For example, to maintain 90% genetic diversity 20 years after the proposed release of four males and four females into the Bangweulu wetlands in Zambia, it would be necessary, at a minimum, to reinforce the initial population with another four males and four females every other year for 4 years (Marker, 2010). The optimum number and scheduling of cheetahs to be reintroduced is likely to vary with characteristics of the release site or metapopulation (e.g., prey availability, resident cheetah population) and, therefore, needs to be part of future site feasibility studies.

Group Composition of Cheetahs for Release

Female cheetahs are primarily solitary, and while they can be released with dependent cubs, this is likely to put extra demands upon the female when trying to establish a new home range. Male cheetahs occur either solitary or in coalitions of two or three, rarely four males (Chapter 9). Coalitions of males are more likely to hold a territory and to keep that territory for longer than singletons both in the wild (Caro, 1994) and in reintroduced populations (Hunter, 1998b). Therefore, it is desirable for male cheetahs to be released as coalitions (of two to three males), rather than individuals, to increase their chances of establishment (Fig. 20.1). Given the limited number of cheetahs released when forming a new population of reintroduced animals, it is recommended to use nonrelated individuals (Moritz, 1999). These coalitions can be formed in captive-holding facilities (see Marnewick et al., 2009 for methods used to create artificial coalitions). Relations established during the prerelease period of captivity can be strong and remain stable after release until the death of the animals (Hayward et al., 2007b; Marker et al., in preparation). Females destined for release have also successfully bonded under captive circumstances (Marker et al., in preparation). Full social compatibility of group members

FIGURE 20.1   **Coalition of two male cheetahs released in Botswana.** *Source: Lorraine Boast, Cheetah Conservation Botswana.*

before the release is likely to increase the survival rates of each of its members after the release (Somers and Gusset, 2009).

Method and Timing of Cheetah Release

To reduce intraspecific competition, it has been advised that cheetah male groupings should be released simultaneously in different parts of the reintroduction site (Hayward et al., 2007b). Subsequent cheetah releases should be outside of the established territories of previously released or resident cheetahs (Hunter, 1999), thereby giving cheetahs a chance to establish themselves and recover from the relocation stress before facing territorial conspecifics. In fenced areas, female cheetahs should be released before males (Marnewick et al., 2009). However, in unfenced environments, Marker et al. (in preparation) recommend releasing males first, while female cheetahs remain held in a holding-pen as an "anchor." When this method was tested in the NamibRand Nature Reserve, an introduced male coalition of cheetahs explored and marked the territory but continued to return to the pen holding the females. Once released, females are likely to scent mark in similar areas as the males, keeping all cheetahs in the same general area,

at least during the initial critical postrelease period.

Using a holding-pen to temporarily hold translocated animals while they acclimatize to the release area is known as soft release, as opposed to animals that are released directly into the reintroduction site (hard release). In general, the soft release of carnivores is associated with less postrelease movements and stress than occurs during hard release (Teixeira et al., 2007), subsequently resulting in increased survival (Linnell et al., 1997; Massei et al., 2010; Somers and Gusset, 2009). However, as cheetahs naturally exhibit wide-roaming movements, soft-released cheetahs are still likely to move beyond the boundaries of the release site (e.g., protected area) (Houser et al., 2011; Purchase and Vhurumuku, 2005; Weise et al., 2015) and, therefore, ensuring the suitability of habitat beyond the release site remains crucial for the long-term success of reintroductions.

Postrelease Monitoring of Cheetahs

Determining reproductive and overall success of reintroduction programs requires long-term, targeted, and intensive monitoring of both the reintroduced individuals and their impact on the environment. Monitoring is an essential part of the adaptive management process and is critical to improve the success of the reintroduction program (Gusset, 2009). Cheetahs should ideally be monitored daily during the initial few weeks to ensure they are hunting successfully (Marnewick et al., 2009). Veterinary intervention, if required, is most likely to be necessary during this period. For the release of rehabilitated cheetahs, the need for intensive monitoring (several times a day to several times a week) is likely to be prolonged relative to wild-caught cheetahs, and supplementary feeding will be necessary initially. The intensity of monitoring can be reduced in the long term.

Monitoring has been made easier with the use of satellite GPS collars, which send regular GPS

FIGURE 20.2 **Translocated cheetah in Namibia showing the GPS tracking collar.** *Source: Cheetah Conservation Fund.*

positions of animals (Fig. 20.2) (Chapter 32). However, without visual follow-ups of individuals, it is challenging to determine the health and well-being of the individual and assess the outcome of the translocation (Boast et al., 2016; Wolf et al., 1996). There are also animal-welfare considerations with the use of GPS collars; collars can be fitted only to adult cheetahs, it is advisable the collar weight does not exceed 400 g, the collar must be removed or drop-off at the end of the study, and ultimately the collar should not compromise the cheetah's survival.

Postrelease monitoring substantially increases the costs of any reintroduction program; for example, 56%–60% of the estimated costs to translocate cheetahs relate to monitoring expenses (Boast et al., 2016; Weise et al., 2014). As a result, many programs neglect postrelease monitoring, or limit it to only a few months after the reintroduction of animals (due to both time and financial constraints) (Gusset, 2009). However, the importance of long-term monitoring has been highlighted by the IUCN SSC Reintroduction Specialist Group (Armstong and Seddon, 2007). Weise et al. (2014) found that monitoring expenses were the cost item that the public was most willing to fund, making long-term monitoring feasible as long as it is properly budgeted.

Support from Surrounding Communities

Surrounding communities' attitudes toward the reintroduction of cheetahs are likely to be mixed. Residents farming livestock or game animals are likely to be concerned that such reintroductions will impact their livelihoods, while landowners conducting ecotourism may welcome the reintroductions. Educational workshops, site visits, and involving local residents in the program will be necessary at potential release sites to listen to and ease people's concerns (Hayward et al., 2007b; Weise et al., 2015). Residents may need to be assured that cheetahs are not a threat to human life and should be offered advice on what to do if they see a cheetah. Providing assistance to improve livestock husbandry to protect livestock against predators and offering compensation in cases where it can be proven that a released cheetah has killed livestock, might improve residents' attitudes toward the release (Weise et al., 2015). Communicating the whereabouts of released cheetahs to the owners of the land where they roam has also shown to improve land-owners' interest in and attitudes toward released animals (Weise et al., 2015). If communities can obtain a tangible benefit from the cheetah's presence, for example, through photographic tourism, they are more likely to support its release (Lindsey et al., 2009b). Obtaining the backing of the surrounding communities is necessary for the long-term success of translocations and for the protection of the species and its habitat, and the importance of local human attitudes should not be overlooked (IUCN SSC, 2013).

Financial Costs of Reintroductions

The average cost of translocating a single cheetah in Namibia or South Africa was approximately US $2730 (Marnewick et al., 2009; Weise et al., 2015), excluding fixed costs, such as holding-pens and capture cages. The individual conservation cost of translocating cheetahs in

Namibia, defined as "the cost of one successfully translocated individual [success was defined as a nonhoming individual surviving to at least 1 year postrelease while causing minimal conflict (≤5 livestock per year)] adjusted by costs of unsuccessful events of the same species," was US $6898 (Weise et al., 2014). The authors found that most of these expenses were recovered through fundraising and in-kind donations of veterinary services and vehicles. The cost of translocations substantially increases if extended captive care or the rehabilitation of captive or orphaned cheetahs is necessary (Houser et al., 2011; Marnewick et al., 2009; Weise et al., 2014). The cost of the prey animals consumed by reintroduced predators is an additional factor to be considered (e.g., on private game reserves) (Hayward et al., 2007b). However, the costs of reintroduction programs are unlikely to be a limiting factor; indeed, in some areas the cheetah's presence is also likely to provide a source of revenue from photographic tourism (Lindsey et al., 2007). While these expenses represent a significant proportion of the budget of non-government organizations, it is a small cost to pay if it is a successful conservation action.

FUTURE PLANS FOR THE TRANSLOCATION OF CHEETAHS

In light of a growing need for the strategy, potential release sites for reintroduction have been identified across the cheetah's historical range, in Asia and Africa. In Asia, plans to reintroduce cheetahs into free-ranging environments in Uzbekistan, Turkmenistan, and India have been discussed (Breitenmoser, 2002; Marker, 2012; Ranjitsinh and Jhala, 2010). However, to date, the potential release sites do not meet the habitat requirements for the reintroduction of cheetahs, primarily due to a lack of prey. In Africa, cheetah experts identified approximately 2.7% of the cheetah's historical range in southern Africa and 1.6% of its historical range in north, west,

and central Africa as recoverable land (IUCN/SSC, 2012; RWCP and IUCN/SSC, 2015). These recoverable areas were defined as being sufficiently large with suitable habitat and prey for the reintroduction or natural colonization of cheetahs within 10 years, if reasonable conservation action was taken. No recoverable range was identified in East Africa (IUCN/SSC, 2007). Areas identified as recoverable in north, west, and central Africa included parts of Senegal, Cameroon, Benin, Chad, Democratic Republic of the Congo, and Egypt (IUCN/SSC, 2012). In southern Africa, recoverable areas were identified in Angola, Malawi, Mozambique, South Africa, and Zambia (RWCP and IUCN/SSC, 2015). Plans for cheetah reintroductions into southern Africa are part of a regional project known as "painting the map red," so called because red is the color used to denote resident range in the cheetah status and action plan documents (Cheetah Rangewide Programme, 2011). To date, the only area for which a feasibility study has been conducted is the Bangweulu flood-plains in Zambia; the area was found to be suitable for cheetah reintroduction (Marker, 2010), but cheetah releases have not yet taken place.

CONCLUSIONS

As the cheetah continues to decline in numbers and distribution, the need for cheetah translocation and reintroduction programs becomes stronger. Securing habitat for the cheetah remains the priority both in terms of safeguarding existing habitat and securing new habitat within the species' historical range for recolonization and potential reintroduction. Introducing connector populations to aid natural dispersal is likely to become a major conservation management tool in the future, if cheetah populations continue to become smaller and more fragmented. Cheetahs have been reintroduced into fenced reserves with great success in South Africa. However, fenced reserves require intensive management as a metapopulation and should only be promoted as a model in regions where there is no scope for maintaining a viable free-ranging population long term. The tools that have been developed, and the lessons learned from reintroduction into fenced reserves, can be applied to future metapopulation management and, where applicable, to reintroductions into free-ranging environments.

Potential reintroduction sites for cheetahs need to be evaluated carefully and comprehensively, with an emphasis on existing predator populations and the suitability of the greater landscape for cheetah survival and population connectivity. Feasibility studies need to be conducted for all potential release sites so that reintroduction programs can be prioritized according to the local/regional endangerment of cheetahs and the area's suitability for reintroduction. Investment into postrelease monitoring and techniques to reduce/manage postrelease movements need to be emphasized.

If cheetah populations continue to dwindle, the need for reintroduction programs will increase. Although the emphasis should remain on protecting existing free-ranging populations, it is essential that reintroduction techniques have been, and continue to be, developed to aid the survival of translocated cheetahs.

References

Adamson, J., 1969. The Spotted Sphinx. Harcourt, Brace & World, San Diego, USA.

Anderson, J.L., 1983. A Strategy for Cheetah Conservation in Africa. Endangered Wildlife Trust, Johannesburg, South Africa.

Armstong, D.P., Seddon, P.J., 2007. Directions in reintroduction biology. Trends Ecol. Evol. 23 (1), 20–25.

Bauer, G.B., 2005. Research training for releasable animals. Conserv. Biol. 19, 1779–1789.

Bissett, C., 2004. The Feeding Ecology, Habitat Selection and Hunting Behaviour of Re-Introduced Cheetah on Kwandwe Private Game Reserve, Eastern Cape Province. MSc thesis, Rhodes University, South Africa.

Boast, L., Good, K.M., Klein, R., 2016. Translocation of problem predators: is it an effective way to mitigate conflict between farmers and cheetahs *Acinonyx jubatus* in Botswana? Oryx 50 (3), 537–544.

Breitenmoser, U., 2002. Feasibility study on re-introduction of cheetah in Turkmenistan. Cat News 36, 13–15.

Caro, T.M., 1994. Cheetahs of the Serengeti Plains—Group Living in an Asocial Species. The University of Chicago Press, Chicago, USA.

Castro-Prieto, A., Wachter, B., Melzheimer, J., Thalwitzer, S., Hofer, H., Sommer, S., 2012. Immunogenetic variation and differential pathogen exposure in free-ranging cheetahs across Namibian farmlands. PLoS One 7 (11), e49129.

Charruau, P., Fernandes, C., Orozoo-Terwengel, P., Peters, J., Hunter, L.T.B., Ziaie, H., Jourabchian, A., Jowkar, H., Schaller, G., Ostrowski, S., Vercammen, P., Grange, T., Schlotterer, C., Kotze, A., Geigl, E., Walzer, C., Burger, P., 2011. Phylogeography, genetic structure and population divergence time of cheetahs in Africa and Asia: evidence of long-term geographic isolates. Mol. Ecol. 20 (4), 706–724.

Cheetah Rangewide Programme, 2011. Exploring opportunities for restoration of cheetah within Southern Africa. A Workshop Held at the National Zoological Gardens, 14 June 2011, Pretoria, South Africa.

Chelysheva, E.V., 2011. Cheetah (*Acinonyx jubatus*) Reintroduction—46 years of Translocations (in Russian). Scientific Research at Zoological Parks, Moscow, Russia, pp. 135–179; (Translated to English by Chelysheva, E.V.).

Dickman, A.J., 2008. Key Determinants of Conflict Between People and Wildlife, Particularly Large Carnivores, Around Ruaha National Park, Tanzania. PhD thesis, University College London, United Kingdom.

Du Preez, J.S., 1970. Report on the feeding and release of 30 cheetah in Etosha. Namibian Nature Conservation Report 50/7, Namibia.

Durant, S., Mitchell, N., Groom, R., Pettorelli, N., Ipavec, A., Jacobson, A., Woodroffe, R., Bohm, M., Hunter, L., Becker, M., Broekhuis, F., Bashir, S., Andresen, L., Aschenborn, O., Beddiaf, M., Belbachir, F., Belbachir-Bazi, A., Berbash, A., Brandao de Matos Machado, I., Breitenmoser, C., Chege, M., Cilliers, D., Davies-Mostert, H.T., Dickman, A.J., Ezekiel, F., Farhadinia, M.S., Funston, P.J., Henschel, P., Horgan, J., Hans de longh, H., Houman, J., Klein, R., Lindsey, P.A., Marker, L.L., Marnewick, K., Melzheimer, J., Merkle, J., M'soka, J., Msuha, M., O'Neil, H., Parker, M., Purchase, G., Sahailou, S., Saidu, Y., Samna, A., Schmidt-Küntzel, A., Selebatso, M., Sogbohossou, E., Soultan, A., Stone, E., Van der Meer, E., Van Vuuren, R., Wykstra, M., Young-Overton, K., 2017. Disappearing spots, the global decline of cheetah and what it means for conservation. Proc. Natl. Acad. Sci. USA 114, 528–533.

Ferguson, M., 1993. Introduction of Cheetah Into Mthethomusha Game Reserve, South Africa. International Cheetah Studbook, Washington, DC, USA.

Fitzpatrick, B.M., Ryan, M.E., Johnson, J.R., Corush, J., Carter, E.T., 2015. Hybridization and the species problem in conservation. Curr. Zool. 61 (1), 206–216.

Fontúrbel, F., Simonetti, J.A., 2011. Translocations and human-carnivore conflicts: problem solving or problem creating? Wildl. Biol. 17, 217–224.

Gusset, M., 2009. A framework for evaluating reintroduction success in carnivores: lessons from African wild dogs. In: Hayward, M.W., Somers, M.J. (Eds.), Reintroduction of Top-Order Predators. Wiley-Blackwell, Oxford, UK, pp. 307–320.

Hayward, M.W., Adendorff, J., O'Brien, J., Sholto-Douglas, A., Bissett, C., Moolman, L.C., Bean, P., Fogarty, A., Howarth, D., Slater, R., Kerley, G., 2007a. The reintroduction of large carnivores to the Eastern Cape, South Africa: an assessment. Oryx 41 (2), 205–214.

Hayward, M.W., Adendorff, J., O'Brien, J., Sholto-Douglas, A., Bissett, C., Moolman, L.C., Bean, P., Fogarty, A., Howarth, D., Slater, R., Kerley, G.I.H., 2007b. Practical considerations for the reintroduction of large, terrestrial, mammalian predators based on reintroductions to South Africa's Eastern Cape Province. Open Conserv. Biol. J. 1, 1–11.

Hayward, M.W., Slotow, R., 2016. Management of reintroduced wildlife populations. In: Jachowski, D.S., Millspaugh, J.J., Angermeier, P.L., Slotow, R. (Eds.), Reintroduction of Fish and Wildlife Populations. University of California Press, Oakland, California, USA, pp. 319–340.

Hayward, M.W., Somers, M.J. (Eds.), 2009. Reintroduction of Top-Order Predators. Blackwell Publishing Ltd, UK.

Hofmeyr, M., van Dyk, G., 1998. Cheetah introductions to two north west parks: case studies from Pilanesberg National Park and Madikwe Game Reserve. Symposium on Cheetahs as Game Ranch Animals, 23–24 October 1998. South African Veterinary Association Wildlife Group, Onderstepoort, South Africa, p. 71.

Houser, A., Gusset, M., Bragg, C.J., Boast, L., Somers, M.J., 2011. Pre-release hunting, training and post-release monitoring are key components in the rehabilitation of orphaned large felids. S. Afr. J. Wildl. Res. 41 (1), 11–20.

Hunter, L.T.B., 1998a. The Behavioural Ecology of Reintroduced Lions and Cheetahs in the Phinda Resource Reserve, Kwazulu-Natal, South Africa. PhD thesis, University of Pretoria, South Africa.

Hunter, L.T.B., 1998b. Early Post-Release Movements and Behaviour of Reintroduced Cheetahs and Lions, and Technical Considerations in Large Carnivore Restoration. Wildlife Group of the South African Veterinary Association, Onderstepoort, South Africa.

Hunter, L.T.B., 1999. Large felid restoration: lessons from the Phinda Resource Reserve, South Africa, 1992–1999. Cat News 30, 20–21.

IUCN/SSC, 2007. Regional Conservation Strategy for the Cheetah and African Wild Dog in Eastern Africa. IUCN/SSC, Gland, Switzerland.

IUCN/SSC, 2012. Regional Conservation Strategy for the Cheetah and African Wild Dog in Western, Central and Northern Africa. IUCN/SSC, Gland, Switzerland.

IUCN/SSC, 2013. Guidelines for Reintroductions and Other Conservation Translocations. Version 1.0. IUCN/SSC, Gland, Switzerland.

Johnson, S., Mengersen, K., de Waal, A., Marnewick, K., Cilliers, D., Houser, A., Boast, L., 2010a. Modelling cheetah relocation success in southern Africa using an iterative Bayesian network development cycle. Ecol. Model. 221 (4), 641–651.

Johnson, W.E., Onorato, D.P., Roelke, M.E., Land, E.D., Cunningham, M., Belden, R.C., McBride, R., Jansen, D., Lotz, M., Shindle, D., 2010b. Genetic restoration of the Florida panther. Science 329 (5999), 1641–1645.

Jule, K.R., Leaver, L.A., Lea, S.E.G., 2008. The effects of captive experience on reintroduction survival in carnivores: a review and analysis. Biol. Conserv. 141, 355–363.

Lindsey, P.A., Alexander, R., Mills, M.G.L., Romañach, S.S., Woodroffe, R., 2007. Wildlife viewing preference of visitors to protected areas in South Africa: implications for the role of ecotourism in conservation. J. Ecotour. 6 (1), 19–33.

Lindsey, P.A., Davies-Mostert, H.T., 2009. South African action plan for the conservation of cheetahs and African wild dogs. Report from a national conservation action planning workshop for South Africa. Endangered Wildlife Trust, Johannesburg, South Africa.

Lindsey, P.A., Marnewick, K., Davies-Mostert, H.T., Rehse, T., Mills, M.G.L., Brummer, R., Buk, K., Traylor-Holzer, K., Morrison, K., Mentzel, C., Daly, B. (Eds.), 2009. Cheetah (Acinonyx jubatus) Population Habitat Viability Assessment Workshop Report. Endangered Wildlife Trust, Modderfontein, South Africa.

Lindsey, P.A., Romañach, S.S., Davies-Mostert, H.T., 2009b. A synthesis of early indicators of the drivers of predator conservation on private lands in South Africa. In: Hayward, M.W., Somers, M.J. (Eds.), Reintroduction of Top-Order Predators. Wiley-Blackwell, Oxford, UK, pp. 321–344.

Linnell, J.D.C., Aanes, R., Swenson, J.E., 1997. Translocation of carnivores as a method for managing problem animals: a review. Biodivers. Conserv. 6, 1245–1257.

Marker, L.L., 2002. Aspects of Cheetah (Acinonyx jubatus) Biology, Ecology and Conservation Strategies on Namibian farmlands. PhD thesis, University of Oxford, United Kingdom.

Marker, L., 2010. Reintroduction of cheetahs to Zambia—Feasibility Study, (Trip report). Cheetah Conservation Fund, Otjiwarongo, Namibia.

Marker, L., 2012. Reintroduction of cheetahs to Uzbekistan—Feasibility Study, (Trip report). Cheetah Conservation Fund, Otjiwarongo, Namibia.

Marker, L.L., 2016. 2015—International Cheetah (Acinonyx jubatus) Studbook. Cheetah Conservation Fund, Otjiwarongo, Namibia.

Marker, L.L., Dickman, A.J., Mills, M.G.L., Jeo, R.M., Macdonald, D.W., 2008. Spatial ecology of cheetah on north-central Namibian farmlands. J. Zool. 274 (3), 226–238.

Marnewick, K., Hayward, M.W., Cilliers, D., Somers, M.J., 2009. Survival of cheetahs relocated from ranchland to fenced protected areas in South Africa. In: Hayward, M.W., Somers, M.J. (Eds.), Reintroduction of Top-Order Predators. first ed. Wiley-Blackwell, Oxford, UK, pp. 282–306.

Massei, G., Quy, R.J., Gurney, J., Cowan, D.P., 2010. Can translocations be used to mitigate human-wildlife conflicts? Wildl. Res. 37, 428–439.

McGranahan, D.A., 2008. Managing private, commercial rangelands for agricultural production and wildlife diversity in Namibia and Zambia. Biodivers. Conserv. 17, 1965–1977.

Moritz, C., 1999. Conservation units and translocations: strategies for conserving evolutionary processes. Hereditas 130 (3), 217–228.

Munson, L., Marker, L.L., Dubovi, E., Spencer, J.A., Evermann, J.F., O'Brien, S.J., 2004. Serosurvey of viral infections in free-ranging Namibian cheetahs (*Acinonyx jubatus*). J. Wildl. Dis. 40 (1), 23–31.

O'Brien, S.J., Mayr, E., 1991. Bureaucratic mischief: recognizing endangered species and subspecies. Science 251 (4998), 1187–1188.

Pettifer, H.L., 1981. The experimental release of captive-bred cheetah into the natural environment. In: Chapman, J.A., Pursley, D. (Eds.), Worldwide Furbearer Conference, pp. 1001–1024.

Pettifer, H.L., Muller, P.J., De Kock, J.P.S., Zambatis, N., 1982. The Experimental Relocation of Cheetahs (*Acinonyx jubatus*) from the Suikerbosrand Nature Reserve to the Eastern Transvaal Lowveld. Hans Hoheisen Wildlife Research Station, South Africa.

Purchase, G., Vhurumuku, G., 2005. Evaluation of a Wild-Wild Translocation of Cheetah (*Acinonyx jubatus*) from Private Land to Matusadona National Park, Zimbabwe (1994–2005). Zambesi Society, Harare, Zimbabwe.

Ranjitsinh, M.K., Jhala, Y.V., 2010. Assessing the potential for reintroducing the cheetah in India. Wildlife Trust of India; the Wildlife Institute of India, Noida; Dehradun (India).

Ripple, W.J., Estes, J.A., Beschta, R.L., Wilmers, C.C., Ritchie, E.G., Hebblewhite, M., Berger, J., Elmhagen, B., Letnic, M., Nelson, M.P., 2014. Status and ecological effects of the world's largest carnivores. Science 343 (6167), 1241484.

RWCP and IUCN/SSC, 2015. Regional Conservation Strategy for the Cheetah and African Wild Dog in Southern Africa. RWCP and IUCN/SSC (revised and updated, August 2015).

Seddon, P.J., Armstrong, D.P., Maloney, R., 2007. Developing the science of reintroduction biology. Conserv. Biol. 21, 303–312.

Somers, M.J., Gusset, M., 2009. The role of social behaviour in carnivore reintroductions. In: Hayward, M.W., Somers, M.J. (Eds.), Reintroduction of Top-Order Predators. Wiley-Blackwell, Oxford, UK, pp. 270–281.

Teichman, K.J., Cristescu, B., Darimont, C.T., 2016. Hunting as a management tool? Cougar-human conflict is positively related to trophy hunting. BMC Ecol. 16 (1), 44.

Teixeira, C.P., Schetini de Azevedo, C., Mendl, M., Cipreste, C.F., Young, R.J., 2007. Revisiting translocation and reintroduction programmes: the importance of considering stress. Anim. Behav. 73, 1–13.

Trinkel, M., Ferguson, N., Reid, A., Reid, C., Somers, M.J., Turrelli, L., Graf, J.A., Szykman, M., Cooper, D., Haverman, P., Kastberger, G., Packer, C., Slotow, R., 2008. Translocating lions into an inbred lion population in the Hluhluwe-iMfolozi Park: South Africa. Anim. Conserv. 11, 138–143.

van der Meer, E., 2016. The Cheetahs of Zimbabwe. Distribution and Population Status 2015. Cheetah Conservation Project Zimbabwe, Victoria Falls, Zimbabwe.

Weise, F.J., Stratford, K.J., van Vuuren, R.J., 2014. Financial costs of large carnivore translocations—accounting for conservation. PLoS One 9 (8), e105042.

Weise, F.J., Lemeris, Jr., J.R., Munro, S.J., Bowden, A., Venter, C., van Vuuren, M., van Vuuren, R.J., 2015. Cheetahs (*Acinonyx jubatus*) running the gauntlet: an evaluation of translocations into free-range environments in Namibia. PeerJ 3, e1346.

Wolf, C.M., Griffith, B., Reed, C., Temple, S.A., 1996. Avian and mammalian translocations: update and reanalysis of 1987 survey data. Conserv. Biol. 10 (4), 1142–1154.

Global Cheetah Conservation Policy: A Review of International Law and Enforcement

Kristin Nowell*, Tatjana Rosen**

*Cat Action Treasury and World Conservation Union (IUCN) Red List Programme,
Cape Neddick, ME, United States
**Panthera and IISD Earth Negotiations Bulletin, Khorog, GBAO, Tajikistan

INTRODUCTION

Influence of the World's First Environmental Treaty on National Policies for Cheetahs Today

One of the first attempts at international environmental policy for Africa was drawn up in London in 1900 by colonial powers for African wildlife, the Convention for the Preservation of Wild Animals, Birds, and Fish in Africa (London Convention). Although it never entered into force (as it was not signed by every negotiating party), its principles have resonated through African history and have inspired the establishment of the first nature reserves, and lists of species subject to different levels of protection. Although most large predators (lion *Panthera leo*, leopard *Panthera pardus*, spotted hyena *Crocuta crocuta*, and African wild dog *Lycaon pictus*) were placed on schedule 5 (Harmful animals desirable to be reduced in number, within sufficient limits), the cheetah (*Acinonyx jubatus*) was given a higher level of protection on schedule 4 (Animals to be protected from hunting and destruction, except in limited numbers). The cheetah was included under its older common and scientific name "the Cheetali (*Cynalurus*)" in the original London Convention of 1900, but was omitted from the updated version of 1933 (Mitchell, 2016a,b).

This level of protection is still evident in national policies covering the cheetah, which is fully protected throughout most of its extant range (Durant et al., 2015). However, in practice this protection has not prevented widespread offtake—by people seeking to protect livestock (Chapter 13), by direct hunting for their skins or for illegal trade (Chapter 14), or by indirect capture in snares set for wild meat. Several southern African countries (Botswana, South Africa and Zimbabwe) protect cheetahs but also allow problem animals to be captured or killed by private citizens under a permit system (Purchase et al., 2007). In Namibia, since 1975 cheetahs may be killed or captured in defense of life and livestock "whilst the life of such livestock is actually being threatened" without a permit, but the person doing so must report this to the nearest government authority within 10 days, and must apply for a permit if the skin or live animal is to be retained. In 1996, permits issued for cheetah removals were analyzed for a National Namibia Conservation Strategy (Table 21.1; Nowell, 1996). Table 21.1 shows that the level of reported removals has been very high in the preceding decades, averaging 827 cheetahs per year from 1978 to 1985, and declining to an average of 297 from 1986 to 1995. These figures do not include cheetahs killed illegally without permits, or killed legally without application for a permit to retain the animal or the skin, and thus may under-represent the true level of removals.

Trophy hunting is allowed, in limited numbers under international (the Convention on International Trade in Endangered Species of Wild Fauna and Flora, or CITES) and national law (as envisaged in the London Convention), in just two countries: Namibia and Zimbabwe. Namibia is the main exporter; trophy hunting of cheetah was legalized there under national law in 1982 (Nowell, 1996), and Namibia has argued that legally taking a regulated number of wild cheetahs for export (so that national policy thus enters the realm of international law under the CITES treaty which regulates wildlife trade between nations) benefits their conservation on private lands. "In Namibia the cheetah is viewed as the single most important predator on livestock on both commercial and communal farms[…]. Trophy hunting and export of live cheetah have been encouraged in Namibia in an attempt to curb the number of cheetah shot as predators of livestock, and to change the attitude of the farmers toward the cheetah from 'kill at all cost' to one where cheetah would be tolerated and accepted. By providing some form of financial return for the losses caused, farmers are now encouraged to utilize the cheetah on a sustainable basis, rather than implement total eradication" (Govt. of Namibia, 1992).

The policy framework of southern African countries differs from that of most other African range countries in that landowners are granted legal ownership of wildlife species occurring on their land if certain regulations are met, unlike many other countries where wildlife is the property of the government. The increased economic benefits accruing to landowners from wildlife can also have negative impacts for cheetahs, in that owners may persecute cheetahs perceiving them as a net loss to other more valuable trophy antelope species (Johnson et al., 2013). In Namibia, killing of cheetahs by humans is the top known cause of cheetah mortality outside protected areas (Marker and Dickman, 2004). However, interviews with land owners reveal that approximately 50% see cheetahs as a desirable species to have on their land, even though 87% reported financial losses to cheetah (Lindsey et al., 2013), suggesting that limited legal offtakes for the export trade can help constrain illegal and unlimited offtake of cheetahs.

The Current Role of International Environmental Agreements in Cheetah Conservation

Several aspects of cheetah ecology lead international law to have a particularly important role in conservation of the species. As the most wide-ranging of the big cats (Chapter 8), viable cheetah populations require large areas of habitat, which in many cases are most readily provided when governments put together transboundary protected areas. Such areas require communication and cooperation for their effective management. Cheetahs typically occur at lower densities than sympatric large carnivores, thus lending them an intrinsic rarity. The diplomacy involved in the functioning of international wildlife treaties is an important

TABLE 21.1   Number of Permits Issued to Citizens to Remove Cheetahs by the Namibian Ministry of Environment and Tourism, 1978–95 (Nowell, 1996)

| | No. of cheetah removal permits | | | | |
| | Permit category[a] | | | | |
Year	2	3	4	5	Total
1978	234	0	711[b]	0	945
1979	125[b]	1	711[b]	0	836
1980	125[b]	0	623	0	748
1981	125[b]	0	669	0	794
1982	125[b]	0	907[c]	0	1032
1983	88	0	725[c]	12	825
1984	107	0	633[c]	7	747
1985	117	0	552[c]	21	690
1986	79	0	318	17	414
1987	84[b]	0	317	12[c]	413
1988	95	0	272	20	387
1989	132	21	271	17	441
1990	84[b]	2	301	24[c]	411
1991	54	1	145	40	240
1992	95	0	34	35	164
1993	44	0	105	20	169
1994	32	0	111[d]	20	146
1995	50	0	116	20	186
Total	1795	25	7521	265	9588

[a]Since 1975, the Permit Office of the government agency now known as Namibia's Ministry of Environment and Tourism has issued permits for removal of cheetahs over the years under the following categories:
1. Capture, keeping, and selling of game by game dealers. (Permit category 1 is not included in Table 21.1 to eliminate the possibility of double-counting, as game dealers would more frequently purchase and keep cheetahs captured by farmers rather than run their own capture operations for this species.)
2. Capture, keeping, and selling of game by nongame dealers.
3. Shooting of game in communal areas and other State land.
4. Possession of skins of protected and specially protected game.
5. Trophy hunting of cheetah.
A new system was put in place in 1994. Category 4 is subsumed into category 2, but it is still possible to distinguish between the two as the notation "live" versus "skin" is usually appended. Data on cheetah permits issued under these categories were collected from Permit Office annual reports and computerized databases. For category 5 more permits may have been issued than cheetahs actually removed, but for other categories the number of cheetahs removed likely exceeds the number of permits issued (Nowell, 1996).

[b]Permit category data are not broken down by species in this year's annual government Permit Office report; figure given is an average for the previous, surrounding or consecutive 5-year-period as appropriate.
[c]Number in the Permit Office annual report differed from figure given in Namibia's CITES Appendix I proposal (Govt. of Namibia, 1992); figure given is an average of the two.
[d]Figure given is an average of years 1993 and 1995.

element in motivating range country governments to prioritize conservation actions for rare and threatened species. Because monetary values for cheetahs are highest outside Africa (as trophies, zoo animals, and, illegally, as exotic pets), a functioning international framework is necessary to regulate legal export and import, alongside international law enforcement cooperation to detect and combat illegal trade.

International legal agreements bring signatory national governments (legally described as contracting Parties) together to address global and regional problems. Once national governments ratify an international treaty by officially recognizing their participation in it in their domestic law, its contents become binding upon them (although of all the agreements reviewed here, only CITES is considered binding in that there are consequences for noncompliance, including trade sanctions). At regular meetings (Conferences of the Parties, or COPs), a huge array of annexes, initiatives, decisions, and recommendations have been adopted around the international environmental treaties. Implementation and enforcement of internationally agreed policy depends entirely upon the contracting Parties, which means that international policy solutions are only as effective as national governments make them. National governments are often motivated to action at the behest of a wide range of non-governmental organization (NGO) stakeholders which follow the regular meetings and documents of the international conventions quite closely, and lobby governments to introduce strengthened policy and to enforce these policies. In some cases, national policies take precedence over international ones, such as the United States' stricter domestic measures for cheetah (which is listed as Endangered under the United States' Endangered Species Act) precluding the import of hunted trophies (which for some species listed as Endangered may be allowed if shown to enhance its conservation). Specifically, the US Fish and Wildlife Service has not allowed the import of sport-hunted cheetahs

from Namibia and Zimbabwe because it has not found that current hunting and management programs enhance the survival of cheetahs, although the United States did support the granting of an export quota for cheetah trophies in 1992 under CITES (Nowell, 1996).

As shown in Table 21.2, there are eight international environmental agreements which directly affect cheetah conservation. Five are concerned primarily with *in situ* conservation; the other three exclusively with wildlife trade controls and their enforcement. This chapter describes their general nature briefly, focuses on provisions which pertain to cheetah conservation, and suggests ways they could be employed for greater effect. The most space will be devoted to CITES, the international forum which has devoted the greatest amount of attention to the cheetah specifically, and is currently of the most direct relevance to cheetah conservation, through (1) regulating legal trade in wild cheetahs, thus facilitating national policies of consumptive sustainable use in both Namibia and Zimbabwe; (2) regulating legal trade in captive cheetahs, thus facilitating global *ex situ* conservation; and (3) working to stop illegal trade in cheetahs (a problem discussed in detail in Chapter 14).

INTERNATIONAL AGREEMENTS PRIMARILY RELEVANT TO IN SITU CHEETAH CONSERVATION

The African Convention on the Conservation of Nature and Natural Resources

The African Convention has been described as "the youngest and most modern among the oldest environmental conventions" (IUCN, 2004) because its roots trace back to the 1900 London Convention, while it has been extensively revised to reflect contemporary environmental stewardship practices. As African nations gained independence, the need for a new nature

TABLE 21.2 International Environmental Agreements of Relevance to Cheetahs

Name and primary nature	Known as	Commencement[a]	Cheetah range countries which have ratified or acceded to the agreement[b]	Reference
Conservation agreements				
African Convention on the Conservation of Nature and Natural Resources (Revised, 2003)	African Convention	2017[c]	Angola, Benin, Burkina Faso, Chad, Mali, Niger, and South Africa	AU (2017)
Convention on Biological Diversity	CBD	1992	All cheetah range countries	CBD (2016)
Convention on the Conservation of Migratory Species of Wild Animals	CMS	1983	[d]Algeria, Angola, *Benin*, Burkina Faso, Chad, *Ethiopia*, *Iran*, Kenya, Mali, Mozambique, *Niger*, Somalia, *South Africa*, *Tanzania*, *Uganda*, and Zimbabwe	CMS (2016a,b)
Southern African Development Community Protocol on Wildlife Conservation and Law Enforcement	SADC Protocol	2003	Botswana, Mozambique, Namibia, South Africa, Tanzania, Zambia, and Zimbabwe	Ecolex (2016b)
Convention concerning the Protection of the World Cultural and Natural Heritage	World Heritage Convention	1972	All cheetah range countries except Somalia	Ecolex (2016c)
Trade agreements				
Convention on International Trade in Endangered Species of Wild Fauna and Flora	CITES	1975	All cheetah source, transit, and destination countries except South Sudan	CITES (2016c)
Lusaka Agreement on Co-operative Enforcement Operations Directed at Illegal Trade in Wild Fauna and Flora	Lusaka Agreement Task Force	1994	Kenya, Tanzania, Uganda, and Zambia	Ecolex (2016d)
African Common Strategy on Combatting Illegal Exploitation and Trade in Wild Fauna and Flora in Africa	African Common Strategy	2015	Endorsed by the Executive Council of the African Union at its 27th meeting[e]	TRAFFIC (2016)

[a]*Denotes the year the agreement was accepted by the United Nations as having legal force.*
[b]*Accession has the same legal effect as ratification, and refers to when a country accepts the opportunity to become party to a treaty already signed by other nations.*
[c]*Official confirmation from the African Union regarding the Revised Convention's apparent entry into force in March 2017 had not been received by the time this chapter went to press.*
[d]*Range countries which have included some information about cheetah in their triannual national reports are in italic.*
[e]*The Strategy is a nonbinding Declaration of the African Union not requiring ratification by member countries.*

conservation treaty was recognized, one that moved away from protecting animals and plants for utilitarian purposes and toward a shared African responsibility for sustainable development. Under the auspices of the Organization of African Unity (now the African Union, AU), the African Convention entered into force in 1969 and was ratified by 31 countries as of December 2001 (Van der Linde, 2002). It is credited with motivating the writing and adopting of national environmental legislation in a number of African countries (Lyster, 1993). There were significant omissions in the document; however, among them provisions for institutional structures to facilitate implementation by Party nations, as well as mechanisms to encourage compliance and enforcement. Through a series of meetings and consultations, the treaty was extensively revised, and adopted by the AU in 2003 (Ecolex, 2016a). Among the notable changes is the exclusion of reservations, thus requiring Parties to formulate common solutions to common problems, with no opting out by individual countries.

Article IX of the revised convention requires Parties to adopt legislation ensuring that all forms of taking or harvest of flora and fauna are sustainably managed, and to employ scientific monitoring of populations. Article X compels contracting governments to identify and eliminate the factors causing the depletion of threatened species. These are suggested (but not prescribed) to be defined according to the IUCN Red List criteria, which would include the cheetah. Article XI requires Parties to regulate trade, possession, and transport of these species to ensure that they are taken or obtained in accordance with both domestic legislation and international law, and enact appropriate penal sanctions and confiscation practices. This addresses a weakness of CITES that Party nations often lack robust national implementing legislation. Article XI also calls for Parties to cooperate through bilateral and subregional agreements to control illegal wildlife trade. Article XII encourages Parties to establish protected areas

identified by competent international organizations for the preservation of threatened species.

The Revised Convention is seen as being much stronger than the original and having great potential for African wildlife conservation (Erinosho, 2013; IUCN, 2004; Lubbe, 2015; Van der Linde, 2002). However, after adoption by the African Union in 2003, it has taken much longer than anticipated to receive the minimum number of ratifications by African countries (15) for it to take effect. The Revised Convention appears to have finally awoken from its "sleeping treaty" status in March 2017 with little fanfare: the AU provides this date for adoption of the Revised Convention (AU, 2017), but this likely refers instead to entry into force (A. Lukacs, Ecolex, and N. Lubbe, North-West University, South Africa, personal communication), with 16 signatories including six cheetah range countries as shown in Table 21.1. The conservation community must now partner with and motivate signatory governments to undertake the treaty's long-awaited implementation, including encouraging the participation of the remaining cheetah range countries.

Convention on Biological Diversity

The Convention on Biological Diversity (CBD) aims broadly at the conservation and sustainable use of biological diversity. Although its text does not identify any species of particular concern, many of its obligations are relevant to the cheetah and other large carnivores, but its language has been described as insufficient to establish "in practice a clear boundary between compliance and violation" (Trouwhorst, 2015). Perhaps the chief value of this treaty is in providing a high-profile forum for the development and adoption of nonbinding but authoritative guidance, of which the most relevant for the cheetah are two biodiversity targets adopted by the CBD Strategic Plan for 2011–20 in Aichi, Japan, known as Aichi Targets 11 and 12. Target 11 calls for 17% of terrestrial ecosystems to be protected by 2020; a 2012 analysis of the World Protected Areas Database estimated the coverage at 13% (Woodley et al., 2012). Yet, a recent analysis has found that, even if the 17% target should be achieved, this would likely protect less than 25% of the range of large carnivores, including the cheetah (Di Minin et al., 2016). Target 12 calls for the improvement of the conservation status of threatened species in decline by 2020, and similar to the African Convention suggests (but does not prescribe) that "threatened species" be defined according to IUCN Red List criteria (Critically Endangered, Endangered and Vulnerable) (CBD, 2010). Improvement of conservation status (which could be measured by an improvement in Red List category) will be a particular challenge for large carnivores in sub-Saharan Africa, where human populations and habitat loss are projected to increase substantially by 2050 (Visconti et al., 2015), and where loss of savanna landscapes is accelerating (Riggio et al., 2013). With these challenges of projected range loss, and because it is unlikely that protected status can be conferred on enough of their range to reverse projected population declines (Di Minin et al., 2016), the establishment of a Large Carnivore Initiative under CBD with focus on the development of conservation solutions outside protected areas would be highly beneficial. This would broadly benefit biodiversity, as well as specifically addressing the unique problems of large predator conservation. A volume of the CBD Technical Series should be dedicated to provide guidance and best practices for the conservation of large carnivores (Trouwhorst, 2015).

Convention on the Conservation of Migratory Species of Wild Animals

The Convention on the Conservation of Migratory Species of Wild Animals (CMS) defines "migratory species" as species "whose members cyclically and predictably cross one or more national jurisdictional boundaries" (Article I). As noted by Trouwhorst (2015); however, "the term has subsequently been interpreted by the CMS COP in a remarkably flexible manner,

as actually encompassing any species whose range extends across more than one country," and is evolving into an instrument for the conservation of transboundary populations in addition to those that regularly undertake long-distance movements. In 2009, the cheetah was included in CMS Appendix I, which extends the highest degree of protection to "endangered" species, with the exception of the populations of Botswana, Namibia, and Zimbabwe (cheetah populations in these three countries are not listed under the convention, for reasons discussed later). Range countries, which are contracting Parties to the Convention, agree to conserve and restore the habitat of species listed on Appendix I, as well as prohibit their taking (e.g., hunting or other form of removal from the wild) except under strict circumstances (e.g., scientific).

The CMS provisions against taking and recommendations for habitat conservation generally have no "teeth" or enforcement mechanism. For instance, CMS cheetah range states are obliged to submit national reports every 3 years (6 months before each regular Conference of the Parties) to provide information on protection measures, but similar to many other CMS-listed species, the submission of reports has been spotty, the information in the reports is often of dubious utility, and few have reported specifically on cheetahs (Table 21.2). When they have, it is in an abbreviated format and there is no independent review to verify information submitted (such as whether their population is increasing or decreasing). There is no mechanism for penalties for either failing to report or to fulfill their obligations to prevent cheetah taking, conserve cheetah habitat, and remove obstacles to transboundary movement.

However, CMS should not be viewed as ineffective. It provides a political platform for discussing conservation threats and, in the words of one expert: "the Convention is 'soft' [law], in such a way that no state need be reluctant to ratify it, yet it encourages and guides Parties to undertake practical and effective 'hard' work under specific regional Agreements" (Oster-woldt, 1989). These subsidiary agreements are widely viewed as successful implementation process for CMS. A number of these agreements have matured into "sophisticated regimes in their own right, complete with an effective institutional structure and the political will to adopt measures to protect the species for which they are responsible" (Caddell, 2005). The existence of small quotas for international trade in wild cheetah specimens under CITES for Botswana, Namibia, and Zimbabwe, as discussed later in detail, led to the CMS decision to exclude cheetah populations of these countries from Appendix I. Inclusion of these populations in Appendix II (species with an unfavorable conservation status whose conservation requires international cooperation) is an option that was discussed by the Scientific Council at its 16th meeting in 2010 (CMS, 2010), but no action was taken, as neither Botswana nor Namibia is currently party to CMS (other non-Party range states include the Central African Republic, Sudan, South Sudan, and Zambia). Appendix II listing would have obliged exploration of a subsidiary legal agreement for these populations.

These CMS subsidiary agreements and memoranda of understanding provide a vehicle for multinational and regional cooperation tailored to specific groups of animals, particularly birds and marine species (Caddell, 2005). Although originally envisioned to improve the conservation of species only partially protected under national law (and listed on CMS Appendix II), these agreements and related initiatives may also extend to Appendix I species, and are also open to the participation of non-Party countries. Cheetah populations may be already benefiting from two CMS subsidiary agreements: the Saharo-Sahelian Megafauna Action Plan, focused on gazelles in 14 North African countries, and the Central Eurasian Aridland Mammals Concerted Action, which spurred the Central Asian Mammals Initiative (CAMI), and includes Iran and its cheetah population. The cheetah has been recommended

as a target species for the development of a CMS Agreement and for the Convention to, thus, play a more active role in transboundary cheetah conservation (Trouwhorst, 2015). And in 2017, the formation of a new joint CMS-CITES African Carnivore Initiative was announced, to include the cheetah along with the leopard, lion and African wild dog. The primary focus will be on "promoting coexistence, sustainable land management and maintaining connectivity for all carnivores," with a 3 year budget estimated at US $53 million (CITES, 2017).

Similar to CBD and CITES, when CMS governments meet at COPs, they can adopt recommendations. Unlike the Convention text itself or subsidiary agreements, recommendations are nonbinding, but do carry weight by communicating shared priorities and drawing international attention. The ninth CMS COP, held in Rome in 2008, adopted a recommendation on Tigers and other Asian Big Cats, which included the Asiatic cheetah population in Iran (CMS, 2008). As that population is not transboundary, the most relevant part of this recommendation for the Iranian cheetah is its call for increased financial support from donor countries and organizations. CMS has appointed under the CAMI an Asiatic cheetah focal point, responsible for advising CMS on activities related to the conservation of the species.

Southern African Development Community (SADC) Protocol on Wildlife Conservation and Law Enforcement

The primary objective of the Protocol is to establish, within the framework of the respective national laws of each party, common approaches to the conservation and sustainable use of wildlife resources, and to assist with the effective enforcement of laws governing those resources. There are many measures to be standardized, including species and habitat protections, regulation of taking and trade, and powers granted to wildlife officers, among others (Cirelli and Morgera, 2010). The Protocol does not identify any species of concern and thus does not extend any specific protections to the cheetah. It does allow for its governing Council to determine sanctions to be taken against any Party government which undertakes action that undermine the Protocol, or persistently fails to execute its obligations under the Protocol without good reason (Article 12). Two recent significant developments to emerge from the Protocol for the cheetah include the SADC Program on Transfrontier Conservation Areas (SADC TFCA) and the SADC Law Enforcement and Anti-Poaching Strategy 2016–21 (SADC LEAP, 2015). The TFCA Program has an active infrastructure of support, including a Steering Committee and a membership Network of practitioners; 18 existing and potential transboundary protected areas are in the process of being developed (SADC TFCA, 2016), all representing important landscapes for cheetah conservation, with the most active being the Kavango-Zambezi TFCA (KAZA, 2016) and the Great Limpopo Transfrontier Park (GLTP, 2016). The SADC LEAP Strategy urges every member country to create a national task force to coordinate wildlife-related law enforcement and antipoaching issues, and establishes a SADC Wildlife Crime Prevention and Coordination Unit to coordinate the efforts of the national task forces (WWF, 2015).

Convention Concerning the Protection of the World Cultural and Natural Heritage

Contracting governments are committed to doing everything within their power to ensure the "identification, protection, conservation, presentation, and transmission to future generations" of the natural heritage situated on their territories (Article 4). It should be noted that the Convention defines "natural heritage" to include, but not be limited to, sites on the World Heritage List (Trouwhorst, 2015), currently over 1000. A number of important large protected areas for cheetah are World Heritage sites, including the Okavango Delta (Botswana) and the Serengeti National Park (Tanzania). Some indication of the value of this recognition is that less than 5% of listed sites are currently considered "in danger" according to Article 4 (World Heritage Convention, 2016). One of the selection criteria for sites on the List is that they "contain the most important and significant natural habitats for *in situ* conservation of biological diversity, including those containing threatened species of outstanding universal value from the point of view of science or conservation," and thus for the cheetah the Convention has the potential to assist in the preservation of priority sites through their nomination for World Heritage.

INTERNATIONAL AGREEMENTS EXCLUSIVELY FOCUSED ON WILDLIFE TRADE CONTROL

Convention on International Trade in Endangered Species of Wild Fauna and Flora

CITES is best known for its appendices. No commercial trade is permitted for species listed on Appendix I, regulated commercial trade is permitted for species listed on Appendices II and III, and trade in unlisted species is not regulated under the Convention. Parties must, each, designate a Management Authority (MA) and a Scientific Authority (SA), and before any legal trade of species or specimens listed under Appendix II may take place, the range state Party has to conduct a nondetriment finding (NDF) exercise prior to export. The cheetah was first listed in 1975 on Appendix I, which includes species "threatened with extinction which are or may be affected by trade. Trade in specimens of these species must be subject to particularly strict regulation in order not to endanger further their survival and must only be authorized in exceptional circumstances," according to Article III of the Convention text. In practice, exceptional circumstances have been interpreted as not for primarily commercial purposes; in other words, a transaction not for resale (such as the movement of private household effects), or for scientific or educational purposes (CITES, 2010). CITES requires permits from both the country of import and export which approve any trade in Appendix I species. A key cheetah range country, Namibia, acceded to the Convention in 1990 and entered a reservation for the species, meaning that the country would be treated as a non-Party to CITES concerning cheetah trade. As the Convention does allow for trade with non-Party nations, this allowed Namibia to export wild cheetahs without permits (although CITES encourages the use of alternative permits when trading with non-Parties), which it did from the 1960s up until 1975, when it was the main supplier to zoos (Marker-Kraus, 1997). In 1992, Namibia explained at the 8th meeting of the Conference of the Parties in Kyoto, Japan, that it had entered a reservation intending to continue cheetah exports because attaching a commercial value to their populations was considered "a prime means for the species conservation" (CITES, 1992a). Namibia submitted a proposal to list cheetah populations of southern African countries on Appendix II (Govt. of Namibia, 1992), which would have allowed commercial trade with export permits. The proposal was withdrawn after a compromise was worked out between various Parties to CITES, resulting in an annotation to the Appendix I listing allowing annual export quotas for Namibia, Zimbabwe, and Botswana. As a result, Namibia also withdrew its reservation on the cheetah. The Appendix I quotas allow the annual export of both live animals and hunting trophies with the following limits: Namibia (150), Zimbabwe (50), and Botswana (5) (CITES, 1992b).

Most of the CITES trade under this quota system has been in hunting trophies, mainly to EU Member States and particularly to Germany, with Namibia the primary exporter (90% of total net trophy exports shown in Fig. 21.1). Zimbabwe has exported an average of fewer than 10 trophies per year, and Botswana, which does not permit the

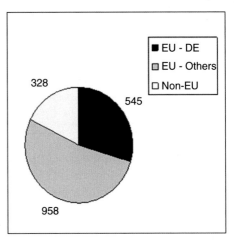

FIGURE 21.1 **Total net imports of cheetah trophies grouped by importer, 1993–2012, from records of the CITES Trade Database** (Nowell, 2014). *DE*, Germany; *EU*, European Union.

trophy hunting of cheetah, has not used its quota at all. CITES Trade Database records indicate that Namibia's exports exceeded its quota of 150 in both 2008 and 2009. There may be recording errors because figures provided by the Namibia government (Nowell, 2014) of trophy exports for 2009 do not correspond to the WCMC-UNEP CITES Trade Database records for that year. Nevertheless, the Namibian government enacted a 1-year moratorium on cheetah trophy hunting in April 2009 (Anon, 2009) to investigate the system, and since then, Namibian exports do not appear to have exceeded its allocated quota.

Cheetah trophy hunting lacks the appeal of "the Big Five" trophy animals although, as just discussed, the CITES quota is mainly trophies and is largely filled annually. One interview survey found 37% of American hunters expressed an interest in hunting cheetahs, but cheetah trophies cannot, currently, be imported into the United States under American national law (Lindsey et al., 2006). South Africa's CITES SA recently undertook a preliminary NDF concerning the possibility of establishing a national trophy hunting quota for cheetah, and found that "there are insufficient data available on population size and trend and inadequate information on the scale of illegal hunting to advise on a sustainable quota" (Nowell, 2014).

The international CITES Appendix I quotas apply solely to wild cheetahs. Wild-born live animals were included in the quota because, in the early 1990s, it was still considered difficult to breed cheetahs in captivity, and most animals in the captive population were wild-caught from Namibia (CITES, 1992c). Since that time; however, improved management and knowledge of the cheetah's unusual mating and reproductive parameters have led to greater success in captive breeding, although only in a few facilities (Chapters 22 and 27). The number of live cheetah imports recorded as captive-bred has increased markedly over the last decade to an average of 88 per year, with South Africa being the major exporter (Nowell, 2014). Namibia exported live wild animals in the 1990s, but has since allocated its quota almost exclusively to hunting trophies; in the 2000s, South Africa's exports of captive cheetahs grew (Fig. 21.2). According to records of the South African CITES MA, 786 live cheetahs were exported from 2002 to 2011 (Nowell, 2014), and now rivals the trophy trade in terms of economic value.

Captive-bred cheetahs are treated differently than wild cheetahs under CITES, in that commercial trade is allowed. Article VII, paragraph 4, of the Convention states that "Specimens of an animal species included in Appendix I bred in captivity for commercial purposes[…] shall be deemed to be specimens of species included in Appendix II." In Resolution Conf. 12.10 (Rev. CoP15) on *Registration of operations that breed Appendix-I animal species in captivity for commercial purposes*, the Conference of the Parties to the Convention has agreed to an interpretation of the provision, as follows: "Parties shall restrict imports for primarily commercial purposes[…] of captive-bred specimens of Appendix-I species to those produced by operations included in the Secretariat's Register and shall reject any [export] document granted[…] if the specimens concerned do not originate from such an operation and if the document does not describe the specific identifying mark applied to each specimen." The same Resolution notes that import of specimens of Appendix I species bred in captivity *not* for commercial purposes, and covered by a certificate of captive breeding, may be authorized for import whether or not the purpose is commercial.

South Africa is the only country which has registered with the CITES Secretariat Appendix I captive breeding operations authorized to export cheetahs for commercial purposes: DeWildt (now Ayn van Dyk) Cheetah and Wildlife

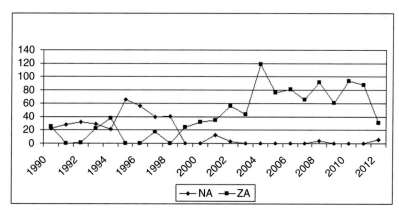

FIGURE 21.2 **Net exports of live cheetahs by major exporter, 1990–2012, according to records of the CITES Trade Database** (Nowell, 2014). *NA*, Namibia—wild cheetahs exported under Appendix I quota; *ZA*, South Africa—cheetahs certified as bred in captivity.

Centre and Hoedspruit Endangered Species Centre. The South Africa CITES MA (Nowell, 2014) notes that "the majority of live cheetah exported from South Africa originate from captive facilities not necessarily registered with CITES," but which are registered according to national legislation with provincial authorities. Less than 5% of South Africa's cheetah exports were coded as commercial in the CITES Trade Database; the purpose of most transactions was reported as Z (zoo), and under Article III of the Convention, this type of trade is considered noncommercial and exporting facilities need not be registered with the CITES Secretariat. Under South African CITES implementation legislation, provinces administer certificates for international trade in listed species, and bred-in-captivity specimens can only be exported by facilities which have registered with their provincial government according to regulations under this law. Although only about 20 facilities are licensed by provincial authorities to breed cheetahs, a survey of captive facilities suggests that more are attempting to breed (Marnewick, 2012). Cheetah experts suspect that some facilities in South Africa may not have mastered the challenge of breeding cheetahs and are instead making false bred-in-captivity declarations for live-captured wild animals, illegally captured in South Africa, as well as neighboring countries (Nowell, 2014).

Concerns raised by conservation NGOs about illegal trade, primarily illicit movement of northeast African wild-caught cheetah cubs to the Gulf states (Chapter 14), but also potential fraudulence in the legal captive live cheetah trade, were instrumental in drawing CITES' attention to the problem. This led to a detailed study (Nowell, 2014) and a recommendation of the CITES Animals Committee to hold an international workshop (CITES, 2014a). Although the Gulf states initially questioned the reliability of information on illegal trade (CITES, 2013a), in 2015 the problem was acknowledged and the government of Kuwait hosted an international workshop convened by the CITES Secretariat

and a CITES intersessional working group composed of governments and NGOs. The Workshop on Illegal Trade in Cheetahs brought together representatives from CITES authorities and enforcement agencies from 13 Parties [Algeria, Bahrain, Botswana, Jordan, Kenya, Kuwait, Qatar, Saudi Arabia, Somalia, South Africa, Sudan, United Arab Emirates (UAE), and Zimbabwe], the Chair of the Animals Committee, and cheetah experts from international and non-governmental organizations (CITES, 2016a,b). The 66th meeting of the CITES Standing Committee in January 2016 adopted the workshop recommendations, and submitted a number of decisions which were adopted at the Convention's 17th COP in Johannesburg in September, 2016. The Decisions include the compilation of a CITES Trade Resource kit for use in law enforcement, including protocols to be followed in case of seizures and guidance on the immediate and long-term disposal of live animals; and the establishment of a Cheetah Forum on the CITES website for Parties, experts, NGOs, and other stakeholders to exchange and share information on cheetahs and illegal trade in cheetah. The COP also called for the Secretariat to report to the Parties on the implementation of all the recommendations, including those adopted by the Standing Committee calling for countries to strengthen their national and regional enforcement actions concerning cheetah (IISD RS, 2016).

At the Kuwait workshop, South Africa presented a number of government interventions being undertaken to strengthen its regulation of cheetah breeding, including development of a DNA database for captive cheetahs to prove parental ancestry, as well as two microchips and photo identification (Tjiane, 2015). After the workshop, a stringent ban on the keeping of exotic animals as pets was announced by the UAE (Anon, 2016), one of the main importers of South African captive cheetahs, and the CITES Party where smuggled cheetah cubs are most frequently confiscated (Nowell, 2014). At the time of this writing, UAE is still working on relevant draft legislation.

Other International Agreements to Combat Illegal Wildlife Trade

Two other international agreements could play a greater role in addressing the illegal trade in cheetahs. This issue was first drawn to international attention (CITES, 2013b) by the Coalition Against Wildlife Trafficking, a public–private partnership with 15 NGO members (including one cheetah-specific NGO, the Cheetah Conservation Fund) and 6 governments (although none in Africa). One prominent government member, the United States, was instrumental in drafting the CITES Animals Committee recommendation that "Parties include cheetahs as a species of priority in their strategies to counter wildlife trafficking" (CITES, 2014a). The Lusaka Agreement Task Force has participated in several coordinated international global enforcement operations resulting in many arrests and seizures of a variety of wildlife, including cheetah skins (CITES, 2014b). With three East African countries as members, the Task Force could play more of a leadership role in helping range states to step up efforts to combat the illegal export to the Gulf States. The newest agreement, the African Common Strategy on Combatting Illegal Exploitation and Trade in Wild Fauna and Flora in Africa, adopted by the AU in 2015, may hold greater potential for effective action because all African countries are members (except Morocco, which is not a cheetah range state). Although it contains nothing specific to cheetahs, it embodies a new and unprecedented level of political will. "There has never been so much high level political momentum in Africa to tackle transnational organized wildlife crime: now there is a plan to turn this into action," declared one NGO observer (TRAFFIC, 2016).

CONCLUSIONS

Environmental treaties are often considered "paper tigers" that do not live up to their full potential. Perhaps their greatest benefit is their convening power: the CITES experience of countries discussing and planning actions against illegal cheetah trade has shown that these fora have the capability for constructive and focused action planning to address specific problems and issues. The NGO cheetah conservation community must continue to motivate CITES and the other environmental bodies described in this chapter, through participation in meetings and provision of researched recommendations, to take targeted actions to protect cheetahs in their natural habitats and undertake intelligence-led enforcement to prevent their smuggling across national borders.

References

Anon, 2009. Moratorium on trophy hunting cheetah permits in Namibia. Namibia Adventure Safaris, 24 April. Available from: http://namibiahuntsafaris.blogspot.com/2009/04/moratorium-for-trophy-hunting-cheetah.html.

Anon, 2016. UAE to ban possession of wild animals, emirates 24│7, 20th January. Available from: http://www.emirates247.com/news/uae-to-ban-possession-of-wild-animals-2016-01-20-1.617991.

AU, 2017. Revised African Convention on the Conservation of Nature and Natural Resources. Available from: https://au.int/en/treaties/african-convention-conservation-nature-and-natural-resources-revised-version.

Caddell, R., 2005. International law and the protection of migratory wildlife: an appraisal of twenty-five years of the Bonn Convention. Colo. J. Int. Environ. Law Policy 16, 113–156.

CBD, 2010. Quick Guide to the Aichi Biodiversity Targets: Extinction Prevented. Available from: www.cbd.int/doc/strategic-plan/targets/T12-quick-guide-en.pdf.

CBD, 2016. Convention on Biological Diversity: List of Parties. Available from: https://www.cbd.int/information/parties.shtml.

Cirelli, M.T., Morgera, E., 2010. Wildlife Law in the Southern African Development Community. Joint Publication of the FAO and CIC, Budapest, 136p.

CITES, 1992a. Summary report of the Committee I meeting, Com I.8.1 (Rev.).

CITES, 1992b. Amendments to Appendices I and II of the Convention adopted by the Conference of the Parties at its eighth meeting in Kyoto, Japan, 2–13 March.

CITES, 1992c. Quotas for trade in specimens of cheetah. Doc. 8.22 (Rev.).

CITES, 2010. Definition of "primarily commercial purposes." Resolution Conf. 5.10 (Rev. CoP 15).

CITES, 2013a. Summary record of the ninth meeting of Committee II. Sixteenth Conference of the Parties, CoP16 Com. II Rec. 9 (Rev. 1).

CITES, 2013b. Illegal trade in cheetahs. Sixteenth Conference of the Parties, CoP16 Doc. 51 (Rev. 1).

CITES, 2014a. Illegal trade in cheetahs (*Acinonyx jubatus*). Agenda item 18, 27th meeting of the CITES Animals Committee. AC27 WG3 Doc. 1.

CITES, 2014b. CITES Secretariat welcomes Operation Cobra II results. Available from: https://cites.org/eng/news/sundry/2014/20140210_operation_cobra_ii.php.

CITES, 2016a. Illegal trade in cheetahs (*Acinonyx jubatus*): Report of the Working Group. SC66 Doc. 32.5.

CITES, 2016b. Summary record of the Sixty-sixth meeting of the Standing Committee, Geneva (Switzerland), 11–15 January.

CITES, 2016c. Convention on International Trade in Endangered Species of Wild Fauna and Flora: Member Countries. Available from: https://cites.org/eng/disc/parties/index.php.

CITES, 2017. Joint CMS-CITES African Carnivores Initiative. AC29 Doc. 29 Annex.

CMS, 2008. COP 9 Proceedings. UNEP/CMS Recommendation 9.3: Tigers and other Asian Big Cats, pp. 149–150.

CMS, 2010. The listing of the cheetah in Appendix II. 16th meeting of the CMS Scientific Council, Bonn, Germany, 28–30 June. UNEP/CMS/ScC16/Doc.20 Agenda Item 14.3.

CMS, 2016a. Convention on the Conservation of Migratory Species of Wild Animals: Parties and Range States. Available from: http://www.cms.int/en/parties-range-states.

CMS, 2016b. Convention on the Conservation of Migratory Species of Wild Animals: National Reports. Available from: http://www.cms.int/en/documents/national-reports.

Di Minin, E., Slotow, R., Hunter, L.T., Pouzols, F.M., Toivonen, T., Verburg, P.H., Leader-Williams, N., Petracca, L., Moilanen, A., 2016. Global priorities for national carnivore conservation under land use change. Sci. Rep. 6, 23814.

Durant, S., Mitchell, N., Ipavec, A., Groom, R., 2015. *Acinonyx jubatus*. The IUCN Red List of Threatened Species 2015: e.T219A50649567. Available from: http://dx.doi.org/10.2305/IUCN.UK.2015-4.RLTS.T219A50649567.en.

Ecolex, 2016a. African Convention on the Conservation of Nature and Natural Resources (Revised Version) (July 11, 2003): Country/Participant. Available from: http://www.ecolex.org/details/african-convention-on-the-conservation-of-nature-and-natural-resources-revised-version-tre-001395/participants/?q=African+Convention+on+the+Conservation+of+Nature+and+Natural+Resources.

Ecolex, 2016b. Protocol on Wildlife Conservation and Law Enforcement (August 18, 1999): Country/Participant. Available from: http://www.ecolex.org/details/protocol-on-wildlife-conservation-and-law-enforcement-tre-001348/participants/?q=Protocol+on+Wildlife+Conservation+and+Law+Enforcement&xdate_min=&xdate_max.

Ecolex, 2016c. Convention concerning the Protection of the World Cultural and Natural Heritage (November 23, 1972): Country/Participant. Available from: http://www.ecolex.org/details/convention-concerning-the-protection-of-the-world-cultural-and-natural-heritage-tre-155235/participants/?q=World+Heritage+Convention.

Ecolex, 2016d. Lusaka Agreement on Co-operative Enforcement Operations directed at Illegal Trade in Wild Fauna and Flora (September 8, 1994): Country/Participant. Available from: https://www.ecolex.org/details/treaty/lusaka-agreement-on-co-operative-enforcement-operations-directed-at-illegal-trade-in-wild-fauna-and-flora-tre-001197/?q=Lusaka+Agreement+on+Co-operative+Enforcement+Operations+directed+at+Illegal+Trade+in+Wild+Fauna+and+Flora+%28September+8%2C+1994%29%3A&xdate_min=&xdate_max=.

Erinosho, B.T., 2013. The revised African convention on the conservation of nature and natural resources: prospects for a comprehensive treaty for the management of Africa's natural resources. Afr. J. Int. Comp. Law 21, 378–397.

GLTP, 2016. Great Limpopo Transfrontier Park. Available from: https://www.sanparks.org/conservation/transfrontier/great_limpopo.php.

Govt. of Namibia, 1992. Proposal to transfer *Acinonyx jubatus* (populations of Botswana, Malawi, Namibia, Zambia, Zimbabwe) from Appendix I to Appendix II. CITES Ref. Doc. 8.46 no. 9. CITES Secretariat, Geneva.

IISD RS, 2016. Summary and Analysis of CITES COP17. Available from: http://www.iisd.ca/cites/cop17/.

IUCN, 2004. An introduction to the African Convention on the Conservation of Nature and Natural Resources. IUCN Environmental Policy and Law Paper No. 56. Gland, Switzerland.

Johnson, S., Marker, L., Mengersen, K., Gordon, C.H., Melzheimer, J., Schmidt-Küntzel, A., Nghikembua, M., Fabiano, E., Henghali, J., Wachter, B., 2013. Modeling the viability of the free-ranging cheetah population in Namibia: an object-oriented Bayesian network approach. Ecosphere 4 (7), 1–19.

KAZA, 2016. Kavango Zambezi Transfrontier Conservation Area. Available from: http://www.kavangozambezi.org/.

Lindsey, P.A., Alexander, R., Frank, L.G., Mathieson, A., Romanach, S.S., 2006. Potential of trophy hunting to create incentives for wildlife conservation in Africa where alternative wildlife-based land uses may not be viable. Anim. Conserv. 9 (3), 283–291.

Lindsey, P.A., Havemann, C.P., Lines, R., Palazy, L., Price, A.E., Retief, T.A., Rhebergen, T., Van der Waal, C., 2013. Determinants of persistence and tolerance of carnivores on Namibian ranches: implications for conservation on southern African private lands. PLoS One 8 (1), e52458.

Lubbe, W.D., 2015. We need a new convention to protect Africa's environment. The Conversation, 7 May. Available from: http://theconversation.com/we-need-a-new-convention-to-protect-africas-environment-40648.

Lyster, S., 1993. International Wildlife Law: An Analysis of International Treatise Concerned With the Conservation of Wildlife. Cambridge University Press, Cambridge.

Marker, L., Dickman, A., 2004. Human aspects of cheetah conservation: lessons learned from the Namibian farmlands. Hum. Dimens. Wildl. 9, 297–305.

Marker-Kraus, L., 1997. History of the cheetah: *Acinonyx jubatus* in zoos 1829–1994. Int. Zoo Yearbook 35, 27–43.

Marnewick, K., 2012. Captive breeding, keeping and trade of captive cheetahs in South Africa. Endangered Wildlife Trust, Johannesburg, South Africa.

Mitchell, R.B., 2016a. Convention designed to ensure the conservation of various species of wild animals in Africa, which are useful to man or inoffensive. International Environmental Agreements (IEA) Database Project. Available from: http://iea.uoregon.edu/treaty-text/1900-PreservationWildAnimalsBirdsFishAfricaENtxt.

Mitchell, R.B., 2016b. Convention Relative to the Preservation of Fauna and Flora in Their Natural State. International Environmental Agreements (IEA) Database Project. Available from: http://iea.uoregon.edu/treaty-text/1900-PreservationWildAnimalsBirdsFishAfricaENtxt.

Nowell, K., 1996. Namibian cheetah conservation strategy. Ministry of Environment and Tourism, Windhoek, Namibia, 89 pp. Available from: http://www.catsg.org/cheetah/05_library/5_2_strategies-&-action-plans/Nowell_1996_Namibia_cheetah_conservation_strategy.pdf.

Nowell, K., 2014. An assessment of conservation impacts of legal and illegal trade in cheetahs *Acinonyx jubatus*. IUCN SSC Cat Specialist Group report prepared for the CITES Secretariat, 65th meeting of the CITES Standing Committee, Geneva, 7–11 July. SC65 Doc. 39 (Rev. 2).

Osterwoldt, Ralph, 1989. Implementation and enforcement issues in the protection of migratory species. Nat. Resour. J. 29, 1017–1049.

Purchase, G., Marker, L., Marnewick, K., Klein, R., Williams, S., 2007. Regional assessment of the status, distribution

and conservation needs of the cheetah in southern Africa. Cat News (Special Issue 3), 44–46.

Riggio, J., Jacobson, A., Dollar, L., Bauer, H., Becker, M., Dickman, A., Funston, P., Groom, R., Henschel, P., de Iongh, H., Lichtenfeld, L., 2013. The size of savannah Africa: a lion's (*Panthera leo*) view. Biodivers. Conserv. 22 (1), 17–35.

SADC LEAP, 2015. SADC Law Enforcement and Anti-Poaching Strategy, 2016–2021. Available from: http://www.gaborone.diplo.de/contentblob/4715602/Daten/6225475/SADC_LEAP_FINAL.pdf.

SADC TFCA, 2016. SADC Transfrontier Conservation Areas. Available from: http://www.sadc.int/themes/natural-resources/transfrontier-conservation-areas/.

Tjiane, M., 2015. South Africa country presentation at the Workshop on Illegal Trade in Cheetahs, Kuwait, City, Kuwait, 3–5 November.

TRAFFIC, 2016. Africa steps up fight against wildlife crime, 20th April. Available from: http://www.traffic.org/home/2016/4/21/africa-steps-up-fight-against-wildlife-crime.html.

Trouwhorst, A., 2015. Global large carnivore conservation and international law. Biol. Conserv. (24), 1567–1588.

Van der Linde, M., 2002. Review of the African convention on nature and natural resources. Afr. Hum. Rights Law J. 2 (1), 33–59.

Visconti, P., Bakkenes, M., Baisero, D., Brooks, T., Butchart, S.H., Joppa, L., Alkemade, R., Di Marco, M., Santini, L., Hoffmann, M., Maiorano, L., 2015. Projecting global biodiversity indicators under future development scenarios. Conserv. Lett. 9 (1), 5–13.

Woodley, S., Bertzky, B., Crawhall, N., Dudley, N., Londoño, J.M., MacKinnon, K., Redford, K., Sandwith, T., 2012. Meeting Aichi Target 11: what does success look like for protected area systems? Parks 18 (1), 23–36.

World Heritage Convention, 2016. List of World Heritage in Danger. Available from: http://whc.unesco.org/en/danger/.

WWF, 2015. Bold new strategy to tackle wildlife crime in southern Africa, 6th November. Available from: http://wwf.panda.org/wwf_news/?255930/Bold-new-strategy-to-tackle-wildlife-crime-in-southern-Africa.

CAPTIVE CHEETAHS

History of Cheetahs in Zoos and Demographic Trends Through Managed Captive Breeding Programs

Laurie Marker, Kate Vannelli*, Markus Gusset**,
Lars Versteege†, Karen Ziegler Meeks‡, Nadja Wielebnowski§,
Jan Louwman¶, Hanneke Louwman¶, Laurie Bingaman Lackey***

*Cheetah Conservation Fund, Otjiwarongo, Namibia
**World Association of Zoos and Aquariums, Gland, Switzerland
†Safaripark Beekse Bergen, Hilvarenbeek, The Netherlands
‡White Oak Conservation, Yulee, FL, United States
§Oregon Zoo, Portland, OR, United States
¶Wassenaar Wildlife Breeding Centre, Wassenaar, The Netherlands

INTRODUCTION

With an increasing number of species nearing global extinction, it is necessary to consider both *in situ* (in the wild) and *ex situ* (under human care in zoological institutions) conservation efforts to fully realize the conservation potential for many threatened species (IUCN/SSC, 2014). Captive facilities worldwide hold animals representing 23% of the world's threatened species of fauna (Conde et al., 2013). These facilities, therefore, serve as an excellent resource for conservation efforts and can provide conservation support for dwindling populations in the wild through awareness raising, fundraising, education, and research (Chapter 23; IUCN/SSC, 2014). Through the culmination of knowledge and experience gained since the first keeping of captive wild animals over 2500 years ago (Lyles, 2001), modern-day zoos now play a critical role in conserving species throughout the world (Barongi et al., 2015; Minteer and Collins, 2013).

The cheetah (*Acinonyx jubatus*) is a species whose survival depends on *in situ* and *ex situ* integrated conservation planning (Bartels et al., 2002). The captive population numbers were 1730 in 2014, equivalent to approximately 24% of the wild adult and adolescent cheetah population (Durant et al., 2017). This chapter discusses the history of cheetahs in captivity and the establishment of the International Cheetah Studbook (ICS). Demographic trends in captive cheetahs from 1956 to 2014 are then described both globally and regionally. Lastly, the chapter presents historic facility breeding success and the breeding strategies used at three of the most successful cheetah breeding facilities.

HISTORY OF CHEETAHS IN ZOOS

Early History of Captive Cheetahs

For thousands of years, cheetahs have been highly revered by humans and kept in captivity as pets and hunting companions (Chapter 2). Cheetahs can be difficult to breed. The first record of a captive-born litter was made by the Mogul emperor Akbar the Great in the 16th century (Guggisberg, 1975); however, from then until the 1970s, captive breeding success was extremely rare. The capture of wild cheetahs to be used for hunting or to be placed in captivity was common in both Asia and Africa, and is believed to be one of the primary reasons for the extinction of populations in many countries during the early 1900s (Pocock, 1939). For example, in 1952, the species was declared extinct in India, and by 1956, Israel had a single wild cheetah left (Divyabhanusinh, 1999; Kraus and Marker-Kraus, 1992).

The Early Captive Cheetah Population (1829–1988)

The first record of a captive cheetah in a zoo occurred in 1829 at the Zoological Society of London (Marker-Kraus, 1997); an adult that was wild-caught and lived less than a year in captivity. None were held in zoos again until 1851 at Antwerp Zoo in Belgium and through the 1860s in the Berlin, Frankfurt, and Hamburg zoos in Germany. In North America, the first cheetah was exhibited in 1871 at the Central Park Zoo in New York. During this period, little was known about holding cheetahs, and only 14 facilities worldwide exhibited them between 1851 and 1900. During the early 1900s, holders included Cairo Zoo in Egypt, Basel Zoo in Switzerland, and, in the United States, the Smithsonian Institution's National Zoo in Washington, DC, Lincoln Park Zoo in Illinois, and Saint Louis Zoo in Missouri. Mortality in these early captive cheetahs was high—for those cheetahs that survived their first year in captivity (both cubs and adults), mean longevity was 3 years and 9 months (Marker-Kraus, 1997). Between 1829 and 1954, 184 cheetahs imported from the wild were displayed in 41 facilities worldwide. There was no breeding and there were 148 deaths (Marker, 2015). At the end of 1954, 41 cheetahs were exhibited in 25 institutions (Marker, 2015). By 1975 numbers had risen to 423 cheetahs primarily through wild imports; prior to 1975, 701 cheetahs had been imported from the wild, compared to only 206 births in captivity (Fig. 22.1). Table 22.1 provides an overview of imports, births, and deaths by different time periods and breeding success by regions.

Between 1955 and 2014, a total of 2223 wild cheetahs were imported into zoological institutions (Table 22.1). Cheetahs were initially sourced from East Africa, primarily Kenya and Somalia. However, as cheetahs became scarce in these areas in the 1960s (Marker-Kraus, 1997), Namibia became the main source of wild cheetah exports from the late 1960s onward (Marker-Kraus, 1997). In 1974, the last major importation of wild cheetahs from Namibia occurred—84 individuals, representing 16% of that year's global captive population. Importations of wild cheetahs were drastically reduced thereafter due

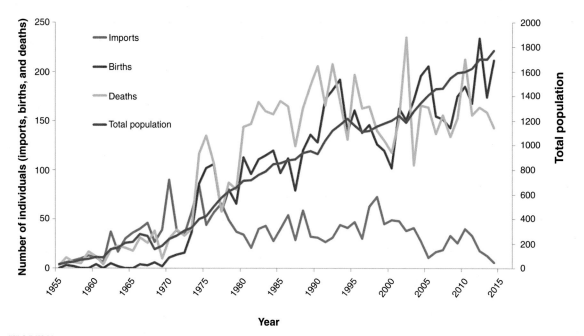

FIGURE 22.1 Changes in the global captive cheetah population from 1955 to 2014, including imports, births, deaths, and total population size.

TABLE 22.1 Summary of Births and Deaths of Captive Cheetahs Across Years and Regions

Global						End of the time period	
Time periods	Imports	Litters[a]	Cubs born[a]	Breeding institutions	Deaths	Animals	Holding institutions
1955–1964	212	4	10	2	108	166	63
1965–1974	489	61	196	24	327	514	112
1975–1984	393	311	1027	48	1035	826	147
1985–1994	452	426	1419	69	1432	1163	198
1995–2004	462	488	1506	90	1581	1327	222
2005–2014	215	623	1918	109	1605	1722	267
Total	2223	1913	6076	—	6088	—	—

Regions (totals up to 2014)	Holding institutions (Countries)	Breeding institutions (Countries)	Years of breeding	Litters[a]	Cubs born[a]	Infant mortality (%)	
						≤1 month	≤ 12 months
Africa	120 (12)	38 (3)	1970–2014	684	1987	17	28
Australia/ New Zealand	19 (2)	5 (2)	1976–2014	38	119	46	49
Europe	235 (36)	79 (19)	1966–2014	577	1994	26	38
Far East	41 (10)	12 (5)	1979–2014	107	348	19	30
India	5 (1)	1 (1)	2011–2014	5	20	50	83
Middle East	17 (4)	5 (4)	1994–2014	106	296	28	29
North America	159 (2)	47 (2)	1956–2014	396	1331	23	28
Total	586 (67)	187 (36)	—	1913	6095	22	28

[a]Births for which the location is unknown are not included.

to the enactment of the Convention of International Trade in Endangered Species (CITES), which restricted the trade of wild cheetahs (Chapter 21). As a result, breeding efforts intensified, and became increasingly successful. By 2014, the captive cheetah population registered in the ICS totaled 1722 individuals: 87% captive born, 12% wild imports, and the remaining 1% of unknown birth type (Fig. 22.1).

The Establishment of the International Cheetah Studbook

By the late 1970s, in partial response to the new restrictions imposed by CITES on collecting species from the wild, *ex situ* population management programs were being instigated for many species, including cheetahs (Chapter 23).

The development of a demographic and genetic profile of the *ex situ* population was an important aspect of these management programs (Marker-Kraus, 1997). Collaborative conservation initiatives for cheetahs began within regional zoo associations (Chapter 23), for example, Australasia (Zoo and Aquarium Association, ZAA), Europe (European Association of Zoos and Aquaria, EAZA), Japan (Japanese Association of Zoos and Aquariums, JAZA), North America (Association of Zoos and Aquariums, AZA), and Africa (Pan-African Association of Zoos and Aquaria, PAAZA). These efforts culminated in the development of regional cheetah studbooks in North America, Japan, and Great Britain in the late 1980s, Europe in 1992, and Australia in the late 1990s. The studbooks recorded all captive cheetahs held in zoological and private facilities that were members of each of the aforementioned associations. To enable greater coordination of the *ex situ* cheetah population, an ICS was established in 1988 under the auspices of the World Association of Zoos and Aquariums (WAZA), with the purpose of recording all captive cheetahs held by zoological and private facilities (Marker-Kraus, 1990). Information is submitted directly by cheetah holders, as well as

from the Species360 database (previously known as the International Species Information System) (Chapter 23) and regional studbook holders coordinated through the ICS. Historical data from pre-1988 is also included in the ICS and was obtained from historic zoo records (Marker-Kraus, 1997). The population of captive cheetahs held in each country is regulated by national wildlife authorities working with registered zoos, and the ICS provides information on their demographic history. Zoos registered with zoo associations are mandated to register with the ICS, and in some countries, the issuing of cheetah ownership permits to private holders is conditional on registration with the ICS.

The ICS provides a database from which information regarding the historical captive cheetah population can been drawn (Marker, 2015). The ICS is published annually and is provided to cheetah facilities and stakeholders. Information as of December 31, 2014 shows the captive population at 1722 (887.829.6) cheetahs: the conventional representation for 1722 total cheetahs, of which 887 were male, 829 female and 6 were of unknown sex; these animals were held in 267 facilities in 44 countries in 7 geographical regions (Tables 22.1 and 22.2) (Marker, 2015). Data for this chapter are provided by the 2014 ICS (Marker, 2015) and the associated software, Single Population Analysis and Record Keeping System, SPARKS v1.66 (Scobie and Bingaman Lackey, 2012) and PMx v1.3 (Ballou et al., 2014). Information is dependent on reporting by international institutions and is subject to change over time as new historical information becomes available. The most up-to-date data at the time of writing are presented here.

DEMOGRAPHIC TRENDS

Population trends are derived from the combination of births, deaths, and imports into the various regional captive populations, as well as exports (Fig. 22.1). Cheetah reproduction has improved over the years. However, death numbers have been nearly equal to the number of births, with imports responsible for the growth of the population, until recently.

Reproduction (1955–2014)

Between 1955 and 2014, of the 563 total facilities that held cheetahs during this time, 187 (33%) reported successful breeding: a total of 6076 (2906.2713.457) cubs were born in 1913 litters. Thirty-two percent (1952) of these cubs died before 1 year of age, of which 1296 (21% of all cubs born) died at less than 1-month-old (Table 22.1). During 2014, 35 facilities reported breeding success for the year, with a total of 67 litters and 213 (84.109.20) cubs born. Of these, 23% died before 1 year of age, 15% at 1 month or less.

Between 1956, when the first litter was born at the Philadelphia Zoo in Pennsylvania, and 1974, 65 litters were born in the USA and Europe at 26 facilities (Table 22.1). After this, the number born in captivity increased yearly. Between 1980 and 2000 (with the exceptions of 1981 and 1987 when there were fewer), at least 30 litters were born annually, increasing to an average of 60 litters annually between 2000 and 2014.

Litters are born most frequently in April, May, and June, with the second highest frequency in September and October. This seasonality of captive litters (most of which being from the northern hemisphere) likely reflects the purposeful timing of breeding opportunities aimed at litters being born under best weather conditions (spring and autumn).

Age at first reproduction was between 18 months and 11 years, averaging 5.5 years for males and 5.0 years for females. Average litter size was 3.3 ($\pm$SD 1.5) cubs (Marker, 2015). As of December 2014, 15% of the captive

TABLE 22.2 Demographic Status of Captive Cheetah Populations Managed by Region and in Total as of December 2014

Regions	Animals overall	Institutions overall (countries)	Regional associations	Animals registered	Institutions registered (countries)	Census λ 1985–2014	Life table λ 1985–1999	Life table λ 2000–2014	N_e/N
Africa	545	39 (7)	Africa (PAAZA)	227	10 (2)	1.02	1.05	1.08	0.14
Australia/ New Zealand	54	14 (2)	Australasia (ZAA)	51	12 (2)	1.05	0.88	0.95	0.12
Europe	453	107 (24)	Europe (EAZA)	425	95 (24)	1.02	0.97	0.96	0.23
Far East	137	19 (5)	Japan (JAZA)	17	2 (1)	1.06	0.91	0.97	0.16
India	9	3 (1)	—	—	—	N/A	N/A	N/A	N/A
Middle East	126	11 (4)	—	—	—	N/A	N/A	N/A	N/A
North America	398	74 (2)	North America (AZA)	295	53 (2)	1.02	0.96	0.98	0.21
Total	1722	267 (45)	N/A	1015	172 (31)	1.02	0.97	1.02	0.20

AZA, Association of Zoos and Aquariums; EAZA, European Association of Zoos and Aquaria; JAZA, Japanese Association of Zoos and Aquariums; PAAZA, Pan-African Association of Zoos and Aquaria; ZAA, Zoo and Aquarium Association.

population was older than the optimal breeding age of 10 years, 55% were of optimal breeding age (3–9 years) (6% of these animals were wild-born), 28% were below optimal breeding age (0–2 years), and 2% of unknown age. Between 1956 and 2014, 1303 (593.710) cheetahs produced cubs at least once in their lifetime (proven breeders), representing 15% of the total captive population. Of cheetahs alive at the end of 2014, 19% were proven breeders. In the 25 years since the first ICS publication, the proportion (N_e/N) of living breeders (effective population size, N_e) in the current population (N) has increased from 11.9% (111.43) to 16.4% (290.85) breeders (Marker, 2015); however, this is still lower than optimally needed for retention of genetic diversity (see section, "Demographic Trends").

Mortality (1955–2014)

Between 1955 and 2014, there were 6088 captive deaths (Table 22.1). In 2014, 23% of deaths occurred under 1 year of age, with 15% occurring under 1 month of age. Of the remaining 77%, 42% of deaths were adults over the age of 8 years, due to diseases commonly afflicting cheetah (Chapter 25).

Demography (1985–2014)

Table 22.2 presents the annual census growth rate [census lambda (λ)], and the life table λ. Census λ is a measure of a population's average annual growth rate based on the total number of living animals in the captive population at the end of each year; in addition to births and deaths it includes importations and exportations. Life table λ is a measure of a population's annual growth rate in the absence of importation/exportation; it is based on age-specific estimates of population survival and fecundity. The difference between these two measures is important: a population with a life table $\lambda < 1$ is not self-sustaining (Table 22.2); if life table $\lambda < 1$ and census $\lambda > 1$, then it means that there is pop-

ulation growth, but that it is dependent on imports. Considering data over the past 30 years, the global captive population has been barely self-sustaining—the life table λ from 1985 to 1999 was 0.97 (corresponding to a 3% annual decline if only considering births and deaths) and from 2000 to 2014 improved to only 1.02 (a 2% annual growth even in the absence of imports) (Table 22.2). Current status of the managed populations is shown in Fig. 22.2.

Imports of wild cheetahs, as well as movement of captive cheetahs between regions, have supported the captive populations. For instance, AZA zoos imported 130 cheetahs between 1985 and 2012, 85% from captive-bred sources in South Africa. However, less than 40% of these imported animals successfully bred. Almost all imports for EAZA zoos were from the wild (vs. captive-born), with 50% of this population being wild-caught in the mid-1980s. However, by 2014, less than 5% were wild-caught, due to deaths and no wild imports. Interestingly, AZA imported a number of captive-bred EAZA stock over the years, while no AZA cheetahs were imported by EAZA, as they did not require additional animals.

Success in breeding cheetahs varies greatly, both between and within regions. In Table 22.2, estimates of λ are based on data between 1985 and 2014. Although most regions have maintained or improved reproduction over this period, only the captive cheetah population of the southern African region (South Africa specifically) can be considered self-sustaining.

Some of the regions are more successful at involving larger proportions of their populations in breeding. However, the ratio of effective living breeders to the overall population size (N_e/N) in cheetahs remains low (12%–23%; Table 22.2) compared with other captive species (30%–40%; L. Bingaman Lackey, unpublished data), which is due in part to the behavioral challenges of breeding cheetahs. Regional differences are due to different levels of institutional expertise in cheetah breeding techniques (Table 22.3).

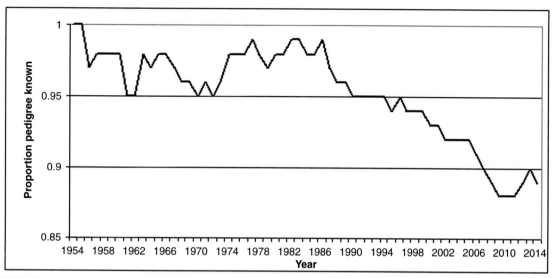

FIGURE 22.2   Changes in the proportion of cheetahs with a known pedigree in the global captive cheetah population managed by regional zoo associations, from 1955 to 2014.

There are significant differences in breeding success among facilities within regions as well. For example, although the North American cheetah population had a life table λ of 0.96 between 1985 and 1999 (Table 22.2), the life table λ for that period for three active breeding facilities (Fossil Rim Wildlife Center in Texas, White Oak Conservation Center in Florida, and San Diego Zoo Safari Park in California) was 1.10. In 1999, it was decided to set up off-exhibit breeding

centers at zoos where larger numbers of cheetahs could be held, thereby allowing a greater potential for mate choice (Chapter 9). Eight of these institutions (the original three plus Cincinnati Zoo and the Wilds in Ohio, Wildlife Safari in Oregon, the Smithsonian Conservation Biology Institute in Virginia, and the National Zoo in Washington, DC) produced 88% of litters within AZA and are today an important part of the breeding consortium called Conservation Centers for

TABLE 22.3   Percent age of Cheetah Holding Institutions Involved in Breeding, and Genetic Status of Populations Managed by Regional Associations as of December 2014

Regional association	1985–1999		2000–2014		2014	
	Holding institutions	Breeding institutions	Holding institutions	Breeding institutions	FGEs	pFGEs
Africa (PAAZA)	10	7 (70%)	14	7 (50%)	13	68
Australasia (ZAA)	7	2 (29%)	14	4 (29%)	8	16
Europe (EAZA)	84	32 (38%)	112	50 (45%)	16	53
North America (AZA)	58	21 (36%)	70	17 (24%)	23	42
Japan (JAZA)	—	—	—	—	3	6

FGEs, Founder genome equivalents; pFGEs, potential founder genome equivalents.

Species Survival (C2S2) (Chapter 23). Life table λ for this group continued at 1.10; however, the North American population as a whole is not yet self-sustaining if looking at the last 15 years combined (Table 22.2).

Population Management to Optimize Genetic Diversity (1985–2014)

The goal of population management programs is to retain representative founder alleles in descendants to maintain genetic diversity in the population. A founder is an individual brought into a population from the wild or from another zoo association that has no known relatives in that population. To date, individuals from the captive cheetah population have not been characterized genetically, therefore substitute variables are used to represent genetic diversity. The measure "founder genome equivalent" (FGE) is a convenient method of expressing gene diversity (GD) in a population. FGE is the number of founders that would produce a population with the same diversity of founder alleles as the pedigree population, assuming all founders contributed equally to each generation of descendants [FGE = 1/ $(2 \times (1 - GD))$]. In other words, FGE estimates how many of the founder animals are represented in the current population to a comparable extent. pFGE is the maximum potential FGE if all extant founder alleles could be brought to the same frequency. Table 22.3 contains the number of FGEs and pFGEs for each regional population. Maximum pFGEs are not actually attainable, as the calculations assume ideal, yet biologically impossible conditions, such as breeding all animals with all other animals, including males with males and females with females, equal litter sizes, and so on. The difference between actual and potential FGEs indicates the extent to which GD could be increased in a particular population by preferentially breeding underrepresented founder lines using Mean Kinship values (= average relatedness of an individual to the whole population, including itself).

Although the number of animals in Table 22.3 can be summed to determine an overall registered captive cheetah number, the same is not true of FGEs, as some founder lines are shared between regional populations. The actual total FGE for the aggregate regional populations is only 34, even though the sum would be 63, indicating considerable overlap of founder lines among regions. To ensure that animals (or reproductive material) imported into a region can help to increase the region's FGE, a careful analysis of the ICS is necessary to identify appropriate individuals.

Morphological traits should not influence breeding decisions if this leads to increased levels of inbreeding therefore mate choice needs to be guided by ICS information (Mean Kinship). To date, only one inheritable morphological trait has been documented in the population. The first captive "king cheetah," a pelage variation, was born in 1981 in South Africa from breeding stock from the De Wildt Centre. King cheetahs are now incorporated into some of the species' management programs in the same manner as all other cheetah, and the unique coat pattern is not specifically bred or selected for in those regions.

Recent developments in the managed population are cause for concern. As animals with unknown history or pedigrees are entering the managed population, the percentage of the overall pedigree known lessens (Fig. 22.2). As relationships between potential breeding animals become more obscured or unknown, sensible management becomes more difficult and existing genetic diversity cannot be maintained optimally. In the EAZA's European Endangered Species Programme (EEP), unknown pedigreed individuals are therefore not included in the breeding population. However, individuals of unknown descent have the potential to contribute new diversity to an existing population. In the AZA Cheetah Species Survival Plan® (SSP), unknown pedigreed individuals are therefore included in the breeding population. Genetic analysis can determine the level of relatedness of unknown animals to the existing breeding stock and should therefore be used by regional programs (Chapter 6).

REGIONAL CAPTIVE POPULATION GROWTH AND LIMITATIONS

Members of ZAA (Australasia), EAZA (Europe), JAZA (Japan), AZA (North America), and PAAZA (Africa) hold 58% of the global captive cheetah population. This subpopulation consists of 1015 cheetahs in 172 different facilities in 31 countries (Table 22.2; Fig. 22.3). The animals in this subpopulation are more likely to be in a managed breeding program, either now or in the near future, than animals not held within a facility registered with one of these regional zoo associations. Each regional breeding program only includes animals from their respective member institutions. The development of and collaboration between successful regional breeding programs, such as AZA's SSP and EAZA's EEP, are

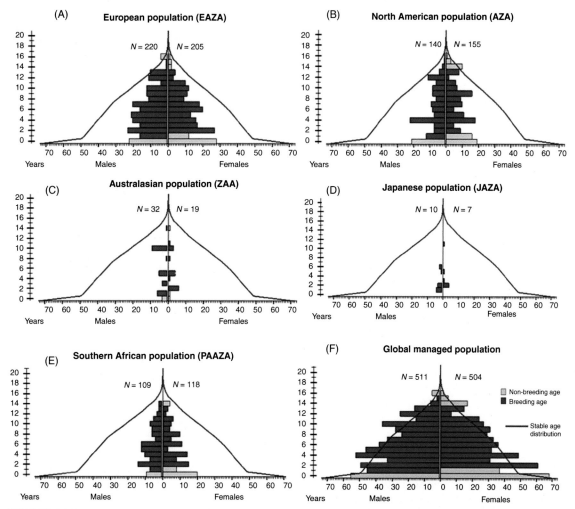

FIGURE 22.3 **Age pyramids of captive cheetah populations managed by regional associations and globally as of December 2014. Note that counts in Fig. 22.3 may differ from Table 22.2, as animals with unknown birth dates cannot be included in the age pyramids.**

critical to contribute to the sustainability of the captive cheetah population worldwide (see Chapter 23 for more details on the regional programs).

The various regional programs have different characteristics which will be described hereafter and are presented in Table 22.2. For instance, the Australian Species Management Program and the SSP utilize ambassador cheetahs for outreach programs and education (Chapter 28), while there are no ambassador cheetahs in the EEP; as of yet, ambassador individuals are not included in any breeding programs.

Europe

In Europe, as of 2014, there were 453 cheetahs at 107 institutions, of which 425 cheetahs at 95 institutions are under EEP management (Table 22.2; Fig. 22.3). Within this region, two populations are managed separately: the southern African cheetah (*Acinonyx jubatus jubatus*), which is managed by most zoos, and the northern African cheetah (*Acinonyx jubatus soemmeringii*). The northern African cheetah subspecies is managed collaboratively by a number of EAZA facilities in Europe as well as registered EAZA facilities in the Middle East, with a total of 155 (90.65) cheetahs. The founders for this population were sourced from the United Arab Emirates as offspring of confiscated animals from the illegal pet trade (Chapter 14).

In the last few years, there have been more births than deaths: growth from 260 to 360 individuals in 4 years demonstrates the EEP's success, as maximum carrying capacity of the holding facilities has almost been reached. In addition, since 2008/2009 more breeding animals are available as a result of captive-bred imports from South Africa. Therefore, the current EEP policy is to be flexible and to observe the growing population without intensive management of breeding decisions. In 2014, there were 29 groups of bachelor males, and 10 male/female groups housed in the same exhibit, which is not conducive to breeding (Chapter 27). Within the

population, 89% of ancestry is known, with a 2% inbreeding coefficient (the proportional decrease in observed heterozygosity relative to the expected heterozygosity of the founder population) and 23% N_e/N.

North America

In 2014, the cheetah SSP population consisted of 295 (144.162) cheetahs, which were distributed across 35 facilities in North America. An additional 21 non-AZA facilities held 103 cheetahs not involved in the SSP program. However, 15% of the SSP population [47 (20.27) animals held at 16 institutions] were ambassador animals and not available for breeding. An additional 15% were too old to breed. Thirty of the 35 institutions held single-sex groups. Approximately 40% of the population was held by the breeding centers and N_e/N was 21%.

Australasia

In Australia and New Zealand in 2014, the population included 54 cheetahs (32.33) at 14 institutions, of which 51 individuals from 12 institutions are members of ZAA. About a third of the animals were too old to breed. Half of the institutions held single-sex groups. Roughly half of the population were ambassador animals. The ZAA population has a N_e/N of 12%.

Japan

In Japan, the population stood at 137 cheetahs in 19 institutions at the end of 2014. However, only 17 (10.7) were held in two JAZA member locations. The JAZA population has a N_e/N of 16%.

Africa

There were 227 (109.118) cheetahs in the 10 PAAZA member institutions in South Africa and

Namibia in 2014. An additional 320 (176.139.5) animals lived in 29 other facilities in Spain's Canary Islands, Morocco, Tunisia, Ethiopia, and South Africa. The N_e/N ratio of the PAAZA population was 14% with a recent life table λ of 1.08, thereby representing the only association with a positive overall life table λ.

HISTORIC FACILITY BREEDING SUCCESS AND BREEDING STRATEGIES

When it comes to the success of captive cheetah breeding, a few facilities have been particularly influential in developing the successful breeding strategies used today. These strategies were mostly developed through trial and error, and strive to mimic natural behavior. Three main facilities, each with a slightly different approach, are presented here as case studies. All three approaches take into account the cheetah's natural behavior and attempt to mimic interactions in the wild (Caro, 1993; Laurenson, 1993).

The Ann van Dyk Cheetah Centre (also known as the De Wildt Cheetah and Wildlife Centre) in South Africa had their first successful breeding in March of 1975. Between 1975 and 2014, they produced 1003 (425.443.135) cubs, with a mortality rate of 34% under 1 year of age, and 22% under 1-month-old. Their breeding strategy was based on the complete separation of males and females (out of both sight and auditory range) to avoid creating familiarity. When a female came into heat, she was briefly brought together, one by one, with a selection of males in a neutral area (later termed "lover's lane") allowing the cheetahs an opportunity for mate choice (van Dyk, 1991).

In the USA, Wildlife Safari had their first successful breeding in 1973. A total of 167 (77.81.9) cubs were born between 1973 and 2014, with a total mortality rate of 35% under 1 year of age, and 24% under 1-month-old. Wildlife Safari used both an off-exhibit breeding area and a large 4 ha drive-through exhibit in which a male coalition and multiple females roamed and bred freely in a natural environment. A few weeks before birthing, pregnant females were moved to one of the adjacent 0.15 ha enclosures from where they could continue to interact with the other cheetahs through the fence. Wildlife Safari later linked the initial off-exhibit breeding center and the main drive-through area with a lover's lane, thus combining three successful strategies.

The Wassenaar Wildlife Breeding Centre was a successful private European zoo run by Jan and Hanneke Louwman. Wassenaar produced 168 (87.81) cubs, with a mortality of 22% under 1 year of age and 12% under 1 month, between 1980 and 2006, when they retired and closed their facility. Their breeding strategy included housing animals separately, away from the main cheetah yards and alternating a male and female daily in a large breeding enclosure. The timing of the breeding was determined based on the male indicating that the female was in heat by sniffing, scent marking, and loud vocalizations (stutter call; Chapter 9) (Beekman et al., 1997).

Concomitant with the emergence of various management strategies applied by a handful of captive breeding facilities, and with the compilation of ICS data, the SSP appointed a research committee to identify factors influencing breeding success (Wildt and Grisham, 1993; Chapter 27). In addition, ongoing long-term studies on wild cheetahs provided important information on natural behaviors and life history (Caro, 1994; Caro and Collins, 1987a,b; Eaton, 1974; Frame and Frame, 1981; Laurenson et al., 1992; Marker et al., 2008a,b; Chapter 9). The research confirmed the importance of enclosure design allowing for: (1) easy movement of individuals between enclosures; (2) intermittent visual and auditory separation of the sexes; and (3) the ability to have multiple individuals to choose from, to accommodate the yet little-understood nature of cheetah mate choice.

CONCLUSIONS

With the introduction of cooperative management programs, the relative success of the captive cheetah population continues to progress. Although the effective population size has increased over the years, large differences in reproductive success remain between regions.

The cheetah is one of the most specialized felids, given its capacity for high speed and associated physical and physiological adaptations (Chapter 7). The species has also been shown to be highly susceptible to stress in a captive setting (Chapter 25). Ultimately, long-term sustainability and welfare of captive cheetahs may depend on the mimicking of wild cheetah behavior and ecology (Chapters 8 and 9).

As the free-ranging population of cheetahs continues to decline (Chapters 4, 5, and 39) and as the genetic diversity of the remaining population decreases (Chapter 6), the captive and wild populations should be managed in cooperation under an integrated species conservation strategy. The development of a Global Species Management Plan for cheetah under the auspices of WAZA may be an important next step for the long-term future of the species and will require the cooperation of all stakeholders. With continued management, the goal of a self-sustaining and genetically diverse captive cheetah population could be achieved.

References

Ballou, J.D., Lacy, R.C., Pollak, J.P., 2014. PMx: software for demographic and genetic analysis and management of pedigreed populations (Version 1.3). Chicago Zoological Society, Brookfield, IL. Available from: http://www.vortex10.org/PMx.aspx.

Barongi, R., Fisken, F.A., Parker, M., Gusset, M. (Eds.), 2015. Committing to Conservation: The World Zoo and Aquarium Conservation Strategy. WAZA Executive Office, Gland.

Bartels, P., Berry, H.H., Cilliers, D., Dickman, A., Durant, S.M., Grisham, J., Marker, L., Munson, L., Mulama, M., Schoeman, B., Tubbesing, U., Venter, L., Wildt, D.E., Ellis, S., Freidmann, Y. (Eds.), 2002. Global Chee-

tah Conservation Action Plan—Final Report from the Workshop. Global Cheetah Conservation Action Plan—Workshop held at Shumba Valley Lodge in South Africa, 27–30 August 2001.

Beekman, S. P. A., Wit, M. D., Louwman, J., Louwman, H., 1997. Breeding and observations on the behaviour of cheetah (Acinonyx jubatus) at Wassenaar Wildlife Breeding Centre. Int. Zoo Yearbook 35 (1), 43–50.

Caro, T.M., 1993. Behavioral solutions to breeding cheetahs in captivity: Insights from the wild. Zoo Biol. 12, 19–30.

Caro, T.M., 1994. Cheetahs of the Serengeti Plains: Group Living in an Asocial Species. University of Chicago Press, Chicago, IL.

Caro, T.M., Collins, D.A., 1987a. Ecological characteristics of territories of male cheetahs (Acinonyx jubatus). J. Zool. 211, 89–105.

Caro, T.M., Collins, D.A., 1987b. Male cheetah social organization and territoriality. Ethology 74, 52–64.

Conde, D.A., Colchero, F., Gusset, M., Pearce-Kelly, Byers, O., Flesness, N., Browne, R.K., Jones, O.R., 2013. Zoos through the lens of the IUCN Red List: a global metapopulation approach to support conservation breeding programs. PLoS One 8, e80311.

Divyabhanusinh, 1999. The End of a Trail: The Cheetah in India. Banyan Books, New Delhi.

Durant, S.M., Mitchell, N., Groom, R., Pettorelli, N., Ipavec, A., Jacobson, A., Woodroffe, R., Bohm, M., Hunter, L., Becker, M., Broekuis, F., Bashir, S., Andresen, L., Aschenborn, O., Beddiaf, M., Belbachir, F., Belbachir-Bazi, A., Berbash, A., Brandao de Matos Machado, I., Breitenmoser, C., Chege, M., Cilliers, D., Davies-Mostert, H., Dickman, A., Fabiano, E., Farhadinia, M., Funston, P., Henschel, P., Horgan, J., de Iongh, H., Jowkar, H., Klein, R., Lindsey, P., Marker, L., Marnewick, K., Melzheimer, J., Merkle, J., Msoka, J., Msuha, M., O'Neill, H., Parker, M., Purchase, G., Saidu, Y., Samaila, S., Samna, A., Schmidt-Küntzel, A., Selebatso, E., Sogbohossou, E., Soultan, A., Stone, E., van der Meer, E., van Vuuren, R., Wykstra, M., Young-Overton, K., 2017. The global decline of cheetah and what it means for conservation. Proc. Natl. Acad. Sci. USA 114, 528–533.

Eaton, R.L., 1974. The Cheetah: The Biology, Ecology and Behavior of an Endangered Species. Van Nostrand Reinhold Company, New York, NY.

Frame, G., Frame, L.H., 1981. Swift and Enduring: The Cheetahs and Wild Dogs of the Serengeti. Dutton, New York, NY.

Guggisberg, C.A.W., 1975. Wild Cats of the World. Taplinger Publishing Co, New York, NY.

IUCN/SSC, 2014. Guidelines on the Use of Ex Situ Management for Species Conservation. Version 2.0. IUCN Species Survival Commission, Gland, Switzerland, 1–15.

Kraus, D., Marker-Kraus, L., 1992. Current Status of the cheetah (Acinonyx jubatus). In: Marker-Kraus, L. (Ed.)

The International Cheetah Studbook. NOAHS Center, National Zoological Park, Smithsonian Institution, Washington, DC.

Laurenson, M.K., 1993. Early maternal behavior of wild cheetahs-implications for captive husbandry. Zoo Biol. 12, 31–43.

Laurenson, M.K., Caro, T.M., Borner, M., 1992. Female cheetah reproduction. Natl. Geogr. Res. Expl. 8, 64–75.

Lyles, A.M., 2001. Zoos and zoological parks. Encyclopedia of Biodiversity, 5, 901–912.

Marker-Kraus, L., 1990. 1988 International Cheetah (*Acinonyx jubatus*) Studbook. NOAHS Center, National Zoological Park, Smithsonian Institution, Washington, DC.

Marker, L., 2015. 2014 International Cheetah (*Acinonyx jubatus*) Studbook. Cheetah Conservation Fund, Otjiwarongo, Namibia.

Marker, L.L., Dickman, A.J., Mills, M.G.L., Jeo, R.M., Macdonald, D.W., 2008a. Spatial ecology of cheetahs (*Acinonyx jubatus*) on North-central Namibian farmlands. J. Zool. 274, 226–238.

Marker, L., Pearks-Wilkerson, A.J., Sarno, R.J., Martenson, J., Breitenmoser-Wursten, C., O'Brien, S.J., Johnson, W.E., 2008b. Molecular genetic insights on cheetah (*Acinonyx jubatus*) ecology and conservation in Namibia. J. Hered. 99 (1), 2–13.

Marker-Kraus, L., 1997. History of the cheetah in zoos 1829–1994. Int. Zoo Yearbook 35, 27–43.

Minteer, B.A., Collins, J.P., 2013. Ecological ethics in captivity: balancing values and responsibilities in zoo and aquarium research under rapid global change. ILAR J. 54 (1), 41–51.

Pocock, R.I., 1939. The Fauna of British India Including Ceylon and Burma. Mammalia. I. Primates and Carnivora. In: Sewell, R.B.S. (Ed) Taylor and Francis LTD, London.

Scobie, P., Bingaman Lackey, L., 2012. SPARKS: Single Population Analysis and Record Keeping System. Version 1.66. International Species Information System (renamed to Species360). Eagan, MN.

van Dyk, A., 1991. The Cheetahs of De Wildt. Struik Publishers (Pty) Ltd, Cape Town.

Wildt, D.E., Grisham, J., 1993. Basic research and the cheetah SSP program. Zoo Biol. 12, 3–4.

The Role of Zoos in Cheetah Conservation: Integrating *Ex Situ* and *In Situ* Conservation Action

Karin R. Schwartz*, Markus Gusset**, Adrienne E. Crosier[†],
Lars Versteege[‡], Simon Eyre[§], Amanda Tiffin[§], Antoinette Kotzé[¶,††]

*Cheetah Conservation Fund, Otjiwarongo, Namibia
**World Association of Zoos and Aquariums, Gland, Switzerland
[†]Smithsonian Conservation Biology Institute, Front Royal, VA, United States
[‡]Safaripark Beekse Bergen, Hilvarenbeek, The Netherlands
[§]Wellington Zoo Trust, Wellington, New Zealand
[¶]National Zoological Gardens of South Africa, Pretoria, South Africa
[††]University of Free State South Africa, Bloemfontein, South Africa

INTRODUCTION

With their sleek build and captivating faces with golden eyes accentuated by black tear marks, cheetahs have long been admired as one of the icons of the wild, fascinating humans with their beauty and speed. Today, wild populations are in decline, and zoos, which maintain this iconic feline species, have taken on an important role in its race against extinction. Wild cheetah populations are threatened with extinction due to a number of anthropogenic threats (Chapters 10–14) combined with low genetic diversity (Chapter 6). Immediate conservation measures are critical to reverse the trend of declining numbers and to sustain viable populations (Chapter 38). The registered global cheetah population maintained in zoological institutions at the end of 2014 totaled 1730 animals in 267 known facilities in 44 countries in 7 geographical regions (Marker, 2015; Chapter 22). This population puts zoos in a position to strive for sustainable populations within their institutions, as well as contribute to conservation of the cheetah in the wild.

The role of zoos has evolved from menageries for entertainment to accredited institutions with an increased emphasis on conservation (Barongi et al., 2015; Rabb, 1994; Chapter 22). Zoological institutions collectively constitute a conservation powerhouse providing environmental education programs to promote awareness for visitors, scientific breeding management programs for maintaining sustainable *ex situ* (under human care in zoological institutions) populations, and contributing to *in situ* (in the wild) conservation through research, funding support, and involvement in conservation action planning. These institutions are organized into international and regional zoological associations that support collaborative conservation programs for threatened species. Through these zoological associations, zoos work together to maintain sustainable *ex situ* populations within their regions and contribute to the conservation of cheetahs in the wild. It is evident that to mitigate the decline of natural cheetah populations, it is necessary to use all the tools available and combine *in situ* and *ex situ* conservation efforts into a One Plan Approach for holistic cheetah conservation (Byers et al., 2013).

This chapter outlines the organization of zoos into zoological associations and the role they play in cheetah conservation through scientific breeding management and funding support for *in situ* conservation programs. It will then discuss further conservation roles through education, awareness raising, participation in conservation action planning through IUCN Species Survival Commission Specialist Groups, and research.

CHEETAH CONSERVATION THROUGH ZOO ASSOCIATIONS

The World Association of Zoos and Aquariums (WAZA) is the international association that serves as an umbrella organization to bring together the world zoo and aquarium community in orchestrated efforts for animal welfare, environmental education and global conservation. WAZA unifies members from over 300 zoos, aquariums, zoo associations, and affiliate and corporate partners around the world. Zoos are organized into regional associations, such as in North America [Association of Zoos and Aquariums (AZA)], Europe [European Association of Zoos and Aquaria (EAZA)], and Australasia [Zoo and Aquarium Association (ZAA)]. These three regions maintain the majority of cheetahs in zoological institutions and participate in collaborative breeding management programs such that the cheetah population is managed throughout the region as a whole. The World Zoo and Aquarium Conservation Strategy, published by WAZA (Barongi et al., 2015), postulates that a holistic approach to species conservation combining both *in situ* and *ex situ* efforts is necessary; that is, the One Plan Approach to species conservation planning (Fig. 23.1). The cheetah represents a prime example for such an integrated approach, as effective conservation action ranges along a continuum of management intensity, from little, if any, human intervention in wild populations all the way to intensively managed populations in some reserves and in zoos (Gusset and Dick, 2013). WAZA promotes healthy *ex situ* populations through conservation breeding, and manages the international studbook program and the WAZA conservation project branding scheme.

Conservation Breeding

The zoological associations utilize breeding management programs to maintain sustainable *ex situ* populations that are genetically diverse, demographically stable, and behaviorally competent. Maintaining healthy, sustainable *ex situ* populations offers conservation benefits by limiting the necessity for further infusion of genetic variability from wild-caught cheetahs and provides the institutions with healthy individuals to educate and engage visitors to care about cheetahs in the wild. These healthy populations could potentially contribute to reintroduction programs in terms of numbers and/or genetic diversity in the advent that a cheetah

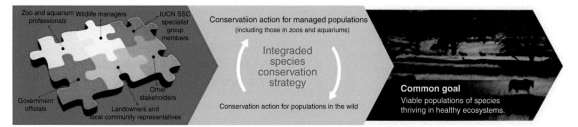

FIGURE 23.1 **The One Plan Approach—a term coined by the IUCN Species Survival Comission (SSC) Conservation Breeding Specialist Group (CBSG)—refers to integrated species conservation planning that considers all populations of the species (inside and outside the natural range), under all conditions of management, and engages all responsible parties and resources from the start of the conservation planning initiative.** *Source: Used with permission from Barongi, R., Fisken, F.A., Parker, M., Gusset, M. (Eds.), 2015. Committing to Conservation: The World Zoo and Aquarium Conservation Strategy. WAZA Executive Office, Gland.*

conservation translocation (reintroduction/supplementation) program using captive-bred animals becomes a feasible and necessary conservation option (Chapter 20). Gusset and Dick (2012) demonstrated that conservation breeding in zoos and subsequent reintroduction has played a role in the recovery of one quarter of the 64 species that showed a reduction in threat status on the IUCN Red List of Threatened Species over a period of 20 years.

International Cheetah Studbook (ICS)

Due to the lack of genetic diversity in the species (Chapter 6), cheetahs are managed particularly carefully within the international and regional zoo associations to preserve as much of the remaining genetic diversity as possible. International and regional studbooks provide the data that facilitate the coordination of conservation breeding efforts across zoological institutions. International studbooks are administered under the auspices of WAZA, with regional studbooks administered by the regional zoo associations.

The International Cheetah Studbook (ICS) maintains the genealogical background of all documented cheetahs in *ex situ* institutions worldwide (Marker, 2015; Chapter 22). The ICS documents wild-born versus captive-born status for each individual, linking the pedigrees of

those in *ex situ* institutions to their wild ancestors. The importance of studbooks in population management has been demonstrated in zoo-held ruminants, as those species with an international studbook have a significantly higher relative life expectancy than those without (Müller et al., 2011). This suggests that the existence and use of a studbook to facilitate scientific breeding management to maintain genetic diversity and reduce inbreeding may positively impact conservation breeding efforts.

The ICS, established in 1988 and first published in 1990 by Dr. Laurie Marker (Marker, 2015), has been maintained annually since the initial publication date (Chapter 22). Most conservation breeding programs are managed at the regional level for logistical and regulatory reasons, each using a regional studbook. The ICS forms the basis for the regional cheetah studbooks used in these breeding programs. A new way of fostering collaboration interregionally for animals maintained in the ICS is being tested through Global Species Management Plans (GSMPs) administered under the auspices of WAZA. A GSMP involves the management of a particular taxon with a globally agreed set of goals, while building upon and respecting existing regional processes (Gusset and Dick, 2011b). The GSMP option to achieve a sustainable population (one of the key emerging issues for zoos globally) (Gusset et al., 2014) needs to be explored for cheetah.

Institutional Records and Studbook Management

Breeding plans are dependent on the use of a studbook for analysis of the genetic and demographic composition of the whole population (Chapter 22). Thus, accurate records on origin and parentage are critical. Based on a combination of genetic and demographic evaluations, breeding recommendations are made annually for individual animals.

Data in a studbook come from the institutions holding the species via information exchange, or record-keeping programs, such as the Species360 database for Species360 members. Species360 (previously International Species Information System) is an organization that governs the centralized database system that compiles animal records from over 1050 zoological institutions in 90 countries worldwide (Species360, 2016). Species360 manages the Zoological Information Management System (ZIMS) application, a global web-based database platform for maintaining animal records, covering information such as life history, physiology, reproduction, behavior, and health to facilitate cooperative animal husbandry and breeding management processes. A new ZIMS Population Management module (deployed April 2017) has added a studbook component that replaces previous studbook software and the system now encompasses husbandry, health, and studbook development within the one information management system. Thanks to this new development, studbooks, such as the ICS can now be updated in real time within ZIMS, and the studbook data exported to population management software, such as PMx (Ballou et al., 2011) for analysis resulting in breeding recommendations. These recommendations facilitate pairings that will result in sustainable populations that maintain genetic diversity,

are demographically stable, and behaviorally competent (Leus et al., 2011).

WAZA-Branded Conservation Projects for Cheetah

Conservation projects are branded by WAZA after consideration of biological, operational, and institutional and partnership endorsement criteria. Seven WAZA-branded projects target the conservation of cheetahs in the wild, including programs run by organizations, such as the Cheetah Conservation Fund (CCF), Cheetah Conservation Botswana (CCB), and Action for Cheetahs in Kenya (ACK) (Gusset and Dick, 2010). The projects focus on research, conservation action, conflict resolution, community outreach, environmental education, and capacity building. All of these projects receive substantial financial, technical, and logistical support from the international zoo community.

THE REGIONAL ZOO ASSOCIATIONS AND CHEETAH CONSERVATION

Association of Zoos and Aquariums (AZA)

AZA is a nonprofit organization representing more than 230 institutions in the United States, Canada, Mexico, and a small number of other locations. AZA institutions collectively draw more than 183 million visitors every year (AZA, 2017a,b). Each AZA-member institution is accredited according to stringent criteria to meet the highest standards in animal management and care, including appropriate habitats, natural social groups, health, and nutrition, as well as involvement in conservation, research, and education programs.

AZA manages ~300 cheetahs in 57 facilities in North America through its *Cheetah Species Survival Plan® (SSP)* (Chapter 22). AZA SSP programs strive to manage and conserve a threatened or endangered species *ex situ* population by working with the AZA Population Management and Reproduction Management Centers. The SSP is responsible for developing a Breeding and Transfer Plan (Crosier et al., 2016) that identifies population management goals and recommendations to ensure the sustainability of a healthy, genetically diverse, and demographically varied population. This Breeding Plan excludes individuals that are used in educational programs (Ambassador Animals); however, each year a certain number of cubs from the breeding program may be designated to the Ambassador Animal program, depending on the success of each of the managed populations (AZA, 2016; Chapter 28).

AZA formed the collaborative *Conservation Centers for Species Survival (C2S2)* in 2005, targeting threatened species with special needs that would benefit from large areas, natural group sizes, minimal human disturbance, and research. The cheetah was the representative carnivore species and 1 of 12 original focal species that would benefit from this management strategy. C2S2 developed and implemented a Cheetah Sustainability Program within the Cheetah SSP in 2013, whereby cheetah holders join a consortium, contribute a participation fee, and commit to cooperative and collaborative management of cheetahs. Breeding is concentrated at nine breeding centers, each with large areas dedicated to cheetahs, a minimum of four breeding females with an appropriate number of males, genetically matched to produce offspring that will maintain the demographic stability and genetic diversity of the North American *ex situ* population (Fig. 23.2). Fourteen other Cheetah SSP member institutions are part of C2S2 and focus on exhibition or education with program animals, but with opportunity to also contribute to the breeding program. Participation fees are utilized to support *in situ* conservation and *ex situ* research through a competitive, peer-reviewed grants program, as well as to support cheetah reproduction in the breeding centers by offsetting a small proportion (<5%) of costs incurred by these facilities.

FIGURE 23.2 **The Smithsonian Conservation Biology Institute (SCBI) in Front Royal, VA participates as one of the nine cheetah breeding centers within the Association of Zoos and Aquariums (AZA) Conservation Centers for Species Survival, dedicating nine acres of land to their cheetah breeding program.** *Source: Lisa Ware, SCBI.*

In 2015, AZA launched a new conservation initiative called *SAFE: Saving Animals From Extinction,* and the cheetah was 1 of 10 inaugural species selected. SAFE is a collaborative initiative of specialists with species-specific expertise including AZA members, scientists, non-government and government partners, and organizations working on field conservation efforts, to identify the threats, develop action plans, obtain new resources, and engage the community (AZA, 2017a). A 2015 cheetah conservation planning workshop, sponsored by AZA SAFE and cohosted by the Range Wide Conservation Program (RWCP) for Cheetah and African Wild Dogs, brought together almost 50 stakeholders representing dozens of governmental agencies, non-governmental and research organizations, as well as AZA representatives, to outline a unified approach to address the highest priority issues facing cheetah conservation. The results of this workshop yielded the development of the 2016–2018 SAFE Cheetah Conservation Action Plan (www.aza.org/SAFE-cheetah). The plan includes several collaborative projects focusing on carnivore-positive landscapes, monitoring and law enforcement, capacity building, distribution and population connectivity, snaring abatement, and the South African managed cheetah metapopulation.

European Association of Zoos and Aquaria (EAZA)

With 377 member institutions in 43 countries throughout Europe and the Middle East, EAZA is the largest zoological association in the world. EAZA's mission is "to facilitate cooperation within the European zoo and aquarium community with the aim of furthering its professional quality in keeping animals and presenting them for the education of the public, and of contributing to scientific research and to the conservation of global biodiversity" (EAZA, 2017). With approximately 140 million visitors a year, the EAZA zoos can have an enormous impact by engaging visitors with their conservation message.

The European Endangered Species Programme (EEP) manages 380 cheetahs in 89 zoological institutions (October 2015; Chapter 22). Historically, the population heavily relied on imports of captive-bred individuals from southern African countries, but as with every intensively managed breeding program, self-sustainability is the goal. Because of the difficulty of breeding cheetahs, in part due to behavioral differences between individuals (Chapter 27), the EEP grants any institution the option to breed. However, the general objective is to place genetically important specimens at the most experienced and successful institutions.

Zoo and Aquarium Association Australasia (ZAA)

The ZAA has a membership of 99 organizations throughout Australasia (Australia, New Zealand, Papua New Guinea, Singapore, and New Caledonia) of which 94 are zoos, aquariums, and museums. Other members include universities, technical and further education organizations (TAFEs), and government departments.

Accreditation requirements include professional standards on animal health and welfare, training and assessment tools, biosecurity, and international collaboration for conservation programs. Each year, ZAA institutions engage over 17 million visitors and provide education programs that reach over 600,000 students (ZAA, 2017).

The aim of the ZAA Cheetah *Australasian Species Management Plan (ASMP)* is to provide a regionally sustainable zoo-held population to meet the conservation advocacy needs of its members. Participants in the ASMP include 14 facilities in Australia and New Zealand with a total cheetah population of 55 animals (33 males and 22 females) by the end of 2014 (Marker, 2015; Chapter 22). The ASMP recognizes two valuable roles for cheetahs in the region: breeding cheetahs and ambassador cheetahs. The movement of animals between these two roles is undertaken through a negotiated process in consultation with the species coordinator and the holders.

The ZAA Cheetah ASMP statistics demonstrate a consistently low level of breeding success, with a reliance on captive-bred imports from southern Africa to sustain the regional population. The 2014 ZAA Captive Management Plan incorporated a multilevel strategy to increase the breeding success within the Australasian region to reduce this program's dependence on importation. First, similarly to AZA (mentioned earlier), the holders with extensive facilities and larger numbers of cheetahs have a proportionally larger commitment to breed cheetahs. These facilities also play the important role of running cheetah-specific husbandry training workshops. Second, ZAA excludes animals from the breeding population when they become postreproductive or when hand-raised animals in the ambassador program that have not yet bred are over 3 years old. Third, breeding recommendations in the Australasian region are made on the basis that any male can mate with any female as long as they are not closely related.

The ASMP has facilities that are not breeding centers but care for and train cheetahs as ambassador animals. ZAA now manages the breeding population and the ambassador animals separately, as the inclusion of ambassador animals in the breeding population gives an inaccurate picture of the genetics and demographics of the breeding population.

FUNDING SUPPORT

The regional zoo associations' member institutions generate substantial funding support for cheetah research and conservation efforts *in situ*. Zoos around the world spend about US $350 million on wildlife conservation every year (Gusset and Dick, 2011a). In relation to major international conservation organizations, the international zoo community is among the main providers of conservation funding.

AZA institutions provide a large amount of logistical and monetary support for *in situ* cheetah conservation. From 2011 to 2015, 58 AZA-accredited zoos reported taking part in 142 instances of cheetah field conservation, investing over US $1.6 million for these conservation projects (AZA, 2017a,b). This provided support to field conservation partners like CCF, ACK, and CCB. In addition, the American Association of Zoo Keepers (AAZK) supports ACK through their Bowling For Rhinos Fund, a yearly event that raises funds and awareness for rhinoceros and habitat conservation (AAZK, 2017). Cheetahs share the same habitat as rhinoceros in Kenya and thus support for ACK helps to save the habitat where both of these endangered species are found.

Many European institutions within EAZA fundraise for, and are actively involved with, cheetah conservation organizations. CCF and Africat in Namibia, and ACK in Kenya are three of the many conservations organizations that have benefited from funds contributed by over 30 European institutions. ZAA institutions involved in the ASMP likewise support CCF, as well as organizations such as CCB and Cheetah Outreach in South Africa. For example, Wellington Zoo Trust has an ongoing program of support for Cheetah Outreach and has contributed over US $10,000 since 2009.

INFORMATION SHARING THROUGH GLOBAL CHEETAH CONSERVATION ACTION PLANNING

Zoo professionals share their expertise in the care and management of *ex situ* cheetah populations through participation in the IUCN Cat Specialist Group and IUCN Conservation Breeding Specialist Group (CBSG) and are actively involved in cheetah conservation action planning. Specialist Groups exist under the umbrella of the IUCN Species Survival Commission, a science-based network of volunteer experts that includes researchers, government officials, wildlife veterinarians, and zoological institution professionals with a common mission to promote the conservation of threatened species. Population and Habitat Viability Assessment (PHVA) workshops, conducted by CBSG, bring together stakeholders from both the *in situ* and *ex situ* communities to evaluate extinction risk using quantitative analyses, and to develop conservation action plans.

The Namibian Cheetah and Lion PHVA, held in 1996 in Namibia, was the first PHVA for cheetahs (CBSG, 1996; Chapter 38). This collaborative workshop hosted by CCF and sponsored by British Airways and 13 AZA institutions, involved personnel from the Namibian Ministry of Environment and Tourism, CCF, the AZA Felid Taxon Advisory Group, the AZA Cheetah and Lion SSPs, and CBSG. The 2002 Global Cheetah Conservation Action Plan (CBSG, 2002) and 2009 South African Cheetah PHVA (CBSG, 2009; Chapter 38) also benefited from funding and involvement by the zoo community (EAZA Cheetah EEP and representatives from the

National Zoological Gardens in Pretoria, South Africa).

EDUCATION AND AWARENESS RAISING THROUGH CLOSE-UP ENCOUNTERS

With more than 700 million visits worldwide every year (Gusset and Dick, 2011a), zoos are major providers of effective environmental education (Gusset and Lowry, 2014; Moss et al., 2015, 2017). To ensure that people care about the survival of cheetahs in the wild, they must have an emotional connection to the animal and a drive to conserve it. However, few people outside of the remaining cheetah range are likely to see a cheetah in the wild. Zoos offer a real-life sensory experience that enables the visitor to form that emotional connection. For example, a close-up view bringing the visitor face-to-face with a cheetah (Fig. 23.3) leaves an indelible impression on the viewer. Some zoos also maintain

FIGURE 23.3 **A subterranean tunnel leads to a viewing window that offers a close up and personal experience to engage visitors in the lives of cheetahs at the Parco Zoo Falconara in Falconara Marittima, Italy.** Parco Zoo Falconara currently hosts 2 cheetah brothers as a member of the European Association of Zoos and Aquaria (EAZA) Southern Cheetah (*Acinonyx jubatus jubatus*) European Endangered Species Programme (EEP). *Source: Parco Zoo Falconara.*

ambassador cheetahs (habituated animals that appear with handlers outside of their exhibits) to further engage visitors (Chapter 28).

Adjacent to cheetah enclosures, zoos usually display educational graphics on species biology, ecology, and conservation status. In addition, some zoos share the information through interactive exhibits or through cheetah talks delivered by zoo docents, volunteers, and educators. The content of the educational messages, which may include suggestions for conservation action that can be taken, can be tailored to a specific audience to make them more effective. The main aim of the written and verbal information-sharing is to create an awareness of the threats cheetahs face in the wild and to help portray a conservation message outlining what visitors can do to support cheetah conservation. The AZA Felid Taxonomy Advisory Group (Felid TAG) provides suggested conservation messages that Cheetah SSP member institutions can incorporate into their programming. The enlightened visitor thereby becomes more engaged in conservation and more receptive to public calls to action concerning the status of cheetahs in the wild.

Zoos provide visitors the opportunity for a personal learning experience by viewing animals in naturalistic habitats engaging in natural behaviors. There are a number of zoos around the world that incorporate the use of a running track equipped with a mechanical pulley device that moves a lure (consisting of a rag, feathers, or meat treat) quickly around the habitat, enticing the cheetah to give chase (Fig. 23.4). Whether with the help of enrichment devices or simply by giving them a lot of space to exercise, behavioral enrichment has many benefits for both the cheetahs and the watching visitors. The cheetahs have the chance to exhibit natural hunting behaviors while gaining a health benefit from the exercise (increasing fitness and reducing stress levels; Chapter 24). The visitors can witness the cheetah's demonstration of speed critical for survival in the

FIGURE 23.4 **Mechanical pulley system for a lure as behavioral enrichment for cheetahs at Ree Park in Elbeltoft, Denmark.** Ree Park is a participant of the European Association of Zoos and Aquaria (EAZA) Southern Cheetah (*Acinonyx jubatus jubatus*) European Endangered Species Programme (EEP), holding several breeding animals of different bloodlines. *Source: Henrik Nordvig.*

wild and learn about the cheetah's unique adaptations as the fastest land mammal on Earth (Chapter 7). "Cheetah runs" are very popular (>100,000 spectators per year at the Cincinnati Zoo; 1200 per day during peak season at the Columbus Zoo; Chapter 28, Table 28.1) and provide the opportunity to share information on cheetah conservation with the guests.

Engaging visitors at zoos to care about cheetah conservation may inspire them to travel to range countries to see these animals in the wild. Zoos often lead tours or encourage visitors to travel to cheetah range countries and this adds another level of connection between animals in a zoo setting with their wild counterparts, underscoring the importance of contributing to conservation efforts. Many cheetah-range countries (e.g., Botswana, Kenya, Namibia, South Africa, and Tanzania) depend on tourism as a source of revenue. Therefore, wildlife-based tourism gives local people an incentive to keep wildlife on their land alive. It has been shown that when local communities witness wildlife-generating tourism dollars, an economic value is instilled on the wildlife; this in turn reinforces local conservation efforts and increases

the cheetah's chances for long-term survival (Marker, 2014).

RESEARCH

Research conducted and/or supported by zoological institutions has become an important part of the conservation of threatened species. A good portion of the knowledge on cheetahs has been generated as a direct result of studies conducted with animals in *ex situ* facilities (Wildt and Grisham, 1993). In addition, research techniques for monitoring cheetahs in the wild have been developed or advanced by studies conducted in zoos (Wharton, 2007). Zoos also support and conduct *in situ* research projects through substantial contributions of funds or direct provision of technical expertise. Research is often conducted via collaborative efforts with and shared among, the broader conservation community (academics, government authorities, local communities, non-governmental organizations). Multiple studies have taken advantage of the link between the *in situ* and *ex situ* populations, investigating comparative health (Munson et al., 2004, 2005), reproduction (Crosier et al., 2007, 2009), and phylogenetics using cheetah samples from museums, zoos, and the wild (Franklin et al., 2016). These rigorous scientific efforts have provided, and continue to provide, a greater understanding of many aspects of cheetah biology, including health (Chapter 25), physiology (Chapter 7), endocrinology (Chapter 27 and 31), behavior (Chapter 9), reproduction (Chapter 27), nutrition (Chapter 26), genetics (Chapter 6), and management (Chapters 22 and 24).

Zoological Conservation and Research Departments

Many zoological institutions maintain dedicated field conservation and research departments. Four major players are noted hereafter as examples.

The Center for Cheetah Conservation of the Saint Louis Zoo WildCare Institute, MO, USA focuses on cheetah conservation programs in Tanzania, Kenya, Botswana, Namibia, and South Africa. The Center partners with other zoo professionals, researchers, and project managers in cheetah range countries to promote cheetah conservation through support of the IUCN Global Cheetah Action Plan (CBSG, 2002). The Center supports (through technical expertise and funding) the Serengeti Cheetah Project in the development and implementation of reliable census techniques to assess status and health of cheetahs in the Serengeti National Park, Tanzania. For *ex situ* research, the Center collaborates with other facilities within the AZA SSP to conduct reproductive and behavioral research.

The Smithsonian Conservation Biology Institute (SCBI) in Front Royal, VA, closely associated with the National Zoo in Washington, DC, USA, hosts scientists whose research focuses on reproductive biology and conservation genetics. SCBI is one of the cheetah breeding centers of C2S2 [section "Association of Zoos and Aquariums (AZA)"]. Among others, work has been done to develop and optimize assisted reproductive techniques, such as artificial insemination (Howard et al., 1992) and sperm cryopreservation. This enables conservationists to collect and bank genetically valuable reproductive samples (Crosier et al., 2007, 2009).

The Conservation and Research Department at Chester Zoo, an EAZA member in the United Kingdom, supports approximately 70 projects for threatened species around the world and maintains close connections with cheetah field conservation organizations. Since 2009, Chester Zoo has contributed technical expertise and funding to a variety of carnivore research projects in cheetah range countries, including the N/a'an ku sê Carnivore Research Project

in Namibia and the Ruaha Carnivore Project in Tanzania. In addition to supporting research, Chester Zoo also sponsors conservation workshops that facilitate research of animals in the wild (e.g., cheetah monitoring using the footprint identification technique) (Wildtrack, 2015). In addition, *ex situ* research projects have been conducted at the zoo and include studies on nutrition (diet and gastrointestinal health), social behavior, and reproductive behavior.

The Centre for Conservation Science at the National Zoological Gardens of South Africa (a designated National Research Facility) has been involved in multidisciplinary projects to investigate cheetah diseases, genetics, nutrition, reproduction, behavior, and physiology. The Centre maintains a database of diseases found in more than 900 cheetah individuals and conducts research to identify biomarkers suitable for the early detection of cheetah diseases. Scientists at the NZG partnered with academic scientists from Austria, France, and Portugal in a 5-year genetic study to analyze DNA samples of living animals in the wild, in zoos, and from museum specimens. Results supported identification of distinct subspecies of the cheetah (Charruau et al., 2011; Chapter 6).

The research supported at these zoos provides a unique opportunity to train a significant number of undergraduate and graduate students, as well as postdoctoral fellows in cheetah-specific areas of research. For example, NZG as a National Research Facility provides a platform for high-quality research in areas of national importance, and for the development of a knowledge economy. From 2008 to 2016, a total of three doctoral, four masters and two bachelors of technology degrees were awarded to students conducting research on aspects of cheetah conservation at NZG. SCBI trained five graduate students and two postdoctoral fellows in cheetah biology from 2006 to 2016. This level of commitment to training the next generation of cheetah biologists, including those from cheetah range countries, is a crucial aspect of developing a selfsustaining research and management program for cheetahs in *in situ* and *ex situ* locations.

SUMMARY: *EX SITU* AND *IN SITU* INTEGRATION FOR HOLISTIC CHEETAH CONSERVATION

All available resources are needed to combat the threats to wild cheetah populations and ensure their future. According to the World Zoo and Aquarium Conservation Strategy (Barongi et al., 2015), animal welfare and conservation must be the primary focus for zoological institutions, all of which are urged to adopt an integrated One Plan Approach to conservation. Thus, partnerships between the global zoo community and those individuals and organizations working on cheetah conservation *in situ* are critical to secure the future of the species. By adopting a One Plan Approach, the *ex situ* and *in situ* communities develop conservation strategies together with the common goal of establishing and maintaining long-term viable cheetah populations in healthy ecosystems.

Modern-day zoos are uniquely positioned to contribute to effective conservation of threatened species and ecosystems on many levels. Cheetahs cared for in accredited zoos play a variety of conservation roles including raising awareness as display animals, connecting with the visitors as ambassadors, and as a backup resource for genetic diversity. Targeted zoo-based research, advances in husbandry management, animal welfare and health care, and an increased knowledge of cheetah biology and physiology, have assisted in a greater understanding of the natural life of cheetahs in the wild. Considerable funding support from zoos and participation of zoo professionals in *in situ* research, as well as in holistic conservation action planning have linked the *ex situ* and *in situ* conservation

communities together to work toward an improved status of the cheetah in the wild. Only through these integrated collaborations will conservation efforts enable the cheetah to win the race against extinction.

References

American Association of Zoo Keepers (AAZK), 2017. Bowling for Rhinos. Available from: https://www.aazk.org/bowling-for-rhinos.

Association of Zoos and Aquariums (AZA), 2016. Cheetah SSP Sustainability Report 2016. Silver Spring, MD.

Association of Zoos and Aquariums (AZA), 2017a. AZA SAFE-Cheetah. Available from: https://www.aza.org/SAFE-Cheetah.

Association of Zoos and Aquariums (AZA), 2017b. About Us. Available from: https://www.aza.org/about-us.

Ballou, J.D., Lacy, R.C., Pollak, J.P., 2011. PMx: Software for Demographic and Genetic Analysis and Management of Pedigreed Populations (Version 1.0). Chicago Zoological Society, Brookfield, IL.

Barongi, R., Fisken, F.A., Parker, M., Gusset, M. (Eds.), 2015. Committing to Conservation: The World Zoo and Aquarium Conservation Strategy. WAZA Executive Office, Gland.

Byers, O., Lees, C., Wilcken, J., Schwitzer, C., 2013. The One Plan Approach: the philosophy and implementation of CBSG's approach to integrated species conservation planning. WAZA Magazine 14, 2–5.

Charruau, P., Fernandes, C., Orozco-ter Wengel, P., Peters, J., Hunter, L., Ziaie, H., Jourabchian, A., Jowkar, H., Schaller, G., Ostrowski, S., Vercammen, P., Grange, T., Schlotterer, C., Kotze, A., Geigle, E.-M., Walzer, C., Burger, P.A., 2011. Phylogeography, genetic structure and population divergence time of cheetahs in Africa and Asia: evidence for long-term geographic isolates. Mol. Ecol. 20, 706–724.

Conservation Breeding Specialist Group (CBSG), 1996. Namibian Cheetah and Lion PHVA. Available from: http://www.cbsg.org/content/namibian-cheetah-and-lion-phva-1996.

Conservation Breeding Specialist Group (CBSG), 2002. Global Cheetah Conservation Action Plan. Available from: http://www.cbsg.org/content/global-cheetah-conservation-plan-2002.

Conservation Breeding Specialist Group (CBSG), 2009. South African Cheetah PHVA. Available from: http://www.cbsg.org/content/south-african-cheetah-phva-2009.

Crosier, A.E., Henghali, J.N., Howard, J.G., Pukazhenthi, B.S., Terrell, K.A., Marker, L.L., Wildt, D.E., 2009. Improved quality of cryopreserved cheetah (*Acinonyx jubatus*) spermatozoa after centrifugation through Accudenz. J. Androl. 30, 298–308.

Crosier, A.E., Marker, L., Howard, J., Pukazhenthi, B.S., Henghali, J.N., Wildt, D.E., 2007. Ejaculate traits in the Namibian cheetah (*Acinonyx jubatus*): influence of age, season and captivity. Reprod. Fertil. Dev. 19, 370–382.

Crosier, A., Moloney, E., Long, S., 2016. Population Analysis & Breeding and Transfer Plan: Cheetah (*Acinonyx jubatus*) AZA Species Survival Plan® Yellow Program. Association of Zoos and Aquariums Population Management Center, Chicago, IL.

European Association of Zoos and Aquaria (EAZA), 2017. About Us. Available from: http://www.eaza.net/about-us.

Franklin, A.D., Schmidt-Küntzel, A., Terio, K.A., Marker, L.L., Crosier, A.E., 2016. Serum amyloid A protein concentration in blood is influenced by genetic differences in the cheetah (*Acinonyx jubatus*). J. Hered. 107, 115–121.

Gusset, M., Dick, G., 2010. 'Building a Future for Wildlife'? Evaluating the contribution of the world zoo and aquarium community to *in situ* conservation. Int. Zoo Yearb. 44, 183–191.

Gusset, M., Dick, G., 2011a. The global reach of zoos and aquariums in visitor numbers and conservation expenditures. Zoo Biol. 30, 566–569.

Gusset, M., Dick, G. (Eds.), 2011. WAZA Magazine 12: Towards Sustainable Population Management. WAZA Executive Office, Gland.

Gusset, M., Dick, G. (Eds.), 2012. WAZA Magazine 13: Fighting Extinction. WAZA Executive Office, Gland.

Gusset, M., Dick, G. (Eds.), 2013. WAZA Magazine 14: Towards Integrated Species Conservation. WAZA Executive Office, Gland.

Gusset M., Fa J.E. and Sutherland W.J., The Horizon Scanners for Zoos and Aquariums, 2014. A horizon scan for species conservation by zoos and aquariums, Zoo Biol. 33, 375–380.

Gusset, M., Lowry, R. (Eds.), 2014. WAZA Magazine 15: Towards Effective Environmental Education. WAZA Executive Office, Gland.

Howard, J., Donoghue, A.M., Barone, M.A., Goodrowe, K.L., Blumer, E.S., Snodgrass, K., Starnes, D., Tucker, M., Bush, M., Wildt, D.E., 1992. Successful induction of ovarian activity and laparoscopic intrauterine artificial insemination in the cheetah (*Acinonyx jubatus*). J. Zoo Wildl. Med. 23, 288–300.

Leus, K., Traylor-Holzer, K., Lacy, R.C., 2011. Genetic and demographic population management in zoos and aquariums: recent developments, future challenges and opportunities for scientific research. Int. Zoo Yearb. 34, 213–225.

Marker, L., 2014. A Future for Cheetahs. Cheetah Conservation Fund, Alexandria, VA.

Marker, L., 2015. 2014 International Cheetah (*Acinonyx jubatus*) Studbook. Cheetah Conservation Fund, Otjiwarongo, Namibia.

Moss, A., Jensen, E., Gusset, M., 2015. Evaluating the contribution of zoos and aquariums to Aichi Biodiversity Target 1. Conserv. Biol. 29, 537–544.

Moss, A., Jensen, E., Gusset, M., 2017. Impact of a global biodiversity education campaign on zoo and aquarium visitors. Front. Ecol. Environ. 15, 243–247.

Müller, D.W.H., Bingaman Lackey, L., Streich, W.J., Fickel, J., Hatt, J.M., Clauss, M., 2011. Mating system, feeding type and *ex situ* conservation effort determine life expectancy in captive ruminants. Proc. R. Soc. B 278, 2076–2080.

Munson, L., Marker, L., Dubovi, E., Spencer, J.A., Evermann, J.F., O'Brien, S.J., 2004. Serosurvey of viral infections in free-ranging Namibian cheetahs (*Acinonyx jubatus*). J. Wildl. Dis. 40, 23–31.

Munson, L., Terio, K.A., Worley, M., Jago, M., Bagot-Smith, A., Marker, L., 2005. Extrinsic factors significantly affect patterns of disease in free-ranging and captive cheetah (*Acinonyx jubatus*) populations. J. Wildl. Disease 41, 542–548.

Rabb, G.B., 1994. The changing roles of zoological parks in conserving biological diversity. Am. Zool. 34, 159–164.

Species360, 2016. Available from: www.Species360.org.

Wharton, D., 2007. Research by zoos. In: Zimmerman, A., Hatchwell, M., Dickie, L.A., West, C. (Eds.), Zoos in the 21st Century-Catalysts for Conservation? Cambridge University Press, New York, pp. 178–191.

Wildt, D.E., Grisham, J., 1993. Basic research and the cheetah SSP program. Zoo Biol. 12, 3–4.

Wildtrack, 2015. The Second Footprint Identification Technique (FIT) For Cheetah Workshop, Namibia. Available from: http://wildtrack.org/wp-content/uploads/2015/07/WildTrack-2nd-International-FIT-workshop-report-s-.pdf.

Zoo and Aquarium Association (ZAA), 2017. Who we are. Available from: www.zooaquarium.org.au.

Clinical Management of Captive Cheetahs

Ana Margarita Woc Colburn, Carlos R. Sanchez**,
Scott Citino†, Adrienne E. Crosier‡, Suzanne Murray§,
Jacques Kaandorp¶, Christine Kaandorp††,
Laurie Marker‡‡*

*Nashville Zoo at Grassmere, Nashville, TN, United States
**Fort Worth Zoo, Fort Worth, TX, United States
†White Oak Conservation Center, Yulee, FL, United States
‡Smithsonian Conservation Biology Institute, Front Royal, VA, United States
§Smithsonian Biology Institute, National Zoological Park, Washington, DC, United States
¶Safaripark Beekse Bergen, Hilvarenbeek, The Netherlands
††GaiaZOO, Kerkrade, The Netherlands
‡‡Cheetah Conservation Fund, Otjiwarongo, Namibia

INTRODUCTION

As wild cheetah populations continue to decline, captive populations in zoological institutions, breeding and conservation centers grow in importance. Though similar in many aspects to other species in the felid family, cheetahs exhibit unique adaptations that can make their captive care more challenging. They are more susceptible to stress induced diseases than other species (Chapter 25), and in particular captive cheetahs appear to be more susceptible to

infectious diseases, some of which was initially attributed to their lack of genetic variability (Heeney et al., 1990; Munson et al., 2005; O'Brien et al., 1985; Chapters 6 and 25). Management requires a comprehensive program to maintain a healthy population of captive cheetahs, including veterinary care, nutrition, housing, exercise, and enrichment. Initial information can be found through the cheetah Animal Care Manual. The manual, published by the cheetah Species Survival Plan® (SSP) and the North American captive cheetah specialists, provides general guidelines for the care of cheetahs, and has been shared with the international cheetah community. These guidelines may need to be adjusted based on each institution's local and regional needs. This chapter covers the aspects of housing, stress management, restraint/handling, anesthesia, preventative medicine (including vaccination, health examinations, preshipment and quarantine examinations, and diagnostics), necropsy, and cub care. Protocols and forms that are relevant to this chapter can be found at https://www.elsevier.com/books-and-journals/book-companion/9780128040881.

SPECIAL HOUSING REQUIREMENTS

Facility design is critical for effective management of the cheetah. Size, barriers, substrate, shelter, transfer chutes, training, and animal management should be carefully considered. How a facility is built and what management practices are undertaken, should depend on the social groupings (single or in groups) and purpose (breeding, display, ambassador) of the cheetahs to be housed in it. Facilities should incorporate enclosures for isolation or separation, as well as having some yards interconnected to allow for ease of transfer. It is also recommended to have a chute or restraint cage so that individuals can be safely handled for procedures without the need for anesthesia (Ziegler-Meeks, 2009). By taking into account management challenges and risk factors, such as disease transmission, fence-line aggression, and ability to separate groups into their housing needs, the well-being and health of captive cheetahs can be greatly improved (Terio and Munson, 2005).

Indoor facilities should have easy access to the outdoor areas. Flooring should provide good traction and be easily disinfected. Adequate platforms should be in place so that the cheetah can be off the floor, with bedding hay or shavings for the cheetah(s) to lay on, especially during the winter months. Proper ventilation

is also needed to prevent accumulation of substances like ammonia from urine, which can result in respiratory problems (Ziegler-Meeks, 2009). The SSP recommends that cheetahs should not be confined indoors, unless it is necessary for medical management or due to inclement weather conditions. Outdoor enclosures for all cheetahs should be as large as possible. The SSP guidelines recommend a minimum enclosure size of ~750 m² for up to 2 cheetahs.

Cheetahs can be housed in open-topped enclosures behind moats, chain-link or wire mesh, solid walls, glass windows, or a combination of these materials (Ziegler-Meeks, 2009). Certain states or countries have more specific regulations for cheetah enclosures, and they should be consulted before the enclosure is constructed. Typically, wire mesh is used for the enclosure and should be no lighter than 25-cm gauge and have spaces no larger than 5 by 10 cm.

Several considerations should be taken when constructing the outdoor enclosure to prevent injury or escapes. Adult cheetahs are generally considered to be poor climbers. However, they have been reported to climb over 3 m of solid wall when no overhang was present (Marker and Schumann 1998), and they are good climbers when immature and are able to jump over a 3.5-m moat (Ziegler-Meeks, 2009). To prevent an escape, a fence made of either solid vertical walls or wire mesh should be at least 2.5 m tall with an additional 60 cm mesh overhang into the enclosure at a 45-degree angle. Overhangs are a critical part of containment and should be made of chain-link or wire mesh. Strands of wire, barbed wire, or hot wire should not be used for the overhang, as cheetahs can easily go through these and get injured in the process (Cheetah SSP manual, in preparation). Fences should be sunk around the perimeter and corners should not be tighter than 90 degrees or contain small spaces that facilitate climbing. If needed, electrified wire can be used as a supplemental deterrent to climbing.

Breeding Facilities

Breeding facilities require additional considerations. Because cheetahs exhibit a high degree of mate selection (Chapters 9 and 27), facilities need to be able to hold a large number of potential breeding animals (four to six females and at least two groups of males) and be subdivided into multiple interconnected enclosures, to increase the chances of reproductive success (Bertschinger et al., 2008; Wielebnowski et al., 2002; Ziegler-Meeks, 2009). Cheetahs used for breeding purposes are currently not recommended to be used concurrently as educational program animals (ambassador cats). The SSP guidelines recommend that all males be maintained in their natural sibling coalitions to improve reproductive success, as well as maintain strong, healthy coalition bonds. Alternatively, captive male cheetahs can be successfully introduced at a young age, to form stable coalitions that simulate natural social groupings of wild male cheetahs (Chadwick et al., 2013; Chapter 27). Koester et al. (2015) reported improved testis function and more normal, motile spermatozoa and androgen production in males held in coalitions compared to those held singly.

The breeding facility should include several maternity enclosures, which are isolated from other enclosures and public viewing, to decrease stress to the dam. These maternity yards should still have easy access to the females' original enclosures and a minimum of two potential dens should be in place. Dens should be constructed with easy access to the cubs by keeper and veterinary staff, while minimizing stress to the dam (Ziegler-Meeks, 2009). Offspring should stay with the dam for at least 1 year and up to 18 months of age. Male and female siblings will have to be separated at around 20–22 months of age. Male siblings can stay together for life. Facilities should be prepared to hold offspring for up to 2 years after being separated from the dam.

STRESS MANAGEMENT

Several cheetah diseases, as well as poor reproductive performance have been linked to stress levels (Chapters 25 and 27); stress management is therefore an important part of clinical management. Many factors, including exhibit design, animal movements, exercise, and enrichment, need to be addressed to minimize stress. Cheetah exhibits should be located far from potential sources of stress, such as other large carnivores (e.g., lions, hyenas), and should be designed to encourage natural behavior. Limiting the need for animal handling, particularly during breeding situations, decreases acute stresses. Exercise is crucial for cheetahs, and should be facilitated through sufficient enclosure size, as well as a lure course system within the enclosure. Mechanical lure coursing equipment pulls a lure (such as a rag, feathers, or meat treat) quickly (80 kph) around the area. The running provides enrichment and health opportunities for the animals. Other enrichment opportunities, such as providing meat on the bone or carcasses (Chapter 26), novel scents and 'safe' toys, and climbing structures, should be provided for the well-being of the cheetah.

RESTRAINT AND HANDLING

Medical Training and Restraint

Cheetahs, whether captive- or wild born, adapt quickly to daily routine procedures in captive settings. Cheetahs can be trained to "station" voluntarily along a fence or in a chute for medical procedures using positive reinforcement. This type of training decreases the stress associated with procedures and allows the use of lower anesthetic drug dosages during induction. Cheetahs are amenable to being trained for multiple routine procedures, such as voluntary venipuncture (usually lateral tail vein or saphenous vein), ultrasounds, radiography, administration of medications, and subcutaneous fluids.

FIGURE 24.1   **Crating a cheetah at the Cheetah Conservation Fund.**

Cheetahs can also be trained to readily shift from their enclosure to a chute, restraint cage, or crate for veterinary procedures (Fig. 24.1). Once restrained, cheetahs can be injected intramuscularly (i.m.) or intravenously (i.v.) via hand- or pole-syringe. The volume of the drug combination for intramuscular injections performed on a physically restrained animal (e.g., in a squeeze crate) should be kept to a minimum, while using a relatively large luer lock syringe (e.g., 5 or 10 cc) and needle size (20 or 18 g) to facilitate administration and reduce the risk of partial drug delivery.

If shifting into a crate or stationing is not possible, and anesthesia is necessary, cheetahs can be darted by blowpipe or with CO_2 powered dart guns inside their enclosures. However, given their leaner muscle mass, care must be taken to avoid spiral fractures along the femoral bone due to penetration of the dart needle into the shaft of the bone (Meltzer, 1999).

Anesthesia

Multiple anesthesia protocols have been used in the cheetah. Selection of anesthesia protocol should be tailored to account for the temperament of the animal, pertinent clinical history, route of administration, procedure to be performed, and clinician expertise. Most protocols use tiletamine–zolazepam or ketamine combined with an α2 agonist (Table 24.1). Protocols that include tiletamine–zolazepam, whether in combination with an α2 agonist or on its own, are often associated with prolonged recoveries and hypertension. However, these protocols are useful when small volumes of anesthetic drugs are needed (e.g., for darting). Several fully reversible anesthetic protocols have also been reported using α2 agonists, such as medetomidine or dexmedetomidine in combination with butorphanol and midazolam (Table 24.1). These protocols provide a fast onset of anesthesia, good muscle relaxation and rapid recovery; however, sudden arousals can be observed at the reported dosages and total injection volume can become large. Varying degrees of hypertension have been observed with α2 agonist combinations (LaFortune et al., 2005; Woc Colburn et al., 2017).

Intravenous anesthetic protocols have also been used in cheetahs trained to accept injections. Either a butterfly infusion set or i.v. catheter can be inserted into the lateral tail vein or medial saphenous vein for drug administration (Fig. 24.2). Alfaxalone can be intravenously titrated to effect and allows general handling of cheetahs, as well as electroejaculation (EEJ). While poikilothermia has been reported in the past (Button et al., 1981), these reports are from a time when alfaxalone was combined with alphadolone and cremophor EL as solvents, as opposed to the current combination of alfaxalone and cyclodextran in water as the solvent (Goodchild et al., 2015). Intravenous propofol-fentanyl protocols have also been reported in the cheetah. These protocols produced rapid induction and stable cardiorespiratory parameters and wore off 15–20 min from induction. Anesthesia can readily be maintained by gas inhalant anesthesia, such as isoflurane. Maintenance of anesthesia with propofol-fentanyl via total intravenous anesthesia is not recommended in cheetahs because, as in domestic cats, it can produce prolonged anesthetic recoveries and episodes of apnea (Woc Colburn et al., 2009).

TABLE 24.1   Previously reported anesthesia protocols in the cheetah

Generic Name	Dosage	Route	Reversal Agents and Dosage	Comment
Tiletamine-Zolazepam	2–4 mg/kg	i.m.		Higher doses may cause apnea
Tiletamine-Zolazepam	4.2 mg/kg	i.m.	Flumazenil 0.031 mg/kg or Sarmazenil 0.1 mg/kg	Improved recovery
Tiletamine-Zolazepam Ketamine	1.9–2.6 mg/kg 1.38–3.38 mg/kg	i.m.		
Tiletamine-Zolazepam Ketamine Medetomidine	1.0–1.4 mg/kg 1.0–1.4 mg/kg 0.010-0.014 mg/kg	i.m.	Atipamezole 5× Medetomidine	
Tiletamine-Zolazepam Ketamine Dexmedetomidine	1.5–1.7 mg/kg 1.5–1.7 mg/kg 0.01 mg/kg	i.m.	Atipamezole 7× Dexmedetomidine	
Tiletamine-Zolazepam Medetomidine	1.5 mg/kg 0.030 mg/kg	i.m.	Atipamezole 0.15 mg/kg	Hypertension
Tiletamine-Zolazepam Medetomidine	2.9 mg/kg 0.027 mg/kg	i.m.	Atipamezole 5× Medetomidine	Hypertension
Tiletamine-Zolazepam Ketamine Xylazine	1.15 mg/kg 1.84 mg/kg 0.46 mg/kg	i.m.	Yohimbine 0.1–0.2 mg/kg	Prolonged recovery
Ketamine Medetomidine	6.9 mg/kg 0.027 mg/kg	i.m.	Atipamezole 5× Medetomidine	Transient seizures, hypertension
Ketamine Medetomidine	2.5 mg/kg 0.04–0.07 mg/kg	i.m.	Atipamezole 0.3mg/kg	
Ketamine Midazolam	6.9 mg/kg 0.4 mg/kg	i.m.		Hypertension
Ketamine Dexmedetomidine	5.0 mg/kg 0.017 mg/kg	i.m.	Atipamezole 10× Dexmedetomidine	Fast and smooth induction, excellent sedation level Cardiovascular parameters similar to medetomidine and ketamine
Medetomidine Butorphanol Midazolam	0.035 mg/kg 0.2 mg/kg 0.15 mg/kg	i.m.	Atipamezole 0.0175 mg/kg + Flumazenil 0.006 mg/kg + Naltrexone 0.25mg/kg	Rapid induction and recovery Hypertension
Dexmedetomidine Butorphanol Midazolam	0.0158 mg/kg 0.22 mg/kg 0.18 mg/kg	i.m.	Atipamezole 0.125 mg/kg + Naltrexone 0.1 mg/kg	Rapid induction Less hypertension than with other protocols
Alfaxalone Alfadolone acetate	54–90 mg 18–30 mg	i.v.		Initial bolus, then add to effect Poikilothermia
Alfaxalone	2 mg/kg	i.v.		Used for EEJ
Propofol Fentanyl	5–6.5 mg/kg 0.002 mg/kg	i.v.		Smooth induction. CRI caused prolonged recovery

CRI, Constant rate of infusion; EEJ, electroejaculation; i.m., intramuscularly; i.v., intravenously.
Modified from Woc Colburn, et al., 2017

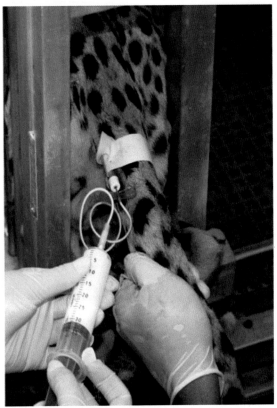

FIGURE 24.2 **Administration of Propofol intravenously through an i.v. catheter placed in the lateral tail vein of a cheetah at Smithsonian Institution National Zoological Park.**

Anesthesia Monitoring

While under general anesthesia, the cheetah's heart rate, respiratory rate, body temperature, oxygen saturation, systemic blood pressure, and end-tidal CO_2, should be monitored closely. Frequency is determined based on the patient and anesthetic protocol, but ideally should be recorded minimally every 5 min. An anesthetized cheetah's heart rate and respiratory rate can vary depending on the anesthetic drug combination used. Lower heart rates are seen with α2 agonists, such as medetomidine and dexmedetomidine. Lower respiratory rates are observed with respiratory depressant anesthetics, such as butorphanol and propofol. Body temperature should be monitored closely. If

hyperthermia ($>40°C/104°F$) is present, the body should be cooled down in a controlled manner to prevent a rapid drop of body temperature. Hypothermia ($<37°C/98°F$) should be addressed aggressively to prevent hemodynamic changes. Both direct and indirect blood pressure monitoring during anesthesia have been performed in the cheetah (Sadler et al., 2013; Sant Cassia et al., 2015), and measurement of blood pressure is recommended as part of overall anesthesia monitoring. Intravenous or subcutaneous fluids should be provided throughout the anesthesia.

The cheetah's eyes will remain open and should be lubricated with tear gel as soon as an ophthalmic examination has been performed. A blindfold or eye cover should be placed. The positions of the eye and pupil size are dependent on the depth of anesthesia, as well as anesthetic agents used.

Recovery

Recovery from anesthesia should be provided in a quiet, dark space. The head should be placed straight, making sure that the throat is not kinked and that the airway is clear. Cheetahs can be given access to water and then food once they are able to stand and walk normally.

PREVENTATIVE MEDICINE

A complete, consistent preventative medicine program should be instituted in all cheetah facilities. The recommendations for cheetah are very similar to those for other exotic felids. Specific testing recommendations exist for health examinations, quarantine, and preshipment protocols (Table 24.2).

Health examination and preventative medicine protocol

Health examinations are performed as part of a complete preventative medicine program. The program should be based on each institution's needs and disease risks, as well as the

TABLE 24.2 Required and Recommended Medical Procedures as Part of the Quarantine, Routine, and Preshipment Examinations for Cheetah

Procedure	Quarantine	Routine exam	Preshipment
SSP REQUIRED MEDICAL PROCEDURES			
Physical exam	X	X	X
CBC, serum chemistry	X	X	X
FeCOV, FIV, FeLV serology	X	X	X
Heartworm testing			X
Serum bank	X	X	X
Urinalysis	X	X	X
Survey radiographs			X
Fecal ova and parasite screening	X	X	X
Fecal FeCoV PCR	X		X
Immunization	If needed	If needed	If needed
Dental prophylaxis	If needed	If needed	If needed
Body weight	X	X	X
Permanent ID	X	X	X
SSP RECOMMENDED MEDICAL PROCEDURES			
FPV, FHV, FCV, Toxoplasma serology	X	X	X
Survey radiographs	X	X	
Heartworm testing	X	X	
Fecal Culture for enteric pathogens	X		X
Fecal FeCoV PCR		X	
Gastroscopy and gastric biopsies	X	X	X
Abdominal ultrasound	X	X	X
Approved research requests	X	X	X

CBC, Complete blood count; FCV, feline calicivirus; FeCoV, feline enteric coronavirus; FeLV, feline leukemia virus; FHV, feline herpesvirus; FIV, Feline immunodeficiency virus; FPV, feline parvovirus/panleukopenia; PCR, polymerase chain reaction, SSP, Species Survival Plan® (SSP) for cheetahs
Data based on Citino, S., Haefele, H., Junge, R., Lamberski, N., McClean, M., Sanchez, C., 2009. Cheetah SSP Health Chapter. In: Ziegler-Meeks, K. (Ed.), Husbandry Manual for the Cheetah (Acinonyx jubatus), White Oak Conservation Center, Yulee, Florida pp. 242–277.

overall health of the animal collection. Health examinations include a complete physical examination, body weight, body condition score (Dierenfeld et al., 2007), complete blood count, serum biochemistry, urinalysis, viral serology, and fecal screening (Citino et al., 2009) (Table 24.2). Other recommended procedures are based on each institution's preventative medicine program. These may include survey radiographs, abdominal ultrasonography, gastric endoscopy for baseline evaluation, and testing for heartworm (antigen and antibody test) or specific infectious disease (Table 24.2).

Normal hematology and serum biochemistry values for both captive and free ranging cheetahs have been reported in the literature (Depauw et al., 2012; Hudson-Lamb et al. 2016; ISIS, 2013; Munson and Marker, 1997). Feces should be examined for ova and parasites, and appropriate deworming and ectoparasitic treatment should be administered. Routine viral screening should include serology tests for feline immunodeficiency virus (FIV), feline leukemia virus (FeLV), and feline enteric coronavirus (FeCov), as well as fecal FeCoV PCR (Gaffney et al., 2012) should be performed. Screening tests should be performed by a standardized laboratory that is familiar with analyzing sera and fecal samples from non-domestic species.

During the physical examination, attention should be paid to lesions caused by viruses, such as feline calicivirus (FCV) and feline herpesvirus (FHV), as cheetahs are highly susceptible to these viruses. Similarly, given the propensity for a wide variety of hepatic, gastric, and renal diseases, special attention should be given to those organ systems (Chapter 25). The oral cavity should be evaluated for lesions, such as papillomatous plaques under the tongue or oral ulcerations. A thorough dental examination should be conducted. Crowding of incisors and palatal depressions are common in captive and wild cheetahs (Marker and Dickman, 2004; Steenkamp et al., 2017; Chapter 7). Palatal depressions can become pathological (focal palatine erosion, FPE; Chapter 25); these crypts can get impacted and infected, leading to abscessations and oronasal fistulas. Gently removing the points from the mandibular M1 (using a Dremel or similar tool) prevents further trauma to the crypts. Tooth abrasions and fractures, particularly those of the canine teeth, are commonly found on oral examination in captive cheetahs (Steenkamp and Boy, 2009). Resorptive lesions have also been documented in both captive and wild Namibian cheetah skulls (Roux et al., 2009).

Vaccinations

Cheetahs are susceptible to similar infectious diseases as seen in domestic cats (Chapter 25) and they should be vaccinated appropriately.

The current recommendation by the Cheetah Veterinary Scientific Advisory Group of the American Association of Zoos and Aquariums (AZA), the SSP, and the European Endangered Species Programme (EEP) are broken down between core vaccines and those that are supplemental, based on each institution's disease risk assessment.

Core vaccines include: rabies, feline parvovirus/panleukopenia (FPV), FHV, and FCV. FHV, FPV, and FCV should be given as a killed vaccine (e.g., Fel-O-Vax from Boehringer Ingelheim, Fevaxyn Pentofel from Zoetis) at 6, 9, 12, and 15–16 weeks and a booster at 6 months (Citino et al., 2009). Rabies vaccine should either be a killed (Imrab 3, Merial) or a canary-pox vectored subunit vaccine, and should be administered at 4–6 months of age and boostered at 1 year of age. Adults should be vaccinated ever 1–3 years for FPV, FHV, FCV, and Rabies. Serum antibody titers can be monitored to evaluate response to the vaccination. Pregnant females are recommended to be revaccinated for FHV, FPV, and FCV with a killed virus vaccine 3 weeks prepartum.

For breeding institutions that have had herpesvirus infections associated with severe and recurrent lesions in their young cheetah cubs (<3 weeks of age), a modified-live virus (MLV) vaccine against FHV (Merial PureVax Feline 3, Merial) has been trialed as a booster. This MLV vaccine is under current investigation and should only be used as a booster in adult females that are going to be bred and that have already received a killed FPV, FHV, FCV vaccine 3 weeks prior, as the calicivirus component could cause the development of pad vesicles (Citino, unpublished data). The MLV herpes vaccine should not be used in kittens, juveniles, or in those not deemed healthy, as disease may be induced (Citino, unpublished data). When using the MLV herpes vaccine, there is a potential for shedding of the vaccine strain virus, therefore recently vaccinated cats should only be in proximity (e.g., sharing fence-lines) with low-risk cheetahs for approximately 3 weeks after vaccination. A thorough risk-based analysis should be performed prior to making the decision to use modified-live vaccines in cheetahs and precautions to prevent disease transmission to other cheetahs should be taken. Vaccinations for FeCoV, FeLV, FIV are not recommended at this time. A risk-based analysis should also be performed before deciding to vaccine for the Canine Distemper Virus (CDV) at the institution.

Quarantine and preshipment protocol

Any newly acquired animal should undergo a quarantine period before being introduced into an existing collection, for both the safety of the individual animal, as well as the collection. Quarantine period and examinations vary among institutions. Factors that affect timing and duration include, but are not limited to, health and disease status of the individual, place of origin, outcome of examination performed prior to its transfer, and history of infectious disease in both the sending and receiving institution. The quarantine period is typically 30 days. During that time period, and depending on preshipment testing, similar procedures to those required for routine health examinations should be completed.

Advanced diagnostics technique

Diagnostics used in domestic cats can readily be applied to the cheetah. Radiographs, ultrasonography, and gastric endoscopy with biopsies, can provide additional information about the health status of the animal. Physiological findings specific to the cheetah are briefly described in Chapter 7. Baseline two-view thoracic and abdominal radiographs are routinely taken by some institutions.

Cheetahs are prone to several renal and hepatic conditions including hepatic veno-occlusive disease, glomerulosclerosis, medullary amyloidosis, pyelonephritis, and suppurative nephritis (Chapter 25) making abdominal ultrasonography a useful diagnostic tool. In normal ultrasound of the liver and the spleen, myelolipomas are commonly seen and are an incidental finding (Chapter 25). Ultrasonography, particularly transrectal ultrasonography, is a useful tool to monitor the reproductive organs and ovarian status in cheetahs (Goeritz et al., 1999; Schulman et al., 2015; Chapter 27).

Gastritis is prevalent in the North American, European, and South African captive cheetah population and is commonly associated with *Helicobacter* infection. The high prevalence was attributed to stress responses (Chapter 25). The only current, definitive antemortem diagnostic test is gastroscopy with gastric biopsies. C-urea breath test can be utilized to see if *Helicobacter* is present, however, this cannot identify the degree of gastritis (Chatfield et al., 2004).

Newer methods for early detection of chronic renal disease are being researched: Glomerular filtration rate, renal plasma flow, endogenous creatinine clearance, and fractional excretion studies have been performed in healthy captive cheetahs (Holder et al., 2004; Sanchez et al., 2007; Terio and Citino, 1997). The use of urine protein-creatinine ratio can be useful for the confirmation of proteinuria. Serum symmetric dimethylarginine, a renal biomarker, can also be used to detect early renal disease. This test appears to be promising, however to date there are no reference values for non-domestic felids, including cheetahs. Newer research into renal biomarkers are being conducted in several exotic felids and can potentially be applied to cheetahs. Serum and acute phase proteins are being used to detect early stages of inflammation, secondary to gastrointestinal and renal diseases. Baseline serum and acute phase proteins are described in captive cheetahs (Depauw et al., 2012). These can be used in nonhealthy cheetahs to aid in the early diagnosis of a variety of gastrointestinal and renal diseases.

POSTMORTEM

In the event of a death, a complete necropsy should be performed following available guidelines (e.g., produced by Cheetah SSP and EEP programs). A complete set of tissues should be submitted in formalin for histopathology and pertinent tissues should be stored frozen (−80°C) for potential additional testing. Findings should be reported to the appropriate studbook keeper and veterinary advisors so that ongoing and emerging disease issues can be identified and researched.

HAND-REARING

Cheetahs have a gestation length of approximately 92 days (Chapter 27). During parturition and cub rearing, it is important to decrease potential stress and disturbance to the dam to prevent potential neglect or trauma to the cubs (Ziegler-Meeks, 2009). Prior to the birth of the cubs, a hand-rearing protocol should be in place along with equipment and supplies, in case they are needed. Every effort should be made to allow the dam to rear the cubs on her own. However, if illness, maternal neglect, abandonment, or trauma occurs, or if only 1 cub is born, a decision to hand-rear has to be made rapidly. Fostering should always be done for single cubs (when another litter is available) as the single cub does not stimulate the dam enough for adequate milk production. Fostering on to a similar aged litter has been successful with cubs that differ by as much as 3 weeks in age.

When cubs are pulled for hand-rearing, they should be evaluated by the veterinary team to assess for health problems, hydration status, and congenital defects. Cubs with failure of passive transfer of antibodies due to the lack of adequate intake of colostrum, are more susceptible to infectious diseases due to their decreased immunity. If failure of passive transfer is present, a plasma transfusion from the dam or another cheetah may be needed. Plasma can be given to the cubs subcutaneously or by mouth for the first 1–2 weeks to ensure that full amount is received. Approximately 19% of hand-reared cubs do not survive, with a large percentage of deaths occurring before 30 days of age (Lombardi et al., 2009). Gastrointestinal disorders, pneumonia, and upper respiratory diseases are common causes of mortality with a variety of potential etiologies (Bell, 2005). To prevent exposure to infectious diseases, hand-rearing should be performed in an isolated room and proper personal protective equipment, such as gloves, gowns, masks, and boots, should be worn. Deworming with pyrantel pamoate should begin at 6 weeks and be administered every 2 weeks until cubs are 16 weeks of age. Facilities in geographic locations that are endemic for heartworm disease (*Dirofilaria immitis* infection) should start prophylactic heartworm prevention as early as 6 weeks of age, when the first set of vaccines is given.

Cubs should be weighed every day at the same time prior to being fed. The average birth weight of cubs is approximately 474 g. Lombardi et al. (2009) reported that the average growth rate in a healthy hand-reared cheetah cub averaged 48 g/day. Those that did not survive averaged much lower daily gains (~5 g/day). Mean growth rate was affected by the type of first solid food given, with ground beef having the lowest growth rate. Cheetah milk is lower in carbohydrates than domestic cat milk and milk replacers; therefore Lombardi et al. (2009) recommends diluting commercial milk formula with distilled water to help decrease carbohydrate load and diminish frequency of diarrhea. Formulas, such as Kitten Milk Replacer (KMR) have been used successfully, mixed either at a 1:2 or 1:3 ratio, and lead to higher survival rates than dog formulas (Esbilac; Lombardi et al., 2009). Zoologic formulas 42/52, 33/40, and 30/55 contain less carbohydrates compared to KMR, and are less frequently associated with diarrhea. Different institutions have added supplemental vitamins, protein sources, or have combined milk replacers to bring the milk composition closer to natural cheetah milk, the constituents of which have been well documented (Lombardi et al., 2009). It is recommended for the first feedings to consist of 5% dextrose mixed with distilled water or pedialyte, and then introduce the formula gradually once cubs are stable and taking several feedings well. Frequency of bottle-feeding depends on the cub's age, and is gradually decreased from every 2 h for newborns to approximately six feedings per day by week 4. Amount of formula usually starts around 10%–12% of body weight per day and is gradually increased to 15%–20%. Feedings greater than 20% have led to loose stool and other gastrointestinal disturbances (Lombardi et al., 2009). Cheetah cubs should be adequately hydrated to prevent impactions. During feeding, care should be taken to prevent aspiration pneumonia.

Meat-based baby foods can be added at 3 weeks of age. If doing well, cooked meat is added to the diet, followed by raw meat, around 4–5 weeks of age. Cubs can then be weaned by 6–7 weeks of age (Citino, unpublished data). Weaning at a younger age leads to less carpal valgal abnormalities, which are commonly seen in hand-reared cheetahs (Bell et al., 2011; Chapter 25).

Diets fed to captive cheetahs, especially supplemented meat, can vary significantly in vitamin and mineral content. These dietary differences alter blood mineral and vitamin levels, and may predispose cheetahs to nutritional disease, such as metabolic bone disease (Beckmann et al., 2013; Depauw et al., 2012). For this reason, diets should be carefully evaluated, particularly in growing cheetahs, to ensure that they not only meet dietary recommendations for felids, but also that circulating levels of nutrients are adequate.

CONCLUSIONS

Cheetahs pose many husbandry and veterinary challenges, in part due to their unique anatomy, physiology (Chapter 7), and propensity to develop disease in the captive setting (Chapter 25). Maintaining the cheetahs' health by providing the proper care in captivity is essential for successful captive propagation, which will become more important as wild populations continue to decline. Providing proper care in captivity depends on a good understanding of their clinical and management needs. The topics of enclosure set-up, stress management and exercise, as well as clinical care are especially pertinent to ensure increased welfare and long-term sustainability of the captive population. Significant progress has been seen over the past decades and management will continue to improve as additional insight is gained from *in situ* and *ex situ* studies, as well as from substantial management experience. Global management programs, such as the SSP, EEP, Japanese Association of Zoos and Aquariums (JAZA), and Australasian Zoo and Aquarium Association (ZAA) continue to work within their regions to promote best-practice guidelines for the captive care of cheetahs.

References

Beckmann, K.M., O'Donovan, D., McKeown, S., Basu, P., Bailey, T.A., 2013. Blood vitamins and trace elements in Northern-East African cheetahs (*Acinonyx jubatus soemmeringii*) in captivity in the Middle East. J. Zoo Wildl. Med. 44 (3), 613–626.

Bell, K.M., 2005. Morbidity and mortality in hand-reared cheetah cubs. Animal Keeper's Forum: the Journal of the American Association of Zoo Keepers, pp. 306–314.

Bell, K.M., Van Zyl, M., Ugarte, C.E., Hartman, A., 2011. Bilateral carpal valgus deformity in hand-reared cheetah cubs (*Acinonyx jubatus*). Zoo Biol. 30 (2), 199–204.

Bertschinger, H.J., Meltzer, D.G.A., van Dyk, A., 2008. Captive breeding of cheetahs in South Africa—30 years of data from the de Wildt cheetah and wildlife centre. Reprod. Domest. Anim. 43 (Suppl. 2), 66–73.

Button, C., Meltzer, D.G.A., Mulders, M.S., 1981. Saffan induced poikilothermia in cheetah (*Acinonyx jubatus*). J. S. Afr. Vet. Assoc. 52 (3), 237–238.

Chadwick, C.L., Rees, P.A., Stevens-Wood, B., 2013. Captive-housed male cheetahs (*Acinonyx jubatus soemmeringii*) form naturalistic coalitions: measuring associations and calculating chance encounters. Zoo Biol. 32 (5), 518–527.

Chatfield, J., Citino, S., Munson, L., Konopka, S., 2004. Validation of the ^{13}C-urea breath test for use in cheetahs (*Acinonyx jubatus*) with *Helicobacter*. J. Zoo Wildl. Med. 35 (2), 137–141.

Citino, S., Haefele, H., Junge, R., Lamberski, N., McClean, M., Sanchez, C., 2009. Cheetah SSP health chapter. In: Ziegler-Meeks, K. (Ed.), Husbandry Manual for the Cheetah (*Acinonyx jubatus*). White Oak Conservation Center, Yulee, FL, pp. 242–277.

Depauw, S., Hesta, M., Whitehouse-Tedd, K., Stagegaard, J., Buyse, J., Janssens, G.P.J., 2012. Blood values of adult captive cheetahs (*Acinonyx jubatus*) fed either supplemented beef or whole rabbit carcasses. Zoo Biol. 31 (6), 629–641.

Dierenfeld, E., Fuller, L., Meeks, K., 2007. Development of a standardized body condition score for cheetahs (*Acinonyx jubatus*). Proceedings of the Seventh conference on Zoo and Wildlife Nutrition, AZA Nutrition Advisory Group, Knoxville, TN 202–204.

Gaffney, P.M., Kennedy, M., Terio, K., Gardner, I., Lothamer, C., Coleman, K., Munson, L., 2012. Detection of feline coronavirus in cheetah (*Acinonyx* jubatus) feces by reverse transcription-nested polymerase chain reaction in cheetahs with variable frequency of viral shedding. J. Zoo Wildl. Med. 43 (4), 776–796.

Goeritz, F., Maltzan, J., Hermes, R., Wiesner, H., Spelman, L.H., Blottner, S., Fritsch, G., Hildebrandt, T.B., 1999. Transrectal ultrasound evaluation of cheetahs. Proceedings of the American Association of Zoo Veterinarians, 194–195.

Goodchild, C.S., Serrao, J.M., Kolosov, A., Boyd, B.J., 2015. Alphaxalone reformulated: a water-soluble intravenous anesthetic preparation in sulfobutyl-ether- ß-Cyclodextrin. Anesth. Analg. 120, 1025–1031.

Heeney, J.L., Evermann, J.F., McKeirnan, A.J., Marker-Kraus, L., Roelke, M.E., Bush, M., Wildt, D.E., Meltzer, D.G., Colly, L., Lukas, J., Manton, V.J., Caro, T., O'Brien, S.J., 1990. Prevalence and implications of feline coronavirus infections of captive and free-ranging cheetahs (*Acinonyx jubatus*). J. Virol. 64 (5), 1964–1972.

Holder, E.H., Citino, S.B., Businga, N., Cartier, L., Brown, S.A., 2004. Measurement of glomerular filtration rate, renal plasma flow, and endogenous creatinine clearance in cheetahs (*Acinonyx jubatus*). J. Zoo Wildl. Med. 35 (2), 175–178.

Hudson-Lamb, G.C., Schoeman, J.P., Hooijberg, E.H., Heinrich, S.K., Tordiffe, A.S., 2016. Reference intervals for selected serum biochemistry analyses in cheetahs (*Acinonyx jubatus*). J. S. Afr. Vet. Assoc. 87 (1), 1–6.

International Species Information System (ISIS), 2013. In: Teare, J.A., (Ed.), ISIS physiological reference intervals for captive wildlife. Eagan, Minnessotta (A CD-ROM resource).

Koester, D.C., Freeman, E.W., Wildt, D.E., Terrell, K.A., Franklin, A.D., Meeks, K., Crosier, A.E., 2015. Group

management influences reproductive function of the male cheetah (*Acinonyx jubatus*). Reprod Fertil. Dev., 29, 496–508.

LaFortune, M., Gunkel, C., Valverde, A., Klein, L., Citino, S.B., 2005. Reversible anesthetic combination using medetomidine-butorphanol-midazolam (MBMz) in cheetahs (*Acinonyx jubatus*). Procedure American Association of Zoo Veterinarian, p. 270.

Lombardi, C., McFerron, K., Bates, S., 2009. Collection and analysis of hand-reared cheetah (*Acinonyx jubatus*) records in the captive North American population. In: Ziegler-Meeks, K. (Ed.), Husbandry Manual for the Cheetah. White Oak Conservation Center, Yulee, FL, pp. 130–230.

Marker, L., Dickman, A.J., 2004. Dental anomalies and incidence of palatal erosions in Namibian cheetahs (*Acinonyx jubatus*). J. Mammal. 85 (1), 19–24.

Marker, L., Schumann, B.D., 1998. Cheetahs as problem animals: management of cheetahs on private land in Namibia. In: Penhorn, B.L., (Ed.), Symposium of Cheetahs as Game Ranch Animals, Onderstepoort, South Africa, pp. 90–99.

Meltzer, D.G.A., 1999. Medical management of a cheetah breeding facility in South Africa. In: Fowler, M.E., Miller, R.E. (Eds.), Zoo and Wildlife Medicine, Vol. 4. W.B. Saunders, Philadelphia, Pennsylvania, pp. 415–417.

Munson, L., Marker, L., 1997. The impact of capture and captivity on the health of Namibian farmland cheetahs (*Acinonyx jubatus*). Proceedings of the 50th Namibian Vet Congress.

Munson, L., Terio, K.A., Worley, M., Jago, M., Bagot-Smith, A., Marker, L., 2005. Extrinsic factors significantly affect patterns of disease in free-ranging and captive cheetah (*Acinonyx jubatus*) populations. J. Wildl. Dis. 41 (3), 542–545.

O'Brien, S.J., Roelke, M.E., Marker, L., Newman, A., Winkler, C.A., Meltzer, D., Colly, L., Evermann, J.F., Bush, M., Wildt, D.E., 1985. Genetic basis for species vulnerability in the cheetah. Science 227 (4693), 1428–1434.

Roux, P., Berger, M., Stich, H., Schawalder, P., 2009. Oral examination and radiographic evaluation of the dentition in wild cats from Namibia. J. Vet. Dent. 26 (1), 16–22.

Sadler, R.A., Hall, N.H., Kass, P.H., Citino, C.B., 2013. Comparison of noninvasive blood pressure measurements techniques via the coccygeal artery in anesthetized cheetahs (*Acinonyx jubatus*). J. Zoo Wildl. Med. 44 (4), 928–935.

Sanchez, C.R., Murray, S., Brown, S., Marker, L., Citino, S., 2007. Single-injection inulin clearance for routine determination of glomerular filtration rate in cheetahs (*Acinonyx jubatus*). Proceedings of American Association of Zoo Veterinarians, 212–213.

Sant Cassia, E.V., Boswood, A., Tordiffe, A.S.W., 2015. Comparison of high-definition oscillometric and direct arterial blood pressure measurement in anesthetized cheetahs (*Acinonyx jubatus*). J. Zoo Wildl. Med. 46 (3), 506–516.

Schulman, M.L., Kirberger, R.M., Tordiffe, A.S.W., Marker, L.L., Schmidt-Küntzel, A., Hartman, M.J., 2015. Ultrasonographic and laparoscopic evaluation of the reproductive tract in older captive female cheetahs (*Acinonyx jubatus*). Theriogenology 84 (9), 1611–1619.

Steenkamp, G., Boy, S.C., 2009. Dental and oral health in captive cheetahs in southern Africa. 23rd Veterinary Dental Forum, Phoenix, Arizona.

Steenkamp, G., Boy, S.C., van Staden, P.J., Bester, M.N., 2017. How the cheetah's specialized palate accommodates its abnormally large teeth. J. Zool. 301 (4), 290–300.

Terio, K., Citino, S.B.,1997. The use of fractional excretion for early diagnosis of renal damage in cheetahs (*Acinonyx jubatus*). Proceedings of American Association of Zoo Veterinarians, 266–267.

Terio, K.A., Munson, L., 2005. Linking stress with altered gastric immune responses in captive cheetahs (*Acinonyx jubatus*). Proceedings of American Association of Zoo Veterinarians, 51–52.

Wielebnowski, N.C., Ziegler, K., Wildt, D.E., Lukas, J., Brown, J.L., 2002. Impact of social management on reproductive, adrenal, and behavioural activity in the cheetah (*Acinonyx jubatus*). Anim. Conserv. 5, 291–301.

Woc Colburn, A.M., Murray, S., Hayek, L.C., Marker, L., Sanchez, C.R., 2017. Cardiorespiratory effects of dexmedetomidine-butorphanol-midazolam (DBM): a fully reversible anesthetic protocol in captive and semi-free ranging cheetahs (*Acinonyx jubatus*). J. Zoo Wildl. Med. 48 (1), 40–47.

Woc Colburn, A.M., Sanchez, C.R., Murray, S., 2009. Comparison of two fentanyl/propofol anesthesia protocols in cheetahs (*Acinonyx jubatus*). Proceedings of American Association of Zoo Veterinarians, 97.

Ziegler-Meeks, K. (Ed.), 2009. Husbandry Manual for the Cheetah (*Acinonyx jubatus*). White Oak Conservation Center, Yulee, Florida.

Diseases Impacting Captive and Free-Ranging Cheetahs

Karen A. Terio, Emily Mitchell**, Chris Walzer†,
Anne Schmidt-Küntzel‡, Laurie Marker‡, Scott Citino§*

*University of Illinois, Brookfield, IL, United States
**National Zoological Gardens of South Africa, Pretoria, South Africa
†University of Veterinary Medicine, Vienna, Austria
‡Cheetah Conservation Fund, Otjiwarongo, Namibia
§White Oak Conservation Center, Yulee, FL, United States

INTRODUCTION

Captive cheetahs have been affected by a variety of diseases limiting the sustainability of these populations. To better understand these captive cheetah diseases on a global scale and determine if they impact wild cheetahs, several long-term studies have been conducted (Munson, 1993; Munson et al., 1999, 2005). As a species, cheetahs are considered more susceptible to disease than other carnivores and are carefully managed in captivity (Chapter 24). Thoughts over the causes of disease susceptibility in captive cheetahs have changed over time as new research has been conducted. This chapter will discuss the evolution in our understanding of cheetah diseases and provide details on some of the most significant diseases affecting captive and wild cheetahs. Protocols and forms that are

relevant to this chapter can be found at https://www.elsevier.com/books-and-journals/book-companion/9780128040881.

Research from the 1980s had suggested that reduced genetic diversity (Chapter 6) may be the cause of poor health in captivity and high neonatal mortality rates. It was hypothesized that the lack of heterogeneity at major histocompatibility complex (MHC) loci (loci which encode peptides that mediate immune responses to pathogens) was an important factor in a feline infectious peritonitis (FIP) epidemic in one captive population, and it was assumed that cheetahs could be particularly vulnerable to infectious diseases (Evermann et al., 1988). However, in the 1990s, aspects of captive management were shown to affect juvenile mortality (Wielebnowski, 1996), and pathology studies indicated that morbidity and mortality in captive cheetah populations were predominately due to degenerative rather than infectious diseases (Munson, 1993; Munson et al., 1999). Study of wild Namibian cheetahs, a primary source for many founders of captive populations, indicated that these diseases were absent or extremely rare in the wild (Munson et al., 2005). Serology and pathology surveys showed that wild cheetahs were exposed to a wide range of viral pathogens without evident associated disease (Munson et al., 2004a,b, 2005; Thalwitzer et al., 2010). Together, these studies suggest that environmental factors, such as captivity-induced stress, are of equal or greater importance than genetic factors in the development of disease in captive cheetahs (Munson et al., 2005).

Multiple studies have now documented both physiological and behavioral (e.g., stereotypies, such as pacing) chronic stress responses in captive cheetahs (Jurke et al., 1997; Terio et al., 2004; Wells et al., 2004; Wielebnowski, 1999). Captive cheetahs have larger adrenal glands and higher baseline levels of corticoids than wild cheetahs (Terio et al., 2004; Chapter 7). These persistently elevated corticoids can have profound negative effects on many physiological functions, including metabolism, reproductive health, and immune responses. Corticoids can alter immune responses by affecting transcription of mediators that coordinate and determine the type of immune responses (e.g., cytokines and inflammatory mediators). Comparison of these mediators between captive and free-ranging cheetahs has demonstrated that captive, but not free-ranging, cheetahs have decreased expression of T_H1-type cytokines (IL-1, IL-2, IFNγ) consistent with elevated corticoids (Terio and Munson, 2005). Because T_H1-type cytokines are important in cell-mediated immune responses, stress-induced immune modulation may be associated with the unusual inflammatory reactions observed in captive cheetahs in response to common pathogens.

Corticoid levels differ among institutions and among individual cheetahs residing at the same institution, suggesting that captive management and individual temperament may affect the stress response (Terio et al., 2004; Wells et al., 2004). Gastritis, one of the major cheetah diseases in captivity (see section, "Helicobacter-Associated Gastritis"), has been associated with increased stress levels and has served as a model to study stress in captivity. In cheetahs, severe gastritis correlates with high individual temperament scores for the adjectives "eccentric" and "easy to work with." Cheetahs with severe gastritis are also more likely to have been exposed to suspected environmental stressors, such as small enclosures, high density of cheetahs, movement to multiple institutions over an individual's lifetime, and high amounts of exposure to the public. In contrast, gastritis severity is negatively correlated with "excitable" and "aggressive" temperaments, a behavioral phenotype more often seen in wild-caught cheetahs. A negative correlation with "aggressive" is also noted with corticoid concentrations. Cheetahs with opportunities for exercise (running on a lure) have lower fecal glucocorticoid concentrations (Terio et al., 2014), supporting the importance of regular exercise for captive cheetahs.

Understanding the cheetah stress response is important not only for the health and well-being of captive cheetahs, but for management of free-ranging cheetahs. Wild-caught cheetahs temporarily held in captivity develop diseases like those noted in captive-born cheetahs, which may impact the cheetahs' survival when released back into the wild. One wild cheetah that had previously been in captivity had evidence of "captive cheetah" diseases at the time of death (Munson et al., 2005). As capture, holding, and translocation (Chapter 20) are tools that can be used for mitigating human-wildlife conflict and habitat fragmentation (Chapters 10 and 13), research aimed at reducing stress in these situations is essential to minimize the impact of capture and captivity on the health of cheetahs worldwide (Teixeira et al., 2007).

IMPORTANT DISEASES IN THE WILD

Infectious Disease

Anthrax

There have been multiple published and anecdotal cases of anthrax in wild and captive cheetahs (Ebedes, 1976; Good et al., 2008; Jäger et al., 1990; de Pienaar, 1961). Most outbreaks can be linked to contaminated carcasses fed to captive animals or concurrent outbreaks in other wildlife. Affected cheetahs may have increased respiratory rate and effort and can vomit (Jäger et al., 1990). Similar to other carnivores, cheetahs may have ventral and cervical edema, bloody nasal discharge, scleral congestion, and pulmonary edema (Good et al., 2008; Jäger et al., 1990). Serosurveys after an anthrax outbreak in Botswana found that only 1 of 16 potentially exposed cheetahs had antibodies (Good et al., 2008) and cheetahs in Namibia also do not have antibodies to anthrax (Switzer et al., 2016). These results support claims that cheetahs are susceptible because they have low levels of exposure (as they generally do not scavenge), thus infection is often fatal (Lindeque et al., 1998). Vaccination may be of benefit for certain high-risk populations as cheetahs do mount a potentially protective response to a commercially available anthrax vaccine (Turnbull et al., 2004).

Mycobacterium bovis

Spillover infection of *Mycobacterium bovis* occurs in cheetahs in areas with infected hoofstock (Keet et al., 1996). Typical granulomatous lesions occur in the lungs of affected cats but the lower airways can also be affected, suggesting that environmental contamination and or direct spread from affected cheetahs is possible. It is unlikely that cheetahs would become maintenance hosts for the disease as their contact rates are relatively low. However, sporadic spill over both to and from cheetahs is a possibility, especially in endemic areas (De Vos et al., 2001).

Sarcoptic Mange

Wild cheetahs in the Masai Mara in East Africa have been diagnosed with mange associated with *Sarcoptes scabei* infection that is a significant infectious cause of morbidity and mortality (Mwanzia et al., 1995). One study found the prevalence in cheetahs to be 12.8% (Gakuya et al., 2012). Lesions in affected cheetahs include hair loss and thickening of the skin. Infection in cheetahs was associated with the dry season and in particular *S. scabei* infection in Thompson's gazelles.

Other Health Problems

Ocular Trauma

Traumatic eye injuries including scratches and lacerations to the lid, nictitans, and cornea, with penetration and secondary uveitis and cataracts, have been documented in free-ranging cheetahs in Namibia (Bauer, 1998). In rare cases, portions of thorns have been found within affected eyes leading to the hypothesis that the trauma is caused by encroachment of thorn bushes due to habitat degradation.

IMPORTANT DISEASES IN CAPTIVITY

Infectious Diseases

Helicobacter-Associated Gastritis

The majority of captive and wild cheetahs worldwide are infected with *Helicobacter* sp., spiral bacteria that colonize the stomach. In wild cheetahs there are numerous bacteria, but little to no associated inflammation (Terio et al., 2005). In contrast, captive cheetahs commonly have some degree of inflammation (gastritis; Fig. 25.1) that may be asymptomatic or associated with

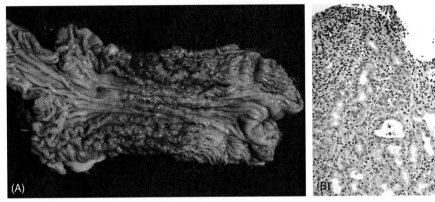

FIGURE 25.1   (A) Stomach from a captive cheetah with gastritis. Note the thickened, reddened mucosa. (B) Histologic section of stomach from a captive cheetah with gastritis with large numbers of inflammatory cells in the mucosa, superficial epithelial erosion, and loss of gastric parietal cells. Inset: Normal stomach from a wild cheetah.

regurgitation, vomiting, passage of undigested feed, and weight loss (Eaton et al., 1993; Terio et al., 2011). Although gastric colonization with *Helicobacter* sp. can be confirmed using the urea breath test, antemortem diagnosis of gastritis requires histologic evaluation of endoscopic biopsies (Chapter 24). It should be noted that the appearance of the stomach on endoscopy does not always correlate with the degree of histologic inflammation and damage.

The pathogenesis of gastritis is not completely understood. The absence of gastritis in the few cheetahs without evidence of ever being colonized by *Helicobacter* suggests that the bacteria play a role, but *Helicobacter* are not the only factor in disease development [Association of Zoos and Aquaria (AZA) Cheetah Species Survival Plan® (SSP) Pathology Archives, unpublished data]. In cheetahs, the distribution of T lymphocytes (T cells) and diffuse upregulation of MHC II expression on gastric epithelium are qualitatively similar to that in other species infected with *Helicobacter* (Terio et al., 2011). What distinguishes cheetahs with severe gastritis from those with mild disease are large numbers of activated B cells (CD79a+ CD21−) and plasma cells (Fig. 25.1B) which may be due to

alterations in the Th1:Th2 cytokine balance and stress-induced modulation of the immune system (Terio et al., 2011).

Longitudinal studies of gastritis in captive cheetahs have documented that while inflammation can wax and wane, in general, it worsens over time irrespective of treatment (Citino and Munson, 2005). Some lesions progress to fibrosis (scarring) and gastric atrophy that limits the ability of cheetahs to eat and digest a complete meal in one sitting (Citino and Munson, 2005; Terio et al., 2011). Although treatment with antibiotic triple therapy may provide short-term decrease in inflammation and eradication of *Helicobacter*, none of the treatments evaluated to date have had a significant effect on long-term *Helicobacter* infection or the lifelong progression of inflammation (Citino and Munson, 2005; Lane et al., 2004; Wack et al., 1997). The use of triple therapies may also impact the gut microbiome and potentially predispose to other conditions. Therefore, antibiotic treatment is recommended only in severely affected or symptomatic cheetahs. Diet may contribute to and or ameliorate signs of gastritis, but there are contradictory results on the impact of diet on gastritis severity (Lane et al., 2012; Whitehouse-Tedd et al., 2015).

Herpesvirus

Captive cheetahs are commonly infected with feline herpesvirus and can develop mild sneezing, nasal discharge, and ocular lesions similar to other felids (van Vuuren et al., 1999). In some cases, severe corneal ulcers and/or keratitis and proliferative skin lesions on the face and/ or forelimbs are noted (Flacke et al., 2015; Junge et al., 1991; Munson et al., 2004a,b; Witte et al., 2013) (Fig. 25.2). Disease has not been noted in wild cheetahs despite evidence of viral exposure (Munson et al., 2004a,b, 2005; Thalwitzer et al., 2010). Studies of the virus have found genetic and antigenic similarity with feline herpesvirus type 1 (FHV-1) (Scherba et al., 1988).

Diagnosis of FHV-1 infection relies on characteristic clinical signs and/or specific diagnostic tests, such as virus isolation, PCR, histopathology with characteristic herpesviral inclusions, or immunohistochemistry (Witte et al., 2013). Serology alone cannot be used to confirm infection (Witte et al., 2013). Biopsy is essential to differentiate proliferative skin lesions from neoplasms (tumors). Histologically, dense infiltrates of eosinophils and mast cells are present in the dermis, and the intact epithelium adjacent to the ulcer may be hyperplastic with viral inclusions (Munson et al., 2004a,b). Treatment for mild FHV-1 is supportive. Administration of L-lysine

and famcyclovir have been helpful in some cases, while skin lesions often require cryotherapy. Vaccination is recommended, although variable responses have been noted in cheetahs and natural exposure is not uncommon (Spencer and Burroughs, 1992; Wack et al., 1993; Chapter 24).

Immune responses to FHV-1 in cheetah have been studied *in vitro* (Miller-Edge and Worley, 1992). In these studies, cheetah cells (lymphocytes and monocytes) as a group had a significantly lower proliferative response to FHV-1 than domestic cats, although individual responses varied widely. Interestingly, cheetah cellular responses improved when they were first cultured with interleukin-2 (IL-2), a TH-1 cytokine critical to an effective cell-mediated immune response. As previously discussed, low levels of IL-2 have been found in captive but not wild cheetahs (Terio et al., 2005), suggesting that ineffective immune responses, possibly due to stress-induced suppression, may play a role. Disease tends to be more severe if cubs are infected by their dams at a young age (<3 weeks). As cubs at this age should still have maternal antibodies, this finding suggests inadequate colostral antibodies against herpesvirus. Often, these cubs are also more likely to have recurrent lesions throughout their lives.

Feline Infectious Peritonitis/Feline Enteric Coronavirus

Feline enteric coronavirus (FCoV) infection can cause mild enteritis, as well as fatal FIP. Lesions and clinical signs of FIP in cheetahs are similar to those described in domestic cats and include high protein effusions within the abdominal and or thoracic cavities with pyogranulomatous inflammation centered on vessels. In 1983, there was an epidemic of FIP in a captive cheetah population with >60% mortality that resulted in stringent testing and quarantine measures to limit the spread of disease within the North American captive cheetah population (Heeney et al., 1990). Complicating attempts to control spread of the disease was a lack of

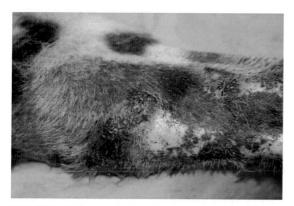

FIGURE 25.2 Plaque-like proliferative and ulcerated skin lesion on the forelimb of a cheetah with herpesviral dermatitis.

concordance between serology and PCR identi-fication of virus from feces (Kennedy et al., 2001; Munson and Citino, 2005). Some cheetahs have been found to shed the virus only intermittent-ly to rarely; therefore, captive populations are now managed as endemically infected (Gaffney et al., 2012; Munson and Citino, 2005).

Despite evidence for continued presence of FCoV in the population, fatalities to FIP have been rare (<2.9% total deaths from 1991 to 2016, AZA SSP Pathology Archives). The role of FCoV in enteritis is less clear and an area of needed study. Some cheetahs appear to be persistently infected with FCoV and persistently shed the vi-rus, and could be a source for viral replication and possible mutation to more virulent strains (Kennedy et al., 2006). Some persistently infect-ed cheetahs appear to develop ulcerative colitis with weight loss and lymphoid proliferation.

Fungal Disease

Although fungal disease is generally rare in cheetahs, there have been several individual case reports of *Cryptococcus* infections in cap-tive cheetahs (Beehler, 1982; Berry et al., 1997; Bolton et al., 1999; Illnait-Zaraozí et al., 2011; Millward and Williams, 2005). Infections pres-ent with nasal discharge, neurological deficits, and shortness of breath; lesions are similar to those noted in domestic cats. In these cases, fun-gi have been classified using varying methods as either *Cryptococcus neoformans, C. neoformans* var *gattii*, or *C. gattii. C. neoformans* var *gattii* has been recently reclassified as *C. gattii*, so it is uncertain whether these cases truly repre-sent *C. neoformans* and *C. gattii* infections or just *C. gattii*. There is no evidence to suspect inher-ent or acquired susceptibility to this infection in cheetah as immune responses are similar to those noted in domestic cats; there has not been any evidence of concurrent infection with im-munosuppressive viruses (e.g., feline immuno-deficiency virus or feline leukemia virus) (Berry et al., 1997; Bolton et al., 1999; Miller-Edge and Worley, 1992).

Dermatophytosis is common in some cap-tive populations of cheetahs. Affected cheetahs have patchy to generalized alopecia and *Micros-porum canis* can be cultured from affected hair. Young cubs can have more severe infections with involvement of tail tips and nail beds that, if not treated aggressively, can lead to slough-ing of nails, digits, and/or the tail tip. Adverse reactions have been noted in cats treated with griseofulvin, including lethargy, diarrhea, bone marrow depletion, and death (Wack et al., 1992). Therefore, topical treatments rather than sys-temic antifungal therapies are recommended.

Other Infectious Skin and Oral Mucosal Conditions

There have been multiple outbreaks of ortho-poxvirus infection resulting in ulcerated skin and oral lesions and, in rare cases, severe acute respiratory disease (Baxby et al., 1982; Maren-nikova et al., 1977). Histologically, poxviruses were suspected due to large eosinophilic intra-cytoplasmic inclusions and virus characteristics consistent with cowpox. Rodents used as intend-ed food items, as well as wild rodents caught by cheetahs are the source of infection (Baxby et al., 1982; Marennikova and Shelukhina, 1976).

Sublingual plaques are not uncommon and may be related to papillomavirus infection. A potential cause of oral ulceration is calicivirus. In contrast to herpesvirus, caliciviral infections tend to be self-limiting and more closely fol-low the course seen in domestic and other non-domestic felids. Treatment is supportive. One notable and important caveat is that caliciviral disease has developed after vaccination of some naïve cheetahs with modified live virus (MLV) vaccine. Similarly, although unproven due to similarities in viral strains, some cheetahs have developed herpesviral lesions after vaccination with MLV vaccines (see also Chapter 24).

Other Infectious Diarrheal Conditions

There are multiple potential causes of diarrhea in cheetah, including infection with parvoviruses, *Salmonella* spp., *Clostridium perfringens,* and *Plesiomonas shigelloides.* Parvoviruses indistinguishable from feline panleukopenia (FPLV) have been reported to cause diarrhea in cheetahs (Steinel et al., 2000; Valícek et al., 1993). Canine parvovirus type 2b has also been identified in cheetahs with chronic diarrhea due to chronic necrotizing enteritis (Steinel et al., 2000). None of these cheetahs had concurrent leukopenia. Molecular phylogeny of outbreaks of FPLV in South Africa has indicated the presence of more than one strain of the virus in captive cheetahs and other felids (Lane et al., 2016). As non-domestic species may be reservoirs of parvoviruses and these viruses readily change host specificity, the risks of FPLV transmission between captive bred and free-ranging carnivores, and domestic cats and dogs, warrants further research. Salmonella infection, including *Salmonella typhimurium,* is another potential cause of diarrhea, especially in cheetahs fed poultry or horsemeat (Venter et al., 2003). Complicating interpretation of positive fecal cultures are high rates of fecal shedding in seemingly healthy felids fed raw diets (Clyde et al., 1997).

Pancreatitis

In addition to infectious causes, diarrhea and vomiting can be noted in cheetahs with pancreatitis. Pancreatitis is not uncommon in captive cheetahs, especially if fed offal or diets high in fat. There is no association between pancreatitis and pancreatic duct ectasia, a common incidental finding.

Degenerative Disease

Amyloidosis

Cheetahs in captivity can develop amyloidosis when insoluble amyloid fibrils deposit within tissues where they disrupt organ function. Amyloidosis can be due to a genetic predisposition (protein sequence that is more likely to precipitate) or secondary to increased production of SAA which occurs in chronic inflammation or neoplasia. Previous studies identified concurrent inflammation in 100% of the cheetahs with amyloidosis; however, not all cheetahs with inflammation develop amyloidosis. Studies of the amyloid fibrils suggest that these deposits are due to deposition of the full serum amyloid A (SAA) protein, as well as truncated fragments (Bergström et al., 2006; Johnson et al., 1997). Studies of the cheetah AA protein sequence have found it to be similar to those of domestic cats with familial amyloidosis, suggesting a possible genetic predisposition (Chen et al., 2012; Johnson et al., 1997). A genetic role for increased production of the SAA protein was proposed (Zhang et al., 2008b). Subsequent studies, however, have determined that while specific genetic variants were associated with differing SAA serum concentrations, there was no significant association with disease (Franklin et al., 2016). To date, amyloidosis has not been described in a free-ranging cheetah (Munson et al., 2005; Terio and Mitchell, unpublished data).

In affected cheetahs, amyloid deposits most commonly within the kidney, specifically the medullary interstitium, leading to renal failure (Papendick et al., 1997). Amyloid fibrils may also be deposited within the liver, contributing to decreases in hepatic function. Clinical signs are similar to those in other types of renal and hepatic disease and require biopsy for definitive diagnosis. There is no specific treatment for amyloidosis, other than supportive therapy.

In some species AA amyloidosis is thought to be transmissible, and studies have identified AA fibrils within cheetah feces, suggesting a possible mechanism for transmission (Zhang et al., 2008a). If true, this would help explain the high prevalence of the disease in certain collections. As cheetahs are not typically coprophagic and enclosures are cleaned daily in most captive facilities, it is unclear whether this represents a true risk. Ongoing epidemiological studies of amyloidosis that controlled for the presence of concurrent inflammation have not yet found evidence for transmission within North American captive facilities (McLean and Garabed, 2015).

In addition to systemic amyloidosis, beta amyloid deposits have been described in the brains of some cheetahs (Serizawa et al., 2012). Although amyloid deposits can develop in aged animals of many species, some affected cheetahs had concurrent neurofibrillary tangles similar to the pattern identified in humans with Alzheimer's disease. Declines in cognitive function are difficult to assess in wild animals, and the relative advanced age of affected cats (all were >10 years) suggests that while this may impact the well-being of select captive cheetahs, it is unlikely to be of conservation significance.

Glomerulosclerosis

Glomerulosclerosis is a primary renal disease of cheetahs and, as with many other diseases, occurs primarily in captive cheetahs (Munson, 1993; Munson et al., 1999, 2005). Damage to renal glomeruli consists of thickening of basement membranes due to deposition of collagen, glycoproteins, and advanced glycosylation end products (Bolton and Munson, 1999). Glomerular damage leads to renal inflammation and scarring. The severity and prevalence of glomerulosclerosis increases with age, but there is no sex predilection. Clinically, affected cheetahs have signs of renal disease and, eventually, failure with proteinuria due to leakage through damaged glomeruli. Treatment is supportive. Feeding diets formulated for renal disease or commercial cat diets rather than raw meat may provide some clinical improvement, particularly in serum urea nitrogen (Lane et al., 2012). The cause of glomerulosclerosis is not known. Histologically, lesions resemble that of a diabetic nephropathy, although most affected cheetahs are not diabetic. The higher prevalence in captivity may be due to stress-induced hyperglycemia or another aspect of the captive environment.

Veno-Occlusive Disease

Veno-occlusive disease (VOD) is an unusual liver disease described in cheetahs and snow leopards. Scarring (subendothelial fibrosis) occurs in the central and sublobular hepatic veins with subsequent collapse and loss of the liver tissue surrounding affected vessels. Cirrhosis occurs in severe cases. Affected cheetahs can be asymptomatic or have signs of hepatic failure, including hypoproteinemia, abdominal effusion, elevated liver enzymes (aspartate aminotransferase, alanine aminotransferase), icterus, and in some cases neurologic signs due to hepatic encephalopathy (AZA Cheetah SSP Pathology Archives, unpublished data; Gosselin et al., 1988). In rare cases, effusions have been chylous (Terrell et al., 2003). A few cases of severe acute VOD have occurred in captive cheetahs with centrilobular to massive hepatic necrosis and hemorrhage (AZA Cheetah SSP Pathology Archives).

No specific cause for VOD has been proven in cheetahs. Estrogens are one cause for VOD in humans, and captive cheetah diets may contain high levels of phytoestrogens (Setchell et al., 1987). However, other lesions of excess dietary estrogen are not noted and the disease is seen in North American and South African captive cheetahs fed different diets (Munson et al., 1999). Interestingly, VOD has not been reported in the European captive population (Kotsch et al., 2002). Elevated dietary levels of vitamin A may cause VOD, but findings have been inconsistent (Gosselin et al., 1988). A few wild cheetahs have been diagnosed with VOD which seems to refute the hypothesis that captive diets are causative (Munson et al., 2005). Clinical pathology studies of trapped wild cheetahs have documented elevations in hepatic enzymes (alanine aminotransferase) within the first week of captivity, which could be the result of hepatocellular damage associated with VOD (Munson and Marker, 1997). One hypothesis is that catecholamine release due to the severe acute stress of capture leads to shunting of blood from the liver and hypoxic damage. The potential health impact of stress upon wild cheetahs during capture or periods of temporary captivity warrants further study.

Myelopathy

Myelopathy in cheetahs is a distinct neurological disorder characterized by degenerative lesions of the spinal cord, causing ataxia and paresis (Walzer and Kübber-Heiss, 1995). The disease emerged in the last 20 years and has limited growth of the European captive cheetah population (Walzer et al., 2003). To date, more than 100 cases have been identified in at least 16 different locations in Europe and the United Arab Emirates, resulting in euthanasia of many cheetahs in breeding programs. There is no apparent sex predilection. All affected cheetahs have been captive bred in European, Middle Eastern, or South African institutions from captive-born or wild-caught parents, belonging to the southern African subspecies (*Acinonyx jubatus jubatus*) or northeastern African subspecies (*Acinonyx jubatus soemmeringii*). Onset of ataxia ranges from 2.5 months to 12 years, is usually acute, and can occur spontaneously or following a stressful experience for the individual or for the litter. In cubs, clinical signs are often temporally associated with sneezing and ocular discharge typical of FHV-1 infection. The course of the disease is variable; ataxia and paresis may develop rapidly to hind limb dragging or recumbency, or progress slowly and stabilize for several months or years, with possible acute relapsing episodes. Pathologically, the disease is characterized by bilateral symmetrical degeneration of the white matter of the spinal cord, with loss of myelin exceeding axonal loss, suggesting a primary myelin disorder (Robert and Walzer, 2009). The etiology is unknown, but several causes have been considered (Burger et al., 2004; Palmer et al., 2001; Shibly et al., 2006; Walzer et al., 2003). Interestingly, the incidence of myelopathy has decreased, with only a single reported case since 2006.

Leukoencephalopathy (Leukoencephalomyelopathy)

In 1996, a novel neurodegenerative disease was seen in cheetahs at multiple captive breeding facilities throughout North America (Brower et al., 2014). Affected cheetahs were older (median and mean age of 12 years) and presented with ataxia and suspected blindness, leading to inability to follow keepers and locate food. Neurological signs included progressive altered behavior, disorientation, proprioceptive deficits, and, in some cases, seizures. Anesthesia in affected cheetahs was complicated by the need for increased doses of anesthetic agents and prolonged recoveries. Magnetic resonance imaging identified areas with white matter loss and hydrocephalus. Corresponding lesions were observed at necropsy. Testing for possible viruses and epidemiological evaluations failed to identify any potential etiologies. Cases increased with a peak of 22 cases in 1998 after which case numbers declined. No cases have been identified since 2005.

Other Renal Diseases

Cheetahs can also develop kidney disease due to oxalate nephrosis. Although oxalate nephrosis is commonly associated with exposure to ethylene glycol (antifreeze) in other species, exposure to ethylene glycol has not been confirmed despite extensive toxicological evaluations (AZA Cheetah SSP Pathology Survey; Mitchell, unpublished data). Furthermore, this disease has been diagnosed in cheetahs from multiple geographic locations, including North America, Europe, the Middle East, and southern Africa. Affected cheetahs develop acute renal failure due to tubular damage. Crystals can be noted in the urine antemortem. No correlation with other common cheetah diseases has been noted, so genetic and/or dietary causes are targets of ongoing investigations. Pyelonephritis is also noted in cheetahs, commonly females, and associated with coliform bacteria.

OTHER CONDITIONS OF WILD AND CAPTIVE CHEETAHS

In addition to the earlier specifically mentioned diseases and those in Tables 25.1 and 25.2, cheetahs can be affected by other degenerative and infectious diseases common to the felid taxa. Exposure to environmental toxins (intentional and otherwise), ingestion of foreign bodies, as well as injuries from snake bites and insect bites/stings are possible with signs and lesions similar to those in other species.

Focal Palatine Erosion

Cheetahs have a set of depressions in the hard palate; the most pronounced one being medial to the first upper molar to accommodate the mandibular molar (carnassial tooth) (Steenkamp et al., 2017; Chapter 7). In some wild and captive cheetahs, corresponding depressions are accompanied with mucosal lesions (Fitch and Fagan, 1982). In severe cases, the palatine bone is perforated resulting in communication between the oral and nasal cavities. Both mild and severe lesions were initially coined Focal Palatine Erosion (FPE). When first identified, the soft nature of captive diets and absence of exposure to tougher dietary items like skin and bones were presumed to play a role

in the pathogenesis (Phillips et al., 1993). However, studies in Namibia found that over 40% of wild cheetahs had deep palatine depressions with perforation of the palate in 15.3%, suggesting that captive diets are not causative (Marker and Dickman, 2004). In this study, wild cheetahs with FPE were also more likely to have crowded lower incisors and absent upper premolars. Studies of museum skulls identified only rare cases in free-ranging East African cheetahs, suggesting that FPE may be more common in the Namibian population (Zordan et al., 2012).

Neoplasia

Tumors of any organ system are possible in cheetahs, but generally uncommon with the exception of myelolipomas. Myelolipomas are benign, often multiple, tumors of the spleen and occasionally the liver, composed of mature adipocytes (fat cells) mixed with hematopoietic cells (Munson, 1993; Munson et al., 1999). These tumors can be identified by ultrasound and have been mistaken for more malignant tumors resulting in unnecessary removal of the spleen (Walzer et al., 1996). Myelolipomas also occur in wild cheetahs (Mitchell, unpublished data). Cutaneous mast cell tumors and systemic mastocytosis are also noted with some frequency (AZA SSP Pathology Archives).

TABLE 25.1   Common Incidental Findings in Cheetahs

Condition	Morphology	Occurrence	Reference
Kinked tails	Kink in posterior vertebra of tail	Wild and captive cheetahs	Marker-Kraus (1997)
Foam cell foci	Subpleural aggregates of foamy macrophages; appear as 1–2-mm yellow foci on lungs	Wild and captive cheetahs	Munson (1993), Munson and Terio (unpublished data)
Pancreatic duct ectasia	Dilation of pancreatic ducts without inflammation; can look like small cysts on the pancreas	Common in captive cheetahs; uncertain in wild cheetahs	Munson (1993)
Telangiectasia	Dilation and pooling of blood of hepatic sinusoids; red areas on liver	Wild and captive cheetahs	Munson (1993), Munson and Terio (unpublished data)
Parovarian cysts	Cystic structures adjacent to the ovary associated with remnant Wolffian ducts	Wild and wild-caught captive held cheetahs	Munson (1993), Munson and Terio (unpublished data)

TABLE 25.2   Uncommon Conditions of Wild and Captive Cheetahs

Causes	Clinical signs/findings	Outcome	Reference
VIRAL			
Astrovirus	Diarrhea, regurgitation	Recover with treatment	Atkins et al. (2009)
Canine Distemper Virus	Seropositivity in wild cheetahs	Unknown	Munson et al., 2004a,b; Munson et al., 2005; Thalwitzer et al., 2010
Feline leukemia virus (FeLV)	Lethargy, anorexia, lymphadenopathy	Multicentric T-cell lymphoma	Marker et al. (2003)
Feline immunodeficiency virus/cheetah immunodeficiency virus	Appears to be rare, incidental infection	None reported	
Papillomavirus	None	Incidental plaque-like lesions on tongue and oral cavity	
PARASITIC			
Babesia spp. (multiple species)	RBC parasite in southern African cheetah	Uncertain, may be incidental finding	
Coccidia (Isospora)	Mild diarrhea	Recover with treatment	Penzhorn et al. (1994)
Cytauxzoon felis	Fever, lameness, severe lethargy	Potentially fatal; blood vessels filled with macrophages containing schizonts	
Giardia sp.	Mild diarrhea	Recover with treatment	
Haemoplasma felis	Rare infection	Incidental finding	Krengel et al. (2013)
Ollulanus tricuspis	Vomiting, gastritis	Recover with treatment	Collett et al. (2000)
Otodectes cynotis	Ear mite	Recover with treatment	
Theileria-like	Occasionally seen in RBC	Incidental finding	
Toxocara sp.	Intestinal parasite	Incidental finding except with heavy burdens in young cubs	
Toxoplasma sp.	Fever, respiratory distress, abdominal effusion	Multicentric organ necrosis with *Toxoplasma* zoites	Lloyd and Stidworthy (2007)
MISCELLANEOUS			
Diabetes mellitus	Elevated blood sugar, fructosamine	Managed with treatment	Sanchez et al. (2005)
Lower esophageal sphincter dysfunction	Prolapse of stomach into esophagus and acquired diaphragmatic hernias	Associated with gastritis leading to gastric esophageal reflux, may be associated with acquired hiatal hernia	Teunissen et al. (1978); Kimber et al. (2001)
Prion/Spongiform encephalopathy	Behavioral changes, tremors, ataxia; PrP protein within brain	Fatal, associated with eating contaminated meat (e.g., beef infected with BSE) or vertical transmission	Peet et al. (1992); Kirkwood et al. (1995); Baron et al. (1997); Lezmi et al. (2003); Bencsik et al. (2009); Eiden et al. (2010)
Soft tissue mineralization	Renal failure if kidney involved	If renal involvement, can be fatal. No apparent exposure to vitamin D	

BSE, Bovine Spongiform Encephalopathy; RBC, red blood cell.

DISEASES OF NEONATAL AND JUVENILE CHEETAHS

Many of the important noninfectious diseases of neonatal and juvenile cheetahs are nutritionally based (Chapters 24 and 26). Bilateral carpal valgus deformities that may require surgical repair have been noted in captive cheetahs at multiple facilities (Allan et al., 2008; Bell et al., 2011). In these cases, cubs presented with lameness around 6 months of age or during weaning. Keeping hand-reared kittens on formula too long may predispose to this condition; adding meat early (e.g., 3 weeks) and weaning by 5-6 weeks can be preventative. Excess dietary calcium can inhibit cartilage maturation and contribute to osteochondrosis, an underlying cause of these deformities (Allan et al., 2008). In some cases, however, affected cheetahs were fed the same diet as unaffected littermates, suggesting that other factors may also be involved (Bell et al., 2011). Fractures have also been reported in some cheetahs, especially in king cheetahs, after minimal trauma, raising concerns about inappropriate dietary mineral content.

CONCLUSIONS

Much of our understanding of disease in cheetahs comes from long-term pathology surveys of cheetahs in captivity and in the wild. As many of the diseases of cheetahs occur primarily in captivity, it is incumbent upon us to better understand and reduce risk factors that lead to this higher prevalence of disease in captive settings. Using death codes reported in the International Studbook from 1980 to 2009, the relative prevalence of reported infectious disease decreased from 50% to less than 25%, suggesting improvements in infectious disease management (Schmidt-Küntzel et al., in preparation). Excluding infectious, accidental, and congenital causes, diseases impacting the urinary system have been increasing from less than 20% in 1980–84 to over 45% of cases from 2005 to 2009. This likely reflects increasing longevity of the captive cheetah population, with a corresponding increase in chronic renal disease, a common cause of death in many domestic and wild aged felids. Unfortunately, some diseases continue to be intractable. Continued research into management is important not only for improved well-being and medical management of captive populations, but also in limiting development of disease in wild-caught cheetahs that are held in captive settings, either temporarily or permanently, due to human-wildlife conflict (Chapter 13) or other reasons. Although previous studies of wild cheetahs have found them to be generally healthy, it is known that they have exposure to many of the infectious agents that can cause problems in captive cheetahs. It is also possible that pressures including reduced size and quality of free-ranging cheetah habitats (Chapter 10), increased contact with domesticated animals, and climate change (Chapter 12) could change disease dynamics. Continued monitoring of both captive and free-ranging cheetah populations, especially small geographically isolated populations and an improved understanding of normal clinical values, is needed for understanding and management of disease risks to species survival.

References

Allan, G., Portas, T., Bryant, B., Howlett, R., Blyde, D., 2008. Ulnar metaphyseal osteochondrosis in several captive bred cheetahs (Acinonyx jubatus). Vet. Radiol. Ultrasound 49, 551–556.

Atkins, A., Wellehan, Jr., J.F., Childress, A.L., Archer, L.L., Fraser, W.A., Citino, S.B., 2009. Characterization of an outbreak of astroviral diarrhea in a group of cheetahs (Acinonyx jubatus). Vet. Microbiol. 136, 160–165.

Baron, T., Belli, P., Madec, J.Y., Moutou, F., Vitaud, C., Savey, M., 1997. Spongiform encephalopathy in an imported cheetah in France. Vet. Rec. 141, 270–271.

Bauer, G.A., 1998. Cheetah—running blind. In: Penzhorn, B.L. (Ed.), Proceedings of a Symposium on Cheetahs as Game Ranch Animals. Onderstepoort, pp. 106–108.

Baxby, D., Ashton, D.G., Jones, D.M., Thomsett, L.R., 1982. An outbreak of cowpox in captive cheetahs: virological and epidemiological studies. J. Hyg. (Lond.) 89, 365–372.

Beehler, B.A., 1982. Oral therapy for nasal cryptococcosis in a cheetah. J. Am. Vet. Med. Assoc. 181, 1400–1401.

Bell, K.M., van Zyl, M., Ugarte, C.E., Hartman, A., 2011. Bilateral carpal valgus deformity in hand-reared cheetah cubs (Acinonyx jubatus). Zoo Biol. 30, 199–204.

Bencsik, A., Debeer, S., Petit, T., Baron, T., 2009. Possible case of maternal transmission of feline spongiform encephalopathy in a cheetah. PLoS One 4, e6929.

Bergström, J., Ueda, M., Une, Y., Sun, X., Misumi, S., Shoji, S., Ando, Y., 2006. Analysis of amyloid fibrils in the cheetah (Acinonyx jubatus). Amyloid 13, 93–98.

Berry, W.L., Jardine, J.E., Espie, I.W., 1997. Pulmonary cryptococcoma and cryptococcal meningoencephalomyelitis in a king cheetah (Acinonyx jubatus). J. Zoo Wildl. Med. 28, 485–490.

Bolton, L.A., Lobetti, R.G., Evezard, D.N., Picard, J.A., Nesbit, J.W., van Heerden, J., Burroughs, R.E., 1999. Cryptococcocsis in captive cheetah (Acinonyx jubatus): two cases. J. S. Afr. Vet. Assoc. 70, 35–39.

Bolton, L.A., Munson, L., 1999. Glomerulosclerosis in captive cheetahs (Acinonyx jubatus). Vet. Pathol. 36, 14–22.

Brower, A.I., Munson, L., Radcliffe, R.W., Citino, S.B., Bingaman Lackey, L., Van Winkle, T.J., Stalis, I., Terio, K.A., Summers, B.A., de Lahunta, A., 2014. Leukoencephalomyelopathy of mature captive cheetahs and other large felids: a novel neurodegenerative disease that came and went? Vet. Pathol. 51, 1013–1021.

Burger, P.A., Steinborn, R., Walzer, C., Petit, T., Mueller, M., Schwarzenberger, F., 2004. Analysis of the mitochondrial genome of cheetahs (Acinonyx jubatus) with neurodegenerative disease. Gene, 111–119.

Chen, L., Une, Y., Higuchi, K., Mori, M., 2012. Cheetahs have 4 serum amyloid A genes evolved through repeated duplication events. J. Hered. 103, 115–129.

Citino, S.B., Munson, L., 2005. Efficacy and long-term outcome of gastritis therapy in cheetahs (Acinonyx jubatus). J. Zoo Wildl. Med. 36, 401–416.

Clyde, V.L., Ramsay, E.C., Bemis, D.A., 1997. Fecal shedding of Salmonella in exotic felids. J. Zoo Wildl. Med. 28, 148–152.

Collett, M.G., Pomroy, W.E., Guilford, W.G., Johnstone, A.C., Blanchard, B.J., Mirams, S.G., 2000. Gastric Ollulanus tricuspis infection identified in captive cheetahs (Acinonyx jubatus) with chronic vomiting. J. S. Afr. Vet. Assoc. 71, 251–255.

de Pienaar, U.V., 1961. A second outbreak of anthrax amongst game animals in the Kruger National Park. Koedoe 4, 4–16.

De Vos, V., Bengis, R.G., Kriek, N.P., Michel, A., Keet, D.F., Raath, J.P., Huchzermeyer, H.F., 2001. The epidemiology of tuberculosis in free-ranging African buffalo (Syncerus caffer) in the Kruger National Park, South Africa. Onderstepoort J. Vet. Res. 68, 119–130.

Eaton, K.A., Radin, M.J., Kramer, L.W., Wack, R.F., Sherding, R., Krakowka, S., Fox, J.G., 1993. Epizootic gastritis associated with gastric spiral bacilli in cheetahs (Acinonyx jubatus). Vet. Pathol. 30, 55–63.

Ebedes, H., 1976. Anthrax epizoonotics in Etosha National Park. Madoqua 10, 99–118.

Eiden, M., Hoffmann, C., Balkema-Buschmann, A., Müller, M., Baumgartner, K., Groschup, M.H., 2010. Biochemical and immunohistochemical characterization of feline spongiform encephalopathy in a German captive cheetah. J. Gen. Virol. 91, 2874–2883.

Evermann, J.F., Heeney, J.L., Roelke, M.E., McKeirnan, A.J., O'Brien, S.J., 1988. Biological and pathological consequences of feline infectious peritonitis virus infection in the cheetah. Arch. Virol. 102, 155–171.

Fitch, H.M., Fagan, D.A., 1982. Focal palatine erosion associated with dental malocclusion in captive cheetahs. Zoo Biol. 1, 295–310.

Flacke, G.L., Schmidt-Küntzel, A., Marker, L., 2015. Treatment of chronic herpesviral dermatitis in a captive cheetah (Acinonyx jubatus) in Namibia. J. Zoo Wildl. Med. 46, 641–646.

Franklin, A.D., Schmidt-Küntzel, A., Terio, K.A., Marker, L., Crosier, A.E., 2016. Serum amyloid A protein concentration in blood is influenced by genetic differences in the cheetah (Acinonyx jubatus). J. Hered. 117, 115–121.

Gaffney, P.M., Kennedy, M., Terio, K., Gardner, I., Lothamer, C., Coleman, K., Munson, L., 2012. Detection of feline coronavirus in cheetah (Acinonyx jubatus) feces by reverse transcription-nested polymerase chain reaction in cheetahs with variable frequency of viral shedding. J. Zoo Wildl. Med. 43, 776–786.

Gakuya, F., Ombui, J., Maingi, N., Muchemi, G., Ogara, W., Soriquer, R.C., Alassad, S., 2012. Sarcoptic mange and cheetah conservation in Masai Mara (Kenya): epidemiological study in a wildlife/livestock system. Parasitology 139, 1587–1595.

Good, K.M., Houser, A., Antzen, L., Turnbull, P.C.B., 2008. Naturally acquired anthrax antibodies in a cheetah (Acinonyx jubatus) in Botswana. J. Wildl. Dis. 44, 721–723.

Gosselin, S.J., Loudy, D.L., Tarr, M.J., Balisteri, W.F., Setchell, K.D., Johnston, J.O., Kramer, L.W., Dresser, B.L., 1988. Veno-occlusive disease of the liver in captive cheetah. Vet. Pathol. 25, 48–57.

Heeney, J.L., Evermann, J.F., McKeirnan, A.J., Marker-Kraus, L., Roelke, M.E., Bush, M., Wildt, D.E., Meltzer, D.G., Colly, L., Lukas, L., Manton, V.J., Caro, T., O'Brien, S.J., 1990. Prevalence and implications of feline coronavirus infections in captive and free-ranging cheetahs (Acinonyx jubatus). J. Virol. 64, 1964–1972.

Illnait-Zaraozí, M.T., Hagen, F., Fernández-Andreu, C.M., Martinez-Machin, G.F., Polo-Leal, J.L., Boekhout, T., Klaassen, C.H., Meis, J.F., 2011. Reactivation of a Cryptococcus gattii infection in a cheetah (Acinonyx jubatus) held in the National Zoo, Havanna, Cuba. Mycoses 54, e889–e892.

Jäger, H.G., Booker, H.H., Hubschle, O.J., 1990. Anthrax in cheetahs (*Acinonyx jubatus*) in Namibia. J. Wildl. Dis. 26, 423–424.

Johnson, K.H., Sletten, K., Munson, L., O'Brien, T.D., Papendick, R., Westermark, P., 1997. Amino acid sequence analysis of amyloid protein (AA) from cats (captive cheetahs: *Acinonyx jubatus*) with a high prevalence of AA amyloidosis. Amyloid 4, 171–177.

Junge, R.E., Miller, R.E., Boever, W.J., Scherba, G., Sundberg, J., 1991. Persistent cutaneous ulcers associated with feline herpesvirus 1 infection in a cheetah. J. Am. Vet. Med. Assoc. 198, 1057–1058.

Jurke, M.H., Czekala, N.M., Lindburg, D.G., Millard, S.E., 1997. Fecal corticoid metabolite measurement in the cheetah (*Acinonyx jubatus*). Zoo Biol. 16, 133–147.

Keet, D.F., Kriek, N.P., Penrith, M.L., Michel, A., Huchzermeyer, H., 1996. Tuberculosis in buffaloes (*Synverus caffer*) in the Kruger National Park: spread of the disease to other species. Onderstepoort J. Vet. Res. 63, 239–244.

Kennedy, M., Citino, S., Dolorico, T., McNabb, A.H., Moffat, A.S., Kania, S., 2001. Detection of feline coronavirus infection in captive cheetahs (*Acinonyx jubatus*) by polymerase chain reaction. J. Zoo Wildl. Med. 32, 25–30.

Kennedy, M.A., Moore, E., Wilkes, R.P., Citino, S.B., Kania, S.A., 2006. Analysis of genetic mutations in the 7a7b open reading frame of coronavirus of cheetahs (*Acinonyx jubatus*). Am. J. Vet. Res. 67, 627–632.

Kimber, K., Pye, G.W., Dennis, P.M., Citino, S.B., Heart, D.J., Bennett, R.A., 2001. Diaphragmatic hernias in captive cheetahs (*Acinonyx jubatus*). In: Proceedings of the AAZV, AAWV, ARAV, NAZWV Joint Conference 151.

Kirkwood, J.K., Cunningham, A.A., Flach, E.j., Thornton, S.M., Wells, G.A.H., 1995. Spongiform encephalopathy in another captive cheetah (Acinonyx jubatus): Evidence for variation in susceptibility or incubation periods between species? J. Zoo Wildl. Med. 26, 577–582.

Kotsch, V., Kübber-Heiss, A., Url, A., Walzer, C., Schmidt, P., 2002. Diseases of captive cheetahs (*Acinonyx jubatus*) within the European Endangered Species Program (EEP)—a 22-year retrospective histopathological study. Vet. Med. Austria 89, 341–350.

Krengel, A., Meli, M., Cattori, V., Wachter, B., Willi, B., Thalwitzer, S., Melzheimer, J., Hofer, H., Lutz, H., Hofmann-Lehmann, R., 2013. First evidence of hemoplasma infection in free-ranging Namibian cheetahs (*Acinonyx jubatus*). Vet. Microbiol. 162, 972–976.

Lane, E.P., Brettschneider, H., Caldwell, P., Oosthuizen, A., Dalton, D.L., du Plessis, L., Steyl, J., Kotze, A., 2016. Feline panleukopaenia virus in captive non-domestic felids in South Africa. Onderstepoort J. Vet. Res. 83, a1099.

Lane, E., Lobetti, R., Burroughs, R., 2004. Treatment with omeprazole, metronidazole, and amoxicillin in captive South African cheetahs (*Acinonyx jubatus*) with spiral bacterial infection and gastritis. J. Zoo Wildl. Med. 35, 15–19.

Lane, E.P., Miller, S., Lobetti, R., Caldwell, P., Bertschinger, H.J., Burroughs, R., Kotze, A., van Dyk, A., 2012. Effect of diet on the incidence of and mortality owing to gastritis and renal disease in captive cheetahs (*Acinonyx jubatus*) in South Africa. Zoo Biol. 31, 669–682.

Lezmi, S., Benscik, A., Monks, E., Petit, T., Baron, T., 2003. First case of feline spongiform encephalopathy in a captive cheetah born in France: PrP9sc analysis in various tissues revealed unexpected targeting of kidney and adrenal gland. Histochem. Cell Biol. 119, 415–422.

Lindeque, P.M., Nowell, K., Preisser, T., Brain, C., Turnbull, P.C.B., 1998. Anthrax in wild cheetahs in the Etosha national Park, Namibia. In: Proceedings ARC-Onderstepoort OIE International Anthrax Congress, Onderstepoort, pp. 9–15.

Lloyd, C., Stidworthy, M.F., 2007. Acute disseminated toxoplasmosis in a juvenile cheetah (*Acinonyx jubatus*). J. Zoo Wildl. Med. 38, 475–478.

Marennikova, S.S., Maltseva, N.N., Korneeva, V.I., Garanina, N.M., 1977. Outbreak of pox disease among carnivore (*Felidae*) and Edentata. J Infect. Dis. 135, 358–366.

Marennikova, S.S., Shelukhina, E.M., 1976. White rats as a source of pox infection in Carnivora of the family *Felidae*. Acta Virol. 20, 442.

Marker-Kraus, L., 1997. Morphological abnormalities reported in Namibian cheetahs (*Acinonyx jubatus*). In: Proceedings 50th Anniversary Congress of the Veterinary Association of Namibia and the 2cd Africa Congress of the World Veterinary Association, 9.

Marker, L.L., Dickman, A.J., 2004. Dental anomalies and incidence of palatal erosion in Namibian cheetahs (*Acinonyx jubatus jubatus*). J. Mammol. 85, 19–24.

Marker, L., Dickman, A.J., Mills, M.G.L., Macdonald, D.W., 2003a. Aspects of the management of cheetahs trapped on Namibian farmlands. Biol. Conserv. 114, 401–412.

Marker, L., Munson, L., Basson, P.A., Quackenbush, S., 2003b. Multicentric T-cell lymphoma associated with feline leukemia virus infection in a captive Namibian cheetah (*Acinonyx jubatus*). J. Wildl. Dis. 39, 690–695.

McLean, K., Garabed, R., 2015. Amyloidosis in cheetahs (*Acinonyx jubatus*), transmissible? In: Proceedings Annual Meeting of the American Association of Zoo Veterinarians. Portland, Oregon.

Miller-Edge, M.A., Worley, M.B., 1992. In vitro responses of cheetah mononuclear cells to feline herpesvirus-1 and *Cryptococcus neoformans*. Vet. Immunol. Immunopathol. 30, 261–274.

Millward, I.R., Williams, M.C., 2005. Cryptococcus neoformans granuloma in the lung and spinal cord of a free-ranging cheetah (*Acinonyx jubatus*). A clinical report and literature review. J. S. Afr. Vet. Assoc. 76, 228–232.

Munson, L., 1993. Diseases of captive cheetahs (*Acinonyx jubatus*): results of the Cheetah Research Council pathology survey 1989–1992. Zoo Biol. 12, 105–124.

Munson, L., Citino, S., 2005. Major health concerns of cheetahs. Infectious diseases: feline enteric coronavirus/feline infectious peritonitis virus. In: Proceedings of the AZA Cheetah SSP Disease Management Workshop, Yulee Florida, pp. 18–20.

Munson, L., Marker, L., 1997. The impact of capture and captivity on the health of Namibian farmland cheetahs (*Acinonyx jubatus*). In: Proceedings of the 50th Namibian Vet Congress.

Munson, L., Marker, L., Dubovi, E., Spencer, J.A., Evermann, J.F., O'Brien, S.J., 2004a. Serosurvey of viral infections in free-ranging Namibian cheetahs (*Acinonyx jubatus*). J. Wildl. Dis. 40, 23–31.

Munson, L., Nesbit, J.W., Meltzer, D.G.A., Colly, L.P., Bolton, L., Kriek, N.P.K., 1999. Diseases of captive cheetahs (*Acinonyx jubatus*) in South Africa: a 20-year retrospective survey. J. Zoo Wildl. Med. 30, 342–347.

Munson, L., Terio, K.A., Worley, M., Jago, M., Bagot-Smith, A., Marker, L., 2005. Extrinsic factors significantly affect patterns of disease in free-ranging and captive cheetah (*Acinonyx jubatus*) populations. J. Wildl. Dis. 41, 542–548.

Munson, L., Wack, R., Duncan, M., Montali, R.J., Boon, D., Stalis, I., Crawshaw, G.J., Cameron, K.N., Mortenson, J., Citino, S., Zuba, J., Junge, R.E., 2004b. Chronic eosinophilic dermatitis associated with persistent feline herpes virus infection in cheetahs (*Acinonyx jubatus*). Vet. Pathol. 41, 170–176.

Mwanzia, J.M., Kock, R., Wambua, J., Kock, N., Jarret, O., 1995. An outbreak of Sarcoptic mange in the free-living cheetah (*Acinonyx jubatus*) in the Mara region of Kenya. In: Proceedings of American Association of Zoo Veterinarians and American Association of Wildlife Veterinarians Joint Conference, pp. 105–112.

Palmer, A.C., Callanan, J.J., Guerin, L.A., Sheanan, B.J., Stronach, N., Franklin, R.J.M., 2001. Progressive encephalomyelopathy and cerebellar degeneration in 10-captive bred cheetahs. Vet. Rec. 149, 49–54.

Papendick, R.E., Munson, L., O'Brien, T.D., Johnson, K.H., 1997. Systemic AA amyloidosis in captive cheetahs (*Acinonyx jubatus*). Vet. Pathol. 34, 549–556.

Peet, R.L., Curran, J.M., 1992. Spongiform encephalopathy in an imported cheetah (*Acinonyx jubatus*). Aust. Vet. J. 69, 171.

Penzhorn, B.L., Booth, L.M., Meltzer, D.G., 1994. *Isospora rivolta* recovered from cheetahs. J. S. Afr. Vet. Assoc. 65, 2.

Phillips, J.A., Worley, M.B., Morsbach, D., Williams, T.M., 1993. Relationship between diet, growth, and occurrence of focal palatine erosion in wild-caught captive cheetahs. Madoqua 18, 79–83.

Robert, N., Walzer, C., 2009. Pathological disorders in captive cheetahs (Patologías de guepardos en cautividad). In: Vargas, A., Breitenmoser-Würsten, C., Breitenmoser, U. (Eds.), Iberian Lynx *Ex Situ* Conservation: An Interdisciplinary Approach (Conservación Ex Situ del Lince

Ibérico: Un Enfoque Multidisciplinar). Fundación Biodiversidad in collaboration with: IUCN Cat Specialist Group, Madrid, pp. 265–272.

Sanchez, C., Bronson, E., Deem, S., Viner, T., Pereira, M., Saffoe, C., Murray, S., 2005. Diabetes mellitus in a cheetah: attempting to treat the untreatable? In: Proceedings of the AAZV, AWV, AZA/NAG Joint Conference 101.

Scherba, G., Hajjar, A.M., Pernikoff, D.S., Sundberg, J.P., Basgall, E.J., Leon-Monzon, M., Nerukar, L., Reichmann, M.E., 1988. Comparison of a cheetah herpesvirus isolate to feline herpesvirus type 1. Arch. Virol. 100, 89–97.

Serizawa, S., Chambers, J.K., Une, Y., 2012. Beta amyloid deposition and neurofibrillary tangles spontaneously occur in the brains of captive cheetahs (*Acinonyx jubatus*). Vet. Pathol. 49, 304–312.

Setchell, K.D., Gosselin, S.J., Welsh, M.B., Johnston, J.O., Balistreri, W.F., Kramer, L.W., Dresser, B.L., Tarr, M.J., 1987. Dietary estrogens—a probable cause of infertility and liver disease in captive cheetahs. Gastroenterology 93, 225–233.

Shibly, S., Schmidt, P., Robert, N., Walzer, C., Url, A., 2006. Immunohistochemical screening for viral agents in cheetahs (*Acinonyx jubatus*) with myelopathy. Vet. Rec. 159, 557–561.

Spencer, J.A., Burroughs, R., 1992. Decline in maternal immunity and antibody response to vaccine in captive cheetah (*Acinonyx jubatus*) cubs. J. Wildl. Dis. 28, 102–104.

Steenkamp, G., Boy, S.C., van Staden, P.J., Bester, M.N., 2017. How the cheetahs' specialized palate accommodates its abnormally large teeth. J. Zool. 301, 290–300.

Steinel, A., Munson, L., van Vuuren, M., Truyen, U., 2000. Genetic characterization of feline parvovirus sequences from various carnivores. J. Gen. Virol. 81, 345–350.

Switzer, A.D., Munson, L., Beesley, C., Wilkins, P., Blackburn, J.K., Marker, L., 2016. Free-ranging Namibian farmland cheetahs (*Acinonyx jubatus*) demonstrate immunologic naivety to anthrax (*Bacillus anthracis*). Afr. J. Wildl. Res. 46, 139–143.

Teixeira, C.P., Schetini de Azevedo, C., Mendl, M., Cipreste, C.F., Young, R.J., 2007. Revisiting translocation and reintroduction programmes: importance of considering stress. Anim. Behav. 73, 1–13.

Terio, K.A., Munson, L., 2005. Linking stress with altered gastric immune responses in cheetahs. In: Proceedings Annual Meeting of the American Association of Zoo Veterinarians. Omaha, Nebraska.

Terio, K.A., Marker, L., Munson, L., 2004. Evidence for chronic stress in captive but not wild cheetahs (*Acinonyx jubatus*) based on adrenal morphology and function. J. Wildl. Dis. 40, 259–266.

Terio, K.A., Munson, L., Marker, L., Aldridge, B.M., Solnick, J.V., 2005. Comparison of *Helicobacter* spp. in cheetahs (*Acinonyx jubatus*) with and without gastritis. J. Clin. Microbiol. 43, 229–234.

Terio, K.A., Munson, L., Moore, P.F., 2011. Characterization of the gastric immune response in cheetahs (*Acinonyx jubatus*) with *Helicobacter*-associated gastritis. Vet. Pathol. 49, 824–833.

Terio, K., Whitham, J., Chosy, J., Sanchez, C., Marker, L., Wielebnowski, N., 2014. Associations between gastritis, temperament and management risk factors in captive cheetahs (*Acinonyx jubatus*). In: Proceedings of the American Association of Zoological Veterinarians Annual Conference. Orlando, Florida.

Terrell, S.P., Fontenot, D.K., Miller, M.A., Weber, M.A., 2003. Chylous ascites in a cheetah (*Acinonyx jubatus*) with venoocclusive liver disease. J. Zoo Wildl. Med. 34, 380–384.

Teunissen, G.H., Happe, R.P., Van Toorenburg, J., Wolvekamp, W.T., 1978. Esophageal hiatal hernia. Case report of a dog and a cheetah. Tijdschr Diergeneeskd 103, 742–749.

Thalwitzer, S., Wachter, B., Robert, N., Wibbelt, G., Müller, T., Lonzer, J., Meli, M.L., Bay, G., Hofer, H., Lutz, H., 2010. Seroprevalences to viral pathogens in free-ranging and captive cheetahs (*Acinonyx jubatus*) on Namibian farmland. Clin. Vaccine Immunol. 17, 232–238.

Turnbull, P.C.B., Tindall, B.W., Coetzee, J.D., Conradie, C.M., Bull, R.L., Lindeque, P.M., Huebschle, O.J.B., 2004. Vaccine-induced protection against anthrax in cheetah (*Acinonyx jubatus*) and black rhinoceros (*Diceros bicornis*). Vaccine 2, 3340–3347.

Valícek, L., Smíd, B., Váhala, J., 1993. Demonstration of parvovirus in diarrhoeic African cheetahs (*Acinonyx jubatus jubatus*, Schreber 1775). Vet. Med. (Praha) 38, 245–249.

van Vuuren, M., Goosen, T., Rogers, P., 1999. Feline herpesvirus infection in a group of semi-captive cheetahs. J. S. Afr. Vet. Assoc. 70, 132–134.

Venter, E.H., van Vuuren, M., Carstens, J., van der Walt, M.L., Nieuwoudt, B., Steyn, H., Kriek, N.P.J., 2003. A molecular epidemiologic investigation of *Salmonella* from a meat source to the feces of captive cheetah (*Acinonyx jubatus*). J. Zoo Wildl. Med. 34, 76–81.

Wack, R.F., Eaton, K.A., Kramer, L.W., 1997. Treatment of gastritis in cheetahs (*Acinonyx jubatus*). J. Zoo Wildl. Med. 28, 260–266.

Wack, R.F., Kramer, L.W., Cupps, W., 1992. Griseofulvin toxicity in four cheetahs (*Acinonyx jubatus*). J. Zoo Wildl. Med. 23, 442–446.

Wack, R.F., Kramer, L.W., Cupps, W.L., Clawson, S., Hustead, D.R., 1993. The response of cheetahs (*Acinonyx jubatus*) to routine vaccination. J. Zoo Wildl. Med. 24, 109–117.

Walzer, C., Hittmayer, C., Wagner, C., 1996. Ultrasonographic identification and characterization of splenic lipomatosis or myelolipomas in cheetahs (*Acinonyx jubatus*). Vet. Radiol. Ultrasound 37, 267–273.

Walzer, C., Kübber-Heiss, A., 1995. Progressive hind limb paralysis in adult cheetahs (*Acinonyx jubatus*). J. Zoo Wildl. Med. 26, 430–435.

Walzer, C., Url, A., Robert, N., Kübber-Heiss, A., Nowotny, N., Schmidt, P., 2003. Idiopathic acute onset myelopathy in cheetah (*Acinonyx jubatus*) cubs. J. Zoo Wildl. Med. 34, 36–46.

Wells, A., Terio, K.A., Ziccardi, M.H., Munson, L., 2004. The stress response to environmental change in captive cheetahs (*Acinonyx jubatus*). J. Zoo Wildl. Med. 35, 8–14.

Whitehouse-Tedd, K.M., Lefebvre, S.L., Janssens, G.P., 2015. Dietary factors associated with faecal consistency and other indicators of gastrointestinal health in the captive cheetah (*Acinonyx jubatus*). PLoS One 10, e0120903.

Wielebnowski, N., 1996. Reassessing the Relationship between Juvenile Mortality and Genetic Monomorphism in Captive Cheetahs. Zoo Biology 15, 353–369.

Wielebnowski, N., 1999. Behavioral differences as predictors of breeding status in captive cheetahs. Zoo Biol. 18, 335–349.

Witte, C.L., Lamberski, N., Rideout, B.A., Fields, V., Shields Teare, C., Barrie, M., Haefele, H., Junge, R., Murray, S., Hungerford, L.L., 2013. Development of a case definition for clinical feline herpesvirus infection in cheetahs (*Acinonyx jubatus*) housed in zoos. J. Zoo Wildl. Med. 44, 634–644.

Zhang, B., Une, Y., Fu, X., Yan, J., Ge, F., Yao, J., Sawashita, J., Mori, M., Tomozawa, H., Kametani, F., Higuchi, K., 2008a. Fecal transmission of AA amyloidosis in the cheetah contributes to high incidence of disease. Proc. Natl. Acad. Sci. USA 105, 7263–7268.

Zhang, B., Une, Y., Ge, F., Fu, X., Qian, J., Zhang, P., Sawashita, J., Higuchi, K., Mori, M., 2008b. Characterization of the cheetah serum amyloid A1 gene: critical role and functional polymorphism of a cis-acting element. J. Hered. 99, 355–363.

Zordan, M., Deem, S.L., Sanchez, C.R., 2012. Focal palatine erosion in captive and free-living cheetahs (*Acinonyx jubatus*) and other felid species. Zoo Biol. 31, 181–188.

Nutritional Considerations for Captive Cheetahs

Katherine Whitehouse-Tedd, Ellen S. Dierenfeld**,
Anne A.M.J. Becker†, Geert Huys†,
Sarah Depauw‡, Katherine R. Kerr§, J. Jason Williams¶,
Geert P.J. Janssens†*

*Nottingham Trent University, Southwell, United Kingdom
**Ellen S. Dierenfeld, LLC, St. Louis, MO, United States
†Ghent University, Ghent, Belgium
‡Odisee College University, Sint Niklaas, Belgium
§San Diego Zoo Global, San Diego, CA, United States
¶Indianapolis Zoological Society, Indianapolis, IN, United States

INTRODUCTION

Due to the role of nutrition in physiological and psychological well-being, providing a nutritionally complete and balanced diet to captive cheetahs (*Acinonyx jubatus*) is among the top priorities for maintaining healthy individuals, and consequently a healthy captive population. In captivity, cheetahs are typically fed a diet either comprised of a range of muscle meats (e.g., beef, chicken, horse), partial or whole prey carcasses, a minced meat commercial preparation, or a combination of these diet types, and are usually supplemented with a vitamin and mineral premix

(Whitehouse-Tedd et al., 2015). However, the optimal diet for cheetahs is largely unknown, such that, when managing the dietary provision of captive cheetahs, we must either extrapolate knowledge from domestic felid studies, or utilize studies of free-ranging cheetahs to estimate a typical "wild-type" diet.

The study of diets fed to captive collections that have a proven history of reproductive success and long-term health may offer some additional insights, although findings are highly dependent on the level of veterinary vigilance and accuracy of dietary records, as well as other potentially influencing environmental or management-related factors. Interspecific differences in physiology, metabolism, and behavior are key concerns when considering the suitability of other species as a model species upon which to base nutritional decisions for captive cheetah diets. Alternatively, items comprising a free-ranging cheetah's diet, while fairly well documented in terms of the "what and when," have so far not been comprehensively documented from a nutritional perspective (i.e., which nutrients are consumed rather than which prey species), making its application to the diet of captive cheetahs challenging. To add complexity, free-ranging individuals are limited by what is available and attainable in the environment, leading to their nutritional status not necessarily being ideal and harsh environmental factors may influence nutritional needs.

In captivity, cheetahs are typically fed on a regular basis without any foraging costs, are under veterinary care, and do not encounter competitors. However, simply providing sufficient nutrients to sustain life and permit reproduction, while crucial, may be insufficient to ensure animal welfare, as cheetah behavioral and gastrointestinal health are influenced by factors, such as diet form, meal frequency, and feeding schedule. Captive cheetahs are known to suffer from a range of ailments and diseases that are not seen (or are far less prevalent) in wild populations, often live well beyond the life expectancy of their wild conspecifics, and are subject to different environmental and physiological stressors. Therefore, the characteristics of a truly optimal diet may be highly situation dependent, and our approach to nutritional provision must be appropriately case and context specific. This chapter reviews our current understanding of the cheetah's nutritional needs in terms of nutrients and gastrointestinal environment, relevant associated diseases and disorders, and highlights knowledge gaps that should form the basis of future research priorities.

LESSONS FROM THE WILD

Cheetahs in the wild preferentially prey upon medium-sized ungulates (23–56 kg) (Hayward et al., 2006), although larger prey may be targeted by cheetah coalitions (Caro, 1994; Chapter 8). Animal tissues are typically high in protein relative to physiological requirements for protein synthesis in felids, and low in carbohydrates, but the nutritional composition of prey items can vary widely. For example, factors, such as prey age, species, diet, and season are known to influence prey composition (Clum et al., 1996; Dierenfeld et al., 2002). Likewise, nutrient intake will vary depending on the carcass component(s) consumed (Table 26.1).

Predators are known to selectively target specific macronutrient ratios, indicating that the seasonal and geographic variation in prey species and carcass component consumed may serve as important mechanisms by which free-ranging predators balance their nutrient intake (Kohl et al., 2015). Captive carnivores have also been shown to demonstrate such nutrient-seeking behavior (Hewson-Hughes et al., 2011, 2013; Kohl et al., 2015), therefore the potential exists to improve captive dietary provision by providing the animals with a degree of choice.

In comparison to the wild, foraging behaviors are restricted in captivity, thereby potentially causing frustration and reduced welfare. For example, carcass-feeding (compared to a minced meat diet) resulted in increased frequency of natural behaviors, such as possessiveness over food, improved appetite, and increased the time spent investigating the food (e.g., sniffing), number of bites per swallow, duration of time spent using the molar teeth, and total time spent feeding (Bond and Lindburg, 1990). Likewise, feeding regime or meal schedule are known to influence behavior and digestive function in captive carnivores. In particular, randomization of feeding times has been shown to have positive physiological and behavioral outcomes in lions (*Panthera leo*) (Altman et al., 2005) and cheetahs

TABLE 26.1 Nutrient Ranges in Adult Whole Prey and Meats Eaten by/fed to Cheetahs

Animal Descriptors	Farmed antelope[a] Whole, small, and medium	Beef[b] Lean, meat only	Beef[b] Whole	Chicken[c] Meat and skin	Chicken[c] Light meat only	Deer[b] Whole, ingesta free	Horse[b] Meat only	Rabbit[d] Whole	Rodents[e] Whole	Quail[f] Whole	Venison[b] Meat only
Water (%)	58–65	72–75	68–73	66	75		73	70–73	66–71	65–70	73
ME (kcal/g)	3.3–4.4	4.2–4.4	5.3–5.7	6.3	4.5		4.9	5.0–5.4	4.7–6.3	5.2–6.1	4.5
Total dietary fiber[g] (%)		1	2					3–4	2–3	1	
Crude fat (%)	4–20	9–10	34–38	44	7		17	24–26	28–29	19–32	9
Crude protein (%)	63–66	79–86	42–55	55	92		78	61–62	59–66	70–72	87
Taurine (%)		0.9			0.06		0.11	0.6			
Vitamin A (IU/g)			36	4	1			6	16–500+	70	
Vitamin E (mg/kg)			51	9					100–139	67	8
Calcium (%)	4	0.1–0.2	2–3	0.03	0.05	3	0.02	2–6	2–3	1–3	0.02
Copper (mg/kg)	2	3–4	4	1	2	26	5	5–35	6–10	3–5	10
Iron (mg/kg)	38	73–192	83–122	27	29	165	140	100–175	123–198	75–148	129
Magnesium (%)	0.1	0.08–0.09	0.07	0.06	0.11	0.2	0.09	0.1–0.2	0.1–0.2	0.06–0.07	0.09
Manganese (mg/kg)	0.3	0.5–2	3–10	0.6	0.7	29	0.7	2–6	8–37	3–6	1.6
Phosphorus (%)	3	0.6–0.8	1–2	0.4	0.7	2	0.8	1–3	2	1	0.8
Potassium (%)	0.9	1.2–1.3	0.4–0.5	0.6	1.0	1	1.3	0.6–0.7	0.7–0.9	0.7	1.2
Sodium (%)	1	0.2	0.2	0.2	0.3	0.4	0.2	0.3–0.4	0.3–0.4	0.6	0.2
Zinc (mg/kg)	16	100–201	57–116	39	39	68	106	72–94	57–104	53–81	79

Nutrients, except water, presented on a dry matter basis. ME, metabolizable energy.

[a]Farmed antelope (blesbok, impala, and springbok) (Van Zyl and Ferreira, 2004). Note: mineral data (in italics) extrapolated from domestic Holstein calves with similar proximate composition (House and Bell, 1993).

[b]Meats: USDA compositional database (Bechert et al., 2002; Depauw et al., 2012b).

[c]Chicken (Dierenfeld et al., 2002; Kerr et al., 2014, 2013a; Kerr et al., 2013c).

[d]Rabbits (Depauw et al., 2012b; Dierenfeld et al., 2002; Kerr et al., 2014).

[e]Rodents (guinea pigs, mice, and rats) (Dierenfeld et al., 2002; Kerr et al., 2014; Kremen et al., 2013).

[f]Quail (Dierenfeld et al., 2002; Kerr et al., 2014).

[g]Total dietary fiber concentration as determined using standard laboratory methods developed for use in determining plant fiber. This method has not been validated for use in carnivore diets.

(Quirke and O'Riordan, 2011). Overall, management changes aimed at providing a more naturalistic diet in terms of composition, schedule, and texture, are likely to offer significant health and welfare benefits to captive cheetahs.

NUTRITIONAL ADAPTATIONS AND CONSEQUENCES OF EVOLUTIONARY SPECIALIZATION FOR A CARNIVOROUS DIET

The feeding strategy of cheetahs, termed obligate carnivory, refers to the fact that cheetahs, like other felids, are dependent on animal tissues as their main source of nutrition. The short and simple gastrointestinal tract (GIT) of carnivores, including cheetahs (Chapter 7), reflects their dietary reliance on animal tissues that require minimal processing compared to plant material. As felids consume animal tissues containing many nutrients in their biologically active forms, the pathways for the biosynthesis of essential compounds from plant-derived precursors are inactive or insufficient to meet nutritional requirements (Morris, 2002).

The domestic cat is currently the only available well-researched nutritional model for cheetahs (and other felids) and it is likely that the two species share many dietary specializations for carnivory. Recent exploration of the cheetah genome suggested that some expanded gene families differed compared to the feline common ancestor and were likely to be related to the species' carnivorous lifestyle (Dobrynin et al., 2015). Of note, two families showed marked expansion in gene numbers in three felids (cheetahs, domestic cats, and lions) compared to other investigated mammals (Dobrynin et al., 2015), indicating potentially similar nutritional adaptations. Indeed, fatty acid metabolism in cheetahs is considered identical to that in domestic cats (Bauer, 1997; Bauer et al., 1996). The majority of other specializations to carnivory (e.g., requirement for

preformed vitamin A, taurine and arginine, and the metabolic capacity of various glucose and amino acid (AA) pathways; reviewed in Depauw, 2012, Fig. 26.1) are untested in cheetahs, and assumed to be common to both cheetahs and domestic cats. Recent advances in our understanding of cheetah nutritional physiology warrants caution in the exclusive use of the domestic cat model for predicting nutritional requirements for cheetahs.

However, some physiological differences between the domestic cat and cheetah are now known, such as distinct intestinal microbiota (Becker et al., 2014), as well as a higher brain glucose requirement in the domestic cat compared to larger carnivores (e.g., Weddell seals and dogs, and as predicted for tigers, mountain lions, and polar bears; Eisert, 2011). Moreover, domestic cats do not have the feast-famine lifestyle exhibited by many of the larger carnivores (Fitzgerald and Karl, 1979), including the cheetah (reported to make a kill once every 2–7 days; Caro, 1994; Mills et al., 2004) and the wolf.

The wolf does not exhibit the same metabolic adaptations to a carnivorous lifestyle as documented in the domestic cat (Bosch et al., 2015). Specifically, AA-catabolising enzymes are much less adaptable in domestic cats than wolves, such that domestic cats can only survive short periods of fasting due to their constantly high rate of AA catabolism (Bosch et al., 2015). The extended fasting periods experienced by wolves may potentially be tolerated due to their efficient protein sparing capacity via the downregulation of AA catabolic enzymes (Bosch et al., 2015). Other species living a feast-famine lifestyle (i.e., cheetahs) may therefore use similar protein sparing and biosynthetic metabolic pathways as the wolf. In cheetahs, signatures of natural selection on four genes associated with negative regulation of catabolic processes were documented (Dobrynin et al., 2015). However, more studies are warranted to determine the relevance of this finding in regards to the cheetah's metabolic capacities

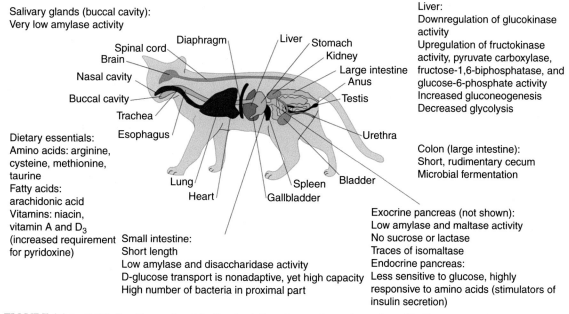

Salivary glands (buccal cavity):
Very low amylase activity

Liver:
Downregulation of glucokinase activity
Upregulation of fructokinase activity, pyruvate carboxylase, fructose-1,6-biphosphatase, and glucose-6-phosphate activity
Increased gluconeogenesis
Decreased glycolysis

Dietary essentials:
Amino acids: arginine, cysteine, methionine, taurine
Fatty acids: arachidonic acid
Vitamins: niacin, vitamin A and D_3 (increased requirement for pyridoxine)

Small intestine:
Short length
Low amylase and disaccharidase activity
D-glucose transport is nonadaptive, yet high capacity
High number of bacteria in proximal part

Colon (large intestine):
Short, rudimentary cecum
Microbial fermentation

Exocrine pancreas (not shown):
Low amylase and maltase activity
No sucrose or lactase
Traces of isomaltase
Endocrine pancreas:
Less sensitive to glucose, highly responsive to amino acids (stimulators of insulin secretion)

FIGURE 26.1 **Felid digestive and metabolic adaptations for carnivory (as reviewed in Depauw, 2012).**

to cope with extended intervals between meals. Differences or similarities between species are likely to vary depending on which aspect of nutritional physiology is being evaluated.

Energy

Daily energy expenditure has been estimated at approximately 580 kJ/kg $BW^{0.75}$ (i.e., per unit of metabolic body weight) in free-ranging cheetahs (Scantlebury et al., 2014). Feed intake in free-ranging cheetahs (0.5–4.0 kg/day) varies by reproductive state, sex and social system; amounts of around 1–1.5 kg appear reasonable for nonreproducing solitary animals (Caro, 1994; Caro et al., 1987; Mills et al., 2004). Using data for antelope energy content (Van Zyl and Ferreira, 2004), this equates to a daily metabolizable energy (ME) intake of roughly 850 kJ/kg $BW^{0.75}$, which is (unsurprisingly, given the lower activity rate of captive animals), at the higher end of the range reported

for captive cheetahs. Maintenance ME intakes of captive cheetahs consuming raw meat-based diets ranged from 355 to 875 kJ/kg $BW^{0.75}$/day (Depauw et al., 2012b; Kerr et al., 2013a), or 750 kJ/kg $BW^{0.75}$/day when fed whole rabbits (Depauw et al., 2012b). These values correspond with practical recommendations of 418 kJ ME/kg $BW^{0.67}$ established empirically for domestic felids (National Research Council, 2006).

Proteins and Amino Acids

The cat's high protein requirement [or secondarily increased protein requirement induced by a high glucose demand as proposed by Eisert (2011)] is a consequence of the nonadaptive metabolic processes that occur in the cat's liver (Morris, 2002). Excessively high protein diets, such as purely muscle meat-based diets may contribute to the chronic renal disease frequently observed in cheetahs kept in captivity (Bechert et al., 2002). Specific AA metabolism may also underlie the elevated blood creatinine and/or urea values observed in muscle meat-fed captive cheetahs, relative to free-ranging cheetahs (Bechert et al., 2002; Beckmann et al., 2013; Depauw et al., 2012b). However, AA balance has not been examined in detail for any non-domestic felid.

Sulfur Amino Acids

Taurine biosynthesis is low in cats, which compounds their dietary needs for this sulfur AA, as they are also unique in their obligatory use of taurine in bile salt conjugation (Morris, 2002). Taurine deficiency is associated with retinal degeneration, coronary disease, reproductive failure, and growth abnormalities in both the domestic cat and cheetah (Hedberg et al., 2007; Ofri et al., 1996). Relative to felid requirements (0.3%–0.4% of dry matter, DM), meat and whole prey diets consumed by captive cheetahs appear to range widely (e.g., 0.1%–0.6% in rabbits and 6.6% in turkey; Bechert et al., 2002; Depauw et al., 2012b; Dierenfeld, 1993). Prey species and/or form (particularly inclusion of indigestible components) may impact taurine status.

Lipids and Fatty Acids

Fats, lipids, and fatty acids are important sources of energy and fat-soluble vitamins for animals. Both protein and fat are preferentially used by felids to meet their energy needs (Bechert et al., 2002). However, dietary fat content is often restricted to minimize obesity issues in captive animals (Bechert et al., 2002), even though optimal dietary protein to fat ratios have not yet been determined for cheetahs. One report has indicated that ~60% of the large felids' total energy needs may be obtained from dietary fats, and that diets containing up to 67% fat were well tolerated in captive animals (Scott, 1968). A dietary fat content of 37% (farmed antelope contains only 4%–20% fat), resulted in loose fecal

consistency in cheetahs, which were also shown to digest fat with reduced efficiency compared to other exotic felids (Vester et al., 2008). In contrast, cheetahs fed a whole prey (rabbit) diet containing 26% fat exhibited improved fecal consistency compared to a supplemented beef diet with an even lower fat content (10%) (Depauw et al., 2013). Confounding differences in fiber content may also have contributed to the observed differences in fecal consistency. However, the whole rabbit diet also produced significantly higher serum cholesterol levels in these cheetahs (Depauw et al., 2012b) which could lead to hyperlipidemia (Backues et al., 1997), further stressing the need to better define optimum macronutrient profiles in cheetahs.

Recent research has demonstrated a preferred macronutrient profile in domestic cats under laboratory conditions, which is similar to the calculated intake of feral cats (Hewson-Hughes et al., 2011; Plantinga et al., 2011). The near-identical values for protein and fat intake support the use of the wild-diet as a proxy for nutritional requirements in carnivores (Kohl et al., 2015), and suggest that captive cheetah diets could be formulated to mimic natural dietary intake of nutrients. Whole prey typically consumed by wild cheetahs contain protein to fat ratios of ~2.5–3.5:1 (%DM; see Table 26.1), whereas various muscle meat-based diets fed to captive cheetahs range from about 3:1 up to 9:1 (e.g., turkey, horse, deer, or beef: Bechert et al., 2002; beef: Depauw et al., 2012b). This indicates a potential oversupply of protein, and/or inadequate fat supply in captive diets. The total amount, distribution (visceral, subcutaneous, or intramuscular), and composition of dietary lipids consumed by wild cheetahs depends upon the feeding habits of both predator and prey, seasonality, and species. Cheetahs likely have limited access to the lipids in bone marrow or brain tissues due to their skull morphology, and instead rely on adipose deposits. Differences between serum fatty acid profiles in captive and free-ranging cheetahs have recently

TABLE 26.2 Mean Concentrations of Circulating Nutrients and Associated Biochemistry Values

| | Unit (SI) | Captive cheetahs | | Wild cheetahs | Felid reference range |
		Commercial diet or supplemented meat	Whole prey		
Vitamin A (retinol)[a,b,c,d,e,f]	µmol/L	1.59–4.4	22	1.35–1.36	0.7–3.35
Vitamin A (retinol + esters)[b,e]	µmol/L	3.44–3.74			0.7–2.1
25-Hydroxyvitamin D[a,b,f]	nmol/L	30.45–109.4		77.6	14.95–20.02
1,25-Dihydroxyvitamin D[b]	pmol/L	132			
Beta-carotene[b,f]	µmol/L	0.448–1.97		0.43	
Vitamin E (α-tocopherol)[a,b,c,d,e,f]	µmol/L	14.85–44.9	35.6	16.67	13.9–22.88
Copper[a,c,d,f]	µmol/L	11.62–17.03	12.53	20.3	6.28–18.84
Cobalt[f]	nmol/L	3.56		3.05	
Iron[a,c,f]	µmol/L	7.57–21.84		5.94	12.53–39.38
Manganese[f]	nmol/L	63.52		58.79	
Magnesium[a,c,g]	nmol/L	0.96–1.01	1	1	0.75–1.0
Molybdenum[f]	nmol/L	23.97		18.55	
Selenium[a,d,f,h]	µmol/L	5.72–7.62	3.1	6.22	4.45–6.35
Zinc[a,c,d,f,h]	µmol/L	9.64–14.08	10.87	9.95	8.26–14.69
Calcium (ionized)[f]	mmol/L	1.36		1.22	
Calcium[a,c,i,j,k]	mmol/L	2.45–2.85		1.22	1.98–2.92
Phosphorus[a,h,i,j,k]	mmol/L	1.32–2.49			1.61–2.26
Sodium[a,c,g,i,j,k]	mmol/L	164.13–155.78	154	155	158.73–161.00
Potassium[a,g,i,j,k]	mmol/L	4.55–4.96	4.5	4.5	3.5–5.1
Chloride[a,g,i,j,k]	mmol/L	118–126	122	119	117–129
Taurine[a,c,l]	µmol/L	128–159	157		60–160
Triglycerides[b,l]	mmol/L	0.32–0.59	0.51		0.11–0.43
Saturated fatty acids[m]	µmol/L	3398		3727	
Monounsaturated fatty acids[m]	µmol/L	874.3		221.1	
Polyunsaturated fatty acids[m]	µmol/L	1183		363	
Cholesterol[a,b,l]	mmol/L	3.37–4.97	5.08		3.37–10.80
Glucose[a,l]	mmol/L	6.27–6.78			3.37–6.89
Leptin[l]	µg/L	2.4	2.3		1.7–5.0
Thyroxine (T4)[l]	nmol/L	0.14	0.12		0.08–0.30
Creatinine[a,g,l]	µmol/L	180–274.04	163–184	169	24–264
Urea[a,g,l]	mmol/L	12.71–16.1	12.8–15.9	12.3	5.5–22.5

[a]Bechert et al. (2002); [b]Crissey et al. (2003); [c]Dierenfeld (1993); [d]Beckmann et al. (2013); [e]Schweigert et al. (1990); [f]Williams et al. (unpublished); [g]Hudson-Lamb et al. (2016); [h]Puls (1988); [i]Wack (2003); [j]Plumb (2002); [k]Anderson and Brown (1979); [l]Depauw et al. (2012b); [m]Tordiffe et al. (2016).

been reported (Tordiffe et al., 2016; Table 26.2). Free-ranging cheetahs exhibited higher serum concentrations of saturated fat and lower concentrations of most unsaturated fatty acids (except arachidonic and hypogeic acids) compared to captive conspecifics (Tordiffe et al., 2016). The higher utilization of ruminant tissues by free-ranging cheetahs, compared to the nonruminant prey typically offered to captive cheetahs, is likely a contributing factor since ruminants have a markedly lower unsaturated to saturated fat ratio (Wood et al., 2008). Tordiffe et al.

(2016) propose that preferential oxidation of polyunsaturated fatty acids (PUFA) in periods of increased exercise or fasting in free-ranging cheetahs may explain their low concentration. However, freshly killed meat is rich in saturated fat (and hence more oxidatively stable); this may indicate that the species lacks extensive antioxidant mechanisms (Tordiffe et al., 2016).

Investigations of lipid metabolism confirm that cheetahs, similar to domestic felids, have limited delta-6-desaturase enzyme activity for elongation of long-chain fatty acids, and thus require dietary sources of arachidonic acid (Bauer, 1997; Bauer et al., 1996); clinical cases of fatty acid deficiency have been documented in captive cheetahs (Bauer et al., 1996; Davidson et al., 1986a,b). Chemical analyses of supplemented raw muscle-meat (beef, horse, elk, and bison) and farmed whole rabbits has determined these diets to be deficient in linoleic and eicosapentaenoic acids, relative to the domestic cat requirements (Depauw et al., 2012b; Kerr et al., 2013a). Moreover, farmed whole rabbit was also deficient in arachidonic acid, compared to other meats that met felid requirements (Depauw et al., 2012b). The important role of prey diet and rearing conditions in determining their subsequent nutrient composition (Hoffman and Wiklund, 2006; Li et al., 2012; Szendrő and Zotte, 2011) may explain the relatively poor fatty acid profiles determined in farmed animals compared to nutrient recommendations. The more desirable ratio of omega-6 to omega-3 in wild versus farmed prey (Cordain et al., 2002; Depauw et al., 2012b; Hoffman and Wiklund, 2006) may significantly impact lipid metabolism, inflammatory diseases, and oxidative status of felids consuming those meats. Combined, these findings have important anti-inflammatory and immunological implications for cheetah health based on our understanding of their role in human nutrition and health (Calder, 2001). Preliminary evidence indicates that stored tissue fatty acid profiles may also vary according to diet; differences were detected in hepatic fatty acid concentrations of two free-ranging cheetahs, compared to a single captive animal (Davidson et al., 1986b).

Carbohydrates

Like other obligate carnivores, felids do not have a dietary requirement for carbohydrate (National Research Council, 2006), and natural diets typically comprise only 1%–2% carbohydrate (Table 26.1; Plantinga et al., 2011; Bosch et al., 2015). The loss of enzymatic pathways involved in the metabolism of carbohydrates is considered another felid specialization linked with its carnivorous feeding habits. Although the majority of studies in this area have excluded non-domestic felids, it has been shown that cheetahs are deficient in circulating α-amylase concentrations (Langley and Carney, 1976), and lack the behavioral preference for, or genetic ability to taste sugars (Li et al., 2005). Lions have reduced maltase and sucrose activities in the jejunal mucosa compared to the domestic cat (Hore and Messert, 1968), and are assumed to have poor lactase activity (Wittmeyer Mills, 1990); similar assumptions could be extended to cheetahs.

Although diabetes mellitus has been reported (Sanchez et al., 2005) and is known to occur anecdotally in captive cheetahs, a pathology study of 87 captive cheetah kidneys failed to reveal any diabetic lesions, despite highly prevalent glomerulosclerosis resembling diabetic nephropathy (Bolton and Munson, 1999). Hyperglycemia from nondiabetic metabolic conditions were hypothesized, while diet was considered potentially etiological in terms of the glomerulosclerosis (Bolton and Munson, 1999).

Fiber, Fermentation, and the Microbiota

The anaerobic degradation of carbohydrates and protein complexes (from dietary fiber) by microbiota in the large intestine is known conventionally as fermentation, during which various metabolites are produced. The health-promoting impact of dietary modulation on the composition of the feline gut microbiota and its functional pathways has been illustrated in domestic cats (Minamoto et al., 2012), but as gut microbiomes of domestic cats and cheetahs differ to a larger extent than previously thought, generalizations across species is unfounded (Becker, 2015; Becker et al., 2014). The key microbial groups residing in the captive cheetah's gut represent a typical carnivore-like microbiota with an enrichment of lineages from the phylum Firmicutes, especially *Clostridium*, *Blautia*, *Coprococcus*, and *Enterococcus*, and a dearth of Bacteroidetes and *Bifidobacteriaceae* (Becker et al., 2014; Ley et al., 2008). The most abundant carbohydrate- and protein-degrading key players belong to *Clostridium* clusters I, XI, and XIVa (Becker et al., 2014, 2015). The composition of the predominant fecal microbial communities remained relatively stable over 3 years, and compromised host health (e.g., idiopathic diarrhea and chronic renal failure) was considered likely to have reflected perturbations of phylogenetic core groups (Becker et al., 2015). Additionally, fecal samples from free-ranging cheetahs revealed the same set of key microbial groups, though differing in relative proportion between the two conspecific populations (Becker, 2015). One pronounced difference in this regard concerned the lower abundance of *Clostridium* cluster XI members in free-ranging cheetahs compared to captive cheetahs (Becker, 2015). Potentially, this can be attributed to different protein content and quality of respective diets as *Clostridium* cluster XI has been positively correlated with dietary protein content (Schwab et al., 2011). Although diet appears to be a key driver in the host–microbiota homeostasis, additional studies are needed to elucidate the multifactorial interactions with host genetic and environmental influences.

Studies investigating fecal consistency and fermentation metabolites in cheetahs have determined that the type of diet significantly affects the source, amount, and ratio of fermentation metabolites in cheetahs (Depauw et al., 2013; Kerr et al., 2013b; Vester et al., 2010), and have therefore confirmed a role for dietary fiber in the intestinal function of cheetahs. *In vitro* fermentation of various animal fibers with cheetah fecal inoculum revealed substrate-dependent differences in fermentation profiles and kinetics (Depauw et al., 2012a). Rabbit skin, hair, and bone appear to be poorly fermentable substrates, eliciting similar fermentation profiles and metabolite production to cellulose (Depauw et al., 2012a). In a separate study, cheetahs consuming whole rabbits exhibited preferable fecal fermentation profiles and improved fecal consistency, compared to when fed muscle meat alone (Depauw et al., 2013). Feeding trials with plant fibers added to muscle-meat based diets have produced similar results; fecal consistency, output and DM content were improved in cheetahs when cellulose was used as the fiber source (compared to a moderately fermentable fiber source, beet pulp) (Kerr et al., 2013b). However, moderately fermentable fiber sources were suggested as optimal to elicit desired effects on nutrient digestibility in domestic cats (Rochus et al., 2014) and cheetahs (Kerr et al., 2013b), as well as benefiting colonic morphology in companion animals (Rochus et al., 2014). It is likely that a blend of poorly and moderately fermentable fibers is optimal for gut health, but will be highly dependent on inclusion level, species (Kerr et al., 2013b), and even body condition (Rochus et al., 2014). Pending further confirmation in cheetahs, these data highlight the importance of fiber fermentation for the general well-being of this species.

Vitamins

Fat-soluble vitamins A, D, and E are stored primarily in the liver, kidney, and adipose tissues; hence consumption of internal organs provides a major dietary source of these nutrients for carnivores in nature. Various studies with cheetahs fed whole prey, vitamin/mineral-supplemented mixed carcass meats, and/or commercial raw meat blends have evaluated plasma levels of vitamins A, D, and/or E (Bechert et al., 2002; Beckmann et al., 2013; Dierenfeld, 1993, 1994; Kaiser et al., 2014). Compared with normal ranges for domestic felids, whole prey is typically shown to provide adequate vitamin A and E nutrition (Kaiser et al., 2014) without supplementation, although excessive vitamin A intake may occur when consuming whole prey containing vitamin A–rich organs, such as the liver, on a daily basis (Depauw et al., 2012b). However, commercial meat-based diet blends containing higher vitamin levels than supplemented meat (3–4 times felid requirements) resulted in higher plasma concentrations of nutrient metabolites without identified negative health effects (Bechert et al., 2002). Unsupplemented or lean meats are likely to be deficient in vitamins A and E, but the requirement for vitamin E is also dependent on dietary PUFA content (Raederstorff et al., 2015).

As a fat-soluble vitamin, retinol (i.e., preformed vitamin A) is readily stored in the body and is therefore not required on a daily basis. Excessive amounts of vitamin A can accumulate, causing toxicity, and have been associated with liver and kidney disease, as well as possibly reproductive failure, in captive cheetahs (Bechert et al. 2002). Furthermore, hepatic fibrosis (venoocclusive disease, VOD; Chapter 25), seen in captive cheetahs, has been linked to vitamin A over-supplementation, but a definitive causative effect has not been confirmed (Allen et al., 1996; Bechert et al., 2002; Gosselin et al., 1998).

Physiological concentrations of circulating retinyl esters bound to lipoproteins are high in carnivores (Table 26.2) and would indicate signs of severe hypervitaminosis A in other mammals (Schweigert et al., 1990). The unique urinary secretion of vitamin A by carnivores may explain their reduced susceptibility to excess vitamin A in the context of a predictably (or sporadically) high intake from a wild diet (Crissey et al., 2003), and their tolerance may

have evolved in response to variable dietary sources (Bechert et al., 2002).

In contrast, hypovitaminosis A has caused ataxia and hind limb paresis, as well as abnormal skull morphology in lions and cheetahs (Bartsch et al., 1975; Kaiser et al., 2014; O'Sullivan et al., 1977). The current minimum hematologic reference for vitamin A (retinol) in felids has been suggested at 1.05 µM/L (Rucker et al., 2008), however reports of cheetahs exhibiting no apparent clinical abnormalities despite having vitamin A levels well below this mark suggest that minimum requirements based on domestic felids may overestimate cheetah requirements.

Cheetahs, like other felids (How et al., 1994), are assumed to be unable to synthesize adequate vitamin D_3 (cholecalciferol) simply through sunlight exposure. However, cheetahs and lions appear to maintain high serum levels of both metabolites of vitamin D [25(OH)D and 1,25(OH)$_2$D], even despite low levels of dietary intake (Crissey et al., 2003), and further studies with alternative diets are required.

Vitamin E deficiencies have been reported in non domestic felids, especially those fed high-fat diets (Ashton and Jones, 1979). Given the antioxidant function of this vitamin, the consumption of partially oxidized animal muscle and fat tissue (i.e., after prolonged storage) may increase the vitamin E requirement of captive cheetahs. Circulating vitamin E concentrations in cheetahs fed fresh and frozen whole prey indicate that this dietary strategy was more than sufficient for the cheetah (Beckmann et al., 2013; Williams, Schmidt-Küntzel, Marker, unpublished. Select serum nutrient, biochemistry, and hematologic values from captive and free-ranging cheetah (*Acinonyx jubatus*) in southern Africa). Cheetahs consuming commercial meat blends with dietary vitamin E at concentrations lower than recommendations for felids were shown to have circulating plasma α-tocopherol concentrations within the normal reference range (Dierenfeld, 1993). Homeostatic mechanisms cannot be ignored and research including different meat types is required to determine optimal dietary (and circulating) vitamin E concentrations in cheetahs. Dietary concentrations of the fat-soluble vitamins should be evaluated as a group, given their synergistic interactions (Bechert et al., 2002).

Thiamine deficiency has previously been reported as the most likely cause of illness and death in a group of cheetahs (Christie, 1998), although the actual dietary requirement is unknown. Lethargy, ventroflexion of the neck, and hind limb ataxia are typical symptoms of the acute stage of thiamine deficiency in the cheetah, but can be successfully treated with parenteral thiamine (Christie, 1998).

Minerals

Essential minerals, such as calcium (Ca) must be obtained in the wild through consumption of skeletal elements (Table 26.1), thus whole prey and/or diets based on whole prey composition provide the basis for nutritional balance. Insufficient Ca, or inappropriate ratios of Ca and phosphorus (P), can lead to developmental bone malformation, and metabolic bone disease in cheetahs fed inappropriately supplemented meat (Allan et al., 2008; Bell et al., 2010). Conversely, excessive supplementation of Ca is also known to contribute toward enlarged joints, splayed feet, angular limb deformities, and stunted growth in growing animals (Bell et al., 2010). However, whole prey and supplemented diets containing up to 10-fold higher Ca than felid requirements (but with appropriate Ca:P ratios in most cases) resulted in normal circulating Ca concentrations in adult cheetahs (Bechert et al., 2002; Depauw et al., 2012b). Captive cheetahs held in their natural habitat and fed a naturalistic diet had higher ionized Ca serum concentrations than wild cheetahs (Williams et al., unpublished).

Copper (Cu) deficiency has been reported in cheetahs, and confirmed through low plasma and hepatic Cu concentrations, as well as spinal cord demyelination, or via response to treatment (Downes, 1998; Kaiser et al., 2014). Captive cheetahs fed a ruminant-based meat diet demonstrated significantly elevated Cu values compared to those fed a poultry-based diet (Beckmann et al., 2013), the former of which resulted in serum values more similar to free ranging cheetahs (Williams et al., unpublished). Likewise, Cu status of captive cheetahs held in their natural habitat and fed ruminant and horse meat (Dierenfeld, 1993; Williams et al., unpublished) was within the range for domestic felids (Rucker et al., 2008), albeit at the upper end.

Poultry products are notoriously poor sources of Cu (Depauw et al., 2012b; Dierenfeld, 1993; Kaiser et al., 2014) and yet are common staples in captive cheetah diets. Although previous reports suggest that guinea fowl carcasses (in wild cheetah diets) contain higher levels of dietary Cu than domestic chickens (Downes, 1998), current research appears to refute this, whereby farmed guinea fowl were shown to be particularly poor sources of dietary Cu (Tlhong, 2008) with prey diet known to influence body composition (Clum et al., 1996).

Nutrient Digestibility

Similar to other carnivores, cheetahs efficiently digest high protein and fat diets (Clauss et al., 2010), with published digestion coefficients for various commercial, meat, and whole prey diets ranging from 76% to 89% (DM), 89% to 93% (energy), 90% to 96% (protein), and 89% to 96% (fat) (Bechert et al., 2002; Kerr et al., 2012, 2013b; Vester et al., 2008, 2010). Although species-specific differences in digestive and/or fermentative end-products and microbial populations have been reported when cats and cheetahs were fed the same diets (Vester et al., 2008, 2010), general patterns associated with digestibility, gut health or function, and fecal characteristics in domestic felids (Rochus et al., 2014) are also seen in cheetahs.

NUTRITION THROUGH THE LIFESTAGES

Nutrition of Cheetahs During Growth

Nutrition of mother and cubs are intimately linked. Maternal nutrition and body stores provide the nutrients required for fetal growth and development during gestation (90–95 days), and is equally important during lactation for the nursing cubs (0–4-months-old) (Caro, 1994; Oftedal, 1984). Wild cubs also depend on the mother to supply prey during the weaning process (starting at approximately 5-weeks-old), and post weaning, until they become independent hunters (12–18-months-old) (Caro, 1994). Key nutritional factors for formulating growth diets in kittens include absolute concentrations of Ca, P, vitamins A and D_3, and taurine, as well as, the Ca:P ratio (National Research Council, 2006). Maternal cheetah milk is similar in macronutrient profile to that of the domestic cat, although some differences have been reported in whey-to-casein ratio (being higher in cheetahs), and fatty acid profile (Osthoff et al., 2006).

Nutrition of Cheetahs During Pregnancy and Lactation

In late pregnancy and lactation, energy expenditure can increase two- to fivefold over maintenance in wild cheetahs, due to gestational cub growth, milk production during lactation, increased time spent obtaining prey and water, and investments in care and defense of cubs (Laurenson, 1995a). Based on data from domestic cats, maternal energy requirements peak in midlactation, and are 2–3 times the maintenance energy needs (National Research Council, 2006). In the wild, maternal food intake >1.5 kg/days was necessary to maintain adequate milk production for cub growth, whereby decreased cub growth occurred at <1.5 kg/days, and no added benefit was seen in instances of higher intake (up to 5 kg/days) (Laurenson, 1995b). To sustain higher intake levels, lactating cheetahs alter their hunting behavior, including increasing the hunting attempts targeting large prey, and increasing the proportion of large prey killed (Cooper et al., 2007; Laurenson, 1995b). Lactating female cheetahs eat 65%–97% more than nonreproducing females (Laurenson, 1995b). Thus, increased dietary quantities and/or nutrient concentrations may be needed to support lactation and cub growth in captive cheetahs.

Nutrition of Mature Cheetahs in Managed Populations

No specific data are available on the nutrition of mature cheetahs (e.g., >~10 years of age). However, based on data from domestic carnivores (National Research Council, 2006), maintenance energy requirements are expected to decrease by approximately 25%. The aging process is influenced by an individual's genetics, nutrition, environment, and disease state. As a result, meeting nutritional requirements and recommendations in practice must be highly individualized (i.e., targeted dietary management). In this regard, body condition scoring alongside dietary evaluation and comprehensive record keeping should be considered fundamental aspects of captive cheetah management (Chapter 24).

NUTRITIONAL DISORDERS AND DIET-RELATED DISEASES

While many nutritional disorders affecting captive carnivores are considered to be of only historical significance (Slusher et al., 1965) due to improved nutritional management, a number of nutritionally-relevant disorders are still of concern in captive cheetahs. Specific nutrient deficiencies have been discussed previously and consequences of inappropriate provision range from abnormal bone formation to gastrointestinal disease and behavioral disorders, and are summarized in Table 26.3.

TABLE 26.3 Nutrition and Diet-Related Health Concerns Reported in Captive Cheetahs

Disease/disorder	Nutritional involvement	Health consequences
Metabolic bone disease[a,b,c,d]	Imbalance of dietary calcium and phosphorus. Typically involves other interacting factors, such as vitamin D_3 intake, and growth rate	Ca deficiency: developmental bone malformation (osteodystrophy), spontaneous fractures, bowed legs, and compromised mobility
		Ca excess: developmental deformity of the forelegs (osteochondrosis dissecans), enlarged joints, splayed feet, angular limb deformities, and stunted growth in slow growing carnivores
Oral health (general)[e,f,g,h,i,j]	Natural diets require significantly greater chewing action than soft diets. However, the size of chewing bones is important	Plaque build-up, poor development or maintenance of masticatory apparatus of the skull
	A link between diet texture with altered jaw and skull morphology has been suggested in captive felids	Altered jaw and skull morphology
	The unnatural scenario of a cheetah spending a significant proportion of its time gnawing large skeletal components in the absence of a meal may result in abnormal behaviors	Stimulation of appetitive responses, including hypersalivation, gastric acid release and unrewarded feeding anticipation (hypothesized)
Focal palatine erosion (FPE)[j,k,l]	Prior to 2004 the occurrence of FPE in cheetahs was considered to be an artifact of captivity and most likely related to a soft captive diet	Erosion of the upper palate, resulting in infection and bone loss
	However, evidence of FP and other dental anomalies in free-ranging cheetahs has subsequently questioned this notion and the disease is now considered multifactorial	
Gastrointestinal disease[m,n,o,p,q,r,s,t,u]	Animal fiber	Improved fecal consistency, fermentation metabolite profile, and reduced biomarkers of intestinal inflammation
	Diet type and ingredients	Risk of gastrointestinal disease and gastritis was reduced when muscle meat and carcass components were fed. Risk of disease increased when horsemeat was fed
	Supplementation with feathers of a commercial diet (captive foxes)	Increased gut microbiota diversity was hypothesized to have potential beneficial effects for host immunity and/or physiological performance
	Dietary fiber (domestic carnivores)	Wide range of benefits for gut function, morphology and health
	Plant-based fiber sources (beet pulp, cellulose) have been tolerated well in cheetahs	Impacts on fecal consistency, nutrient digestibility and fermentation metabolite profiles
Hepatic disease[v,w,x]	Excessive vitamin A due to over-supplementation	Veno-occlusive disease (hepatic fibrosis), although further research is necessary

(Continued)

4. CAPTIVE CHEETAHS

TABLE 26.3   Nutrition and Diet-Related Health Concerns Reported in Captive Cheetahs (*cont.*)

Disease/disorder	Nutritional involvement	Health consequences
Renal disease[y,z]	High protein and/or amino acid imbalance	Glomerulosclerosis, diabetes mellitus and other renal disease
Behavioral abnormalities[e,aa,bb,cc]	Reduced time spent foraging, and lack of ability to express behaviors associated with prey capture and manipulation (compared to free-ranging cheetahs)	Feeding-related behavioral abnormalities represent a significant concern for captive cheetah welfare
		Boredom and stereotypic behaviors
	Naturalistic diet or feeding regime	Positive effects on cheetah psychological well-being
Reproductive failure[w,dd,ee,ff,gg]	Potential roles have been hypothesized for vitamin A and phytoestrogens, although the latter appears unlikely given findings in domestic cats	Altered estrous cycles and teratogenic effects
	Reproductive failure may occur secondary to other systemic diseases with dietary involvement (e.g., gastrointestinal disease, renal, or hepatic disease)	Morbidity arising from other diseases is likely to reduce reproductive capacity (hypothesized)

[a]Allan et al., 2008; [b]Bell et al., 2010; [c]Hedhammer et al. (1974); [d]Hazewinkel et al. (1991); [e]Bond and Lindburg (1990); [f]Vosburgh et al. (1982); [g]Fagan (1980); [h]Kapoor et al. (2016); [i]Hartstone-Rose et al. (2014); [j]Phillips (1993); [k]Fitch and Fagan (1982); [l]Marker and Dickman (2004); [m]Depauw et al. (2014); [n]Depauw et al. (2013); [o]Depauw et al. (2012b); [p]Depauw et al. (2012a); [q]Whitehouse-Tedd et al. (2015); [r]Zhang et al. (2014); [s]De Godoy et al. (2013); [t]Fahey et al. (2004); [u]Kerr et al. (2013c); [v]Allen et al. (1996); [w]Bechert et al. (2002); [x]Gosselin et al. (1998); [y]Bolton and Munson (1999); [z]Sanchez et al. (2005); [aa]Bashaw et al. (2003); [bb]Carlstead (1996); [cc]Quirke and O'Riordan (2011); [dd]Setchell et al. (1987a); [ee]Setchell et al. (1987b); [ff]Whitehouse-Tedd et al. (2013); [gg]Bell et al. (2007).

CURRENT CHALLENGES/GAPS IN OUR KNOWLEDGE

The nutrient requirements, especially their nutritional idiosyncrasies, have been well defined for domestic cats (Morris, 2002; National Research Council, 2006), while species-specific requirements for captive exotic felids have not been adequately explored. The nutritional requirements of carnivores are likely to align to the nutrient composition of their naturally available prey (Kohl et al., 2015; Raubenheimer et al., 2009). However, detailed analytical data pertaining to the nutritional composition of wild prey for cheetahs are surprisingly lacking in the published literature. The use of a "wild-type" diet as a proxy for evolved nutritional requirements in cheetahs, and other predator species, holds great potential. Gaining a better understanding of the nutrient intake of free-ranging cheetahs should be considered a research priority, with direct and immediate benefits to captive diet evaluation, formulation and improvement. Moreover, elucidating nutrient intake according to carcass component, temporal and spatial variation in prey, carcass component utilization, or targeting by free-ranging cheetahs is likely to challenge the use of the relatively monotonous dietary provision offered in many captive situations.

Fundamental aspects of captive cheetah management in need of further scientific attention include identifying the role of diet, specific nutrients (and their interactions), nonnutritive dietary components, dietary format and feeding schedules in captive cheetahs' GIT, behavioral and overall health. Although findings in other species may prove to be applicable in cheetahs, the unusually high incidence of GIT disorders and poor reproductive output of this species compared to other large captive carnivores warrants a species-specific focus in current and future nutritional research.

Of particular current interest is the role of dietary fiber. Decreased putrefactive compounds arising from intestinal fermentation have been produced in captive cheetahs when fed a diet containing either plant or animal fibers. These studies indicate that whole prey diets, or meat-diets supplemented with an appropriate fiber source are likely to exert important, positive, metabolic outcomes relevant for cheetah health (Depauw et al., 2012b, 2013; Kerr et al., 2013b; Vester et al., 2008). However, while research to date appears promising, we are a long way from defining the mechanisms by which dietary components influence the gastrointestinal environment or metabolism in order to beneficially impact host health. Advances in our understanding of the cheetah's gastrointestinal microbiome (Becker et al., 2014, 2015) have laid the necessary foundation for these types of future investigations. Identifying and understanding the dietary factors involved will require the testing of a variety of hypotheses postulated in recent publications, and a holistic approach appears essential to the assessment of health and welfare in this species.

References

Allan, G., Portas, T., Bryant, B., Howlett, R., Blyde, D., 2008. Ulnar metaphyseal osteochondrosis in seven captive bred cheetahs (*Acinonyx jubatus*). Vet. Radiol. Ultrasound 49 (6), 551–556.

Allen, M.E., Oftedal, O., Baer, D.J., 1996. The feeding and nutrition of carnivores. In: Kleiman, D.G. et al., (Ed.), Wild Mammals in Captivity: Principles and Techniques. The University of Chicago Press, Chicago, pp. 139–147.

Altman, J.D., Gross, K.L., Lowry, S.R., 2005. Nutritional and behavioral effects of gorge and fast feeding in captive lions. J. Appl. Anim. Welf. Sci. 8 (1), 47–57.

Anderson, J.H., Brown, R.E., 1979. Serum thyroxine (T4) and triiodothyronine (T3) uptake values in normal adult cats as determined by radioimmunoassay. Am. J. Vet. Res. 40, 1493–1494.

Ashton, D.G., Jones, D.M., 1979. Veterinary aspects of the management of wild cats. In: Barzdo, J. (Ed.), Proceedings of the 4th Symposium on Management of Wild Cats in Captivity.

Backues, K.A., Hoover, J.P., Bauer, J.E., Campbell, G.A., Barrie, M.T., 1997. Hyperlipidemia in four related male cheetahs (*Acinonyx jubatus*). J. Zoo Wildl. Med. 28 (4), 476–480.

Bartsch, R.C., Imes, G.D., Smit, J.P.J., 1975. Vitamin A deficiency in the captive African lion cub (*Panthera leo*; Linnaeus, 1978). Onderstepoort J. Vet. Res. 42 (2), 43–54.

Bashaw, M.J., Bloomsmith, M.A., Marr, M.J., Maple, T.L., 2003. To hunt or not to hunt? A feeding enrichment experiment with captive large felids. Zoo Biol. 22, 189–198.

Bauer, J.E., 1997. Fatty acid metabolism in domestic cats (*Felis catus*) and cheetahs (*Acinonyx jubatus*). Proc. Nutr. Soc. 56, 1013–1024.

Bauer, J.E., Backues, K., Dunbar, B.L., Hoover, J.P., Barrie, M.T., Citino, S.B., Wallace, R., 1996. Serum lipid fatty acids in captive cheetahs: evidence of N-6 and N-3 fatty acid chain elongation and omega-5 desaturation. In: Proceedings of the 14th ACVIM Forum. San Antonio, Texas.

Bechert, U., Mortenson, J., Dierenfeld, E.S., Cheeke, P., Keller, M., Holick, M., Chen, T.C., Rogers, Q., 2002. Diet composition and blood values of captive cheetahs (*Acinonyx jubatus*) fed either supplemented meat or commercial food preparations. J. Zoo Wildl. Med. 33 (1), 16–28.

Becker, A.A.M.J., 2015. Diversity and dynamics of gut microbiota in captive cheetahs (*Acinonyx jubatus*): a baseline for dietary interventions in a strict carnivore with vulnerable status. PhD thesis, University of Ghent, Belgium.

Becker, A.A.M.J., Hesta, M., Hollants, J., Janssens, G.P.J., Huys, G., 2014. Phylogenetic analysis of faecal microbiota from captive cheetahs reveals underrepresentation of *Bacteroidetes* and Bifidobacteriaceae. BMC Microbiol. 14 (1), 43.

Becker, A.A.M.J., Janssens, G.P.J., Snauwaert, C., Hesta, M., Huys, G., 2015. Integrated community profiling indicates long-term temporal stability of the predominant faecal microbiota in captive cheetahs. PLoS One 10 (4), e0123933.

Beckmann, K.M., O'Donovan, D., McKeown, S., Wernery, U., Basu, P., Bailey, T., 2013. Blood vitamins and trace elements in northern-east African cheetahs (*Acinonyx jubatus soemmeringii*) in captivity in the Middle East. J. Zoo Wildl. Med. 44 (3), 613–626.

Bell, K., Ugarte, C.E., Tucker, L.A., Thomas, D.G., 2007. Genistein and daidzein do not affect puberty onset or oestrus cycle parameters in the domestic cat (*Felis catus*). Asia Pac. J. Clin. Nutr. 16 (Suppl. 3), S72.

Bell, K., van Zyl, M., Ugarte, C.E., Hartman, A., 2010. Bilateral carpal valgus deformity in hand-reared cheetah cubs (*Acinonyx jubatus*). Zoo Biol. 30 (2), 199–204.

Bolton, L., Munson, L., 1999. Glomerulosclerosis in captive cheetahs (Acinonyx jubatus). Vet. Pathol. 36 (1), 14–22.

Bond, J.C., Lindburg, D.G., 1990. Carcass feeding of captive cheetahs (Acinonyx jubatus): the effects of a naturalistic feeding program on oral health and psychological well-being. Appl. Anim. Behav. Sci. 26, 373–382.

Bosch, G., Hagen-Plantinga, E., Hendriks, W.H., 2015. Dietary nutrient profiles of wild wolves: insights for optimal dog nutrition? Br. J. Nutr. 113 (S1), S40–S54.

Calder, P.C., 2001. Polyunsaturated fatty acids, inflammation, and immunity. Lipids 36, 1007–1024.

Carlstead, K., 1996. Effects of captivity on the behavior of wild mammals. In: Kleiman, D.G. (Ed.), Wild Mammals in Captivity: Principles and Techniques. University of Chicago Press, Chicago, pp. 317–333.

Caro, T.M., 1994. Cheetahs of the Serengeti Plains: Group Living in an Asocial Species. The University of Chicago Press, Chicago.

Caro, T., Holt, M., FitzGibbon, C., Bush, M., Hawkey, C., Kock, R., 1987. Health of adult free-living cheetahs. J. Zool. 212, 573–584.

Christie, P., 1998. Thiamine deficiency in cheetah. International Cheetah Studbook 1997/1998. Cheetah Conservation Fund, Otjiwarongo, Namibia, pp. S1–S7.

Clauss, M., Kleffner, H., Kienzle, E., 2010. Carnivorous mammals: nutrient digestibility and energy evaluation. Zoo Biol. 29 (6), 687–704.

Clum, N.J., Fitzpatrick, M.P., Dierenfeld, E.S., 1996. Effects of diet on nutritional content of whole vertebrate prey. Zoo Biol. 15 (5), 525–537.

Cooper, A.B., Pettorelli, N., Durant, S.M., 2007. Large carnivore menus: factors affecting hunting decisions by cheetahs in the Serengeti. Anim. Behav. 73 (4), 651–659.

Cordain, L., Watkins, B., Florant, G.L., Kelher, M., Rogers, L., Li, Y., 2002. Fatty acid analysis of wild ruminant tissues: evolutionary implications for reducing diet-related chronic disease. Eur. J. Clin. Nutr. 56 (3), 181–191.

Crissey, S.D., Ange, K.D., Jacobsen, K.L., Slifka, K., Bowen, P.E., Stacewicz-Sapuntzakis, M., Langman, C.B., Sadler, W., Kahn, S., Ward, A., 2003. Serum concentrations of lipids, vitamin D metabolites, retinol, retinyl esters, tocopherols and selected carotenoids in twelve captive wild felid species at four zoos. J. Nutr. 133 (1), 160–166.

Davidson, B.C., Cantrill, R.C., Varaday, D., 1986a. The reversal of essential fatty acid deficiency symptoms in the cheetah. S. Afr. J. Zool. 21 (2), 161–164.

Davidson, B.C., Morsbach, D., Cantrill, R.C., 1986b. The fatty acid composition of the liver and brain of Southern African cheetahs. Prog. Lipid Res. 25, 97–99.

De Godoy, M., Kerr, K., Fahey, G., 2013. Alternative dietary fiber sources in companion animal nutrition. Nutrients 5, 3099–3117.

Depauw, S., Bosch, G., Hesta, M., Whitehouse-Tedd, K., Hendriks, W.H., Kaandorp, J., Janssens, G.P.J., 2012a. Fermentation of animal components in strict carnivores: a comparative study with cheetah fecal inoculum. J. Anim. Sci. 90 (8), 2540–2548.

Depauw, S., Heilmann, R.M., Whitehouse-Tedd, K., Hesta, M., Steiner, J.M., Suchodolski, J.S., Janssens, G.P.J., 2014. Effect of diet type on serum and faecal concentration of S100/calgranulins in the captive cheetah. J. Zoo Aquar. Res. 2 (2), 33–38.

Depauw, S., Hesta, M., Whitehouse-Tedd, K., Stagegaard, J., Buyse, J., Janssens, G.P.J., 2012b. Blood values of adult captive cheetahs (Acinonyx jubatus) fed either supplemented beef or whole rabbit carcasses. Zoo Biol. 31 (6), 629–641.

Depauw, S., Hesta, M., Whitehouse-Tedd, K., Vanhaecke, L., Verbrugghe, A., Janssens, G.P.J., 2013. Animal fibre: the forgotten nutrient in strict carnivores? First insights in the cheetah. J. Anim. Physiol. Anim. Nutr. 97 (1), 146–154.

Dierenfeld, E., 1993. Nutrition of captive cheetahs: food composition and blood parameters. Zoo Biol. 12 (1), 143–152.

Dierenfeld, E.S., 1994. Vitamin E in exotics: effects, evaluation and ecology. J. Nutr. 124, 2579S–2581S.

Dierenfeld, E.S., Alcorn, H.L., Jacobsen, K.L., 2002. Nutrient Composition of Whole Vertebrate Prey (Excluding Fish) Fed in Zoos. National Agricultural Library, Beltsville, Maryland, (p. 20).

Dobrynin, P., Liu, S., Tamazian, G., Xiong, Z., Yurchenko, A.A., Krasheninnikova, K., Kliver, S., Schmidt-Küntzel, A., Koepfli, K.-P., Johnson, W., Kuderna, L.F.K., García-Pérez, R., Manuel, M. de, Godinez, R., Komissarov, A., Makunin, A., Brukhin, V., Qiu, W., Zhou, L., et al., 2015. Genomic legacy of the African cheetah, Acinonyx jubatus. Genome Biol. 16 (1), 277.

Downes, S.J., 1998. A case report of suspected copper deficiency in cheetah (Acinonyx jubatus). International Cheetah Studbook 1997/1998. Cheetah Conservation Fund, Otjiwarongo, Namibia.

Eisert, R., 2011. Hypercarnivory and the brain: protein requirements of cats reconsidered. J. Comp. Physiol. B 181, 1–17.

Fagan, D.A., 1980. Diet consistency and periodontal disease in exotic carnivores. In: Proceedings of the 1980 Conference of the American Association of Zoo Veterinarians, pp. 34–37.

Fahey, G.C., Flickinger, E.A., Grieshop, C.M., Swanson, K.S., 2004. The role of dietary fiber in companion animal diets. In: van der Kamp, J.W. (Ed.), Dietary Fibre: Bioactive Carbohydrates for Food and Feed. Wageningen Academic Publishers, Wageningen, The Netherlands, pp. 295–328.

Fitch, H.M., Fagan, D.A., 1982. Focal palatine erosion associated with dental malocclusion in captive cheetahs. Zoo Biol. 1, 295–310.

Fitzgerald, B.M., Karl, B.J., 1979. Foods of feral house cats (*Felis catus* L.) in forest of the Orongorongo Valley, Wellington. N. Z. J. Zool. 6 (1), 107–126.

Gosselin, S.J., Setchell, K.D.R., Harrington, G.W., Welsh, M.B., Pylypiw, H., Kozeniauskas, R., Dollard, D., Tarr, M.J., Dresser, B.L., 1998. Nutritional considerations in the pathogenesis of hepatic veno-occlusive disease in captive cheetahs. Zoo Biol. 8, 339–347.

Hartstone-Rose, A., Selvey, H., Villari, J.R., Atwell, M., Schmidt, T., 2014. The three-dimensional morphological effects of captivity. PLoS ONE 9 (11), e113437.

Hayward, M.W., Hofmeyr, M., O'Brien, J., Kerley, G.I.H., 2006. Prey preferences of the cheetah (*Acinonyx jubatus*) (Felidae: Carnivora): morphological limitations or the need to capture rapidly consumable prey before klepto-parasites arrive? J. Zool. 270 (4), 615–627.

Hazewinkel, H.A.W., van den Brom, W.E., van T'Klooster, A.T., Voorhout, G., van Wees, A., 1991. Growth and skeletal development calcium metabolism in Great Dane dogs fed diets with various calcium and phosphorus levels. J. Nutr. 121, S99–S106.

Hedberg, G.E., Dierenfeld, E.S., Rogers, Q.R., 2007. Taurine and zoo felids: considerations of dietary and biological tissue concentrations. Zoo Biol. 26 (6), 517–531.

Hedhammer, A.F., Wu, F., Krook, L., Schryver, H.F., DeLahunta, A., Whalen, J.P., Kallfelz, F.A., Nunez, E.A., Hintz, H.F., Sheffy, B.E., Ryan, G.D., 1974. Overnutrition and skeletal disease: an experimental study in growing Great Dane dogs. Cornell Vet. 64 (Suppl. 5), S1–S160.

Hewson-Hughes, A.K., Hewson-Hughes, V.L., Colyer, A., Miller, A.T., McGrane, S.J., Hall, S.R., Butterwick, R.F., Simpson, S.J., Raubenheimer, D., 2013. Geometric analysis of macronutrient selection in breeds of the domestic dog, *Canis lupus familiaris*. Behav. Ecol. 24 (1), 293–304.

Hewson-Hughes, A.K., Hewson-Hughes, V.L., Miller, A.T., Hall, S.R., Simpson, S.J., Raubenheimer, D., 2011. Geometric analysis of macronutrient selection in the adult domestic cat, *Felis catus*. J. Exp. Biol. 214, 1039–1051.

Hoffman, L.C., Wiklund, E., 2006. Game and venison—meat for the modern consumer. Meat Sci. 74 (1), 197–208.

Hore, P., Messert, M., 1968. Studies on disaccharidase activities of the small intestine of the domestic cat and other carnivorous mammals. Comp. Biochem. Physiol. A 24, 717–725.

House, W.A., Bell, A., 1993. Mineral accretion in the fetus and adnexa during late gestation in Holstein cows. J. Dairy Sci. 76, 2999–3010.

How, K.L., Hazewinkel, H.A.W., Mol, J.A., 1994. Dietary vitamin D dependence of cat and dog due to inadequate cutaneous synthesis of vitamin D. Gen. Comp. Endocrinol. 96 (1), 12–18.

Hudson-Lamb, G.C., Schoeman, J.P., Hooijberg, E.H., Heinrich, S.K., Tordiffe, A.S.W., 2016. Reference intervals for selected serum biochemistry analytes in cheetahs (*Acinonyx jubatus*). J. S. Afr. Vet. Assoc. 87 (1), e-1–e-6.

Kaiser, C., Wernery, U., Kinne, J., Marker, L., Liesegang, A., 2014. The role of copper and vitamin A deficiencies leading to neurological signs in captive cheetahs (*Acinonyx jubatus*) and lions (*Panthera leo*) in the United Arab Emirates. Food Nutr. Sci. 5, 1978–1990.

Kapoor, V., Antonelli, T., Parkinson, J.A., Hartstone-Rose, A., 2016. Oral health correlates of captivity. Res. Vet. Sci. 107, 213–219.

Kerr, K., Beloshapka, N., Morris, C.L., Parsons, C.M., Burke, S.L., Utterback, P.L., Swanson, K.S., 2013a. Evaluation of four raw meat diets using domestic cats, captive exotic felids, and cecectomized roosters. J. Anim. Sci. 91, 225–237.

Kerr, K., Kappen, K.L., Garner, L.M., Swanson, K.S., 2014. Commercially available avian and mammalian whole prey diet items targeted for consumption by managed exotic and domestic pet felines: macronutrient, mineral, and long-chain fatty acid composition. Zoo Biol. 33 (4), 327–335.

Kerr, K., Morris, C.L., Burke, S.L., Swanson, K.S., 2013b. Influence of dietary fiber type and amount on energy and nutrient digestibility, fecal characteristics, and fecal fermentative end-product concentrations in captive exotic felids fed a raw beef-based diet. J. Anim. Sci. 91, 2199–2210.

Kerr, K., Morris, C.L., Burke, S.L., Swanson, K.S., 2013c. Apparent total tract macronutrient and energy digestibility of 1- to- 3-day-old whole chicks, adult ground chicken, and extruded and canned chicken-based diets in African wildcats (*Felis silvestris lybica*). Zoo Biol. 32 (5), 510–517.

Kerr, K., Vester Boler, B.M., Morris, C.L., Liu, K.J., Swanson, K.S., 2012. Apparent total tract energy and macronutrient digestibility and fecal fermentative end-product concentrations of domestic cats fed extruded, raw beef-based, and cooked beef-based diets. J. Anim. Sci. 90 (2), 515–522.

Kohl, K.D., Coogan, S.C.P., Raubenheimer, D., 2015. Do wild carnivores forage for prey or for nutrients? BioEssays 37 (6), 701–709.

Kremen, N.A., Calvert, S.C.P., Larsen, J.A., Baldwin, R.A., Hahn, T.P., Fascetti, A.J., 2013. Body composition and amino acid concentrations of select birds and mammals consumed by cats in northern and central California. J. Anim. Sci. 91 (3), 1270–1276.

Langley, D.J., Carney, J.A., 1976. Comparative effects of antisera to human pancreatic alpha-amylase on serum amylases of several mammalian species. Comp. Biochem. Physiol. B 55, 563–565.

Laurenson, M.K., 1995a. Behavioural costs and constraints of lactation in free-living cheetahs. Anim. Behav. 50, 815–826.

Laurenson, M.K., 1995b. Cub growth and maternal care in cheetahs. Behav. Ecol. 6 (4), 405–409.

Ley, R.E., Hamady, M., Lozupone, C., Turnbaugh, P.J., Ramey, R.R., Bircher, J.S., Schlegel, M.L., Tucker, T., Schrenzel, M.D., Knight, R., Gordon, J.I., 2008. Evolution of mammals and their gut microbes. Science 320 (5883), 1647–1651.

Li, X., Li, W., Wang, H., Cao, J., Maehashi, K., Huang, L., Bachmanov, A., Reed, D.R., Legrand-Defretin, V., Beauchamp, G.K., Brand, J.G., 2005. Pseudogenization of a sweet-receptor gene accounts for cats' indifference toward sugar. PLoS Genet. 1 (1), 27–35.

Li, R.G., Wang, X.P., Wang, C.Y., Ma, M.W., Li, F.C., 2012. Growth performance, meat quality and fatty acid metabolism response of growing meat rabbits to dietary linoleic acid. Asian-Australasian J. Anim. Sci. 25 (8), 1169–1177.

Marker, L.L., Dickman, A.J., 2004. Dental anomalies and incidence of palatal erosion in Namibian cheetahs (Acinonyx jubatus jubatus). J. Mammal. 85 (1), 19–24.

Mills, M.G.L., Broomhall, L.S., du Toit, J.T., 2004. Cheetah Acinonyx jubatus feeding ecology in the Kruger National Park and a comparison across African savanna habitats: is the cheetah only a successful hunter on open grassland plains? Wildl. Biol. 10 (3), 177–186.

Minamoto, Y., Hooda, S., Swanson, K., Suchodolski, J., 2012. Feline gastrointestinal microbiota. Anim. Health Res. Rev. 13, 64–77.

Morris, J.G., 2002. Idiosyncratic nutrient requirements of cats appear to be diet-induced evolutionary adaptations. Nutr. Res. Rev. 15 (1), 153–168.

National Research Council, 2006. Nutrient Requirements of Dogs and Cats. National Academy of Sciences, Washington, DC, USA.

O'Sullivan, B.M., Mayo, F.D., Hartley, W.J., 1977. Neurologic lesions in young captive lions associated with vitamin A deficiency. Austr. Vet. J. 53, 187–189.

Ofri, R., Barishak, R.Y., Eshkar, G., Aizenberg, I., 1996. Feline central retinal degeneration in captive cheetahs (Acinonyx jubatus). J. Zoo Wildl. Med. 27 (1), 101–108.

Oftedal, O.T., 1984. Milk composition, milk yield and energy output at peak lactation: a comparative review. Symp. Zool. Soc. Lond. 51, 33–85.

Osthoff, G., Hugo, A., de Wit, M., 2006. The composition of cheetah (Acinonyx jubatus) milk. Comp. Biochem. Physiol. B 145, 265–269.

Phillips, J.A., 1993. Bone consumption by cheetahs at undisturbed kills: evidence for a lack of focal-palatine erosion. J. Mammal. 74 (2), 487–492.

Plantinga, E., Bosch, G., Hendriks, W.H., 2011. Estimation of the dietary nutrient profile of free-roaming feral cats: possible implications for nutrition of domestic cats. Br. J. Nutr. 106, S35–S48.

Plumb, D., 2002. Laboratory values. Veterinary Drug Handbook. Iowa State Press, Iowa, USA.

Puls, R., 1988. Mineral Levels in Animal Health. Diagnostic data. Sherpa International, Clearbrook, BC, Canada.

Quirke, T., O'Riordan, R.M., 2011. The effect of a randomised enrichment treatment schedule on the behaviour of cheetahs (Acinonyx jubatus). Appl. Anim. Behav. Sci. 135 (1–2), 103–109.

Raederstorff, D., Wyss, A., Calder, P.C., Weber, P., Eggersdorfer, M., 2015. Vitamin E function and requirements in relation to PUFA. Br. J. Nutr. 114 (8), 1113–1122.

Raubenheimer, D., Simpson, S.J., Mayntz, D., 2009. Nutrition, ecology and nutritional ecology: toward an integrated framework. Funct. Ecol. 23, 4–16.

Rochus, K., Janssens, G.P.J., Hesta, M., 2014. Dietary fibre and the importance of the gut microbiota in feline nutrition: a review. Nutr. Res. Rev. 27 (2), 295–307.

Rucker, R.B., Morris, J., Fascetti, A.J., 2008. Clinical biochemistry of domestic animals. In: Kaneko, J.J. et al., (Ed.), Vitamins. sixth ed. Academic Press, Burlington, Minnesota, pp. 695–730.

Sanchez, C., Bronson, E., Deem, S., Viner, T., Pereira, M., Saffoe, C., Murray, S., 2005. Diabetes mellitus in a cheetah: attempting to treat the untreatable? In: Proceedings of the AAZV, AAWV, AZA/NAG Joint Conference, pp. 101–103.

Scantlebury, D.M., Mills, M.G.L., Wilson, R.P., Wilson, J.W., Mills, M.E.J., Durant, S.M., Bennett, N.C., Bradford, P., Marks, N.J., Speakman, J.R., 2014. Flexible energetics of cheetah hunting strategies provide resistance against kleptoparasitism. Science 346 (6205), 79–82.

Schwab, C., Cristescu, B., Northrup, J., Stenhouse, G., Gänzle, M., 2011. Diet and environment shape fecal bacterial microbiota composition and enteric pathogen load of grizzly bears. PLoS One, 6e27905.

Schweigert, F.J., Ryder, O.A., Rambeck, W.A., Zucker, H., 1990. The majority of vitamin A is transported as retinyl esters in the blood of most carnivores. Comp. Biochem. Physiol. A 95 (4), 573–578.

Scott, P.P., 1968. The special features of nutrition of cats, with observations on wild Felidae nutrition in the London Zoo. Symp. Zool. Soc. Lond. 21, 21–36.

Setchell, K.D.R., Gosselin, S.J., Welsh, M.B., Johnston, J.O., Balistreri, W.F., Dresser, B.L., 1987a. Dietary factors in the development of liver disease and infertility in the captive cheetah. In: Edney, A.T.B. (Ed.), Nutrition, Malnutrition and Dietetics in the Dog and Cat: Proceedings of An International Symposium. British Veterinary Association in Collaboration with the WALTHAM Centre for Pet Nutrition, Hanover, Germany, pp. 97–100.

Setchell, K.D.R., Gosselin, S.J., Welsh, M.B., Johnston, J.O., Balistreri, W.F., Kramer, L.W., Dresser, B.L., Tarr, M.J., 1987b. Dietary estrogens—a probable cause of infertility and liver disease in captive cheetahs. J. Gastroenterol. 93, 225–233.

Slusher, R., Bistner, S.I., Kirchner, C., 1965. Nutritional secondary hyper-parathyroidism in a tiger. J. Am. Vet. Med. Assoc. 147, 1109–1115.

Szendrő, Z., Zotte, A.D., 2011. Effect of housing conditions on production and behaviour of growing meat rabbits: a review. Livest. Sci. 137 (1–3), 296–303.

Tlhong, T.M., 2008. Meat Quality of Raw and Processed Guinea Fowl (*Numeda meleagris*). MSc thesis, Stellenbosch University, South Africa.

Tordiffe, A.S.W., Wachter, B., Heinrich, S.K., Reyers, F., Mienie, .L.J., 2016. Comparative serum fatty acid profiles of captive and free-ranging cheetahs (*Acinonyx jubatus*) in Namibia. PLoS ONE 11, e0167608.

Van Zyl, L., Ferreira, A.V., 2004. Physical and chemical carcass composition of springbok (*Antidorcas marsupialis*), blesbok (*Damaliscus Dorcas phillipsi*) and impala (*Aepyceros melampus*). Small Rumin. Res. 53 (1), 103–109.

Vester, B.M., Burke, S.L., Dikeman, C.L., Simmons, L.G., Swanson, K.S., 2008. Nutrient digestibility and fecal characteristics are different among captive exotic felids fed a beef-based raw diet. Zoo Biol. 27 (2), 126–136.

Vester, B.M., Beloshapka, A.N., Middelbos, I.S., Burke, S.L., Dikeman, C.L., Simmons, L.G., Swanson, K.S., 2010. Evaluation of nutrient digestibility and fecal characteristics of exotic felids fed horse- or beef-based diets: use of the domestic cat as a model for exotic felids. Zoo Biol. 29 (4), 432–448.

Vosburgh, K.M., Barbiers, R.B., Sikarskie, J.G., Ullrey, D.E., 1982. A soft versus hard diet and oral health in captive timber wolves (*Canis lupus*). J. Zoo Wild Anim. Med. 13 (3), 104–107.

Wack, R.F., 2003. Felidae. In: Fowler, M.E., Miller, R.E. (Eds.), Zoo and Wild Animal Medicine. Current Therapy. fifth ed. W.B. Saunders Company, Philadelphia, USA, pp. 491–501.

Whitehouse-Tedd, K.M., Cave, N.J., Roe, W.D., Ugarte, C.E., Thomas, D.G., 2013. Preliminary investigation of the influence of long-term dietary isoflavone intake on reproductive tract histology and sex steroid receptor expression in female domestic cats (*Felis catus*). J. Appl. Anim. Nutr. 1, e8.

Whitehouse-Tedd, K.M., Lefebvre, S.L., Janssens, G.P.J., 2015. Dietary factors associated with faecal consistency and other indicators of gastrointestinal health in the captive cheetah (*Acinonyx jubatus*). PLoS One 10 (4), e0120903.

Wittmeyer Mills, A., 1990. A comparative study of the digestibility and economy of three feline diets when fed to lions and tigers in confinement. In: Montali, R.J., Migaki, G. (Eds.), The Comparative Pathology of Zoo Animals: The Symposia of the National Zoological Park. Smithsonian Institution Press, Washington DC, USA, pp. 87–91.

Wood, J.D., Enser, M., Fisher, A.V., Nute, G.R., Sheard, P.R., Richardson, R.I., Hughes, S.I., Whittington, F.M., 2008. Fat deposition, fatty acid composition and meat quality: a review. Meat Sci. 78, 343–358.

Zhang, L., Yang, S., Xu, Y., Dahmer, T.D., 2014. The influence of dietary feathers on the fecal microbiota in captive arctic fox: do dietary hair or feather play a role in the evolution of carnivorous mammals? Integr. Zool. 9, 583–589.

Reproductive Physiology of the Cheetah and Assisted Reproductive Techniques

Adrienne E. Crosier*, Bettina Wachter**,
Martin Schulman[†], Imke Lüders[‡], Diana C. Koester[‡‡],
Nadja Wielebnowski[§], Pierre Comizzoli[¶],
Laurie Marker[††]

*Smithsonian Conservation Biology Institute, Front Royal, VA, United States
**Leibniz Institute for Zoo and Wildlife Research, Berlin, Germany
[†]University of Pretoria, Pretoria, South Africa
[‡]GEOlife's Animal Fertility and Reproductive Research, Hamburg, Germany
[§]Oregon Zoo, Portland, OR, United States
[¶]Smithsonian Conservation Biology Institute, Washington, DC, United States
[††]Cheetah Conservation Fund, Otjiwarongo, Namibia
[‡‡]Cleveland Metroparks Zoo, Cleveland, OH, United States

INTRODUCTION

Reproduction of cheetahs (*Acinonyx jubatus*) in captivity has been very challenging and regular successful breeding only started after the 1970s (Chapter 22). Few individuals were successfully mated and historically only 15% of all wild-caught animals contributed to the captive gene pool (Marker-Kraus and Grisham, 1993). Today, approximately 20% of cheetahs included in the North American Species Survival Plan® (SSP) have reproduced (Cheetah SSP®, 2016), with a similar situation, seen globally (Chapter 22). It was previously suggested that the reproductive impairments observed in *ex situ* populations (Wildt et al., 1983) may be linked to low levels of genetic diversity (Chapter 6). However, despite similarly low-levels of genetic diversity in free-ranging cheetahs, reproduction in the wild does not appear to face the same challenges as seen in captivity (Caro, 1994; Eaton and Craig, 1973; Laurenson et al., 1992; Marker et al., 2003; Wachter et al., 2011). In the wild, cheetah females have high fecundity (Kelly et al., 1998; Laurenson et al., 1992) with up to 95% of observed-females successfully producing cubs (Laurenson et al., 1992).

The reduced fecundity of captive cheetah populations led scientists to initiate an extensive study of the North American population to identify factors influencing breeding success (Wildt et al., 1993), and in 1989, the North American cheetah SSP conducted a systematic survey of physiology and health status (Munson 1993; Wildt et al., 1993). These early studies revealed no significant differences in health, genetics, nutritional status or reproductive, and endocrine characteristics among proven and unproven breeders (Wildt et al., 1993). Very few pathologies, related to reproduction were detected in males. While males were found to have poor quality semen, they were reported to be able to produce pregnancies (Lindburg et al., 1993). Management was subsequently found to play a major role in captive breeding, and a number of factors associated with *ex situ* management, such as opportunity of mate choice, public exposure, exhibit space, and exercise, may significantly impact reproductive success. These early studies contributed to the most extensive biological database on an endangered species, including endocrine evaluation via radio immunoassays (RIA), laparoscopy, seminal analysis, and sperm function *in vitro*. Since then, a plethora of research has been conducted on the cheetah's reproductive biology, primarily in *ex situ* populations. This chapter provides information related to male and female cheetah reproductive physiology, endocrinology, and reproductive management, including assisted reproduction technology (ART). Data are compiled from peer-reviewed publications and from unpublished, newly emerging research from the scientific community. Protocols and forms that are relevant to this chapter can be found at https://www.elsevier.com/books-and-journals/book-companion/9780128040881.

FEMALE REPRODUCTIVE PHYSIOLOGY

Puberty in Females

Using enzyme immunoassays (EIAs), *ex situ* female cheetahs monitored from 6 to 36 months old exhibited a significant increase in mean fecal estrogen metabolite concentrations from 18 to 24 months of age compared to 25–30 months of age. In addition, the number of both estrogen peaks and cycles was greatest for females at 25–30 months of age (Maly et al., 2015). These data suggest that captive cheetah females reach puberty at 25–30 months of age (Maly et al., 2015). This age range is based on physiological data only, and does not account for the behavioral component of breeding interest/receptivity that must be in place, before successful mating can occur. Nevertheless, this approximate age concurs with data from free-ranging females which presumably become pregnant soon after puberty and have their first litter on average at 29 months of age, after a 3-month gestation period (Kelly et al., 1998).

Reproductive Behavior

Cheetah females are induced ovulators. A physiological evaluation for a minimum of 60 days on 24 females without physical contact to males, showed that an increase in fecal estradiol values did not result in a corresponding increase in fecal progesterone (the hormone released following ovulation) (Brown et al., 1996). In two females, an increase in progesterone without physical contact (but with auditory and visual contact) with a male cheetah was noted, although the concentrations were lower than levels documented following natural breeding. In cheetahs several follicles mature simultaneously leading to litter sizes of up to 6 cubs, with the potential of multiple paternities if females mate with multiple males during estrus (Gottelli et al., 2007; Chapter 6).

Cheetahs are nonseasonal breeders (Chapter 9). Free-ranging cheetah females are either pregnant, lactating or raising young (Laurenson et al., 1992; Wachter et al., 2011). They readily become pregnant, and resume cyclicity quickly when a litter is lost (Laurenson et al., 1992). Female cheetahs lactate for several months and remain with their cubs for approximately 18 months (Caro, 1994; Chapter 9). Once independent, the young female cheetahs remain with their siblings until they reach puberty and then roam solitarily (Caro, 1994). Therefore, housing practices in zoological institutions primarily house females individually. Placing unrelated females in pairs or groups in captivity can suppress ovarian cyclicity, likely due to the associated stress (Jurke et al., 1997), particularly in subordinate animals (Wielebnowski and Brown, 1998; Wielebnowski et al., 2002). Separating the females by a fence resulted in the re-initiation of ovarian activity in the reproductively suppressed females (Brown et al., 1996). In contrast, housing closely bonded females, such as siblings together resulted in synchronous and uninterrupted ovarian cyclicity (Koester et al., 2017; Terio et al., 2003; Wielebnowski and Brown, 1998).

Estrus

Unlike many other felid species, behavioral signs of estrus are difficult to detect in female cheetahs and require considerable observation and experience (Wielebnowski, 1999; Wielebnowski and Brown, 1998). Many *ex situ* females never display overt signs of behavioral estrus, such as rolling, rubbing, sniffing, vocalizing, and urine spraying and/or breeding receptivity to male cheetahs (Wielebnowski and Brown, 1998). In captivity, the duration of estrus, or receptivity to breeding, varies from 2 to 6 days, with estrogen elevation for 4.1 ± 0.8 days (mean $\pm$ SEM, standard error of the mean; Brown et al., 1996). Behavioral observations of free-ranging females exhibiting behavioral signs of estrus revealed a 2-week interval between cycles (Eaton and Craig, 1973). However, cycle length of cheetah females in captivity

shows a wide variation among and within individuals from 5 to 30 days, as detected with fecal estrogen metabolites (Fig. 27.1) assessed by RIA (Brown et al., 1996), EIA (Crosier et al., 2017), as well as vaginal cytology (Asa et al., 1992; Khwaja et al., in preparation). These studies also determined that estrogen peaks are not always associated with breeding receptivity (Fig. 27.1), making estrus prediction using this hormone difficult. However, fluctuating fecal estrogen metabolite concentrations reflect follicular growth and regression in this species (Crosier et al., 2017) and can thus establish whether a female is cycling or acyclic. The lack of regular patterns of cyclicity in female cheetahs is unique in that the ovaries are actively producing continuous waves of estradiol, but exhibit sporadic intervals of shut-down (Fig. 27.1; Crosier et al., 2017).

Anestrus

Acyclic phases for captive cheetahs have been confirmed through longitudinal fecal hormone metabolite monitoring (Brown et al., 1996; Crosier et al., 2017). During a 1-year monitoring period, none of the seven observed females cycled continuously, but rather showed anestrus periods regardless of season or age (Brown et al., 1996). These periods of ovarian inactivity are not influenced by previous breeding history or reproductive success (Fig. 27.1; Crosier et al., 2017), and are not caused by chronic stress (Wachter et al., 2011). Females that were observed to have periods of anestrus were able to mate naturally and gave birth to litters in the future, indicating that these females were not infertile and there appeared to be no long term negative effects of prolonged acyclic periods (Fig. 27.1; Crosier et al., 2017).

Monitoring Ovarian Activity and Reproductive Organs

Hormonal Assessments

The original work on cheetah ovarian activity was conducted by blood hormone monitoring

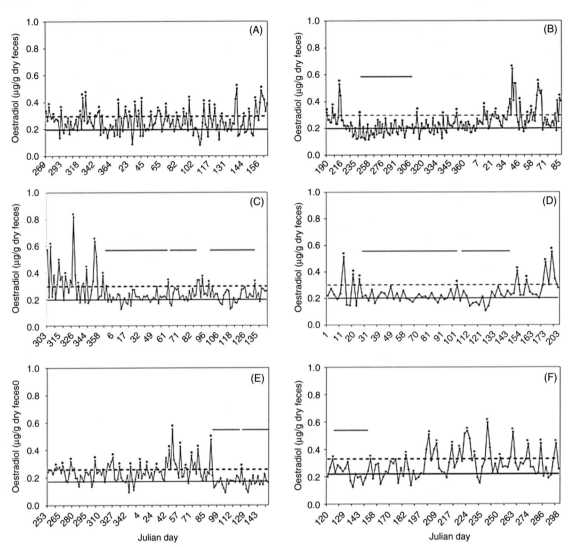

FIGURE 27.1 **Fecal estradiol metabolite patterns in adult, female cheetahs.** (A) 8 year old, parous female. (B) 3 year old nulliparous female that subsequently produced three litters. (C) 5 year old nulliparous female that never reproduced. (D) 3 year old nulliparous cheetah that subsequently produced one litter. (E) 4 year old nulliparous female that subsequently produced three litters. (F) 7 year old parous female. Each asterisk indicates an estradiol peak, and each orange line represents an acyclic period (>30 days with no estradiol peaks). For each individual female, the estradiol baseline is in solid black, and the peak line is broken. *Source: Crosier et al., 2017.*

of a significant proportion ($n = 68$; ~40%) of the North American captive female population, in conjunction with laparoscopy (Wildt et al., 1993). Following anesthesia, serial blood samples ($n = 12$ samples per female) were collected to confirm the level of ovarian activity observed via laparoscopy. This extensive endocrine evaluation using RIAs found that in most

TABLE 27.1 Basal Serum Concentrations for Luteinizing Hormone (LH), Follicle-Stimulating Hormone (FSH), Estradiol-17β, and Progesterone in Captive Adult Female Cheetahs Without Detectable Luteal Tissue

	LH (ng/mL)	FSH (ng/mL)	Estradiol-17β (pg/mL)	Progesterone (ng/mL)
Overall (n = 66)[a]	0.91 ± 0.06	8.99 ± 0.05	7.49 ± 0.09	0.210 ± 0.01
Overall (n = 66)[b] range	0.10 – 2.07	2.50 – 25.90	2.50 – 48.09	0.10 – 0.81
Minimum individual value	0.01	2.0	2.5	0.1
Maximum individual value	2.4	27.6	61.7	2.87
Proven breeders (n = 13)	0.71 ± 0.12	9.09 ± 0.87	6.46 ± 0.49	0.23 ± 0.05
Unproven breeders (n = 28)	1.00 ± 0.11	9.55 ± 0.88	8.84 ± 1.52	0.24 ± 0.03

Unless indicated otherwise, concentrations are given in mean ± standard error of the mean.
[a] All females with no discernible luteal tissue.
[b] Mean lowest to mean highest value among all females with no discernible luteal tissue.
From (Wildt, et al., 1993).

females, circulating estradiol-17β and progesterone concentrations were basal, corresponding to the acyclic status observed with laparoscopy (Wildt et al., 1993; Table 27.1). Circulating concentration of progesterone can be accurately used to determine luteal status and if the female had been mated. Fecal hormone metabolite monitoring can establish whether a female is cycling or is acyclic, and it can also be used to detect and monitor pregnancy.

Vaginal Cytology

Assessments of cheetah cyclicity can also be conducted using vaginal cytology. However, this technique should only be performed by experienced staff using aseptic technique to prevent infection of the vagina. Vaginal swabs can be obtained from anesthetized females (Asa et al., 1992; Schulman et al., 2015); however, the technique has also been successfully performed on nonanesthetized females that were hand-raised (Khwaja et al., in preparation). Clinical monitoring of vaginal cytology is a practical method for assigning reproductive status for facilities without ready access to hormonal monitoring capabilities (Schulman et al., 2015).

Laparoscopy

The first laparoscopy studies on cheetahs allowed visual examination of the female reproductive organs and produced the first detailed photographic records of cheetah uterine and ovarian morphology and associated structures: follicles, corpora hemorrhagica (CH), corpora lutea (CL), and luteal scars (Wildt et al., 1993). CH/CL were not observed in any nonpregnant or nonlactating female, supporting that cheetahs are induced ovulators.

Follicles were categorized into three size groups: <2 mm (immature), 2 to <4 mm, and ≥4 mm (fully mature). This early research classified 33.3% of females as having only immature follicles, and 22.7% of females as having fully mature follicles. Among cheetahs with ovaries producing any size of follicles, or between proven and unproven breeders, there were no differences in mean circulating concentrations of luteinizing hormone (LH), follicle-stimulating hormone (FSH), or estadiol-17β [before administering gonadotrophin releasing hormone (GnRH)]. The average diameter of the uterine horns was 8.4 ± 0.4 mm (range, 4–17 mm), and the mean oviductal diameter was 3.8 ± 0.1 mm (range,

2–7 mm). In a more recent study, ovarian volumes for those categorized as cyclically active (i.e., proestrus or estrus; $n = 9$) were significantly greater (1322 mm^3, IQR: 1024–1363 mm^3; $P = 0.005$) than those in anestrus ($n = 7$; 906 mm^3, IQR: 387–964 mm^3) (Schulman et al., 2015).

Ultrasonography

Measurements of ovarian and uterine morphology, as well as horn width and thickness have been evaluated for female cheetahs using transabdominal ultrasound (Crosier et al., 2011). The uterine body and cervix can be visualized using transrectal ultrasonography, while it may be obstructed by the pelvic canal during transabdominal ultrasonography. Precise differentiation of ovarian structures, especially differentiation of the various follicular developmental stages or luteal structures, is possible when employing high-resolution ultrasonography, either transabdominally, or transrectally. Preovulatory follicles measure >6.0 mm in the ultrasound image (Schulman et al., 2015). The estrus cycle stage may be accurately determined when combining ultrasound with other techniques, such as vaginal cytology (Schulman et al., 2015) or serum estradiol and progestogen assays (Bertschinger et al., 2002a; Crosier et al., 2011; Schulman et al., 2015). Structural and functional effects of estrogen in carnivores include tissue edema, hyperemia, and keratinization of the vaginal epithelial (Schulman et al., 2015). Anestrus can be differentiated from active reproductive cycles (e.g., nonpregnant luteal phase, proestrus, estrus) via identification of follicles or luteal structures, as well as measurement of ovarian volumes (Schulman et al., 2015; Wachter et al., 2011).

Impact of Stress on Reproductive Activity

Several reports have suggested that the reproductive health and performance of cheetahs were negatively affected by chronic physiological stress (Jurke et al., 1997; Terio et al., 2004, 2014). Stress indicators, such as elevated fecal glucocorticoid metabolites (fGCM; Wielebnowski et al., 2002) have been suggested to lead to reproductive quiescence (Jurke et al., 1997; Munson et al., 2005; Terio et al., 2004). Stressors suggested to suppress ovarian activity were environmental and management conditions, such as overcrowding and unnatural social groupings (Brown et al., 1994, 1996; Jurke et al., 1997; Wielebnowski et al., 2002). However, in a recent study, neither being housed on- or off-exhibit or the number of adult conspecifics (male and female) in the same facility, had an impact on estrogen or glucocorticoid metabolite excretion in female cheetahs (Koester et al., 2017). Interestingly, estrogen and glucocorticoid peaks were correlated within individuals (Koester et al., 2017).

Pathologies of the Reproductive Tract and Asymmetric Reproductive Aging

Of 68 captive female cheetahs surveyed in the North American population, 11 (16.2%) were reproductively unsound (Wildt et al., 1993). The abnormalities included degenerate/fibrous ovaries, having only a single ovary, unilateral ovarian adhesion to adjacent omentum, infantile reproductive tract, mass within uterine horn(s), and an abnormally small vaginal opening (Wildt et al., 1993). A nulliparous female with an infantile horn and ovaries, as well as one multiparous female with a unilateral ovarian adhesion to adjacent omentum have also been documented (Crosier et al., unpublished data). Numerous captive females, both multiparous and nulliparous, have been reported to have paraovarian cysts as single or multiple fluid-filled pockets (Crosier et al., 2011; Schulman et al., 2015; Wildt et al., 1993). These cysts are histologically not indicative of anatomical unsoundness or disease (Munson, 1993), they are remnants of the mesonephric duct, usually located near the attachment of the broad ligament, and are variable in size, as documented in the domestic cat (Schlafer and Gifford, 2008). Twenty-three of the thirty-five females (66%) that had paraovarian cysts had them on both sides (bilateral), with an average range in cyst diameter of 2.0–35.0 mm, with an overall mean ± SEM of 9.4 ± 1.2 mm (Wildt et al., 1993). Paraovarian cysts were present in all age groups, although less prevalent in females, 48 months of age or less (Munson 1993). In free-ranging cheetahs, paraovarian cysts are rare. In 13 Namibian free-ranging cheetahs only a single paraovarian cyst was observed in one female by ultrasound imaging (Wachter et al., 2011).

In aged captive cheetahs (≥9 years of age), endometrial hyperplasia is the most common uterine pathology (Crosier et al., 2011), whereas fibroleiomyoma has been described in only one individual (Walzer et al., 2003). Muco- or pyometra may also occur, but with low incidence; only one of 21 aged, nulliparous females showed these pathologies on transabdominal ultrasound and laparoscopy (Schulman et al., 2015). Exogenous hormone administration, particularly the prolonged progestin supplementation in captive cheetahs used as a contraceptive, is a risk factor for development of severe endometrial hyperplasia (Munson et al., 2002). However, the administration of GnRH agonists, such as deslorelin (Suprelorin) in captive cheetah females for downregulation of ovarian activity does not appear to be associated with obvious adverse effects on uterine integrity (Bertschinger et al., 2002b, 2008; Schulman et al., 2015).

In the absence of pregnancies and lactation, the reproductive tract of nonreproducing captive females is subjected to frequent fluctuations in estrogen concentration, which can have a negative impact on the uterus and lead to an increase in genital pathologies and lesions with age (Crosier et al., 2011; Wachter et al., 2011). It has been speculated that if estrogen fluctuations persist for years, the reproductive tract of females could be affected such that reproduc-tion can no longer be established, as observed in select hoofed-stock species [elephants (*Elephantidae* spp.) and rhinoceros (*Rhinocerotidae* spp.)] (Hermes et al., 2004). This "asymmetric reproductive aging" is a nonreversible process and this phenomenon has been observed in captive cheetahs (Wachter et al., 2011). Free-ranging cheetah females are not affected by asymmetric reproductive aging because as observed by ultrasound examination females in the wild are either cycling, pregnant, or lactating (Namibia, N = 13; Wachter et al., 2011).

Genital lesions were documented in nulliparous captive populations in Namibia, North America, and Europe; hydrosalpinx and dense connective tissue in the ovarian stroma was observed on ultrasound at a mean age of 5.6 years (Namibia, N = 14; Wachter et al., 2011). In European zoos similar symptoms were detected (Europe, N = 12; Ludwig, et al., unpublished data). Severe uterine pathologies which reduce the likelihood that females can establish a pregnancy (e.g., cysts within the endometrium, fibrosis, adenomyosis, pyometra, and endometrial atrophy) were detected in over 50% of aged females (North America, N = 105; Namibia, N = 13; Crosier et al., 2011). Uterine hyperplasia was observed on necropsy in 90% of females of more than 9 years (compared to only 50% of females aged 6–8 years). The likelihood of females developing uterine hyperplasia increased with the length of time from her last litter (Crosier et al., 2011). Vice versa, parity has been associated with a decreased risk of developing endometrial hyperplasia (Munson et al., 2002). Additionally, lactation until natural weaning further reduces the frequent fluctuation of estrogen concentration and thus is likely to extend the reproductive lifespan of captive cheetah females (Ludwig et al., unpublished data; Wachter et al., 2011). Therefore, it seems advisable to breed cheetah females at an early age, and to establish pregnancy regularly (every 2–3 years) as would occur in the wild.

MALE REPRODUCTIVE PHYSIOLOGY

Puberty in Males

Based on semen collection data via electro-ejaculation (EE), captive cheetah males start producing sperm at 14 months of age (Crosier et al., 2007) and can sire offspring before 48 months of age (Marker, 2015). However, in free-ranging populations males usually do not repro-duce until they are prime adults (48–96 months), at an age at which they are able to obtain and de-fend a territory from conspecifics (Caro, 1994). In *ex situ* managed cheetahs, captive males had a significant rise in baseline fecal androgen metabolite concentrations at 12–18 months of age (Maly et al., 2015). This coincides with the time free-ranging cheetah adolescents separate from their mother as a sibling group and even-tually males separate from the female siblings (Caro, 1994; Chapter 9).

Male Social Structure

Male cheetahs are either solitary or live in co-alition groups usually consisting of littermates (Caro and Collins, 1987). In captivity, their so-cial living structure has been shown to affect reproduction. Ejaculate quality was higher in group-housed males compared to that of males housed singly (Koester et al., 2015b). Numbers of other conspecifics housed at the same insti-tution; however, did not influence either fGCM concentrations, testosterone concentrations, or ejaculate quality (Koester et al., 2015a). Monitor-ing an entire year of fecal androgen and gluco-corticoid concentrations from males housed in the North American SSP revealed no fluctua-tions among seasons (Koester et al., 2015a). Fecal androgen metabolite concentrations were posi-tively correlated with superior sperm quality in group-housed males but not in males kept singly (Koester et al., 2015b). However, there was no correlation between seminal metrics and fGCM

concentrations, indicating that the mechanism limiting seminal quality is not related to gluco-corticoid-associated stress (Koester et al., 2015a).

Semen Collection and Quality

Semen Collection

Prior to semen collection, the reproductive organ health of male cheetahs should be evalu-ated. Testes volume and tone should be assessed to ensure the absence of abnormalities. Total testes volume ranged from 8700 ± 700 mm^3 to 10200 ± 1600 mm^3 for wild-born Namibian males ($N = 97$; Crosier et al., 2007). The penis should also be examined for abnormalities, injuries, and the presence of a normal complement of spines. To collect semen, three primary methods have been described in cheetahs: artificial vagina, transrectal EE, and urethral catheterization.

The *artificial vagina* (AV) has been success-fully used in cheetahs and other felid species (Durrant et al. 2001; Zambelli and Cunto, 2006). Tame (hand-raised), nonsedated male cheetahs have been trained to ejaculate into an AV, and samples collected had similar quality to ejacu-lates collected by other methods (Durrant et al., 2001). A longitudinal study of a single male cheetah providing an average of 26 ejaculate samples via AV per year from 2 to 15 years of age, revealed that semen volume increased with age, and motility and concentration peaked at 8–10 years of age (Durrant et al., 2001).

Transrectal EE has been the most commonly applied method to obtain semen from chee-tahs (Wildt et al. 1983, 1988; Table 27.2). Under anesthesia, an appropriately sized EE probe (1.6–1.9 cm diameter) is used to deliver low elec-tric stimuli (2–6 volts, 50–200 mA). Up to 100 stimuli of increasing voltage in three sets during an approximately 30–40 min stimulation period are commonly used to produce ejaculates (Cro-sier et al., 2006, 2007; Wildt et al., 1987). With this technique, urine contamination may occur due to stress during anesthesia induction causing urination and presence of urine in the urethra.

TABLE 27.2 Summary of Basic Cheetah Ejaculate Traits Collected by Different Researchers From Captive and Free-Ranging Populations

Cheetah origin	Number of males examined	Method of semen collection	Ejaculate volume (mL)	Sperm concentration (×10⁶ mL⁻¹)	Total motility (%)	Abnormal sperm (%)	Normal sperm (%)	Reference
Semi-wild, South Africa	18	EE	2.1 ± 0.2	14.5 ± 1.8	54.0 ± 3.0	71.0 ± 0.9	~29	Wildt et al. (1983)
Captive, USA	23	EE	~1.0	25.1 ± 4.4	70.7 ± 3.5	70.6 ± 3.3	~28.4	Wildt et al. (1987)
Wild, East Africa	8	EE	~1.4	36.4 ± 12.2	63.1 ± 3.9	75.9 ± 4.4	~24.1	Wildt et al. (1987)
Captive, USA and South Africa	3	EE	1.8 ± 0.3	27.3 ± 5.8	69.0 ± 5.8	64.6 ± 4.9	~35.4	Wildt et al. (1988)
Captive, USA	60	EE	1.5 ± 0.1	29.3 ± 5.6	67.0 ± 2.0	~78.7	21.3 ± 2.0	Wildt et al. (1993)
Captive, wild-born, Namibia	13	EE	3.7 ± 0.4	20.4 ± 3.1	78.0 ± 1.4	~78.3	21.7 ± 2.4	Crosier et al. (2006)
Captive, South Africa	32	EE	0.7 ± 0.5	32.7 ± 36.1	nr	~59.7	40.3 ± 17.5	Bertschinger et al. (2008)
Captive, on exhibit, USA	116	EE	~1.0	19.3 ± 7.6	75.1	~75	~25	Koester et al. (2015a)
Captive, off exhibit, USA	43	EE	~1.0	63.8 ± 16.2	~69.3	~75	~25	Koester et al. (2015a)
Captive, UAE, (partly South African origin)	13	EE	0.4 ± 0.4	189.0 ± 317.6	56.7 ± 20.7	nr	nr	Marker (unpublished data)
Captive, Europe (South African origin)	8	UC	0.2 ± 0.1	15.1 ± 7.3	44.4 ± 25.7	63.5 ± 17.4	~36.5	Lueders (unpublished data)

EE, Electroejaculation; nr, not recorded; UC, urethral catheterization; UAE, United Arab Emirates. Values given in mean ± standard deviation for Bertschinger et al., 2008; Marker, unpublished; and Lueders, unpublished or ± standard error of the mean for Wildt et al., 1983, 1987, 1988, 1993; Crosier et al., 2006; Koester et al., 2015a.

In addition, anesthetic drugs with a low ratio of α2:α1 binding affinity or α2 adrenergic agonists (causing retrograde sperm flow), inexperience of the collector, too deep placement of electrodes in the rectum (induces contraction of bladder), and age of the animal all may contribute to urine contamination (Bertschinger et al., 2002b; Marrow et al., 2015; Virtanen, 1989). Urine contamination may be determined by measurement of pH (<8.7 likely to be contaminated), ejaculate appearance (yellowish), and microscopic observation of urine crystals (Bertschinger et al., 2008; Crosier et al., 2007).

The third method is semen collection by *urethral catheterization* (UC). This newest technique is currently rarely used for cheetahs, despite requiring less experience or technical equipment compared to EE. Preliminary results suggest that application of the UC method is limited to cheetah fertility assessment (Lueders, unpublished data), as it tends to yield less volume and lower concentrations of sperm than EE (Table 27.2), and therefore may not be appropriate if the aim is to collect semen for reproductive purposes. This technique was first described for domestic cats and has been successfully used in lions (*Panthera leo*) and other large, non-domestic felids (Lueders et al., 2012; Zambelli et al., 2008). α2 agonists (i.e., medetomidine) administered as an anesthetic agent induces sperm release into the urethra. The sperm can be collected by capillary force with a sterilized urinary catheter (500 mm length, 2.0–2.6 mm diameter) inserted via the urethra to the prostate or as deep as 20 cm; a transrectal ultrasound can provide guidance.

Sperm Cryopreservation

Sperm cryopreservation has provided an important tool for the banking of genetic material from hundreds of cheetah males living in zoos in North America, from >150 wild born males at the Cheetah Conservation Fund (CCF) in Namibia, (Comizzoli et al., 2009; Crosier et al., 2007; Wildt et al., 1993) and from north African cheetahs held in the United Arab Emirates (CCF, unpublished

data). Samples from *in situ* Namibian males and *ex situ* Emirati males were cryopreserved using "field friendly" techniques that can be easily applied for variable working conditions (Crosier et al., 2007).

Semen Evaluation

A common finding across all studied cheetah populations of captive and free-ranging males was the high percentage (approximately 75%) of structurally deformed spermatozoa in the ejaculate (Table 27.2; Crosier et al., 2007; Wildt et al., 1993). The reduced genetic diversity in cheetahs (Chapter 6) is thought to contribute to this observed teratospermia (defined as the production of ≥60% structurally abnormal spermatozoa in an ejaculate) and genetic changes, such as sequence variants were identified to affect spermatogenesis (Dobrynin et al., 2015). However, sperm quality does not appear to be influenced by increased homozygosity (Terrell et al., 2016). It is reassuring that breeding activity and conception do occur in cheetahs with more than 75% of structurally abnormal sperm in an ejaculate (Lindburg et al., 1993; Wildt et al., 1993).

Standard metrics evaluated for every semen sample include semen volume, sperm concentration, and total motility, forward progressive motility, and sperm motility index (SMI) (Crosier et al., 2007; Howard 1993). Most adult male cheetahs in both North American and Namibian facilities produce motile sperm (Crosier et al., 2007; Wildt et al., 1993). No difference in testes size, semen quality, or sperm morphology has been documented for proven versus unproven breeders in the North American population (Wildt et al., 1993). Overall, cheetahs produce comparatively lower sperm concentration than other felids (Wildt et al. 1983, 1988). Testes volumes, ejaculate volume, sperm concentration, and sperm motility of young males less than 2 years of age were lower than those of adult (2–10 years of age) and aged males (>12 years old; Crosier et al., 2007). Testosterone concentrations do not reach adult levels until approximately

2 years of age, whereas male cheetahs as old as 15 years produce spermic ejaculates, even in the presence of low testosterone (Wildt et al., 1993).

Sperm Function

Two assays have been used to characterize the viability and functionality of cheetah spermatozoa: (1) longevity of sperm motility *in vitro*, and (2) ability to penetrate the oocyte's intact zona pellucida (ZP) of domestic cat oocytes in an *in vitro* fertilization (IVF) system (Howard et al., 1993).

Ejaculate quality and the ability of cheetah sperm to perform *in vitro* bioassays (Donoghue et al., 1992) were considerably inferior to the results from normospermic felid species (Andrews et al., 1992). Even structurally normal sperm from teratospermic cheetahs are compromised in the ability to penetrate the oocyte ZP and achieve embryo cleavage *in vitro* (Roth et al., 1995). Also, sperm viability and IVF rates in the cheetah are vastly inferior to that observed for other felid species (Donoghue et al., 1992). Both captive and free-ranging cheetah males have equal percentages of teratospermia (Crosier et al., 2007); however, this does not hinder reproductive success in the wild. This suggests that challenges of breeding *ex situ* are not primarily related to their unique sperm phenotypes (Crosier et al., 2007; Wildt et al., 1993). Rather, the cause is likely to be due to behavior or libido dysfunction, both of which may be associated with husbandry, endocrine imbalance, or some other unknown problem (Wildt et al., 1993).

EX SITU BREEDING AND ASSISTED REPRODUCTION TECHNOLOGY (ART)

ARTs comprise of methods developed to achieve pregnancy when natural breeding is not possible or when assisted techniques are necessary to enhance the genetic variation of a population. These methods include semen collection and cryopreservation (covered under "Male Reproductive Physiology"), hormonal estrus and ovulation induction, oocyte aspiration and *in vitro* maturation, artificial insemination (AI) with fresh or frozen-thawed spermatozoa, *in vitro* embryo production, and embryo transfer. Cryorepositories of systematically collected gametes, embryos, blood, tissue, and DNA for defined conservation programs, termed "genome resource banks" (GRB), provide a genetic insurance policy to mitigate against a potential reduction of free-ranging populations, and have provided the ability to infuse the captive population with genetic material without requiring removal of animals from the wild (Wildt et al., 1997).

Ovarian Stimulation and Ovulation Induction

Hormonal treatment is required to induce follicle growth, oocyte maturation, and follicular ovulation, followed by timed AI or oocyte aspiration (Wildt et al., 1981). Usually, these protocols include exogenous gonadotropins (equine and human chorionic-gonadotropins; eCG and hCG; Conforti et al., 2013), particularly hCG, which has sufficient LH-like properties to induce ovulation in felids (Swanson et al., 1997). Although successful in other felid species, comparatively little is known about the use of other exogenous hormones, such as GnRH analogs or porcine LH (*p*LH) to induce ovulation in cheetahs (Conforti et al., 2013; Pelican et al., 2006). It has been advised that administration of exogenous CG only be repeated after 6–9 months, because of the potential for antibody production (Roth et al., 1997; Swanson et al., 1997), and its only moderate to good success in cheetahs (Howard et al., 1992, 1997). Ovulation induction by injection of a single dose of the GnRH analog buserelin acetate on the day of AI during natural heat has been successful in several felid species (Asiatic golden cats, *Catopuma temmincki* and Persian leopard, *Panthera pardus saxicolor*;

Lueders et al., 2015). However, the lack of overt estrus behaviors in cheetahs decreases the feasibility of this approach.

The occurrence of natural estradiol peaks in cheetah females is important for the effectiveness of exogenous gonadotropin protocols. When the treatment occurred at least 3 days (72 h) after a detected estradiol peak, more viable oocytes were recovered and more normative progestogen patterns were produced after aspiration compared with females given gonadotropins ≤2 days after an estradiol peak (Crosier et al., 2017). Exogenous progestin supplementation (e.g., Altrenogest) has been used to first achieve ovarian suppression and quiescence for improved timing of subsequent administration of exogenous gonadotropins (Crosier et al., 2017).

Artificial Insemination (AI)

Due to the poor reproductive success in captive-held cheetahs, development of AI techniques has been pursued more for cheetahs than for other large felids. There are three main components important for AI success: (1) correct timing to induce follicular growth and ovulation by hormonally-induced ovarian activity (exogenous gonadotropin administration induction of follicular growth and ovulation; section "Ovarian Stimulation and Ovulation Induction"), (2) availability of acceptable quality semen (section "Semen Collection and Quality"), and (3) placement of the semen sample within the appropriate location of the female reproductive tract at the optimal time (hereafter).

Sperm Requirements for AI

Freshly collected ejaculates with high concentration of sperm (10^7–10^8 cells/mL) are suitable for intravaginal deposition (Pelican et al., 2006), whereas ejaculates with small volumes, low-sperm concentration, or frozen-thawed semen may warrant intraoviductal (Lambo et al., 2013) or deep intrauterine placement (Howard

et al., 1997). AI with freshly collected sperm has been successfully applied, and live births were reported in multiple species of felids (Howard and Wildt, 2009). Generally, fresh or chilled spermatozoa are preferred over frozen-thawed semen because of the substantial damage to acrosomal membranes and loss of motile cells during the cryopreservation process (Crosier et al., 2006). However, in cheetahs, fresh (Howard et al., 1992, 1997) and frozen-thawed semen AI (Howard and Wildt, 2009) have resulted in successful pregnancies following intrauterine deposition via a laparoscopic approach.

The number of sperm has a strong influence on the success of the insemination procedure. Cheetah females that established pregnancy were inseminated with an average of $15.8 \pm 3.8 \times 10^6$ fresh motile sperm compared with $8.0 \pm 2.4 \times 10^6$ for females that did not establish pregnancy (Howard et al., 1997). When cryopreserved samples were used for intrauterine AI, no pregnancies were achieved for the 5 females inseminated with $<4 \times 10^6$ motile sperm compared with 3 of 6 females becoming pregnant following insemination with 6–16×10^6 motile sperm (Howard et al., 2002).

Deposition of Sperm

Tame females could potentially be trained for intravaginal deposition of sperm, however, any deeper placement of the AI catheter warrants full anesthesia. Nonanesthetized methods require the AI to take place prior to ovulation to allow passage of the cervix and for uterine sperm transport. A study in domestic cats suggested that if the female is preovulatory, the anesthesia induction agents interfere with ovulation and transuterine sperm transport, which prompted AI research in felids directed toward postovulatory timing of insemination (Howard and Wildt, 2009).

A relatively new approach in felid AI is intraoviductal sperm deposition, with pregnancies achieved in domestic cats (Conforti et al., 2013; Lambo et al., 2013) and clouded leopards

(Comizzoli and Crosier, unpublished data). The laparoscopic intraoviductal AI is a potentially appropriate technique for the cheetah, especially with consideration of the small semen dose/concentration needed and relatively poor semen quality in this species. Domestic cat pregnancies ($n = 3$), and the birth of five live offspring, have been achieved using as few as 1×10^6 total motile sperm per oviduct (Conforti et al., 2013).

Success of AI in Cheetahs

Early AI attempts in cheetahs yielded good results with approximately 45% of inseminated ovulatory females establishing pregnancy (Howard et al., 1997). Based on compiled historical data, a total of 74 female cheetahs were stimulated with a combination of 200 IU eCG and one of three hCG doses (100, 125, or 150 IU) (Crosier et al., 2017). Of these, a total of 49 females (66.2%) had fresh corpora lutea on their ovaries at the time of insemination. These 49 females were inseminated with an average of 14.6×10^6 total motile sperm, and 11 litters (22.4% of females inseminated) were produced. Despite continued attempts, the last living cheetah cub produced by AI was in 2003 (Crosier et al., 2017).

In vitro Embryo Production and Oocyte Quality

Early research into IVF focused on recovering oocytes from ovarian tissues that were collected following ovariohysterectomy or from deceased females (Johnston et al., 1991). Because no gonadotropin stimulation was used, only ~8% of these oocytes became mature following in vitro maturation, and none successfully fertilized when exposed to conspecific spermatozoa in culture. More recently, a large-scale collaborative study was conducted including animals from the North American population and cheetahs housed at CCF in Namibia (Crosier et al., 2011). Females were stimulated with eCG and hCG and oocytes aspirated laparoscopically

for assessment of morphological quality, maturation (proportion at metaphase II), and ability to become fertilized and form embryos. There was no difference among age groups in the proportion of mature (metaphase II) oocytes recovered, or in the ability of oocytes to support fertilization and cleavage to early embryonic stages. Approximately 41% of cleavage-stage embryos progressed to at least the 8-cell stage of development in vitro, with 24% of all cleaved embryos developing to the morula stage, and 5% forming blastocysts. This was the first report of blastocyst stage cheetah embryos produced in vitro. Noteworthy, only cryopreserved cheetah sperm were utilized for IVF of in vivo matured cheetah oocytes, resulting in 70% fertilization of all metaphase II oocytes cocultured with sperm. These results indicate that oocyte quality doesn't diminish with female age, and that it is feasible to harvest oocytes and generate embryos from older females for potential transfer to younger surrogates (Crosier et al., 2011). This is especially important for older cheetah females that have never reproduced. Results also support the importance of genome resource banking and the use of cryopreserved spermatozoa for ART.

Pregnancy Diagnosis and Pseudopregnancy

Currently there is no practical, reliable test for diagnosing pregnancy in the cheetah until the second half of gestation. The most commonly used method is fecal sample analysis for progestagen metabolite concentrations using EIA or RIA (Table 27.3; Adachi et al., 2011; Brown et al., 1996). In general, progesterone concentrations remain at baseline levels until an ovulation is induced via mating or exogenous hormonal stimulation (Adachi et al., 2011; Brown et al., 1996; Crosier et al., 2011; Graham et al., 1995). Regardless of whether fertilization occurs, luteal activity will maintain the elevated progestogen concentrations for approximately 55 days (range, 45–65 days) following ovulation.

TABLE 27.3   Concentrations of Fecal Steroid Hormone Metabolites Quantified by Radioimmunoassay in Captive Adult Female Cheetahs

	Estradiol (ng/g dry feces)	Progesterone (ng/g dry feces)
Baseline nonmated values (range)	25–60	0.7–6.0
Peak estrus value, conceived (mean ± SEM)	284.3 ± 45.5	
Peak estrus value, did not conceive (mean ± SEM)	314.8 ± 41.9	
Luteal phase, pregnant females (mean ± SEM)		202.9 ± 15.3
Luteal phase, nonpregnant females (mean ± SEM)		240.6 ± 26.4

SEM, Standard error of the mean.
Compiled from (Brown, et al, 1996).

This physiological state, without the development of a fetus, is termed a "pseudopregnancy" or a nonpregnant luteal phase. Differentiation between pregnancy and pseudopregnancy by progestogens measurement only becomes possible after levels return to baseline in nonpregnant females (Brown et al., 1996).

The utility of measuring prostaglandin F2α metabolite (PGFM) in cheetah feces has been demonstrated for accurate, noninvasive pregnancy determination (Dehnhard et al., 2012). The PGFM levels rise to 47–49 after fertile mating, but remain at baseline in a nonconceptive luteal phase. Although ultrasonography may be used to determine pregnancy in cheetahs, compliance is rare without the assistance of anesthesia, which is contraindicated during a potential pregnancy. Alternatively, radiography can be used in nonanesthetized cheetahs (by confining the cheetah to a modified crate or chute; Fig. 27.2A) for definitive pregnancy assessment and accurate determination of cub number (Ware et al., 2016). Accurate pregnancy determination has been documented as early as

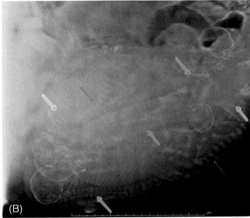

FIGURE 27.2   **Crate to perform radiography of nonanesthetized cheetah females.** (A) Crate with radiolucent X-ray tabletop material incorporated into one side (*) for pregnancy diagnosis and cub number determination. Plexiglas on the alternate side allows viewing of cheetah position in the crate (B) Radiograph of a cheetah female pregnant with 7 cubs (*arrows pointing to spinal columns*) at day 73 postcopulation. *Source: Part A, SCBI Department of Animal Medicine; part B, Adrienne Crosier, SCBI.*

day 55 and accurate cub number by approximately day 65 postcopulation using this technique (Fig. 27.2B; Crosier, unpublished data). Early determination of cub number assists husbandry planning, especially in the case of a single cub, which does not stimulate consistent milk production, making hand raising necessary (Ziegler-Meeks, 2009; Chapter 24).

New research focusing on noninvasive methods to identify fecal proteins (biomarkers) that distinguish pregnancy from pseudopregnancy is under way. Proteomic techniques (e.g., two-dimensional gel electrophoresis and tandem mass spectrometry) have been successfully applied to identify relevant biomarkers in the polar bear (*Ursus martimus*) (Curry et al., 2012). Using similar techniques, significant progress has been made investigating the protein profile present in the fecal material of pregnant and nonpregnant cheetah females to identify and reliably detect biomarker candidates associated with pregnancy establishment in cheetahs (Koester et al, in preparation).

CONCLUSIONS

Future reproductive research in cheetahs should be focused on three primary areas: improvement of ovarian synchronization for AI, development of methods for early pregnancy diagnosis and improvement of husbandry conditions to encourage successful breeding. A major challenge for successful AI in cheetahs is proper timing of exogenous gonadotropin treatments relative to the females' natural ovarian cycles. Early pregnancy diagnosis would improve reproductive management of females and improve overall breeding success. Identification of pregnancy biomarkers will enable researchers to better understand whether females are breeding and not conceiving, or are aborting a pregnancy following early embryo/fetal development. Rapid, early pregnancy diagnosis techniques developed in cheetahs have the potential to be applied to other threatened and endangered felid species. Husbandry management should be improved to simulate the natural social and mating system of cheetahs. This includes the early and regular breeding of cheetah females to prevent the development of pathologies of the reproductive tract.

References

Adachi, I., Kusuda, S., Kawai, H., Ohazama, M., Taniguchi, A., Kondo, N., Yoshihara, M., Okuda, R., Ishikawa, T., Kanda, I., Doi, O., 2011. Fecal progestagens to detect and monitor pregnancy in captive female cheetahs (*Acinonyx jubatus*). J. Reprod. Dev. 57, 262–266.

Andrews, J.C., Howard, J.G., Bavister, B.D., Wildt, D.E., 1992. Sperm capacitation in the domestic cat (*Felis catus*) and leopard cat (*Felis bengalensis*) as studied with a salt-stored zona pellucida penetration assay. Mol. Reprod. Dev. 31, 200–207.

Asa, C.S., Junge, R.E., Bircher, J.S., Noblem, G.A., Sarri, K.J., Plotka, E.D., 1992. Assessing reproductive cycles and pregnancy in cheetahs (*Acinonyx jubatus*) by vaginal cytology. Zoo Biol. 11, 139–151.

Bertschinger, H.J., Meltzer, D.G.A., van Dyk, A., 2008. Captive breeding of cheetahs in South Africa- 30 years of data from the de Wildt Cheetah and Wildlife Centre. Reprod. Domest. Anim. 43, 66–73.

Bertschinger, H.J., Nöthling, J.O., Nardini, R.M., Hemmelder, S., Broekhuisen, M.H., 2002b. Collection of semen in cheetahs (*Acinonyx jubatus*) using electro-ejaculation: attempts to avoid urine contamination. Adv. Ethol. 37, 122.

Bertschinger, H.J., Trigg, T.E., Jochle, W., Human, A., 2002a. Induction of contraception in some African wild carnivores by downregulation of LH and FSH secretion using the GnRH analogue deslorelin. Reprod. Suppl. 60, 41–52.

Brown, J.L., Wasser, S.K., Wildt, D.E., Graham, L.H., 1994. Comparative aspects of steroid-hormone metabolism and ovarian activity in felids, measured noninvasively in feces. Biol. Reprod. 51, 776–786.

Brown, J.L., Wildt, D.E., Wielebnowski, N., Goodrowe, K.L., Graham, L.H., Wells, S., Howard, J.G., 1996. Reproductive activity in captive female cheetahs (*Acinonyx jubatus*) assessed by faecal steroids. J. Reprod. Fertil. 106, 337–346.

Caro, T.M., 1994. Cheetahs of the Serengeti Plains: Group Living in an Asocial Species. The University of Chicago Press, Chicago, IL.

Caro, T.M., Collins, D.A., 1987. Male cheetah social organization and territoriality. Ethol. 74, 52–64.

Cheetah Species Survival Plan (SSP®), 2016. Association of Zoos and Aquariums Cheetah Breeding and Transfer Plan Report. Silver Spring, MD.

Comizzoli, P., Crosier, A.E., Songsasen, N., Gunther, M.S., Howard, J.G., Wildt, D.E., 2009. Advances in reproductive science for wild carnivore conservation. Reprod. Domest. Anim. 44, 47–52.

Conforti, V.A., Bateman, H.L., Schook, M.W., Newsom, J., Lyons, L.A., Grahn, R.A., Deddens, J.A., Swanson, W.F., 2013. Laparoscopic oviductal artificial insemination improves pregnancy success in exogenous gonadotropin-treated domestic cats as a model for endangered felids. Biol. Reprod. 89, 1–9.

Crosier, A.E., Comizzoli, P., Baker, T., Davidson, A., Munson, L., Howard, J., Marker, L.M., Wildt, D.E., 2011. Increasing age influences uterine integrity, but not ovarian function or oocyte quality, in the cheetah (Acinonyx jubatus). Biol. Reprod. 85, 243–253.

Crosier, A.E., Comizzoli, P., Koester, D.C., Wildt, D.E., 2017. Circumventing the natural, frequent oestrogen waves of the female cheetah (Acinonyx jubatus) using oral progestin (Altrenogest). Reprod. Fertil. Dev 29, 1486–1498.

Crosier, A.E., Marker, L., Howard, J., Pukazhenthi, B.S., Henghali, J.N., Wildt, D.E., 2007. Ejaculate traits in the Namibian cheetah (Acinonyx jubatus): influence of age, season and captivity. Reprod. Fertil. Dev. 19, 370–382.

Crosier, A.E., Pukazhenthi, B.S., Henghali, J.N., Howard, J.G., Dickman, A.J., Marker, L., Wildt, D.E., 2006. Cryopreservation of spermatozoa from wild-born Namibian cheetahs (Acinonyx jubatus) and influence of glycerol on cryosurvival. Cryobiology 52, 169–181.

Curry, E., Stoops, M.A., Roth, T.L., 2012. Non-invasive detection of candidate pregnancy protein biomarkers in the feces of captive polar bears (Ursus maritimus). Theriogenology 78, 308–314.

Dehnhard, M., Finkenwirth, C., Crosier, A., Penfold, L., Ringleb, J., Jewgenow, K., 2012. Using PGFM (13,14-dihydro-15-keto-PGF2α) as a non-invasive pregnancy marker for felids. Theriogenology 77, 1088–1099.

Dobrynin, P., Liu, S., Tamazian, G., Xiong, Z., Yurchenko, A.A., Krasheninnikova, K., Kliver, S., Schmidt-Küntzel, A., Koepfli, K.P., Johnson, W., Kuderna, L.F., García-Pérez, R., Manuel, M., Godinez, R., Komissarov, A., Makunin, A., Brukhin, V., Qiu, W., Zhou, L., Li, F., Yi, J., Driscoll, C., Antunes, A., Oleksyk, T.K., Eizirik, E., Perelman, P., Roelke, M., Wildt, D., Diekhans, M., Marques-Bonet, T., Marker, L., Bhak, J., Wang, J., Zhang, G., O'Brien, S.J., 2015. Genomic legacy of the African cheetah, Acinonyx jubatus. Genome Biol. 16, 277.

Donoghue, A.M., Howard, J.G., Byers, A.P., Goodrowe, K.L., Bush, M., Blumer, E., Lukas, J., Stover, J., Snodgrass, K., Wildt, D.E., 1992. Correlation of sperm viability with gamete interaction and fertilization in vitro in the cheetah (Acinonyx jubatus). Biol. Reprod. 46, 1047–1056.

Durrant, B.S., Millard, S.E., Zimmerman, D.M., Lindburg, D.G., 2001. Lifetime semen production in a cheetah (Acinonyx jubatus). Zoo Biol. 20, 359–366.

Eaton, R.L., Craig, S.J., 1973. Captive management and mating behavior of the cheetah. In: Eaton, R.L. (Ed.), The World's Cats I: Ecology and Conservation. World Wildlife Safari, Winston, OR, pp. 217–254.

Gottelli, D., Wang, J., Bashir, S., Durant, S.M., 2007. Genetic analysis reveals promiscuity among female cheetahs. Proc. R. Soc. Lond. 274, 1993–2001.

Graham, L.H., Goodrowe, K.L., Raeside, J.I., Liptrap, R.M., 1995. Non-invasive monitoring of ovarian function in several felid species by measurement of fecal estradiol-17β and progestins. Zoo Biol. 14, 223–237.

Hermes, R., Hildebrandt, T.B., Göritz, F., 2004. Reproductive problems directly attributable to long-term captivity—asymmetric reproductive aging. Anim. Reprod. Sci. 82–83, 49–60.

Howard, J.G., 1993. Semen collection and analysis in carnivores. In: Fowler, M.E. (Ed.), Zoo and Wild Animal Medicine Current Therapy III. W.B. Saunders Co, Philadelphia, PA, pp. 390–399.

Howard, J., Donoghue, A.M., Barone, M.A., Goodrowe, K.L., Blumer, E.S., Snodgrass, K., Starnes, D., Tucker, M., Bush, M., Wildt, D.E., 1992. Successful induction of ovarian activity and laparoscopic intrauterine artificial insemination in the cheetah (Acinonyx jubatus). J. Zoo Wildl. Med. 23, 288–300.

Howard, J.G., Donoghue, A.M., Johnston, L.A., Wildt, D.E., 1993. Zona pellucida filtration of structurally abnormal spermatozoa and reduced fertilization in teratospermic cats. Biol. Reprod. 49, 131–139.

Howard, J.G., Roth, T.L., Byers, A.P., Swanson, W.F., Wildt, D.E., 1997. Sensitivity to exogenous gonadotropins for ovulation induction and laparoscopic artificial insemination in the cheetah and clouded leopard. Biol. Reprod. 56, 1059–1068.

Howard, J.G., Wildt, D.E., 2009. Approaches and efficacy of artificial insemination in felids and mustelids. Theriogenology 71, 130–148.

Howard, J.G., Marker, L., Pukazhenthi, B.S., Roth, T.L., Swanson, W.F., Grisham, J., Wildt, D.E., 2002. Genome resource banking and successful artificial insemination with cryopreserved sperm in the cheetah. In: Proceedings of the 9th International Symposium on Spermatology. 70, (PL15).

Johnston, L.A., Donoghue, A.M., O'Brien, S.J., Wildt, D.E., 1991. Rescue and maturation in vitro of follicular oocytes collected from nondomestic felid species. Biol. Reprod. 45, 898–906.

Jurke, M.H., Czekala, N.M., Lindburg, D.G., Millard, S.E., 1997. Fecal corticoid metabolite measurement in the cheetah (Acinonyx jubatus). Zoo Biol. 16, 133–147.

Kelly, M.J., Laurenson, M.K., FitzGibbon, C.D., Collins, D.A., Durant, S.M., Frame, G.W., Bertram, B.C.R., Caro, T.M., 1998. Long-term demography of the Serengeti Cheetah population: the first 25 years. J. Zool. 244, 473–488, London.

Koester, D.C., Freeman, E.W., Brown, J.L., Wildt, D.E., Terrell, K.A., Franklin, A.D., Crosier, A.E., 2015a. Motile sperm output by male cheetahs (*Acinonyx jubatus*) managed *ex situ* is influenced by public exposure and number of care-givers. PLoS One 10, e0135847.

Koester, D.C., Freeman, E.W., Wildt, D.E., Terrell, K.A., Franklin, A.D., Meeks, K., Crosier, A.E., 2015b. Group management influences reproductive function of the male cheetah (*Acinonyx jubatus*). Reprod. Fertil. Dev. 29, 496–508.

Koester, D.C., Wildt, D.E., Brown, J.L., Meeks, K., Crosier, A.E., 2017. Public exposure and number of conspecifics have no influence on ovarian and adrenal activity in the cheetah (*Acinonyx jubatus*). Gen. Comp. Endo. 243, 120–129.

Lambo, C.A., Bateman, H.L., Swanson, W.F., 2013. Application of laparoscopic oviductal artificial insemination for conservation management of Brazilian ocelots and Amur tigers. Reprod. Fertil. Dev. 26, 116.

Laurenson, K.M., Caro, T.M., Borner, M., 1992. Female cheetah reproduction. Natl. Geogr. Res. Explor. 8, 64–75.

Lindburg, D.G., Durrant, B.S., Millard, S.E., Oosterhuis, J.E., 1993. Fertility assessment of cheetah males with poor quality semen. Zoo Biol. 12, 97–103.

Lueders, I., Ludwig, C., Weber, H., 2015. Nonsurgical artificial insemination in felids: Asiatic golden cat (*Catopuma temmincki*) and Persian leopard (*Panthera pardus saxicolor*). In: Proceedings of the International Conference on Diseases of Zoo and Wild Animals. Barcelona, Spain.

Lueders, I., Luther, I., Scheepers, G., van der Horst, G., 2012. Improved semen collection method for wild felids: Urethral catheterization yields high sperm quality in African lions (*Panthera leo*). Theriogenology 78, 696–701.

Maly, M.A., Edwards, K.L., Whisnant, C.S., Koester, D.C., Farin, C.E., Crosier, A.E., 2015. Assessing hormonal onset of puberty in captive cheetahs (*Acinonyx jubatus*) using fecal steroid metabolites. In: Goff, D., Morris, C. (Eds.), Proceedings of the Felid Taxon Advisory Group Annual Meeting. Pittsburgh, PA.

Marker, L.L., 2015. International cheetah (*Acinonyx jubatus*) studbook. Cheetah Conservation Fund, Namibia.

Marker, L.L., Dickman, A.J., Jeo, R.M., Mills, M.G.L., Macdonald, D.W., 2003. Demography of the Namibian cheetah, *Acinonyx jubatus jubatus*. Biol. Conserv. 114, 413–425.

Marker-Kraus, L., Grisham, J., 1993. Captive breeding of cheetahs in North American zoos: 1987–1991. Zoo Biol. 12, 5–18.

Marrow, J.C., Woc-Colburn, M., Hayek, L.A.C., Marker, L., Murray, S., 2015. Comparison of two α2-adrenergic agonists on urine contamination of semen collected by electrojaculation in captive and semi-free ranging cheetah (*Acinonyx jubatus*). J. Zoo Wildl. Med. 46, 417–420.

Munson, L., 1993. Disease of captive cheetahs (*Acinonyx jubatus*): results of the cheetah research council pathology survey, 1989–1992. Zoo Biol. 12, 105–124.

Munson, L., Gardner, A., Mason, R.J., Chassy, L.M., Seal, U.S., 2002. Endometrial hyperplasia and mineralization in zoo felids treated with melengestrol acetate contraceptives. Vet. Pathol. 39, 419–427.

Munson, L., Terio, K.A., Worley, M., Jago, M., Bagot-Smith, A., Marker, L., 2005. Extrinsic factors significantly affect patterns of disease in free-ranging and captive cheetah (*Acinonyx jubatus*) populations. J. Wildl. Dis. 41 (3), 542–548.

Pelican, K.M., Wildt, D.E., Pukazhenthi, B., Howard, J., 2006. Ovarian control for assisted reproduction in the domestic cat and wild felids. Theriogenology 66, 37–48.

Roth, T.L., Swanson, W.F., Blumer, E., Wildt, D.E., 1995. Enhancing zona penetration by spermatozoa from a teratospermic species, the cheetah (*Acinonyx jubatus*). J. Exp. Zool. 271, 323–330.

Roth, T.L., Wolfe, B.A., Long, J.A., Howard, J.G., Wildt, D.E., 1997. Effects of equine chorionic gonadotropin, human chorionic gonadotropin, and laparoscopic artificial insemination on embryo, endocrine, and luteal characteristics in the domestic cat. Biol. Reprod. 57, 165–171.

Schlafer, D.H., Gifford, A.T., 2008. Cystic endometrial hyperplasia, pseudo-placentational endometrial hyperplasia, and other cystic conditions of the canine and feline uterus. Theriogenology 70, 349–358.

Schulman, M.L., Kirberger, R.M., Tordiffe, A.S.W., Marker, L.L., Schmidt-Küntzel, A., Hartman, M.J., 2015. Ultrasonographic and laparoscopic evaluation of the reproductive tract in older captive female cheetahs (*Acinonyx jubatus*). Theriogenology 84, 1611–1619.

Swanson, W.F., Wolfe, B.A., Brown, J.L., Martin-Jimenez, T., Riviere, J.E., Roth, T.L., Wildt, D.E., 1997. Pharmacokinetics and ovarian-stimulatory effects of equine and human chorionic gonadotropins administered singly and in combination in the domestic cat. Biol. Reprod. 57, 295–302.

Terio, K.A., Marker, L., Munson, L., 2004. Evidence for chronic stress in captive but not free-ranging cheetahs (*Acinonyx jubatus*) based on adrenal morphology and function. J. Wildl. Dis. 42 (2), 259–266.

Terio, K., Marker, L., Overstrom, E., Brown, J., 2003. Analysis of ovarian and adrenal activity in Namibian cheetahs. S. Afr. J. Wildl. Res. 33, 71–78.

Terio, K.A., Whitham, J.C., Chosy, J., Sanchez, C., Marker, L., Wielebnowski, N., 2014. Associations Between Gastritis, Temperament and Management Risk Factors in Captive Cheetahs (*Acinonyx jubatus*). In: Proceedings of the Annual AAZA, American Association for Zoo Veterinarians. Conference, October 2014, Orlando, FL.

Terrell, K.A., Crosier, A.E., Wildt, D.E., O'Brien, S.J., Anthony, N.M., Marker, L., Johnson, W.E., 2016. Continued decline in genetic diversity among wild cheetahs (*Acinonyx jubatus*) without further loss of semen quality. Biol. Conserv. 200, 192–199.

Virtanen, R., 1989. Pharmacological profiles of medetomidine and its antagonist, atipamezole. Acta Vet. Scand. Suppl. 85, 29–37.

Wachter, B., Thalwitzer, S., Hofer, H., Lonzer, J., Hildebrandt, T.B., Hermes, R., 2011. Reproductive history and absence of predators are important determinants of reproductive fitness: the cheetah controversy revisited. Conserv. Lett. 4, 47–54.

Walzer, C., Kübber-Heiss, A., Bauder, B., 2003. Spontaneous uterine fibroleiomyoma in a captive cheetah. J. Vet. Med. A. Physiol. Pathol. Clin. Med. 50, 363–365.

Ware, L.H., Crosier, A.E., Braun, L., Lang, K., Joyner, P., Lockhart, S., Aitken-Palmer, C., 2016. Use of a novel radiographic technique for pregnancy determination and fetal count in cheetahs (*Acinonyx jubatus*) and maned wolves (*Chrysocyon brachyurus*). 35th AZVT Annual Conference. Point Defiance, September 9–12, 2015, Tacoma, WA.

Wielebnowski, N., 1999. Individual behavioral differences in captive cheetahs as predictors of breeding status. Zoo Biol. 18, 335–349.

Wielebnowski, N., Brown, J.L., 1998. Behavioral correlates of physiological estrus in cheetahs. Zoo Biol. 17 (3), 193–209.

Wielebnowski, N.C., Ziegler, K., Wildt, D.E., Lukas, J., Brown, J.L., 2002. Impact of social management on reproductive, adrenal and behavioural activity in the cheetah (*Acinonyx jubatus*). Anim. Conserv. 5, 291–301.

Wildt, D.E., Brown, J.L., Bush, M., Barone, M.A., Cooper, K.A., Grisham, J., Howard, J.G., 1993. Reproductive status of cheetahs (*Acinonyx jubatus*) in North American zoos: the benefits of physiological surveys for strategic planning. Zoo Biol. 12, 45–80.

Wildt, D.E., Bush, M., Howard, J.G., O'Brien, S.J., Meltzer, D., Van Dyk, A., Ebedes, H., Brand, D.J., 1983. Unique seminal quality in the South African cheetah and a comparative evaluation in the domestic cat. Biol. Reprod. 29 (4), 1019–1025.

Wildt, D.E., Obrien, S.J., Howard, J.G., Caro, T.M., Roelke, M.E., Brown, J.L., Bush, M., 1987. Similarity in ejaculate-endocrine characteristics in captive versus free-ranging cheetahs of two subspecies. Biol. Reprod. 36, 351–360.

Wildt, D.E., Phillips, L.G., Simmons, L.G., Chakraborty, P.K., Brown, J.L., Howard, J.G., Teare, A., Bush, M., 1988. A comparative analysis of ejaculate and hormonal characteristics of the captive male cheetah, tiger, leopard, and puma. Biol. Reprod. 38 (2), 245–255.

Wildt, D.E., Platz, C.C., Seager, S.W.J., Bush, M., 1981. Induction of ovarian activity in the cheetah (*Acinonyx jubatus*). Biol. Reprod. 24, 217–222.

Wildt, D.E., Rall, W.F., Critser, J.K., Monfort, S.L., Seal, U.S., 1997. Genome resource banks. Bioscience 47, 689–698.

Zambelli, D., Cunto, M., 2006. Semen collection in cats: techniques and analysis. Theriogenology 66 (2), 159–165.

Zambelli, D., Prati, F., Cunto, M., Iacono, E., Merlo, B., 2008. Quality and in vitro fertilizing ability of cryopreserved cat spermatozoa obtained by urethral catheterization after medetomidine administration. Theriogenology 69 (4), 485–490.

Ziegler-Meeks, K., 2009. Husbandry Manual for the Cheetah (*Acinonyx jubatus*). White Oak Conservation Center, Yulee, FL.

Communicating the Conservation Message—Using Ambassador Cheetahs to Connect, Teach, and Inspire

Suzi Rapp*, Kate Vannelli**, Linda Castaneda†, Annie Beckhelling‡, Susie Ekard§, Cathryn Hilker†, Janet Rose-Hinostroza§, Alicia Sampson†, Michelle Lloyd¶, Linda Stanek*

*Columbus Zoo and Aquarium, Powell, OH, United States
**Cheetah Conservation Fund, Otjiwarongo, Namibia
†Cincinnati Zoo & Botanical Garden, Cincinnati, OH, United States
‡Cheetah Outreach, Cape Town, South Africa
§San Diego Zoo Safari Park, San Diego, CA, United States
¶Monarto Zoo, Monarto, SA, Australia

INTRODUCTION

With collections of living representatives of a variety of species, zoos are uniquely positioned to share a conservation message with their visitors. The Association of Zoos and Aquariums (AZA) states that conservation education helps enhance the public's understanding of wildlife and the need to conserve the places where these animals live (https://www.aza. org/conservation-education). They define an ambassador animal as a representative of the wild members of their species that "interacts with the public in support of institutional education and conservation goals" (CEC Ambassador Animal Position Statement) to inspire individuals to embark on conservation actions (Manion, 2013; Pooley and O'connor, 2000; Povey and Spaulding, 2005). Studies show that ambassador animals help foster empathy, as well as emotional development among learners (Daly and Suggs, 2010).

The cheetah is an excellent candidate species for ambassador programs; it rates highly on the Program Animal Rating and Information System (PARIS), an evaluation tool used by the AZA Ambassador Animal Scientific Advisory Group to recommend the suitability of species to be program (ambassador) animals (PARIS, n.d.). The high scores are due to cheetahs' nonaggressive temperament and ease of trainability and handling when raised by experienced and qualified handlers. Public encounters with ambassador cheetahs vary between facilities, and include cheetah running demonstrations, educational presentations, media appearances, and up-close encounters. Encounters can create personal connections between humans and animals (Povey and Rios, 2002), with the potential for fostering understanding for the species and concern for the threats wild cheetahs face. In range countries in Africa, ambassador cheetahs can have a positive impact on the attitudes of local people by helping them understand the role that predators play in keeping their ecosystems healthy.

Most ambassador cheetahs are found in the United States (US) and South Africa, however some facilities in New Zealand, Australia, and Namibia also hold ambassador cats. There are currently no ambassador cheetahs in Europe (Marker, 2015).

THE HISTORY OF CHEETAH AMBASSADOR PROGRAMS

Although revered for thousands of years by nobles in many countries (Chapter 2), few cheetahs lived in zoos until the 1970s, due to poor captive reproductive success (Chapter 27). Wildlife Safari, in Oregon, was one of the few facilities worldwide to successfully breed cheetahs (Chapter 22). In 1976, Laurie Marker, the nursery and clinic supervisor and curator of the cheetah-breeding program, hand-raised a 1-month-old female cheetah cub, which was born there. She named the cub Khayam, and unintentionally began the concept of ambassador cheetah. Marker paired Khayam with her 4-year-old dog, a female Labrador mix, as a companion animal. Khayam had access to 2 ha of fenced land, learned to chase a mechanical lure, and later she ran behind a truck that pulled a lure. This provided her with the exercise to keep her healthy. Once grown, Marker took Khayam to Namibia, the birthplace of Khayam's parents, as part of a television documentary. The aim was to determine whether a cheetah's ability to hunt was instinctual or learned, and what steps were involved in teaching a captive-born cheetah to hunt in the wild. While Khayam learned to successfully hunt, she remained habituated to humans. After several months, Marker and Khayam returned to the United States.

Back in Oregon, Khayam became an ambassador for her species. She traveled with Marker throughout the United States for the next 10 years and assisted Marker in raising awareness of threats faced by wild cheetahs. Khayam met hundreds of thousands of people including school children, former US Presidents, members of the United Kingdom's Royal Family, famous artists, and Neil Armstrong, the first man to walk on the moon. "The American Sportsman," a television series, further increased awareness when it shared Khayam's story with a national audience (Marker, personal communication).

Khayam's popularity and reach demonstrated the potential of ambassador cheetahs, and in 1981, Wildlife Safari raised two more cheetah ambassadors. The female, Damara, remained at Wildlife Safari and the male, Arusha, was placed with the San Diego Zoo. Following Marker's advice, the zoo's Animal Training Supervisor, Kathy Marmack, matched the cheetah cub with a golden retriever puppy and the duo's popularity soared. Together, they interacted with millions of visitors in California. At about that same time, Cathryn Hilker at the Cincinnati Zoo & Botanical Garden,. received Angel, a cheetah

FIGURE 28.1 **Cathryn Hosea Hilker, founder of the Cat Ambassador Program at the Cincinnati Zoo & Botanical Garden, educating and inspiring people about the need to help save the cheetah with the help of an ambassador cheetah.** *Source: Cincinnati Zoo & Botanical Garden.*

cub from the Columbus Zoo and Aquarium (Fig. 28.1). She raised Angel with the help of her Great Dane. Later, Hilker bonded a young cougar with Angel. The cats became lifelong companions, and their story also reached millions.

In 1995, after Marker established the Cheetah Conservation Fund (CCF) in Namibia, she rescued an orphaned cheetah cub. She called the cub Chewbaaka, and paired him with an Anatolian shepherd puppy, a breed of dog typically used to guard livestock from predators. The duo laid the foundation for a story told today through many zoos, about how these dogs live and successfully protect livestock, in turn protecting cheetahs and other large predators by reducing human-wildlife conflict (Chapter 15; Fig. 28.2). Today CCF engages visitors at their Research and Education Center through cheetah run demonstrations.

In South Africa, Cheetah Outreach began an ambassador cheetah program in 1998 after Annie Beckhelling visited Namibia and learned more about the conflict with cheetahs and farmers. Cheetah Outreach, a public education facility near Cape Town, uses ambassador cheetahs to educate people on the plight of wild cheetahs, and the funds raised by these encounter programs support their livestock guarding dog program

FIGURE 28.2 **Orphaned as a young cub at the Cheetah Conservation Fund in Namibia, Chewbaaka, was raised with, Koya, an Anatolian shepherd livestock guarding dog.** Many zoos tell the story about using these dogs to protect livestock and chase off cheetahs, thus protecting cheetahs from being killed. Chewbaaka and Koya were lifelong companions. *Source: Laurie Marker, Cheetah Conservation Fund.*

for farmers in South Africa (Chapter 15). In 2000, Cheetah Outreach partnered with the DeWildt Cheetah and Wildlife Centre (now called the Anne vanDyke Cheetah Centre) to hand raise captive-born cubs selected for placement in zoos as ambassador cheetahs.

In 1999, Suzi Rapp started an ambassador cheetah program at the Columbus Zoo. She received 5 cubs from the Oak Hill Center for Rare and Endangered Species in Oklahoma.

Two of the cubs stayed at the Columbus Zoo, the other three went to Cincinnati Zoo, San Diego Zoo, and Wildlife Safari. All but one became educational ambassadors.

In 2002, the National Zoo and Aquarium in Canberra, Australia, received 3 cheetahs, which had been born at the DeWildt Cheetah and Wildlife Centre in South Africa and raised at Cheetah Outreach, beginning their cheetah ambassador program, as well as the ambassador cheetah movement in Australasia's zoos. In 2004, a cheetah at the Monarto Zoo in South Australia gave birth to a litter of cubs that required hand-rearing and started their ambassador program. Two of these cubs eventually went to the Australia Zoo in Queensland, continuing the spread of cheetah ambassador programs in Australia. In 2005, Michelle Lloyd initiated an ambassador program at Monarto Zoo, after learning about the plight of the wild cheetah, while volunteering in Namibia.

New Zealand's cheetah ambassador programs first began at Orana Park, located in Christchurch, in 2007 with 2 cubs from South Africa. Other cheetah ambassador programs later expanded throughout the country to include programs at Auckland Zoo, Hamilton Zoo, and Wellington Zoo. While New Zealand's ambassadors initially came from South Africa, New Zealand–bred cheetahs are now used.

The methods used to raise and care for today's ambassador cheetahs are based on Marker's original strategy with Khayam. These early cheetah ambassadors, along with the committed work of their trainers, paved the way for today's ambassador cheetah programs.

AMBASSADOR CHEETAHS' IMPACT ON CHEETAH CONSERVATION

Reaching the Public and Fostering Connections

The reach of ambassador cheetahs is considerable (Table 28.1). Not only do these ambassadors reach people at zoos, but also many "outreach" programs and special events take place off zoo grounds as well. Ambassador outreach programs visit local schools, public libraries, fairs, religious institutions, events, and media programs.

Between 2013 and 2015, 10 zoos with ambassador cheetah programs (9 in the United States and 1 in Australia) reached more than 6 million people through on- and off-site education programs and fundraisers for conservation partners (Table 28.1). The off-site programs disseminate educational materials and extend the conservation message beyond visitors to the zoo, thereby broadening each zoo's reach.

Within zoos, visitors vary widely in the range and breadth of their knowledge about wildlife and conservation issues (Moss et al., 2014). Therefore, people who encounter ambassador cheetahs bring a wide variety of interest levels, understanding of cheetahs and their struggles, and personal involvement in conservation issues. Studies show that ambassador animals effectively engage audiences and nurture emotional connections between the visitors and the animals (Povey and Spaulding, 2005), thus making the plight of the species tangible (McGuire, 2015). Creating an emotional connection is the most effective means of inspiring people to take actions that benefit cheetahs, as well as fostering a long-term stewardship ethic (Manion, 2013; Pooley and O'connor, 2000; Povey and Spaulding, 2005).

Some zoos have facilities specially designed to demonstrate cheetahs' running speed. There, an ambassador cheetah runs while an educator interprets the experience and conveys a conservation message. The Cincinnati Zoo's cheetah run draws over 100,000 visitors a year, and the Columbus Zoo reports about 1,200 visitors attending their cheetah runs each day from April to October (Table 28.1).

Local and national news and media outlets provide an opportunity to reach an even broader audience. Ambassador cheetahs at the

TABLE 28.1 Average Annual Number of Ambassador Cheetah Outreach Programs Between 2013 and 2015 at the Main Cheetah Ambassador Facilities in the United States and one Facility in Australia

Facility	Zoo programs	Special event programs	School programs	Fundraisers for conservation partners
Columbus Zoo, Ohio, USA	365 (75,167)	737 (96,333)	13 (2,755)	9 (4000)
San Diego Global, California, USA	1426 (270,400)	332 (34,690)	23 (1,004)	2 (400)
Cincinnati Zoo, Ohio, USA	250 (133,333)	2 (400)	20 (6,240)	1 (800)
Wildlife Safari, Oregon, USA	1095 (54,750)	15 (7,500)	10 (1,000)	5 (150)
Cat Haven, California, USA	1414 (28,030)	4 (600)	33 (12,967)	1 (337)
Busch Gardens, Florida, USA	2000 (300,000)	10 (1 million[a])	0 (N/A)	1 (300)
Wild Cat Conservation and Education Fund, California, USA	N/A (N/A)	N/A (N/A)	75 (37,750)	3 (300)
Wild Wonders, California, USA	187 (2,067)	? (?)	N/A (N/A)	6 (100)
Zoo Miami, Florida, USA	534 (14,000)	13 (1,667)	N/A (N/A)	3 (625)
Monarto Zoo, South Australia	450 (3,600)	N/A (N/A)	N/A (N/A)	4 (400)
Total	7721 (881,347)	1113 (1,141,190[a])	174 (61,716)	35 (7412)

For each event type the average number of events per year between 2013 and 2015 is indicated in the table; the average number per year of people attending the events is indicated in brackets.
[a] Directed media including TV appearances.

Columbus Zoo travel year-round across the United States to appear at programs with the Columbus Zoo Director Emeritus, Jack Hanna (Fig. 28.3).

Ambassador cheetahs are important in range countries as well—especially in areas where wild cheetahs are threatened because of negative perceptions about the species. For instance, at Cheetah Outreach, almost 60,000 people visited in 2015, and more than 71,000 in 2016. Cheetah Outreach provides the general public with hands-on experiences, allowing visitors to touch a cheetah, and either feel or hear their purrs. In Namibia, CCF's Field Research & Education Center welcomed more than 10,000 tourists in 2016. An additional 1000 school children and 300 farmers from throughout the country also traveled to CCF to learn about cheetahs during that year. While the majority of CCF's cheetahs are not habituated to humans, the orphan cheetahs that are viewable by the public instill a sense of awe in visitors who witness their feeding and social interactions. Seeing cheetahs in this light helps motivate local people to take responsibility of their own natural resources. In these high-priority conservation areas, animal encounters can help shift attitudes and create a new respect and appreciation of the species.

FIGURE 28.3 Jack Hanna, Columbus Zoo Director Emeritus, on a television show with a cheetah cub and puppy. *Source: Grahm S. Jones, Columbus Zoo.*

As a flagship species, a cheetah ambassador also represents the struggles faced by predators throughout the world. Even in areas where no cheetahs live, ambassador cheetahs can help the public understand and connect with the challenges faced by their own native predators. Utilizing these ambassadors, educators foster a connection between people and predators to teach the value of a predator's role in keeping an ecosystem healthy and productive for the benefit of all species there.

Enhancing Learning

Evidence indicates that public engagement is lengthened when people view a live animal in an interpretive setting (Povey and Rios, 2002). A study of adults at the Brookfield Zoo showed memory retention rates as high as 83% on some of the educational messages 6 weeks after a live animal show (note that the study lacked a control group; Heinrich and Birney, 1992). Povey and Rios (2002), showed that 69% of visitors who watched presentations that included ambassador animals said they learned something, versus 9% for the visitors who viewed the same animal in its exhibit without being accompanied by an educational message. The study also showed that visitors had an 11 times higher ability to answer questions about an ambassador animal versus one on exhibit. Similarly, a 2015 study done at The Ohio State University looked at the relationship between having live animals in the classroom and learning. Results showed higher student retention of information when animals were in the room, further supporting the use of ambassador cheetahs as powerful allies in teaching conservation (George and Cole, 2016). Another study looked at the transmission of conservation messages from children to their parents, and found that under certain circumstances teaching young people creates an increase in knowledge in both their parents and the community, supporting the use of ambassador cheetahs with people of all ages for a positive effect on conservation (Vaughan et al., 2003).

Cheetah ambassadors create experiential learning opportunities, which Kolb et al. (2000) defines as "the process by which knowledge is created through the transformation of experience." This happens as visitors go through the process of observing a cheetah (a concrete experience), to abstract conceptualization (this is what a cheetah is like), to reflective observation (this is what I now know), and finally to active experimentation (this is what I will do) (Kolb, 1984). Therefore, ambassador cheetahs provide a greater opportunity for public education as compared to exhibit animals or educational graphics (Heinrich and Birney, 1992; Manion, 2013; Povey and Spaulding, 2005; Yerke and Burns, 1991, 1993 as cited in AZA, 2015).

Inspiring Conservation

Ambassador cheetahs are also useful in inspiring conservation actions. To be effective, cheetah conservation groups and other non-governmental organizations (NGOs) need support and funding from the public for vital aspects of integrated conservation, such as research into health and reproductive challenges, finding and implementing strategies for habitat restoration, and mitigation of human-wildlife conflict, as well as expanding *in situ* education opportunities. To drive these donations, the public needs to gain an awareness of the threats and the stark realization of what is at stake if measures are not taken to conserve cheetahs. People who encounter ambassador cheetahs become inspired to tell others about the challenges these cats face, and some become donors to support conservation efforts. Guests can impact the species' conservation both directly, by donating to or engaging in cheetah conservation, and indirectly, by visiting cheetah range countries and thus contributing to the local economy (Chapter 23).

Ambassador cheetahs also appear with conservationists when they give lectures or hold fundraising events (Table 28.1). In addition to increasing attendance, donors at CCF's conservation events in the United States gave 65% more in donations at events that included an ambassador cheetah as compared to events without one, demonstrating the impact of having a live cheetah present. Following an event with a cheetah, a single donor gave US $1,000,000 to fund field conservation programs for 5 years. At another event with a cheetah, a donor gave US $250,000 for the purchase of land in Africa. Cheetah ambassadors help conservationists drive their points home by giving attendees an up-close and personal look at the species, illustrating what might be lost.

Cheetah ambassador programs also create opportunities for zoos to raise funds for *in situ* conservation. The Columbus Zoo donates 100% of their profits from the sale of cheetah merchandise, as well as a percentage of cheetah program revenues to *in situ* conservation organizations. These combined revenues added up to more than US $250,000 between 2000 and 2014 (AZA, 2015). The Cincinnati Zoo's Angel Fund, named after their first ambassador cheetah, supports a variety of *in situ* cheetah conservation projects and also benefits cheetahs *ex situ*. In 15 years, the Angel Fund donated over US $1,000,000 to cheetah conservation (Cincinnati Zoo & Botanical Garden, n.d.). The Monarto Zoo ambassador program's proceeds support CCF's programs. CCF in Namibia, Cheetah Conservation Botswana (CCB), Cheetah Outreach in South Africa, and Action for Cheetahs in Kenya (ACK) have all benefited through direct funding and enhanced zoo partnerships. In addition to the funding, ambassador cheetahs inspire people to develop careers in conservation, animal research, and zoo keeping.

CREATING A CHEETAH AMBASSADOR PROGRAM

SSP and Government Guidelines

The AZA Cheetah Species Survival Plan's® (SSP) (Chapter 23) position statement supports the use of ambassador program cheetahs, labeling them as a powerful tool for conservation education, public relations, and fundraising. It also outlines the conditions required for their use. These conditions include appropriate acquisition policies working through the Cheetah SSP and compliance with its individual Institutional Program Animal Policies for the management and training of ambassador cats (Ziegler–Meeks, 2009). AZA's accreditation standards further require that "the conditions and treatment of animals in education programs must meet standards set for the remainder of the animal collection, including species-appropriate shelter, exercise, appropriate environmental enrichment, access to veterinary care, nutrition, and other related standards" (AZA, 2015). Guidelines for cheetah ambassador programs also cover hand-raising. These guidelines can be obtained through the AZA website (SSP Program).

Working with a cheetah in public is never done without careful thought and planning. Cheetah handlers follow specific safety protocols when a cheetah is presented, in order to maintain public, animal, and staff safety. When traveling, it is recommended that cheetahs are transported in an appropriate-sized crate, inside a fully enclosed, temperature-controlled vehicle, in accordance with AZA rules. Cheetah trainers always evaluate the site where the cheetah will appear before bringing the cheetah into an event. Although not the case in all countries, the United States Department of Agriculture regulations dictate that the public is not allowed direct contact with an ambassador cheetah (USDA, 2012), and a sufficient distance and/or barrier between the public and the ambassador cheetah is required. When working with any wild animal, relevant government agencies' rules apply. Zoo association accreditation ensures that an institution's ambassador program is following zoo association guidelines and working in accordance with regional legislation.

Ambassador Cubs—Circumstances and Considerations

Many factors are vital to the successful raising of a cheetah as an ambassador. Cheetahs are not pets; they are wild animals and should be treated as such.

When selecting a cub for use in an ambassador program, habituation to humans must start at a very early age so the cub imprints on a caregiver, ideally before 3 weeks of age. Cubs may be selected from a litter in a zoo or other registered facility for hand raising, or they may be hand raised because they were orphaned or pulled from the mother due to absent or insufficient maternal care. Hand raised cubs require intensive care in order to survive (Chapter 24) and usually become ambassador cheetahs. In rare cases, wild orphan cubs are rescued and require bottle feeding and extensive care, which habituates them to humans. However the cub is acquired, the individual temperament determines whether or not it will become a good ambassador. Each cheetah's suitability must be assessed early on and before the animal is brought into a public setting.

Hand-Rearing, Socializing, and Training

Once the decision is made to raise a cheetah cub as an ambassador, bonding between the cheetah and handler is essential. The cheetah needs to be extremely comfortable with people and unfamiliar situations to ensure future success.

It is important to consider the health and safety of the cheetah, as well as the natural history of the species when embarking on hand-rearing an ambassador. Cubs are very active from an early age, and their bone development is dependent on a good diet and appropriate activity. Dietary requirements must be followed strictly to maintain good health (Chapters 24 and 26). As a cub grows and develops, exercise and play are essential for its physical and mental development.

A cub should never be allowed to engage in behavior that would not be tolerated as an adult (e.g., chasing and tripping people), and positive reinforcement training is the recommended method (Povey, 2017). As cubs get older, a lure course for running is a good form of exercise and helps them maintain physical and mental health.

Early on, cubs should be habituated, desensitized, and socialized to the world around them. It is important to habituate the cheetah to potentially intimidating situations, such as doors shutting, televisions, crowds of people, or anything it may come in contact with throughout its life. Program cheetahs need to be socialized to behavioral constraints, such as sitting during programs, and need to learn that being on-leash means that it's time to work.

Handlers work long and hard to train their animals to behave gently and appropriately. The importance of consistency cannot be overstated. Learning species' precursors and each individual's cues can help handlers redirect potentially bad behavior by recognizing the warning signs. Most cubs are raised with one or two primary handlers. As the ambassadors get older, they can be introduced to additional handlers, but consistency in care and training, and familiarity with the handlers is necessary to provide the cheetah with security. While it is important to adhere to best practices in rearing and training, it is equally important to remember that each cheetah is an individual, and so the care and training of each must be individually tailored. It is critical that all ambassador staff members work closely with their program's veterinary staff, as well as with outside mentors, to determine what is best for each cheetah. They must also keep in mind that not every cheetah will be a successful ambassador, even with appropriate training.

Working with cheetahs requires extensive previous experience with animal training and handling. There are both courses and books that teach training practices that use positive reinforcement. Additionally, trainers new to cheetahs must begin their hands-on work with a mentor or with experienced cotrainers.

Companion Dogs

Species-appropriate exercise and enrichment for a young cheetah, mandated by the AZA, includes the ability to run and play with a sibling, much like it would in the wild. Pairing an ambassador cheetah with a companion dog accomplishes many of these goals. The cheetah can engage in natural play behavior with a dog without learning to stalk, grab, or chase humans, which is the concern if it plays with its handlers. Pairing a cheetah with a dog fosters confidence in the cheetah. Cheetahs have a natural tendency to flee versus fight, but when raised with a puppy they tend to take their cues from, and stay with, their dog (Fig. 28.2). A cheetah will assess its dog's reaction to unfamiliar situations and read the dog's body language to gauge whether a situation is safe. The dog can also act as a barrier between the cheetah and people, helping the cheetah to feel safe. Overall, rearing a cheetah with a dog often increases its chance for success as an ambassador and its overall quality of life. This practice has been so successful, that the AZA recommends it to their zoos.

Though not every cheetah will have a dog companion its whole life, building a relationship between a cheetah and a dog should begin at a very young age. The ideal age for both is between 6 and 8 weeks, once both are vaccinated and through their quarantine periods. Cubs have also been successfully paired with older dogs that have the right temperament, energy, and patience. An ideal dog companion is willing to play, and is easily and quickly calmed down.

LOOKING TO THE FUTURE

Cheetahs that are designated as ambassadors in the United States have been excluded from the AZA Cheetah SSP's population management analyses, which means that they are not recommended for, and thus do not become a part of, the breeding program. In 2014, 15% of the SSP population were ambassadors, and thus not available for breeding. Projections state that under current conditions, the core SSP population will not reach its target population size to retain sustainability for the next 100 years (AZA, 2013; Crosier et al., 2016). In the Australasian Species Management Program (ASMP) (Chapter 23), half of the cheetah population is involved in the breeding program, while the other half works strictly as ambassadors. So while the ambassador cheetah populations are not large, they have the potential of boosting population numbers. That being said, breeding decisions are always predicated on genetic suitability to protect the viability of the population, which would be determined in the United States by the SSP (Chapters 22 and 23), and ultimately on cheetah mate choice.

Breeding success of hand raised cheetahs has not yet been examined, however, Hampson and Schwitzer (2016) found that cheetahs that were hand raised and successfully bred have a lower rate of infant mortality in their litters than mother-raised females, possibly because they are less stressed following a birth as their handlers reassure them.

CONCLUSIONS

Zoos often identify cheetahs as one of the top species visitors come to see. International cheetah educators regularly witness remarkable connections made between their animals and the public. Excited and engaged people telling others about the critical nature of the cheetah's struggles is invaluable, as generating awareness of the problem is the first step to finding and funding help. When people have encounters with ambassador cheetahs, they are likely to learn more, remember more, and make a personal connection that can spark interest, understanding, and compassion for these beautiful cats and their wild counterparts. Cheetah Outreach's motto perfectly illustrates the vital impact ambassador cheetahs can have in terms of conservation: "See it, sense it, save it." As the wild cheetah population struggles to survive in an ever-changing landscape, zoos with ambassador programs continue to evolve, exploring best practices to make the critical connection between the public and the cheetah. Every zoo's fervent aim is that this connection will, in turn, contribute to saving cheetahs from extinction.

References

AZA, 2013. Cheetah (*Acinonyx jubatus*) AZA Animal Program Population Viability Analysis Report. AZA, Silver Spring, MD.

AZA, 2015. CEC Ambassador Animal Position Statement. Available from: https://www.aza.org/cec-ambassador-animal-position-statement.

Cincinnati Zoo & Botanical Garden, n.d. Cheetah Conservation Available from: http://cincinnatizoo.org/conservation/field-projects/cheetah-conservation-angel-fund/.

Columbus Zoo's Cheetah Conservation. Association of Zoos and Aquariums, n.d. Available from: http://s3.goeshow.com/aza/annual/2015/profile.cfm?profile_name(exhibitor&master_key(F6559FDC-DA0E-E511-BD45-04DA201B3FF&inv_mast_key(EF1B2158-E2E8-613A-6542-50E13AEFD3A7&xtemplate.

Crosier, A., Moloney, E., Long, S., 2016. Population Analysis and Breeding and Transfer Plan.Cheetah (*Acinonyx jubatus*) AZA Species Survival Plan Yellow Program. AZA, Silver Spring, MD.

Daly, B., Suggs, S., 2010. Teachers' experiences with humane education and animals in the elementary classroom: implications for empathy development. J. Moral Educ. 39, 101–112.

George, K.A., Cole, K., 2016. Educational Value of Human-Animal Interactions. Poster session presented at the meeting of the North American Colleges and Teachers of Agriculture, Mānoa, HI.

Hampson, M.C., Schwitzer, C., 2016. Effects of Hand-Rearing on Reproductive Success in Captive Large Cats *Panthera tigris altaica, Uncia uncia, Acinonyx jubatus* and *Neofelis nebulosa*. PLoS One 11, e0155992.

Heinrich, C.J., Birney, B.A., 1992. Effects of live animal demonstrations on zoo visitors' retention of information. Anthrozoös 5, 113–121.

Kolb, D.A., 1984. Experiential Learning: Experience as a Source of Learning and Development. Prentice-Hall, Englewood Cliffs, New Jersey.

Kolb, D.A., Boyatzis, R.E., Mainemelis, C., 2000. Experiential learning theory: previous research and new directions. Perspect. Think. Learn. Cogn. Styles 1, 227–247.

Manion, K., 2013. The Effect of Different Learning Experiences with African Penguins on Visitor Knowledge, Attitude, and Behavior. Unpublished research project. George Mason University. Fairfax, VA.

Marker, L., 2015. 2014 International Cheetah (*Acinonyx jubatus*) Studbook. Cheetah Conservation Fund, Otjiwarongo.

McGuire, N.M., 2015. Environmental Education and behavioral change: an identity-based environmental education model. Int. J. Environ. Sci. Educ. 10, 695–715.

Moss, A., Jensen, E., Gusset, M., 2014. A Global Evaluation of Biodiversity Literacy in Zoo and Aquarium Visitors. WAZA Executive Office, Gland, p 37.

PARIS n.d. Program Animal Rating and Information System, AZA Available from: http://zooparis.wikispaces.com/.

Pooley, J.A., O'Connor, M., 2000. Environmental education and attitudes: emotions and beliefs are what is needed. Environ. Behav. 32, 711–723.

Povey, K., 2017. The Management of Ambassador Animals: A Handbook for Wildlife Professionals. Association of Zoos and Aquariums, Silver Spring.

Povey, K.D., Rios, J., 2002. Using interpretive animals to deliver affective messages in zoos. J. Interpret. Res. 7, 19–28.

Povey, K.D., Spaulding, W., 2005. Message design for animal presentations: a new approach. Presented at the Association of Zoos and Aquarium National Conference.

United States Department of Agriculture, 2012. Animal and Plant Health Inspection Services. Animal Care. USDA, Available from: https://www.aphis.usda.gov/ animal_welfare/hp/downloads/strategic_plan/AC-Strategic-Plan-2016-2020_092716.pdf.

Vaughan, C., Gack, J., Solorazano, H., Ray, R., 2003. The effect of environmental education on schoolchildren, their parents, and community members: a study of intergenerational and intercommunity learning. J. Environ. Educ. 34, 12–21.

Ziegler–Meeks, K., 2009. Husbandry Manual for the Cheetah (*Acinonyx jubatus*). White Oak Conservation Center, Yulee.

TECHNIQUES AND ANALYSES

The Use of Remote Camera Trapping to Study Cheetahs: Past Reflections and Future Directions

Ezequiel Fabiano*, Lorraine K. Boast**,
Angela K. Fuller[†,a], Chris Sutherland[‡]

*University of Namibia, Katima Mulilo, Namibia
**Cheetah Conservation Botswana, Gaborone, Botswana
[†]Cornell University, Ithaca, NY, United States
[‡]University of Massachusetts-Amherst, Amherst, MA, United States

Remote camera trapping and associated advances in ecological statistics (i.e., capture–recapture [C–R] modeling) provide effective and efficient means to study wild animal populations (O'Connell et al., 2011). Photographic C–R has been applied across a diverse range of taxa to investigate many types of ecological questions, for example, to assess the impacts of land-use on species diversity (Kauffman et al., 2007), to estimate and assess trends in species distribution, abundance, and density (e.g., tigers *Panthera tigris*; Karanth et al., 2006), and to examine species' behavior and social interactions (e.g., Eurasian lynx *Lynx lynx*; Vogt et al., 2014). Camera trapping provides a non-invasive approach especially useful for studying low-density and elusive species, such as

the cheetah (*Acinonyx jubatus*). In addition, cheetahs are individually recognizable from photographs by their unique spot patterns, which remain unaltered throughout the lifetime of an individual (Caro and Durant, 1991; Chapter 32), providing natural marks that can be used to estimate abundance using C–R methods (O'Connell et al., 2011).

Here, we provide a brief overview of previous applications of camera trapping of cheetahs, focusing on spatial capture–recapture (SCR) approaches to estimate cheetah abundance and density. We discuss the challenges faced when surveying cheetahs using camera traps and provide some suggestions that can improve the success of future cheetah camera trapping studies.

[a]Any use of trade, firm, or product names is for descriptive purposes only and does not imply endorsement by the U.S. Government.

Cheetahs: Biology and Conservation
http://dx.doi.org/10.1016/B978-0-12-804088-1.00029-0

SURVEYING CHEETAHS: APPLICATIONS OF CAMERA TRAPPING

The most common application of camera trap surveys in cheetah research has been to estimate abundance and density using C–R methods. These studies have been conducted in a variety of environments, ranging from the hyperarid Saharan desert in Algeria to thornbush and woodland savannah habitats in farmland/Botswana, Kenya, Namibia, and South Africa (Table 29.1). In addition to abundance and density, camera trapping has been used to investigate other aspects of cheetah ecology, such as confirming cheetah presence in remote habitats of the Sahara desert (Sillero-Zubiri et al., 2015), assessing annual survival on farmlands (Cheetah Conservation Fund, unpublished data), evaluating temporal activity patterns (Belbachir et al., 2014; Fabiano, 2013), investigating the role of scent marking posts for social communication (Fig. 29.1) (Fabiano, 2013; Marnewick et al., 2006), and monitoring distribution patterns (Andresen et al., 2014).

ESTIMATING CHEETAH ABUNDANCE AND DENSITY WITH CAMERA TRAPS

Conducting camera trap surveys to estimate abundance or density requires that multiple cameras are placed throughout a study area. The number of cameras deployed, their exact location, the distance between cameras, and the duration of study should be determined by the specific ecological question of interest, the ecology of the species, knowledge gained through pilot or previous surveys, and the availability of resources. To capture both flanks of an individual for individual identification, cameras are often, although not always (e.g., Brassine and Parker, 2015), placed opposite each other in pairs. Camera placement depends strongly on the movement patterns of the species; ideal locations are those areas that are likely to be utilized by all classes of individuals (i.e., males and females of all ages). Typically for cheetahs, cameras are placed along roads or trails or at scent marking sites. To obtain quality photographs of the entire body of a passing cheetah, and to maximize the likelihood of individual identification, cameras should be placed at approximately shoulder height of an adult cheetah (e.g., 75 cm; Boast et al., 2013), and to reflect daily activity patterns, should be operational for 24 h a day (Cozzi et al., 2012). Individual identification has been manually performed (e.g., Marnewick et al., 2008), but spot recognition software is available (Kelly, 2001).

It is well understood that when surveying a population, not all individuals are likely to be encountered (i.e., surveys rarely result in a complete census). As such, statistical models are required to estimate the unobserved proportion of the population and thus the total population size. When a population is sampled repeatedly and individuals are identifiable by natural markings or otherwise, closed population C–R models can be used to analyze the resulting capture histories to estimate detectability and abundance (Royle et al., 2014; Williams et al., 2001). Capture histories represent a temporal sequence of individual encounters over multiple sampling occasions, which are commonly (though not a requirement) defined as a single day in camera trap studies. The period of time over which a survey is conducted is assumed to be closed (i.e., no additions or losses to the population), and depends on the biology, movement, and life history of the focal species. For large carnivores this period has typically been approximately 90 days (Hedges et al., 2016) including for cheetahs (Table 29.1). One criticism of traditional C–R approaches is that they are nonspatial, that is, capture histories indicate only the occasion during which an individual was detected and ignore the location of the detection (Chapter 1 in Royle et al., 2014).

TABLE 29.1 A Summary of the Study Design and Calculated Estimates of Cheetah Abundance and Density in Camera Trapping Studies

		Marnewick et al. (2008)	Marker et al. (2008)	O'Brien and Kinnaird (2011)	Fabiano (2013)[a]	Belbachir et al. (2014)	Boast et al. (2015)	Brassine and Parker (2015)[b]
Survey design	Study site/country (number of surveys)	Fenced reserve/South Africa (n = 1)[c]	Conservancy/Namibia (n = 1)[d]	Conservancy/Kenya (n = 1)[d]	Conservancy/Namibia (n = 10)[d]	Unfenced protected area/Algeria (n = 2)	Botswana farmland/Botswana (n = 1)[d]	Fenced reserve/reserve/Botswana (n = 2)[c]
	Survey area size (km²)	240	277	200	341 ± 41	2551	475	240
	Study duration (days)	30	90	84	90 ± 0	60; 90	84	90; 130
	No. of cameras stations	12	13	21	16 ± 1	40	26	60; 60
	Space between camera stations (km ± SD)	—	17.0 ± 9.2	1.3 ± 0	17.0 ± 9.2	10.0 ± 0	4.3 ± 0.8	3.7; 3.1
	Survey design[e]	Static	Static	Block	Static	Static	Block	Block
	Statistical analysis method[f]	Nonspatial	Nonspatial	Nonspatial	Nonspatial; B-SCR	Nonspatial	B-SCR	B-SCR
Results	No. of camera trap nights	120	1170	1764	1377 ± 120	1862; 3367	1063	2660; 3750
	No. of days to first cheetah capture	3	1	—	38 ± 28	—	6	9
	Days until all identified individuals are detected	9	84	—	45 ± 20	35; 82	45	—
	No. of adult cheetahs detected (male: female)	5:0	11:0	3:0	32:7	4:0; 3:0	4:1	5:2; 5:2
	No. of independent captures of adult cheetahs	12	72	4	332 ± 231	15; 17	17	18; 31

(Continued)

5. TECHNIQUES AND ANALYSES

TABLE 29.1 A Summary of the Study Design and Calculated Estimates of Cheetah Abundance and Density in Camera Trapping Studies (cont.)

	Marnewick et al. (2008)	Marker et al. (2008)	O'Brien and Kinnaird (2011)	Fabiano (2013)[a]	Belbachir et al. (2014)	Boast et al. (2015)	Brassine and Parker (2015)[b]
Capture probability ± SD (95% HPD)	0.17	0.29	0.04	0.32 ± 0.21 (0.12 – 0.78)	0.20; 0.21	NA	NA
Baseline encounter[g] ± SD (95% HPD)	—	—	NA	0.34 ± 0.33 (0.09 – 1.03)	—	0.04 ± 0.03 (0.002 – 0.091)	0.03 ± 0.03 (0.01 – 0.07); 0.01 ± 0.01 (0.00 – 0.02)
Spatial scale (mean ± SD (km, 95% HPD)	—	—	NA	5.03 ± 1.70 (1.63 – 7.26)	—	8.83 (±13.55, 3.60 – 15.49)	5.10 (±1.02, 3.27 – 7.14); 6.29 (±1.57, 3.78 – 9.48)
FMMDM (km)	—	12.4	0	9.67 ± 4.7	44.9; 44.9	11.9	28.0
Density/1000 km² ± SD (95% CI/HPD)[h]	NA	9.4 ± 2.1	22.5	Nonspatial 6.0 ± 4.0 (1.0 – 15.0); B-SCR 11.0 ± 4.0 (1.0 – 19.0)	0.3; 0.2	3.2 (0.40 – 7.7)[g]	6.1 ± 1.8 (3.0 – 9.0) 5.8 ± 2.0 (2.4 – 9.0)

Measures of precision are the standard deviation (SD), the 95% confidence intervals (95% CI) or highest posterior distribution (95% HPD).

[a] Average of individual surveys (mean ± SD, where applicable).

[b] Results presented are those based on the 90 and 130 trapping days with cameras placed at focal points.

[c] Fenced but does not restrict cheetah movement.

[d] Unfenced cattle and wildlife conservancy.

[e] Block (cameras are relocated to a new survey block which is part of the total area to be surveyed) and static (1 survey block, cameras are not relocated).

[f] Nonspatial, density estimation determined by adding a buffer around the camera trapping grid [based on the full mean maximum distance moved (FMMDM) of individuals captured at different camera stations]; B-SCR, Bayesian Spatial Capture Recapture.

[g] Baseline encounter = probability of detection if the camera trap was exactly at the activity center of an individual.

[h] The assumption that all individuals have independence movements was only upheld in Boast et al. (2015).

FIGURE 29.1 **A male cheetah scent marking a tree by spraying urine, captured by a remote camera trap in north-central Namibia.** *Source: Cheetah Conservation Fund.*

Estimating absolute density becomes problematic in nonspatial C–R methods because individuals may move outside the area delimited by a camera trap array inducing a form of individual heterogeneity in detectability which makes it difficult to define precisely the effective area being sampled by an array of cameras. Various methods have been proposed to calculate this area in nonspatial C–R (Karanth and Nichols, 2002; Wilson and Anderson, 1985a,b). These methods include creating buffers around each camera station or around the minimum convex polygon of the outer camera locations. Buffer size can be based on the mean maximum distance moved (MMDM) of individuals captured at different locations, the half MMDM, or using the mean home range radius (HRR) calculated from a representative sample of collared cheetahs from the same area and time. The full MMDM and the HRR are considered the most accurate for estimating density using nonspatial closed population models in large carnivores (Sharma et al., 2010), including cheetahs (Marker et al., 2008). Although these methods are commonly applied they are ad hoc, and fail

to explicitly model species movement parameters during density estimation.

In addition, nonspatial C–R fails to explicitly accommodate the individual encounter heterogeneity (i.e., patterns of detection that vary by individual) due to the between–individual variability in trap exposure. It is logical to assume that individuals with activity centers (home range centers) closer to a camera trap will have more detections than that of an individual located farther from the same camera trap. Because the density of camera traps will be lowest at the edges of a trapping array, individuals that spend most of their time on the periphery or off the trapping array will have a lower overall detection probability than those located within the center. Failure to account for detection heterogeneity can result in biased estimates of abundance (Otis et al., 1978).

A preferable alternative to nonspatial C–R is SCR (Borchers and Efford, 2008; Efford, 2004; Royle et al., 2009a). The major advance of SCR models is the inclusion of a spatially explicit encounter model that relates the detection of individuals to the distance between an individual's activity center and a camera trap, and a spatial point process model that describes the distribution of activity centers within a prescribed area. Key to estimating the spatial scale of detection (i.e., movement range of an individual) are the unique spatial locations where individuals are detected, that is, spatial encounter histories that acknowledge when, and importantly where, individuals were encountered. The objective in SCR is to estimate the number of unobserved individuals and thus the total number of individuals (activity centers) within the prescribed area, providing an estimate of absolute density. Thus, SCR directly addresses the two major concerns of nonspatial C–R models: individual heterogeneity in detectability due to the juxtaposition of individuals and traps (Efford et al., 2009; Royle et al., 2009b), and the explicit definition of the sampled area included as part of the model.

TABLE 29.2   A Review of Core Nonspatial and Spatial Capture Recapture (SCR) Model Assumptions and an Indication of Potential Camera Study Design Aspects That can Reduce the Degree of Bias in Estimates Resultant When These Assumptions are Violated

Assumptions	Nonspatial models	SCR model	Aspect of camera study design to consider to reduce risk of violating assumption
1. All individuals are correctly identified from photographs	✓	✓	Use multiple individual observers to identify individuals
2. No unmodeled variation in the probability of detection	✓	✓	Adequate camera trap spacing, density and placement of cameras at focal points that maximize both sex detection
3. The population is demographically closed	✓	✓	Length of study, time of year survey conducted (e.g., avoid birth peaks)
4. The population is geographically closed	✓	✓	Survey areas include multiple individuals' complete home ranges. Avoid sampling when dispersal is likely
5. Each capture is an independent event	✓	✓	Include covariates (e.g., social groups); or by inflating variance (e.g., c-hat, an estimate of overdispersion); further analysis needed
6. All individual's home ranges are circular		✓	Apply, if necessary and data permits, the nonstationary home range SCR model using ecological distances
7. Distribution of individuals follows a Poisson or Binomial distribution		✓	
8. Individuals have independent activity centers	—	✓	Apply, if data allow, SCR models that allow for dependence between locations
9. Home ranges are fixed during the survey period	—	✓	Conduct surveys within a single season; apply Markovian SCR models or include a resource selection function as a density covariate
10. Probability of detection declines with distance of the activity center from the trap in a Euclidean manner	—	✓	Apply the nonstationary home range SCR model using ecological distances
11. Areas around cameras is homogenous in terms of habitat suitability	✓	✓	Selection of nonsuitable habitat in SCR models, large study area

Adapted from Boast L., 2014.

SCR models can be analyzed using maximum likelihood methods (Borchers and Efford, 2008; Efford, 2004) or Bayesian methods (Royle et al., 2014). Both approaches are equally suitable (Gerber and Parmenter, 2015), and the choice of analysis depends on the study objectives (e.g., Fuller et al., 2016; Royle et al., 2014). Sutherland and Royle (2016) provide an overview of the software available to conduct either analysis. Additional core assumptions for nonspatial and SCR models are presented in Table 29.2.

CHALLENGES FACED WHEN SURVEYING CHEETAHS USING CAMERA TRAPS

Although the application of camera traps has some distinct advantages for studying elusive species, such as the cheetah (see section, Surveying Cheetahs: Applications of Camera Trapping), the method is not without challenges. The primary challenge is that cheetahs occur naturally at low density throughout their range, making it difficult to obtain sufficient sample sizes (i.e., number of unique individuals and spatial recaptures) to ensure precise density estimates. Previous cheetah surveys have captured between 3 and 11 adult cheetahs (median = 5; Table 29.1). These sample sizes are less than the 30 individuals with 20 recaptures recommended for SCR by Tobler and Powell (2013). These low sample sizes contributed to the typically low precision in density estimates in all previous cheetah studies (Table 29.1).

In an attempt to increase sample sizes, cheetah surveys have often placed camera stations at scent marking sites. However, marking sites are predominately used by territorial males (Chapter 9). This camera placement can, therefore, result in a substantial sex bias toward males (in particular territorial males) as observed in all previous cheetah camera surveys (i.e., 56 males compared to 10 females;

Table 29.1). Small sample sizes limit the ability to estimate sex-specific detection probabilities. A consequence of this limitation is that the averaged estimates of detectability will be weighted toward the more frequently observed sex. Nonetheless, despite low sample sizes of females, estimation of a sex-specific spatial scale parameter is still possible, provided females are detected at multiple locations. Indeed, accounting for sex is vital if intersex overall detection probabilities differ largely (Efford and Mowat, 2014; Sollmann et al., 2011).

An additional challenge when surveying cheetahs with cameras is the sociality of males (i.e., male cheetahs can occur alone or in coalitions of two or more individuals), which violates the assumption that all captures are independent and that individuals have single, independent activity centers (Table 29.2). This violation tends to inflate the apparent precision of density estimates (Head et al., 2013; Efford, personal communication). Head et al. (2013) accounted for this violation by including a categorical social group covariate (solitary vs. group individuals) when working with group living species (i.e., chimpanzees *Pan troglodytes troglodytes*, gorillas Gorilla *gorilla gorilla*, and forest elephants *Loxodonta cyclotisii*). Development of models that account for social grouping and efforts to increase the sample size of cheetahs to enable the inclusion of covariates are thus needed.

In low-density, wide-ranging species, a large sample size is difficult to achieve, and few published camera trap surveys of felids have reached the general recommendations on sample size (Foster and Harmsen, 2012). However, efforts can be made to optimize cheetah camera trap survey design to maximize sample size while maintaining model assumptions (Table 29.2) using accrued knowledge from previous cheetah camera surveys and research conducted on cheetah behavior and space use (Chapters 8 and 9).

LESSONS LEARNED TO MAXIMIZE CHEETAH DETECTION AND OPTIMIZE CHEETAH CAMERA SURVEYS

Timing of Camera Trap Study

To date, no peer-reviewed camera trap study has compared cheetah detection rates between seasons. However, a long-term camera trapping study, involving five surveys in winter (June–October) and summer (September–April), in north-central Namibia showed no significant difference in cheetah detection rates between seasons (captures = 849 in summer and 970 in winter, $U = 17.5$, $P = 0.29$) (Fabiano, 2013). However, females showed a trend for increased capture probability at scent marking sites in the summer (10 out of 15 capture events; Fabiano, 2013). We encourage that sampling be conducted across seasons in other cheetah populations, as density and detectability may be driven by seasonal factors.

The median study duration of previous cheetah camera surveys was 90 days (range 30–130 days; Table 29.1). Extending the survey length increases the opportunity of capture as was observed in a cheetah survey in Botswana (e.g., up to 130 days; Brassine and Parker, 2015) and in a leopard survey in China (up to 123 days; Hedges et al., 2016). However, increased precision in density estimates due to higher recapture rates may be countered by the inclusion of transient individuals, potentially inflating density estimates (Larrucea et al., 2007). SCR models, however, can be robust to the inclusion of transient individuals (Royle et al., 2016), therefore longer survey lengths should be considered.

Selection of Camera Placement

Cheetah captures are greater when cameras are selectively placed at sites associated with known cheetah presence (e.g., scent marking sites), as opposed to random placement (Brassine

and Parker, 2015; O'Brien and Kinnaird, 2011). Although this targeted placement of cameras may induce heterogeneity in capture probabilities (e.g., due to differential use of scent marking sites between sexes), this approach is often used in camera survey studies of large carnivores (e.g., Sharma et al., 2010). We suggest that local knowledge of the study area and information from previous studies on space-use and behavior, particularly of female and nonterritorial male cheetahs, be utilized to select camera station locations. Additional considerations include adaptive sampling approaches that represent two-phase sampling, such as combining random sampling followed by targeted sampling in areas of documented animal presence, based on explicit criteria, such as counts of individuals detected in phase-one sampling (Conroy et al., 2008).

Maximize Survey Area Size With Appropriate Camera Spacing

When designing a camera trap survey, it is important to consider the balance between spatial coverage to increase the number of individuals encountered, trap density to increase the number of spatial recaptures of individuals, and attempts to sample all habitat types (Foster and Harmsen, 2012; Royle et al., 2014). Cheetahs do not use space homogeneously (Broekhouis and Gopalaswamy, 2016); therefore, surveys should sample the various habitat types available in proportion to their occurrence in the landscape of interest.

Camera spacing and the size of the survey area are often constrained by practical considerations (e.g., the number of available cameras, personnel, and logistics). The median survey area size in previous studies was 277 km^2 (range 200–2551 km^2; Table 29.1). Therefore, study areas were generally smaller than a cheetah home range size, which is reported to vary between 125 and 2161 km^2 depending on sex and habitat (Chapter 8). To maximize survey area size, it has been suggested that cameras be placed up to a home range radius apart without introducing bias (Sollmann et al., 2012; Sun et al., 2014). The home range radius used should be that of the smaller ranging sex (Kelly et al., 2013). Based on cheetah home range estimates, intercamera distance could be between 7 and 26 km, depending on habitat. These values would fall within the MMDM of cheetahs in previous camera trap surveys (9.7–45.0 km; Table 29.1). Hence, information on cheetah movements, gathered at the camera survey site (or if unavailable from a population occurring under similar natural systems), should be used to inform and determine camera trap spacing.

An additional way to increase survey area size is to use only one camera at each station, as implemented by Brassine and Parker (2015). Although recently developed models exist to account for partial identity of individuals (i.e., photographing only one flank) in nonspatial (McClintock, 2015) and spatial (Augustine et al., 2016) C–R models, pairs of cameras are preferred to minimize the risk of data loss (e.g., due to camera malfunction). Alternatively, survey designs in which pairs of cameras are rotated within a larger sampling area, known as a block design, can increase spatial coverage without sacrificing the double camera deployment (Karanth and Nichols, 2002). A variation on the block design recommends placing cameras in clusters of two to four, with at least two clusters per home range (Sun et al., 2014). The advantage is that clusters of cameras can be spread farther apart from regular camera placement, increasing the area sampled and hence the number of individuals exposed to cameras stations (Sun et al., 2014). Design considerations and associated parameter sensitivity to trap spacing and array configuration are provided by Efford and Fewster (2013), in Chapter 10 in Royle et al. (2014), Sollmann et al. (2012), and Sun et al. (2014).

Data analysis

SCR methods offer a powerful statistical framework for estimating density. These models allow ecologists and conservation biologists to explicitly investigate hypotheses about species' spatial ecology (Royle et al., 2014, 2017), which is vital to the conservation of species and not available using nonspatial C–R models. Previous camera studies of cheetahs have used the R package SPACECAP (Bayesian analysis; Gopalaswamy et al., 2012); however, the maximum likelihood estimation R packages secr (Efford, 2015) and oSCR (Sutherland et al., 2016) are also applicable. Camera trap survey design is an important consideration to minimize the risks of violating model assumptions (Table 29.2), which continues to be an active area of research. Pilot surveys and computer simulations can be informative about design considerations, such as number of cameras, distance between camera stations, and trapping array size. For example, the R package secrdesign (Efford, 2016) can be used to assess the potential bias and predicted precision of density estimates under specific survey designs (Chapter 10 in Royle et al., 2014).

CONCLUSIONS

Surveying cheetahs to estimate abundance and density is particularly challenging because the species occurs at low densities across large areas, commonly resulting in small sample sizes. Nevertheless, monitoring cheetahs with camera traps and analyzing the resulting data using SCR methods provide a promising solution for studying the spatial ecology of cheetahs while simultaneously estimating absolute density, and ultimately improving the conservation of this highly threatened species. Researchers are encouraged to perform simulations prior to conducting camera trapping surveys to determine sampling effort and trap spacing. The conservation, management, and ultimate survival of cheetahs can benefit greatly from well-conceived camera trap studies combined with analytical tools as has been described in this chapter.

References

Andresen, L., Everatt, K.T., Somers, M.J., 2014. Use of site occupancy models for targeted monitoring of the cheetah. J. Zool. 292, 212–220.

Augustine, B.C., Royle, J.A., Kelly, M.K., Satter, C.B., Alonso, R.S., Boydston, E.E., Crooks, K.R., 2016. Spatial capture–recapture with partial identity: an application to camera-traps. J. Am. Stat. Assoc, doi: https://doi.org/10.1101/056804.

Belbachir, F., Pettorelli, N., Wacher, T., Belbachir-Bazi, A., Durant, S.M., 2014. Monitoring rarity: the critically endangered Saharan cheetah as a flagship species for a threatened ecosystem. PLoS One 10, e0115136.

Boast, L., Houser, A.M., Good, K., Gusset, M., 2013. Regional variation in body size of the cheetah (*Acinonyx jubatus*). J. Mammal. 94, 1293–1297.

Boast L., 2014. Exploring the causes of and mitigation options or human-predator conflict on game ranches in Botswana. How is coexistence possible? PhD thesis, University of Cape Town, South Africa.

Boast, L., Reeves, H., Klein, R., 2015. Camera-trapping and capture–recapture models for estimating cheetah density. Cat News 62, 34–37.

Borchers, D.L., Efford, M.G., 2008. Spatially explicit maximum likelihood methods for capture–recapture studies. Biometrics 64, 377–385.

Brassine, E., Parker, D., 2015. Trapping elusive cats: using intensive camera trapping to estimate the density of a rare African felid. PLoS One 10, e0142508.

Broekhouis, F., Gopalaswamy, A.M., 2016. Counting cats: spatially explicit population estimates of cheetah (*Acinonyx jubatus*) using unstructured sampling data. PLoS One 11, e0153875.

Caro, T.M., Durant, S.M., 1991. Quantitative analyses of pelage characteristics to reveal family resemblances in genetically monomorphic cheetahs. J. Hered. 82, 8–14.

Conroy, M.J., Runge, J.P., Barker, R.J., Fonnesbeck, C.J., 2008. Efficient estimation of abundance for patchily distributed populations via two-phase, adaptive sampling. Ecology 89, 3362–3370.

Cozzi, G., Broekhuis, F., McNutt, J.W., Turnbull, L.A., Macdonald, D.W., Schmid, B., 2012. Fear of the dark or dinner by moonlight? Reduced temporal partitioning among Africa's large carnivores. Ecology 93, 2590–2599.

Efford, M.G., 2004. Density estimation in live-trapping studies. Oikos 106, 598–610.

Efford, M.G., 2015. secr: Spatially Explicit Capture-Recapture Models. R package version 2.10.0.

Efford, M.G., 2016. secrdesign: Sampling Design for Spatially Explicit Capture–Recapture. Available from: http://cran.r-project.org/.

Efford, M.G., Borchers, D.L., Byrom, A.E., 2009. Density estimation by spatially explicit capture–recapture: likelihood-based methods. In: Thomson, D.L., Cooch, E.G., Conroy, M.J. (Eds.), Modeling Demographic Processes in Marked Populations. Springer, New York, NY, pp. 255–269.

Efford, M.G., Fewster, R.M., 2013. Estimating population size by spatially explicit capture–recapture. Oikos 122, 918–928.

Efford, M.G., Mowat, G., 2014. Compensatory heterogeneity in spatially explicit capture–recapture data. Ecology 95, 1341–1348.

Fabiano, E., 2013. Historical and Contemporary Demography of Cheetahs (*Acinonyx jubatus*) in Namibia, Southern Africa. PhD thesis, Pontifícia Universidade Católica do Rio Grande do Sul, Brazil.

Foster, R.J., Harmsen, B.J., 2012. A critique of density estimation from camera-trap data. J. Wildl. Manage. 76, 224–236.

Fuller, A.K., Sutherland, C.S., Royle, J.A., Hare, M.P., 2016. Estimating population density and connectivity of American mink using spatial capture–recapture. Ecol. Appl. 6, 1125–1135.

Gerber, B.D., Parmenter, R.R., 2015. Spatial capture–recapture model performance with known small-mammal densities. Ecol. Appl. 25, 695–705.

Gopalaswamy, A.M., Royle, A.J., Hines, J.E., Singh, P., Jathanna, D., Kumar, N.S., Karanth, K.U., 2012. Program SPACECAP: software for estimating animal density using spatially explicit capture–recapture models. Methods Ecol. Evol. 3, 1067–1072.

Head, J.S., Boesch, C., Robbins, M.M., Rabanal, L.I., Makaga, L., Kühl, H.S., 2013. Effective sociodemographic population assessment of elusive species in ecology and conservation management. Ecol. Evol. 3, 2903–2916.

Hedges, L., Lam, W.Y., Campos-Arceiz, A., Rayan, D.M., Laurance, W.F., Latham, C.J., Saaban, S., Clements, G.R., 2016. Melanistic leopards reveal their spots: Infrared camera traps provide a population density estimate of leopards in Malaysia. J. Wildl. Manage. 79, 846–853.

Karanth, K.U., Nichols, J.D. (Eds.), 2002. Monitoring Tigers and Their Prey: A Manual for Researchers Managers and Conservationists in Tropical Asia. Centre for Wildlife Studies, Bangalore, India.

Karanth, K.U., Nichols, J.D., Kumar, N.S., Hines, J.E., 2006. Assessing tiger population dynamics using photographic capture–recapture sampling. Ecology 87, 2925–2937.

Kauffman, M.J., Sanjayan, M., Lowenstein, J., Nelson, A., Jeo, R.M., Crooks, K.R., 2007. Remote camera-trap methods and analyses reveal impacts of rangeland management on Namibian carnivore communities. Oryx 41, 70–78.

Kelly, M.J., 2001. Computer-aided photograph matching in studies using individual identification: an example from Serengeti cheetahs. J. Mammal. 82, 440–449.

Kelly, M.J., Tempa, T., Wangdi, Y., 2013. Camera trapping protocols for wildlife studies (with emphasis on tiger density estimation). In: Mills, L.S., Tshering, E.,

Cheng (Eds.), Wildlife Research Techniques in Rugged Mountainous Asian Landscapes. Ugyen Wangchuck Institute for Conservation and Environment, Bhutan, pp. 92–113.

Larrucea, E.S., Serra, G., Jaeger, M.M., Barrett, R.H., 2007. Censusing bobcats using remote cameras. West. North Am. Nat. 67, 538–548.

Marker, L.L., Fabiano, E., Nghikembua, M., 2008. The use of remote camera traps to estimate density of free ranging cheetahs in north-central Namibia. Cat News 49, 22–24.

Marnewick, K.A., Bothma, J. du P., Verdoorn, G.H., 2006. Using camera-trapping to investigate the use of a tree as a scent-marking post by cheetahs in the Thabazimbi district. S. Afr. J. Wildl. Res. 36, 139–145.

Marnewick, K., Funston, P.J., Karanth, K.U., 2008. Evaluating camera trapping as a method for estimating cheetah abundance in ranching areas. S. Afr. J. Wildl. Res. 38, 59–65.

McClintock, B.T., 2015. multimark: an R package for analysis of capture–recapture data consisting of multiple "noninvasive" marks. Ecol. Evol. 5, 4920–4931.

O'Brien, T.G., Kinnaird, M.F., 2011. Density estimate of sympatric carnivores using spatially explicit capture–recapture methods and standard trapping grid. Ecol. Appl. 21, 2908–2916.

O'Connell, A.F., Nichols, J.D., Karanth, K.U. (Eds.), 2011. Camera Traps in Animal Ecology. Methods and Analysis. Springer, New York, NY.

Otis, D.L., Burnham, K.P., White, G.C., Anderson, D.R., 1978. Statistical inference from the capture data on closed animal populations. Wildl. Monogr. 62, 1–135.

Royle, A.J., Chandler, R.B., Sollmann, R., Gardner, B., 2014. Spatial Capture–Recapture. Academic Press, San Diego, CA.

Royle, J.A., Fuller, A.K., Sutherland, C., 2016. Spatial capture–recapture models allowing Markovian transience or dispersal. Popul. Ecol. 58, 53–62.

Royle, J.A., Fuller, A.K., Sutherland, C., 2017. Unifying population and landscape ecology with spatial capture-recapture. Ecography doi: 10.1111/ecog.03170.

Royle, A.J., Karanth, K.U., Gopalaswamy, A.M., Kumar, S.N., 2009a. Bayesian inference in camera trapping studies for a class of spatial capture–recapture models. Ecology 90, 3233–3244.

Royle, A.J., Nichols, J.D., Karanth, K.U., Gopalaswamy, A.M., 2009b. A hierarchical model for estimating density in camera-trap studies. J. Appl. Ecol. 46, 118–127.

Sharma, R.K., Jhala, Y.V., Qureshi, Q., Vattakaven, J., Gopal, R., Nayak, K., 2010. Evaluating capture–recapture population and density estimation of tigers in a population with known parameters. Anim. Conserv. 13, 94–103.

Sillero-Zubiri, C., Rostro-García, S., Burruss, D., Matchano, A., Harouna, A., Rabeil, T., 2015. Saharan cheetah *Acinonyx jubatus hecki*, a ghostly dweller on Niger's Termit massif. Oryx 49, 591–594.

Sollmann, R., Furtado, M.M., Gardner, B., Hofer, H., Jácomo, A.T.A., Tôrres, N.M., Silveira, L., 2011. Improving density estimates for elusive carnivores: accounting for sex-specific detection and movements using spatial capture–recapture models for jaguars in central Brazil. Biol. Conserv. 144, 1017–1024.

Sollmann, R., Gardner, B., Belant, J., 2012. How does spatial study design influence density estimates from spatial capture–recapture models? PLoS One 7, e34575.

Sun, C.C., Fuller, A.K., Royle, J.A., 2014. Trap configuration and spacing influences parameter estimates in spatial capture–recapture models. PLoS One 9, e88025.

Sutherland, C., Royle, J.A., 2016. Estimating abundance. In: Dodd, C.K. (Ed.), Reptile Ecology and Conservation: A Handbook of Techniques. Oxford University Press, New York, USA, pp. 388–399.

Sutherland, C., Royle, J.A., Linden, D., 2016. oSCR: Multi-Session Sex-Structured Spatial Capture–Recapture Models. R package version 0.30.0.

Tobler, M., Powell, G.V.N., 2013. Estimating jaguar densities with camera traps: problems with current designs and recommendations for future studies. Biol. Conserv. 159, 109–118.

Vogt, K., Zimmermann, F., Kölliker, M., Breitenmoser, U., 2014. Scent-marking behavior and social dynamics in a wild population of Eurasian lynx *Lynx lynx*. Behav. Processes 106, 98–106.

Williams, B.K., Nichols, J.D., Conroy, M.J., 2001. Analysis and Management of Animal Populations. Academic Press, New York, NY.

Wilson, K.R., Anderson, D.R., 1985a. Evaluation of a density estimator based on a trapping web and distance sampling theory. Ecology 66, 1185–1194.

Wilson, K.R., Anderson, D.R., 1985b. Evaluation of a nested grid approach for estimating density. J. Wildl. Manage. 49, 675–678.

Spoor Tracking to Monitor Cheetah Populations

Lorraine K. Boast*, *Linda van Bommel***,†,
Leah Andresen‡, *Ezequiel Fabiano*§

*Cheetah Conservation Botswana, Gaborone, Botswana
**University of Tasmania, Hobart, TAS, Australia
†Australian National University, Canberra, ACT, Australia
‡Nelson Mandela University, Port Elizabeth, South Africa
§University of Namibia, Katima Mulilo, Namibia

The cheetah (*Acinonyx jubatus*) is a cryptic species that generally occurs at low densities (Chapter 8). For such species, the use of direct research methods, such as visual observation (Chapter 32), or mark and recapture, can be invasive, time consuming, and expensive (Wilson and Delahay, 2001). Therefore, indirect techniques using, for example, motion cameras (Chapter 29) or signs of the animal [e.g., spoor (*also known as footprints or tracks*) or feces (Chapter 31)] are more common in cheetah research. This chapter will focus on spoor and how it is used to assess cheetah populations.

Tracking an animal using its spoor is a long-practiced technique used by many traditional hunter-gatherer societies. In Africa, this skill is exemplified by the San bushmen whose skills have been utilized for many spoor surveys (Bothma and Le Riche, 1984; Stander et al., 1997). In cheetah research, tracks are used to analyze behavior, to locate individuals for further study, and to estimate species occupancy, abundance, and density. The introduction of hand-held field computers and Cybertracker data recording software (http://www.cybertracker.org/) has enabled people from any language, educational, or cultural background to use pictorial icons to enter biological field data when an animal track is found (Liebenberg et al., 1998). In addition, to complement and advance the ability of humans to track cheetahs, computer software has been developed to identify individual cheetahs from their footprints (Jewell et al., 2016).

Locating and identifying spoor can be done opportunistically or during standardized surveys. Surveys are conducted on specific roads or animal trails either of predetermined lengths (transects) or on numerous small sections (track stations). Detailed information about the recommended methodology for these techniques can be found in Long and MacKay (2008) and a description of the cheetah's footprint can be found in Box 30.1. Study design and the specific data collected during surveys differ depending on the practitioner's goal, that is, behavioral data, presence–absence, occupancy, density, or individual identification.

PRESENCE–ABSENCE STUDIES OF CHEETAHS

Presence–absence data can be used to determine the distribution of populations on a large spatial scale, or to identify habitats of high value (e.g., Winterbach et al., 2015). For example, the presence of the critically endangered northwest African cheetah (*A. jubatus hecki*) was determined in Niger's Termit massif using spoor transect surveys (Sillero-Zubiri et al., 2015). However, it is difficult to determine if a species is absent as opposed to undetected (a false-absence). The probability of detecting a species depends on its population density and on the efficiency of the sampling approaches and the sampling effort (Gu and Swihart, 2004). Species that occur at low densities, such as the cheetah, are particularly prone to being misclassified as absent. Therefore, caution should be exercised when interpreting cheetah absence data, as potential misclassification can alter habitat models and potentially affect management decisions (Gu and Swihart, 2004). An alternative to presence–absence studies is to use occupancy modeling.

OCCUPANCY MODELING OF CHEETAH SPOOR

Site occupancy models present a relatively new methodological and statistical approach in the use of traditional presence–absence data. Occupancy models use replicated detection/ nondetection histories to estimate detection probability. They can account for false-absences and, therefore, produce an unbiased estimate of occurrence (MacKenzie et al., 2002). Occupancy modeling allows conservation practitioners to provide reliable estimates of the area of occurrence (i.e., the proportion of area occupied) and to investigate the factors influencing resource selection across landscapes (e.g., Barber-Meyer et al., 2013; Sunarto et al., 2012; Chapter 36).

An occupancy approach can be applied to spoor surveys by dividing spoor transects into multiple segments (e.g., 1–4 km) and considering each segment as a unique sampling occasion on which cheetah spoor is detected or nondetected. Replication can be performed temporally (Andresen, 2013) or spatially (Hines et al., 2010), making large-scale occupancy surveys logistically feasible for wide-roaming terrestrial carnivores (e.g., Karanth et al., 2011). Site-specific variables, such as prey, sympatric predators, anthropogenic factors, and habitat structure can be recorded along transects to provide additional information (Hebblewhite et al., 2011). For example, the identification of sites used by bush-meat hunters during spoor surveys in Mozambique revealed that habitat selection by cheetahs was most strongly determined by an avoidance of these sites (Andresen, 2013).

Many of the site-specific variables will influence both occurrence and detectability (Gu and Swihart, 2004). As occupancy modeling examines occurrence and detectability simultaneously, the approach allows reliable inferences to be made about the factors that most affect the area of occurrence and habitat selection, without being confounded by factors influencing

BOX 30.1

Cheetah foot structure is characteristic of cats, with two lobes on the anterior edge of the main pad and three lobes on the posterior edge of the main pad. Other distinguishing features of cheetah prints are the tear-drop shape of the toes and the overall elongated oval shape of the track (Stuart and Stuart, 2001) (Fig. 30.1). Cheetahs most often walk in an overstep, so that the hind print lands in front of the fore print in each track pair on either side of the body (Fig. 30.2). Cheetah tracks are smaller and more elongated than those of lion (*Panthera leo*) or leopard (*Panthera pardus*), but larger than those of serval (*Felis serval*) or caracal (*Caracal caracal*). They are the only big cat species that are unable to fully retract their claws; therefore, their footprints are often identifiable from other cat species by the presence of nail marks. Due to the nail marks, cheetah tracks can be confused with African wild dog (*Lycaon pictus*), domestic dog (*Canis lupus familiaris*), or hyena (*Hyaenidae* spp.) tracks. However, they can be differentiated from these species by the variation in the main pad impression: main pads of dogs and hyenas are roughly triangular in shape, lacking the lobes on the anterior part of the main pad, and having only two lobes on the posterior part of the main pad.

Front foot

Measurements
Length: 70–85 mm
Width: 65–75 mm

Hind foot

Length: 75–90 mm
Width: 60–70 mm

FIGURE 30.1 **Cheetah footprints: anatomy of the prints.**

FIGURE 30.2 **Cheetah tracks (circled) in the sand; the foot sequence from foreground to background is rear left, front right, rear right.** *Source: Lorraine Boast.*

the probability of detecting cheetah spoor (MacKenzie et al., 2006).

The effect of survey-specific covariates on species detectability and sampling efficacy—such as substrate, survey method, trail type, and different trackers—can also be included in the analysis, providing information for the improvement of monitoring protocols. Other potentially useful applications to cheetahs are two-species occupancy models (MacKenzie et al., 2004), which can be used to investigate competition and predation. In addition, if surveys are repeated over time, factors influencing the dynamic processes of local extinction and colonization can be explored through metapopulation studies (MacKenzie et al., 2006).

As with all studies, careful consideration needs to be given to the design of occupancy surveys (MacKenzie and Royle, 2005; MacKenzie et al., 2006). The size of the sample unit should depend on the scale of interest (Johnson, 1980), and surveys should be conducted to both maximize spatial coverage and the probability of detecting cheetahs (Andresen, 2013; Karanth et al., 2011; MacKenzie et al., 2006). The survey effort (e.g., the number of transect segments within each sample unit) has to be adequate to achieve a probability of detecting the target species >0.15, as estimates are unreliable below that value (MacKenzie et al., 2006). The amount of survey effort required to achieve this goal should be estimated before the start of the research (Andresen et al., 2014; Kéry, 2002; MacKenzie and Royle, 2005), and incorporated in the study design to ensure adequate data are collected for analysis. For low density species, such as the cheetah, the general rule of thumb is to survey a larger number of grid cells less intensively rather than a smaller number more intensively. Unequal sampling across sites can be accommodated within occupancy models (MacKenzie and Royle, 2005; MacKenzie et al., 2006). Sample design trade-offs can be explored using simulations performed in programs, such as PRESENCE (Hines, 2006) or GENPRES (Bailey et al., 2007). For example, simulations could be used to determine the optimal number of sites required to maintain an adequate power of detecting a decline in cheetah occurrence while reducing sampling intensity (e.g., Sewell et al., 2012).

DENSITY ESTIMATION: CALIBRATION FACTORS FOR CHEETAH

Spoor surveys, like other indirect sampling techniques, generate an index of animal density. In this case, the number of kilometers surveyed to view one footprint is referred to as the "spoor encounter rate" or "spoor frequency." Two conversion approaches have been applied to estimate true cheetah density from the spoor encounter rate: general linear models (Funston et al., 2001; Funston et al., 2010; Houser et al., 2009) and the Formozov–Malyshev–Pereleshin (FMP) formula (Keeping, 2014; Kuzyakin, 1983), also referred to as the Formozov formula (Mirutenko, 1986). Using these techniques, cheetah density has been calculated to range between 0.3 and 2.7 cheetahs per 100 km^2 in various habitats (Boast and Houser, 2012; Funston et al., 2010; Houser et al., 2009; Keeping, 2014) (where applicable data were standardized using the calibration equation recommended; Winterbach et al., 2016).

The general linear model approach is based on surveying several discrete animal populations with both a direct technique, such as mark and recapture, and a spoor survey. By double sampling the populations, the relationship between the two results can be examined and, ultimately, a calibration equation can be determined to be applied to future spoor surveys. The technique has been applied to a variety of African large carnivores (e.g., Funston et al., 2010; Stander, 1998). Initially, calibration equations were species-specific and, in the case of cheetahs, season-specific (Houser et al., 2009; Stander, 1998), however, a general equation for large African carnivores was published in 2001

(Funston et al., 2001). This equation was based on lion populations and was then extrapolated to all large African carnivores, including cheetahs (based on predicted cheetah density from a photographic survey). The equation was refined in 2010 by comparing known carnivore densities with their respective spoor densities at 18 study sites, 7 of which included data on cheetahs (Funston et al., 2010). This equation was specific to sandy soils that dominate cheetah habitat. Although African large carnivores exhibit variety in their daily movements (Keeping, 2014), this calibration equation (observed track density = 3.15 × carnivore density + 0.4) was found to account for 96% of variation in species track densities (Funston et al., 2010). It has therefore been applied to subsequent cheetah spoor surveys (Boast and Houser, 2012). However, the equation is inappropriate for low density populations (Winterbach et al., 2016), as the linear equation does not intercept zero, and therefore spoor densities of less than 0.4 spoor per 100 km^2 result in negative density estimates.

To address this limitation, Winterbach et al. (2016) showed that the calibration equation "observed track density = 3.26 × carnivore density" based on linear regression through the origin (i.e., zero spoor encounter rate equals zero species density) is more appropriate for the dataset used by Funston et al. (2010) than linear regression through an intercept other than zero. This equation is suitable for species with densities of 0.27 per 100 km^2 or higher. Lower densities are outside of the sample range upon which the model was based, and using the model for these densities should only be undertaken with caution. This multispecies calibration equation is recommended for use until a cheetah-specific equation based on numerous populations is developed.

The second approach, the FMP formula, requires an estimate of the average daily distances animals travel (day range) in order to estimate the probability of an individual animal crossing a transect within the 24-h period before the survey (Kuzyakin, 1983; Stephens et al., 2006). To accurately determine a species' day range within the study area, additional surveying, such as collaring of individuals or following an individual's spoor must be performed, increasing the cost and labor intensity of the surveys. Alternatively, published data from other study sites can be utilized, based on the assumption that a species' day range will be similar in comparable habitats (Keeping and Pelletier, 2014). For example, Keeping (2014) used empirical data on cheetah day range estimates from the Kgalagadi Transfrontier Park to estimate cheetah density over a wider area of the southern Kalahari.

The applicability of this method to cheetahs is currently limited due to the paucity of data available on cheetah day range and the temporal and environmental factors influencing these movements (Keeping, 2014). As an alternative, the established relationship between carnivore body mass and day range (Carbone et al., 2005; Garland, 1983) can be used to estimate the average daily movements of cheetahs (Keeping, 2014). Nevertheless, utilizing data on daily movements from within the study area would be preferable (Keeping, 2014), and efforts to increase our knowledge of cheetah movement patterns is required.

Both approaches rely on the assumption that a species' day range, and therefore, the linear relationship between spoor counts and a "true" density estimate, remains constant between the populations to which the approach is applied (Keeping and Pelletier, 2014; Stander, 1998). The approaches also assume that all spoor is detected and correctly identified, and that the reliability of trackers to detect and identify spoor remains constant within and between surveys. It is, therefore, imperative to use trained and experienced spoor trackers for all surveys.

In addition, the current applications of general linear models to cheetahs require that each individual is only recorded once per day (Stander, 1998). Potentially subjective assumptions as to whether the same or a new individual has been detected have to be made. This subjectivity can be compounded in social or semisocial species, such as the cheetah, where multiple individuals travel together, but may be detected on transects at different times. The development of new calibration equations based on detecting all recrossings of spoor, as is required by the FMP approach (Keeping 2014; Keeping and Pelletier, 2014), would reduce this subjectivity. It is recommended that future studies record all recrossings of spoor, allowing future spoor survey data to be analyzed with either technique (Keeping, personal communication).

Cheetahs naturally occur at low densities and are wide ranging, therefore a substantial sample effort is likely to be required when surveying for cheetah spoor. To obtain precise density estimates, defined as having a coefficient of variation (CV) set arbitrarily at below 20%, Funston et al. (2010) determined that 19–30 large carnivore spoor need to be detected. At low densities—for example, 0.27 per km^2, equivalent to 0.88 spoor per 100 km, the minimum density at which the Winterbach et al. (2016) linear equation is valid—this will require the sampling of 2200–3400 km. However, it should be noted that cheetahs do not use the landscape in a uniform fashion. They are likely to be heterogeneously spread across an area, resulting in hotspots of cheetah density (Broekhuis and Gopalaswamy, 2016; Muntifering et al., 2006). Combining hot-spots and lesser-used areas within one survey are likely to result in higher CV values (Quinn and Keough, 2002). As a result, CV values of less than 20% may be difficult to achieve in cheetah surveys. For example, despite observing >50 cheetah spoor in the wet season and >120 spoor in the dry season, CV values in a spoor survey in Botswana were always above 20% (Houser et al., 2009). Due to cheetahs' heterogeneous use of the landscape, sampling the population in a stratified fashion, that is, sampling from multiple subpopulations within the overall area, will be necessary

to enable extrapolation to the wider landscape (Winterbach, personal communication).

FOOTPRINT IDENTIFICATION METHODS IN RELATION TO CHEETAH

Footprint identification methods have been developed which allow information, such as species, sex, age-class, and individual identity of an animal to be ascertained from the footprints it has left behind (Sharma et al., 2005). Many different protocols have been used to identify individual large carnivores from their footprints, ranging from simple visual assessment and shape descriptions to complex morphometrics using statistical analysis of measurements taken from the prints (see Sharma et al. (2005) for an overview of past research. Not all methods have been equally successful, depending on the rigor of data collection and analysis, proper method validation, and sample sizes used in the models. However, a few studies have reported more than 90% accuracy in assigning individual identity or sex based on footprint analysis (e.g., Alibhai et al., 2008; Gu et al., 2014; Sharma et al., 2003).

None of the footprint identification methods have originally been designed for cheetahs, although they could be applied to them. However, one method that was originally developed for black rhinocerous (*Diceros bicornis*) has been now adapted for cheetahs (Jewell et al., 2016) and several other species (Jewell and Alibhai, 2013). This method, the Footprint Identification Technique (FIT) from the JMP Division of the SAS Institute (www.jmp.com) is also the only method that has customized software available for the full analysis of footprint data (for information on the software see www.wildtrack.org).

To use FIT, a standard photo protocol is used to take a series of digital images along several trails of footprints of the species of interest (Jewell and Alibhai, 2013). These images are then imported into FIT, and for cheetah

136 measurements are automatically generated from the prints through the manual placement of landmarks (Jewell and Alibhai, 2013; Jewell et al., 2016). The best measurements are selected for analysis through a stepwise procedure, and the footprints are classified with a customized model based on pair-wise discriminant analysis with a Ward's clustering analysis (Jewell et al., 2016). The model's algorithm is initially constructed from a training set of footprints of known individual animals, and can then be applied to classify tracks belonging to unknown individuals (Jewell and Alibhai, 2013). It can also determine the sex of the individual.

FIT is more than 90% accurate in identifying individual cheetahs (Jewell et al., 2016). Similar classification accuracy has been found for a range of species, including black and white (*Ceratotherium simum*) rhinoceros and mountain lions (*Puma concolor*), and FIT was found to be 98% accurate in assigning sex of Amur tigers (*Panthera tigris altaica*) (Alibhai et al., 2008; Gu et al., 2014; Jewell et al., 2001, 2014). FIT seems to perform equally well classifying footprints collected from wild animals in the field as it does with tracks collected from captive individuals (Alibhai et al., 2008; Jewell et al., 2001).

Due to the standardized photo protocol used for FIT, many different recorders can photograph footprints without compromising classification accuracy. The placement of landmarks for measurement extraction requires expert positioning, and is therefore probably best done by one or a few experienced researchers. FIT also allows the depth of the substrate to be estimated, and adjusts measurements accordingly (Jewell et al., 2016). Therefore, tracks from many different substrates can be used for classification, provided the outline of the track is clear.

Footprint identification can be used for noninvasive monitoring of wildlife populations, including population estimates, detection of problem individuals or following an individual of interest. If the study is designed correctly, it can also be used to investigate habitat preferences, home ranges, social interactions, and animal behavior.

CONCLUSIONS

Spoor surveys remain a commonly-used method for the study of cheetah populations. They are relatively affordable, require little equipment, do not require the capture of the animal, and are easy to conduct compared to direct research techniques. Spoor surveys can provide information about species density, occupancy, behavior, and data for habitat modeling. In addition, multiple species can be studied in one survey. The use of spoor tracking, however, is limited by the availability of suitable substrate and the substantial survey effort required to attain a reasonable level of accuracy and precision for a low-density species. Despite this, the technique remains a valuable addition to the cheetah biologist's toolbox, and with the implementation of new technologies, including Cybertracker and footprint identification technology, potentially much more information can be gained from spoor tracking.

References

Alibhai, S., Jewell, Z.C., Law, P.R., 2008. A footprint technique to identify white rhino *Cerototherium simum* at individual and species levels. Endanger. Species Res. 4, 205–218.

Andresen, L., 2013. Site Occupancy and Habitat Selection of Cheetahs *Acinonyx jubatus* (Shreber, 1775) in a Human-Influenced Landscape in Mozambique. MSc thesis, University of Pretoria, South Africa.

Andresen, L., Everatt, K.T., Somers, M.J., 2014. Use of site occupancy models for targeted monitoring of the cheetah. J. Zool. 292 (3), 212–220.

Bailey, L.L., Hines, J.E., Nichols, J.D., MacKenzie, D.I., 2007. Sampling design trade-offs in occupancy studies with imperfect detection: examples and software. Ecol. Appl. 17 (1), 281–290.

Barber-Meyer, S., Jnawali, S., Karki, J., Khanal, P., Lohani, S., Long, B., MacKenzie, D., Pandav, B., Pradhan, N., Shrestha, R., Subedi, N., Thapa, G., Thapa, K., Wikramanayake, E., 2013. Influence of prey depletion and human disturbance on tiger occupancy in Nepal. J. Zool. 289 (1), 10–18.

Boast, L., Houser, A., 2012. Density of large predators on commercial farmland in Ghanzi, Botswana. S. Afr. J. Wildl. Res. 42 (2), 138–143.

Bothma, J.D.P., Le Riche, E., 1984. Aspects of the ecology and the behaviour of the leopard *Panthera pardus* in the Kalahari Desert. Koedoe 27 (2), 259–279.

Broekhuis, F., Gopalaswamy, A.M., 2016. Counting cats: spatially explicit population estimates of cheetah (*Acinonyx jubatus*) using unstructured sampling data. PLoS ONE 11 (5), e0153875.

Carbone, C., Cowlishaw, G., Isaac, N.J., Rowcliffe, J.M., 2005. How far do animals go? Determinants of day range in mammals. Am. Nat. 165 (2), 290–297.

Funston, P.J., Frank, L.G., Stephens, T., Davidson, Z., Loveridge, A.J., Macdonald, D.W., Durant, S.M., Packer, C., Mosser, A., Ferreira, S.M., 2010. Substrate and species constraints on the use of track incidences to estimate African large carnivore abundance. J. Zool. 281, 56–65.

Funston, P.J., Herrmann, E., Babupi, P., Kruiper, A., Kruiper, H., Jaggers, H., Masule, K., Kruiper, K., 2001. Spoor frequency estimates as a method of determining lion and other large mammal densities in the Kgalagadi Transfrontier Park. In: Funston, P.J. (Ed.), Kalahari Transfrontier Lion Project. Endangered Wildlife Trust, Johannesburg, South Africa, pp. 36–52.

Garland, Jr., T., 1983. Scaling the ecological cost of transport to body mass in terrestrial mammals. Am. Nat. 121 (4), 571–587.

Gu, J., Alibhai, S.K., Jewell, Z.C., Jiang, G., Ma, J., 2014. Sex determination of Amur tigers (*Panthera tigris altaica*) from footprints in snow. Wildl. Soc. Bull. 38 (3), 495–502.

Gu, W., Swihart, R.K., 2004. Absent or undetected? Effects of non-detection of species occurrence on wildlife–habitat models. Biol. Conserv. 116 (2), 195–203.

Hebblewhite, M., Miquelle, D.G., Murzin, A.A., Aramilev, V.V., Pikunov, D.G., 2011. Predicting potential habitat and population size for reintroduction of the Far Eastern leopards in the Russian Far East. Biol. Conserv. 144 (10), 2403–2413.

Hines, J.E., 2006. PRESENCE 4—Software to Estimate Patch Occupancy and Related Parameters. USGS-PWRC, Laurel, USA.

Hines, J., Nichols, J., Royle, J., MacKenzie, D., Gopalaswamy, A., Kumar, N.S., Karanth, K., 2010. Tigers on trails: occupancy modeling for cluster sampling. Ecol. Appl. 20 (5), 1456–1466.

Houser, A., Somers, M.J., Boast, L., 2009. Spoor density as a measure of true density of a known population of free-ranging wild cheetah in Botswana. J. Zool. 278, 108–115.

Jewell, Z., Alibhai, S., 2013. Identifying Endangered Species From Footprints. The International Society for Optics and Photonics (SPIE) Newsroom, Bellingham, Washington, USA.

Jewell, Z.C., Alibhai, S.K., Evans, J.W., 2014. Monitoring mountain lion using footprints: a robust new technique. Wild Felid Monit. 7 (1), 26–27.

Jewell, Z.C., Alibhai, S.K., Law, P.R., 2001. Censusing and monitoring black rhino (*Diceros bicornis*) using an objective spoor (footprint) identification technique. J. Zool. 254, 1–16.

Jewell, Z.C., Alibhai, S., Weise, F., Munro, S., van Vuuren, M., van Vuuren, R., 2016. Spotting cheetahs: identifying individuals by their footprints. J. Vis. Exp. 111, e54034.

Johnson, D.H., 1980. The comparison of usage and availability measurements for evaluating resource preference. Ecology 61 (1), 65–71.

Karanth, K.U., Gopalaswamy, A.M., Kumar, N.S., Vaidyanathan, S., Nichols, J.D., MacKenzie, D.I., 2011. Monitoring carnivore populations at the landscape scale: occupancy modelling of tigers from sign surveys. J. Appl. Ecology 48, 1048–1056.

Keeping, D., 2014. Rapid assessment of wildlife abundance: estimating animal density with track counts using body mass–day range scaling rules. Anim. Conserv. 17 (5), 486–497.

Keeping, D., Pelletier, R., 2014. Animal density and track counts: understanding the nature of observations based on animal movements. PLoS ONE 9 (5), e96598.

Kéry, M., 2002. Inferring the absence of a species: a case study of snakes. J. Wildl. Manage. 66 (2), 330–338.

Kuzyakin, V.A., 1983. Results of Modelling Winter Transect Counts. Central Research Laboratory of Glavokhota, Moscow, Russia.

Liebenberg, L., Blake, E., Steventon, L., Benadie, K., Minye, J., 1998. Integrating traditional knowledge with computer science for the conservation of biodiversity. In: 8th International Conference on Hunting and Gathering Societies and Post-Foraging Societies: History, Politics and Future. Osaka, Japan, pp. 26–30.

Long, R.A., MacKay, P. (Eds.), 2008. Noninvasive Survey Methods for Carnivores. Island Press, USA.

MacKenzie, D.I., Bailey, L.L., Nichols, J., 2004. Investigating species co-occurrence patterns when species are detected imperfectly. J. Anim. Ecol. 73 (3), 546–555.

MacKenzie, D.I., Nichols, J.D., Lachman, G.B., Droege, S., Andrew Royle, J., Langtimm, C.A., 2002. Estimating site occupancy rates when detection probabilities are less than one. Ecology 83 (8), 2248–2255.

MacKenzie, D.I., Nicholls, J.D., Royle, A.J., Pollock, K.H., Bailey, L.L., Hines, J.E., 2006. Occupancy Estimation and Modeling: Inferring Patterns and Dynamics of Species Occurrence. Academic Press, London, UK.

MacKenzie, D.I., Royle, J.A., 2005. Designing occupancy studies: general advice and allocating survey effort. J. Appl. Ecol. 42 (6), 1105–1114.

Mirutenko, V.S., 1986. Reliability of the Analysis of Route Surveys of Game Animals. Central Research Laboratory of Glavokhota, Moscow, Russia.

Muntifering, J., Dickman, A., Perlow, L., Hruska, T., Ryan, P., Marker, L., Jeo, R., 2006. Managing the matrix for large carnivores: a novel approach and perspective from cheetah (*Acinonyx jubatus*) habitat suitability modelling. Anim. Conserv. 9 (1), 103–112.

Quinn, G.P., Keough, M.J., 2002. Experimental Design and Data Analysis for Biologists. Cambridge University Press, Cambridge, UK.

Sewell, D., Guillera-Arroita, G., Griffiths, R.A., Beebee, T.J., 2012. When is a species declining? Optimizing survey effort to detect population changes in reptiles. PLoS ONE 7 (8), e43387.

Sharma, S., Jhala, Y., Sawarkar, V.B., 2003. Gender discrimination of tigers by using their pugmarks. Wildl. Soc. Bull. 31 (1), 258–264.

Sharma, S., Jhala, Y.V., Sawarkar, V.B., 2005. Identification of individual tigers (*Panthera tigris*) from their pugmarks. J. Zool. 267, 9–18.

Sillero-Zubiri, C., Rostro-García, S., Burruss, D., Matchano, A., Harouna, A., Rabeil, T., 2015. Saharan cheetah *Acinonyx jubatus hecki*, a ghostly dweller on Niger's Termit massif. Oryx 49 (4), 591–594.

Stander, P.E., 1998. Spoor counts as indices of large carnivore populations: the relationship between spoor frequency, sampling effort and true density. J. Appl. Ecol. 53, 378–385.

Stander, P.E., Ghau, II., Tsisaba, D., OMA, II. and VI, I.I. 1997. Tracking and the interpretation of spoor: a scientifically sound method in ecology. J. Zool. 242, 329–341.

Stephens, P., Zaumyslova, O.Y., Miquelle, D., Myslenkov, A., Hayward, G., 2006. Estimating population density from indirect sign: track counts and the Formozov–Malyshev–Pereleshin formula. Anim. Conserv. 9 (3), 339–348.

Stuart, C., Stuart, T., 2001. Southern, Central, and East African Mammals: A Photographic Guide. Struik, Cape Town, South Africa.

Sunarto, S., Kelly, M.J., Parakkasi, K., Klenzendorf, S., Septayuda, E., Kurniawan, H., 2012. Tigers need cover: multi-scale occupancy study of the big cat in Sumatran forest and plantation landscapes. PLoS ONE 7 (1), e30859.

Wilson, G.J., Delahay, R.J., 2001. A review of methods to estimate the abundance of terrestrial carnivores using field signs and observation. Wildl. Res. 28, 151–164.

Winterbach, H.E., Winterbach, C.W., Boast, L., Klein, R., Somers, M.M., 2015. Relative availability of natural prey versus livestock predicts landscape suitability for cheetahs *Acinonyx jubatus* in Botswana. Peer J 3, e1033.

Winterbach, C.W., Ferreira, S.M., Funston, P.J., Somers, M.J., 2016. Simplified large African carnivore density estimators from track indices. PeerJ 4, e2662.

Mining Black Gold—Insights From Cheetah Scat Using Noninvasive Techniques in the Field and Laboratory: Scat-Detection Dogs, Genetic Assignment, Diet and Hormone Analyses

Anne Schmidt-Küntzel, Claudia Wultsch**,
Lorraine K. Boast†, Birgit Braun‡, Leanne Van der Weyde†,
Bettina Wachter§, Rox Brummer¶, Eli H. Walker*,
Katherine Forsythe††, Laurie Marker**

*Cheetah Conservation Fund, Otjiwarongo, Namibia
**American Museum of Natural History, New York, NY, United States
†Cheetah Conservation Botswana, Gaborone, Botswana
‡Action Campaign for Endangered Species, Korntal-Münchingen, Germany
§Leibniz Institute for Zoo and Wildlife Research, Berlin, Germany
¶Green Dogs Conservation, Alldays, South Africa
††Anchor Environmental Consultants, Tokai, South Africa

INTRODUCTION

Noninvasive techniques are sometimes the only method available to study elusive carnivores. They are also the method of choice for scientists who elect to address research questions without imposing capture (Chapter 33) and handling stress (Munson and Marker, 1997) for the individual. Cheetahs (*Acinonyx jubatus*) rarely rub against features for territorial and social communication; therefore it is difficult to collect hair in sufficient quantity for it be used for analysis. As a result, the most applicable noninvasive biological sample source for cheetah is scat (feces). Scat samples can be used as a geophysical landmark for cheetah presence or as a source of material for a wide variety of downstream analyses.

Due to the dry environmental conditions throughout most of the cheetah's range, fecal samples can remain in the field for several weeks, thus tens of fecal samples can be found for any given individual at any point of time. However, the cheetah's relatively low densities (Chapter 8) make it challenging to locate those samples. The chance of finding scat samples in the wild can be increased by following cheetah spoor (Cheetah Conservation Botswana, unpublished data) or targeting conspicuous land marks that cheetahs use as marking sites (e.g., marking trees, termite mounds). The use of professionally-trained scent detection dogs has been effectively applied in the field to help locate marking sites and cheetah fecal samples (section "Scat-Detection Dogs"). Scat samples found in the field are usually of unknown individual identity, and sometimes even of unconfirmed species. Identification of species, individual, and sex of the animal that has deposited the scat sample can be determined using molecular genetic techniques (section "Assigning Species and Individual Identify of Noninvasive Samples Using Genetics"), allowing for a variety of follow-up questions.

Geophysical locations can be used as presence points for geospatial analyses (Chapters 8 and 36), to guide other surveys, such as camera trap placement (Chapter 29), or to study marking behavior (Chapter 9). Collectible scat samples can also be used to conduct traditional diet analysis based on indigestible prey remains (e.g., hair; section "Dietary Analysis") regardless of the age of the sample. If samples are not decayed, fecal DNA can be extracted and applied for a wide range of genetic analyses (e.g., genetic diversity, phylogeography; Chapter 6). Hormone levels can be measured in fresh scat samples to assess stress levels and reproductive status (section "Hormone Analysis"). Samples found within 24 h can be assessed for presence of parasites via fecal egg counts. Protocols and forms that are relevant to this chapter can be found at https://www.elsevier.com/books-and-journals/book-companion/9780128040881.

SCAT-DETECTION DOGS

Scat-detection dogs are sniffing dogs similar to those used for detection of narcotics, weapons, explosives, human remains, or wildlife contraband. They differ from tracking (or trailing) dogs in that they do not search for a continuous scent trail, but rather scan the air for a scent cone dissipating from the scat sample. With over 220 million olfactory receptors, dogs have an excellent sense of smell (e.g., Wasser et al., 2004; de Oliveira et al., 2012) and scat-detection dogs can detect scat over large distances. Under ideal circumstances a dog can detect a cheetah scat sample from over 500 m, but distances of 50–100 m are more common (Brummer, personal observation).

Detection dogs are a powerful tool to locate scat samples of low-density target species, such as the cheetah (e.g., Maruping, 2011). Former studies showed that scat dogs can be more successful at detecting the presence of target species than remote-cameras, hair traps, and scent stations (e.g., Long et al., 2007a,b). Scat-detection dogs can be trained to identify multiple species/target scents. While dogs indicating on multiple species can be taught different signals for each species, more often molecular genetic or scent-matching techniques (using additional scat-detection dogs) are applied to assign a species to the samples (e.g., Wasser et al., 2009). Scat-detection dogs are also able to identify specific individuals based on fecal samples (e.g., Amur tigers, *Panthera tigris altaica*; Kerley and Salkina, 2007), although this has not yet been attempted for cheetahs.

Considerations for Scat-Detection Dog Work

A detection dog for cheetah scat can be hired as part of a professional dog-handler team (e.g., Working Dogs for Conservation, Conservation Canines), purchased or leased from a professional trainer (e.g., Green Dogs for Conservation, Steve Austin Canine Training), or trained in house. Given the low-density occurrence of cheetahs and their scat samples, it is critical that the detection dog has a high motivation and endurance level to search for the target scent for several hours under potentially harsh field conditions. The physical characteristics of the dog (e.g., size, coat density, and color) must be appropriate for the field conditions in cheetah range countries, which are mostly hot and arid, with habitat ranging from wide-open spaces to thick thorny vegetation.

To compensate for the fact that scat samples may not be detected during every search effort, it is important to provide sufficient training opportunities with known samples. Additional co-occurring species can be included as target scents to increase encounter of target samples, and thus successful detections and rewards, which in turn is a motivating factor for the dog. Including additional target species in a survey also increases the cost-efficiency per species. To avoid preference toward one of the target scents (e.g., scat samples from species with stronger

odor or more frequent occurrence), known scat samples of the target species (training aids) need to be included during training and search work. Species identity of field-collected scat samples should be verified genetically on a regular basis to assess the specificity of the detection dog.

Scat-detection dogs are usually taught a passive alert (e.g., sitting next to target scent; Fig. 31.1) upon detection of the target scat sample. This ensures that the fecal sample is not disturbed or contaminated with dog saliva, which may falsify the results of downstream analyses. Barking can be used as an active alert to locate the dog in thick vegetation, however, it is not recommended to leave the dog out of sight. Most field projects use a toy reward.

Hiring professional scat dog teams may be more cost effective for pilot studies or short-term/infrequent survey efforts despite travel expenses (e.g., study conducted by the Zambian Carnivore Project), whereas long-term studies or monitoring programs benefit from renting or buying a scat-detector dog and getting their own dog handler trained [e.g., Cheetah Conservation Fund (CCF), Action for Cheetah Kenya]. Locally trained dogs have the advantage that they do not require time for adaptation to local environmental conditions (e.g., climate, flora and fauna) prior to initiating a field survey. Regardless of the route chosen, the associated cost is generally justified by the increased number of scat samples detected in the field.

Considerations for Search Protocols/ Survey Designs

Study Design

During the dry season scat samples desiccate within less than a day, making them stable for genetic analyses, but nonsuited to evaluate parasite eggs, which are particularly susceptible to desiccation. During the rainy season, dung beetle activity is high and may remove all signs of a scat sample within hours. Fire and flooding can also contribute to the destruction of scat samples

FIGURE 31.1   **Scat-detection dog indicating on a cheetah scat sample found next to a termite mound.** Dog is fitted with a working harness and GPS collar to track work efforts. *Source: Birgit Braun, AGA/CCF.*

in the field and in addition put scat dogs and their handlers at risk. While dogs may detect burned and highly degraded samples that are not visible to human investigators, the data that can be extracted from such samples may be compromised (e.g., low DNA amplification success).

Search techniques can mostly be assigned to one of two categories: spot searches of points of interest, such as marking sites or systematic searches along transects selected randomly within grid cells. If the aim is to locate as many cheetah scat samples as possible within a short period of time then the former category is advisable, while for a systematic survey the latter category is required. Spot searches are time efficient; however, they require prior knowledge of the points of interest present in the area, and samples found at scent-marking sites will be biased toward males (Chapter 9). For a systematic search, it is important to track the search effort of the dog/handler team in the field (e.g., fitting a GPS unit on the dog or handler; Fig. 31.1)

to infer the area covered by the search and determine the success rate of sample detection. Training aids placed along transects provide a way to calibrate the search, as well as to motivate the dog by providing additional successful detections.

Points of Interest—Marking Sites

Marking sites (often trees or termite mounds) are used by male cheetahs to deposit scent in the form of feces and urine and are the best-known point of interest for cheetahs. During 63 searches conducted at 20 known marking sites in dense habitat in Namibia, 30 cheetah fecal samples were detected, corresponding to a search effort of 2 site searches/sample (CCF, unpublished data). During the pilot study a 50 m radius was searched around each marking tree, however, no scat sample was retrieved further than 10 m from the tree and most (approx. 90%) were located on the tree or within a 3 m radius; therefore an extensive search is not recommended.

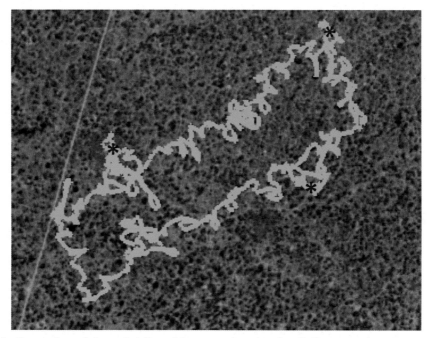

FIGURE 31.2 **Screen shot of the path followed by a scat-detection dog during a search on a rectangular transect.** *Red* stars indicate the location of known samples placed as training aids. *Source: CCF.*

Randomized Transects—Linear Features or Geometric Circuits

Survey transects are randomly selected within grid cells superimposed on the study area (e.g., Wasser et al., 2004; Wultsch et al., 2014). Transects along linear features can be conducted following roads, trails, or fence lines. Terrain along fences and roads is more passable, which presents a relevant advantage in bush encroached areas. In Namibia, 1 confirmed sample location (with 4 cheetah scat samples) was found during 40 1-km road searches, corresponding to a search effort of 40 km/sample location; the location was identified as a marking tree, which was not known prior to the study (CCF, unpublished data). An additional 181 carnivore samples were opportunistically collected during those searches, emphasizing the potential of including additional target species in cheetah scat surveys. In south-western Mozambique, 1434 carnivore scat samples, corresponding to 4 carnivore species

including cheetah, were found on 686 km of transects. Transects were initiated from a random start location and performed along trails, roads, and riverbeds to include as many landscape features of potential interest (e.g., water points, marking areas) as possible (Andresen, 2015).

Transects following a geometric circuit are selected irrespective of proximity to roads or other features and are commonly of rectangular (Fig. 31.2) or diamond shape. They are the least biased type of search but can be challenging in terms of terrain, in particular in areas with high levels of bush encroachment, as found in some parts of the cheetah's range. In Namibia, no cheetah fecal samples were detected during 18 1-km rectangular-transect searches performed in dense habitat, although the scat dog found 73% of the training aids placed up to 50 m from the transect (CCF, unpublished data). In open habitat in Zambia, 35 suspected cheetah scat samples were found at 17 locations during searches of 74 3-km transects and 15 at places of interest searched opportunistically (Becker et al., 2017). However, only 27 of the 50 total samples and 8 of 17 sample locations (as seen on Figure 2 in Becker et al., 2017) could be genetically confirmed to have originated from cheetahs, corresponding to a search effort of 27.8 km/ sample location. Two of the sample locations were scent-posts and contained multiple scat samples.

Considerations for Scat Sample Processing and Storage

Once collected, scat samples need to be kept dry, frozen, or in storage liquid (e.g., buffer, ethanol) to preserve the quality of the sample for downstream analysis. Dry samples and samples placed in storage buffer can usually be kept at room temperature. Frozen samples should be kept at a constant temperature, as temperature variation (and in particular freeze-thaw cycles) affects the quality of the samples. If samples are to be used for genetic work, particular attention needs to be paid during handling and processing of scat to avoid contamination with DNA of other species (e.g., collector, detection dog, or other species present in the sample collection) as the basis of any genetic work includes an amplification step, which will also amplify any contaminant DNA.

ASSIGNING SPECIES AND INDIVIDUAL IDENTITY OF NONINVASIVE SAMPLES USING GENETICS

With the advancement in genetic techniques over the past decades, it is now feasible to obtain genetic information from fecal samples. Inhibitors found in the diet need to be removed during the DNA extraction process, and amplification protocols need to be adapted to the generally low amounts and quality of target DNA found in scat samples, as well as concurrent presence of prey and microbial DNA.

Genetic Assignment of Species Identity

Mitochondrial markers are the marker of choice to verify the species identity of a scat sample. The shorter the sequence segment targeted the more likely it is to amplify. Two published mtDNA fragments have been used in combination to identify cheetah samples in Algeria (Busby et al., 2009) and a single nucleotide fragment of short length has been trialed in poor quality scat samples in southern Africa (CCF, unpublished data). Species assignment is conducted by comparing DNA sequences with available reference sequences.

Genetic Assignment of Sex

The sex of an individual can also be determined from scat samples using molecular markers. For low quality samples, it is important to use a test that is not based on presence/absence of a given marker (i.e., SRY), as it is not possible to distinguish between lack of amplification due to the absence of the markers or due to insufficient sample quality. A difference in amplification size in a fragment of the zinc-finger gene between the X and Y chromosomes (Pilgrim et al., 2005) has been applied to identify the sex of individuals from cheetah fecal samples (Becker et al., 2017).

Genetic Assignment of Individual Identity

Microsatellite (or short tandem repeat) markers are the markers of choice to identify individuals within a species. Despite the low genetic diversity of the cheetah, microsatellite markers are sufficiently variable to distinguish individuals (Chapter 6). Individual identity is assigned to the scat samples using fragment size polymorphism analysis. Several microsatellite markers have been used in cheetah genetics (Chapter 6). Markers designed in the domestic cat (*Felis catus*) were used on cheetah samples in the Serengeti (Gottelli et al., 2007) and Zambia to identify individuals (Becker et al., 2017). Published microsatellite markers can be adapted to low quality/quantity DNA to increase amplification success rates when using field-collected scat samples (Schmidt-Küntzel et al., in preparation). Using published marker information (i.e., allele frequency; Marker et al., 2008), it was determined that for Namibian cheetahs three microsatellite markers were sufficient to assign a sample to a given individual (as opposed to an unrelated individual) with a 1999/2000 likelihood (Mhuulu, 2015). Identification of individual cheetahs based on scat samples provides researchers with the opportunity to address various genetic, biomedical, and ecological questions. For instance, identifying recaptures (i.e., multiple scat samples collected from the same individual) provides information, such as density estimation for demographic studies.

RESEARCH QUESTIONS APPLICABLE TO FECAL SAMPLES OF KNOWN SPECIES AND/OR INDIVIDUAL IDENTITY

Dietary Analysis

Understanding the diet of free-ranging carnivores is an important part of carnivore ecology and conservation. Carnivore diet can be assessed through direct observation of kills or carcasses, or indirect analysis of stomach or scat contents, as well as stable isotope and genetic analysis. Direct observations are often biased toward large prey, as these kills are more likely to be observed. Stomach contents can be challenging to analyze as carcasses of carnivores are rarely obtained fresh. Stable isotopic analysis is considered a useful method for determining diet, but it requires the collection

of muscle, blood, hair, or breath from predator species, and reference muscle samples of prey species. Genetic studies can be performed on stomach content or scat but are more expensive than structural analysis and similarly to isotope analysis require adequate laboratory facilities. Traditional diet analyses based on scat content are relatively cheap and logistically the most feasible analysis method to study carnivore diet under field conditions.

Traditional scat content analysis methods quantify undigested prey remains, such as hair, feathers, bones, teeth, hooves, and claws (Geffen et al., 1992). To separate out the prey remains, scat samples are washed or crushed. The species identity is then determined by comparing the unique scale patterns of the cuticle (Fig. 31.3) or the shape and structure of the cross-section of the hair to a reference guide (e.g., Keogh, 1983; Figure 31.3).

The frequency of occurrence of scat samples containing a given species provides an indication of the diet in terms of species content (Breuer, 2005). This method, however, is biased toward the detection of small prey animals, as they have a higher fur-to-meat ratio than larger animals. To account for the different digestibility of prey species, biomass models were developed using feeding trials with captive cheetahs. A linear correction factor between prey mass and excreted scat numbers was determined (Marker et al., 2003). More recently, using an increased range of prey sizes, nonlinear relationships were shown to be more predictive (Wachter et al., 2012; Khorozyan et al., 2017). Furthermore, a second correction factor to determine the number of consumed individuals relative to the amount of prey mass was proposed (Wachter et al., 2012). Fecal analyses have contributed to the knowledge of the cheetah's diet composition in the wild in numerous regions (Boast et al., 2016; Craig et al., 2017; Farhadinia et al., 2012; Marker et al., 2003; Wachter et al., 2006; Chapter 8).

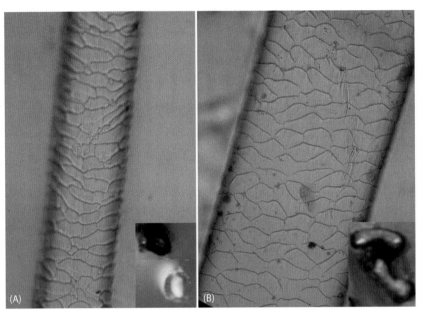

FIGURE 31.3   Imprint of the cuticle scale pattern with inset of a cross-section of a hindquarter hair from a cheetah (A) and a steenbok (*Raphicerus campestris*) (B) as seen under the microscope. *Source: CCF.*

Hormone Analysis

Fecal (or urine) samples can provide endocrine information on reproductive (i.e., progestogens, estrogens, or testosterone) and stress-related hormones (i.e., cortisol). These fecal hormones can be efficiently extracted, and assays to measure hormone metabolite concentrations have been validated for cheetahs (Brown et al., 1994; Ludwig et al., 2013; Pribbenow et al., 2016; Terio et al., 1999). Obtained information can provide health and physiological information about specific individuals and can also be used at the population level to examine questions, such as the influences of anthropogenic activities on stress levels, which can be of relevance for conservation efforts (Kersey and Dehnhard, 2014).

In free-ranging cheetahs, the use of noninvasive endocrine studies is limited due to difficulties in finding fecal samples of sufficient freshness. Hormones degrade when exposed to high temperature or precipitation, therefore only a fraction of fecal samples found in the wild

qualify for hormone studies (Millspaugh and Washburn, 2004) and caution should be used when interpreting the data. In general, for free-ranging cheetahs, population studies are more feasible than longitudinal studies of a specific individual overtime, due to the difficulties in collecting fresh scat from the same individual multiple times. However, if territorial male cheetahs are targeted specifically, repeated samples can be collected at marking trees, which are regularly visited by the same individual(s). Population studies can assess hormone levels relative to sex, group, location, or season. However, results need to be interpreted with caution because hormone concentrations differ for different age classes and reproductive status, which are usually unknown in fecal samples and thus might confound analysis outcome (Millspaugh and Washburn, 2004).

In captive cheetahs, longitudinal hormone evaluations are often performed on fecal samples to avoid having to immobilize the animal. The results can improve our understanding on the health and reproductive performance of the animals and thus help guide better husbandry conditions (Chapters 24, 25, and 27). Several studies have compared fecal metabolite concentrations of captive and free-ranging cheetahs (Terio et al., 2004), linking stress to ovarian activity (Jurke et al., 1997; Terio et al., 2003) and captive management (Koester et al., 2017). Importance of stress in relation to behavior, social status, and anthropogenic factors has been assessed in other carnivore species (e.g., spotted hyenas, *Crocuta crocuta*, Van Meter et al., 2009; African wild dogs, *Lycaon pictus*, Creel et al., 1997), but similar studies in cheetahs are currently lacking.

SUMMARY

Using scat-based techniques opens up a suite of possible uses and questions. Noninvasive techniques allow the researcher to gather important biomedical data that can often otherwise only be gained through live capture and anesthesia, which is more expensive, requires the presence of a veterinarian (Chapter 33), and is stressful for the animal. Being able to gather large sample sizes of field-collected scat is particularly valuable in species, such as cheetah, where low densities and cryptic behavior can severely hamper other research techniques.

Scat analysis techniques are sometimes used in conjunction with other noninvasive study approaches. For instance, genetic and hormone data from scat samples found at marking sites can be analyzed in conjunction with remote camera trap footage (Chapter 29) to combine information on visible health status (e.g., injuries, level of nutrition), social grouping (e.g., male coalitions), with information on relatedness and stress. Integration of well-trained scat-detection dogs into field projects is of substantial benefit to increase the number of scat samples available for research. Detection dogs can be trained on multiple target scents (e.g., scat samples of co-occurring carnivore species), which increases cost-efficiency and likely serves as motivating factor for the dog. Studies based on marking sites are biased toward males, whether they are based on invasive (e.g., capture) or noninvasive (e.g., remote camera traps, scat) techniques. Systematic searches performed with scat-detection dogs are expected to minimize this bias by detecting similar scat numbers for female cheetahs and thus increase the knowledge about female biology, prey ecology, and marking behavior.

References

Andresen, L., 2015. Cheetah distribution, threats and landscape connectivity in south-western Mozambique annual progress report. University-Centre for African Conservation Ecology, Nelson Mandela University.

Becker, M.S., Durant, S.M., Watson, F.G.R., Parker, M., Gottelli, D., M'soka, J., Droge, E., Nyirenda, M., Schuette, P., Dunkley, S., Brummer, R., 2017. Using dogs to find cats: detection dogs as a survey method for wide-ranging cheetah. J. Zool. 302 (3), 184–192.

Boast, L.K., Houser, A.M., Horgan, J., Reeves, H., Phale, P., Klein, R., 2016. Prey preferences of free-ranging cheetahs on farmlands: scat analysis versus farmer perceptions. Afr. J. Ecol. 54, 424–433.

Breuer, T., 2005. Prey choice of large carnivores in northern Cameroon. Afr. J. Ecol. 43, 97–106.

Brown, J.L., Wasser, S.K., Wildt, D.E., Graham, L.H., 1994. Comparative aspects of steroid hormone metabolism and ovarian activity in felids, measured noninvasively in feces. Biol. Reprod. 51, 776–786.

Busby, G.B., Gottelli, D., Wacher, T., Marker, L., Belbachir, F., De Smet, K., Belbachir-Bazi, A., Fellous, A., Belghoul, M., Durant, S.M., 2009. Genetic analysis of scat reveals leopard *Panthera pardus* and cheetah *Acinonyx jubatus* in southern Algeria. Oryx 43, 412–415.

Craig, C.A., Brassine, E.I., Parker, D.M., 2017. A record of cheetah (*Acinonyx jubatus*) diet in the Northern Tuli Game Reserve, Botswana. Afr. J. Ecol, Doi: 10.1111/aje.12374.

Creel, S., Creel, N.M., Mills, M.G.L., Monfort, S.L., 1997. Rank and reproduction in cooperatively breeding African wild dogs: behavioral and endocrine correlates. Behav.l Ecol. 8, 298–306.

de Oliveira, M.L., Norris, D., Ramírez, J.F.M., de, F., Peres, P.H., Galetti, M., Duarte, J., 2012. Dogs can detect scat samples more efficiently than humans: an experiment in a continuous Atlantic forest remnant. Zoologia 29, 183–186.

Farhadinia, M.S., Hosseini-Zavarei, F., Nezami, B., Harati, H., Absalan, H., Fabiano, E., Marker, L., 2012. Feeding ecology of the Asiatic cheetah *Acinonyx jubatus venaticus* in low prey habitats in northeastern Iran: Implications for effective conservation. J. Arid Environ. 87, 1–6.

Geffen, E., Hefner, R., Macdonald, D.W., Ucko, M., 1992. Diet and foraging behavior of Blanford's foxes *Vulpes cana*, in Israel. J. Mammal. 73, 395–402.

Gottelli, D., Wang, J., Bashir, S., Durant, S.M., 2007. Genetic analysis reveals promiscuity among female cheetahs. Proc. R. Soc. B 274, 1993–2001.

Jurke, M.H., Czekala, N., Lindburg, D.G., Millard, S.E., 1997. Fecal corticoid metabolite measurement in the cheetah (*Acinonyx jubatus*). Zoo Biol. 16, 133–147.

Keogh, H.J., 1983. A photographic reference system of the microstructure of the hair of the Southern African Bovids. S. Afr. J. Wildl. Res. 13, 89–131.

Kerley, L.L., Salkina, P.G., 2007. Using scent-matching dogs to identify individual Amur tigers from scats. J. Wildl. Manage. 71, 1349–1356.

Kersey, D.C., Dehnhard, M., 2014. The use of noninvasive and minimally invasive methods in endocrinology for threatened mammalian species conservation. Gen. Comp. Endocrinol. 203, 296–306.

Khorozyan, I., Lumetsberger, T., Ghoddousi, A., Soofi, M., Waltert, M., 2017. Global patterns in biomass models describing prey consumption by big cats. Mam. Rev. 47, 124–132.

Koester, D.C., Wildt, D.E., Brown, J.L., Meeks, K., Crosier, A.E., 2017. Public exposure and number of conspecifics have no influence on ovarian and adrenal activity in the cheetah (*Acinonyx jubatus*). Gen. Comp. Endocrinol. 243, 120–129.

Long, R.A., Donovan, T.M., Mackay, P., Zielinski, W.J., Buzas, J.S., 2007a. Comparing scat-detection dogs, cameras, and hair snares for surveying carnivores. J. Wildl. Manage. 71 (6), 2018–2025.

Long, R.A., Donovan, T.M., Mackay, P., Zielinski, W.J., Buzas, J.S., 2007b. Effectiveness of scat-detection dogs for detecting forest carnivores. J. Wildl. Manage. 71 (6), 2007–2017.

Ludwig, C., Wachter, B., Silinski-Mehr, S., Ganswindt, A., Bertschinger, H.J., Hofer, H., Dehnhard, M., 2013. Characterisation and validation of an enzyme-immunoassay for the non-invasive assessment of faecal glucocorticoid metabolites in cheetahs (*Acinonyx jubatus*). Gen. Comp. Endocrinol. 180, 15–23.

Marker, L.L., Muntifering, J.R., Dickman, A.J., Mills, M.G.L., Macdonald, D.W., 2003. Quantifying prey preferences of free-ranging Namibian cheetah. S. Afr. J. Wildl. Res. 33, 45–53.

Marker, L.L., Pearks Wilkerson, A.J., Sarno, R.J., Martenson, J., Breitenmoser-Würsten, C., O'Brien, S.J., Johnson, W.E., 2008. Molecular genetic insights on cheetah (*Acinonyx jubatus*) ecology and conservation in Namibia. J. Hered. 99, 2–13.

Maruping, Nkabeng Thato, 2011. The re-introduction of captive bred cheetah into a wild environment, Makulu Makete Wildlife Reserve, Limpopo province. PhD thesis, University of Pretoria, South Africa.

Mhuulu, L., 2015. Identification of Individual Cheetahs (*Acinonyx jubatus*) Represented in a Sample Collection, Combining Non-Invasive Genetic and Camera-Trapping Techniques. MSc thesis, Cheetah Conservation Fund and University of Namibia, Namibia.

Millspaugh, J.J., Washburn, B.E., 2004. Use of fecal glucocorticoid metabolite measures in conservation biology research: considerations for application and interpretation. Gen. Comp. Endocrinol. 138, 189–199.

Munson, L., Marker, L., 1997. The impact of capture and captivity on the health of Namibian farmland cheetahs (*Acinonyx jubatus*). Proceedings of the 50th Namibian Veterinary Congress.

Pilgrim, K.L., McKelvey, K.S., Riddle, A.E., Schwartz, M.K., 2005. Felid sex identification based on noninvasive genetic samples. Mol. Ecol. Notes 5, 60–61.

Pribbenow, S., Wachter, B., Ludwig, C., Weigold, A., Dehnhard, M., 2016. Validation of an enzyme-immunoassay for the non-invasive monitoring of faecal testosterone metabolites in male cheetah (*Acinonyx jubatus*). Gen. Comp. Endocrinol. 228, 40–47.

Terio, K.A., Citino, S.B., Brown, J.L., 1999. Fecal cortisol metabolite analysis for noninvasive monitoring of adrenocortical function in the cheetah (*Acinonyx jubatus*). J. Zoo Wildl. Med. 30, 484–491.

Terio, K.A., Marker, L., Munson, L., 2004. Evidence for chronic stress in captive but not free-ranging cheetahs (*Acinonyx jubatus*) based on adrenal morphology and function. J. Wildl. Dis. 40, 259–266.

Terio, K.A., Marker, L.L., Overstrom, E.W., Brown, J.L., 2003. Analysis of ovarian and adrenal activity in Namibian cheetahs. S. Afr. J. Wildl. Res. 33, 71–78.

Van Meter, P.E., French, J., Dloniak, S.M., Watts, H.E., Kolowski, J.M., Holekamp, K.E., 2009. Fecal glucocorticoids reflect socio-ecological and anthropogenic stressors in the lives of wild spotted hyenas. Horm. Behav. 55, 329–337.

Wachter, B., Blanc, A.-S., Melzheimer, J., Honer, O.P., Jago, M., Hofer, H., 2012. An advanced method to assess the diet of free-ranging large carnivores based on scats. PLoS One 7, e38066.

Wachter, B., Jauernig, O., Breitenmoser, U., 2006. Determination of prey hair in faeces of free-ranging Namibian cheetahs with a simple method. CAT News 44, 8–9.

Wasser, S.K., Davenport, B., Ramage, E.R., Hunt, K.E., Parker, M., Clarke, C., Stenhouse, G., 2004. Scat-detection dogs in wildlife research and management: application to grizzly and black bears in the Yellowhead Ecosystem, Alberta, Canada. Can. J. Zool. 82, 475–492.

Wasser, S.K., Smith, H., Madden, L., Marks, N., Vynne, C., 2009. Scent-matching dogs determine number of unique individuals from scat. J. Wildl. Manage. 73, 1233–1240.

Wultsch, C., Waits, L., Kelly, M., 2014. Noninvasive individual and species identification of jaguars (*Panthera onca*), pumas (*Puma concolor*) and ocelots (*Leopardus pardalis*) in Belize, Central America using cross-species microsatellites and fecal DNA. Mol. Ecol. Resour. 14, 1171–1182.

Field Methods for Visual and Remote Monitoring of the Cheetah

Femke Broekhuis*,**, Charlene Bissett†,
Elena V. Chelysheva‡

*Mara Cheetah Project, Kenya Wildlife Trust, Nairobi, Kenya
**University of Oxford, Tubney, Abingdon, United Kingdom
†South African National Parks, Kimberley, South Africa
‡Mara-Meru Cheetah Project, Kenya Wildlife Service, Nairobi, Kenya

INTRODUCTION

Carnivore studies are very diverse, and can include anything from estimating densities, to studying individuals and populations remotely or interactively, recording presence and absence, and understanding habitat, movements, and physiology (Boitani and Powell, 2012). In recent years, studies have changed with the development of new technology and techniques which aid the collection of data on elusive large carnivores, such as cheetahs (*Acinonyx jubatus*). Although the classic approach of direct observations is still frequently used in collecting behavioral data (Altmann, 1974), the ongoing development of technology has now allowed researchers to achieve this remotely (Altmann and Altmann, 2003; Cooke, 2008; Wilmers et al., 2015).

The remote collection of data can occur through both noninvasive and invasive techniques. Noninvasive methods include fecal studies (Chapter 31) and using devices, such as camera traps and microphone arrays, which are placed within the environment usually according to a predefined study design. Camera traps, for example, have been used to estimate cheetah densities (Brassine and Parker, 2015) and to investigate cheetah scent-marking behavior (Marnewick et al., 2006; Chapter 29). Invasive techniques generally involve devices that are placed on or in an animal to collect data on that specific individual and its environment, for example, collars and dataloggers. As invasive methodologies are an extension of, and complementary to, the classic methods of collecting behavioral data, we will briefly discuss the original methods and then focus on the use of dataloggers in cheetah research.

DIRECT OBSERVATIONS

Direct observations, through sightings or continuous follows, are classic methods used to study carnivores, such as cheetahs. Data can be collected on a species' habitat use, hunting behavior, intraguild and interguild interactions, and activity patterns (Bissett and Bernard, 2007; Bissett et al., 2015; Mills, 1992; Pettorelli et al., 2009). In some instances, and depending on the research question, knowing the identity of a cheetah is essential. For example, data based on individual identification of cheetahs have been used to determine relatedness (Kelly et al., 1998) and reproductive success (Pettorelli and Durant, 2007) and to estimate densities within a mark-recapture framework (Brassine and Parker, 2015; Broekhuis and Gopalaswamy, 2016). Consequently, individual identification of cheetahs is dependent on the overall objective of the study.

Cheetah Identification

Each individual cheetah can be recognized by its unique spot pattern (Caro, 1994; Chelysheva, 2004; Kelly, 2001) and the spot patterns remain constant throughout their life (Caro and Durant, 1991). In the 1970s, researchers mostly used facial markings for cheetah identification, but in the last decades researchers have used spot patterns from all parts of the cheetah's body, including the face, chest, limbs, and tail (Caro and Durant, 1991; Chelysheva, 2004; Kelly, 2001).

Using these markings there are two basic ways of identifying individual cheetahs based on a dataset of known individuals (i.e., a reference library): (1) by visually comparing photographs and (2) by using a computer-aided identification system. Manual identification can be time-consuming, so in some cases it is useful to speed up this process with the help of an identification software. The first three-dimensional computer-matching system for cheetah identification was designed for the Serengeti Cheetah Project (Kelly, 2001). Later, the HotSpotter software for individual animal identification and the Wildbook ecological information management systems were developed for striped and spotted animals. The Wildbook servers feed into an Image-Based Ecological Information System which turns massive collections of images into a high-resolution information database about animals (Crall et al., 2013).

Undoubtedly, computer-aided identification software saves time for the researcher. However, there are also some disadvantages to these systems. Each new image requires substantial preprocessing (e.g., cropping or enhancing brightness and contrast), which can be time-consuming, requires training, and can introduce subjectivity and error. In addition, computer-made comparisons of photographs at skewed camera angles have a tendency to reduce the coefficients of similarity (Kelly, 2001). Furthermore, individual features (e.g., spots and stripes) of natural patterns often vary in shape and size, yet many computer-aided identification systems ignore this variability and focus on the arrangement of these features instead (Anderson et al., 2010).

Challenges

Direct observations can provide long-term, detailed data, which are essential for the species conservation in the wild and improving their husbandry in captivity (Caro, 1993; Laurenson et al., 1992). Data from direct observations can also be useful for studying the rewilding of cheetahs, as in-depth behavioral information can assist with refining rewilding techniques (Houser et al., 2011). However, to be able to collect data using direct observations, the study animal needs to be habituated to ensure they can be followed and observed (Caro, 1994). Habituation of cheetahs is more feasible in protected areas, making it difficult to collect data on individuals outside protected areas. If behavioral observations were limited to protected areas, this could bias data to traits of cheetahs in these areas, a particular problem as most cheetah populations reside outside protected areas (Durant et al., 2017). Observations of cheetahs also tend to be biased toward favorable conditions (i.e., in open habitat and during the day) and the presence of an observer may disturb or alter behavior (Caro, 1994; Tomkiewicz et al., 2010). Remote techniques remove some of these biases while still providing important behavioral data on free-ranging wild cheetahs.

DATALOGGERS

Due to the difficulties of observing marine animals in their natural environment, the development of remote data collection technology has been led by research in this field. More recently, these techniques have been applied to terrestrial species (Wilmers et al., 2015). These developments in biologging are allowing researchers to study elusive and wide-ranging species, such as the cheetah, in their natural environment (Hebblewhite and Haydon, 2010; Wilmers et al., 2015).

Very High Frequency Collars

Very High Frequency (VHF) collars/transmitters emit a pulsed radio signal which allows a person to physically locate the cheetah by homing into the signal using a receiver and directional antenna. VHF collars have been used in cheetah studies since the late 1970s (Pettifer, 1981) and have helped give insights into feeding behavior (e.g., Bissett and Bernard, 2007), cub mortality (e.g., Mills and Mills, 2014), habitat selection (e.g., Bissett and Bernard, 2007) and home ranges (e.g., Marker et al., 2008).

Geographical Positioning System Collars

Geographical Positioning System (GPS) collars remotely record the location of a collared individual based either on a time schedule that is set by the researcher (for more information about GPS technology, see Tomkiewicz et al., 2010) or that is triggered by changes in an individual's activity (Brown et al., 2012). Depending on the times when locations are recorded, GPS data can be used to investigate an array of different behaviors in a variety of different ways. For example, locations can be used to identify clusters (locations in close proximity over a specified length of time) or to examine movement by taking into account both the temporal and spatial relationship between fixes. For cheetahs, spatial data have been used to investigate habitat use and avoidance behavior in relation to other predators, such as lions *Panthera leo* and spotted hyenas *Crocuta crocuta* in the Okavango Delta in Botswana (Broekhuis et al., 2013) and in a number of reserves in South Africa (Bissett et al., 2015; Rostro-García et al., 2015). Clusters of fixes can be used to identify site-fidelity behavior. Clusters of nonconsecutive points spread out over a long period of time (weeks–months) can be used to locate den sites (Ciarniello et al., 2005; Krofel et al., 2013) and marking sites, whereas clusters of consecutive points over a short period of time (hours–days) have been used to identify kill sites. This has given insight into predation rates and feeding behavior for various species, including cougars *Puma concolor* (Anderson and Lindzey, 2003), leopards *Panthera pardus* (Cristescu et al., 2014; Martins et al., 2011), and lions (Tambling et al., 2010). Clusters of fixes have also been used to try to locate cheetah kills in South Africa (Bissett, personal communication); however, due to a short kill retention time for cheetahs, that is, no longer than 2 h (Broekhuis et al., 2014), this method had a low success rate. Fine scale temporal resolution is required to increase the success of finding cheetah kills (<2 hourly fixes) and may need to be even finer for detecting small prey, such as hares or birds. Consecutive GPS points can be transformed into movement data (Gurarie et al., 2009). Various movement parameters, including distance traveled, tortuosity, speed can be related to different behavioral states such as walking, hunting, and resting (Nams, 2014). Very fine-scale GPS data, for example, in combination with other sensors, have been used to investigate cheetah hunting behavior (Wilson et al., 2013a,b).

Animal-Borne Cameras

Animal-borne cameras, duped "Crittercams" by National Geographic, can record stills, video, or both, allowing the observer to see and hear what an animal is likely to see and hear in the wild (Moll et al., 2007). Although it was originally used as an educational rather than a research tool, it has since been adapted to investigate behaviors, such as foraging and interactions with other individuals (both intra- and interspecific), for a variety of different species, including domestic cats *Felis catus* (Loyd et al., 2013), bottlenose dolphins *Tursiops aduncus* (Heithaus et al., 2002), and Adélie penguins *Pygoscelis adeliae* (Watanabe and Takahashi, 2013). Crittercam collars have been tested on cheetahs in Botswana to determine the feasibility of their use to gain insights into the hunting behavior of cheetahs on farmlands (Van der Weyde, personal communication). Animal-borne video cameras can give some unique insights into animal behavior, but are still limited because of a short lifespan (3–4 days) due to data storage capacity (Hays, 2015) and battery consumption.

Audio Recording Devices

Audio recording can be used to investigate behavior through acoustic signal processing; for example, it has been used to investigate the feeding and vigilance behavior of mule deer *Odocoileus hemionus* (Lynch et al., 2013, 2015). It could similarly be applied to cheetah research to determine when an individual is feeding and how long they are feeding for, and to study social interactions through communication. In addition, audio recording devices

could be used to record environmental noise, such as tourism vehicles, livestock, and people, to indicate the presence of anthropogenic centered factors.

Activity Sensors

Activity sensors, often in the form of accelerometers, collect data on the body movement of an animal which can be used to characterize and quantify behavior. Most modern sensors collect data along three axes, X, Y, and Z. Raw activity data can be used to see whether an animal is active or not. A study by Cozzi et al. (2012) used accelerometer data for cheetahs, wild dogs (*Lycaon pictus*), lions, and spotted hyenas to investigate temporal avoidance, and found that both cheetahs and wild dogs were more nocturnal than was previously thought and that this was positively correlated with moonlight intensity. The same cheetah accelerometer data were then transformed into three distinct behavioral states: walking, resting, and feeding (Grünewälder et al., 2012; also see section, "Data Analysis"). These data revealed that the observed nocturnal activity could be explained by feeding, and therefore hunting, behavior. In addition, the behavioral data also provided information on feeding bouts and illustrated how the length of feeding bouts differed seasonally (Broekhuis et al., 2014).

Other Sensors

There are various other sensors that can be deployed on animals to collect data, such as body temperature and heart rate. These data, also known as biospatial data, can be used as covariates for behavioral analysis (Wall et al., 2014). For example, Hetem et al. (2013) placed both temperature- and movement-sensitive implants in cheetahs, and found that cheetahs do not abandon their hunts due to overheating, as was previously thought. Sensors can also be fitted to record information about the animal's surroundings, such as ambient air temperature or the proximity to other (tagged) individuals, which can give insights into social relationships and reproductive behavior (Ralls et al., 2013).

Challenges

Although dataloggers are widely used for behavioral studies on cheetahs, there are certain shortfalls with the technology. Dataloggers are often expensive and are therefore generally only deployed on a small number of individuals in a population. The resolution of the data is often limited by battery life which in turn is restricted to the weight of the datalogger. In addition, collar failures have been experienced which negatively affect data collection and analysis (Houser et al., 2009; Weise et al., 2014). However, technology is consistently being improved making the collars lighter and extending the battery life.

FIGURE 32.1 **A female cheetah fitted with a GPS collar that collects data on the individual's location and activity through accelerometers.** *Source: Femke Broekhuis.*

DEVICE DEPLOYMENT AND ETHICAL CONSIDERATIONS

There are various ways that dataloggers can be deployed on an animal depending on the battery and satellite requirements. Larger devices, such as GPS units that need both large quantities of power and a clear view of the sky for satellite communication, are generally fitted on a collar which is then placed around the animal's neck (Fig. 32.1). The Guidelines of the American Society of Mammalogists state that external devices can weigh between 5% and 10% of an animal's body weight (Sikes and Gannon, 2011); however, for cheetahs this is likely to be too heavy. It is advisable that collars do not exceed 400 g and collars that weigh <1% of a cheetah's body weight have succesfully been used in the past (Broekhuis et al., 2014; Marker et al., 2008; Wilson et al., 2013a). Smaller devices, such as temperature loggers, can be fitted as implants, usually in the abdomen of an animal; however, implants are more invasive as they need a surgical procedure for placement (Hetem et al., 2013). Similarly, when collars are fitted on a wild animal, it requires that the animal is immobilized.

As with any animal study, researchers should adhere to strict ethical considerations such as the impact of devices on animal behavior and well-being and external access to the data to prevent misuse (i.e., cyber poaching). Devices should be tried and tested before being deployed on free-ranging animals (Tomkiewicz et al., 2010) and fitted or placed by professionals, and removed once they are no longer working or on completion of the study. The number of immobilizations should be minimized and for collars this can be done by adding a drop-off mechanism to a collar which triggers a time-based release. However, drop-off mechanisms add extra weight to the collar and are not always reliable due to triggering prematurely or not at all (McKenna, personal communication; Marker, personal communication). Therefore, before deploying any device on cheetahs, researchers must anticipate the possibility of negative effects on the well-being of the study individuals and correlate it with the potential outcome of the proposed study.

DATA ANALYSIS

The last couple of decades has seen not only the rapid development of devices to collect data remotely, but also the development of processing and analytical methods needed to interpret and understand the large and complex datasets that such devices produce.

One of the challenges of collecting data remotely is the interpretation of those data and what those data represent in a behavioral context. Using algorithms, there are two ways that data collected remotely can be translated, or classified, into different behaviors: supervised and unsupervised classification (Bishop, 2008). Supervised classification uses direct observations, either in a wild or in a captive situation, to interpret the data collected by the dataloggers (Bom et al., 2014; Grünewälder et al., 2012). Unsupervised classification is when there is no prior knowledge about the behavioral state of the animal, and behavioral states are determined based on significant differences within the data. If an animal can be observed, then the results from the unsupervised classification can be validated with direct observations (Grünewälder et al., 2012). Both GPS and activity data can be transformed into behavioral states or used to determine transition probabilities between states using methods, such as Support Vector Machines (Grünewälder et al., 2012), Hidden Markov Models (Patterson et al., 2009), Random Forest (Bom et al., 2014), and Behavioral Change Point Analysis (Gurarie et al., 2009). Some examples of packages that can be used for data analysis in statistical software R (R Development Core Team, 2016) are highlighted in Table 32.1; however, it is important to note that because of the rapidly evolving nature of technology (both in terms of dataloggers and computational power) packages are continuously being updated and new packages created.

It is important that the behavioral data collected remotely are put into context. By only using remotely sensed data there are concerns that ecologist are moving away from being in the field (Hebblewhite and Haydon, 2010). However, there is still a place for behavioral ecologists to be in the field to collect not only behavioral data for classification or validation purposes, but also environmental data, such as prey abundance, grass height, and so on. In addition, a visual of the tagged animals on a regular basis is necessary to ensure that devices are not harming the individual and that they are still in good condition.

TABLE 32.1   Examples of Packages Developed for Statistical Software R That can be Used for Analyzing Data From Dataloggers

Package	Description	References
moveHMM	For statistical modeling of animal movement data using hidden Markov models	Michelot et al. (2016)[a]
Ctmm	For analyzing animal relocation data as a continuous-time stochastic process	Calabrese et al. (2016)[a]
randomForest	Classification and regression based on forest of trees using random inputs	Bom et al. (2014)[b]
adehabitatHR	For analysis of animal home range	Calenge (2006)[a]
KNN	For analysis of classifying animal behavior modes	Bidder et al. (2014)[b]

[a]Publication about the package.
[b]Publication using the package.

APPLICATION TO CHEETAH BEHAVIOR AND THE FUTURE

In cheetah research, dataloggers have been used to investigate behavioral budgets (Broekhuis et al., 2014), feeding behavior (Broekhuis et al., 2014), hunting behavior (Hetem et al., 2013; Wilson et al., 2013a,b) and interspecific avoidance behavior (Broekhuis et al., 2013; Rostro-García et al., 2015). However, the use of these technologies in cheetah behavioral research is lagging compared with those in other terrestrial mammals (Wall et al., 2014). This is partially related to the cheetah's size as data collection devices are limited by weight which limits the available battery power. Although this is slowly changing as more advanced technology is being developed that is suitable for smaller species (Wilson et al., 2013b). As technology develops, allowing for higher resolution data, the scope of behaviors that can be investigated in cheetahs will increase.

References

Altmann, J., 1974. Observational study of behavior: sampling methods. Behaviour 49, 227–266.

Altmann, S.A., Altmann, J., 2003. The transformation of behaviour field studies. Anim. Behav. 65, 413–423.

Anderson, C.J.R., Da Vitoria Lobo, N., Roth, J.D., Waterman, J.M., 2010. Computer-aided photo-identification system with an application to polar bears based on whisker spot patterns. J. Mammal. 91, 1350–1359.

Anderson, J.C.R., Lindzey, F.G., 2003. Estimating cougar predation rates from GPS location clusters. J. Wildl. Manage. 67, 307–316.

Bidder, O.R., Campbell, H.A., Gómez-Laich, A., Urgé, P., Walker, J., Cai, Y., Gao, L., Quintana, F., Wilson, R.P., 2014. Love thy neighbour: automatic animal behavioural classification of acceleration data using the K-nearest neighbour algorithm. PLoS One 9, e88609.

Bishop, C.M., 2008. Pattern Recognition and Machine Learning. Springer, New York, NY.

Bissett, C., Bernard, R.T.F., 2007. Habitat selection and feeding ecology of the cheetah (Acinonyx jubatus) in thicket vegetation: is the cheetah a savanna specialist? J. Zool. 271, 310–317.

Bissett, C., Parker, D.M., Bernard, R.T., Perry, T.W., 2015. Management-induced niche shift? The activity of cheetahs in the presence of lions. Afr. J. Wildl. Res. 45, 197–203.

Boitani, L., Powell, R.A., 2012. Introduction: research and conservation of carnivores. In: Boitani, L., Powell, R.A. (Eds.), Carnivore Ecology and Conservation: A Handbook of Techniques. Oxford University Press, New York, NY.

Bom, R.A., Bouten, W., Piersma, T., Oosterbeek, K., Van Gils, J.A., 2014. Optimizing acceleration-based ethograms: the use of variable-time versus fixed-time segmentation. Mov. Ecol. 2, 6.

Brassine, E., Parker, D., 2015. Trapping elusive cats: using intensive camera trapping to estimate the density of a rare African felid. PLoS One 10, e0142508.

Broekhuis, F., Cozzi, G., Valeix, M., Mcnutt, J.W., Macdonald, D.W., 2013. Risk avoidance in sympatric large carnivores: reactive or predictive? J. Anim. Ecol. 82, 1098–1105.

Broekhuis, F., Gopalaswamy, A.M., 2016. Counting cats: spatially explicit population estimates of cheetah (Acinonyx jubatus) using unstructured sampling data. PLoS One 11, e0153875.

Broekhuis, F., Grünewälder, S., Mcnutt, J.W., Macdonald, D.W., 2014. Optimal hunting conditions drive circalunar behavior of a diurnal carnivore. Behav. Ecol. 25, 1268–1275.

Brown, D.D., Lapoint, S., Kays, R., Heidrich, W., Kümmeth, F., Wikelski, M., 2012. Accelerometer-informed GPS telemetry: reducing the trade-off between resolution and longevity. Wildl. Soc. Bull. 36, 139–146.

Calabrese, J.M., Fleming, C.H., Gurarie, E., 2016. ctmm: an r package for analyzing animal relocation data as a continuous-time stochastic process. Methods Ecol. Evol. 7, 1124–1132.

Calenge, C., 2006. The package "adehabitat" for the R software: a tool for the analysis of space and habitat use by animals. Ecol. Model. 197, 516–519.

Caro, T.M., 1993. Behavioral solutions to breeding cheetahs in captivity: insights from the wild. Zoo Biol. 12, 19–30.

Caro, T.M., 1994. Cheetahs of the Serengeti: Group Living in an Asocial Apecies. University of Chicago Press, Chicago, IL.

Caro, T.M., Durant, S.M., 1991. Use of quantitative analyses of pelage characteristics to reveal family resemblances in genetically monomorphic cheetahs. J. Hered. 82, 8–14.

Chelysheva, E., 2004. A new approach to cheetah identification. Cat News 41, 27–29.

Ciarniello, L.M., Boyce, M.S., Heard, D.C., Seip, D.R., 2005. Denning behavior and den site selection of grizzly bears along the Parsnip River, British Columbia, Canada. Ursus 16, 47–58.

Cooke, S.J., 2008. Biotelemetry and biologging in endangered species research and animal conservation: relevance to regional, national, and IUCN Red List threat assessments. Endanger. Species Res. 4, 165–185.

Cozzi, G., Broekhuis, F., Mcnutt, J.W., Turnbull, L.A., Macdonald, D.W., Schmid, B., 2012. Fear of the dark or dinner by moonlight? Reduced temporal partitioning among Africa's large carnivores. Ecology 93, 2590–2599.

Crall, J.P., Stewart, C.V., Berger-Wolf, T.Y., Rubenstein, D.I., Sundaresan, S.R., 2013. HotSpotter—Pattern Species Instance Recognition. In: IEEE Workshop on Applications of Computer Vision. Tampa, Florida.

Cristescu, B., Stenhouse, G.B., Boyce, M.S., 2014. Predicting multiple behaviors from GPS radiocollar cluster data. Behav. Ecol. 26, 452–464.

Durant, S., Mitchell, N., Groom, R., Pettorelli, N., Ipavec, A., Jacobson, A., Woodroffe, R., Bohm, M., Hunter, L., Becker, M., Broekhuis, F., Bashir, S., Andresen, L., Aschenborn, O., Beddiaf, M., Belbachir, F., Belbachir-Bazi, A., Berbash, A., Brandao De Matos Machado, I., Breitenmoser, C., Chege, M., Cilliers, D., Davies-Mostert, H.T., Dickman, A.J., Ezekiel, F., Farhadinia, M.S., Funston, P.J., Henschel, P., Horgan, J., Hans, H., De Longh, Houman, J., Klein, R., Lindsey, P.A., Marker, L.L., Marnewick, K., Melzheimer, J., Merkle, J., M'soka, J., Msuha, M., O'neil, H., Parker, M., Purchase, G., Sahailou, S., Saidu, Y., Samna, A., Schmidt-Küntzel, A., Selebatso, M., Sogbohossou, E., Soultan, A., Stone, E., Van Der Meer, E., Van Vuuren, R., Wykstra, M., Young-Overton, K., 2017. Disappearing spots, the global decline of cheetah and what it means for conservation. Proc. Natl. Acad. Sci. USA 114, 528–533.

Grünewälder, S., Broekhuis, F., Macdonald, D.W., Wilson, A.M., Mcnutt, J.W., Shawe-Taylor, J., Hailes, S., 2012. Movement activity based classification of animal behaviour with an application to data from cheetah (Acinonyx jubatus). PLoS One 7, e49120.

Gurarie, E., Andrews, R.D., Laidre, K.L., 2009. A novel method for identifying behavioural changes in animal movement data. Ecol. Lett. 12, 395–408.

Hays, G.C., 2015. New insights: animal-borne cameras and accelerometers reveal the secret lives of cryptic species. J. Anim. Ecol. 84, 587–589.

Hebblewhite, M., Haydon, D.T., 2010. Distinguishing technology from biology: a critical review of the use of GPS telemetry data in ecology. Philos. Trans. R. Soc. Lond. B Biol. Sci. 365, 2303–2312.

Heithaus, M.H., Dill, L.D., Marshall, G.M., Buhleier, B.B., 2002. Habitat use and foraging behavior of tiger sharks (Galeocerdo cuvier) in a seagrass ecosystem. Mar. Biol. 140, 237–248.

Hetem, R.S., Mitchell, D., De Witt, B.A., Fick, L.G., Meyer, L.C.R., Maloney, S.K., Fuller, A., 2013. Cheetah do not abandon hunts because they overheat. Biol. Lett. 9, 20130472.

Houser, A., Gusset, M., Bragg, C.J., Boast, L., Somers, M.J., 2011. Pre-release hunting, training and post-release monitoring are key components in the rehabilitation of orphaned large felids. S. Afr. J. Wildl. Res. 41, 11–20.

Houser, A., Somers, M.J., Boast, L.K., 2009. Home range use of free-ranging cheetah on farm and conservation land in Botswana. S. Afr. J. Wildl. Res. 39, 11–22.

Kelly, M.J., 2001. Computer-aided photography matching in studies using individual identification: an example from Serengeti cheetahs. J. Mammal. 82, 440–449.

Kelly, M.J., Laurenson, M.K., Fitzgibbon, C.D., Collins, D.A., Durant, S.M., Frame, G.W., Bertram, B.C., Caro, T.M., 1998. Demography of the Serengeti cheetah (Acinonyx jubatus) population: the first 25 years. J. Zool. 244, 473–488.

Krofel, M., Skrbinšek, T., Kos, I., 2013. Use of GPS location clusters analysis to study predation, feeding, and maternal behavior of the Eurasian lynx. Ecol. Res. 28, 103–116.

Laurenson, K.M., Caro, T.M., Borner, M., 1992. Female cheetah reproduction. Natl. Geogr. Res. Explor. 8, 64–75.

Loyd, K.a.T., Hernandez, S.M., Carroll, J.P., Abernathy, K.J., Marshall, G.J., 2013. Quantifying free-roaming domestic cat predation using animal-borne video cameras. Biol. Conserv. 160, 183–189.

Lynch, E., Angeloni, L., Fristrup, K., Joyce, D., Wittemyer, G., 2013. The use of on-animal acoustical recording devices for studying animal behavior. Ecol. Evol. 3, 2030–2037.

Lynch, E., Northrup, J.M., Mckenna, M.F., Anderson, C.R., Angeloni, L., Wittemyer, G., 2015. Landscape and anthropogenic features influence the use of auditory vigilance by mule deer. Behav. Ecolo. 26, 75–82.

Marker, L.L., Dickman, A.J., Mills, M.G.L., Jeo, R.M., Macdonald, D.W., 2008. Spatial ecology of cheetahs on north-central Namibian farmlands. J. Zool. 274, 226–238.

Marnewick, K.A., Bothma, J.D.P., Verdoorn, G.H., 2006. Using camera-trapping to investigate the use of a tree as a scent-marking post by cheetahs in the Thabazimbi district. S. Afr. J. Wildl. Res. 36, 139–145.

Martins, Q., Horsnell, W., Titus, W., Rautenbach, T., Harris, S., 2011. Diet determination of the Cape Mountain leopards using global positioning system location clusters and scat analysis. J. Zool. 283, 81–87.

Michelot, T., Langrock, R., Patterson, T.A., 2016. moveHMM: an R package for the statistical modelling of animal movement data using hidden Markov models. Methods Ecol. Evol. 7, 1308–1315.

Mills, M.G.L., 1992. A comparison of methods used to study food habits of large African carnivores. In: Mccullough, D.R., Barrett, R.H. (Eds.), Wildlife 2001: Populations. Elsevier Applied Science, London/New York, NY.

Mills, M., Mills, M., 2014. Cheetah cub survival revisited: a re-evaluation of the role of predation, especially by lions, and implications for conservation. J. Zool. 292, 136–141.

Moll, R.J., Millspaugh, J.J., Beringer, J., Sartwell, J., He, Z., 2007. A new 'view' of ecology and conservation through animal-borne video systems. Trends Ecol. Evol. 22, 660–668.

Nams, V.O., 2014. Combining animal movements and behavioural data to detect behavioural states. Ecol. Lett. 17, 1228–1237.

Patterson, T.A., Basson, M., Bravington, M.V., Gunn, J.S., 2009. Classifying movement behaviour in relation to environmental conditions using hidden Markov models. J. Anim. Ecol. 78, 1113–1123.

Pettifer, H.L., 1981. Aspects on the ecology of cheetah (*Acinonyx jubatus*) on the Suikerborand Nature Reserve. In: Chapman, J.A., Pursely, D. (Eds.), Worldwide Furbearer Conference Proceedings. R.R. Donnelly & Sons, Falls Church, VA.

Pettorelli, N., Durant, S.M., 2007. Family effects on early survival and variance in long-term reproductive success of female cheetahs. J. Anim. Ecol. 76, 908–914.

Pettorelli, N., Hilborn, A., Broekhuis, F., Durant, S.M., 2009. Exploring habitat use by cheetahs using ecological niche factor analysis. J. Zool. 277, 141–148.

R Development Core Team, 2016. R: A Language and Environment for Statistical Computing. R Foundation for Statistical Computing, Vienna, Austria.

Ralls, K., Sanchez, J.N., Savage, J., Coonan, T.J., Hudgens, B.R., Cypher, B.L., 2013. Social relationships and reproductive behavior of island foxes inferred from proximity logger data. J. Mammal. 94, 1185–1196.

Rostro-García, S., Kamler, J.F., Hunter, L.T.B., 2015. To kill, stay or flee: the effects of lions and landscape factors on habitat and kill site selection of cheetahs in South Africa. PLoS One 10, e0117743.

Sikes, R.S., Gannon, W.L., 2011. Guidelines of the American Society of Mammalogists for the use of wild mammals in research. J. Mammal. 92, 235–253.

Tambling, C.J., Cameron, E.Z., Du Toit, J.T., Getz, W.M., 2010. Methods for locating African lion kills using global positioning system movement data. J. Wildl. Manage. 74, 549–556.

Tomkiewicz, S.M., Fuller, M.R., Kie, J.G., Bates, K.K., 2010. Global positioning system and associated technologies in animal behaviour and ecological research. Philos. Trans. R. Soc. B Biol. Sci. 365, 2163–2176.

Wall, J., Wittemyer, G., Klinkenberg, B., Douglas-Hamilton, I., 2014. Novel opportunities for wildlife conservation and research with real-time monitoring. Ecol. Appl. 24, 593–601.

Watanabe, Y.Y., Takahashi, A., 2013. Linking animal-borne video to accelerometers reveals prey capture variability. Proc. Natl. Acad. Sci. 110, 2199–2204.

Weise, F.J., Stratford, K.J., Van Vuuren, R.J., 2014. Financial costs of large carnivore translocations—accounting for conservation. PLoS One 9, e105042.

Wilmers, C.C., Nickel, B., Bryce, C.M., Smith, J.A., Wheat, R.E., Yovovich, V., 2015. The golden age of bio-logging: how animal-borne sensors are advancing the frontiers of ecology. Ecology 96, 1741–1753.

Wilson, A., Lowe, J., Roskilly, K., Hudson, P., Golabek, K., McNutt, J., 2013a. Locomotion dynamics of hunting in wild cheetahs. Nature 498, 185–189.

Wilson, J.W., Mills, M.G.L., Wilson, R.P., Peters, G., Mills, M.E.J., Speakman, J.R., Durant, S.M., Bennett, N.C., Marks, N.J., Scantlebury, M., 2013b. Cheetahs, *Acinonyx jubatus*, balance turn capacity with pace when chasing prey. Biol. Lett. 9, 20130620.

Capture, Care, Collaring, and Collection of Biomedical Samples in Free-Ranging Cheetahs

Laurie Marker, Anne Schmidt-Küntzel*,
Ruben Portas**, Amy Dickman[†], Kyle Good[‡],
Axel Hartmann[§], Bogdan Cristescu*, Joerg Melzheimer***

*Cheetah Conservation Fund, Otjiwarongo, Namibia
**Leibniz Institute for Zoo and Wildlife Research, Berlin, Germany
[†]University of Oxford, Tubney, Abingdon, United Kingdom
[‡]Cheetah Conservation Botswana, Bulawayo, Zimbabwe
[§]Bicornis Veterinary Consulting, Bicornis Conservation Trust, Otjiwarongo, Namibia

INTRODUCTION

Wildlife capture and handling allow veterinary care and help in answering health and genetic questions that require biomedical samples. In addition, ecological questions based on spatial data can be addressed by collaring individuals with radio very high frequency (VHF) or global positioning system (GPS) tracking collars. However, capturing and anesthetizing free-ranging carnivores is a controversial task, which entails potential risks for the animal. Adverse effects may be manifested a few days after capture, sometimes leading to death (Putman, 1995). The ethical aspect is particularly relevant when the target is an endangered or threatened species, such as the cheetah (*Acinonyx jubatus*). Repeated captures of the same individual within a short time frame should be avoided. Physiological risks to the cheetah increase when the captured individual is old and/or in poor physical condition. The benefit of the knowledge gained must be weighed against the possibility of injury and/or physiological stress to the animal (Caulkett and Shury, 2007). Cheetahs caught in capture cages are highly stressed, which is reflected in increased liver enzymes and biochemical stress markers (Munson and Marker, 1997). Therefore, research projects should consider whether necessary data could be collected using noninvasive methods (Chapters 29–31) prior to making the decision to capture.

Capturing techniques and anesthetic procedures must be detailed in field protocols and regularly reviewed to ensure that they are optimal in terms of current best practices and of comparability with previous studies. Captures should be undertaken by experienced field biologists and veterinarians. Research plans and field protocols should be reviewed by an Animal Care and Use Ethical Committee, as well as other relevant authorities (e.g., Department of Environment/Conservation, National Parks System, and Research Advisory Board). Capture and handling procedures should follow the highest standards in safety and animal welfare (Sikes et al., 2011). Careful thought should be given to the choice of capture methodology, bearing in mind welfare, safety, effectiveness, and selectivity. Several guidelines and articles on animal capture and welfare for research are available and should be consulted prior to capture being attempted [e.g., Caulkett and Shury, 2007; Iossa et al., 2007; Kock and Burroughs, 2014; Powell and Proulx, 2003; The American Society of Mammalogists (http://www.mammalogy.org/)]. This chapter covers some of the basic techniques used for such procedures. Protocols and forms that are relevant to this chapter can be found at https://www.elsevier.com/books-and-journals/book-companion/9780128040881.

FACTORS TO CONSIDER BEFORE CAPTURE

With legal and research ethics clearances in effect, the trained capture personnel must be familiar with:

- Method of capture to be used.
- Method of drug administration and choice of anesthetic agent, its properties, effects, hazards, and legal regulation.

- Rapid assessment of animal's health condition and suitability for anesthesia (i.e., diagnosis of potential dehydration, undernourishment, or evidence of disease).
- Physiological monitoring of the anesthetized cheetah.
- Handling and proper placement of collars.
- First aid for cardiopulmonary resuscitation and emergency treatment for cheetahs in case of preexisting conditions or capture- and anesthesia-related issues.
- First aid for humans in the event of accidental exposure to any of the drugs on hand or other injury during the capture process.
- Euthanasia as a last resort, using the methodology approved by the relevant institutional committees that reviewed the capture protocol.

Materials necessary for capture include specialized capture equipment, drugs and drug delivery devices, monitoring equipment, sample collection supplies (e.g., syringes, needles, and vials for blood/tissue samples), identification devices (e.g., ear tags, microchip, and respective applicators), and, where necessary, equipment for GPS collar fitting (e.g., wrenches and punches for adjusting collar belting). In the event of a severe injury the animal may have to be transferred to a specialized facility, so contingency plans should be made for this in advance.

CHEETAH CAPTURE METHODS

The method most commonly used for physical capture of cheetahs is a cage trap, also known as capture cage or box trap (Marker et al., 2003a; Marnewick and Somers, 2015; Wachter et al., 2011; Fig. 33.1A). Foot restraining devices are usually not recommended and free-darting is only feasible in areas where cheetahs are not used to human presence. Cheetahs are shy by nature and difficult to approach, especially in areas where they are harassed. Trapping should be avoided during extremely low or high temperatures and/or during the rainy season when road conditions make it difficult to check traps and to attend captured animals. Capture or anesthesia should preferentially be done early in the day to avoid the hottest hours and to allow for recovery and release before nightfall, when cheetahs are most vulnerable to other predators.

Captured cheetahs should be worked on at the place of capture in the presence of a veterinarian. If more extensive medical attention is necessary, cheetahs can be transferred to a holding facility, veterinary clinic, or research facility for examination. Specially designed transport crates provide a dark and secure environment.

Cage Traps

Cage traps are the most commonly used method to live capture cheetahs. The functional parts include the cage with two self-closing gates at either end (Fig. 33.1A), a door locking mechanism, and a trigger plate (treadle) in the middle [Fig. 33.1B (1)]. When the cheetah enters the trap and steps on the floor treadle, the trigger releases the trap gates, closing the trap. Cage sizes vary but measure approximately $2.40 \times 0.95 \times 0.80$ m^3, with a minimum recommended length of 2.20 m. Due to the cheetah's speed, short cage traps allow for escape and can cause potentially severe injuries if the gate falls on the cheetah's back as it attempts to run out.

Cage traps should be strongly built with appropriate mesh wire walls, ceiling, and floor [Fig. 33.1B (2)]. The gates should have a locking mechanism to prevent being opened by a trapped animal, and the trap should include "tail savers," which ensure that the gates leave a small gap when fully closed [Fig. 33.1B (3)].

If cage traps are used properly, the risk of injury to the cheetah is relatively low. Responding quickly to a capture event and reducing stress to the cheetah after capture minimize the chance of injury. Trapped cheetahs are easily stressed, therefore only essential personnel should approach the cage trap while the animal is conscious, and all noise should be kept to a minimum. Domestic animals must be kept

FIGURE 33.1 **Typical metal cage trap used for catching cheetahs.** (A) Cage with open trap doors. (B) Close up showing (1) the trigger plate or treadle (which triggers the trap door closure), (2) smooth wire mesh, and (3) a tail saver (which prevents the trap door from closing completely). *Source: IZW–Ruben Portas.*

away from the area. Some nontarget species are susceptible to self-injury when trapped or/and suffer a high amount of stress that can be lethal (e.g., warthog and honey badger) (Powell and Proulx, 2003). Cage traps should allow for easy release of captured, nontarget animals, including potentially dangerous ones, by remotely raising the cage door via a cable (e.g., winch) fed to a vehicle. Cage traps should be checked at a minimum twice daily, or more frequently if government regulations require it. Whenever possible, the trap should be placed in the shade (i.e., under a tree or near a boulder; Fig. 33.2A). If the trapping location does not provide shade, the trap should be covered with a tarp, shade net, or bushes (Fig. 33.2B).

While a central plate on the trap's floor is still the most common trigger, computerized systems exist (e.g., an array of light beams monitoring the trap and electronically activating a release mechanism). By adapting the height of the laser array and programming the activity times of the target species, bycatch is significantly reduced (Melzheimer and Portas, unpublished data).

Foot Snares

Foot snares may be considered in very special circumstances and by very experienced trappers. They have been used successfully to capture a small number of cheetahs in the remote mountains of Iran (Hunter et al., 2007). Snares have to be set in exposed locations to prevent entanglement and injury of the animal. Snares should be equipped with a shock absorption device that can extend when the animal attempts to pull its foot out of the trap (e.g., metal spring or rubber bungee cord with suitable tension) and at least two solid in-line swivels and a stopper to prevent full closure of the snare on to the animal's foot. Some studies have shown that animals captured with foot-restraining devices had higher stress levels than those captured with cage traps (Cattet et al., 2008; Iossa et al., 2007).

Free-Darting

In areas where cheetahs are relatively habituated to vehicles, such as some national parks and game reserves, free-darting is the most commonly selected method of immobilization (Caro, 1994). The method requires excellent long-range darting skills and knowledge of ballistics, and cheetah anatomy, to avoid accidental debilitating injuries to the animal caused by an ill-placed dart. Once darted, the animal will attempt to escape and needs to be continuously monitored to avoid losing contact. The surrounding environment plays an important role not only to initially sight the animal and follow it, but also for its safety. A cheetah darted in a dense, bushed area may not be found before the drugs have metabolized and worn off and, in the worst case situation, it might die whether from the consequences of an unmonitored anesthesia complication or from the inability to defend itself against encountered threats. Radiotelemetry darts can be used to locate the animal after darting.

CAGE TRAP SETUP

Placement of Cheetah Traps

The most effective locations for cheetah trap placement are those that show signs of repeated use. Known locations of interest for cheetahs are scent-marking sites, such as trees (Fig. 33.2A–B), bushes, rocks, termite mounds (Marnewick et al., 2006; Chapter 9), and—especially in areas of dense vegetation—walk-through areas, such as roadways, trails, and fence lines (Fig. 33.2C). In arid open areas, water points, such as natural waterholes, dams, or troughs are also potential locations. It is noteworthy that capturing cheetahs at scent-marking sites introduces a gender bias, as these are primarily visited by adult male cheetahs (Marker et al. 2003b; Marnewick et al., 2006; Chapter 9). Remote cameras (Chapter 29) can

FIGURE 33.2 **Cage trap setups.** (A) Next to a scent-marking tree, with a ring of thorny bushes. A solar panel is placed inside the boma to power electronic devices. A second trap is within the thorn bush circle with either another trapped cheetah or a goat, which acts as bait. (B) Next to a scent-marking tree, with a second cage to catch a second cheetah set next to the cage containing the first cheetah. A tarp provides shade to the cheetah in the trap cage. (C) Along a fence line. Thorn bush on the sides acts as a natural funnel. *Source: Parts A and C, IZW–Joerg Melzheimer; part B, CCF–Laurie Marker.*

reveal detailed information on cheetah frequency of use and behavior (e.g., where they mark and direction of travel).

To capture cheetahs at a scent-marking site, a closed or open setup can be followed: access to the scent-marking site can be blocked by a ring of thorny bushes with the exception of a passage leading through the cage trap (Fig. 33.2A). For trap-shy animals or when thornbushes are not available in the area, the trap can be placed immediately next to the trunk of the scent-marking tree or under bent trees where cheetahs mark. More than one trap can be set (Fig. 33.2A–B).

Baits and Lures

Although not necessary under most circumstances, it is possible to attract the cheetah into the trap using bait or lures. An animal, such as a goat or lamb can be placed in a secure cage to call and attract the predator to that spot. It is vital that the bait animal has adequate shade, food, water, and bedding; is checked regularly; and is not left in the lure cage for extended periods. The use of live bait should be carefully thought through, as the animal will experience considerable amounts of stress; it is not allowed in some countries, and does not comply with the ethics requirements of some scientific journals. Imitation animals could be an effective alternative, such as a robot goat used in Kenya to lure cheetahs to cage traps (Hermsen, 2013). Audio call-in systems can attract cheetahs (Melzheimer and Portas, unpublished data) and other carnivores (Cozzi et al., 2013), and urine from females in estrus attracts trap-shy males to traps (Divyabhanusinh, 1995; Mills, 1996).

Catching a Group of Cheetahs

Adult males are either found alone or in coalitions of two to three, rarely four individuals, while adult females are typically on their own or accompanied by dependent cubs (Caro, 1994; Marker et al. 2003b; Wachter et al., 2011; Chapter 9). The strong family bond facilitates the capture of the whole group once one member is trapped (Marker et al., 2003b). To ensure that the entire group of cheetahs is captured, traps should be set adjacent to the one containing the trapped individual (Fig. 33.2A–B). While captured cheetahs should be kept in the cage for a minimal amount of time, it is even more important not to disrupt a social group if captured for translocation, by ensuring that all animals are caught. With a large group, trapping may take several days; under such circumstances only the most recently caught cat should be left as a lure and the first ones moved to a holding facility until the entire group is captured. The trapped cheetah(s) must have adequate shade and water. When targeting a social group (coalition of males, sibling group, or mother with cubs), at least two traps should be set from the beginning to speed up the process. Setting a motion-activated camera at the site and inspecting tracks will help in determining the group size, such that enough traps can be placed to capture all the animals promptly.

Electronic Devices to Remotely Monitor Traps

Modern technology allows monitoring of traps to minimize the time an animal spends in a trap (thereby minimizing stress and risk of injury). Animal presence can be detected through photographic evidence using cellular cameras (such as Bushnell, Doerr, and Seissiger), which send pictures via email or GSM (Global System for Mobile communication) message to the user. Alternatively, a "capture signal" can be triggered.

If terrain allows for radio signals to reach the researcher's position, radio collars can be used as "homemade trap monitors." The magnet that activates the VHF collar can be attached to the trigger of the trap (Johansson et al., 2011), so that it is pulled when the trap is triggered, thereby activating the radio beacon. Other radio devices have also been used to monitor traps (Hayes, 1982; Marks, 1996; Nolan et al., 1984).

Devices with satellite communication can be useful in remote areas without network coverage (Marks, 1996). The pin that activates the satellite transmitter can be attached to the trap gate so that it gets pulled when the trap is triggered, thereby initiating an e-mail alert immediately. A daily status e-mail confirms the general functioning of the device. Cell phone or radio triggers that call a phone or trigger a radio signal when the trap is sprung can be utilized.

In areas with GSM coverage, a magnet switch can be used to monitor trap status and the information sent via a GSM modem (Larkin et al., 2003; O'Neill et al., 2007). Alternatively, CCTV cameras can provide live streaming of the trapping location. The information can be used to monitor trap activity or to remotely trigger cage traps when the target animal is inside, thereby allowing for immediate release of non-target species. It is recommended to check cage traps regularly even when these devices are in place in case of failure.

CHEMICAL IMMOBILIZATION

Cheetahs must be chemically immobilized to enable the collection of samples or deployment of a GPS/VHF collar. The time of anesthesia and recovery should be kept to a minimum; however, any anesthetic event should be optimally utilized to maximize the output with regard to sample and data collection. Chemical immobilization is covered in detail in Chapter 24; however, practical information specific to field anesthesia is presented here.

If trapping is used to restrain cheetahs, minimizing the time between capture and immobilization reduces stress-associated risks, including the potential for self-injury, hypo- or hyperthermia, dehydration, and other complications. Monitoring (from a distance) the movements of the captured individual until the chemical immobilization has taken effect, as well as postrelease routine monitoring, are critical. Following chemical immobilization, the movements of large carnivores can be affected for several weeks after release (Cattet et al., 2008), but more typically postcapture effects on movement rates are shorter in duration (Rode et al., 2014; Teräväinen 2016).

Choosing an Anesthetic Agent

Several anesthesia protocols providing smooth immobilization in the cheetah are presented in Chapter 24. In choosing an anesthetic agent for a field workup, several aspects should be taken into consideration. For cheetahs darted in the wild, the drug combination and dosage should ensure a quick induction to avoid the animal traveling long distances postinjection. In most cases, anesthetized cheetahs will show signs of sedation within 5–10 min and will be lying down within 10–15 min. Preference should be given to drugs that have an antagonist (e.g., medetomidine is reversed with atipamezole) to shorten recovery time, and as an emergency tool in case of complications. Certain drugs can have adverse effects on cheetahs (e.g., ketamine alone can cause seizures). For hand injection in particular, small drug volumes should be favored and special consideration should be given to the choice of syringe and needle to reduce the risk of partial drug delivery (Chapter 24). For anesthesia performed from a distance with a dart syringe, volumes need to be appropriate for the drug chamber of the dart syringe; volumes may be adjusted with sterile water for ballistic purposes, if needed.

Dosage

Dosage depends on the anesthetic agent used and the animal's body weight and condition. The correct dose is critical for achieving a safe anesthesia. Therefore, veterinarians, paraveterinarians, and researchers must be able to accurately estimate the weight of the cheetah to be immobilized, often without close examination of the animal. Due to increased stress levels, wild animals (particularly females with dependent cubs) require higher doses compared to captive and habituated animals (Hartmann, personal observation). The dose given should be slightly lower for young and old animals, or those that are dehydrated or in poor condition (Hartmann, personal observation).

Administration of the Anesthetic Agent

Most anesthetics are administered intramuscularly. Administration of the entire prescribed dose at one time as a bolus is preferred over incremental dosing, as otherwise the level of anesthesia achieved may be lighter, with the cheetah taking more time to be sedated and to recover. The two preferred ways of administering anesthetic agents intramuscularly to cheetahs are: hand syringe or darting. Pole syringes can be used if no transport box or darting equipment are available.

Hand Syringe in a Squeeze Crate

The transport/squeeze crate is designed for easy administration of the anesthetic and works well in conjunction with a cage trap (Marker, 2002). When ready to "squeeze," metal poles are attached to the inside panel, which is used to push the cheetah against the side of the crate (small slots in the crate allow for hand injection). Successful drug delivery can be visually verified.

Darting

Blow pipes and pistols are useful when the animal is relatively close, such as in a cage trap situation. Dart rifles are the only effective and safe method for darting at longer distances. Cheetahs do not have a large amount of body mass, and a badly placed dart or one delivered at great velocity could break bones. Barbed or collared darts should be restricted to free-range darting situations whereas nonbarbed darts are typically used in a cage trap.

Stabilizing the Cheetah

The cheetah should be monitored quietly until the drug has taken full effect, as well as during the procedure. To reduce external stimuli, eyes should be covered with a blindfold and cotton placed in the ears. Eyes should be lubricated with oily eye solution, as they stay open during anesthesia. The dart wound should be treated with a topical antibiotic, and if a barded dart is used, the animal should be injected with long-acting penicillin to reduce risk of infection. Body mass, health, and physical status of the cheetah should be assessed and any life-threatening conditions attended to immediately. The body mass information will indicate whether the animal was potentially over- or underdosed.

To provide access to veins, an intravenous catheter should be placed. An endotracheal tube can be placed to provide access to the airways. To address possible dehydration (e.g., if animal was caught in a cage trap for several hours), the cheetah should be placed on an intravenous (or subcutaneous) drip; fluids also assist with the elimination of the drugs. Fluid temperature should be considered, for example, cold fluids should not be administered if the animal exhibits signs of hypothermia (35°C/95°F).

As part of a routine monitoring plan, the cheetah's heart rate, respiratory rate, and body temperature should be measured regularly (Chapter 24). In the absence of electricity, blankets and ethanol spray bottles (for evaporative cooling when sprayed on pads of feet) are important tools for temperature control.

Recovery From Anesthesia

Once the sampling and collaring have been completed, standard anesthesia reversal and recovery procedures should be followed (Chapter 24). It is advisable to place the cheetah in a safe crate or a padded/covered cage trap or transport crate, where it can recover until fully awake and ready to be safely released. Alternatively, the cheetah can be transported to a shaded spot and monitored to protect the animal against natural elements and potential predators until fully recovered. Recovery time depends on the amount and kind of anesthetic administered.

BIOMEDICAL EXAMINATION, SAMPLE AND DATA COLLECTION

Maximizing data and sample collection on free-ranging cheetahs that are being anesthetized should be a standard component to any capture. Samples and data should not be solely collected for ongoing studies, but also banked (inventoried and stored) for potential future research; banked samples reduce the need for future animal captures and should therefore be considered an ethical obligation. Data recording should include health, morphometric, and biological data. A detailed protocol for examinations of free-ranging cheetahs and associated data sheets were developed and agreed upon regionally during an international Population Viability Assessment workshop for cheetahs (Berry et al., 1996). There are four main components of the biomedical examination: (1) unique identification and marking of the animal and its samples; (2) physical assessment, aging, and morphometric data collection (including observation of abnormalities); (3) collection of biomedical samples; and (4) placement of monitoring device (e.g., collar).

Unique Identification

Each cheetah examined should be assigned a unique accession code to identify it and associated samples. Reference photographs should be taken from both sides of the body, and tail, as well as face. Those images can serve as a future reference to identify the individual from images, such as those taken on camera traps (Chapter 29) or by tourists (Chapters 32 and 34). Depending on the research question, two methods of marking are commonly used: ear tagging or microchip insertion.

Ear-Tagging

Small metal smallstock (goat and sheep) or pet ear tags are inexpensive, visible externally, and can be used on cheetahs of all ages. However, they are subject to being ripped or torn out. Care must be taken to place them correctly above the ear fold, so that the edge of the ear is not bunched up and there is minimal space for objects to snag between the ear and the tag. Ear tags can have a unique identifying number and a reference to the research project; some ear tags allow insertion of contact details. If a cheetah is recaptured, killed, or found dead, researchers can be contacted to identify the animal, collect it, and obtain samples and other valuable information.

Microchip Insertion

Microchips are unique identification devices that are inserted subcutaneously (e.g., Trovan Electronic Identification Systems). They cannot be torn off and can be used on cheetahs of any age, but are not visible externally, are relatively expensive, and require a compatible microchip reader. Microchips are recommended to be inserted near the base of the tail beside the spine to minimize migration (Berry et al., 1996). The use of microchips is advisable, especially when sampling young animals that do not yet have a visible coat pattern that may serve for identification. It is crucial to maintain consistency in the identification and recording system, as well as to input the data and identification codes in a dedicated backed-up database.

Physical Assessment, Aging, and Morphometric Data Collection

As part of the physical assessment, the gender and any physical characteristics (both intrinsic and injury related) are recorded, and the age of the cheetah is estimated. Ageing is done by examination of tooth wear and discoloration, gum recession, pelage (coat) condition, body size, the social grouping of the animal at the time of capture, and reproductive condition (Caro, 1994; Marker et al., 2003b).

Body mass is measured using a stretcher on a hanging scale and is recorded to the nearest kilogram (deducting the weight of the stretcher). Vernier calipers are used to record skull length, skull width, muzzle length, canine tooth lengths (Fig. 33.3A), gum recession, footpad measurements (Fig. 33.3B), and testicle lengths and widths. Measurements should be recorded to the nearest 0.1 cm. All other measurements are taken using a flexible measuring tape and are recorded to the nearest 1 cm. Leg measurements are taken with the legs positioned as if the cheetah were standing. The status of traits commonly observed in the cheetah should be assessed and levels recorded (i.e., presence/absence of upper first premolars, crowded lower incisors, palatal indentations formerly classified as mild focal palatine erosion, and kinked tails) (Chapter 7). The wear of teeth should also be noted.

Biomedical Sample Collection

Biomedical samples can provide important information regarding the overall health and genetic diversity of the free-ranging cheetah population, as well as the prevalence of various diseases (Krengel et al., 2013, 2015; Marker et al., 2010; Thalwitzer et al., 2010). Commonly collected samples are: skin, hair, blood (stored as whole blood, as well as separated into fractions of serum, plasma, buffy coat, and red blood cells), feces, swabbing for vaginal cytology, and sperm. Blood and tissue samples are

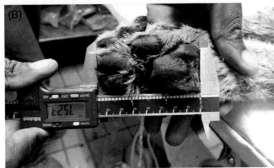

FIGURE 33.3   Digital calipers are used to measure several parts of the cheetah's body, including (A) canine teeth, and (B) paws. (C) Placement of a GPS collar on the cheetah. Leaving space for two fingers perpendicularly to the neck is a usual guide to ensure that collars are no too tight. *Source: CCF–Laurie Marker.*

stored frozen (liquid nitrogen or freezer) or in stabilization solutions (e.g., RNA-later and ethanol). Hair samples can be stored in paper envelopes (with root vs. shaved), ethanol, or dimethylsulfoxide. Sperm should be banked viably to preserve current genetic diversity (Chapter 27).

Collaring for Future Collection of Geocoordinates

VHF and GPS collars provide information regarding movements and home ranges (Chapter 8) with no detected impact on cheetah survival (Marker et al., 2003b). For the collar choice, animal welfare considerations should be prioritized over desired research output. Selection criteria include size (particularly weight); data collection, transmission, and storage capabilities; as well as battery life. Although it is generally accepted that collars for mammals weigh up to 3% of the animal's body weight (Kenward, 2001), we recommend lighter collars, (e.g., <400 g for cheetahs, approximately 1% of animal's body weight) due to the high *g*-forces during the cheetahs' famous acceleration. Collars are usually attached with a ±3.8 cm wide and 30–45 cm long adjustable neoprene belt. Internal antennas are advisable despite the decreased radio-transmitting range, as external antennas get regularly chewed off during allogrooming or fighting.

Correct fitting of the collar is crucial; the collar should be fitted so that two fingers can be inserted perpendicularly between the collar and the neck of the cheetah when the collar is properly fastened (Fig. 33.3C). Collars cannot be used on young animals that are not yet fully grown (<18 months), as the neck grows rapidly. Also adult males grow considerably after becoming territorial, thus this needs to be considered when collaring males. Depending on the research question, it is often sufficient to only collar one member of a social group.

Studies should endeavor to maximize data acquisition by using GPS rather than VHF collars, although the final choice should be made as a trade-off between study questions and unit price (Hebblewhite and Haydon, 2010). Compared to GPS collar data, VHF data is laborious to collect requiring radio tracking and triangulation of the animal, and does not typically allow for advanced movement analyses (Hebblewhite and Haydon, 2010; Latham et al., 2015). Imprecise triangulation can lead to location errors (Fuller and Fuller, 2012) and inaccurate home range estimates (Walter et al., 2015). GPS collars acquire date- and time-specific positions automatically according to preprogrammed schedules by the researcher.

Activity data can also be collected, and such information can be highly detailed if new technology, such as collars embedded with biaxial accelerometers, are used (Grünewälder et al. 2012; Chapter 32). The technology used to transmit the data to the researcher varies. Options include: Iridium or Globalstar satellites, mobile data network (GPRS), GSM network (SMS), or VHF/UHF radio uplink, which normally requires an aircraft to locate and get close enough to the cheetah to download the data. More basic collars that store the data on the unit itself and require collar retrieval for data download are not recommended, as the risk to not recover the data outweighs the cost benefits (Table 33.1).

To ensure that the collar does not remain on the animal indefinitely, it should be fitted with an automatic "drop-off" mechanism and a back-up "rot-off" fabric in the event of drop-off failure. If individuals must be recaptured for deployment of collars with new batteries, recapturing males is easier, as they return to the same marking sites regularly. Collars with an accelerometer-informed GPS (Brown et al., 2012) save up to 70% of power by switching off if the animal is not moving, thus effectively extending battery life. Chapter 32 provides more information on collars and dataloggers.

TABLE 33.1 Some of the Major Companies That Produce Collars (GPS and VHF), With Select Models That can be Used on Cheetahs Provided as Examples

Company	Country	Collar Type	Weight (g)
Africa Wildlife Tracking	South Africa	VHF radio	180
Advanced Telemetry Systems	USA	GPS	410
Eobs	Germany	GPS, remote download	330
Followit	Sweden	GPS, iridium	270
Lotek	Canada	GPS	350
SirTrack	New Zealand	GPS, iridium	385
Telonics	USA	VHF radio	105/175
Telonics	USA	GPS, store-on-board	145
Vectronic	Germany	GPS	370

Life span of VHF collars is typically considerably longer than that of GPS collars in the same battery size class. Life expectancy of GPS collar batteries depends on GPS and radio beacon scheduling scheme, remote data transmission type and frequency, environment (such as terrain, vegetation, and temperature), as well as animal behavior.

CONCLUSIONS

Capturing cheetahs is an invasive procedure, which can provide a large amount of data that would otherwise be difficult or impossible to acquire. While new and developing noninvasive techniques reduce the need for animal capture, live captures will still sometimes be needed to collect specific data. Sample and data collection should be maximized while complying with the highest standards in animal welfare and care.

Live capturing of cheetahs is safe and effective, provided protocols are strictly followed. Cage traps should be the preferred method of capture, and agents for anesthesia should be carefully selected.

Whenever possible, GPS collars should be used for monitoring animal movements. Collaring and sample collection should be undertaken in the framework of a research proposal that addresses specific research objectives. Proposals should be scrutinized by Animal Care and Use Committees prior to initiation of capture efforts. Standardization of protocols and data recording are crucial and facilitate collaborations between various teams, as well as metaanalyses for the common goal of informing cheetah conservation.

References

Berry, H., Bush, M., Davidson, B., Forge, O., Fox, B., Howe, M., Hurlbut, S., Marker-Kraus, L., Martenson, J., Munson, L., Nowell, K., Schumann, M., Shille, T., Stander, F., Venzke, K., Wagener, T., Wildt, D., Ellis, K., Seal, U. (Eds.), 1996. Population Habitat Viability Assessment for the Namibian Cheetah and Lion. Zoo Animal and Wildlife Immobilization and Anesthesia. IUCN/SSC Conservation Breeding Specialist, Apple Valley, Minnesota.

Brown, D.D., LaPoint, S., Kays, R., Heidrich, W., Kümmeth, F., Wikelski, M., 2012. Accelerometer-informed GPS telemetry: reducing the trade-off between resolution and longevity. Wildl. Soc. Bull. 36, 139–146.

Caro, T.M., 1994. Cheetahs of the Serengeti Plains: Group Living in an Asocial Species (Wildlife Behavior and Ecology series). University of Chicago Press, Chicago, IL.

Cattet, M., Boulanger, J., Stenhouse, G., Powell, R.A., Reynolds-Hogland, M.J., 2008. An evaluation of long-term capture effects in Ursids: implications for wildlife welfare and research. J. Mammal. 89, 973–990.

Caulkett, N., Shury, T., 2007. Human safety during wildlife capture. In: West, G., Heard, D., Caulkett, N. (Eds.), Zoo Animal and Wildlife Immobilization and Anesthesia. Blackwell Publishing, USA.

Cozzi, G., Broekhuis, F., McNutt, J.W., Schmid, B., 2013. Density and habitat use of lions and spotted hyenas in northern Botswana and the influence of survey and ecological variables on call-in survey estimation. Biodivers. Conserv. 22, 2937–2956.

Divyabhanusinh, 1995. On trapping, training, treatment of and hunting with cheetahs. In: Divyabhanusinh (Ed.), The End of a Trail—The Cheetah in India. Banyan Books, New Delhi.

Fuller, M.R., Fuller, T.K., 2012. Radio-telemetry equipment and applications for carnivores. In: Boitani, L., Powell, R.A. (Eds.), Carnivore Ecology and Conservation. A Handbook Of Techniques. Oxford University Press, Oxford.

Grünewälder, S., Broekhuis, F., Macdonald, D.W., Wilson, A.M., McNutt, J.W., Shawe-Taylor, J., Hailes, S., 2012. Movement activity based classification of animal behaviour with an application to data from cheetah (*Acinonyx jubatus*). PLoS One 7, e49120.

Hayes, R.W., 1982. A telemetry device to monitor big game traps. J. Wildl. Manage. 46, 551–553.

Hebblewhite, M., Haydon, D.T., 2010. Distinguishing technology from biology: a critical review of the use of GPS telemetry data in ecology. Philos. Trans. R. Soc. 365, 2303–2312.

Hermsen, E., 2013. Using Camera-Traps to Test the Efficacy of Different Bait Types in Luring Cheetahs (*Acinonyx jubatus*) in Kenya, Africa. MSc thesis, Antioch University, United States.

Hunter, L., Jowkar, H., Ziaie, H., Schaller, G., Balme, G., Walzer, C., Ostrowski, S., Peter Zahler, P., Robert-Charrue, N., Kashiri, K., Christie, S., 2007. Conserving the Asiatic Cheetah in Iran: launching the first radio-telemetry study. Cat news 46, 8–11.

Iossa, G., Soulsbury, C.D., Harriset, S., 2007. Mammal trapping: a review of animal welfare standards of killing and restraining traps. Anim. Welf. 16, 335–352.

Johansson, A.T., Johansson, O., McCarthy, T., 2011. An Automatic VHF transmitter monitoring system for wildlife research. Wildl. Soc. Bull. 35 (4), 489–493.

Kenward, R., 2001. A Manual for Wildlife Radio-Tagging. Academic Press, San Diego; London.

Kock, M.D., Burroughs, R, 2014. Chemical and Physical Restrain of Wild Animals: A Training and Field Manual for African Species. Africa, IWVS.

Krengel, A., Cattori, V., Meli, M.L., Wachter, B., Böni, J., Bisset, L.R., Thalwitzer, S., Melzheimer, J., Jago, M., Hofmann-Lehmann, R., Hofer, H., Lutz, H.y., 2015.

Gammaretrovirus-specific antibodies in free-ranging and captive Namibian cheetahs. Clin. Vaccine Immunol. 22, 611–617.

Krengel, A., Meli, M.L., Cattori, V., Wachter, B., Willi, B., Thalwitzer, S., Melzheimer, J., Hofer, H., Lutz, H., Hofmann-Lehmann, R., 2013. First evidence of hemoplasma infection in free-ranging Namibian cheetahs (*Acinonyx jubatus*). Vet. Microbiol. 162, 972–976.

Larkin, R.P., VanDeelen, T.R., Sabick, R.M., Gosselink, T.E., Warner, R.E., 2003. Electronic signaling for prompt removal of an animal from a trap. Wildl. Soc. Bull. 31, 392–398.

Latham, A.D.M., Latham, M.C., Anderson, D.P., Cruz, J., Herries, D., Hebblewhite, M., 2015. The GPS craze: six questions to address before deciding to deploy GPS technology on wildlife. N. Z. J. Ecol. 39, 143–152.

Marker L.L., 2002. Aspects of Cheetah (*Acinonyx Jubatus*) Biology, Ecology and Conservation Strategies on Namibian Farmlands. PhD thesis, University of Oxford, United Kingdom.

Marker, L., Dickman, A.J., Jeo, R.M., Mills, M.G.L., Macdonald, D.W., 2003a. Demography of the Namibian cheetah. Biol. Cons. 114 (3), 413–425.

Marker, L., Dickman, A.J., Mills, M.G.L., Macdonald, D.W., 2003b. Aspects of the management of cheetahs trapped on Namibian farmlands. Biol. Cons. 114 (3), 401–412.

Marker, L., Dickman, A.J., Mills, M.G.L., Macdonald, D.W., 2010. Cheetahs and ranches in Namibia: a case study. In: Macdonald, D.W., Loveridge, J. (Eds.), Biology and Conservation of Wild Felids. Oxford University Press, Oxford, p. 353.

Marks, C.A., 1996. A radiotelemetry system for monitoring the treadle snare in programmes for control of wild canids. Wildl. Res. 23, 381–386.

Marnewick, K.A., Bothma, J.D.P., Verdoorn, G.H., 2006. Using camera-trapping to investigate the use of a tree as a scent-marking post by cheetahs in the Thabazimbi district. S. Afr. J. Wildl. Res. 36, 139–145.

Marnewick, K., Somers, M.J., 2015. Home ranges of cheetahs (*Acinonyx jubatus*) outside protected areas in South Africa. S. Afr. J. Wildl. Res. 45, 223–232.

Mills, M.G.L., 1996. Methodological advances in capture, census, and food habits studies of large African carnivores. In: Gittleman, J.L. (Ed.), Carnivore Behavior, Ecology and Evolution Volume II. Cornell University Press, Ithaca, New York, NY, pp. 223–242.

Munson L. and Marker L., 1997. The impact of capture and captivity on the health of Namibian farmland cheetahs (*Acinonyx jubatus*). Proceedings of the 50th Namibian Vet Congress. September, Namibia.

Nolan, J.W., Russell, R.H., Anderka, F., 1984. Transmitters for monitoring Aldrich snares set for grizzly bears. J. Wildl. Manage. 48, 942–945.

O'Neill, L., De Jongh, A., Ozolinš, J., De Jong, T., Rochford, J., 2007. Minimizing leg-hold trapping trauma for otters with mobile phone technology. J. Wildl. Manage. 71, 2776–2780.

Powell, R.A., Proulx, G., 2003. Trapping and marking terrestrial mammals for research: integrating ethics, performance criteria, techniques, and common sense. ILAR J. 44, 259–276.

Putman, R.J., 1995. Ethical considerations and animal welfare in ecological field studies. Biodivers. Conserv. 4, 903–915.

Rode, K.D., Pagano, A.M., Bromaghin, J.F., Atwood, T.C., Durner, G.M., Simac, K., Amstrup, S.C., 2014. Effects of capturing and collaring on polar bears: findings from long-term research on the southern Beaufort Sea population. Wildl. Res. 41, 311–322.

Sikes, R.S., Gannon, W.L., 2011. The Animal Care and Use Committee of the American Society of Mammalogists. Guidelines of the American Society of Mammalogists for the use of wild mammals in research. J. Mammal. 92 (1), 235–253.

Teräväinen M., 2016. Short Term Effects of Capture on Movements in Free-Ranging Wolves (*Canis lupus*) in Scandinavia. MSc thesis, Hedmark University College, Norway.

Thalwitzer, S., Wachter, B., Robert, N., Wibbelt, G., Müller, T., Lonzer, J., Meli, M.L., Bay, G., Hofer, H., Lutz, H., 2010. Seroprevalences to viral pathogens in free-ranging and captive cheetahs (*Acinonyx jubatus*) on Namibian farmland. Clin. Vac. Immunol. 17, 232–238.

Wachter, B., Thalwitzer, S., Hofer, H., Lonzer, J., Hildebrandt, T.B., Hermes, R., 2011. Reproductive history and absence of predators are important determinants of reproductive fitness: the cheetah controversy revisited. Conserv. Lett. 4, 47–54.

Walter, W.D., Onorato, D.P., Fischer, J.W., 2015. Is there a single best estimator? Selection of home range estimators using area-under-the-curve. Mov. Ecol. 3, 10.

Citizen Science in Cheetah Research

Esther van der Meer*,**, Femke Broekhuis[†,‡],
Elena V. Chelysheva[§], Mary Wykstra[¶],
Harriet T. Davies-Mostert[††,‡‡]

*Cheetah Conservation Project Zimbabwe, Victoria Falls, Zimbabwe
**National University of Science and Technology, Bulawayo, Zimbabwe
[†]Mara Cheetah Project, Kenya Wildlife Trust, Nairobi, Kenya
[‡]University of Oxford, Tubney, Abingdon, United Kingdom
[§]Mara-Meru Cheetah Project, Kenya Wildlife Services, Nairobi, Kenya
[¶]Action for Cheetahs in Kenya Project, Nairobi, Kenya
[††]Endangered Wildlife Trust, Johannesburg, South Africa
[‡‡]University of Pretoria, Pretoria, South Africa

INTRODUCTION

Citizen science is defined as the voluntary participation of members of the public (citizen scientists) in research projects directed by professional scientists (Dickinson et al., 2010; Wiggins and Crowston, 2011). This participation most commonly involves the collection of data (Rotman et al., 2012), but can also entail contributions to the design of a project, data analyses, and dissemination of results (Bonney et al., 2009a). Citizen science is particularly useful when addressing research questions that have a large spatial and/or temporal scope (Dickinson et al., 2010, 2012; Silvertown, 2009), for example, changes in species distribution and abundance over time and space. Since such large-scale monitoring programs can be constrained by the costs of employing adequate numbers of trained professionals (Lovett et al., 2007), the involvement of volunteers has enabled scientists to address research questions on scales far larger than would have been feasible with traditional field research models (Dickinson et al., 2010, 2012; Silvertown, 2009).

The involvement of members of the public in scientific research dates back as far as the 19th century and started in the fields of astronomy and ornithology (Dickinson et al., 2010). One of the most famous and longest running ornithology citizen science project, the Christmas Bird Count, was launched in the USA in 1900 (Dickinson et al., 2010). Currently, more than 50,000 volunteers participate in this annual bird count, and the collected data have been used in hundreds of scientific publications (Dunn et al., 2005). Citizen scientists collect data either opportunistically, for example, during chance encounters with a species, or systematically by following specific survey protocols (Lewandowski and Specht, 2015). The use of internet technology and mobile devices to disseminate project information and electronically submit data has facilitated implementation of citizen science projects and has expanded the scope of data collection (Dickinson et al., 2012; Silvertown, 2009). As a result, citizen science has gained ground and now covers a wide range of disciplines and taxonomic groups (Dickinson et al., 2010).

One of the first documented cases of citizen science in carnivore research in Africa is a study by Reich (1981), who collected African wild dog (*Lycaon pictus*) sightings from tourists by providing sighting sheets at tourist offices in the Kruger National Park (KNP), South Africa. In 1988, the Endangered Wildlife Trust introduced a photo competition in the KNP to collect African wild dog sightings and photographs from tourists (Maddock and Mills, 1994) which was extended to cheetah (*Acinonyx jubatus*) in 1990 (Marnewick et al., 2014). These citizen science surveys have been repeated at 5-year intervals ever since and have assisted researchers with the long-term monitoring of the KNP's African wild dog and cheetah populations (Marnewick et al., 2014).

For species with unique coat patterns, such as African wild dogs and cheetahs, it is possible to identify individuals from photographs (Maddock and Mills, 1994; Marnewick et al., 2014). This ability to identify individuals, enables professional scientists to generate information on demography and behavior from photographic sighting data

(Maddock and Mills, 1994), which can in turn be used to estimate population sizes via capture–recapture modeling (Marnewick et al., 2014) and conduct population viability assessments (Kelly and Durant, 2000). The information collected is also used to map local, national, and international distribution and derive information on movement patterns (IUCN/SSC, 2007).

THE ROLE OF CITIZEN SCIENCE IN CHEETAH RESEARCH

To determine the current role of citizen science in cheetah research, we asked researchers to complete a short questionnaire on their organization's use of citizen science in cheetah research (Box 34.1). Cheetah researchers from 21 projects responded to our request we would sincerely like to thank our colleagues for taking the time to complete our questionnaire and provide us with an insight in the use of citizen science in cheetah research, 11 of these projects were cheetah projects and 9 were general carnivore projects with a cheetah component. All 11 of the cheetah projects and 4 of the general carnivore projects make use of citizen science in cheetah research. The projects primarily use citizen science to collect cheetah sightings (location data) and photographs. These sightings and photographs are collected opportunistically based on chance encounters between citizen scientists and cheetahs (Fig. 34.1). The majority of the citizen scientists are tourists, safari guides, and rangers, but cheetah sightings have also been successfully collected by others, for example, trophy hunters, commercial farmers, and local communities.

Involving local stakeholders in citizen science projects can be an important conservation tool as it encourages people to be involved in the monitoring of their wildlife resources and connects them with conservation issues (Devictor et al., 2010; Gaidet et al., 2003). In addition, participation in citizen science projects improves the participant's knowledge (Bonney et al., 2009a),

BOX 34.1

QUESTIONNAIRE-BASED SURVEY TO DETERMINE THE ROLE OF CITIZEN SCIENCE WITHIN CHEETAH RESEARCH.

To investigate whether cheetah researchers use citizen science as a tool within their research, we approached cheetah and carnivore projects with the following open-ended questions:

How do you use citizen science in cheetah research?

- Methods?
- Type of data collected?
- Target audience (who are your citizen scientists?)
- For how many years have you used this method?
- What are the positive sides of this method?
- What are the negative sides of this method?
- Are there ways in which your method can be improved?

Does your citizen science project include an educational and/or awareness component?

- Description of educational or awareness component.
- Target audience?
- Do you feel this method is effective to educate or raise awareness, why?

How do you engage people, how do you make sure they know about the project, and how do you keep them interested in the project?

- Methods used to engage members of the public in your project?
- Target audience?
- Do you feel this method is effective to engage people for the duration of the project, why?

Have you used members of the public (citizen scientists) for purposes other than data collection, for example, to analyze data or design projects, what was your experience with this?

- Purpose and method?
- Target audience?
- What was your experience with this?

Have you published any research results that included citizen science data?

Do you have any concluding remarks about the use of citizen science in cheetah research, the pro's and con's, future use, etc.?

---

which can positively affect attitudes toward carnivore species and their conservation (Ericsson and Heberlein, 2003). The Cheetah Conservation Fund assists both commercial and communal conservancy members in conducting annual game counts, this informs farmers of trends in their cheetah and wildlife populations and allows for a sustainable utilization of game. Action for Cheetahs in Kenya involves community members in cheetah monitoring and receives 2–10 cheetah sightings from the community each day. Although these sightings have to be verified as there

is regular confusion between cheetah, leopard (*Panthera pardus*), and serval (*Leptailurus serval*), they provide the researchers with a search area to collect proof of cheetah presence often resulting in the identification of individual cheetahs. In many countries cheetahs are heavily persecuted by commercial and communal farmers (Marker and Dickman, 2004); involving these stakeholders in citizen science projects can help to raise awareness, provide education, and build relationships based on trust, all of which benefit cheetah conservation.

FIGURE 34.1   **A chance encounter between citizen scientists and a cheetah.** *Source: Mark Butcher.*

The cheetah and carnivore projects we surveyed encourage citizen scientists to collect cheetah sightings and photographs through the use of sighting sheets/books at tourist offices and camps, posters, stickers, and flyers, and in some cases through local radio adverts. In addition, projects in areas with cell phone coverage and internet connection have started to use more advanced technologies, for example, online sighting forms, social media (e.g., Facebook and Twitter), various web-based platforms (Table 34.1), census hotline numbers, and monitoring applications for mobile devices (e.g., Spot-a-Cat and Carnivore Tracker). The Spot-a-Cat application, developed by the Mara Cheetah Project in Kenya, was the first cheetah monitoring application (maracheetahs.org/Spot-a-Cat). It enables citizen scientists to record cheetah sightings, upload cheetah photographs,

view sightings on an interactive map, and read through a cheetah fact file. In 2 years since its launch in June 2014, ca. 500 individuals have downloaded the application on their mobile devices, resulting in 54 cheetah sightings. Uploaded records are verified by professional scientists and used to monitor individual cheetahs in Kenya and Tanzania, and to determine overall cheetah range and distribution.

The number of cheetah sightings and photographs collected via citizen science depends on the tourism activity in the research area and the efforts made to engage citizen scientists in cheetah research. According to our survey, cheetah and carnivore projects that encourage citizen scientists to collect sightings opportunistically collect between 10 and 800 sightings a year (median = 155), and most of these projects (80%) collect 150 or more sightings annually.

TABLE 34.1 Examples of National, Regional, and Global Web-Based Platforms via Which Cheetah Sightings and Photographs are Being Collected (Information About Registered Users and Collected Cheetah Sightings as per January 2017)

Websites	Forms of submission	Scope	Launch	Project leaders	Descriptions	Data verification	Contributors	Submitted cheetah records
the-eis.com/atlas	Atlasing in Namibia website	National/all species	2011	Several local stakeholders and partners, hosted by the Environmental Information Service	Atlasing in Namibia's objective is to increase the knowledge and understanding of the presence and distribution of species in Namibia. This platform encourages members of the public and professional scientists not only to submit sightings, photographs, camera trap photographs, spoor, and telemetry data but also incorporates historical datasets	Professional scientists	181	690
mammalmap.adu.org.za	Mammalmap website and Africa live safari sightings application	Regional/all species	2010	Animal Demography Unit of the University of Cape Town and the Mammal Research Institute of the University of Pretoria	Mammalmap aims to update distribution maps of African mammal species. This platform facilitates the submission of sightings and camera trap photographs of African mammals by both professional scientists and members of the public. Mammalmap is part of a virtual museum which covers a range of taxa	Professional scientists	465[a]	189[b]

(Continued)

TABLE 36.1   Examples of National, Regional, and Global Web-Based Platforms via Which Cheetah Sightings and Photographs are Being Collected (Information About Registered Users and Collected Cheetah Sightings as per January 2017) (cont.)

Websites	Forms of submission	Scope	Launch	Project leaders	Descriptions	Data verification	Contributors	Submitted cheetah records
iNaturalist.org	iNaturalist website and application	Global/all species	2008	California Academy of Sciences	iNaturalist allows members of the public to share observations of species of various taxa and help each other to learn about nature. This platform is used by various (scientific) specialist groups, for example, the Range Wide Conservation Programme for Cheetah and Wild dog, to incorporate crowd sourced data into biodiversity surveys	Other iNaturalist users	86,850	287
iSpotnature.org	iSpot website and application	Global/all species	2009	Open University	iSpot was created to give members of the public an opportunity to learn about nature. iSpot allows users to upload species sightings and aims to help anyone, anywhere to identify anything in nature. iSpot has a special Southern Africa community with ca. 8,000 registered members: ispotnature.org/communities/southern-africa	Other iSpot users	61,188	38

[a] The virtual museum has a total number of 1682 contributors.

[b] In addition to these submissions, Mammalmap has collected 1844 cheetah distribution records from various other sources.

Opportunistic data collection requires minimal efforts from scientists to familiarize their citizen scientists with the survey method (Lewandowski and Specht, 2015); however, data collection is biased toward areas with better infrastructure and higher visitation rates. The involvement of citizen scientists in systematic data collection, for example, by allowing safari guides and their tourists to radio track rehabilitated cheetahs in a small game reserve (20,000 ha) (AfriCat), or by involving farmers and volunteers in game counts (Cheetah Conservation Fund), has enabled projects to increase the number of cheetah sightings they receive to more than a thousand a year. Systematic data collection ensures data is gathered over a representative time period and geographic range but requires standardized survey protocols and the training of volunteers (Bonney et al., 2009b; Lewandowski and Specht, 2015).

Data collected through citizen science are not limited to sighting data. Citizen scientists can be involved in a wide range of other activities, for example, camera trapping (Chapter 29), the collection of scats (Chapter 31), and recording of spoor (Chapter 30). Within cheetah research, only a few projects engage citizen scientists in activities other than the collection of sightings and photographs: five cheetah projects have involved rangers, volunteers, commercial and communal farmers in the design and/or execution of camera trap surveys; one project involved farmers and volunteers in game counts, and two projects involved farmers in the collection of marking tree locations and cheetah scats. Additionally, some cheetah projects involve commercial farmers in the collection of tissue samples from cheetah carcasses and in assisting in the capture of cheetahs to deploy collars.

An area of citizen science that, according to our survey, remains largely unexplored in cheetah research is data mining (i.e., ordering and examining large datasets). Although some projects have involved volunteers in the identification of cheetahs from photographs and simple data analyses, none of the projects that participated in our survey used citizen scientists for these purposes on a large scale. Tiger Nation (tigernation.org), a web-based citizen science project that promotes the conservation of wild tigers in the world, not only encourages citizen scientists to submit tiger photographs but also engages them in analysis of these photographs by playing a tiger matching game. The game is used to verify results of stripe recognition software and motivates users by providing visual rewards (e.g., points and badges) (Mason et al., 2012). Another prime example of citizen science-based data mining is the web-based platform Zooniverse (zooniverse.org), which is used by projects to engage volunteers in data mining ranging from the identification and classification of species from camera trap pictures (wildcamgorongosa.org), the marking of animal behavior on videos (chimpandsee.org) to the identification of sounds (batdetective.org). The Serengeti Lion Project used Zooniverse to classify camera trap photographs (snapshotserengeti.org); ca. 68,000 volunteers contributed 10.8 million classifications of species, number of individuals, presence of young, and behavior for the animals in the 1.2 million sets of photographs. To ensure accuracy of these classifications each set of two photographs was classified by multiple citizen scientists, after which an algorithm was used to produce a consensus dataset of final classifications. Validation of these citizen science consensus datasets against more than 4000 photograph sets classified by experts, showed a 96.6% accuracy for species identification and 90.0% accuracy for species counts (Swanson et al., 2015).

CHALLENGES AND BENEFITS OF CITIZEN SCIENCE

Data Quality

One of the main challenges of citizen science is to ensure data quality is of a high standard for scientific use. Although some studies find no difference in data quality, others show that data collected by professional scientists are more accurate than by citizen scientists (Lewandowski and Specht, 2015). Variability in the quality and reliability of citizen science data can result in artificial biological patterns or mask existing ones (Gardiner et al., 2012; Lewandowski and Specht, 2015). In a study by Gardiner et al. (2012) variability in quality and reliability of citizen science data resulted in an underestimation of common species and an overestimation of rare species, species richness and diversity. However, if citizen science data were verified by professional scientists and the number of citizen science observations was increased to the number needed to attain the same degree of accuracy as data collected by professional scientists, the costs of verified citizen science would still be much lower than that of traditional science (Gardiner et al., 2012).

Apart from variations in observer skills, bias in citizen science datasets can also result from variations in sampling effort (time) and geographic range (Dickinson et al., 2010; Lewandowski and Specht, 2015; Van Strien et al., 2013). The probability of recording a cheetah sighting is affected by differences in tourist visitation rates and infrastructure, as well as ecological factors, such as prey and vegetation density (Gu and Swihart, 2004; Marnewick et al., 2014). A cheetah is, for example, likely to be less conspicuous in dense vegetation. Such biases in sighting probability influence the outcome of models used to estimate population sizes, distribution, and wildlife–habitat relationships (Gu and Swihart, 2004; Marnewick et al., 2014). In addition, the cheetah sightings and photographs received are not always accurate and useful, it can be difficult to assess the reliability of sightings that are not accompanied by photographs and citizen scientists regularly send in photographs that are not properly time stamped or geo referenced, or photographs that are suboptimal for identification (e.g., low resolution and

face shots). Obtaining complete information can be time consuming, requiring follow up to collect missing information. The transmission of information can also provide challenges. Limited cell phone network and internet access in remote wildlife areas can make it difficult for citizen scientists to submit cheetah sightings and photographs electronically. Once network and internet become available, despite good intentions, citizen scientists might forget to submit their sightings.

A citizen science project consists of several stages (Fig. 34.2), within these stages there are a number of opportunities to enhance data quality. Generating high-quality data starts with the provision of pilot tested, clear, simple and standardized survey protocols, and data collection forms (Bonney et al., 2009b). Additionally, observer errors can be minimized via close supervision by scientists, long-term retention of volunteers, and training (Lewandowski and Specht, 2015). Apart from providing sighting sheets, where applicable, the cheetah and carnivore projects we surveyed tried to improve data quality with training and by providing their citizen scientists with adequate equipment, for example, data loggers and cameras with built in global positioning systems. This not only enhances data quality, but also encourages participation and promotes long-term retention of citizen scientists, which in turn improves data quality. To address complicated research questions, scientists can partner with existing specialist groups, or select a subset of very committed citizen scientists out of a group of existing volunteers (Bonney et al., 2009b; Dickinson et al., 2010). Specialist groups involved in citizen science in cheetah research may include safari guides, hunters, commercial farmers, and rangers.

Regardless of the measures taken to avoid bias and error, it is advisable for citizen science projects to include means of data validation and develop criteria to identify and eliminate data with systematic errors (Bonney et al., 2009b). One can, for example, test whether different

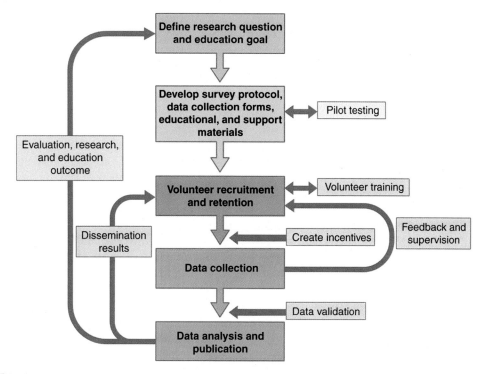

FIGURE 34.2 **Illustration of the various stages within citizen science projects.**

citizen science surveys show consistent population trends or crossvalidate citizen science datasets with reference data collected by professional scientists (Devictor et al., 2010; Riesch and Potter, 2014). Some citizen science projects use data quality filters to flag unexpected electronic entries. If the citizen scientist confirms the flagged entry as correct, he or she is contacted by a regional expert to verify the observation (Dickinson et al., 2010). In addition, bias and error can be accounted for in the statistical analyses of citizen science datasets (Bird et al., 2014), for example, by using occupancy models that take the imperfect detection of species into account (Van Strien et al., 2013; Chapter 37). Geographical bias of cheetah sightings can be corrected in capture–recapture analyses by categorizing sightings according to search effort, for example, infrastructure and visitation rates (Marnewick

et al., 2014). Although citizen science data often require a trade off between data quality and quantity, noisy datasets are not by definition useless. The large sample size of most citizen science datasets enable a level of statistical power and robustness that make it possible to detect biological patterns even when data contain noise (Bonney et al., 2009b; Devictor et al., 2010). Also, as long as sources of bias and error are kept constant over time, it is feasible to monitor trends based on noisy datasets (Devictor et al., 2010).

Availability of Resources

Before setting up a citizen science project, it is important to realize that, although citizen scientists donate their time, there are other costs associated with the project management, volunteer support, data compilation, and validation

(Bonney et al., 2009b). For large-scale citizen science projects, it is often necessary to employ professionals from a variety of disciplines to adequately cover all aspects of the project (Bonney et al., 2009b). Small-scale projects generally do not have the means to employ such a wide range of professionals, instead they can collaborate with other organizations, adapt existing large-scale projects for their own use and use free technologies, such as GoogleMaps, Cybertracker, Flickr, Facebook, and Twitter (Bonney et al., 2009b; Silvertown, 2009). The cheetah and carnivore projects we surveyed make use of those free technologies and consider the collection of sightings and photographs via rangers, safari guides, and tourists; a cost-effective way to collect additional cheetah data over a wider range compared to what they would normally be able to cover. Although financing long-term citizen science projects can be challenging (Wiggins and Crowston, 2011), citizen science can also have unexpected financial benefits, for example, in Namibia commercial farmers have assisted cheetah researchers from the Leibniz Institute for Zoo and Wildlife Research in raising funding by acquiring funding from the farming community and by successfully writing a project proposal to a funding agency.

EDUCATION AND RAISING AWARENESS

The initial decision to volunteer as a citizen scientist is largely driven by personal interest and the desire to learn more about specific topics (Rotman et al., 2012). As such, citizen science projects provide a perfect opportunity to raise awareness, education about nature and science, and promote a connection to nature, all of which ultimately stimulates proenvironmental behavior (Bickford et al., 2012; Chapter 18). Even though citizen science projects do not always include explicit educational objectives (Wiggins and Crowston, 2011), they often use a variety of

educational materials to provide information about the research and explain the data collection protocols (Bonney et al., 2009b). Most of the cheetah and carnivore projects we surveyed (53%) have incorporated an awareness and education component in their citizen science projects by adding information about cheetahs and the threats that the species faces, to recruitment and data collection materials.

Although it is not always evaluated whether educational materials result in the desired learning outcome (Devictor et al., 2010), participation in citizen science projects has been shown to positively affect awareness, knowledge, and an understanding of scientific concepts (Bonney et al., 2009a). Based on the positive feedback and the number of cheetah sightings and photographs received, the cheetah and carnivore projects we surveyed feel that their approach is successful in promoting awareness. Training in particular, seems to be an effective way to raise awareness and educate citizen scientists. Informal tests before and after predator differentiation training sessions conducted by Action for Cheetahs in Kenya show that citizen scientists learn to differentiate cheetahs and their tracks from other carnivore species, which improved the labeling of photographs.

ENGAGEMENT AND CONTINUED INVOLVEMENT OF CITIZEN SCIENTISTS

Providing opportunities to receive badges, prizes, and rewards or take part in contests and challenges can create incentives for citizen scientists to participate in citizen science projects (Dickinson et al., 2012; Mason et al., 2012). Field trips, meetings, and training workshops in which citizen scientists can interact with peers and scientists, create learning opportunities, give a form of recognition, and therefore an incentive for continued participation (Rotman et al., 2012). Cheetah and carnivore projects use a variety of incentives to engage citizen scientists, such as training workshops, distribution of cheetah identification kits to rangers and safari guides, certificates for assistance with the project, regular meetings with rangers, safari guides and lodge managers, photo competitions, and giving citizen scientists the opportunity to name new cheetahs they sight.

Nevertheless, the cheetah and carnivore projects we surveyed indicate that it can be challenging to engage potential citizen scientists in cheetah research and to ensure their continued participation over long periods of time. Factors, such as recognition, appreciation, and usefulness of their contribution to science and the local community, determine whether volunteers continue to participate in a citizen science project (Rotman et al., 2012). As the quality of the collected data is generally positively related to the level of experience of the citizen scientist, it is important to stimulate long-term retention of volunteers (Lewandowski and Specht, 2015). Survey protocols that are too difficult and demanding reduce both the number and long-term retention of participants in citizen science projects (Lewandowski and Specht, 2015). Regular feedback, progress reporting, and acknowledgment all help to foster long-term engagement (Rotman et al., 2012).

The cheetah and carnivore projects we surveyed are aware of the importance of feedback and provide updates via personal communication, talks, newsletters, social media, websites, and popular publications. In addition, they acknowledge the contribution of citizen scientists in their reports, on their websites and Facebook pages, and in some cases in scientific publications. Despite the large contribution of citizen scientists to science, the recognition of this contribution in scientific publications is not a standard practice (Dickinson et al., 2012) and the format in which citizen scientists should be acknowledged in these publications remains part of an ongoing debate (Mason et al., 2012; Riesch and Potter, 2014).

CONCLUSIONS

Citizen science is commonly used in cheetah research and is regarded by cheetah researchers as an efficient way to assist with wide-ranging and long-term monitoring of cheetahs, raising awareness, and providing education. Citizen science in cheetah research predominantly consists of involving tourists, safari guides, and rangers in the collection of cheetah sightings and photographs. However, there are opportunities to expand the scope of these projects to a broader range of citizen scientists and citizen science activities. Citizen science has the ability to connect people to nature, create an understanding of conservation issues, and promote proenvironmental behavior (Bickford et al., 2012; Devictor et al., 2010; Gaidet et al., 2003) and should, therefore, be recognized as an important conservation tool, independently of its contribution to research. The incorporation of a clearly defined awareness and education component in cheetah citizen science projects and evaluation of whether this component results in the desired learning outcome is likely to increase the effectiveness of the materials used. Citizen science in cheetah research, therefore, not only provides a means to collect additional cheetah data at large geographical and temporal scales, but it is also an invaluable means to promote the conservation of cheetahs throughout their range.

References

Bickford, D., Posa, M.R.C., Qie, L., Campos-Arceiz, A., Kudavidanage, E.P., 2012. Science communication for biodiversity conservation. Biol. Conserv. 151, 74–76.

Bird, T.J., Bates, A.E., Lefcheck, J.S., Hill, N.A., Thomson, R.J., Edgar, G.J., Stuart-Smith, R.D., Wotherspoon, S., Krkosek, M., Stuart-Smith, J.F., Wotherspoon, S., Krkosek, M., Stuart-Smith, J.F., Pecl, G.T., Barrett, N., Frusher, S., 2014. Statistical solutions for error and bias in global citizen science datasets. Biol. Conserv. 173, 144–154.

Bonney, R., Ballard, H., Jordan, R., McCallie, E., Phillips, T., Shirk, J., Wilderman, C.C., 2009a. Public Participation in Scientific Research: Defining the Field and Assessing its Potential for Informal Science Education: A CAISE

Inquiry Group Report. Center for Advancement of Informal Science Education, Washington, USA.

Bonney, R., Cooper, C.B., Dickinson, J., Kelling, S., Phillips, T., Rosenberg, K.V., Shirk, J., 2009b. Citizen science: a developing tool for expanding science knowledge and scientific literacy. BioScience 59, 977–984.

Devictor, V., Whittaker, R.J., Beltrame, C., 2010. Beyond scarcity: citizen science programmes as useful tools for conservation biogeography. Divers. Distrib. 16, 354–362.

Dickinson, J.L., Shirk, J., Bonter, D., Bonney, R., Crain, R.L., Martin, J., Phillips, T., Purcell, K., 2012. The current state of citizen science as a tool for ecological research and public engagement. Front. Ecol. Environ. 10, 291–297.

Dickinson, J.L., Zuckerberg, B., Bonter, D.N., 2010. Citizen science as an ecological research tool: challenges and benefits. Ann. Rev. Ecol. Evol. Syst. 41, 149–172.

Dunn, E.H., Francis, C.M., Blancher, P.J., Roney Drennan, S., Howe, M.A., Lepage, D., Robbins, C.S., Rosenberg, K.V., Sauer, J.R., Smith, K.G., 2005. Enhancing the scientific value of the Christmas Bird Count. Auk 122, 338–346.

Ericsson, G., Heberlein, T., 2003. Attitudes of hunters, locals and the general public in Sweden now that the wolves are back. Biol. Conserv. 111, 149–159.

Gaidet, N., Fritz, H., Nyahuma, C., 2003. A participatory counting method to monitor populations of large mammals in non-protected areas: a case study of bicycle counts in the Zambezi Valley, Zimbabwe. Biodivers. Conserv. 12, 1571–1585.

Gardiner, M.M., Allee, L.L., Brown, P.M.J., Losey, J.E., Roy, H.E., Rice Smyth, R., 2012. Lessons from lady beetles: accuracy of monitoring data from US and UK citizen science programs. Front. Ecol. Environ. 10, 471–476.

Gu, W., Swihart, R.K., 2004. Absent or undetected? Effect of non-detection on species occurrence on wildlife-habitat models. Biol. Conserv. 116, 195–203.

IUCN/SSC, 2007. Regional Conservation Strategy for the Cheetah and African Wild Dog in Eastern Africa. International Union for Conservation of Nature and Natural Resources, Gland, Switzerland, Available from: http://www.cheetahandwilddog.org/.

Kelly, M.J., Durant, S.M., 2000. Viability of the Serengeti cheetah population. Conserv. Biol. 14, 786–797.

Lewandowski, E., Specht, H., 2015. Influence of volunteer and project characteristics on data quality of biological surveys. Conserv. Biol. 29, 713–723.

Lovett, G.M., Burns, D.A., Driscoll, C.T., Jenkins, J.C., Mitchell, M.J., Rustad, L., Shanley, J.B., Likens, G.E., Haeuber, R., 2007. Who needs environmental monitoring? Front. Ecol. Environ. 5, 253–260.

Maddock, A., Mills, M.G.J., 1994. Population characteristics of African wild dogs *Lycaon pictus* in the Eastern Transvaal lowveld, South Africa, as revealed through photographic records. Biol. Conserv. 67, 57–62.

Marker, L., Dickman, A., 2004. Human aspects of cheetah conservation: lessons learned from the Namibian farmlands. Hum. Dimens. Wildl. 9, 297–305.

Marnewick, K., Ferreira, S.M., Grange, S., Watermeyer, J., Maputla, N., Davies-Mostert, H.T., 2014. Evaluating the status of and African wild dogs *Lycaon pictus* and cheetahs *Acinonyx jubatus* through tourist-based photographic surveys in the Kruger National Park. PLoS One 9, 1–8.

Mason, A.D., Michalakidis, G., Krause, P.J., 2012. Tiger nation: empowering citizen scientists. Digital Ecosystems Technologies (DEST), 2012 6th IEEE International Conference on Digital Ecosystems and Technologies. Institute of Electrical and Electronics Engineers. Italy.

Reich, A., 1981. The Behaviour and Ecology of the African Wild Dog *Lycaon Pictus* in the Kruger National Park. PhD thesis, Yale University, USA.

Riesch, H., Potter, C., 2014. Citizen science as seen by scientists: methodological, epistemological and ethical dimensions. Public Underst. Sci. 23, 107–120.

Rotman, D., Preece, J., Hammock, J., Procita, K., Hansen, D., Parr, C., Lewis, D., Jacobs, D., 2012. Dynamic changes in motivation in collaborative citizen-science projects. Proceedings of the ACM 2012 conference on Computer Supported Cooperative Work. Association for Computing Machinery, New York, USA, pp. 217–266.

Silvertown, J., 2009. A new dawn for citizen science. Trends. Ecol. Evol. 24, 467–470.

Swanson, A., Kosmala, M., Lintott, C., Simpson, R., Smith, A., Packer, C., 2015. Snapshot Serengeti: high-frequency annotated camera trap images of 40 mammal species in Serengeti National Park. Sci. Data 2, 150026. DOI: 10.1038/sdata.2015.26.

Van Strien, A., van Swaay, C.A.M., Termaat, T., 2013. Opportunistic citizen science data of animal species produce reliable estimates of distribution trends if analysed with occupancy models. J. Appl. Ecol. 50, 1450–1458.

Wiggins, A., Crowston, K., 2011. From conservation to crowdsourcing: a typology of citizen science. In Proceedings of the 44th Hawaii International Conference on System Science. Institute of Electrical and Electronics Engineers, Hawaii, USA, pp. 1–10.

Social Science Methods to Study Human–Cheetah Interactions

Niki A. Rust*,†, Courtney Hughes**,‡

*University of Kent, Canterbury, United Kingdom
**University of Alberta, Edmonton, AB, Canada
†WWF-UK, Woking, United Kingdom
‡Alberta Environment and Parks, Edmonton, AB, Canada

Social science is the study of human society, including its psychological, sociological, organizational, political, and economic aspects (Blaikie, 2004). As humans are both the major cause of cheetah decline and drivers for the species' conservation, social science can help understand the anthropogenic threats toward cheetahs and evaluate methods to mitigate these threats (Dickman et al., 2013). However, many previous human–cheetah coexistence (HCC) studies focused more on how environmental factors, such as seasonality or different habitats, influenced cheetah predation on livestock (Chapter 13), without taking into account people's values, attitudes, and knowledge, along with broader societal issues—hereafter referred to as the human dimensions (Marchini and Macdonald, 2012). Conservationists are becoming increasingly aware of the importance to involve social scientists in research and applied conflict mitigation.

However, to date, there have been limited thorough human dimensions studies of cheetahs conducted, and even fewer in collaboration with trained social scientists.

Social science research methods can assist with understanding complex HCC issues (Manfredo and Dayer, 2004) and have proven essential when improving coexistence with wildlife. Research can include identifying values and attitudes toward cheetahs, rates and types of conflict, and associated economic impacts, efficacy of or receptiveness to new management practices. Examples of previous studies on the impact of social aspects on cheetah conservation have included using quantitative questionnaires with farmers, soliciting cheetah stories from children using qualitative methods, and employing a mixed approach to better understand the complexity of conflict (Boast, 2014; Hughes, 2013; Marker et al., 2003).

Cheetahs: Biology and Conservation
http://dx.doi.org/10.1016/B978-0-12-804088-1.00035-6

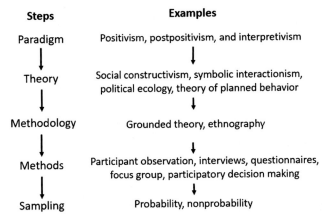

FIGURE 35.1   **Process of designing your social science methods, with examples of each step.**

Due to the limited number of social science studies in cheetah conservation, this chapter aims to set the scene for how one might effectively design and undertake such a study. We briefly describe some potentially useful social science paradigms and theories, followed by an overview of methodologies and methods conservationists can use to study HCC. Fig. 35.1 shows a suggested process in designing a social science study, which we explain in proceeding sections.

RESEARCH AIMS

Clear research aims/questions will direct what paradigm, theories, methodologies, and methods are appropriate for the study. Literature reviews provide an understanding of what has been done in the past, what methods have worked, what ideas you can extract, and where gaps in knowledge exist. Commonly, human dimensions of cheetah conservation seek to understand the cause of conflict and how to address it but it is equally important to understand the efficacy and effect of mitigation, conservation educational, and awareness raising programs.

RESEARCH PARADIGMS

Philosophical perspectives (i.e., belief systems of how the world is perceived) that guide data collection must be considered as part of the research paradigm. Positivism, interpretivism, and postpositivism are arguably the most accepted philosophical perspectives in social science (Grix, 2010). Positivism (e.g., experimental testing) tests a theory or hypothesis about the world via the scientific process using deduction and objectivity, and may be useful when measuring the effectiveness of certain conflict mitigation strategies. Interpretivism infers that the world is socially constructed, and each individual constructs his or her own reality. It could help explain how people's knowledge and experiences with cheetahs affects their values and/or attitudes as researchers seek to interpret and understand relevant perceptions and meanings. Postpositivism can bridge these paradigms, where context and attending to researcher and other biases is important when interpreting the world. Blaikie (2007) provides more information on this topic, including contemporary paradigms, in social research.

THEORETICAL PERSPECTIVES

To date, there has been very limited attention given to the importance of clearly choosing and explaining the theoretical perspective for a HCC study. As such, the following section explains the different theoretical perspectives that could be employed in a HCC study. Theory seeks to explain social reality, such as how people perceive cheetahs and helps to describe or predict a topic or behavior. Initially, a researcher will identify their paradigm and determine what theoretical framework best meets their needs. This is an important reflective process, which should be explicitly stated as part of social research studies but has been lacking in cheetah social research. Some theories potentially useful in HCC studies include:

- *Constructivism, or social constructivism* focuses on culture and context, where knowledge is built through experiences with the world and other people (Young and Collin, 2004). A qualitative study at the Cheetah Conservation Fund employed this approach, to learn how children participating in educational programs constructed their values for cheetahs (Hughes, 2008).
- *Symbolic interactionism*, focusing on interactions with people, objects, and places (Aksan et al., 2009), could be used to help identify symbolic meanings that farmers or others ascribe to cheetahs or livestock, and how this influences behaviors. This theory could be applied to understand perspectives or effectiveness on using cheetahs as a flagship symbol for conservation (e.g., Belbachir et al., 2015).
- *Political ecology* deals with relationships between political, economic, and social aspects of the environment (Robbins, 2004). Research using political ecology has shown that instances of HCC in Namibia is influenced by the country's previous apartheid regime and the associated

inequality in land, wealth, and education (Rust et al., 2016).
- *Theory of planned behavior* assumes human behavior is due to rational choices shaped by a person's attitudes, subjective norms, and perceptions of control over that behavior (Ajzen, 1991). For example, research in South Africa used this theory to identify why farmers adopt or resist cheetah conservation measures (Reijneker, 2014).

Combining paradigms and theory can be challenging, so an example is provided below. To understand if and how people value cheetahs within an interpretivism paradigm, one could seek to understand how people experience cheetahs, and how this influences their behaviors. This paradigm can integrate the theories of social constructivism and symbolic interactionism to theorize that people construct their perceptions of cheetahs through direct, lived experiences, and that behaviors toward cheetahs are influenced by what a predator, or other value objects, such as livestock, means to them.

METHODOLOGIES OF RELEVANCE TO HUMAN–CHEETAH COEXISTENCE

The choice of methodology is influenced by the researcher's paradigm, the context of the research and its aims, and the strengths and weaknesses of the available methods and analytic tools. There are a number of excellent textbooks guiding this process (e.g., Bernard, 2006; Newing et al., 2010). Though various methodological approaches may be useful in HCC studies (e.g., phenomenology or narrative inquiry; Wertz et al., 2011), we will only explain grounded theory (GT) and ethnography, as they have not been widely used in HCC studies, yet have the potential to provide new and detailed insight.

Grounded Theory

GT is an iterative, inductive framework (Glaser and Strauss, 1967) where researchers start data collection before hypotheses are made; the theory is then generated from the collected data. Observations of a particular scenario are collected and immediately analyzed via understanding common themes and relationships, which provide a contextual understanding; this preliminary information guides assumptions and future research direction. Afterward, detailed data is gathered which is immediately examined for possible connections.

GT may uncover previously overlooked findings given its ability to look afresh at problems. It has rarely been used in HCC studies, though Rust (2015a) applied GT to understand underlying drivers of conflict between cheetahs and livestock farmers in Namibia. Qualitative interviews with farmers and their employees indicated that HCC was in part affected by the relationship between farmer and workers. The methodology is, however, time- and labor-intensive, requiring substantial knowledge of the techniques involved in collecting and analyzing data (Backman and Kyngäs, 1999).

Ethnography

Ethnography aims to find cultural understandings of a topic via long-term observation and participation in a specific study site (Atkins and Beschner, 1980). While not common to HCC studies, it can offer a detailed understanding of complex issues related to social, cultural, and symbolic values of wildlife (Hill, 2015). Ethnography generally uses qualitative methods to collect detailed descriptive data, which can be used to test theories and offer new perspectives on study subjects. Indeed, similar to GT, building theory grounded in the data itself can be an outcome (Streubert and Carpenter, 1999). However, the researcher needs to use critical self-examination as it is particularly susceptible to subjective

bias, given the researcher's understanding and potential preconceptions of the topic (Hoholm and Araujo, 2011). It can be difficult to replicate ethnographic studies given context specificity, though mitigations include keeping extensive field notes on the context and reflections of the study. This methodology should be considered by HCC researchers when the aim is to provide a detailed picture of a conflict situation. For instance, Rust and Taylor (2016) used ethnography to understand the relationship between people and predators in Namibia and found that predators had been persecuted due to an "Othering" process that alienated predators by differentiating them from other species.

METHODS FOR STUDYING HUMAN–CHEETAH COEXISTENCE

Theory and methodology should guide method choices, as these will influence data collection and analytics. A wide variety of methods are available to study HCC with choices depending on: the focus of the research question; the extent of prior knowledge about the study's topics; whether conventional thinking is to be challenged; and possible study constraints, e.g., budget, time, cultural, or ethical considerations (Newing et al., 2010). The main methodological approaches are qualitative, quantitative, and mixed methods. To explore sensitive topics in HCC (e.g., poaching) specific quantitative methods, such as randomized response technique (RRT) and unmatched count technique (UCT) are particularly useful tools. These methods are discussed further.

Qualitative Methods

Qualitative methods (Table 35.1) gather rich descriptions of social phenomena typically from direct contact with research participants (e.g., interviews), and sometimes through qualitative content analysis of documents (Elo and

TABLE 35.1   Common Qualitative Methods

Method and description	Strengths	Weaknesses
Informal interview: unplanned conversation between participants and researchers. *Unstructured interview*: prearranged with a set topic; researcher loosely guides conversation. *Semistructured interview*: researcher uses an interview guide that outlines questions or themes to be explored.	Helps design data collection of other methods, for example, questionnaires. Sets initial study scene. Helps triangulate data. Useful for identifying future respondents. Can reveal unexpected information.	Takes time to elaborate on answers. Analysis should include contextual information and nonverbal communication, for example, context and body language.
Focus group: a group interview; researchers set topic(s) in advance.	Highlights interaction between participants; shows issues of contention and consensus.	Less powerful/dominant participants can be uncomfortable participating. Bias toward vocal participants. Complicated to implement.
Participant observation: relatively unstructured; study people in their environment, describing how and why they do things. Researchers live and undertake activities in the community.	Helps explain hidden values, beliefs, attitudes, and intangible motivations that are otherwise unobtainable via structured interviews. Researcher not limited to respondent's recall. Can assist with building positive researcher–participant relationships.	Time and constraints of immersing oneself in study site. Misunderstandings and miscommunication could lead to misrepresentation. External validity is minimal. Conclusions are strongly dependent on the researcher's interpretation.

Kyngäs, 2008). Results are typically analyzed descriptively using themes. These methods can be used to explore values, emotions, cultures, and relationships toward cheetahs. For example, storytelling has been used to understand how children relate to cheetahs (Hughes, 2013; Fig. 35.2). Qualitative methods can also be used to gain in-depth knowledge on rarely explored topics, such as negative consequences that affect cheetahs of poor game-farm management in South Africa (Cousins et al., 2008).

Qualitative data are context-specific, where analysis seeks to identify themes or patterns of socially-constructed meaning. Kent (2011) used informal interviews and participant observation to explore local perceptions of cheetahs and other carnivores in rural Botswana. This research highlighted some respondents' mistrust in the government's ability to effectively manage wildlife, in turn affecting the ability of the government to accurately monitor livestock depredation and lethal predator control.

Quantitative Methods

Quantitative methods (Table 35.2) generate numerical data (e.g., via questionnaires), which are analyzed statistically to explain phenomena under investigation. If data are randomly sampled, findings can be extrapolated. Selebatso et al. (2008) determined farmer perceptions of cheetah conservation in Botswana using questionnaires, identifying that farmers generally tolerated cheetahs.

Changes of specific variables over time (usually decades) can be tracked quantitatively by using a panel/longitudinal study. While preferable to use the same individuals over time, it is not necessary as long as random sampling is used. Panel studies can be useful when aiming to understand changes in attitudes, behavior

FIGURE 35.2   **Using storytelling to understand children's perceptions of cheetahs (Hughes, 2013).**

TABLE 35.2   Common Quantitative Methods

Method and description	Strengths	Weaknesses
Questionnaires: interviews using fixed, structured questions (but can also use open-ended questions).	Used to answer a specific set of questions, or to quantify a problem or solution. Useful toward the end of data collection once the focus of the research topic has been refined, information has been gathered about the topic to produce appropriate questions, and rapport has been built with the study population. Advantageous when recording openly discussed facts, such as husbandry methods or demographic details.	Difficult to determine if questions are relevant or biased without prior research. Limited chance of follow-up questions. Using cross-cultural questionnaires can be problematic. Less helpful when asking "why" questions. Some information is culturally or socially sensitive, rendering this method inappropriate.

or knowledge over time, especially when combined with evaluating the effect of a conflict mitigation intervention in this time frame. As an example, a study of nearly 1900 individuals in the United States showed that attitudes toward wolves became less positive over an 8-year period, with respondents reporting an increase in fear and inclination to poach wolves (Treves et al., 2013).

Mixed Methods

Mixed methods (Table 35.3) combine elements of both techniques in data collection and analysis, but it is not simply an addition of qualitative to quantitative techniques. Rigorously-designed mixed approaches yield both internal and external validity by combining statistics with rich narratives, with the aim to result in a deeper understanding of the research problem. Using mixed methods can be useful when researching the efficacy of HCC strategies. For example, Rust (2016) used a mixed-methods approach in a novel decision-making exercise to determine whether farmers and conservationists could agree on cheetah and other predator management strategies on Namibian farmland. The study built on data previously collected from a content analysis of newspapers (Rust, 2015b), along with interviews, participant observation, and focus groups.

Randomized Response Technique and Unmatched Count Technique

The frequency of illegal killing of cheetahs is challenging to assess, as respondents may be unwilling to share sensitive information (Tourangeau and Yan, 2007). One way to circumvent this problem is to use methods specifically designed to illicit information on illegal behavior. Two common indirect questioning methods that quantify sensitive information are RRT and UCT.

RRT enhances the accuracy of estimating sensitive behavior by in effect anonymizing the response of the participant through a randomization process. Randomizing devices, such as dice or spinners are often used (Warner, 1965). RRT relies on the principle that fixed responses (regardless of their veracity) are given by the respondent for certain outcomes of the randomizing device, while truthful responses are given for other outcomes: for example, the respondent must say "yes" to the statement "I have illegally killed a cheetah in the last 12 months" if a die lands on 1 or 2, "no" if it lands on 3 or 4; and responds truthfully if it lands on 5 or 6. Crucially, the interviewer is unaware of the number on which the die lands, thereby giving privacy to the respondent. Because the randomizing device has a known probability of landing on each number, the probability of answering truthfully can be quantified. This method increases

TABLE 35.3 Common Mixed Methods

Method and description	Strengths	Weaknesses
Participatory research: researchers work with study site individuals to collaboratively design and analyze research.	Coproduces knowledge. Useful for marginalized groups, as power is placed with the community.	Participants guide research, which may not be in the researchers' desired direction. Susceptible to oversimplifying challenges with participation and power.
Multicriteria analysis: differentiates, ranks or analyses trade-offs and scenarios.	Helps understand the effect(s) decisions have on outcomes or impacts. Systematically allows decision-makers to transparently critique options. Integrates social values and local knowledge.	Can lack transparency. Problems with weighting responses to ensure participant views are accurately represented. Challenges with power and bias, which influences fairness and representativeness.

response rates to sensitive questions (Lensvelt-Mulders et al., 2005) but requires a large sample size as response variance is high (i.e., high standard error and deviations), and it is particularly difficult to estimate relationships between the sensitive behavior and other variables (Tourangeau and Yan, 2007). RRT has been used to estimate illegal killing of cheetahs in South Africa (St John et al., 2012) and Botswana (Boast, 2014), and to understand the extent of carcass poisoning to kill predators in Namibia (Santangeli et al., 2016).

UCT was developed to overcome some pitfalls of RRT, including respondents feeling tricked or confused by RRT. UCT presents two very similar lists of activities, both of which contain an identical list of unthreatening activities, and one contains an additional sensitive activity, such as cheetah poaching. Two groups of comparable respondents each receive one of the lists and are asked how many (but not which) items apply to them. By comparing answers from each group, a rough estimate of the frequency of illegal behaviors within a population can be calculated. The challenge is that this method does not always yield more honest answers about the sensitive behavior than direct questions (e.g., Tourangeau and Yan, 2007). UCT also requires complex data analysis and is unable to precisely quantify illegal behaviors due to large standard errors. UCT has been employed to estimate poaching of wildlife in the Serengeti (Nuno et al., 2013). For a more thorough discussion on asking sensitive conservation questions, see Nuno and St John (2014).

PRACTICAL CONSIDERATIONS

There are a number of practical considerations that need to be taken into account when designing a social science study. Quantitative techniques can be administered in person, via telephone, mail, SMS, or online, while qualitative techniques are usually administered in person or via telephone. The most appropriate format depends entirely on the study's context (e.g., if a mail-out is appropriate or Internet is available) and constraints (e.g., budget, time, research permits). Response rates influence the best administration method. Face-to-face questionnaires usually provide the best response rates but can be more time-consuming and expensive. If the study participants speak another language, can you learn the new language or will you use a translator? Both approaches have strengths and weaknesses (Ervin and Bower, 1952) that should be considered. Learning a language requires time but may allow for better integration into the community and help to understand cultural nuances. Using a translator can lead to information being mistranslated. Nonverbal body language cannot be captured remotely, yet is important to note in data collection as it may indicate emotions of the respondent. What is your budget? Collecting social science data can be very costly, particularly if you need to pay field assistants' salaries or for mail-out surveys to hundreds or thousands of participants. Is the study culture different to yours? If so, will this affect your integration into the community and potentially influence the results? In a new area, it can be useful to spend a number of months in the local community learning about the culture and context to better understand the norms and ensure questions are relevant. How will interviewer and participant education levels affect data collection? Participants with very limited education may not be able to complete written questionnaires and field assistants may not be able to conduct them. Also, while many academic institutions have rigorous ethics training and protocols for research involving human subjects, other agencies or organizations may not. It would be beneficial for researchers to familiarize themselves with ethical considerations (Sieber, 1982) when undertaking HCC studies if they involve human subjects.

Sampling

Sampling considerations of face-to-face recruitment for HCC studies include physical location and time of day, community layout, anonymity and confidentiality, informed consent and, in some cases, if participant incentives are necessary. Ensuring representative participation, particularly among marginalized individuals, is vital when the aim is to engage the entire community. Challenges include finding enough participants, particularly in remote locations with low population density or in communities suspicious of researchers. Therefore, it is recommended that key local informants are used to help recruit participants, though researchers must be aware that the respondents can be biased. Larger samples tend to be more representative, but are not always required for qualitative research. Smaller samples can be useful for in-depth interviewing but one must take care not to generalize findings, as data are likely to be context-specific. Sample size might vary dramatically from 10 to 100 for an in-depth qualitative study, and 100–1000s for quantitative questionnaires. Choice of sampling depends on research aims and fulfilling criteria, such as repeatability, validity, credibility, and so on. Ensuring personal data are anonymized—for example, using unique codes (e.g., P1, P2...)—helps protect participants' identity in case there are concerns that identification may put them at risk (e.g., when admitting to illegal killing of cheetahs). In cases where identifying individuals may be necessary, participants must be told in advance so they can make an informed decision to participate. For more information on sampling in social science, including reducing bias, see Newing et al. (2010).

CONCLUSIONS

We have described some of the more common social science paradigms, theories, methodologies, and methods to be used in HCC studies. However, there are many other options available (Outhwaite and Turne, 2007). Designing a social science wildlife study requires detailed considerations to ensure the approach accurately answers the study aims. Other important factors briefly described, but requiring more thorough examination, include the explicit consideration of context, ethics, and researcher reflexivity, discussed in more detail in various textbooks, for example, Anyansi-Archibong (2015) and Bernard (2006). Close collaboration with expert social scientists is strongly advised to ensure that approaches are appropriately chosen, designed, undertaken, and analyzed; the choice of appropriate approaches is a prerequisite for obtaining reliable results.

As human populations grow, expanding into more wild land where conflict between humans and predators such as cheetahs is likely to become more prevalent, it has never been more important to find ways in which people and large carnivores can coexist. Social science studies are vital to finding paths to coexistence.

References

Ajzen, I., 1991. The theory of planned behavior. Organ. Behav. Hum. Decis. Process. 50, 179–211.

Aksan, N., Kısac, B., Aydın, M., Demirbuken, S., 2009. Symbolic interaction theory. Procedia-Social Behav. Sci. 1, 902–904.

Anyansi-Archibong, C.B., 2015. Contemporary Issues Surrounding Ethical Research Methods and Practice. IGI Global, Hershey.

Atkins, C., Beschner, G., 1980. Ethnography: A Research Tool for Policymakers in the Drug & Alcohol Fields. US Department of Health & Human Services, Rockville.

Backman, K., Kyngäs, H.A., 1999. Challenges of the grounded theory approach to a novice researcher. Nurs. Health Sci. 1, 147–153.

Belbachir, F., Pettorelli, N., Wacher, T., Belbachir-Bazi, A., Durant, S.M., 2015. Monitoring rarity: the critically endangered saharan cheetah as a flagship species for a threatened ecosystem. PLoS One 10 (1), e0115136.

Bernard, H.R., 2006. Research Methods in Anthropology: Qualitative and Quantitative Approaches, fourth ed. AltaMira Press, Lanham.

Blaikie, N., 2004. Philosophy of social science. In: Lewis-Beck, M.S., Bryman, A., Liao, T.F. (Eds.), The SAGE Encyclopedia of Social Science Research Methods. Sage Publications, Inc, Thousand Oaks, pp. 821–822.

Blaikie, N., 2007. Designing Social Research, second ed. Polity Press, Cambridge: UK.

Boast, L.K., 2014. Exploring the Causes of and Mitigation Options for Human-Predator Conflict on Game Ranches in Botswana: How is Coexistence Possible? PhD thesis, University of Cape Town, South Africa.

Cousins, J.A., Sadler, J.P., Evans, J., 2008. Exploring the role of private wildlife ranching as a conservation tool in South Africa: stakeholder perspectives. Ecol. Soc. 13, 43.

Dickman, A., Marchini, S., Manfredo, M., 2013. The human dimension in addressing conflict with large carnivores. In: Macdonald, D.W., Wills, K.J. (Eds.), Key Topics in Conservation Biology 2. John Wiley & Sons, Oxford, pp. 110–126.

Elo, S., Kyngäs, H., 2008. The qualitative content analysis process. J. Adv. Nurs. 62, 107–115.

Ervin, S., Bower, R.T., 1952. Translation problems in international surveys. Public Opin. Q. 16, 595–604.

Glaser, B.G., Strauss, A.L., 1967. The Discovery of Grounded Theory: Strategies for Qualitative Research. Aldine Publishing Company, Aldine, Chicago.

Grix, J., 2010. The Foundations of Research. Palgrave Macmillan, Basingstoke.

Hill, C.M., 2015. Perspectives of "Conflict" at the Wildlife–Agriculture Boundary: 10 Years On. Hum. Dimens. Wildl. 20, 1–6.

Hoholm, T., Araujo, L., 2011. Studying innovation processes in real-time: the promises and challenges of ethnography. Ind. Mark. Manage. 40, 933–939.

Hughes, C., 2008. Environmental Education and the Cheetah Conservation Fund: Exploring Children's Value-Based Relationships With Cheetahs. MSc thesis, Lakehead University, Canada.

Hughes, C., 2013. Exploring children's perceptions of cheetahs through storytelling: implications for cheetah conservation. Appl. Environ. Educ. Commun. 12, 173–186.

Kent, V.T., 2011. The Status and Conservation Potential of Carnivores in Semi-Arid Rangelands, Botswana The Ghanzi Farmlands: A Case Study. PhD thesis, Durham University, United Kingdom.

Lensvelt-Mulders, G.J.L.M., Hox, J., van der Heijden, P.G.M., Maas, C.J., 2005. Meta-analysis of randomized response research. Sociol. Methods Res. 33, 319–348.

Manfredo, M.J., Dayer, A.A., 2004. Concepts for exploring the social aspects of human–wildlife conflict in a global context. Hum. Dimens. Wildl. 9, 1–20.

Marchini, S., Macdonald, D.W., 2012. Predicting ranchers' intention to kill jaguars: case studies in Amazonia and Pantanal. Biol. Conserv. 147, 213–221.

Marker, L.L., Mills, G., Macdonald, D.W., 2003. Factors influencing perceptions of conflict and tolerance toward cheetahs on Namibian farmlands. Conserv. Biol. 17, 1290–1298.

Newing, H., Eagle, C., Puri, R., Watson, C., 2010. Conducting Research in Conservation: Social Science Methods and Practice. Routledge, Abingdon.

Nuno, A., Bunnefeld, N., Naiman, L.C., Milner-Gulland, E.J., 2013. A novel approach to assessing the prevalence and drivers of illegal bushmeat hunting in the Serengeti. Conserv. Biol. 27, 1355–1365.

Nuno, A., St John, F.A.V., 2014. How to ask sensitive questions in conservation: a review of specialised questioning techniques. Biol. Conserv. 189, 5–15.

Outhwaite, W., Turne, S., 2007. The Sage Handbook of Social Science Methodology. Sage Publications, Ltd., London.

Reijneker, M.S., 2014. Elephants, Predators & Politics: Understanding Socio-Political Landscape Influences on Farmer-Wildlife Conflict Around Alldays, South Africa. PhD thesis, Wageningen University, Netherlands.

Robbins, P., 2004. Political Ecology. Blackwell Publishing, Malden.

Rust, N.A., 2015a. Understanding the Human Dimensions Affecting Coexistence Between Carnivores and People: a Case Study in Namibia. PhD thesis, University of Kent, United Kingdom.

Rust, N.A., 2015b. Media framing of financial mechanisms for resolving human-predator conflict in Namibia. Hum. Dimens. Wildl. 20, 440–453.

Rust, N.A., 2016. Can stakeholders agree on how to reduce human–carnivore conflict on Namibian livestock farms? A novel Q-methodology and Delphi exercise. Oryx 50, 339–346.

Rust, N.A., Taylor, N., 2016. Qualitative case study on the intersectional conflict: a qualitative case study on the intersectional persecution of predators and people in Namibia. Anthrozoös 29, 653–667.

Rust, N.A., Tzanopoulos, J., Humle, T., MacMillan, D.C., 2016. Why has human-carnivore conflict not been resolved in Namibia? Soc. Nat. Resour. 29, 1079–1094.

Santangeli, A., Arkumarev, V., Rust, N., Girardello, M., 2016. Understanding, quantifying and mapping the use of poison by commercial farmers in Namibia – Implications for scavengers' conservation and ecosystem health. Biol. Conserv. 204, 205–211.

Selebatso, M., Moe, S.R., Swenson, J.E., 2008. Do farmers support cheetah Acinonyx jubatus conservation in Botswana despite livestock depredation? Oryx 42, 430–436.

Sieber, J.E., 1982. The Ethics of Social Research: Fieldwork, Regulation, and Publication. Springer-Verlag, New York.

St John, F.A.V., Keane, A.M., Edwards-Jones, G., Jones, L., Yarnell, R.W., Jones, J.P.G., 2012. Identifying indicators of illegal behaviour: carnivore killing in human-managed landscapes. Proc. R. Soc. B Biol. Sci. 279, 804–812.

Streubert, H., Carpenter, D., 1999. Qualitaitve research: advancing the humanistic imperative, second ed. Lippincott, Philadelphia, PA.

Tourangeau, R., Yan, T., 2007. Sensitive questions in surveys. Psychol. Bull. 133, 859–883.

Treves, A., Naughton-Treves, L., Shelley, V., 2013. Longitudinal analysis of attitudes toward wolves. Conserv. Biol. 27, 315–323.

Warner, S., 1965. Randomized response: a survey technique for eliminating evasive answer bias. J. Am. Stat. Assoc. 60, 63–69.

Wertz, F.J., Charmaz, K., McMullen, L.M., Josselson, R., Anderson, R., McSpadden, E., 2011. Five Ways of Doing Qualitative Analysis: Phenomenological Psychology, Grounded Theory, Discourse Analysis, Narrative Research, and Intuitive Inquiry. Guilford Press, New York.

Young, R.A., Collin, A., 2004. Introduction: Constructivism and social constructionism in the career field. J. Vocat. Behav. 64, 373–388.

Spatial and Landscape Analysis: Applications for Cheetah Conservation

Richard M. Jeo*, Leah Andresen**

*The Nature Conservancy in Montana, Helena, MT, United States
**Nelson Mandela University, Port Elizabeth, South Africa

INTRODUCTION

Quantitative spatial analyses can provide powerful insight into the design of conservation management strategies. Technological advancements and the development of new tools and statistical modeling techniques along with the availability of high-resolution spatial data sets that cover the entire earth, have advanced the ease of spatial analysis. Well-designed quantitative analyses can help target conservation strategies and identify critical gaps in knowledge, which is especially important for a species, such as cheetah (*Acinonyx jubatus*) that survives largely in vast, remote areas, which are difficult to access, throughout its range. In this chapter, we provide a general review of available quantitative spatial conservation modeling approaches, with an emphasis on cheetah conservation applications.

FRAMING THE PROBLEM: WHAT CAN BE GAINED FROM SPATIAL AND LANDSCAPE ANALYSIS?

Imagine you are working for a government wildlife agency in one of the cheetah range countries and you were asked to prioritize management efforts for cheetah in your region (and we hope that at least a few of the readers of this book are in this position!). How would you go about answering this question? There are many approaches for designing and prioritizing conservation efforts and smart managers recognize that quantitative spatial analyses are not always necessary. But for those cases where conservation objectives for cheetahs would benefit from spatial and landscape analysis, these techniques can quantify and rank threats, identify core habitats, corridors, and help prioritize key areas where conservation interventions are required.

Cheetahs: Biology and Conservation
http://dx.doi.org/10.1016/B978-0-12-804088-1.00033-2

Methods for wildlife habitat modeling can be classified into a few primary approaches (for reviews, see Franklin and Miller, 2009; Guisan and Zimmermann, 2000). In this chapter, we discuss essentially two types of models: (1) those used for inference about population status (e.g., occupancy and occurrence modeling and habitat suitability mapping) and (2) those used to project the consequences of management actions or other scenarios on focal populations (e.g., landscape connectivity modeling and populations viability modeling). We provide a general overview of the modeling techniques that can be a practical starting point for users considering the myriad of current options and explore research priorities for cheetah conservation.

Predicting Presence/Absence and Probability of Occurrence

Predicting presence and absence is perhaps the most fundamental problem that spatial analysis can address. This is especially important for a species like the cheetah, which occurs largely outside of protected areas, across vast remote landscapes spanning political boundaries and multiple-use landscapes (Durant et al., 2017). Predicting presence/absence or probability of occurrence relates presence data and absence data (if available) to landscape and habitat features. Presence models can be based simply on expert knowledge (Yamada et al., 2003) or presence recorded from field-data (e.g., from camera trap surveys; O'Connell et al., 2011; Chapter 29, spoor surveys; Houser et al., 2009; Chapter 30, or citizen science; Dickinson et al., 2012; Chapter 34).

Alternatively, more advanced models relate field observations to landscape features to define habitat quality and predict presence. How landscape areas are categorized is perhaps the crux of using these habitat quality thresholds to predict presence. Categorizations are often based on land cover classification, but an appropriate

understanding and utilization of knowledge of the biology of the species in question remains the most important step in any habitat assessment process.

Estimating the probability that cheetahs will occur in a particular habitat (i.e., habitat use or the probability of occurrence), utilizes statistical models fitted to georeferenced detection/nondetection data. The resulting maps are useful for cheetah management in a number of ways. First, this approach can provide quantitative insight into what landscape or habitat features relate to cheetah habitat use and distributions. Second, these maps can provide insight into the extent, quality, and configuration of habitat available in a way that is not always immediately obvious from the observation data alone and therefore can directly inform management practices (MacKenzie, 2006).

Several techniques have been used to predict the statistical likelihood of presence from spatial data, including resource selection function (RSF) models and occupancy models (MacKenzie et al., 2006; Manly et al., 2007). RSF models are functions (e.g., logistic regression or logistic discriminant) that are proportional to the probability of use of a spatial resource unit. When used with typical detection/nondetection data, they assume that the focal species is detected on every spatial unit that it occupies. Occupancy models are essentially RSF models that were developed to deal with nondetection; that is, the focal species is not always detected at the sampled location, even when at least one individual of the species occupies the site (MacKenzie et al., 2006). Distinguishing between true absence and nondetection is particularly important for rare and elusive species, such as the cheetah, which may be present but remain undetected. Occupancy models have been used to generate probability of occurrence maps from footprint (spoor), scat, camera-trap, and interview survey data (Andresen et al., 2014; MacKenzie, 2006; O'Connell et al., 2011; Chapters 29–31, and 35).

Habitat Quality—Models of Habitat Suitability

What is cheetah habitat—that is, what combinations of structural and functional characteristics of landscapes drive survival and reproduction? Habitat suitability models can range from a set of mechanistic needs of an animal based on general physiology and ecological requirements to quantitative predictions of the likelihood of a species' presence based on a set of environmental variables (Guisan and Zimmermann, 2000). However, functional characteristics of habitat (e.g., mortality drivers or prey density) are notoriously difficult to map because comprehensive data are largely absent. A commonly utilized approach relates habitat to *surrogates* for habitat where there are spatial data available. Use of spatial land-cover classifications based on satellite imagery has been a common approach to developing habitat surrogates for a variety of species. Surrogates have been developed from a wide variety of data ranging from environmental conditions, such as temperature regime to landscape features, such as hollow-bearing trees as a surrogate for arboreal mammals (Lindenmayer et al., 2014). Surrogates can be easily derived from spatial data that are widely available, such as digital elevation models, general vegetation type or other layers derived from remote-sensing data, and are often focused on the structural aspects of cheetah habitat. There are two levels of assumptions here: (1) that such surrogates reflect habitat and (2) that habitat is relevant to species distributions or vital rates. In general, surrogates should have predictive value on key aspects of cheetah biology, such as availability of plant forage for prey species or mortality risk associated with anthropogenic features.

A general assumption is that high-quality habitats are used more intensely than poor-quality ones, but a potential pitfall here is that multiple factors can lead to an animal avoiding rich habitats and selecting for poor quality habitats (Railsback et al., 2003). Examples where cheetahs select poor habitat over rich habitat include ecological traps, site fidelity, and avoidance of dominant competitors (Battin, 2004; Durant, 2000; Marker et al., 2003). Identification of mortality sinks and knowledge of factors leading to these sinks could facilitate cheetah conservation efforts (van der Meer et al., 2015).

Occupancy modeling, RSF and other probabilistic approaches to evaluating habitat suitability can provide insight beyond simple maps of habitat and nonhabitat (Andresen et al., 2014). For example, an important use of habitat modeling approaches would be to predict key areas where conflict might occur, in order to implement proactive conservation measures that improve human tolerance (Chapter 13).

Landscape Connectivity Modeling—Least Cost Path and Graph Theory Approaches

Methods to assess habitat fragmentation and mitigate its impacts through identification and protection of landscape corridors have been promoted broadly (Crooks and Sanjayan, 2006). Wildlife corridors were originally viewed as linear strips of habitat that allow organisms to move through the landscape, but the corridor concept has been broadened to encompass landscape linkages that are not necessarily linear or even structurally continuous (Bennett, 1999). For the cheetah, a functional corridor can be a broad area of sufficient permeability where prey and human tolerance are adequate for movement between higher quality habitats (Chapter 10).

There are several GIS-based tools available to assess habitat connectivity and provide insight into the landscape's resistance to movement (Zeller et al., 2012). Least-cost path analysis identifies a route that incurs a minimum cost as an animal moves from one location to another. This assumes that the "cost" metric reflects actual energy, mortality risk, or some other measure of fitness. In addition, it also assumes animal knowledge of, or ability to assess cost, is perfect. A variant of the method is factorial least-cost path analysis, which integrates a vast number of paths or corridors between many points across the landscape. Densities of paths are shown in gradients—representing putative areas that are predicted to be more important in the landscape for connectivity between multiple points. Another similar technique uses circuit theory, where the landscape is treated as if it was an electrical circuit, and uses that analogy to delineate connectivity patterns using relative voltage, resistance, and current as metrics.

Correctly assigning the costs is of course the key to the usefulness of least-cost or other resistance-based connectivity tools. Analytical approaches, like occupancy modeling, offer quantitative methods that can inform least-cost path models. For example, the inverse of the model can be used to generate a cost surface and explore potential linkage areas and predict probability of human conflicts (Chetkiewicz et al., 2006).

Graph theory is another tool that offers insight into landscape permeability. This application was originally derived from analysis of transportation and computer networks and has been applied to ecological systems and landscapes (Bunn et al., 2000). Resource patches (known as "nodes" in graph theory) are defined in a lattice framework and distance between pairwise combinations are computed (i.e., "edges" in graph theory). While similar to least cost path (and other methods that employ a resistance surface), the power of a graph theory approach is that it provides a standard framework that allows evaluation of different pieces of the landscape for their relative importance for overall connectivity. Nodes and edges can be added or subtracted from the habitat model to simulate loss or maintenance of landscape connectivity and various scenarios can be explored and visualized (Minor and Urban, 2008).

Application of landscape genetic techniques has shown recent promise in improving our understanding of functional connectivity. For example, landscape genetics have recently been used to inform cost layers and/or validate connectivity models (Mateo-Sánchez et al., 2015) and have been used to evaluate source-sink dynamics and identify ecological traps. Other recent techniques use Bayesian clustering methods to identify populations genetically and then estimate recent asymmetrical rates of gene flow between subpopulations (Andreasen et al., 2012). However, it is important to acknowledge that level of connectivity differs for genetic and demographic functions, where considerably more movement is needed for demographic effects (Sanderlin et al., 2012).

POPULATION VIABILITY ANALYSIS (PVA)

Population viability analyses (PVA) are models that provide a quantitative assessment of extinction risk, represented by a probability of extinction in a certain time frame (Chapter 38). Perhaps a more insightful question for cheetah management, rather than probability of extinction in 10 or 100 years, is: what are the best kind of management actions or conservation interventions to decrease the extinction probability? How would various management actions compare? PVA techniques that can address such questions are discussed in the following sections.

Spatially Explicit PVA

Population viability can be impacted by loss of habitat, changes in the suitability of remaining habitat areas, and by increasing isolation through changes in land use or human activities in and around habitat areas. Spatially explicit PVA modeling quantifies the impact of changes in habitat quality or quantity and directly links changes in habitat to demographic rates.

Measuring the availability of high-quality habitat is perhaps the simplest form of spatially explicit PVA. This approach combines assumptions about habitat requirements and home range size (Roloff et al., 2002) and estimates whether or not an area can encompass a minimum number of home ranges. Population growth rates can also be estimated using habitat quality data. A potential advantage to this approach is that source-sink dynamics and ecological traps can be identified as habitat changes over time (Foppen et al., 2000). Multiseason occupancy dynamic modeling approaches (e.g., MacKenzie et al. 2006) can also be used to perform spatially-explicit PVA and explore management scenarios. These models can relate probability of local extinction and colonization to habitat. Probability of extinction of an entire area is a function of probability of extinction for all occupied patches (all go extinct) and of colonization for all unoccupied patches (no colonization). The utility of this approach is that it can be computed for any current occupancy state and set of habitat relationships. However, these analyses are quite data intensive, and the data required for spatially explicit models are often unavailable.

Bayesian Networks

Bayesian networks have also been used to assess the statistical links between habitat and the likelihood of different viability states (Uusitalo, 2007). Statistical predictions are based on linking habitat (or any other spatial data) with demographic trends or other proxies for population growth. Bayesian approaches can utilize sparse demographic data and provide a framework that explicitly quantifies both uncertainty and risk. For example, Bayesian approaches have been used to assess risk under the Endangered Species Act in the USA (Taylor et al., 2002).

PVA Summary

Whether it is done with Bayesian statistics or multiseason occupancy modeling, the assumed direct linkage between habitat quality or habitat status and demographic rates are of paramount importance. Obtaining primary data is often difficult, time consuming, and expensive, as it requires mortality information from labor-intensive techniques, such as capture-recapture and collar data (radio or GPS) (Chapter 32). A cautionary note here is that even large amounts of monitoring data are often insufficient to develop PVAs with strong predictive power and high confidence (Reed et al., 2002).

CONSIDERATIONS FOR CHEETAH

Building predictive models that are both reflective of biological reality and useful for conservation and management is a difficult endeavor. Fortunately, methodological advances have improved the techniques available for acquiring primary demographic data needed for modeling (Chapters 29–35). For example, camera-traps coupled with spatially explicit capture-recapture models allow estimation of population density even when sample size is small (Broekhuis and Gopalaswamy, 2016). Similarly, statistical advances in the use of detection/nondetection data obtained from track surveys make it more logistically feasible to obtain occurrence data across vast landscapes (Hines et al., 2010; Midlane et al., 2014). Large-scale occupancy surveys are becoming widely used for quantifying cheetah occurrence and threats across less-known landscapes in Africa (Andresen et al., 2014).

These advances could provide new information that lends itself directly to spatial modeling techniques that can address key conservation questions. Perhaps the most urgent conservation questions for the cheetah involves the relationship between population status and the underlying forces that drive survival (i.e., fitness-related variables) discussed in the following sections.

CONSERVATION INSIGHT INTO FITNESS-RELATED VARIABLES

Following definitions of habitat, based on ecological niche theory, we recognize that any habitat model represents a *fitness landscape* (Mitchell and Hebblewhite, 2012). Perhaps the bottom line here is that cheetah population trajectory is largely driven by only a few variables, primarily: prey availability, interspecific competition with other carnivores, and human-caused mortality (Winterbach et al., 2013, 2014 also see Chapter 10). In the following sections we discuss these fitness-related variables and the potential application of spatial models that can provide direct and immediate insight into the consequences of cheetah management and conservation actions.

Prey Availability

Wild prey resources are a key fitness-related variable for cheetahs (Chapter 11). Vegetation communities, physical land structure, and land-cover classification layers have all been used as surrogates for estimating prey density and/or distribution. However, studies on other carnivores have shown that the predictive value of habitat surrogates for prey availability, are often weak or absent (McLoughlin et al., 2004). Indeed, assumptions of how forage availability relates to prey density can rarely be justified in landscapes where humans deplete prey, particularly in the context of the bushmeat crisis (Lindsey et al., 2013). By contrast, the predictive value between prey abundance and predator density have been more successful (Karanth et al., 2004). Thus, while some measures of prey are available, their use in habitat modeling should be critically evaluated.

Ability to access prey, including the influence of landscape structure (e.g., vegetative cover, edge density, and openness) on prey capture, is a component of this fitness-related variable. Cheetahs have a highly specialized rapid

pursuit hunting strategy (Wilson et al., 2013) and they require both cover for stalking and suitable terrain for short high-speed chases (Chapters 8 and 9).

We suggest that future modeling efforts that provide insight into prey and access to prey represent a high-priority need. For example, carefully designed landscape level occupancy surveys for cheetah prey species would elucidate key limiting factors for prey species (e.g., bush meat harvest or land-cover alterations), as well as interactions with human-caused mortality (another key fitness variable). Such information would lend itself directly to habitat connectivity modeling and conflict reduction strategies—all in a spatially-explicit manner.

Interspecific Competition: Lions and Hyenas

Lions (*Panthera leo*), spotted hyenas (*Crocuta crocuta*) and, to a lesser extent, leopards (*Panthera pardus*) are the cheetah's dominant competitors in Africa and restrict the cheetah's distribution and density (Durant, 2000; Mills and Mills, 2014). Interspecific competition is a primary driver of fitness within protected areas but generally weak or absent outside of protected areas where dominant competitors are largely absent (Durant, 2000). Spatial modeling studies are needed that provide new insight into protected area management that can balance the needs of cheetahs (and other subordinate carnivore species) with those of lions and spotted hyenas. For example, models that are used to design and evaluate new management interventions ranging from altering permanent water delivery to changing the ungulate/prey composition within a protected area could have a profound impact on cheetah numbers. Dynamic multispecies occupancy models provide a promising means of assessing influence of one species on extinction/colonization of another and of projecting associated management consequences (Yackulic et al., 2014).

Land Cover Conversion

Over the next decades, cheetah will be faced with ongoing changes in land cover as the African human population continues to develop and grow. Cheetahs can occur in a wide range of habitats and demonstrate flexibility in their diet (Hayward et al., 2006) and the cheetah's ability to adapt to land cover changes (e.g., conversion to cropland) is another important area of research.

General land cover categorizations (e.g., cropland) represent variation in underlying local factors, such as prey availability and amount of protection and there is a high degree of variability within the same land use categorizations across the cheetahs range. For example, a National Park with high tourism revenue is likely to afford greater protection for cheetahs and their prey than an underfunded one, and livestock grazing areas that employ herders and guarding dogs (Chapter 15) are likely to have less conflict than locations where these measures are not taken. Spatial modeling can help tear apart the relative impact of these factors, which may be critical to the design of conservation management, and models that combine occupancy and land-use dynamics can provide a useful insight (Miller et al., 2012).

Human Conflict

Human-caused mortality is clearly a key fitness-related variable (Winterbach et al., 2013), with factors including persecution and conflict with livestock production, accidental snaring, illegal trafficking, and road mortality (Chapters 5, 13 and 14). Persecution in defense of livestock may be reasonably estimated using surrogates, such as proximity to livestock areas but in many cases, more granular insights are needed for conservation management insight (e.g., livestock species, management practices, local knowledge, governance systems and so on).

Other human impacts, such as accidental snaring and illegal trafficking are more difficult to predict spatially and the underlying drivers (e.g.,

access to calories, economic conditions, global markets) have not been well studied. Primary data on locations of bushmeat snares and origins of trafficked cheetah skins could be utilized where available, but these data are largely absent, although genetic techniques hold some promise here.

Effective conflict reduction strategies have been recently advanced and spatial modeling could provide additional insights. Interview data can be incorporated into spatial analysis to better understand how cultural attitudes impact cheetah distributions (Thorn et al., 2011). Another recent study used countrywide aerial surveys in Botswana to predict landscape suitability for cheetah using prey and livestock as proxy for conflict and found that areas with at least 20% natural prey had below average conflict levels (Winterbach et al., 2015). Nevertheless, there is a lack of data on the underlying sociological drivers of human-caused cheetah mortality (Dickman, 2010) and additional research is sorely needed across the cheetah's range.

Fitness-Based Variables—Summary

Quantitative habitat models have been developed for tigers (*Panthera tigris*; Treves and Karanth, 2003) and wolves (*Canis lupus*; Mech, 1995). The conclusion for both of these studies is that wolf or tiger habitat could be essentially anywhere where there is sufficient prey and a level of human tolerance. A similar conclusion has been drawn for cheetahs (Winterbach et al., 2013, 2014, 2015; Chapter 10). New applications of analytical modeling should focus on providing quantitative insights that are relevant for management actions regarding the underlying drivers of cheetah survival—that is, fitness related variables.

CONCLUSIONS

Quantitative spatial analysis can be a powerful tool for informing management strategies and conservation management that benefit cheetahs.

Models are useful in that they compel us to formalize our hypotheses and organize the existing knowledge and highlight critical gaps in information. The best models for cheetahs will provide information necessary for effective conservation management by providing insight into the fitness-related variables that directly influence survival and reproduction. Moreover, the information can be displayed in a visually compelling format that is useful to managers, that is, maps. Maps can convene debate and inspire discussions and help us build a case for conservation among disparate parties. We hope that this summary will aid managers in assessing and utilizing new mapping and modeling techniques and help keep pace with the challenges that the cheetah faces in the decades to come.

References

Andreasen, A.M., Stewart, K.M., Longland, W.S., Beckmann, J.P., Forester, M.L., 2012. Identification of source-sink dynamics in mountain lions of the *Great Basin*. Molec. Ecol. 21, 5689–5701.

Andresen, L., Everatt, K.T., Somers, M.J., 2014. Use of site occupancy models for targeted monitoring of the cheetah: cheetah occupancy and detectability. J. Zool. 292, 212–220.

Battin, J., 2004. When good animals love bad habitats: ecological traps and the conservation of animal populations. Conserv. Biol. 18, 1482–1491.

Bennett, A.F., 1999. Linkages in the Landscape: The Role of Corridors and Connectivity in Wildlife Conservation. IUCN, Switzerland.

Broekhuis, F., Gopalaswamy, A.M., 2016. Counting Cats: spatially explicit population estimates of cheetah (*Acinonyx jubatus*) using unstructured sampling data. PloS One 11, e0153875.

Bunn, A., Urban, D., Keitt, T., 2000. Landscape connectivity: a conservation application of graph theory. J. Environ. Manage 59, 265–278.

Chetkiewicz, C.-L.B., St. Clair, C.C., Boyce, M.S., 2006. Corridors for conservation: integrating pattern and process. Annu. Rev. Ecol. Evol. Syst. 37, 317–342.

Crooks, K.R., Sanjayan, M., 2006. Connectivity Conservation. Cambridge University Press, Cambridge.

Dickinson, J.L., Shirk, J., Bonter, D., Bonney, R., Crain, R.L., Martin, J., Phillips, T., Purcell, K., 2012. The current state of citizen science as a tool for ecological research and public engagement. Front. Ecol. Environ. 10, 291–297.

Dickman, A., 2010. Complexities of conflict: the importance of considering social factors for effectively resolving human–wildlife conflict. Anim. Conserv. 13, 458–466.

Durant, S.M., 2000. Living with the enemy: avoidance of hyenas and lions by cheetahs in the Serengeti. Behav. Ecol. 11, 624–632.

Durant, S., Mitchell, N., Groom, R., Pettorelli, N., Ipavec, A., Jacobson, A., Woodroffe, R., Bohm, M., Hunter, L., Becker, M., Broekhuis, F., Bashir, S., Andresen, L., Aschenborn, O., Beddiaf, M., Belbachir, F., Belbachir-Bazi, A., Berbash, A., Brandao de Matos Machado, I., Breitenmoser, C., Chege, M., Cilliers, D., Davies-Mostert, H.T., Dickman, A.J., Ezekiel, F., Farhadinia, M.S., Funston, P.J., Henschel, P., Horgan, J., Hans de longh, H., Houman, J., Klein, R., Lindsey, P.A., Marker, L.L., Marnewick, K., Melzheimer, J., Merkle, J., M'soka, J., Msuha, M., O'Neil, H., Parker, M., Purchase, G., Sahailou, S., Saidu, Y., Samna, A., Schmidt-Küntzel, A., Selebatso, M., Sogbohossou, E., Soultan, A., Stone, E., Van der Meer, E., Van Vuuren, R., Wykstra, M., Young-Overton, K., 2017. Disappearing spots, the global decline of cheetah and what it means for conservation. Proc. Natl. Acad. Sci. USA 114, 528–533.

Foppen, R.P., Chardon, J.P., Lietveld, W., 2000. Understanding the role of sink patches in source-sink metapopulations: Reed Warbler in an agricultural landscape. Conserv. Biol. 14, 1881–1892.

Franklin, J., Miller, J.A., 2009. Mapping Species Distributions: Spatial Inference and Prediction. Cambridge University Press, Cambridge; New York.

Guisan, A., Zimmermann, N.E., 2000. Predictive habitat distribution models in ecology. Ecol. Model. 135, 147–186.

Hayward, M.W., Hofmeyr, M., O'Brien, J., Kerley, G.I.H., 2006. Prey preferences of the cheetah (*Acinonyx jubatus*) (Felidae: Carnivora): morphological limitations or the need to capture rapidly consumable prey before kleptoparasites arrive? J. Zool. 270, 615–627.

Hines, J., Nichols, J., Royle, J., MacKenzie, D., Gopalaswamy, A., Kumar, N., Karanth, K., 2010. Tigers on trails: occupancy modeling for cluster sampling. Ecol. Appl. 20, 1456–1466.

Houser, A.M., Somers, M.J., Boast, L.K., 2009. Spoor density as a measure of true density of a known population of free-ranging wild cheetah in Botswana. J. Zool. 278, 108–115.

Karanth, K.U., Nichols, J.D., Kumar, N.S., Link, W.A., Hines, J.E., 2004. Tigers and their prey: predicting carnivore densities from prey abundance. PNAS 101 (14), 4854–4858.

Lindenmayer, D.B., Barton, P.S., Lane, P.W., Westgate, M.J., McBurney, L., Blair, D., Gibbons, P., Likens, G.E., 2014. An empirical assessment and comparison of species-based and habitat-based surrogates: a case study of forest vertebrates and large old trees. PLoS One 9, e89807.

Lindsey, P.A., Balme, G., Becker, M., Begg, C., Bento, C., Bocchino, C., Dickman, A., Diggle, R.W., Eves, H., Henschel, P., Lewis, D., Marnewick, K., Mattheus, J., McNutt, W., McRobb, R., Midlane, N., Milanz, J., Morley, R., Murphree, M., Opyene, V., Phadima, J., Purchase, G., Rentsch, D., Roche, C., Shaw, J., Van der Westhuizen, H., Van Vliet, N., Zisadza-Gandiwa, P., 2013. The bushmeat trade in African savannas: Impacts, drivers, and possible solutions. Biol. Conserv. 160, 80–96.

MacKenzie, D.I., 2006. Modeling the probability of resource use: the effect of, and dealing with, detecting a species imperfectly. J. Wildl. Manage. 70, 367–374.

MacKenzie, D.I., Nichols, J.D., Royle, J.A., Pollock, K.H., Bailey, L.L., Hines, J.E., 2006. Occupancy Estimation And Modeling: Inferring Patterns and Dynamics of Species. Elsevier, Amsterdam; Boston.

Manly, B., McDonald, L., Thomas, D., McDonald, T.L., Erickson, W.P., 2007. Resource Selection by Animals: Statistical Design and Analysis for Field Studies. Springer Science & Business Media, Dordrecht, Netherlands.

Marker, L., Dickman, A., Mills, M.G., Macdonald, D., 2003. Aspects of the management of cheetahs, Acinonyx jubatus jubatus, trapped on Namibian farmlands. Biol. Conserv. 114, 401–412.

Mateo-Sánchez, M.C., Balkenhol, N., Cushman, S., Pérez, T., Domínguez, A., Saura, S., 2015. Estimating effective landscape distances and movement corridors: comparison of habitat and genetic data. Ecosphere 6, 1–16.

McLoughlin, P.D., Walton, L.R., Cluff, H.D., Paquet, P.C., Ramsay, M.A., 2004. Hierarchical habitat selection by tundra wolves. J. Mammal. 85, 576–580.

Mech, L.D., 1995. The challenge and opportunity of recovering wolf populations. Conserv. Biol. 9, 270–278.

Midlane, N., O'Riain, M.J., Balme, G., Robinson, H.S., Hunter, L.T.B., 2014. On tracks: a spoor-based occupancy survey of lion Panthera leo distribution in Kafue National Park, Zambia. Biol. Conserv. 172, 101–108.

Miller, D.A.W., Brehme, C.S., Hines, J.E., Nichols, J.D., Fisher, R.N., 2012. Joint estimation of habitat dynamics and species interactions: disturbance reduces co-occurrence of non-native predators with an endangered toad. J. Anim. Ecol. 81(6), 1288–1297.

Mills, M.G.L., Mills, M.E.J., 2014. Cheetah cub survival revisited: a re-evaluation of the role of predation, especially by lions, and implications for conservation: cheetah cub survival and predation. J. Zool. 292, 136–141.

Minor, E.S., Urban, D.L., 2008. A graph-theory framework for evaluating landscape connectivity and conservation planning. Conserv. Biol 22, 297–307.

Mitchell, M.S., Hebblewhite, M., 2012. Carnivore habitat ecology: integrating theory and application. In: Boitani, L., Powell, R.A. (Eds.), Carnivore Ecology Conservation Handbook of the Techniques. Oxford University Press, London, United Kingdom, pp. 218–255.

O'Connell, A.F., Nichols, J.D., Karanth, K.U., 2011. Camera traps in animal ecology: methods and analyses. Springer, Tokyo.

Railsback, S.F., Stauffer, H.B., Harvey, B.C., 2003. What can habitat preference models tell us? Tests using a virtual trout population. Ecol. Appl. 13, 1580–1594.

Reed, J.M., Mills, L.S., Dunning, J.B., Menges, E.S., McKelvey, K.S., Frye, R., Beissinger, S.R., Anstett, M.-C., Miller, P., 2002. Emerging issues in population viability analysis. Conserv. Biol. 16, 7–19.

Roloff, G., Haufler, J., Scott, J., Heglund, P., Morrison, M., Haufler, J., Raphael, M., Wall, W., Samson, F., 2002. Modeling habitat-based viability from organism to population. Island Press, Washington, DC, USA, pp. 673–686.

Sanderlin, J.S., Waser, P.M., Hines, J.E., Nichols, J.D., 2012. On valuing patches: estimating contributions to metapopulation growth with reverse-time capture-recapture modelling. Proc. R. Soc. B Biol. Sci. 279, 480–488.

Taylor, B.L., Wade, P.R., Ramakrishnan, U., Gilpin, M., Akcakaya, H.R., 2002. Incorporating uncertainty in population viability analyses for the purpose of classifying species by risk. University Chicago Press, Chicago, pp. 239–252.

Thorn, M., Green, M., Keith, M., Marnewick, K., Bateman, P.W., Cameron, E.Z., Scott, D.M., 2011. Large-scale distribution patterns of carnivores in northern South Africa: implications for conservation and monitoring. Oryx 45, 579–586.

Treves, A., Karanth, K.U., 2003. Human-carnivore conflict and perspectives on carnivore management worldwide. Conserv. Biol. 17, 1491–1499.

Uusitalo, L., 2007. Advantages and challenges of Bayesian networks in environmental modelling. Ecol. Model. 203, 312–318.

van der Meer, E., Rasmussen, G.S.A., Fritz, H., 2015. Using an energetic cost-benefit approach to identify ecological traps: the case of the African wild dog. Anim. Conserv. 18, 359–366.

Wilson, A.M., Lowe, J., Roskilly, K., Hudson, P.E., Golabek, K., McNutt, J., 2013. Locomotion dynamics of hunting in wild cheetahs. Nature 498, 185–189.

Winterbach, H.E.K., Winterbach, C.W., Boast, L.K., Klein, R., Somers, M.J., 2015. Relative availability of natural prey versus livestock predicts landscape suitability for cheetahs Acinonyx jubatus in Botswana. PeerJ 3, e1033.

Winterbach, H.E.K., Winterbach, C.W., Somers, M.J., 2014. Landscape suitability in Botswana for the conservation of its six large African carnivores. PLoS One 9, e100202.

Winterbach, H.E.K., Winterbach, C.W., Somers, M.J., Hayward, M.W., 2013. Key factors and related principles in the conservation of large African carnivores: factors and principles in carnivore conservation. Mammal Rev. 43, 89–110.

Yackulic, C.B., Reid, J., Nichols, J.D., Hines, J.E., Davis, R., Forsman, E., 2014. The roles of competition and habitat in the dynamics of populations and species distributions. Ecology 95, 265–279.

Yamada, K., Elith, J., McCarthy, M., Zerger, A., 2003. Eliciting and integrating expert knowledge for wildlife habitat modelling. Ecol. Model. 165, 251–264.

Zeller, K.A., McGarigal, K., Whiteley, A.R., 2012. Estimating landscape resistance to movement: a review. Landsc. Ecol. 27, 777–797.

Now You See Them, Soon You Won't: Statistical and Mathematical Models for Cheetah Conservation Management

Sandra Johnson*, Bogdan Cristescu**,
Jacqueline T. Davis[†], Douglas W. Johnson[‡],
Kerrie Mengersen*

*Queensland University of Technology, Brisbane, QLD, Australia
**Cheetah Conservation Fund, Otjiwarongo, Namibia
[†]University of Cambridge, Cambridge, United Kingdom
[‡]University of Queensland, Brisbane, QLD, Australia

INTRODUCTION

The cheetah is a priority species for conservation planning due to its documented decline and IUCN vulnerable status (Chapters 4 and 5). To become effective, conservation management of cheetahs should be informed by rigorous quantitative methodologies. Statistical and mathematical modeling techniques enable estimation of population parameters and population trajectories, identification of geographic regions of importance to the species, assessment of potentially adverse situations and ecological thresholds, and optimization of conservation decisions, such as translocations and reintroductions (Moilanen et al., 2009). The power of these techniques could potentially allow for much more effective cheetah conservation planning and management.

Similar to many carnivores, observational datasets are generally sparse for cheetahs. Reasons include the cheetah's cryptic behavior, low densities, and large home ranges (Chapter 8). Nonetheless, traditional observational data are increasingly complemented by digital images obtained from camera traps

(Chapter 29), biologging (Chapter 32), satellite data, and aerial surveys from manned and unmanned aircraft (Gonzalez et al., 2016). Therefore, a lack of data should not necessarily be an obstacle to using statistical and mathematical modeling. However, the variety of analytical methods combined with the rapid growth in diversity and complexity of software platforms and development environments can be an intimidating environment for the cheetah conservationist. There is typically a substantial learning curve to choose, acquire, set up, and use modeling software, which can contribute to a lack of understanding about how these methods can inform conservation and management decisions (Moilanen et al., 2009).

In this chapter, we review key statistical and mathematical modeling techniques for use in cheetah conservation management, to aid future endeavors in this field. Our review is guided by an examination of research literature in the ISI Web of Science. Such reviews of digital publication repositories have been used to summarize the state of knowledge on specific conservation topics (Driscoll et al., 2014) and to better align ecological research with conservation needs (Cristescu and Boyce, 2013).

MODELING APPROACHES FOR CHEETAH CONSERVATION

A wide range of mathematical and statistical methods are used in biological conservation, and many of these are suited to cheetah conservation. A survey of the ISI Web of Science online database in October 2016, using the search terms [(*mathematical OR statistical*) *AND* (*ecology OR conservation*) *AND model**] revealed 6881 publications. A word cloud constructed from abstracts of the 200 most cited publications, is shown in Fig. 37.1.

Words, such as "model(s)" and "statistical" are the search terms used to target these publications, and are of less interest. The word cloud in conjunction with collocated terms indicate that species distribution, population, and regression are recurrent themes. Some other concepts worth noting are spatial representation, identification of variables, patterns, change, community, and diversity.

We selected six modeling techniques to review in more detail, on the basis of the relative popularity of these methods in conservation biology, while ensuring that classic models are complemented by some newly emerging techniques. The total number of publications for each tech-

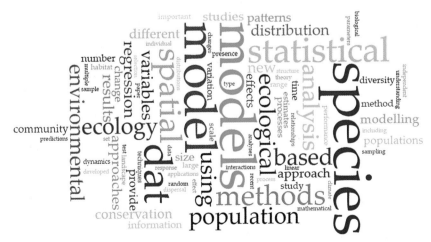

FIGURE 37.1 **Word cloud of abstracts from the 200 most cited publications containing the search words [(*mathematical OR statistical*) *AND* (*ecology OR conservation*) *AND model**] in ISI Web of Science.**

nique were: regression (5827), followed by, occupancy modeling (1280), population viability analysis (719), Marxan (201), Bayesian networks (166), and stochastic dynamic programming (73). There is a degree of overlap in the publication counts, because more than one modeling technique may have been described in a single publication.

When the ISI Web of Science database search described earlier was restricted to relevant mathematical or statistical modeling studies of cheetahs, a total of 50 publications were found using the search terms [*cheetah* AND model* AND (conservation OR population)*]. This is a relatively small number of publications compared to using the same search terms and substituting "cheetah" for another large carnivore; for example, we found 344 for "lion" (sea lion excluded) and 288 for "leopard." This brief review of quantitative research effort on cheetah highlights a great opportunity to use some of the approaches discussed in this chapter for cheetah conservation.

REVIEW OF STATISTICAL AND MATHEMATICAL METHODS

We present a review of common analytical perspectives, how these are related to each other, and how they can be used for particular conservation aims. Mathematical and statistical methods can be characterized according to their conservation aims and the analytical focus (Fig. 37.2), the available data (Ferson and Burgman, 2000), and also in terms of the key modeling objective, and the degree to which they are driven by processes or data (Fig. 37.3).

Conservation Aim and Analytical Focus

Mathematical and statistical techniques require a clear statement of the conservation aim (Fig. 37.2). This overarching aim is typically the culmination of a series of smaller steps, with each step having a more focused conservation aim. For example, if the conservation aim is to ensure the viability of a cheetah population, the first step may be to identify the key areas to conserve (A1), followed by estimating the current population, distribution, and the trajectory within the identified key areas (A2). Thereafter it would be important to identify any adverse situations that may impact the target cheetah population and the thresholds at which this population becomes unviable (A3). Using the insights gained from the previous steps, the ultimate aim would be to optimize the conservation management decisions and strategies (A4).

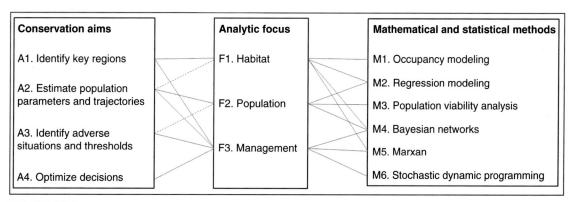

FIGURE 37.2 Connections between the three frameworks describing conservation aims, analytic focus, and mathematical and statistical methods reviewed in this chapter.

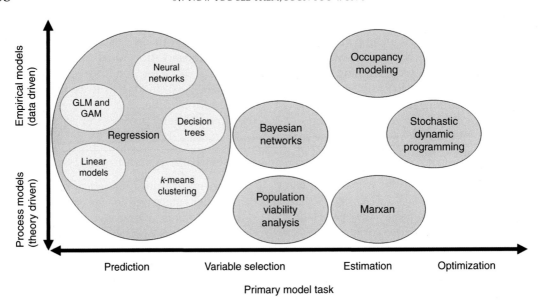

FIGURE 37.3   **Graphical representation of the methods being reviewed, with respect to the primary aim of the technique (x-axis) and model characteristics (y-axis).** Note: this is a general framework useful to characterize the methods reviewed in this chapter, and is not comprehensive. Moreover, methods are typically a blend of several modeling tasks. *GAM*, Generalized additive models; *GLM*; generalized linear models.

These conservation questions broadly fall within three analytic contexts: *habitat-focused*, *population-focused*, and *management-focused*, and often a combination of all three. Habitat-focused models (F1) may, for example, aim to map cheetah habitat and determine important environmental variables influencing cheetah distribution, as was done using species distribution models and occupancy estimation models in Kenya and Namibia (Kuloba et al., 2015; Mwendera, 2015). These and other habitat-focused models can be undertaken using the popular software MaxEnt (Phillips et al., 2006). Population-focused modeling (F2) ranges from description of metapopulations to population abundance estimation. For example, this focus was taken to underpin a cheetah metapopulation project in South Africa (Akçakaya et al., 2007) and to obtain spatial population density estimates and density "hotspots" of cheetah distribution in the Maasai Mara, Kenya (Broekhuis and Gopalaswamy, 2016). Finally, management-

focused modeling (F3) may evaluate a management technique as was done during a cheetah population habitat viability analysis carried out by Lindsey et al. (2009) in South Africa, and for a cheetah case study to evaluate adaptive management methods (Canessa et al., 2016).

Process- and Empirical-Driven Methods

Statistical and mathematical modeling methods range from those that are based strongly on *process* to those that are strongly *empirical*. Process-based approaches are theoretical representations of an underlying biological and physical process, while empirical approaches are based first on representations of the data, and are then evaluated in light of their informativeness about the processes.

Very broadly, mostly process-driven mathematical models have been used in the context of cheetah conservation management; for example, Lubben et al. (2008) extended matrix population models to population projection matrices to assess survival of the Serengeti cheetah. Empirical-focused approaches are typically used for machine-learning models, which use data from known scenarios to construct a model which can then infer the outcome of other scenarios. Statistical models use a combination of process-driven and empirical approaches. For example, Grünewälder et al. (2012) employed support vector machines and hidden Markov models to describe the behavior of cheetahs in Botswana using movement data. Increasingly, mathematical, statistical, and machine learning methods are being blended to provide more informative data-based descriptions of the processes of interest.

Modeling Tasks

Modeling techniques can also be viewed in terms of the primary modeling tasks they perform (Fig. 37.3). Four key modeling tasks in cheetah conservation are *prediction, variable selection, estimation,* and *optimization*. Prediction tasks aim to predict responses inferred from observed data. For example, the future size of a cheetah population 5-years later might be predicted from an existing dataset. Variable selection tasks aim to identify the most important variables that influence the problem or management task, and includes activities, such as clustering (grouping similar variables together, without relation to a response) and dimension reduction (reduce the number of variables to only the most important variables). For example, clustering might identify characteristics that can collectively describe groups of individual cheetahs within a population, and dimension reduction might identify the most important spatial variables in a habitat model and eliminate those variables that have a smaller influence on the result. Estimation tasks aim to estimate parameters, such as population size, within the spatial, temporal, and covariate domain of the observed data. For example, the abundance of a particular cheetah population

might be estimated. Finally, optimization tasks aim to find the optimal solution given a set of constraints within which the problem needs to be solved. For example, the optimal set of management actions to facilitate the growth or establishment of a cheetah population in a particular area. The choice of task is important because it determines which statistical or mathematical method will best address the conservation focus (Fig. 37.3).

REVIEW OF SPECIFIC METHODS

The six selected modeling techniques are reviewed in more detail here, and an integrated framework is discussed in section "Integrating the Frameworks."

Bayesian Networks

A Bayesian network (BN) is a graphical probabilistic model with nodes representing factors of interest in the system, and arrows between the nodes to represent dependencies between them. Once quantified with the available information the BN can be used for estimation, scenario analysis, and optimization. Model extensions include dynamic Bayesian networks (DBNs), to describe time-varying systems and object oriented Bayesian networks (OOBNs), in which related nodes are grouped into submodels. BNs can model complex processes and systems, engage diverse stakeholders in model development and quantification, integrate qualitative and quantitative data, and can be applied when data are scarce or plentiful (Aguilera et al., 2011; Barton et al., 2012; Johnson and Mengersen, 2012; Uusitalo, 2007). They can be applied as habitat-focused, population-focused, or management-focused approaches. For example, Johnson et al. (2013) used OOBN modeling to integrate subnetworks of ecological, biology, and human factors in an assessment of the viability of the free-ranging cheetah population in Namibia.

Marxan

Marxan is an optimization tool that is used to define areas that meet conservation goals with minimum cost. The method is well established in the domain of reserve planning (Watts et al., 2009). Marxan is specifically targeted to management focused questions. For example, Chaminuka (2012) employed Marxan in a comprehensive evaluation of multiple land use options in transfrontier conservation areas in South Africa, with a particular focus on the interface between wildlife and livestock. Advantages of Marxan software include its ability to incorporate spatial planning, capture a variety of activities in different geographic areas, and include multiple objectives (Carwardine et al., 2010; Watts et al., 2009). Alternatives to Marxan have also been proposed, for example, the Zonation framework can prioritize areas in a landscape according to their biodiversity features (Lehtomäki and Moilanen, 2013).

Occupancy Modeling

Occupancy modeling is typically an estimation task designed to estimate the probability of presence of a species across a spatial and temporal domain (MacKenzie et al., 2006), with a particular focus on identifying false absences: where a species is present but undetected. Occupancy modeling is popular for population focused questions, particularly in combination with camera or track surveys (Chapters 29 and 30). For example, Andresen et al. (2014) evaluated the use of occupancy modeling in combination with replicate camera trap and track surveys to estimate cheetah occurrence in the Limpopo National Park, Mozambique. If detectability were perfect, occupancy models would provide similar information to habitat suitability and ecological niche models, although the latter are often more process focused. Occupancy modeling is discussed in more detail in Chapter 36.

Population Viability Analysis

PVA is arguably one of the most established techniques in conservation biology. PVA focuses on estimating the risk of extinction, or conversely the health of a population. It is flexible and can be used for habitat, population, and management focused questions. It is primarily a process-based estimation task, and can be used to estimate population patterns and trends, determine which input variables have the largest effect on the outputs while accounting for other factors, and predict the impact of changes in inputs, such as climate or management interventions on extinction risk. This technique was used by Kelly and Durant (2000) to study the viability of the Serengeti cheetah population. PVA is discussed in more detail in Chapter 38.

Regression

Regression is a well-known statistical modeling approach that is well suited to model many cheetah conservation problems or objectives. Regression techniques aim to estimate or predict one or more responses (also known as outputs or target variables), given a set of covariates (also known as inputs, features, or predictors). In addition to predicting responses, the regression model can identify which covariates are most important in providing good estimates or predictions. Specializations of regression modeling abound and include among others generalized linear (GLM) and additive (GAM) models, maxEnt, as well as machine learning techniques, such as k-nearest neighbors, neural networks, and decision trees.

The following list provides examples of the use of different types of regression for cheetah conservation:

- *Logistic regression*: Gros and Rejmánek (1999) constructed a habitat suitability model for cheetahs in Uganda based on characteristics extracted from a vegetation map; Durant et al. (2004) identified environmental and social factors affecting reproductive success and estimated survival time; Muntifering et al. (2006) discriminated between high and low use areas by cheetahs based on radio-telemetry data; and Lindsey et al. (2013) examined land use characteristics and socioeconomic conditions that influenced farmers' tolerance toward large carnivores, including cheetahs, in Namibia.

- *Linear regression*: Walker et al. (2016) aimed to identify physical characteristics that influenced the preferences of cheetahs for scent-marking trees.
- *Generalized linear models*: zero-inflated (Royle et al., 2007) and hurdle models (Marnewick et al., 2006) have been used to allow for excess zeros in presence–absence and count models; Swanson et al. (2016) employed hurdle models and time-to-event modeling of camera trap data in Tanzania to evaluate patterns of coexistence between cheetahs and other big cats.
- *Nearest neighbor estimation*: This technique was used by Bidder et al. (2014) to classify animal behavior using accelerometer data, and by Marnewick and Somers (2015) to identify home ranges of cheetahs outside protected areas in South Africa.
- *Neural networks*: Dalziel et al. (2008) used neural networks to analyze animal movement trajectories.
- *Decision trees, random forests, boosted regression trees*: Elith et al. (2008) provide a seminal application of boosted regression trees in ecology. This technique has not yet been widely applied to cheetah data, but it has been used for orangutan conservation to combine interview, Geographic Information System (GIS) and media data to obtain spatial quantification of abundance (Meijaard et al., 2011), killing (Davis et al., 2013), and perceived species' threats and population trends (Abram et al., 2015).

Stochastic Dynamic Modeling

Stochastic programming aims to determine an optimal set of decisions, acknowledging complexity, and uncertainty in the associated processes. Conservation problems are represented as a set of decisions with a formally defined relationship, in which all of the possible states of the problem and solution are identified. Fuller et al. (2008) used stochastic programming to select the most cost-effective set of sites for biodiversity conservation. An extension to this modeling approach is stochastic dynamic programming, which allows for sets of decisions to be made over time (Marescot et al., 2013). Chadès et al. (2014) released a software toolbox for stochastic dynamic programing. Variations, extensions, and alternatives to stochastic programming approaches have also been proposed; for example, Nicol and Chadès (2011) suggest a more scalable approach that might be more applicable to cheetah conservation.

INTEGRATING THE FRAMEWORKS

A canonical example of the interplay between the frameworks described in Figs. 37.2 and 37.3 is a species recovery plan. For example, one of the stated objectives in the Rangewide Conservation Planning Process for Cheetah and African Wild Dog is "to collate information on wild dog and cheetah distribution and abundance on an ongoing basis, in order to direct conservation efforts and evaluate the success or failure of these efforts in future years" (IUCN and SSC, 2007). Mathematical and statistical modeling and analysis of the collated data can support habitat-focused analyses by identifying regions critical to the survival of the species (e.g., through occupancy models); population-focused analyses by providing population estimates and predicting population trajectories (e.g., through regression models) or evaluating the effect of social and economic impacts on populations (e.g., through

BNs); and management-focused analyses by identifying adverse situations (e.g., through PVA), defining areas that meet conservation goals (e.g., through Marxan), or optimizing conservation decisions (e.g., through stochastic dynamic programming). The potential and actual successes of the conservation efforts arising from the species recovery plan can also be statistically evaluated.

DISCUSSION

In this chapter we briefly outlined conservation perspectives, modeling objectives, and a selection of techniques that are well suited to cheetah conservation. However, single techniques are typically no panacea and there is rarely one "correct" choice. A pragmatic strategy is to integrate different modeling approaches that complement each other, compensate for shortcomings or overly restrictive assumptions and collectively provide meaningful insights and predictions based on the available information (Guisan and Zimmermann, 2000). We encourage the reader to explore different modeling techniques, but to be wary of inadvertently using the methods inappropriately (due to model assumptions not being satisfied or other modeling considerations), and to present the model, interpret output, and draw conclusions carefully and transparently (Reed et al., 2002). Established, trusted software, such as the statistical programming language R (R Core Team, 2014), has many packages available to ease the implementation of different statistical methods, and good textbooks are available to introduce R. For example, Bolker (2008) is suited to ecological modeling and Logan (2010) to biostatistical modeling.

When choosing a modeling technique, "classic" methods should not automatically be discarded because of past or perceived limitations. For example, modern regression models have the advantage of strong mathematical foundations that enhance evidence-based inference, as well

as advanced features of flexibility, robustness, and computation that may not be sufficiently appreciated or utilized in practice. Traditional robust modeling approaches allow comparisons across regions and a global perspective which may often be preferable to overly complex site-specific models. However, while simpler models are generally more desirable than complicated approaches, and reproducibility of results is certainly essential, these should not be used as an "excuse" for poor modeling choices when more complex models are required. An overriding strategy is to always be transparent with data cleansing, modeling inputs, and assumptions. This ensures that model, data, and distribution assumptions are understood by others, and hence enables the refinement and improvement of the approach used, especially as data collection and analysis capabilities improve and new modeling techniques emerge.

Although traditional refereed journals remain valuable sources of authoritative information, faster and more open communication forums can be invaluable and dissemination of "what not to do" can contribute substantially to the knowledge base. For this reason, we encourage the exploration of new methods and the reporting of both good and poor outcomes.

It has long been acknowledged that expert knowledge is important in creating more robust and biologically sensible models, identifying and quantifying human-related influences, and in potentially reducing uncertainty in model predictions. Methods for eliciting this knowledge and incorporating it in mathematical and statistical models remain a topic of active research (Martin et al., 2012). Human-related influences are at the core of many cheetah conservation challenges and should be incorporated in the modeling process. Moreover, integration of ecological, statistical, mathematical, and social science skillsets and collaboration between experts across disciplines are likely to deliver more effective solutions to complex questions and achieve conservation objectives.

Finally, statistical and mathematical modeling of cheetahs has been surprisingly underutilized despite the species' IUCN Vulnerable status. These tools are invaluable to cheetah conservation managers who have to operate in a highly uncertain set of circumstances. Modeling can allow for early recognition of failed or suboptimal solutions and thus lead to more effective interventions which can ameliorate otherwise potentially disastrous outcomes. It is hence critical that the consequences of modeling-derived management actions are monitored to assess whether they are consistent with predicted outcomes, or whether there are some unintended consequences or knock on effects (Chauvenet et al., 2011).

References

Abram, N.K., Meijaard, E., Wells, J.A., Ancrenaz, M., Pellier, A.S., Runting, R.K., Gaveau, D., Wich, S., Nardiyono, Tjiu, A., Nurcahyo, A., Mengersen, K., 2015. Mapping perceptions of species' threats and population trends to inform conservation efforts: the Bornean orangutan case study. Divers. Distrib. 21, 487–499.

Aguilera, P.A., Fernández, A., Fernández, R., Rumí, R., Salmerón, A., 2011. Bayesian networks in environmental modelling. Environ. Model. Softw. 26, 1376–1388.

Akçakaya, H.R., Mills, G., Doncaster, C.P., 2007. The role of metapopulations in conservation. In: Macdonald, D.W., Service, K. (Eds.), Key Topics in Conservation Biology. Blackwell Publishing by TJ International, Padstow, UK.

Andresen, L., Everatt, K.T., Somers, M.J., 2014. Use of site occupancy models for targeted monitoring of the cheetah. J. Zool. 292, 212–220.

Barton, D.N., Kuikka, S., Varis, O., Uusitalo, L., Henriksen, H.J., Borsuk, M., de la Hera, A., Farmani, R., Johnson, S., Linnell, J.D.C., 2012. Bayesian networks in environmental and resource management. Integr. Environ. Assess. Manage. 8, 418–429.

Bidder, O.R., Campbell, H.A., Gómez-Laich, A., Urgé, P., Walker, J., Cai, Y., Gao, L., Quintana, F., Wilson, R.P., 2014. Love thy neighbour: automatic animal behavioural classification of acceleration data using the K-nearest neighbour algorithm. PLoS One 9, e88609.

Bolker, B.M., 2008. Ecological Models and Data in R. Princeton University Press, Princeton.

Broekhuis, F., Gopalaswamy, A.M., 2016. Counting cats: spatially explicit population estimates of cheetah (*Acinonyx jubatus*) using unstructured sampling data. PLoS One 11, e0153875.

Canessa, S., Guillera-Arroita, G., Lahoz-Monfort, J.J., Southwell, D.M., Armstrong, D.P., Chades, I., Lacy, R.C., Converse, S.J., 2016. Adaptive management for improving species conservation across the captive-wild spectrum. Biol. Conserv. 199, 123–131.

Carwardine, J., Wilson, K.A., Hajkowicz, S.A., Smith, R.J., Klein, C.J., Watts, M., Possingham, H.P., 2010. Conservation planning when costs are uncertain. Conserv. Biol. 24, 1529–1537.

Chadès, I., Chapron, G., Cros, M.J., Garcia, F., Sabbadin, R., 2014. MDPtoolbox: a multi-platform toolbox to solve stochastic dynamic programming problems. Ecography 37, 916–920.

Chaminuka, P., 2012. Evaluating Land Use Options at the Wildlife/Livestock Interface: An Integrated Spatial Land Use Analysis. PhD thesis, Wageningen University, Netherlands.

Chauvenet, A.L.M., Durant, S.M., Hilborn, R., Pettorelli, N., 2011. Unintended consequences of conservation actions: managing disease in complex ecosystems. PLoS One 6, e28671.

Cristescu, B., Boyce, M.S., 2013. Focusing ecological research for conservation. Ambio 42, 805–815.

Dalziel, B.D., Morales, J.M., Fryxell, J.M., 2008. Nicolas, P., Michael, C.W. (Eds.), Fitting Probability Distributions to Animal Movement Trajectories: Using Artificial Neural Networks to Link Distance, Resources, and Memory, vol. 172, The University of Chicago Press for The American Society of Naturalists, United States, pp. 248–258.

Davis, J.T., Mengersen, K., Abram, N.K., Ancrenaz, M., Wells, J.A., Meijaard, E., 2013. It's not just conflict that motivates killing of orangutans. PloS One 8, e75373.

Driscoll, D.A., Banks, S.C., Barton, P.S., Ikin, K., Lentini, P., Lindenmayer, D.B., Smith, A.L., Berry, L.E., Burns, E.L., Edworthy, A., Evans, M.J., Gibson, R., Heinsohn, R., Howland, B., Kay, G., Munro, N., Scheele, B.C., Stirnemann, I., Stojanovic, D., Sweaney, N., Villaseñor, N.R., Westgate, M.J., 2014. The trajectory of dispersal research in conservation biology. systematic review. PLoS One 9, e95053.

Durant, S.M., Kelly, M., Caro, T.M., 2004. Factors affecting life and death in Serengeti cheetahs: environment, age, and sociality. Behav. Ecol. 15, 11–22.

Elith, J., Leathwick, J.R., Hastie, T., 2008. A working guide to boosted regression trees. J. Anim. Ecol. 77, 802–813.

Ferson, S., Burgman, M., 2000. Quantitative Methods for Conservation Biology. Springer New York, New York, NY.

Fuller, T., Morton, D.P., Sarkar, S., 2008. Planning for Biodiversity Conservation Using Stochastic Programming. Birkhäuser, Boston, MA.

Gonzalez, L., Montes, G., Puig, E., Johnson, S., Mengersen, K., Gaston, K., 2016. Unmanned aerial vehicles (UAVs) and artificial intelligence revolutionizing wildlife monitoring and conservation. Sensors 16, 97.

Gros, P.M., Rejmánek, M., 1999. Status and habitat preferences of Uganda cheetahs: an attempt to predict carnivore occurrence based on vegetation structure. Biodivers. Conserv. 8, 1561–1583.

Grünewälder, S., Broekhuis, F., Macdonald, D.W., Wilson, A.M., McNutt, J.W., Shawe-Taylor, J., Hailes, S., 2012. Movement activity based classification of animal behaviour with an application to data from cheetah (*Acinonyx jubatus*). PLoS One 7, e49120.

Guisan, A., Zimmermann, N.E., 2000. Predictive habitat distribution models in ecology. Ecol. Model. 135, 147–186.

IUCN and SSC, 2007. Strategies and National Action Plans. Gland, Switzerland. Available from: http://cheetahandwilddog.org/regional-strategies-national-action-plans/.

Johnson, S., Marker, L., Mengersen, K., Gordon, C.H., Melzheimer, J., Schmidt-Küntzel, A., Nghikembua, M., Fabiano, E., Henghali, J., Wachter, B., 2013. Modeling the viability of the free-ranging cheetah population in Namibia: an object-oriented Bayesian network approach. Ecosphere 4, 1–19, art90.

Johnson, S., Mengersen, K., 2012. Integrated Bayesian network framework for modeling complex ecological issues. Integr. Environ. Assess. Manage. 8, 480–490.

Kelly, M.J., Durant, S.M., 2000. Viability of the Serengeti cheetah population. Conserv. Biol. 14, 786–797.

Kuloba, B.M., Van Gils, H., Van Duren, I., Muya, S.M., Ngene, S.M., 2015. Modeling *Cheetah Acinonyx jubatus* fundamental niche in Kenya. Int. J. Environ. Monit. Anal. 3, 317–330.

Lehtomäki, J., Moilanen, A., 2013. Methods and workflow for spatial conservation prioritization using Zonation. Environ. Model. Softw. 47, 128–137.

Lindsey, P.A., Havemann, C.P., Lines, R., Palazy, L., Price, A.E., Retief, T.A., Rhebergen, T., Waal, C.V., 2013. Determinants of persistence and tolerance of carnivores on Namibian ranches: implications for conservation on southern African private lands. PLoS One 8, e52458.

Lindsey, P.A., Marnewick, K., Davies-Mostert, H.T., Rehse, T., Mills, M.G.L., Brummer, R., Buk, K., Traylor-Holzer, K., Morrison, K., Mentzel, C., Daly, B., 2009. Population and Habitat Viability Assessment for Cheetahs in South Africa. IUCN Conservation Breeding Specialist Group and Endangered Wildlife Trust, Johannesburg, South Africa.

Logan, M., 2010. Biostatistical Design and Analysis Using R: A Practical Guide. Wiley-Blackwell, Hoboken, NJ; Chichester, West Sussex.

Lubben, J., Tenhumberg, B., Tyre, A., Rebarber, R., 2008. Management recommendations based on matrix projection models: the importance of considering biological limits. Biol. Conserv. 141, 517–523.

MacKenzie, D.I., Nichols, J.D., Royle, J.A., Pollock, K.H., Bailey, L.L., Hines, J.E., 2006. Occupancy Estimation and Modeling: Inferring Patterns and Dynamics of Species Occurrence. Elsevier/Academic Press, Burlington, MA, USA.

Marescot, L., Chapron, G., Chadès, I., Fackler, P.L., Duchamp, C., Marboutin, E., Gimenez, O., Freckleton, R., 2013. Complex decisions made simple: a primer on stochastic dynamic programming. Methods Ecol. Evol. 4, 872–884.

Marnewick, K.A., Bothma, J.D.P., Verdoorn, G.H., 2006. Using camera-trapping to investigate the use of a tree as a scent-marking post by cheetahs in the Thabazimbi district. S. Afr. J. Wildl. Res. 36, 139–145.

Marnewick, K., Somers, M.J., 2015. Home ranges of cheetahs (*Acinonyx jubatus*) outside protected areas in South Africa. S. Afr. J. Wildl. Res. 45, 223–232.

Martin, T.G., Burgman, M.A., Fidler, F., Kuhnert, P.M., Low-Choy, S., McBride, M., Mengersen, K., 2012. Eliciting expert knowledge in conservation science. Conserv. Biol. 26, 29–38.

Meijaard, E., Buchori, D., Hadiprakarsa, Y., Utami-Atmoko, S.S., Nurcahyo, A., Tjiu, A., Prasetyo, D., Nardiyono, Christie, L., Ancrenaz, M., Abadi, F., Antoni, I.N.G., Armayadi, D., Dinato, A., Ella, Gumelar, P., Indrawan, T.P., Kussaritano, Munajat, C., Priyono, C.W.P., Purwanto, Y., Puspitasari, D., Putra, M.S.W., Rahmat, A., Ramadani, H., Sammy, J., Siswanto, D., Syamsuri, M., Andayani, N., Wu, H., Wells, J.A., Mengersen, K., 2011. Quantifying killing of orangutans and human-orangutan conflict in Kalimantan, Indonesia. PLoS One 6, e27491.

Moilanen, A., Wilson, K.A., Possingham, H.P., 2009. Spatial Conservation Prioritization: Quantitative Methods and Computational Tools. Oxford University Press, Oxford.

Muntifering, J.R., Dickman, A.J., Perlow, L.M., Hruska, T., Ryan, P.G., Marker, L.L., Jeo, R.M., 2006. Managing the matrix for large carnivores: a novel approach and perspective from cheetah (*Acinonyx jubatus*) habitat suitability modelling. Anim. Conserv. 9, 103–112.

Mwendera, N.Y., 2015. Modelling the Distribution of the Cheetah (*Acinonyx jubatus*) in Namibia. MSc thesis, University of Twente, Netherlands.

Nicol, S., Chadès, I., 2011. Beyond stochastic dynamic programming: a heuristic sampling method for optimizing conservation decisions in very large state spaces. Methods Ecol. Evol. 2, 221–228.

Phillips, S.J., Anderson, R.P., Schapire, R.E., 2006. Maximum entropy modeling of species geographic distributions. Ecol. Model. 190, 231–259.

R Core Team, 2014. R: A Language and Environment for Statistical Computing. R Foundation for Statistical Computing, Vienna, Austria. Available from: http://www.R-project.org/.

Reed, J.M., Mills, L.S., Dunning, J.B., Menges, E.S., McKelvey, K.S., Frye, R., Beissinger, S.R., Anstett, M.-C., Miller, P., 2002. Emerging issues in population viability analysis. Conserv. Biol. 16, 7–19.

Royle, J.A., Dorazio, R.M., Link, W.A., 2007. Analysis of multinomial models with unknown index using data augmentation. J. Comput. Graph. Stat. 16, 67–85.

Swanson, A., Arnold, T., Kosmala, M., Forester, J., Packer, C., 2016. In the absence of a "landscape of fear": how lions, hyenas, and cheetahs coexist. Ecol. Evol. 6, 8534–8545.

Uusitalo, L., 2007. Advantages and challenges of Bayesian networks in environmental modelling. Ecol. Model. 203, 312–318.

Walker, E.H., Nghikembua, M., Bibles, B., Marker, L., 2016. Scent-post preference of free-ranging Namibian cheetahs. Glob. Ecol. Conserv. 8, 55–57.

Watts, M.E., Ball, I.R., Stewart, R.S., Klein, C.J., Wilson, K., Steinback, C., Lourival, R., Kircher, L., Possingham, H.P., 2009. Marxan with zones: software for optimal conservation based land- and sea-use zoning. Environ. Model. Softw. 24, 1513–1521.

A Review of Population Viability Analysis and its use in Cheetah Conservation

Bogdan Cristescu*, Anne Schmidt-Küntzel*,
Karin R. Schwartz*, Carl Traeholt**, Laurie Marker*,
Ezequiel Fabiano†, Kristin Leus‡, Kathy Traylor-Holzer§

*Cheetah Conservation Fund, Otjiwarongo, Namibia
**Copenhagen Zoo, Frederiksberg, Denmark
†University of Namibia, Katima Mulilo, Namibia
‡IUCN SSC Conservation Planning Specialist Group—Europe, Copenhagen Zoo,
Merksem, Belgium
§IUCN SSC Conservation Planning Specialist Group, Apple Valley,
MN, United States

USE OF PVA AS A CONSERVATION TOOL

Wildlife populations face a range of threats that can alter their survival rates and reproductive success, including habitat degradation, loss and fragmentation, off-take (e.g., hunting), and catastrophic events. As populations become smaller, they become increasingly vulnerable and sensitive to additional random (stochastic) processes. Such compounding effects on dwindling populations exacerbate the decline and can cause a spiral (vortex) toward extinction, at which point recovery is extremely difficult (Gilpin and Soule, 1986; Shaffer, 1987). Recent reviews of wild vertebrate species status have highlighted important numeric or range declines in many mammals; in the case of the cheetah (*Acinonyx jubatus*), both numbers and habitat have declined significantly in the last century (Durant et al., 2017; Chapters 4, 5, and 39). Assessing and managing the extinction risk of species like the cheetah, is essential for conservation success (Shaffer, 1990), especially in light of such alarming declines.

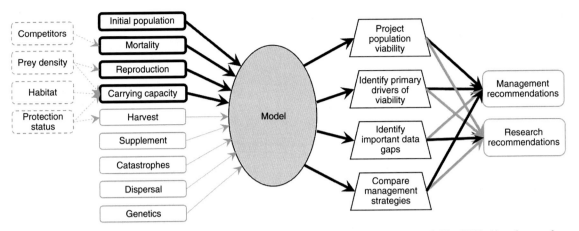

FIGURE 38.1   **PVA workflow.** Schematic including, to the *left* of the model, mandatory *(bold solid black)* and several optional *(solid gray)* input parameters, as well as examples of useful underlying factors that are modeled indirectly via input parameters *(dashed gray)*. Output options, as well as recommendations are provided to the right of the model. For detailed subcategories of inputs (where applicable) refer to Table 38.1.

The viability of a population can be assessed using a variety of measures, such as population trend, genetic diversity, and probability of population persistence over time. Population viability analysis (PVA) is an analytical modeling tool to project future population trends and estimate extinction risk, and can be used to evaluate various management interventions aimed at mitigating population declines (Fig. 38.1). PVA is commonly used in wildlife management and the International Union for Conservation of Nature (IUCN) Species Survival Commission (SSC) Conservation Planning Specialist Group (CPSG) has undertaken PVAs for a range of species, including cheetah, as part of a broader species conservation planning process. These PVAs and associated planning workshop reports are freely available from http://www.cpsg.org/document-repository.

In this chapter we provide a brief overview of the factors affecting PVA outcomes, review PVA applications to cheetah conservation, and discuss the importance of PVA as a tool for understanding cheetah population trajectories and informing conservation decisions.

FACTORS AFFECTING POPULATION PERSISTENCE

Certain factors affecting the survival of a population can be estimated and represented numerically; they are referred to as deterministic (e.g., maximum breeding age, harvest rate). Variation in survival and reproductive rates due to stochastic demographic, environmental, and genetic processes can also affect the decline/increase of a population (Boyce, 1992). All PVAs incorporate deterministic demographic factors such as rates of birth and death in the population (Fig. 38.1). PVAs also often incorporate (directly or indirectly) ecological conditions, such as habitat quality and size and food availability, as well as human impacts (e.g., harvesting or supplementation; Fig. 38.1). Less often, PVAs incorporate stochasticity, which simulates variation in these rates due to chance and annual fluctuations in the environment.

Stochastic variation has a disproportionately greater impact on small populations relative to large populations and can lead to decline (Lacy, 1987). Annual environmental fluctuations (e.g., due to disease, predation pressure, food availability, rainfall), even if localized, can significantly impact a population if the total area of habitat that the population occupies is small (Lacy, 2000). Extreme events, whether they are natural or anthropogenic catastrophes, can lead to severe decline and decimate isolated populations (Reed et al., 2003). Small populations with low genetic diversity, such as the cheetah (Chapter 6), are especially vulnerable to these effects.

Small populations are especially susceptible to rapid changes in gene frequencies due to chance (genetic drift), which decreases the genetic diversity of the population (Lacy, 1997). Genetic drift and a reduction in the availability of mates lead to an increase in inbreeding depression, which has been shown to impact juvenile survival (O'Grady et al., 2006), adult survival (Jiménez et al., 1994), reproduction (Ballou, 1997), and the potential to adapt to changes in the environment (Frankham, 1995).

Large populations are buffered from such stochastic processes and can generally rebound from their effects, while small populations are more likely to be drawn into a feedback loop of decline and potential extinction. Stochasticity should therefore be considered in all PVAs for small populations in order to simulate real life scenarios and to reliably project future population trends.

PVA PROCESS AND SOFTWARE

A variety of approaches and software modeling packages can be used to perform PVA. One of the most widely used PVA modeling tools is VORTEX (Lacy and Pollak, 2017), which was used in the majority of cheetah PVAs. VORTEX is well suited to project the viability of small populations of threatened vertebrates, such as the cheetah, because it can incorporate both deterministic and stochastic processes. The program is flexible, with options to incorporate complex functional relationships and a wide range of input and output options. Output parameters include projected population size, stochastic growth rate, probability of extinction, mean time to extinction, and genetic diversity. An attractive feature of VORTEX is its ability to model a population's genetic characteristics, including the ability to use molecular and pedigree data. VORTEX is provided free of charge by the Species Conservation Toolkit Initiative (http://www.cpsg.org/new-initiatives/species-conservation-toolkit-initiative), making it accessible for conservation scientists and managers worldwide.

PVAs can be carried out as population-level or individual-based models. VORTEX has both capabilities but is usually run as an individual-based model that tracks the fate of all individuals (and their genetic make-up) throughout their lifetime in each population; it can also incorporate variation in demographic rates (e.g., year-to-year variation). VORTEX does not model individuals across a spatial landscape but can model interacting populations, providing outputs for each population, as well as the metapopulation (see Chapter 36 for spatially explicit PVAs). Population-based modeling approaches that ignore stochastic processes are not recommended for cheetahs, because many existing populations are small and because the species' genetic makeup can play an important role in its conservation (O'Brien et al., 1985; Chapter 6).

Since stochastic processes are random, a stochastic model will produce a different result each time it is run. Each new iteration of the model thus produces a different trajectory. An example of possible population trajectories carried out in VORTEX (Lacy and Pollak, 2017) is provided in Fig. 38.2. Means and variation across hundreds of iterations provide insight into future population trends. Simulation time frames should be long enough (ideally ≥10 generations) to detect population trends and risks; however, the uncertainty of viability projections increases over time and with changing future conditions and management. Resulting population status and trends can describe long-term demographic

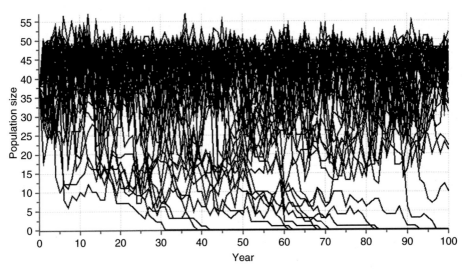

FIGURE 38.2 **Projected population trajectories for the cheetah population in W-Arly-Pendjari (WAP) National Park complex, western Africa.** Initial population size was set to 38 individuals (extrapolated from 25 adults and adolescents described in Durant et al., 2017). Graph shows the possible trajectories over 100 years, based on 50 iterations of the simulated population implemented in *VORTEX*.

and genetic health of the population. Details regarding model structure and format of input parameters can be found in the software manual (Lacy et al., 2017).

Other programs, such as *ALEX* (Possingham and Davies, 1995), *GAPPS* (Harris et al., 1986), *INMAT* (Mills and Smouse, 1994), *RAMAS* (Akçakaya, 2005), and *STELLA* (High Performance Systems, 2001) are also used for PVA. Regardless of the program, it is important that users understand if and how various aspects of population demography and dynamics are modeled in their software package of choice. Advanced users can also code program for viability analyses, using existing scripts as starting points (e.g., in *R*, Durant et al., 2017; in *MATLAB*, Morris and Doak, 2003). *R* subroutines can also be incorporated into *VORTEX* for added complexity. Individual programming allows the user to develop uniquely tailored models specifically for the species or population of focus.

PVAs FOR CHEETAHS

PVAs carried out for cheetahs are listed chronologically in Table 38.1. Analyses include PVAs and broader conservation planning efforts, such as Population and Habitat Viability Assessments (PHVAs), which are multistakeholder species conservation planning workshops that incorporate PVA as a tool to assist evaluation of management options and inform conservation action decisions. Main input parameters used in cheetah PVAs carried out in *VORTEX* and the respective parameter values are listed in Table 38.1.

Many of the input parameter values were derived from long-term research on cheetahs, whereas others were based on expert opinion and other data sources. Some input parameters are general for all cheetahs, whereas other parameters are population-specific and are influenced by environmental conditions and threats

TABLE 38.1 Input Parameters for *VORTEX* Used in Published PVA for Various Cheetah Populations

Parameter	Berry et al. (1997)	Hunter (1998)[a]	Broomhall (2001)		Lindsey et al. (2009)[b]		
Study details							
Country	Namibia	South Africa	South Africa	Tanzania	South Africa	South Africa	South Africa
Location	Farmland	Phinda Resource Reserve	Kruger National Park	Serengeti National Park	Farmland	Kruger National Park	Kgalagadi Transfrontier Park
Protected status	Nonprotected	Protected	Protected	Protected	Nonprotected	Protected	Protected
Ecoclassification							
Habitat	Woodland Savanna	Bushveld	Woodland Savanna	Grassland Savanna	Bushveld	Woodland Savanna	Shrubby desert grassland
Prey density	High	High	High	High	High	High	Low
Competitors present (e.g., lions)	No	Yes	Yes	Yes	No	Yes	Yes
Population							
Initial population size	2500	15	250	250	400	100	80
Initial population structure	Stable age distribution	Specified distribution	Stable age distribution	Stable age distribution	Stable age distribution	Stable age distribution	Stable age distribution
Carrying capacity of habitat (no. of individuals)	1500, 2500, 4000, 6000 (including Botswana)	50 (arbitrary)	500 (arbitrary)	500 (arbitrary)	800	160	90
Connectivity to other populations	No[c]	No	No	No[d]	Not modeled	Not modeled	Yes[e]
Reproduction							
Breeding system	(Short-term) polygyny	(Short-term) polygyny	(Short-term) polygyny	(Short-term) polygyny	(Short-term) polygyny	(Short-term) polygyny	(Short-term) polygyny
Definition of "reproduction" used	First observation (~3–4 months)	Emergence from lair (~6–8 weeks)	?	?	Emergence from lair	Emergence from lair	Emergence from lair
Mean age at first reproduction F/M (years)	3/5	3/3 or 5	2/2	2/2	3/3	3/3	3/3
Maximum breeding age (years)	10	12	12	12	12	12	12
Density dependent reproduction	No	?	No	No	No	No	No
% Males that are potential breeders	66	?	100	100	95	90	100
% Adult females that produce offspring (per year)	60 or 75	33.3	68	87.4	55.4	68	40

(Continued)

TABLE 38.1 Input Parameters for *VORTEX* Used in Published PVA for Various Cheetah Populations (*cont.*)

Parameter	Berry et al. (1997)	Hunter (1998)[a]	Broomhall (2001)	Lindsey et al. (2009)[b]				
EV for reproduction (as CV)	20.8%, 16.7%	?	10%	10%	20%	20%	20%	20%
Maximum number of litters per year	n/a	n/a	1	3	1	1	1	1
Mean litter size (±SD)	3.5 ± ?	4.4 ± ?	3.5 ± ?	3.5 ± ?	3.4 ± 0.4	3.4 ± 0.4	3.4 ± 0.4	3.33 ± 0.8
Maximum litter size	4	6	6	6	6	6	6	5
EV correlation (reproduction and survival EVs)	Yes	?	Yes	Yes	Yes	Yes	Yes	Yes
Genetics								
Inbreeding depression included	No	Yes	No	No	Yes	Yes	Yes	No
Lethal equivalents	n/a	3.14	n/a	n/a	1.57	1.57	1.57	n/a
% Lethal equivalents due to recessive alleles	n/a	50	n/a	n/a	50	50	50	n/a
Mortality rates								
Mortality 0–1 years F/M (%)	46/46	25/25	50/25	90 (85–95)[f]	25/25	25/25	50/50	52/61
Mortality 1–2 years F/M (%)	10/10	10/12.5	15 (15–65)[f]	35[g] (30–40)[f]	18/25	18/25	20/30	25/25
Mortality 2–3 years F/M (%)	10/10	10/12.5	15 (15–65)[f]	35[g] (30–40)[f]	22/30	22/30	25/35	15/34
Mortality > 3 years (annual) F/M (%)	10/10	12.5/17.85	15 (15–25)[f]	15[g] (10–20)[f]	15/15	15/15	15/15	15/15
EV for survival (as CV)	?	?	0	0	20%	20%	20%	20%
Catastrophes								
Catastrophe frequency (%)	5 or 10	5	0	0	2.5	2.5	2.5	2.5
Catastrophe severity on reproduction (proportion of normal value)	1 or 0.8	1 (no effect)	n/a	n/a	1 (no effect)	1 (no effect)	1 (no effect)	1 (no effect)
Catastrophe severity on survival (proportion of normal value)	0.8 or 0.65 or 0.5	0.75	n/a	n/a	0.5	0.5	0.5	0.5
Deterministic growth rate (r)								
Baseline r	0.156	?	?			0.142	0.093	0.02

Harvest

Animals removed annually through harvest function	No	No	No	No	Annually (conflict, illegal hunting)	No	No
Removed 1–2 years F/M (#)	n/a	n/a	n/a	n/a	4/4	n/a	n/a
Removed 2–3 years F/M (#)	n/a	n/a	n/a	n/a	2/3	n/a	n/a
Removed >3 years F/M (#)	n/a	n/a	n/a	n/a	5/13	n/a	n/a

Supplementation

Unrelated animals added annually through supplementation function	No	No	No	No	No	No	Yes
Added F/M (#)	n/a	n/a	n/a	n/a	n/a	n/a	1/1 (migrants from adjacent wild pops)

Model

Number of iterations	200	200	500	500	500	500	500
Number of years	100	100	100	100	100	100	100
VORTEX Version[h]	7.1/7.2	?	8	8	9.94	9.94	9.94

?, Unspecified; CV, coefficient of variation, EV, environmental variation; F, female; M, male; PVA, population viability analysis; SD, standard deviation.
[a] Baseline input parameters largely derived from small (N = 15) reintroduced population monitored during 40 months.
[b] Metapopulations of managed reserves in South Africa were not included as these were not considered free-ranging, but can be inspected in Lindsey et al. (2009).
[c] Potential connectivity to Botswana indirectly modeled via larger initial population size.
[d] Population was considered isolated and immigration not included because no data were available. However, emigration was included via adult and subadult mortality rates.
[e] Connectivity with adjacent populations in other countries.
[f] Brackets denote values used in sensitivity analysis to assess the influence of varying age-specific mortality rates on population persistence. Mortality rates were varied by 5% while holding all other parameters constant.
[g] Includes emigrations.
[h] More recent versions have increased capability to perform complex analyses, affecting several input parameters and their use within the program.

under which the population lives. Some earlier PVAs did not explicitly incorporate rates based on specific ecological conditions, despite the importance of habitat, predation, and competition in driving wildlife population dynamics (Boyce, 1992). Even today, lack of empirical data for many cheetah populations and software limitations typically mean that the modeler needs to find ways to indirectly include ecological aspects, such as prey density or habitat, into the model (Fig. 38.1).

Namibian PHVA

The IUCN CBSG (now CPSG) facilitated a PHVA Workshop for Namibian Cheetah and Lion in 1996, using *VORTEX* for the analytical component of the workshop (Berry et al., 1997). Without human-caused mortality, the model projected a high potential for an increase in the Namibian cheetah population, with an estimated annual growth rate of 10%–15%—that is, doubling in 5–7 years. Adult female survival emerged as a critical parameter that could affect the probability of extinction. High adult female mortality, in combination with catastrophes, led to projected extinction within 100 years under certain scenarios. In contrast, mortality rate of males (over the range tested) had little effect on the population growth rate under a polygynous mating system. Based on these results, it was concluded that conservation efforts for this population should focus on measures that decrease human-caused female cheetah mortality (Table 38.2). In addition, the PHVA showed the need for more research, encouraging the launch of additional research projects and the use of shared protocols.

PVA for a Reintroduced Cheetah Population

Another early cheetah PVA using *VORTEX* was undertaken for a population in a fenced private game reserve in South Africa (Hunter, 1998).

A small number of cheetahs (*n* = 15) and lions (*n* = 13) reintroduced to Phinda Resource Reserve were monitored during 40 months through direct observations and radio tracking, and demographic data were recorded for each individual. This study presented the rare situation of a known starting population size and age distribution; other parameter values were derived from *in situ* monitoring, as well as from the literature (Berry et al., 1997). The PVA suggested that the risk of extinction for this cheetah population was very low, although increased juvenile and subadult mortality increased extinction risk to 19.5% over 100 years. If lions are reintroduced first, cheetahs could be suppressed. Therefore, it was recommended to reintroduce cheetahs into small reserves prior to lion reintroductions to allow the cheetah population to get established through high-reproductive output in the years preceding lion presence (Table 38.2).

PVA for a Grassland Savanna

Cheetahs inhabiting the Serengeti Plains in Tanzania constitute one of the most intensively studied populations of cheetahs in the wild. The dynamics of this population have been modeled by Crooks et al. (1998) and Kelly and Durant (2000). Crooks et al. (1998) used age-structured population models with postbirth censusing to model the demography of female cheetahs, and found that adult female survival had a considerably larger effect on population growth rate than juvenile survival (Table 38.2). Kelly and Durant (2000) used *POPGEN*, an individual-based model that incorporates demographic and environmental stochasticity and found that, in model scenarios with medium- and high-lion density, cheetahs were projected to go extinct within 50 years. They concluded that cheetah populations outside of protected areas (where lions have been largely removed) were therefore important for cheetah survival (Table 38.2). Immigration was excluded from all models, although it is probably an important factor in the viability of the Serengeti cheetah population (Broomhall, 2001).

PVA for Woodland Versus Grassland Savanna

Broomhall (2001) modeled cheetah population viability in two different savanna types. Data for woodland savanna came from Kruger National Park, South Africa (Broomhall, 2001), and literature (Berry et al., 1997; Hunter, 1998; Purchase, 1998), whereas for grassland savanna, data from Serengeti Plains, Tanzania were used (Kelly and Durant, 2000). The grassland savanna population had a negative stochastic growth rate with 3% chance of extinction over 100 years, whereas the woodland savanna population had a positive growth rate with no chance of extinction. The greater resilience of the woodland savanna population could be attributed to higher average growth rates than in the savanna population due to lower juvenile and subadult mortality. Mortality of juveniles in particular is likely lower in woodland savanna because of increased refugia facilitated by habitat cover, such as more effective den concealment and less visibility of the adult female (and thus cubs) by lions.

Namibian Bayesian PVA

Johnson et al. (2013) used object-oriented Bayesian network modeling to integrate expert knowledge and data on cheetahs in Namibia for the purposes of modeling population viability, assessing factors influencing persistence, as well as identifying knowledge gaps. To improve estimates of cheetah population resilience, the analysis identified the need to focus on quantifying mortality and recruitment. The probability that the cheetah population would persist under current conditions was predicted to be 52.4%. When a scenario of decreased cheetah removal by farmers was considered, the probability of cheetah persistence increased to 58.1%,

suggesting the need to focus efforts on facilitating coexistence between cheetahs and landowners on livestock and game farms.

South African PHVA

In 2009, the IUCN CPSG facilitated a cheetah PHVA workshop for South Africa (Lindsey et al., 2009). The workshop served as a foundation for the development of a metapopulation management strategy for cheetahs in South Africa. VORTEX was used to simulate various ecological conditions for South African cheetahs over a 100-year timeframe. For farmland with good prey base and relative absence of lions and other large carnivores, a High-Prey Density/No Competitors scenario was used. PVA results suggested that cheetah populations on such farmland have a high probability of persistence, provided that the incorporated data on illegal harvest of cheetahs (which is rarely reported) were not underestimated and that habitat was not lost in the future. For Kruger National Park a High-Prey Density/Competitors Present scenario was generated to reflect a good prey base and the presence of lions. The Kruger cheetah population was projected to be relatively stable, except under conditions of high juvenile mortality. For the South African section of Kgalagadi Transfrontier Park, a Low-Prey Density/Competitors Present scenario was devised. This population was projected to suffer a negative growth rate if considered in isolation, but regular immigration from adjacent populations could increase the probability of a long-term resilient population.

Within South Africa, cheetahs also occur on multiple small, intensively managed fenced reserves. A metapopulation model of these reserves was developed to explore potential population management strategies. The viability of such a metapopulation was dependent on reserve size, number of reserves, presence of lions, cheetah population structure within reserves, and rate of translocation among reserves.

Sensitivity analyses of these models predicted that adult female breeding rate and adult female mortality influenced projections substantially, highlighting the importance of adult females to the reproductive potential of the population. This also emphasized the importance of collecting reproductive data (e.g., interbirth interval), in concert with juvenile mortality, to improve modeling accuracy.

Rangewide, With Special Reference to Protected Areas

Using data acquired primarily at regional planning workshops for cheetah conservation, Durant et al. (2017) estimated that 67% of the extant cheetah subpopulations are located outside protected areas and that cheetahs only occupy 9% of their historical range. Population simulation modeling in *R* suggested that the risk of extinction was greatest if the percentage of protected land were to decrease, and where cheetah population growth rates were inhibited, which is common for populations outside protected areas. For small populations of 200 individuals, extinction risk modeled over 50 years was greater than 50%, if unprotected populations had a 10% annual decline, relatively high migration rates between unprotected and protected land (10% of the population per year), and land protection corresponded to current levels. For projections regarding extinction risk of the wild cheetah population as a whole, all populations were considered to be connected for modeling purposes. The cheetah population was projected to decline by 50%–70% within three generations if annual declines on unprotected land varied between 10%–20% (lambda = 0.8–0.9) and migration rates were limited to 5% per year. Based on sensitivity analyses, cheetah population decline outside

TABLE 38.2   Main Management Recommendations of Published Population Viability Analyses for Various Cheetah Populations

Study	Subsection heading	Main recommendation
Berry et al. (1997)	Namibian PHVA	Decrease human-caused mortality especially of adult females
Hunter (1998)	PVA for a reintroduced cheetah population	Establish cheetahs before lion reintroduction into small reserves
Crooks et al. (1998); Kelly and Durant (2000)	PVAs for a grassland savanna	Decrease cheetah mortality of adult females in particular[a] Conserve cheetahs outside protected areas, not only in parks/reserves
Broomhall (2001)	PVA for woodland versus grassland savanna	Ensure connectivity, especially for woodland savannas to act as "source populations" for grassland savannas[b] Decrease adult mortality[a,b]
Johnson et al. (2013)	Namibian Bayesian PVA	Facilitate cheetah coexistence with farmers
Lindsey et al. (2009)	South African PHVA	Decrease persecution and human-cheetah conflict Minimize habitat fragmentation and loss Promote connectivity of cheetah populations
Durant et al. (2017)	Rangewide, with special reference to protected areas	Facilitate cheetah coexistence with farmers Elevate IUCN Red List conservation status of cheetah to "Endangered"

Input for metapopulations of managed reserves in South Africa were not included because these were not considered free-ranging cheetahs, but can be inspected in Lindsey et al. (2009).

IUCN, International Union for Conservation of Nature; PHVA, population and habitat viability assessment.

[a] *Cheetah mortality was linked in part to the presence of lions; however, it is natural for lions and cheetahs to share habitat.*

[b] *Management recommendations inferred from study conclusions.*

protected areas could only be compensated if growth rates were high within protected areas; this scenario was dependent on the absence of impermeable fences that prevent cheetah migration. Consequently, the authors suggested that conservation efforts should facilitate cheetah coexistence with local communities outside protected areas, and recommended that, due to the high risk of extinction, the species' IUCN Red List conservation criteria be reassessed and the status of the cheetah be elevated to "Endangered" (Chapter 39; Table 38.2).

DISCUSSION OF BENEFITS AND CAUTIONS IN USING PVA

PVA is a complex process involving the synthesis of information about the target species and population, and development of the best model with the information available (Boyce, 1992). For the cheetah, as for many other rare or endangered species, data availability is limited. The robustness of inferred conclusions can be critically influenced by unreliable or missing data (Coulson et al., 2001). However, estimates can provide valuable input if needed. In the absence of data, parameters for PVAs can be set based on assumptions and extrapolations from other populations or closely related species, or based on expert opinion and elicitation. The level of precision needed for a given parameter can be assessed with sensitivity testing (i.e., comparing modeling output for a range of values of a given parameter). For example, variation of litter size (based on statistical distribution of probabilities) generally has little effect on modeling extinction risk for species with multiparous females (Devenish-Nelson et al., 2013). Sensitivity testing and model validation also determine the degree of uncertainty in the PVA results and assess the model's usefulness and limitations. Like all models, a PVA model is a simplification of reality (Reed et al., 2002) and the assumptions made should be clearly stated.

PVAs may also provide alternative useful outputs in the absence of precise parameter information. While detailed biological and threat data are required for accurate viability projections, less robust models can be used to identify important data gaps and assess the relative impact of alternative management strategies (Miller, 2006).

Overall, PVAs are useful and powerful tools to understand the effects of specific ecological variables, random events, and human-caused pressures on a species in a given system. They help explore the mechanisms of population trajectory and how specific factors might contribute to it. While there is some criticism of PVA, primarily because of lack of validation in realistic systems (Ludwig, 1999; Taylor, 1995), an analysis by Brook et al. (2000) found that the projections from PVAs can be surprisingly accurate irrespective of the analytical software used for PVA implementation. In addition, the PVA process has the potential benefit of bringing together stakeholders with diverse expertise and to identify priority areas where data are missing.

CONCLUSIONS AND FUTURE DIRECTIONS FOR CHEETAH PVAs AND CONSERVATION EFFORTS

Through careful model construction, validation, sensitivity testing, and interpretation, PVAs can provide valuable guidance regarding the likely future of a population, the primary factors affecting its viability, and important research and management priorities to ensure the long-term survival of viable cheetah populations. Such quantitative assessments are essential tools in species conservation planning processes and careful application can contribute to more effective conservation intervention.

Even for a flagship species, such as the cheetah, relevant information needed for PVA is often not available. For example, demographic information (e.g., survival rates), as well as dispersal data, are either insufficient or absent for some ecosystems, especially given different human densities and land use types. Human-caused mortality of cheetahs is important information that currently mostly relies on extrapolation across ecosystems or incomplete data. For many cheetah populations the data on initial population sizes and age structure are unreliable. Knowledge on prey availability and habitat-carrying capacity are of various qualities and largely absent for many cheetah subpopulations. More research on those aspects would certainly improve PVA predictions. Nonetheless, sensitivity analyses within cheetah PVAs have revealed a range of important components that can influence population trajectory.

Past cheetah PVAs focused on a range of cheetah populations, from single small isolated populations to range-wide assessments, from protected areas to human-dominated landscapes, and over a range of habitats and levels of population management. Population projections ranged from a positive growth rate on Namibian farmland (Berry et al., 1997) and Kruger National Park (Broomhall, 2001) to potential extinction within 50 years in the Serengeti plains (Kelly and Durant, 2000) and in small populations (Durant et al., 2017). Differences in projections are likely due to the different conditions in the study areas, as well as differences in the input parameters used.

Overall, the projected outcomes highlighted the importance of adult female survival and reproduction to ensure long-term population resilience. This conclusion corresponds with expectations for a polygynous, relatively long-lived (~12 years in the wild) vertebrate species, in which population growth is driven by female breeders. The loss of adult cheetahs, and in particular adult females, through natural or human-caused mortality will reduce the population's growth rate, and if the rate of adult loss is greater than the rate of replacement, this will inevitably cause population decline (i.e., negative deterministic growth rate). Juvenile

mortality and reduced female reproduction also negatively affected population growth rate, but to a lesser extent than adult female mortality.

Presence of high numbers of predators, as well as increased human-cheetah conflict were shown to have significant impact on the risk of extinction of existing cheetah populations, largely due to their effect on adult cheetah and cub survival. Official protection status of the habitat (i.e., national parks) was considered favorable, as, when well-managed, it may provide the framework for higher cheetah population growth rates than unprotected land due to reduced or absent human-caused mortality. However, interspecific competition with other predators is likely to be greater within protected areas than outside their boundaries (where lions and spotted hyenas have largely been removed); such competition may outweigh the benefits of reduced human-caused mortality within protected areas. With most of the world's free-ranging cheetahs living outside parks and reserves, human-caused mortality is a principal threat to cheetah survival (Chapter 13), which emphasizes the need for conservation programs that promote human–cheetah coexistence.

Extinction risk was further shown to increase as population size decreases and isolation of the population increases. Increased extinction risk of isolated small populations has been attributed to increased vulnerability to changes in the ecological conditions, as well as demographic and genetic random events, resulting in reduced growth rates. This is concerning, as cheetah populations throughout its range are becoming smaller and more isolated due to continued habitat fragmentation (Chapter 10). To improve their chances of survival, small fragmented populations need to be reconnected, whether through metapopulation management (Lindsey et al., 2009) or habitat connectivity (Chapter 10).

While findings of past cheetah studies have provided some conclusions on which conservationists can act upon (Table 38.2), they were based on local scales or, in the case of

Durant et al. (2017), were principally focused on the connection between protected and unprotected land, without taking into account the level of separation between populations from different regions. Given the trend of reduced population sizes and increasing separation between them, a rangewide PVA including population data is needed for cheetahs. The realization of this gap has led to the compilation of information about free-ranging populations for a rangewide cheetah viability analysis (Cristescu et al., in preparation).

While additional data will yield more precise predictions, which in turn will help prioritize specific conservation actions or populations in need of intervention, important recommendations regarding protecting cheetahs (in particular adult females) from persecution and ensuring connectivity between populations have already emerged from existing analyses (Table 38.2). It is therefore important that the knowledge gained from past and future PVAs is taken into account by managers and legislators and translated into conservation action to improve the chances of survival of the cheetah.

References

Akçakaya, H.R., 2005. RAMAS GIS: Linking Spatial Data With Population Viability Analysis. Applied Biomathematics, Setauket, New York.

Ballou, J.D., 1997. Ancestral inbreeding only minimally affects inbreeding depression in mammalian populations. J. Hered. 88, 169–178.

Berry, H., Bush, M., Davidson, B., Forge, O., Fox, B., Grisham, J., Howe, M., Hurlbut, S., Marker-Kraus, L., Martenson, J., Munson, L., Nowell, K., Schumann, M., Shille, T., Venzke, K., Wagener, T., Wildt, D., Ellis, S., Seal, U. (Eds.), 1997. Population and Habitat Viability Assessment for the Namibian Cheetah (*Acinonyx jubatus*) and Lion (*Panthera leo*). Workshop Report. IUCN/SSC Conservation Breeding Specialist Group, Apple Valley, MN.

Boyce, M.S., 1992. Population viability analysis. Annu. Rev. Ecol. System. 23, 481–506.

Brook, B.W., O'Grady, J.J., Chapman, A.P., Burgman, M.A., Akçakaya, H.R., Frankham, R., 2000. Predictive accuracy of population viability analysis in conservation biology. Nature 404, 385–387.

Broomhall, L.S., 2001. Cheetah *Acinonyx jubatus* ecology in the Kruger National Park: a Comparison With Other Studies Across the Grassland-Woodland Gradient in African Savannas. MSc thesis, University of Pretoria, South Africa.

Coulson, T., Mace, G.M., Hudson, E., Possingham, H., 2001. The use and abuse of population viability analysis. Trends Ecol. Evol. 16, 219–221.

Crooks, K.R., Sanjayan, M.A., Doak, D.F., 1998. New insights on cheetah conservation through demographic modeling. Conserv. Biol. 12, 889–895.

Devenish-Nelson, E.S., Stephens, P.A., Harris, S., Soulsbury, C., Richards, S.A., 2013. Does litter size variation affect models of terrestrial carnivore extinction risk and management? PLoS One 8, e58060.

Durant, S.M., Mitchell, N., Groom, R., Pettorelli, N., Ipavec, A., Jacobson, A., Woodroffe, R., Bohm, M., Hunter, L.T.B., Becker, M.S., Broekuis, F., Bashir, S., Andresen, L., Aschenborn, O., Beddiaf, M., Belbachir, F., Belbachir-Bazi, A., Berbash, A., Brandao de Matos Machado, I., Breitenmoser, C., Chege, M., Cilliers, D., Davies-Mostert, H., Dickman, A.J., Ezekiel, F., Farhadinia, M.S., Funston, P., Henschel, P., Horgan, J., de Iongh, H.H., Jowkar, H., Klein, R., Lindsey, P.A., Marker, L., Marnewick, K., Melzheimer, J., Merkle, J., M'soka, J., Msuha, M., O'Neill, H., Parker, M., Purchase, G., Sahailou, S., Saidu, Y., Samna, A., Schmidt-Küntzel, A., Selebatso, E., Sogbohossou, E.A., Soultan, A., Stone, E., van der Meer, E., van Vuuren, R., Wykstra, M., Young-Overton, K., 2017. The global decline of cheetah *Acinonyx jubatus* and what it means for conservation. Proc. Natl. Acad. Sci. USA 114, 528–533.

Frankham, R., 1995. Inbreeding and extinction: a threshold effect. Conserv. Biol. 9, 792–799.

Gilpin, M.E., Soule, M.E., 1986. Minimum viable populations: processes of species extinction. In: Soule, M. (Ed.), Conservation Biology: The Science of Scarcity and Diversity. Sinauer Associates, Sunderland, MA, pp. 19–34.

Harris, R.B., Metzgar, L.H., Bevins, C.D., 1986. GAPPS: generalized animal population projection system. Version 3.0. Montana Cooperative Wildlife Research Unit, University of Montana, Missoula, Montana.

High Performance Systems, 2001. Stella v.7.0.1. High Performance Systems, Lebanon, New Hampshire.

Hunter, L.T.B., 1998. The Behavioural Ecology of Reintroduced Lions and Cheetahs in the Phinda Resource Reserve. PhD thesis, University of Pretoria, South Africa.

Jiménez, J.A., Hughes, A., Alaks, G., Graham, L., Lacy, R.C., 1994. An experimental study of inbreeding depression in a natural habitat. Science 266, 271–273.

Johnson, S., Marker, L., Mengersen, K., Gordon, C.H., Melzheimer, J., Schmidt-Küntzel, A., Nghikembua, M., Fabiano, E., Henghali, J., Wachter, B., 2013. Modeling the viability of the free-ranging cheetah population in Namibia: an object-oriented Bayesian network approach. Ecosphere 4, 1–19.

Kelly, M.J., Durant, S.M., 2000. Viability of the Serengeti cheetah population. Conser. Biol. 14, 786–797.

Lacy, R.C., 1987. Loss of genetic diversity from managed populations: interacting effects of drift, mutation, immigration, selection, and population subdivision. Conserv. Biol. 1, 143–158.

Lacy, R., 1997. Importance of genetic variation to the viability of mammalian populations. J. Mammal. 78 (2), 320–335.

Lacy, R., 2000. Considering threats to the viability of small populations using individual-based models. Ecol. Bull. 48, 39–51.

Lacy, R.C., Miller, P.S., Traylor-Holzer, K., 2017. VORTEX: A Stochastic Simulation of the Extinction Process. Version 10 User's Manual. IUCN SSC Conservation Breeding Specialist Group and Chicago Zoological Society, Apple Valley, MN.

Lacy, R.C., Pollak, J.P., 2017. Vortex: A Stochastic Simulation of the Extinction Process. Version 10.126. Chicago Zoological Society, Brookfield, Illinois, USA.

Lindsey, P., Marnewick, K., Davies-Mostert, H., Rehse, T., Mills, M.G.L., Brummer, R., Buk, K., Traylor-Holzer, K., Morrison, K., Mentzel, C., Daly, B. (Eds.), 2009. Cheetah (Acinonyx jubatus) Population and Habitat Viability Assessment Workshop Report. Conservation Breeding Specialist Group (SSC/IUCN) Southern Africa. Endangered Wildlife Trust.

Ludwig, D., 1999. Is it meaningful to estimate a probability of extinction? Ecology 80, 298–310.

Miller, P.S., 2006. What types of models should I run in my PVA? CBSG internal document. IUCN SSC Conservation Breeding Specialist Group, Apple Valley, MN.

Mills, L.S., Smouse, P.E., 1994. Demographic consequences of inbreeding in remnant populations. Am. Nat. 144, 412–431.

Morris, W.F., Doak, D.F., 2003. Quantitative Conservation Biology: Theory and Practice of Population Viability Analysis. Sinauer Associates, Sunderland, Massachusetts, USA.

O'Brien, S.J., Roelke, M.E., Marker, L., Newman, A., Winkler, C.A., Meltzer, D., Colly, L., Evermann, J.F., Bush, M., Wildt, D.E., 1985. Genetic basis for species vulnerability in the cheetah. Science 227, 1428–1434.

O'Grady, J.J., Brook, B.W., Reed, D.H., Ballou, J.D., Tonkyn, D.W., Frankham, R., 2006. Realistic levels of inbreeding depression strongly affect extinction risk in wild populations. Biol. Conserv. 133, 42–51.

Possingham, H.P., Davies, I., 1995. ALEX: a model for the viability analysis of spatially structured populations. Biol. Conserv. 73, 143–150.

Purchase, G., 1998. An assessment of a Cheetah Re-Introduction Project in Matusadona National Park. MSc thesis, University of Zimbabwe, Zimbabwe.

Reed, J.M., Mills, L.S., Dunning, Jr., J.B., Menges, E.S., McKelvey, K.S., Frye, R., Beissinger, S.R., Anstett, M.-C., Miller, P., 2002. Emerging issues in population viability analysis. Conserv. Biol. 16, 7–19.

Reed, D.H., O'Grady, J.J., Ballou, J.D., Frankham, R., 2003. The frequency and severity of catastrophic die-offs in vertebrates. Anim. Conserv. 6, 109–114.

Shaffer, M., 1987. Minimum viable populations: coping with uncertainty. In: Soule, M. (Ed.), Viable Populations for Conservation. Cambridge University Press, Cambridge, pp. 69–86.

Shaffer, M.L., 1990. Population viability analysis. Conserv. Biol. 4, 39–40.

Taylor, B.L., 1995. The reliability of using population viability analysis for risk classification of species. Conserv. Biol. 9, 551–558.

THE FUTURE

The Conservation Status
of the Cheetah

Sarah M. Durant,**, *Nicholas Mitchell*,**,
Rosemary Groom,**, *Audrey Ipavec*,**, *Rosie Woodroffe**,
Christine Breitenmoser†,‡, *Luke T.B. Hunter*§

*Zoological Society of London, London, United Kingdom
**Wildlife Conservation Society, New York, NY, United States
†IUCN/SSC Cat Specialist Group, Bern, Switzerland
‡KORA, Bern, Switzerland
§Panthera, New York, NY, United States

ASSESSMENT OF CONSERVATION STATUS

The cheetah is a wide-ranging and low-density species (Chapter 8), factors which pose particular challenges for evaluating its conservation status. For cheetah, as for all other species, the IUCN Red list criteria are the primary tool used for assessment of threat status (Mace et al., 2008; Rodrigues et al., 2006). Here, a species is assessed as threatened if it falls within one of three categories that correspond to different estimated risks of extinction: Critically Endangered with an extinction risk of 50%; Endangered with a risk of 20%; and Vulnerable with a risk of 10% (Mace and Lande, 1991). Extinction risk is assessed within a set time frame that relates to a

multiple of the species' generation time. The categories thus reflect actual risk of species extinction and are aligned to the urgency of the threat.

In the IUCN Red List threat assessment framework, extinction risk is usually assessed indirectly through defined criteria using indices of population status, such as population size, range-based measures (extent of occurrence and area of occupancy), and past or projected population trends. An additional criterion allows for direct extinction risk assessment using quantitative analyses (e.g., Population Viability Analysis or PVA; Chapter 38); however, the detailed information necessary for a credible analysis at a broad scale is rarely available. Much of the information on population size and trends used in threat assessment tends to come from relatively well-monitored populations, usually within protected areas (PAs).

Like many species, the cheetah faces spatial variability in the level of threat experienced inside and outside PAs. Yet, most of the data available on cheetah population status comes from sites where the species is likely to be most abundant, either because it is a target of active conservation management (Marker, 2002) or because sites are protected (Belbachir et al., 2015; Chauvenet et al., 2011; Durant et al., 2011; Marnewick et al., 2014). Cheetahs are one of the most wide-ranging predators and occur at some of the lowest densities recorded in free-ranging carnivores (Chapter 8). Finding a rare and well camouflaged cat within a vast landscape is further complicated by its highly secretive habits, particularly in areas where it faces persecution and may flee before being seen by human observers, or may shift into nocturnal activity (Belbachir et al., 2015; Marnewick et al., 2006). These multiple factors make population status assessment particularly difficult (Belbachir et al., 2015).

Despite the challenges in assessing status of cheetah in the wild, there is an international consensus on the known distribution of the species (Durant et al., 2017). In Africa, distributional mapping of cheetah used an expert-based approach established for jaguar and tiger (Dinerstein et al., 2007; Sanderson et al., 2002). This was conducted during IUCN/SSC conservation strategic planning workshops for cheetah and another similarly sparse and wide-ranging species, African wild dog *Lycaon pictus* (IUCN/SSC, 2007a,b, 2012, 2015). Additional map refinements were conducted during National Conservation Action or Management Planning Workshops and from published reports and scientific articles. Mapping in Asia was conducted by a small expert team using information from ongoing survey work in Iran and from the IUCN Red List assessment for the Asiatic subspecies (*Acinonyx jubatus venaticus*; Farhadinia et al., 2016; Jowkar et al., 2008).

In distributional maps, cheetah resident range is defined as land where the species was known to be still resident as recognized by (1) its regular detection in an area within the last 10 years, and over multiple years; and/or (2) evidence of breeding. All land formerly occupied by cheetah was considered to fall inside its historical range. For some areas, detailed historical data on distribution were available; elsewhere, historical distribution was estimated based on the species' broad habitat requirements.

Historically widespread, cheetahs used to occur across Africa through to Asia, favoring a broad range of habitats from thick bush to hyperarid deserts, such as the Sahara. The species was only excluded from dense forest biomes. The consensus on the current distribution is that cheetahs are resident across 2,977,000 km^2 in Africa, representing only 13% of the 23,341,000 km^2 historical range on the continent (Durant et al., 2017; Fig. 39.1). In Asia the situation for cheetah is even worse, with resident range now restricted to 147,000 km^2, entirely within Iran, a figure which amounts to just 2% of a historical range that used to encompass 9,716,000 km^2 (Durant et al., 2017). If we assume cheetahs were widespread across their historical range at the turn of the century, then there has been an annual mean decline in resident range of 2.3% per annum, resulting in a contraction of 30% over the last 15 years or 3 cheetah generations. In reality, contraction in range is likely to have accelerated over time, given increasing rates of human-induced rapid environmental change in the last few decades. It is important to note, however, despite these figures, there remains much uncertainty over current cheetah distribution, particularly in areas of limited access, such as regions subject to recent or ongoing insecurity (IUCN/SSC, 2007a,b, 2012, 2015).

The global population of free-ranging cheetahs is tentatively estimated at 7100 adolescents and adults (Durant et al., 2017; Table 39.1). This estimate was derived using known population estimates (based on surveys and monitoring)

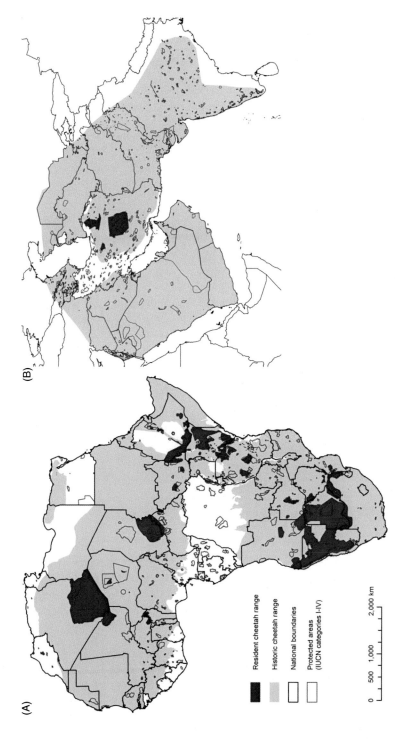

FIGURE 39.1 Known cheetah distribution in (A) Africa and (B) Asia. Gray shading denotes historical range, and red shading range where cheetahs are known to be resident, boundaries of PAs under IUCN categories I–IV are marked in blue. This map was originally published in Durant et al. (2017).

TABLE 39.1 Regional Summary of Known Cheetah Distributional Range and Populations

Area name	Countries	Resident range (km²)	Population size	Overall increase↑[a] / stable↔/ decrease↓	Resident range in PAs	(%) Range in PAs	Population size in PAs	(%) Population in PAs
Africa								
Southern Africa five-country polygon	Angola/Botswana/Mozambique/Namibia/South Africa	1,212,179	4,021	→	283,851	23.4	1,041	25.9
Moxico	Angola	25,717	26	?	0	0.0	0	0.0
Pandmatenga/Hwange/Victoria Falls	Botswana/Zimbabwe	25,926	50	→	15,551	60.0	29	58.0
Banhine	Mozambique	7,266	10	?	0	0.0	0	0.0
Malilangwe/Save/Gonarezhou	Mozambique/Zimbabwe	9,922	46	↔	4,757	47.9	19	41.3
Kafue	Zambia	26,222	65	?	22,185	84.6	55	84.6
Liuwa	Zambia	3,170	20	↑ or ↔	2,921	92.1	18	90.0
Bubiana-Nuanetsi-Bubye Conservancies	Zimbabwe	8,816	40	→	0	0.0	0	0.0
Zambezi valley	Zimbabwe	3,612	12	→	2,102	58.2	7	58.3
Matusadona	Zimbabwe	1,422	3	→	1,422	100.0	3	100.0
Midlands Rhino Conservancy	Zimbabwe	318	4	→	0	0.0	0	0.0
Subtotal southern Africa		*1,324,570*	*4,297*		*332,789*	*25.1*	*1,172*	*27.3*
Afar	Ethiopia	4,480	11	→	1,092	24.4	3	27.3
Blen-Afar	Ethiopia	8,170	20	→	1,856	22.7	5	25.0
Ogaden	Ethiopia	12,605	32	→	0	0.0	0	0.0
Yangudi Rassa	Ethiopia	3,046	8	→	3,046	100.0	8	100.0
Ethiopia/Kenya/South Sudan	Ethiopia/Kenya/South Sudan	191,180	191	?	37,953	19.9	38	19.9
South Turkana	Kenya	3,580	36	?	1,117	31.2	11	30.6
Kidepo/S South Sudan/NW Kenya	Kenya/South Sudan/Uganda	6,694	19	?	1,422	21.2	4	21.1

Site	Country			Trend[a]				
Serengeti/Mara/Tsavo/Laikipia/Samburu	Kenya/Tanzania	280,114	1,362	↓	49,705	17.7	664	48.8
Badingilo NP	South Sudan	8,517	85	?	4,741	55.7	47	55.3
Radom NP	South Sudan	6,821	68	?	0	0.0	0	0.0
Southern NP	South Sudan	14,680	147	?	10,863	74.0	109	74.1
Ruaha ecosystem	Tanzania	30,820	200	↔	25,551	82.9	166	83.0
Maasai Steppe	Tanzania	20,409	51	↓	3,755	18.4	9	17.6
Katavi-Ugalla	Tanzania	23,955	60	?	10,475	43.7	26	43.3
Subtotal eastern Africa		*615,071*	*2,290*		*151,576*	*24.6*	*1,090*	*47.6*
Adrar des Ifoghas/Ahaggar/Ajjer and Mali	Algeria/Mali	762,871	191	?	98,867	13.0	25	13.0
WAP	Benin/Burkina Faso/Niger	25,345	25	?	20,923	82.6	21	82.6
CAR/Chad	CAR/Chad	238,234	238	?	44,396	18.6	44	18.6
Termit Massif	Niger	2,820	1	?	2,820	100.0	1	100.0
Air-T	Niger	8,052	2	?	8,052	100.0	2	100.0
Subtotal western, central, and northern Africa		*1,037,322*	*457*		*175,058*	*16.9*	*93*	*20.3*
Total African		*2,976,963*	*7,044*		*659,423*	*22.2*	*2,355*	*33.4*
Asia								
Central and Eastern Landscapes	Iran	107,566	20	↔	41158	38.3	N/A	N/A
Northern Landscape	Iran	33,445	22	↓	18077	54.04	N/A	N/A
Kavir	Iran	5,856	1	↓	5,856	100	N/A	N/A
Total Asia		*146,867*	*43[b]*		*65,091*	*44.3*	*N/A*	*N/A*
Total global		*3,123,830*	*7,087*		*724,514*	*23.2*	*2,355[c]*	*33.4[c]*

Historical distributional range for cheetah totals 33,057,000 km², comprising 23,341,000 km² African and 9,716,000 km² Asian range (Fig. 39.1). Table originally published in Durant et al. (2017).

[a]Estimates of trend apply to entire polygon thus, for example, populations may increase at specific sites, even though there is an overall decrease across the polygon.
[b]Population estimated as less than 40 adults by Farhadinia et al. (2016).
[c]Does not include Iranian cheetah.

where available, and where these were not available, applying conservative density estimates to areas of cheetah range from comparable habitats and level of protection (IUCN, 2012; IUCN/SSC, 2007a,b, 2012, 2015). The surviving cheetah population is now highly fragmented and distributed across 33 populations, 30 of these populations are on the African continent (Table 39.1; Fig. 39.1). Only 3 populations, collectively estimated to support fewer than 50 adults and adolescents (Durant et al., 2017; Jowkar et al., 2008) (or < 40 adults; Farhadinia et al., 2016), survive in Asia, and are confined to Iran.

Today, the majority of cheetahs occur in a single transboundary population in southern Africa (60%; ca. 4000 adult and adolescent individuals). This population covers one third of known cheetah distributional range and stretches across five countries from southwestern Angola, across Namibia, Botswana, northern South Africa, and western Mozambique (Fig. 39.1). The next largest population of cheetah is in eastern Africa, supporting an estimated 1400 adults and adolescents, encompassing the Serengeti, Tsavo, and Laikipia landscapes in Tanzania and Kenya. All the remaining populations are estimated at below 250 individuals, and most are below 100 individuals (79%). Nearly half number 25 individuals or fewer, with many in the single digits (Table 39.1). These small populations are on the threshold of extinction.

The cheetah was most recently assessed as Vulnerable in the IUCN Red List (Durant et al., 2015b; Box 39.1).

THREATS TO CHEETAH SURVIVAL

Habitat Loss and Fragmentation

The cheetah's low density means that cheetah populations require much larger areas of land to survive than do those of most other large carnivore species, and hence are particularly sensitive to habitat loss and fragmentation (Chapter 10) which represent the overarching threat to cheetah (IUCN/SSC, 2007a,b, 2012). Demographically viable populations of cheetah are estimated to number 300 individuals or more (Durant, 2000), and hence are likely to require areas of land in excess of 10,000 km^2. However, since cheetahs can survive in anthropogenically modified habitats under the right circumstances, the landscapes that cheetahs require for their survival may be protected, unprotected, or a combination of the two. Cheetahs also have excellent dispersal abilities (Boast, 2014), which help to maintain gene flow between populations, and allows recolonization of suitable unoccupied habitat, provided sufficient connectivity is maintained (Chapter 10).

BOX 39.1

The cheetah meets the conditions of Vulnerable under the IUCN Red List Criterion A2acd due to an observed, estimated, inferred, or suspected population size reduction of ≥30% over the last 3 generations where the reduction or its causes may not have ceased OR may not be understood OR may not be reversible, based on direct observation (a); a decline in area of occupancy, extent of occurrence and/or quality of habitat (c); and actual or potential levels of exploitation (d) (IUCN, 2012). The species also qualifies as Vulnerable under criterion C1: population size estimated to number fewer than 10,000 mature individuals and an estimated continuing decline of at least 10% within 3 generations (IUCN, 2012).

Conflict With Livestock and Game Farmers

Cheetahs prefer wild prey to livestock, but they may kill livestock in some circumstances (Boast et al., 2016). As a result, cheetahs may be killed by farmers in retaliation for stock depredation or preemptively due to a perceived threat to stock (Chapter 13). Conflict with game farmers is also widespread in game farming areas as cheetahs are seen as competitors for valuable game offtake. These conflicts may involve subsistence pastoralists or farmers, as well as large-scale commercial ranchers and farmers. The use of traps, to capture and possibly kill cheetah, particularly on farmlands, is an ongoing threat. Nonetheless, cheetahs may persist in a landscape in the face of conflict, because they rarely scavenge (Caro, 1994; Durant et al., 2010), and hence are less susceptible to poisoning than hyenas (*Crocuta* and *Hyaena* spp.), leopards (*Panthera pardus*), and lions (*Panthera leo*). However, increasing pressures on the land and a loss of refuges for cheetahs to escape persecution can make human–cheetah coexistence impossible in some areas.

Loss of Prey

Cheetahs are highly efficient hunters, and can survive in areas of comparatively low-prey density (Belbachir et al., 2015; Caro, 1994; Durant et al., 2010). Nevertheless, loss of prey due to hunting (usually illegal), high-livestock densities and grazing pressure, and/or habitat conversion reduces cheetah densities and directly impacts cheetah populations (Chapter 11). The extraction of bush-meat has reached an industrial scale, and is responsible for massive declines in ungulate species, even within PAs (Benitez-Lopez et al., 2017; Lindsey et al., 2013; Ripple et al., 2016). Cheetahs may also become captured in snares set for bush-meat offtake, even though they may not be the primary targets (Lindsey et al., 2013). Prey loss can also have serious indi-rect effects, since predation by cheetahs on livestock may become more frequent where wild prey are depleted, intensifying conflict with livestock farmers (Chapter 13).

Road Mortality

Roads are increasing across the continent and present a growing threat to cheetahs. This is a particular concern where paved roads cross or adjoin major wildlife areas. In Iran, out of 21 known cheetah mortalities between 2001 and 2012 due to various human-causes, at least 12 were killed on roads through or adjacent to PAs, making it a major cause of anthropogenic mortality, second only to persecution-related killing by people (Iranian Cheetah Society, 2013; Chapter 5). This problem is not confined to Iran. In recent years, there have been a number of incidents of cheetahs being killed by cars on the main road through the Serengeti National Park in Tanzania (Durant, personal observation). Deaths have also been reported in many other countries, including South Africa, Botswana, Namibia, Zambia, Zimbabwe, and Kenya. Such mortality could have a significant impact on the viability of small and isolated populations of cheetah, and may affect dispersal and recolonization.

Illegal Trade

Historical capture and trade in live cheetahs has been reported as a key cause of their disappearance from much of their range in Asia (Chapter 22). There has been recent evidence of an increase in trade in cheetahs (Nowell, 2014), and illegal trade has been identified as a threat by stakeholders at most regional and national conservation action planning workshops. This trade has been of sufficient global concern to gain the attention of CITES (CITES, 2013; Mitchell and Durant, 2017). Live cheetahs are caught and traded illegally to the pet trade and are also hunted for their skin (Chapter 14). An increase in legal trade in cheetahs provides evidence of increasing demand for cheetah. CITES allows a legal quota for "live specimens and hunting trophies" of cheetah in Namibia, Zimbabwe, and Botswana and permits commercial trade in captive-bred animals (Chapter 21). From 2002 to 2011, this legal trade averaged 153 wild cheetah specimens per year (mainly hunting trophies from Namibia), and 88 captive-bred live animals (mainly from South Africa) (Nowell, 2014). Documentation for illegal trade is more problematic. Official records show that, on average, only three confiscations of illegally traded live cheetahs are reported to CITES per year, however, this is likely to be an underestimate of the real trade (Nowell, 2014). Most confiscated animals were destined for a pet trade market in the Gulf States. These consumer countries (Bahrain, Kuwait, Oman, Qatar, Saudi Arabia, and the UAE) are Party to CITES and thus officially prohibit wild cheetah imports. Cheetah skins are traded, often alongside leopard skins, within Africa and to Asia. As most cheetah populations are small, even a low level of illegal trade can threaten wild populations.

Unregulated Tourism

Tourism, when poorly managed, has the capacity to threaten cheetah populations (Roe et al., 1997). Cheetahs are a key attraction for wildlife tourists; in Amboseli National Park in Kenya, tourists spent 12%–15% of their total wildlife viewing time observing cheetahs (Roe et al., 1997). However, large number of tourist vehicles or insensitive tourist behavior can lead to multiple negative effects on cheetahs including interference with hunting, scaring cheetahs away from kills (to which they are unlikely to return), and separating mothers from cubs (Burney, 1980; Henry, 1975, 1980). The death of cheetah cubs as a direct consequence of separation from their mother due to tourist activity has been recorded in the Serengeti National Park (Durant, personal observation). In contrast, well-regulated tourism can make important contributions to

cheetah conservation, not only by the revenue it generates, but also by raising awareness and increasing public support for conservation (Roe et al., 1997).

An Emerging Threat: Infrastructure Development and Barriers to Movement

An emerging threat to cheetahs is increasing resource extraction and the erection of impermeable barriers within the species' range. An increasing proliferation of small scale fencing, such as to protect farmed game and livestock, can remove large areas of habitat from cheetahs (Løvschal et al., 2017). Large infrastructure development, such as mining, oil pipelines, roads, and railways is also increasing, and is often accompanied by fencing and other barriers to movement. Infrastructure development, mainly for the mining and transportation sectors, is thought to be a primary factor in the isolation of Iran's Kavir National Park from the rest of cheetah range in the country (Farhadinia et al., 2016). There are also moves to increase border fencing in several countries (FRONTEX, 2017). Developments such as these may present a death knell to wide-ranging species like cheetah, as they further fragment remaining cheetah populations into smaller and smaller subpopulations, which may no longer be viable.

Regional Variation in Threat and Underlying Drivers

All of the threats identified previously above play a role across cheetah range, although there are regional variations in relative impact. In eastern, southern, central, and western Africa, habitat loss and fragmentation have been identified as a primary threat (IUCN/SSC, 2007a,b, 2012; Chapter 10). Whereas, in northern Africa and Iran, where hyper-arid environments prohibit widespread agriculture, a depleted wild ungulate prey base is a dominant concern (Belbachir et al., 2015; Durant et al., 2014; Farhadinia et al., 2016; Chapter 11). Conflict with livestock farmers due to livestock depredation, either perceived, or real, is a widespread and serious problem in many areas of cheetah range (Chapter 13). Illegal trade in live cheetahs is a particular problem in the Horn of Africa, while trade in skins poses a threat throughout the species' range (Nowell, 2014; Chapter 14).

The multiple direct threats to cheetahs, while being immediate causes of population decline, are a consequence of many ultimate drivers. These include underlying problems, such as insufficient land-use planning, insecurity, and political instability and a lack of awareness or political will to foster cheetah conservation. Many countries harboring cheetahs also suffer from a lack of capacity and financial resources to support conservation and, throughout cheetah

range; there is a lack of sufficient incentives for local people to conserve cheetahs and their prey.

It is impossible to combat the direct threats to cheetahs without understanding and addressing the complex and interconnected drivers of these threats. For this purpose, detailed problem analyses, identifying the full suite of problems confronting cheetah conservation, have been undertaken across cheetah range at regional strategic planning workshops (IUCN/SSC, 2007a,b, 2012, 2015; Fig. 39.2). These analyses show that these problems fall under nine general areas: (1) coexistence, covering problems arising from people and their domestic animals living alongside cheetahs; (2) knowledge and information, covering problems arising from a lack of information about cheetahs, including their distributional range, population

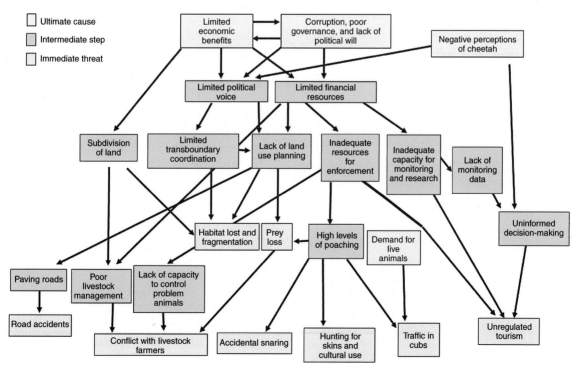

FIGURE 39.2 **Problem tree showing ultimate causes and intermediate steps leading to direct threats to cheetah in eastern Africa.** *Source: Adapted from IUCN/SSC, 2007a.*

status, habitat, and management; (3) sensitization and awareness, covering problems arising from a lack of awareness about the conservation needs and status of cheetahs and management of their habitat; (4) capacity development, covering problems associated with insufficient human resources and a lack of training and equipment; (5) utilization, covering problems arising from unsustainable offtake of cheetah prey and illegal killing of cheetahs; (6) land use, covering problems arising from insufficient or inappropriate land management, including poor rangeland and ecosystem management and loss of connectivity and fragmentation; (7) policy and legislation, covering problems arising from a lack of, or inappropriate, policies, and legal frameworks for wildlife conservation; (8) advocacy, covering problems associated with low importance attached by the public and government to cheetah conservation; (9) National Planning, covering problems associated with a lack of national frameworks for cheetah conservation. Conservation interventions need to address these multiple underlying problems, if they are to successfully combat the direct threats to cheetah survival.

CHEETAH THREAT ASSESSMENT

The majority of current cheetah range in Africa is on unprotected land (78%), which supports an estimated 67% of the cheetah population (Table 39.1). Cheetahs face higher levels of threat outside, compared with inside, PAs (IUCN/SSC, 2007a,b, 2012, 2015). These pressures on populations mean that, although we might expect cheetah populations to be stable in the core of large and well managed PAs, populations on unprotected lands and in small or poorly managed PAs are likely to be in decline. However, because of the considerable survey effort required or because of supplementation from adjacent PAs, such declines are likely to go undetected.

Simulation modeling in R (R Core Team, 2015) has been used to assess how spatial variation in threat across protection gradients in cheetah range affects population viability (Durant et al., 2017). Mean and variance in population growth rate in PAs was taken from the long-term study population of cheetahs living in Serengeti National Park, which implicitly includes the impacts of competitors (such as lion and spotted hyena *Crocuta crocuta*) (Chauvenet et al., 2011; Laurenson, 1995). Growth rate outside PAs was allowed to vary until it was substantially less than replacement and cheetah movement into and out of PAs was allowed within simulations. The global cheetah population was simulated by setting the initial population equal to the estimated population of 7000 individuals, of which 33% occurs in PAs (Table 39.1).

Simulations showed that, when the population growth rate outside PAs was 10% less than replacement, simulated populations declined by 53% over 15 years (Fig. 39.3A). When the growth rate outside PAs was 20% less than replacement, then the decline was 70%. If the growth rate inside PAs was increased until it was above replacement, then this slowed the rate of decline. However, growth rates needed to be high to completely mitigate against declines (Fig. 39.3B). Changing the migration rate had little effect on overall population decline (Durant et al., 2017).

Evidence of recent cheetah population declines is consistent with modeling results. For example, in Zimbabwe, where cheetah distribution is relatively well known, cheetahs were distributed across a contiguous population encompassing 133,000 km^2 in 2007, which contracted to a fragmented population occupying only 49,000 km^2 by 2015 (IUCN/SSC, 2007b, 2015; Van der Meer, 2016). This 63% range contraction over a short period was equivalent to a loss of 11% of distributional range per year, and was largely due to the disappearance of cheetahs outside PAs, associated with major changes in land tenure (Williams et al., 2016). Numbers of cheetahs in Zimbabwe declined even more

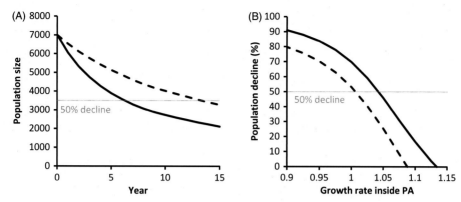

FIGURE 39.3 Simulated (A) population trajectories over 3 generations (15 years) of the global cheetah population and (B) sensitivity analysis to changes in the growth rate within protected areas (PAs). Starting population was the current total estimated global population size of 7000 individuals with 33% of the population on protected lands (Table 39.1). The dashed line depicts results from a multiplicative growth rate (lambda) of 0.9 on unprotected lands, the solid line 0.8. Migration rate was set at 0.05 with standard deviation 0.025. The gray dotted line depicts the 50% threshold for uplisting to endangered status using IUCN Red List criteria A3b (IUCN, 2012). This figure was originally published in Durant et al. (2017).

precipitously, with an estimated decline in excess of 85% between 1999 and 2015 (Van der Meer, 2016), which is equivalent to an annual decline of 13%. Elsewhere, cheetah population trends are largely unknown; however, of the 18 populations, where trends could be assigned by experts in the region, 14 were judged to be in decline, 3 stable, and only 1 stable or increasing (Table 39.1). There have also been recent large-scale extinctions of cheetahs across western and central Africa (Brugiere et al., 2015; de Iongh et al., 2011). Ongoing rapid change is likely across the African continent due to changes in land tenure (Williams et al., 2016); large-scale fencing (Durant et al., 2015a); land grabs (Davis et al., 2014); and political instability (Bouché et al., 2012). However, cheetah status in areas where they are most threatened is usually uncertain, because those areas lack data. On this basis, in line with the precautionary approach and in the absence of alternative information, this analysis supports uplisting cheetah threat status to Endangered under IUCN Red List criterion A3b (a population size reduction of ≥50%, projected or suspected to be met within the next 3 generations, based on an index of abundance; IUCN, 2012).

RANGE WIDE CHEETAH CONSERVATION

The low density of cheetahs throughout their range means they require conservation action on a scale that is seldom seen in terrestrial conservation. This includes transboundary cooperation, land use planning across large landscapes to maintain habitat connectivity, and human-wildlife conflict mitigation across extensive areas (IUCN/SSC, 2007a,b, 2012). Globally, most cheetah range (77%) is on unprotected lands (Table 39.1), where they are particularly vulnerable to anthropogenic pressures. The species is listed on Appendix I of CITES, Appendix 1 of CMS and is protected under national legislation throughout most of its extant and some of its former range (IUCN/SSC, 2007a,b, 2012; Nowell and Jackson, 1996; Chapter 21).

In Africa, nearly all range states are actively involved with the Range Wide Conservation

Program for Cheetah and African Wild Dog (RWCP), which has facilitated the development of regional strategies and national conservation action plans using the IUCN/SSC strategic planning process (IUCN/SSC, 2008). Cheetah and wild dog are combined in this process because of their similarly low densities, large space needs, and ecological requirements. This also increases leverage for conservation action by way of delivering impacts for two threatened species together. There are three regional strategies in place for Africa covering the entire cheetah distributional range on the continent: eastern Africa (IUCN/SSC, 2007a); southern Africa (IUCN/SSC, 2007b, 2015); and western, central, and northern Africa (IUCN/SSC, 2012).

The regional strategies provide a roadmap for developing national conservation action plans within participatory workshops. This allows relatively broad regional commitments to be tailored to the specific policy and legislative environments within each country. National conservation action plans are in place for most cheetah range states (year of the workshop in brackets): Kenya (2007), Botswana (2007), Ethiopia (2010), South Sudan (2009), Zambia (2009), Zimbabwe (2009), South Africa (2009), Benin (2014), Niger (2012), Chad (2015), Tanzania (2013), Mozambique (2010), Malawi (2011), Namibia (2013), Algeria (2015), Angola (2016), and Burkina Faso (2016). In addition, cheetahs are included in Uganda's Large Carnivore National Conservation Action Plan (2010). These action plans cover nearly all of the 30 cheetah populations in Africa and 96% of known African cheetah range. Each national conservation action plan is published by government wildlife authorities and represents each state's commitment to cheetah (and wild dog) conservation.

Regional strategies and national action plans provide a government supported framework to halt or reverse declines in cheetah populations, which uses a comprehensive problem analysis to develop a holistic approach to combat proximate threats together with their ultimate drivers. While there are some differences between individual plans and strategies, they broadly address objectives to improve national capacity for cheetah conservation and management; raise awareness of, and political commitment to, cheetah conservation, promote human-cheetah coexistence, improve land use planning and reduce habitat fragmentation, combat illegal and unsustainable offtake of prey, improve policy and legislation, and address cheetah conservation information needs. Local and national projects, NGOs and governments are key to this planning process. The implementation of the plans and strategies is overseen by three regional coordinators within the Range Wide Conservation Program for Cheetah and African Wild Dog. There is, however, substantial regional variation in implementation. In southern and eastern Africa, a number of projects and/or NGOs dedicated specifically to the conservation and research of cheetahs, or to the wider guild of large carnivores, carry out important site-based conservation activities that benefit cheetahs. However, there is much less NGO or project-based activity in northern, western, or central Africa, despite some recent encouraging developments.

In Iran, the Asiatic cheetah is protected (Chapter 5). The UNDP and Iran's Department of the Environment established a program of work to support conservation of the Asiatic cheetah from 2001 onwards. There is no national action plan (Breitenmoser et al., 2009) although some planning processes have been undertaken (e.g., see CEESP/CENESTA, 2004). In 2009, the Afghan Government placed cheetahs on the country's Protected Species List, meaning all hunting and trading of this species within Afghanistan is now illegal, although it is thought to be extinct in the country. The IUCN SSC Cat Specialist Group maintains a Cheetah Conservation Compendium with a reference library and detailed country information (http://www.catsg.org/cheetah/20_cc-compendium/index.htm), which provides a useful resource for publications relating to all aspects of cheetah ecology and conservation.

HALTING CHEETAH DECLINE

The worsening of the "threat status" of the cheetah should act as a wake-up call. Urgent action is needed if the survival of cheetahs is to be secured. There is an international public support for cheetah and other iconic megafauna, that is beyond doubt. This is clear from the millions of international visitors who travel thousands of kilometers to see cheetahs and other wildlife, and by the millions who avidly watch wildlife programs streaming into their homes. What is missing is the effective means to channel this value into local communities that bear the real costs of living with cheetahs and other potentially problematic species. To halt cheetah decline, the realities of conservation in developing countries that still harbor cheetahs must be confronted. Communities who share their land with cheetahs may face a daily challenge to feed themselves and their families (Middleton et al., 2011). They cannot afford to pay the costs of losing their livestock, even if depredation by cheetahs is a relatively rare event (Boast et al., 2016).

The Range Wide Conservation Program for Cheetah and African Wild Dog has made substantial strides toward developing an approach for cheetah conservation on the large and transboundary scale needed. The program has been working with range state governments for a decade, and the regional strategies and national action plans produced provide a road map for the conservation of cheetahs together with African wild dogs. These strategies and plans have the strong support of range state governments and conservation NGOs and lay out a list of all the actions that need to be undertaken to secure the survival of both species. To implement these road maps, more financial resources are needed, from range state governments, bilateral and multilateral donors, and NGOs, and innovative new mechanisms are needed that allow communities to benefit from the presence of wildlife and to become genuine partners in initiatives to save the cheetah.

Over coming decades Africa faces a critical period for its biodiversity. The continent's human population is predicted to double by 2050 (Population Reference Bureau, 2016). The need to support and feed more people will exert unprecedented pressures on wildlife and the environment. But lessons from Europe show what can be done (Chapron et al., 2014). Here, large carnivores faced imminent extinction toward the end of the 20th century. Yet today, due to protection and restoration programs combined with policies that help foster coexistence between people and wildlife, there has been a resurgence of bears, wolves, and lynx (Chapron et al., 2014). This demonstrates that, in the right circumstances, it is possible for people and large carnivores to live together, even when human densities are relatively high. For cheetahs, we urgently need to find the political will and the financial means to enable people and wildlife to coexist, and for both to prosper. Only then we can be sure that future generations will be able to continue to marvel at the sight of a cheetah at full speed. If we fail, the fate of the cheetah will be in peril.

References

Belbachir, F., Pettorelli, N., Wacher, T., Belbachir-Bazi, A., Durant, S.M., 2015. Monitoring rarity: the critically endangered Saharan cheetah as a flagship species for a threatened ecosystem. PLoS One, 10 (1), e0115136.

Benitez-Lopez, A., Alkemade, R., Schipper, A.M., Ingram, D.J., Verweij, P.A., Eikelboom, J.A.J., Huijbregts, M.A.J., 2017. The impact of hunting on tropical mammal and bird populations. Science 356 (6334), 180–183.

Boast, L., 2014. Exploring the causes of and mitigation options for human-predator conflict on game ranches in Botswana: How is coexistence possible? PhD thesis, University of Cape Town, South Africa.

Boast, L., Houser, A., Horgan, J., Reeves, H., Phale, P., Klein, R., 2016. Prey preferences of free-ranging cheetahs on farmland: scat analysis versus farmers' perceptions. Afr. J. Ecol. 54, 424–433.

Bouché, P., Mange, R.N.M., Tankalet, F., Zowoya, F., Lejeune, P., Vermeulen, C., 2012. Game over! Wildlife collapse in northern Central African Republic. Environ. Monit. Assess. 184, 7001–7011.

Breitenmoser, U., Alizadeh, A., Breitenmoser-Würsten, C., 2009. Conservation of the Asiatic Cheetah, its Natural Habitat and Associated Biota in the I. R. of Iran. Project Number IRA/00/G35. Terminal Evaluation Report, Global Environment Facility. Available from: https://erc.undp.org/evaluation/documents/download/2697.

Brugiere, D., Chardonnet, B., Scholte, P., 2015. Large-scale extinction of large carnivores (lion *Panthera leo*, cheetah *Acinonyx jubatus* and wild dog *Lycaon pictus*) in protected areas of West and Central Africa. Trop. Conserv. Sci. 8, 513–527.

Burney, D.A., 1980. The effects of human activities on cheetah *(Acinonyx jubatus)* in the Mara region of Kenya. MSc thesis, University of Nairobi, Kenya.

Caro, T.M., 1994. Cheetahs of the Serengeti Plains: Group Living in an Asocial Species. University of Chicago Press, Chicago.

CEESP/CENESTA, 2004. A Co-Management Strategy for Cheetah Conservation in Iran. Unpublished Report, IUCN CENESTA, Tehran.

Chapron, G., Kaczensky, P., Linnell, J.D.C., von Arx, M., Huber, D., Andren, H., Lopez-Bao, J.V., Adamec, M., Alvares, F., Anders, O., Balciauskas, L., Balys, V., Bedo, P., Bego, F., Blanco, J.C., Breitenmoser, U., Broseth, H., Bufka, L., Bunikyte, R., Ciucci, P., Dutsov, A., Engleder, T., Fuxjager, C., Groff, C., Holmala, K., Hoxha, B., Iliopoulos, Y., Ionescu, O., Jeremic, J., Jerina, K., Kluth, G., Knauer, F., Kojola, I., Kos, I., Krofel, M., Kubala, J., Kunovac, S., Kusak, J., Kutal, M., Liberg, O., Majic, A., Mannil, P., Manz, R., Marboutin, E., Marucco, F., Melovski, D., Mersini, K., Mertzanis, Y., Myslajek, R.W., Nowak, S., Odden, J., Ozolins, J., Palomero, G., Paunovic, M., Persson, J., Potocnik, H., Quenette, P.Y., Rauer, G., Reinhardt, I., Rigg, R., Ryser, A., Salvatori, V., Skrbinsek, T., Stojanov, A., Swenson, J.E., Szemethy, L., Trajce, A., Tsingarska-Sedefcheva, E., Vana, M., Veeroja, R., Wabakken, P., Wofl, M., Wolfl, S., Zimmermann, F., Zlatanova, D., Boitani, L., 2014. Recovery of large carnivores in Europe's modern human-dominated landscapes. Science 346, 1517–1519.

Chauvenet, A.L.M., Durant, S.M., Hilborn, R., Pettorelli, N., 2011. Unintended consequences of conservation actions: managing disease in complex ecosystems. PLoS One, 6 (12), e28671.

CITES, 2013. Illegal trade in Cheetah. CoP16 Doc. 51 (Rev. 1), Bangkok, Thailand. Earth Negotiation Bull. 21 (79).

Davis, K.F., D'Odorico, P., Rulli, M.C., 2014. Land grabbing: a preliminary quantification of economic impacts on rural livelihoods. Popul. Environ. 36, 180–192.

de Iongh, H.H., Croes, B., Rasmussen, G., Buij, R., Funston, P., 2011. The status of cheetah and African wild dog in the Benoue Ecosystem, North Cameroon. Cat News 55, 29–31.

Dinerstein, E., Loucks, C., Wikramanayake, E., Ginsberg, J., Sanderson, E., Seidensticker, J., Forrest, J., Bryja, G.,

Heydlauff, A., Klenzendorf, S., Leimgruber, P., Mills, J., O'Brien, T.G., Shrestha, M., Simons, R., Songer, M., 2007. The fate of wild tigers. Bioscience 57, 508–514.

Durant, S.M., 2000. Dispersal patterns, social structure and population viability. In: Gosling, L.M., Sutherland, W.J. (Eds.), Behaviour and Conservation. Cambridge University Press, Cambridge.

Durant, S.M., Becker, M.S., Creel, S., Bashir, S., Dickman, A.J., Beudels-Jamar, R.C., Lichtenfeld, L., Hilborn, R., Wall, J., Wittemyer, G., Badamjav, L., Blake, S., Boitani, L., Breitenmoser, C., Broekhuis, F., Christianson, D., Cozzi, G., Davenport, T.R.B., Deutsch, J., Devillers, P., Dollar, L., Dolrenry, S., Douglas-Hamilton, I., Droge, E., FitzHerbert, E., Foley, C., Hazzah, L., Hopcraft, J.G.C., Ikanda, D., Jacobson, A., Joubert, D., Kelly, M.J., Milanzi, J., Mitchell, N., M'Soka, J., Msuha, M., Mweetwa, T., Nyahongo, J., Rosenblatt, E., Schuette, P., Sillero-Zubiri, C., Sinclair, A.R.E., Price, M.R., Zimmermann, A., Pettorelli, N., 2015a. Developing fencing policies for dryland ecosystems. J. Appl. Ecol. 52, 544–551.

Durant, S.M., Craft, M.E., Hilborn, R., Bashir, S., Hando, J., Thomas, L., 2011. Long-term trends in carnivore abundance using distance sampling in Serengeti National Park, Tanzania. J. Appl. Ecol. 48, 1490–1500.

Durant, S.M., Dickman, A.J., Maddox, T., Waweru, M.N., Caro, TM., Pettorelli, N., 2010. Past, present and future of cheetah in Tanzania: from long term study to conservation strategy. In: Macdonald, D.W., Loveridge, A.J. (Eds.), Biology and Conservation of Wild Felids. Oxford University Press, Oxford, pp. 373–382.

Durant, S.M., Mitchell, N., Groom, R., Pettorelli, N., Ipavec, A., Jacobson, A.P., Woodroffe, R., Bohm, M., Hunter, L.T.B., Becker, M.S., Broekhuis, F., Bashir, S., Andresen, L., Aschenborn, O., Beddiaf, M., Belbachir, F., Belbachir-Bazi, A., Berbash, A., Machado, I.B.D., Breitenmoser, C., Chege, M., Cilliers, D., Davies-Mostert, H., Dickman, A.J., Ezekiel, F., Farhadinia, M.S., Funston, P., Henschel, P., Horgan, J., de Iongh, H.H., Jowkar, H., Klein, R., Lindsey, P.A., Marker, L., Marnewick, K., Melzheimer, J., Merkle, J., M'Soka, J., Msuha, M., O'Neill, H., Parker, M., Purchase, G., Sahailou, S., Saidu, Y., Samna, A., Schmidt-Küntzel, A., Selebatso, E., Sogbohossou, E.A., Soultan, A., Stone, E., van der Meer, E., van Vuuren, R., Wykstra, M., Young-Overton, K., 2017. The global decline of cheetah Acinonyx jubatus and what it means for conservation. Proc. Natl. Acad. Sci. USA 114, 528–533.

Durant, S., Mitchell, N., Ipavec, A., Groom, R., 2015b. *Acinonyx jubatus*. The IUCN Red List of Threatened Species 2015: e.T219A50649567. Available from http://dx.doi.org/10.2305/IUCN.UK.2015-4.RLTS.T219A50649567.en.

Durant, S.M., Wacher, T., Bashir, S., Woodroffe, R., De Ornellas, P., Ransom, C., Newby, J., Abaigar, T., Abdelgadir, M., El Alqamy, H., Baillie, J., Beddiaf, M., Belbachir, F., Belbachir-Bazi, A., Berbash, A.A., Bemadjim, N.E.,

Beudels-Jamar, R., Boitani, L., Breitenmoser, C., Cano, M., Chardonnet, P., Collen, B., Cornforth, W.A., Cuzin, F., Gerngross, P., Haddane, B., Hadjeloum, M., Jacobson, A., Jebali, A., Lamarque, F., Mallon, D., Minkowski, K., Monfort, S., Ndoassal, B., Niagate, B., Purchase, G., Samaila, S., Samna, A.K., Sillero-Zubiri, C., Soultan, A.E., Price, M.R.S., Pettorelli, N., 2014. Fiddling in biodiversity hotspots while deserts burn? Collapse of the Sahara's megafauna. Divers. Distrib. 20, 114–122.

Farhadinia, M.S., Akbari, H., Eslami, M., Adibi, M.A., 2016. A review of ecology and conservation status of Asiatic cheetah in Iran. Cat News Special Issue Iran 10, 18–26.

FRONTEX, 2017. Africa-Frontex Intelligence Community Joint Report 2016, Warsaw. Available from: http://frontex.europa.eu/assets/Publications/Risk_Analysis/AFIC/AFIC_2016.pdf.

Henry, W., 1975. A Preliminary Report on Visitor Use in Amboslei National Park. Working paper no. 263. Institute for Development Studies, University of Nairobi, Kenya.

Henry, W., 1980. Patterns of tourist use in Kenya's Amboseli National Park: implications for planning and management. In: Hawkins, D., Shafer, E., Rovelstad, J. (Eds.), Tourism Marketing and Management Issues. George Washington University, Washington, DC.

Iranian Cheetah Society, 2013. More than 40% of Cheetahs Killed on Roads in Iran. Iranian Cheetah Society. Available from: http://www.wildlife.ir/en/2013/01/14/more-than-40-of-cheetahs-killed-on-roads-in-iran/.

IUCN/SSC, 2007a. Regional Conservation Strategy for the Cheetah and African Wild Dog in Eastern Africa. Gland, Switzerland.

IUCN/SSC, 2007b. Regional Conservation Strategy for the Cheetah and African Wild Dog in Southern Africa. Gland, Switzerland.

IUCN/SSC, 2008. Strategic Planning for Species Conservation: An Overview. Version 1.0.

IUCN, 2012. IUCN Red List Categories and Criteria: Version 3.1, second ed. Gland, Switzerland and Cambridge, UK, IUCN.

IUCN/SSC, 2012. Regional Conservation Strategy for the Cheetah and African Wild Dog in Western, Central and Northern Africa. Gland, Switzerland.

IUCN/SSC, 2015. Review of the Regional Conservation Strategy for the Cheetah and African Wild Dog in Southern Africa. Gland, Switzerland.

Jowkar, H., Hunter, L., Ziaie, H., Marker, L., Breitenmoser-Wursten, C., Durant, S., 2008. Acinonyx jubatus ssp. venaticus. The IUCN Red List of Threatened Species 2008, e.T220A13035342. Available from: http://dx.doi.org/10.2305/IUCN.UK.2008.RLTS.T220A13035342.en.

Laurenson, M.K., 1995. Implications of high offspring mortality for cheetah population dynamics. In: Sinclair, A.R.E., Arcese, P. (Eds.), Serengeti II: Dynamics, Management and Conservation of an Ecosystem. University of Chicago Press, Chicago.

Lindsey, P.A., Balme, G., Becker, M., Begg, C., Bento, C., Bocchino, C., Dickman, A., Diggle, R.W., Eves, H., Henschel, P., Lewis, D., Marnewick, K., Mattheus, J., Weldon McNutt, J., McRobb, R., Midlane, N., Milanzi, J., Morley, R., Murphree, M., Opyene, V., Phadima, J., Purchase, G., Rentsch, D., Roche, C., Shaw, J., Westhuizen, H.V.D., Vliet, N.V., Zisadza-Gandiwa, P., 2013. The bushmeat trade in African Savannas: impacts, drivers, and possible solutions. Biol. Conserv. 160, 80–96.

Løvschal, M., Bøcher, P.K., Pilgaard, J., Amoke, I., Odingo, A., Thuo, A., Svenning, J.-C., 2017. Fencing bodes a rapid collapse of the unique Greater Mara ecosystem. Sci. Rep. 7, 41450.

Mace, G.M., Collar, N.J., Gaston, K.J., Hilton-Taylor, C., Akcakaya, H.R., Leader-Williams, N., Milner-Gulland, E.J., Stuart, S.N., 2008. Quantification of extinction risk: IUCN's system for classifying threatened species. Conserv. Biol. 22, 1424–1442.

Mace, G.M., Lande, R., 1991. Assessing extinction threats toward a reevaluation of IUCN threatened species categories. Conserv. Biol. 5, 148–157.

Marker, L., 2002. Aspects of Cheetah (Acinonyx jubatus) Biology, Ecology and Conservation Strategies on Namibian Farmlands. PhD thesis, University of Oxford, United Kingdom.

Marnewick, K.A., Bothma, J.D.P., Verdoorn, G.H., 2006. Using camera-trapping to investigate the use of a tree as a scent-marking post by cheetahs in the Thabazimbi district. S. Afr. J. Wildl. Res. 36, 139–145.

Marnewick, K., Ferreira, S.M., Grange, S., Watermeyer, J., Maputla, N., 2014. Evaluating the status of and African wild dogs Lycaon pictus and cheetahs Acinonyx jubatus through tourist-based photographic surveys in the Kruger National Park. Plos One 9, e86265.

Middleton, N., Stringer, L., Goudie, A., Thomas, D., 2011. The Forgotten Billion: MDG Achievement in the Drylands. United Nations Office at Nairobi, Publishing Services Section, ISO 14001:2004.

Mitchell, N., Durant, S.M., 2017. Steps in tackling the illegal cheetah trade. Cat News 65, 49–50.

Nowell, K., 2014. An assessment of conservation impacts of legal and illegal trade in cheetahs Acinonyx jubatus. IUCN SSC Cat Specialist Group report prepared for the CITES Secretariat, 65th meeting of the CITES Standing Committee, Geneva, 7–11 July. SC65 Doc. 39 (Rev. 2).

Nowell, K., Jackson, P., 1996. Wild Cats: Status Survey and Conservation Action Plan. Burlington Press, Cambridge.

Population Reference Bureau, 2016. 2016 World Population Data Sheet. Population Reference Bureau, Washington, DC. Available from: www.prb.org.

R Core Team, 2015. R: A Language and Environment for statIstical Computing. R Foundation for Statistical

Computing, Vienna, Austria. Available from: http://www.R-project.org/.

Ripple, W.J., Abernethy, K., Betts, M.G., Chapron, G., Dirzo, R., Galetti, M., Levi, T., Lindsey, P.A., Macdonald, D., Machovina, B., Newsome, T.M., Peres, C.A., Wallach, A.D., Wolf, C., Young, H., 2016. Bushmeat hunting and extinction risk to the world's mammals. R. Soc. Open Sci. 3 (10), 160498.

Rodrigues, A.S.L., Pilgrim, J.D., Lamoreux, J.F., Hoffmann, M., Brooks, T.M., 2006. The value of the IUCN Red List for conservation. Trends Ecol. Evol. 21, 71–76.

Roe, D., Leader-Williams, N., Dalal-Clayton, B., 1997. Take only photographs, leave only footprints: the environmental impacts of wildlife tourism. Environmental Planning, IIED Wildlife and Development Series, No.10.

Sanderson, E.W., Redford, K.H., Chetkiewicz, C.B., Medellin, R.A., Rabinowitz, A.R., Robinson, J.G., Taber, A.B., 2002. Planning to save a species: the jaguar as a model. Conserv. Biol. 16, 58–72.

Van der Meer, E., 2016. The Cheetahs of Zimbabwe, Distribution and Population Status 2015. Cheetah Conservation Project Zimbabwe, Victoria Falls, Zimbabwe. Available from: www.cheetahzimbabwe.org.

Williams, S.T., Williams, K.S., Joubert, C.J., Hill, R.A., 2016. The impact of land reform on the status of large carnivores in Zimbabwe. PeerJ. 14 (4), e1537.

What Does the Future Hold for the Cheetah?

Laurie Marker, Lorraine K. Boast**,
Anne Schmidt-Küntzel**

*Cheetah Conservation Fund, Otjiwarongo, Namibia
**Cheetah Conservation Botswana, Gaborone, Botswana

INTRODUCTION

Cheetahs (*Acinonyx jubatus*) have declined in number by more than 90% and have lost more than 91% of their range since 1900 due to human-caused habitat destruction and persecution (Durant et al., 2017; Marker and Dickman, 2004). Today the total free-ranging population is estimated at 7100 adults and adolescents distributed across 33 populations in 19 African countries and Iran (Durant et al., 2017; Chapters 4, 5, and 39). Recent estimates indicate that populations continue to decline (Durant et al., 2017) despite the best efforts of a global network of dedicated researchers and conservationists around the world.

The Iranian and northwest African cheetah subspecies is classified as Critically Endangered with the other African subspecies listed as Vulnerable by the International Union for the Conservation of Nature (IUCN) (Durant et al., 2015). In 2016 a motion was put forward for IUCN to uplist these remaining subspecies to Endangered

(Durant et al., 2017; Chapter 39). The cheetah was ranked 3rd of 36 wild felid species in terms of prioritization of conservation efforts (Dickman et al., 2015). With most population numbers below the inferred threshold of demographic viability (Chapter 39), and all but two below the estimated threshold for long-term viability (Chapter 10), the species' survival across its current range requires action now, and at a much greater scale than is currently underway.

As highlighted in the chapters of this book, the most significant immediate threats to cheetahs are habitat loss and fragmentation (Chapter 10), loss of prey base (Chapter 11), direct persecution due to human-carnivore conflict (Chapter 13), and illegal trade (Chapter 14); with climate change (Chapter 12) and human population growth exacerbating these threats in the future. The species' documented genetic uniformity (Chapter 6) will likely compound these threats in small isolated populations, which will continue to decline and disappear if conservation

Cheetahs: Biology and Conservation
http://dx.doi.org/10.1016/B978-0-12-804088-1.00040-X

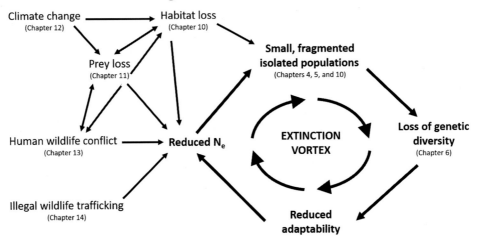

Cheetah Conservation Strategies

Predator-friendly livestock and game-stock management techniques (Chapters 13 and 15)

Improved and alternative livelihood initiatives (Chapter 16)

Conservancies (Chapter 17)

Environmental education (Chapters 18 and 28)

Protected areas (Chapter 19)

Conservation translocation of cheetahs (Chapter 20)

National and international policies and legislation (Chapter 21)

Ex situ cheetah population (Chapters 22 and 23)

FIGURE 40.1 **Summary of the threats to cheetahs and of the conservation actions to prevent the extinction of the cheetah, as presented in the chapters of this book.**

actions are not implemented (Fig. 40.1). It is critical to understand that the majority of cheetahs (67%) live on nonprotected land; and that protected areas, while gaining in importance due to increasing human pressure on cheetah habitat (Chapter 19), are generally too small to support cheetah populations without connectivity to cheetahs outside of their boundaries (Durant et al., 2017). The extinction risk for cheetahs is both real and increasing. Their survival will only be ensured by integrated coexistence with human development.

Cheetah survival is threatened by unsustainable use of natural resources including loss of habitat and prey due to human population pressures. Our species' demand on the planet's resources, as tracked by the Ecological Footprint (Wackernagel et al., 2002), has increased by over 140% from 1961 to 2010 (Galli et al., 2015). This pressure is much greater than what the planet can sustainably support (Galli et al., 2014) and it was estimated that humans will need the equivalent of 2.6 planet's worth of ecological resources by 2050 if we continue "business as usual" (Moore et al., 2012). Despite being able to create new technologies to convert land for human needs and economic development, we have not proven proficient at recognizing or addressing the destruction we are causing, while putting at risk the survival of many species, including our own, as well as that of the cheetah.

WHAT WILL IT TAKE TO SAVE THE CHEETAH?

When the question of "how can we save the cheetah" arises in general discussion, many people expect the response to be a short, simple, declarative, and definitive statement. Sadly, there is no easy solution. A clear lesson from the chapters presented in this book is that conserving cheetahs is an ongoing process involving applied conservation and research through a multifaceted, participatory approach addressing both the needs of the cheetah and of the people who live within the cheetah's range. Interrelated issues of human population expansion, commercialism, and the nonsustainable culture of modern life, poverty, and ignorance are issues that are not easily addressed. To do so will require governmental policies in conjunction with fundamental changes in personal attitudes. It will require all of us to take responsibility for the impact we create if we want to affect the problems cheetahs and other wildlife face today. But these changes will take time, perhaps generations, while the cheetah may not have this much time.

Fortunately, some positive action is occurring. Cheetah conservation has benefited from the presence of a handful of conservation institutions and organizations which have dedicated their efforts for several decades to conserving wild cheetahs (Chapter 1), and understanding their behavioral ecology (Chapters 8 and 9). In 2007, cheetah organizations, partnering with African wild dog (*Lycaon pictus*) organizations, joined their efforts in a range wide program of the IUCN Canid and Cat Specialist Groups. This Range-wide Cheetah and African Wild Dog program has brought together government wildlife authorities, field programs, non-government organizations (NGOs), and other stakeholders

through regional and national meetings, and has developed regional strategies and national action plans, as a "roadmap" for the cheetah's conservation (IUCN/SSC, 2007a,b, 2012; RWCP and IUCN/SSC, 2015). The national action plans cover 96% of known African cheetah range and aim to guide conservation actions of all stakeholders to achieve crucial collective objectives within the next 10–20 years (Chapter 39). As stated in Chapter 39, the Cheetah Range-wide objectives broadly are to: "improve national capacity for cheetah conservation and management, raise awareness of and political commitment to cheetah conservation [sensitize decision makers], promote human–cheetah coexistence, improve land use planning and reduce habitat fragmentation, combat illegal and unsustainable offtake of prey, improve policy and legislation, and address cheetah conservation information needs."

This book summarizes ongoing cheetah conservation strategies (Fig. 40.1) such as: promotion of predator-friendly livestock and gamestock management techniques (Chapters 13 and 15), improved and alternative livelihood initiatives (Chapter 16), development of conservancies (Chapter 17), child and adult environmental education (Chapters 18 and 28), protected areas (Chapter 19), reintroduction and translocation of cheetahs (Chapter 20), national and international policies and legislation (Chapter 21), and maintaining sustainable *ex situ* backup populations (Chapters 22 and 23). But the existing conservation programs are limited in scope, due to primarily financial constraints, while the overwhelming message of the future of cheetah conservation calls for an increase in the scale of conservation efforts across the cheetah's range in Africa and Iran.

CHEETAH CONSERVATION MAGNIFIED

The cheetah is a wide-ranging species that occurs at much lower densities than that recorded for most other large felids (Chapter 8; Fig. 40.2).

of effective conservation programs of critically endangered and undersupported populations is paramount, in particular in northern, western, and central Africa, where cheetah populations are small, fragmented, and understudied.

To achieve the impact and scale needed for cheetah conservation, it is imperative to operate under a broad, multispecies, and interdisciplinary approach. Interdisciplinary approaches can integrate cheetah conservation with grassland management, land use planning, the conservation of other endemic species, and with the economic development of key areas. As the threats are similar for other large African carnivores, in terms of habitat loss and persecution, multispecies approaches enable the expansion of the scale, scope, and efficiency of conservation programs (Macdonald et al., 2012), as was done with the integration of cheetah and African wild dog conservation. A priority for any conservation effort should be to maintain and recover functional ecosystems, promoting conservation at all levels including intact guilds of large carnivores and their prey, along with the structural and functional biodiversity (Mills and Mills 2017; Noss, 1990).

The need for national parks and protected areas will become more critical as human populations increase and affect the health of ecosystems. Partners, such as the African Parks network (https://www.african-parks.org/), have developed strategic agreements with governments to combine sound conservation and park protection with business expertise. Their model combines long-term donor funding with revenue from ecotourism. Such models could enhance the conservation potential of many key cheetah conservation areas. Currently, two protected areas within the African Parks network have cheetahs with another two as possible reintroduction sites (Mills and Mills, 2017). However, the need to feed and employ a growing human population may limit the potential to develop new state- or privately-owned protected areas. The future of the cheetah, therefore, depends on finding innovative ways to support the end-goals of agricultural production,

FIGURE 40.2 **Cheetah observing its surroundings from an elevated location.** *Source: Peter Scheufler.*

Thus, large areas of connected landscape are required for the species' survival, and coordinated, landscape level conservation efforts must be implemented at a much bigger scale than currently being practiced. Transfrontier management will be an important part of this approach and the role of Transfrontier Conservation Areas in conserving biodiversity, socioeconomic development, and promoting a culture of peace (peace parks) has been recognized (Hanks, 2003). As cheetahs are largely found outside of protected areas, their conservation depends on cooperation and coordination between governments and citizens across a matrix of privately owned and rural landscapes. Initiation and expansion in ways that do not negatively affect cheetahs and other wildlife. This book (Chapters 13, 15, 16, and 17) details ways in which the livelihoods of the people who live with cheetahs can be supported thereby reducing conflicts between cheetahs and people. The present dynamics of human population growth suggest that any conservation plan that does not take the need for agricultural production into account may not be effective, even over the next 25–30 years.

CONSERVATION NEEDS

To achieve the grand scale of cheetah conservation needed to reverse the species' decline and reduce the threats the species faces, there is a need for greater attitudinal support from all stakeholders in cheetah conservation (e.g., governments, local, and international communities), and for increased funding and resources. In addition, there is a need for efficient monitoring and research, and a heightened level of cooperation among conservation groups with similar interests.

Support

Governments

To succeed, cheetah conservation needs to be driven from within the range states where cheetahs are resident. Throughout the African regions, concerns over the low priority to wildlife in political decision making and engaging political commitment were highlighted as constraints to alleviating the threats to cheetahs (IUCN/SSC, 2012; RWCP and IUCN/SSC, 2015). Given the need for economic development, health, and other human related issues in most cheetah range countries, this is no surprise, but as the human population is predicted to grow, it is essential that awareness and capacity is raised in governments and plans are put in place now to address wildlife conservation, before it becomes further overlooked.

Capacity building is therefore one of the objectives identified by the regional strategies as being essential to develop the knowledge and skills among decision makers and wildlife professionals, in order to implement policies that can affect the cheetah's survival. Conservation-focused policies around issues such as land use, poverty alleviation, sustainable use of wildlife (e.g., game ranching, hunting, tourism), illegal trade, conservancies, bushmeat hunting, and ecologically friendly livestock and game management should be emphasized. An International Parliamentarian Conservation Caucus Foundation has been developed in several of the cheetah range countries (http://www.internationalconservation.org/). These high profile governmental caucuses can help educate other governmental officials to the benefits of wildlife and conservation.

Communities and Landowners

The cheetah's future is deeply entwined with that of the people who share the land with them. Engaging communities and landowners through education, awareness raising, and human–cheetah coexistence initiatives have been shown to contribute to the establishment of positive attitudes and behaviors that can benefit predator conservation (Chapters 15, 16, and 18). However, the issue of human–carnivore coexistence is complex (Chapter 13), and achieving cheetah conservation in human dominated landscapes will require an integrated approach, which promotes economic development in an environmentally sustainable fashion. For communities to tolerate cheetahs in their "backyards" economic benefits to offset the costs of coexistence must exist (Winterbach et al., 2013). Identifying and developing situations that can both conserve biodiversity, as well as assist communities is difficult (McShane et al., 2011). Engaging individuals to take a greater responsibility and ownership for the environment and the conservation initiatives, through a participatory approach, has been emphasized as an important aspect of current and future conservation programs (Chapter 16). The impact of conservation programs can be further optimized by engaging social science, economic development, and education experts in the programs' planning, development, and implementation (Chapters 18 and 35).

Initiatives, which link cheetah conservation to financial benefits or community resources, are likely to be beneficial (Chapter 16). Examples, such as ecotourism, hunting, and other alternative livelihoods can provide support through employment (e.g., safari guides, camp staff), as well as support services in the wider community. However, economically driven conservation may not incorporate the entire ecosystem, potentially resulting in decreased biodiversity and sustainability, as witnessed in many private game-fenced reserves (Mills and Mills, 2017). Additionally, establishing community benefits can be difficult as dispersing benefits equitably is challenging (Mills and Mills, 2017). Successful community conservation models have been developed, for instance, conservancies in Namibia and Kenya provide large landscapes that can support cheetahs (Chapter 17).

International Community

The international community has an important role to play in all wildlife issues and can indirectly contribute to cheetah conservation by raising funds and awareness toward the conservation goal, through donations, responsible purchasing and ecotourism. While once off donations can significantly contribute to specific conservation projects, predictable income, such as conservation funds generated from market driven profits and initiatives that provide a product or service, have the benefit of providing long-term stability. For example, consumer preference for products and services that reduce environmental impact and promote wildlife friendly practices (e.g., predator friendly products or ecofriendly and community based tourism models) can provide

economic incentives to promote cheetah–human coexistence. Ecotourism, in particular, can provide economic incentives to communities, which can be directly linked to cheetahs, as cheetahs are highly sought after by tourists (Ripple et al., 2014). Consumers should carry the responsibility to inform themselves about wildlife-friendly services and products, and reject nonsustainable or unethical tourism practices (e.g., carnivore petting farms, canned hunting, and the keeping of nonendemic game species, which reduces predator tolerance). Certification schemes can be beneficial in informing consumers and engaging international communities in species conservation while supporting local ecosystem stewardship practices (Chapter 16).

It is also important that all individuals reduce their own individual ecological footprint, through sustainable living. While individual changes toward sustainable living may seem to have a minimal impact, collectively these individual changes could significantly benefit the cheetah's future. Individual changes have even been shown to collectively have the capacity to reduce the extent of global warming, which causes a serious threat to many ecosystems and wildlife species, including cheetahs (Chapter 12).

Funding and Resources

Drastically increased funding over a long time-frame, in a more predictable manner than is currently available, will be required to implement the conservation initiatives at the scale needed to conserve the remaining cheetah populations. However, in the near-term, the available funding and resources to devote to conservation actions are likely to remain limited and highly competitive. Prioritization of conservation efforts may therefore be required to focus resources toward key issues, known as a "triage" approach (ECOS, 2014), which is aimed at preventing scarce resources being spent on what cannot be saved.

Therefore, choices need to be made using a rational set of criteria. Identifying those areas of conservation priority (e.g., key populations or habitat corridors), as well as programs likely to have the greatest impact and likelihood of success will be important steps forward (Chapter 38 for the application of population viability analysis to the prioritization of conservation decisions). In addition to funding conservation initiatives, the importance of funding areas, such as project operating costs, staff retention, and career development of conservationists, needs greater recognition and support. This will increase capacity at an NGO level, which will in turn allow NGOs to work in cooperation with governments to achieve land use needs that will support long-term cheetah survival.

Sharing of limited resources using multispecies and interdisciplinary approaches is one way to maximize available funding. Increasing partnerships with zoos, commercial industries (in particular those which can directly impact cheetahs e.g., mining, tourism, hunting), and international donors is imperative. Zoos in particular have an important role to play in the cheetah's future, not only to host and manage back-up populations, but also to raise awareness, and provide funding, expertise, research opportunities, and technical support (Chapters 22–28).

Information Dissemination and Research

Conservation success for cheetahs requires continuous scientific support for conservation activities while raising awareness for society's role in securing the cheetah's long-term future. Securing support and funding is largely reliant on awareness raising and communication. Disseminating research findings to aid conservation management is an important component of cheetah conservation; and promotes collaboration and communication between conservation personnel, government wildlife officials, and other stakeholders relevant to cheetah conservation. Long-term monitoring, standardization of research techniques, and options for data sharing, as well as collaboration and timely

publication are all important themes emphasized in the "techniques and analyses" section of this book (Chapters 29–38).

It is important that conservation programs include an evaluation step in their initiatives. Evidence of what is and isn't working is necessary to both improve and develop programs, and to prove to funding agencies the success and importance of the work. Further research could aid the cheetah's long-term conservation. For example, developing artificial reproductive techniques (Chapter 27) and improving translocation successes (Chapter 20) could both contribute to the preservation of the existing genetic diversity. Innovative new methods to promote human–cheetah coexistence need to be considered, investigated, and trialed. Saving the cheetah will require looking beyond our current conservation toolbox to find new methods of conflict mitigation and species management.

CONCLUSIONS

As the fastest land mammal, the cheetah is one of the world's most unique species with an interesting evolutionary history, as well as a long history with humans (Chapters 2, 3, and 7). The cheetah has attracted broad scientific interest for decades, of which the resulting data have been summarized in this book. While data gaps have been highlighted in various chapters and some gaps impact the precision of modeling predictions, population viability analysis (PVA) has shown that conclusions regarding the cheetah's chance of extinction can already be drawn and that conservation action is needed immediately if cheetahs are to survive (Chapters 38 and 39).

Action plans for cheetah conservation have been written and the broad requirements of support, funding, resources, and information dissemination have been described, but converting this into appropriate action is difficult. In southern Africa, the only region to have conducted a formal review of the progress made in addressing action plan objectives, most progress has been made in the areas of human–cheetah coexistence and awareness raising, more efforts are needed in the areas of capacity development, information transfer, and policy and legislation (RWCP and IUCN/SSC, 2015). The Durant et al. (2017) publication on cheetah conservation status (discussed in Chapter 39) attracted much international media attention, as did the inclusion of the cheetah as a species of priority at the 2016 CITES conference on illegal trade in wildlife (Chapter 14). The motion to uplist cheetahs to Endangered on the IUCN list of threatened species (Durant et al. 2017), if approved, is likely to attract much needed support for cheetah conservation. Cheetahs have arguably more international attention for their conservation now than they have had before, so now is the time to turn this attention and support into action.

However, unless this support translates into action quickly, conservation efforts are likely to arrive too late for some of the cheetah populations suffering from small numbers or continued habitat loss. If these populations are to be saved, action is required immediately, and should include connection with other populations (via corridors or translocation; Chapters 10 and 20) and gamete banking (Chapter 27). Ideally the level of genetic diversity, rate of its loss, and degree of inbreeding, should be monitored but this quest for knowledge should in no way delay conservation action. Conservation efforts of small isolated populations should not detract efforts of stronghold populations to the point to jeopardize their survival.

Solving the cheetah conservation crisis requires addressing a complex web of social, economic, and environmental issues while stabilizing (and ideally increasing) viable cheetah populations throughout their range. As such, cheetahs are symbolic of many contemporary issues related to conservation and the environment. Today, the cheetah needs everyone's support, as we cannot only depend on government to solve the problems, nor can we place the burden of the cheetah's survival solely on the rural and poor communities living with cheetahs, or with the organizations working toward cheetah conservation. The international community as a whole, has the ability to significantly and positively influence cheetah conservation through their actions.

In Africa, the socioeconomics and economic development of the region will be an integral part of the cheetah's survival strategy. Conserving species in human-dominated landscapes is a complicated venture, time-consuming, and difficult to predict. But through collaboration and commitment, a roadmap to cheetah conservation has been written (IUCN/SSC, 2007a,b, 2012; RWCP and IUCN/SSC, 2015). We must learn from our previous efforts and continue to embrace both ecological and social sciences to inform our decisions. Like the cheetah, humans are running out of space and time. But unlike the cheetah, humans have the power to make global changes to improve the situation, for us and for all other species.

It will be up to the people living today to determine what habitats will exist and what animals will remain to share the planet's resources. Cheetahs are, of course, not the only species of wildlife at risk today. The cheetah shares needs for habitat, space, and conservation protection with other iconic species, such as elephants (*Loxodonta* spp.), rhinoceros (*Rhinocerotidae* spp.), lions (*Panthera leo*), and African wild dogs to name a few; these well known species can be considered "canaries in the coal mine" for ecosystem health. To summarize, we need an army ranging from farmers and consumers to educators and donors to save the cheetah and its ecosystems. As individuals, we must take action and we must do it now.

References

Dickman, A.J., Hinks, A.E., Macdonald, E.A., Burnham, D., Macdonald, D.W., 2015. Priorities for global felid conservation. Conserv. Biol. 29, 854–864.

Durant, S., Mitchell, N., Ipavec, A., Groom, R., 2015. *Acinonyx jubatus*. The IUCN Red List of Threatened Species 2015: e.T219A50649567. Available from: http://dx.doi.org/10.2305/IUCN.UK.2015-4.RLTS.T219A50649567.en.

Durant, S.M., Mitchell, N., Groom, R., Pettorelli, N., Ipavec, A., Jacobson, A.P., Woodroffe, R., Böhm, M., Hunter, L.T.B., Becker, M.S., Broekhuis, F., Bashir, S., Andresen, L., Aschenborn, O., Beddiaf, M., Belbachir, F., Belbachir-Bazi, A., Berbash, A., de Matos Machado, I.B., Breiten-moser, C., Chege, M., Cilliers, D., Davies-Mostert, H., Dickman, A.J., Ezekiel, F., Farhadinia, M.S., Funston, P., Henschel, P., Horgan, J., de Iongh, H.H., Jowkar, H., Klein, R., Lindsey, P.A., Marker, L., Marnewick, K., Melzheimer, J., Merkle, J., M'soka, J., Msuha, M., O'Neill, H., Parker, M., Purchase, G., Sahailou, S., Saidu, Y., Samna, A., Schmidt-Küntzel, A., Selebatso, E., Sogbohos-sou, E.A., Soultan, A., Stone, E., van der Meer, E., van Vuuren, R., Wykstra, M., Young-Overton, K., 2017. The global decline of cheetah *Acinonyx jubatus* and what it means for conservation. Proc. Nat. Acad. Sci. USA 114, 528–533.

ECOS, 2014. More Governments Adopting Controversial "Triage" Approach to Conservation. Available from: http://www.ecosmagazine.com/?paper=EC14126.

Galli, A., Lin, D., Wackernagel, M., Gressot, M., Winkler, S., 2015. Humanity's growing Ecological Footprint: sustainable development implications. Brief for GSDR2015. Available from: https://sustainabledevelopment.un.org/content/documents/5686humanitysgrowingecologicalfootprint.pdf.

Galli, A., Wackernagel, M., Iha, K., Lazarus, E., 2014. Ecological Footprint: implications for biodiversity. Biol. Conserv. 173, 121–132.

Hanks, J., 2003. Transfrontier Conservation Areas (TFCAs) in Southern Africa. J. Sustain. Forest. 17, 127–148.

IUCN/SSC, 2007a. Regional Conservation Strategy for the Cheetah and African Wild Dog in Eastern Africa. IUCN/SSC, Gland, Switzerland.

IUCN/SSC, 2007b. Regional Conservation Strategy for the Cheetah and African Wild Dog in Southern Africa. IUCN/SSC, Gland, Switzerland.

IUCN/SSC, 2012. Regional Conservation Strategy for the Cheetah and African Wild Dog in Western, Central, and Northern Africa. IUCN/SSC, Gland, Switzerland.

Macdonald, D.W., Burnham, D., Hinks, A.E., Wrangham, R., 2012. A problem shared is a problem reduced: seeking efficiency in the conservation of felids and primates. Folia Primatol. 83, 171–215.

Marker, L.L., Dickman, A.J., 2004. Human aspects of cheetah conservation: lessons learned from the Namibian farmlands. Hum. Dimen. Wildl. 9 (4), 297–305.

McShane, T.O., Hirsch, P.D., Trung, T.C., Songorwa, A.N., Kinzig, A., Monteferri, B., Mutekanga, D., Van Thang, H., Dammert, J.L., Pulgar-Vidal, M., Welch-Devine, M., Brosius, J.P., Coppolillo, P., O'Connor, S., 2011. Hard choices: making trade-offs between biodiversity conservation and human well-being. Biol. Conserv. 144 (3), 966–972.

Mills, M.G.L., Mills, M.E.J., 2017. Kalahari Cheetahs: Adaptations to an Arid Region. Oxford University Press, Oxford, UK.

Moore, D., Galli, A., Cranston, G.R., Reed, A., 2012. Projecting future human demand on the Earth's regenerative capacity. Ecol. Indicat. 16, 3–10.

Noss, R.F., 1990. Indicators for monitoring biodiversity—a hierarchical approach. Conserv. Biol. 4, 355–364.

Ripple, W.J., Estes, J.A., Belschta, R.L., Wilmers, C.C., Ritchie, E.G., Hebblewhite, M., Berger, J., Elmhagen, B., Letnic, M., Nelson, M.P., Schmitz, O.J., Smith, D.W., Wallach, A.D., Wirsin, A.J., 2014. Status and ecological effects of the world's largest carnivores. Science 243, 124484.

RWCP and IUCN/SSC, 2015. Regional Conservation Strategy for the Cheetah and African Wild Dog in Southern Africa; Revised and Updated, August 2015.

Wackernagel, M., Schultz, B., Deumling, D., Linares, A.C., Jenking, M., Kapos, V., Monfreda, C., Loh, J., Myers, N., Norgaard, R., Randers, J., 2002. Tracking the ecological overshoot of the human economy. PNAS 99 (14), 9266–9271.

Winterbach, H.E.K., Winterbach, C.W., Somers, M.J., Hayward, M.W., 2013. Key factors and related principles in the conservation of large African carnivores. Mammal Rev. 43, 89–110.

Index

Printed in the United States
By Bookmasters